T0343178

Measuring Ocean Currents

Tools, Technologies, and Data

Measuring Ocean Currents
Tools, Technologies, and Data

Dr. Antony Joseph

Formerly
Chief Scientist
Marine Instrumentation Division
CSIR—National Institute of Oceanography
Dona Paula
Goa – 403 004, India

AMSTERDAM • BOSTON • HEIDELBERG • LONDON • NEW YORK • OXFORD • PARIS
SAN DIEGO • SAN FRANCISCO • SINGAPORE • SYDNEY • TOKYO

ELSEVIER

Elsevier
225 Wyman Street, Waltham, MA 02451, USA
525 B Street, Suite 1900, San Diego, CA 92101-4495, USA

First edition 2014

Copyright © 2014 Elsevier Inc. All rights reserved.

No part of this publication may be reproduced, stored in a retrieval system or transmitted in any form or by any means electronic, mechanical, photocopying, recording or otherwise without the prior written permission of the publisher.

Permissions may be sought directly from Elsevier's Science & Technology Rights Department in Oxford, UK: phone (+44) (0) 1865 843830; fax (+44) (0) 1865 853333; email: permissions@elsevier.com. Alternatively you can submit your request online by visiting the Elsevier web site at http://elsevier.com/locate/permissions, and selecting Obtaining permission to use Elsevier material.

Notice

No responsibility is assumed by the publisher for any injury and/or damage to persons or property as a matter of products liability, negligence or otherwise, or from any use or operation of any methods, products, instructions or ideas contained in the material herein. Because of rapid advances in the medical sciences, in particular, independent verification of diagnoses and drug dosages should be made.

Library of Congress Cataloging-in-Publication Data
Joseph, Antony.
  Measuring ocean currents : tools, technologies, and data / Dr. Antony Joseph. – 1st ed.
    pages cm
  ISBN 978-0-12-415990-7
1. Ocean currents. 2. Ocean currents–Measurement. I. Title.
  GC239.2.J67 2013
  551.46'20287–dc23
                              2013011297

British Library Cataloguing in Publication Data
A catalogue record for this book is available from the British Library

For information on all **Elsevier** publications
visit our web site at store.elsevier.com

ISBN: 978-0-12-415990-7

Working together
to grow libraries in
developing countries

ELSEVIER   Book Aid International

www.elsevier.com • www.bookaid.org

Contents

Acknowledgments ix
Foreword xi
Preface xix

1. Oceanic Currents and Their Implications 1

1.1. Oceans' Thermohaline Conveyor Belt Circulation and Global Climate Change 2
1.2. Meandering Currents, Eddies, Rings, and Hydrographic Fronts 5
    1.2.1. Meandering Currents 5
    1.2.2. Eddies 5
    1.2.3. Rings 9
    1.2.4. Hydrographic Fronts 10
1.3. Influence of Eddies and Fronts on Fishery and Weather 11
1.4. Major Current Systems in the World Oceans 12
    1.4.1. Antarctic Circumpolar Current 13
    1.4.2. Western Boundary Currents in the Atlantic Ocean 15
    1.4.3. Western Boundary Current in the Pacific Ocean: The Kuroshio Current 20
    1.4.4. Western Boundary Currents in the Indian Ocean 22
    1.4.5. Equatorial Undercurrents 24
1.5. Currents of Different Origins 29
    1.5.1. Wind-Driven Current 31
    1.5.2. Inertia Current 33
    1.5.3. Tidal Currents in Open Seas, Estuaries, and Ridge Valleys 34
    1.5.4. Rip Currents 36
1.6. Implications of Ocean Currents 40
References 42

2. The History of Measuring Ocean Currents 51

2.1. Surface Current Measurements 52
    2.1.1. Measurements Based on Motion of Drifting Surface Bodies 52
    2.1.2. Imaging of Surface Water Motion Trajectories and Patterns 57

    2.1.3. Vector Mapping Based on Current-Driven Sea Surface Wave Transport 62
2.2. Subsurface and Abyssal Current Measurements 64
    2.2.1. Early Mariners' Contributions 64
    2.2.2. Spatially Integrated Measurements Based on Earth's Magnetism and Oceanic Sound Speed 65
    2.2.3. Measurements Based on Motion of Drifting Subsurface Floats 69
    2.2.4. Measurements from Fixed Locations at Predetermined Depths 70
2.3. Seafloor Boundary Layer Current Measurements 82
    2.3.1. Mechanical Devices 83
    2.3.2. Nonmechanical Devices 83
2.4. Vertical Profiling of Horizontal Currents 86
References 88
Bibliography 91

3. Lagrangian-Style Surface Current Measurements Through Tracking of Surface Drifters 93

3.1. Radio Buoys 94
    3.1.1. Drifter-Following Radar Transponder 95
    3.1.2. Drifter-Borne Doppler Transponder 96
    3.1.3. Radio Buoys Tracked by Polar-Orbiting Satellites 97
    3.1.4. GPS-Tracked Drifters 100
    3.1.5. Telephonically Tracked Drifters 104
3.2. Limitations of Surface Drifters 105
References 105
Bibliography 107

4. Remote Mapping of Sea Surface Currents Using HF Doppler Radar Networks 109

4.1. Crombie's Discovery 110
4.2. Peculiarities of Pulse Doppler-Radar Echo Spectra 112

4.3. Estimation of Sea Surface Current Using
     the Bragg Resonance Principle   113
4.4. Depth Extent of Doppler Radar-Based
     Sea Surface Current Measurements   113
4.5. Technological Aspects of Doppler
     Radar-Based Surface Current Mapping   114
4.6. Experimental Developments   115
4.7. Instrumentation Aspects   118
4.8. Radial and Total Vector Currents   120
4.9. Developments on Operational Scales   120
     4.9.1. CODAR   121
     4.9.2. OSCR   121
     4.9.3. SeaSonde   123
     4.9.4. WERA   124
     4.9.5. Systems for Special Applications 126
4.10. Intercomparison Considerations   126
4.11. Advantages of Radio-Wave Doppler
      Radar Measurements   129
4.12. Round-the-Clock Coast- and Shelf-
      Observing Role of Doppler Radar   131
4.13. Detection and Monitoring of Tsunami-
      Induced Sea Surface-Current Jets at
      Continental Shelves   132
References   133
Bibliography   135

5. Imaging of Seawater Motion
   Signatures Using Remote
   Sensors   139
5.1. Aerial Photography in the Visible and
     Infrared Bands   140
5.2. Remote Detection by Radiometers   141
     5.2.1. Passive Radiometry in the
            Visible-Wavelength Band   142
     5.2.2. Active Radiometry in the
            Visible-Wavelength Band   145
     5.2.3. Passive Radiometry in the
            Thermal Infrared Band   146
     5.2.4. Microwave Radiometers   149
5.3. Active Microwave Radar Imaging of Sea
     Surface Current Signatures   153
5.4. Active Microwave Radar Imaging
     Technologies   154
     5.4.1. Active Microwave Imaging by the
            RAR Systems   155
     5.4.2. Active Microwave Imaging by
            SAR Systems   158
5.5. Advances in the Development of SAR
     Technology   166
     5.5.1. Interpretation of Image Data   167
5.6. Detection of Seawater Circulation
     Features Using RAR and SAR   168

5.7. Measurement of Sea Surface Currents
     Using Imaging of Ice Floes   170
References   171
Bibliography   174

6. Lagrangian-Style Subsurface Current
   Measurements Through Tracking of
   Subsurface Drifters   177
6.1. Surface-Trackable Subsurface Drifters 177
6.2. Satellite-Recovered Pop-Up Drifters   178
6.3. Swallow Floats Tracked by Ship-Borne
     Hydrophones   180
6.4. Subsurface Floats Transmitting to
     Moored Acoustic Receivers   183
6.5. Subsurface Floats Listening to
     Moored Acoustic Sources   188
6.6. ALACE: Horizontally Displaced and
     Vertically Cycling Subsurface Float   192
6.7. Drifting Profiling Floats (Argo Floats) 195
     6.7.1. Profiling Observations from
            Polar Regions   197
References   198
Bibliography   199

7. Horizontally Integrated Remote
   Measurements of Ocean Currents
   Using Acoustic Tomography
   Techniques   201
7.1. One-Way Tomography   203
7.2. Two-Way Tomography (Reciprocal
     Tomography)   210
7.3. Acoustic Tomographic Measurements
     from Straits   212
7.4. Coastal Acoustic Tomography   215
7.5. River Acoustic Tomography   226
7.6. Acoustic Tomographic Measurements of
     Vorticity   230
7.7. Horizontally Integrated Current
     Measurements Using Space-Time
     Acoustic Scintillation Analysis Technique 232
References   235
Bibliography   237

8. Eulerian-Style Measurements
   Incorporating Mechanical Sensors   241
8.1. Eulerian-Style Measurements   242
8.2. Savonius Rotor Current Meters   242
8.3. Savonius Rotor and Miniature Vane
     Vector-Averaging Current Meters   249

8.4. Propeller Rotor Current Meter (Plessey Current Meter) 250
8.5. Biaxial Dual Orthogonal Propeller Vector-Measuring Current Meters 253
8.6. Calibration of Current Meters 255
8.7. Graphical Methods of Displaying Ocean Current Measurements 258
8.8. Advantages and Limitations of Mechanical Sensors 263
References 263
Bibliography 264

9. Eulerian-Style Measurements Incorporating Nonmechanical Sensors 267
9.1. Electromagnetic Sensors 267
9.1.1. Electromagnetic Current Meters 269
9.2. Acoustic Travel-Time Difference and Phase Difference Sensors 281
9.3. Acoustic Doppler Current Meter 291
References 294
Bibliography 296

10. Vertical Profiling of Horizontal Currents Using Freely Sinking and Rising Probes 297
10.1. Importance of Vertical Profile Measurements of Ocean Currents 297
10.2. Technologies Used for Vertical Profile Measurement of Ocean Currents 300
10.2.1. Freely Sinking and Wire-Guided Relative Velocity Probes 300
10.2.2. Bottom-Mounted, Winch-Controlled Vertical Automatic Profiling Systems 302
10.2.3. Acoustically Tracked Freely Sinking Pingers 305
10.2.4. A Freely Sinking and Rising Relative Velocity Probe (Cyclesonde) 308
10.2.5. A Freely Falling Electromagnetic Velocity Profiler 311
10.2.6. Free-Falling, Acoustically Tracked Absolute Velocity Profiler (Pegasus) 318
10.2.7. Freely Falling, Acoustically Self-Positioning Dropsonde (White Horse) 320
10.2.8. Freely Rising Acoustically Tracked Expendable Probes (Popup) 322
10.3. Technologies Used for Vertical Profile Measurements of Oceanic Current Shear and Fine Structure 325

10.3.1. Free-Fall Shear Profiler (Yvette) 326
10.3.2. Acoustically Tracked Free-Fall Current Velocity and CTD Profiler (TOPS) 326
10.4. Technologies Used for Vertical Profile Measurements of Oceanic Current Shear and Microstructure 329
10.4.1. Towed Acoustic Transducer 329
10.4.2. Free-Fall Probe (PROTAS) 330
10.4.3. Free-Falling Lift-Force Sensitive Probes 332
10.5. Merits and Limitations of Freely Sinking/Rising Unguided Probes 335
References 335
Bibliography 337

11. Vertical Profiling of Currents Using Acoustic Doppler Current Profilers 339
11.1. Basic Assumptions and Operational Issues 340
11.2. Principle of Operation 341
11.3. Profiling Geometries 344
11.3.1. Bottom-Mounted, Upward-Facing ADPs 344
11.4. Trawl-Resistant ADP Bottom Mounts 353
11.5. Horizontal-Facing ADPs 354
11.6. Subsurface Moored ADPs 358
11.7. Downward-Facing Shipboard ADPs 360
11.8. Towed ADPs 363
11.9. Lowered ADCP (L-ADCP) 367
11.10. ADPs for Current Profiling and AUV Navigation 368
11.10.1. AUV-Mounted ADPs for Current Profiling 368
11.10.2. ADPs for AUV Navigation 369
11.11. Calibration of ADPs 371
11.12. Intercomparison and Evaluation 372
11.13. Merits and Limitations of ADPs 374
References 375
Bibliography 378

12. Remote Measurements of Ocean Currents Using Satellite-Borne Altimeters 381
12.1. Oceanic Currents and Associated Features Generated by Sea Surface Slope 382
12.2. Determination of Seawater Motion from Sea Surface Slope Measurements 385

12.3. Technological Intricacies in Realizing Satellite Altimetric Measurements   386

12.4. Correction of Errors in Satellite Altimeter Data   389

    12.4.1. Correction of Satellite Orbit Errors   389

    12.4.2. Correction of Geoid Errors   390

    12.4.3. Null Methods for Obtaining Topographic Height Variability Independent of Geoid   391

12.5. Evolution of Satellite Altimetry   392

References   394

Bibliography   395

13. Conclusions   397

13.1. Progress in Ocean Current Measurement Technologies   397

13.2. Moored Current Meters and Their Limitations   397

13.3. Lagrangian Measurements of Surface Currents   399

13.4. Global Observation of Sea Surface Currents and Their Signatures through Imagery   401

13.5. Real-Time Two-Dimensional Mapping of Sea Surface Current Vectors   402

13.6. Global Observation of Surface Geostrophic Currents and Mesoscale Circulation Features   405

13.7. Current Profile Measurements Using Freely-Moving Sensor Packages and ADPs   406

13.8. Evolution of Acoustic Tomography: Monitoring Water Flow Structure from Open Ocean, Coastal Waters, and Rivers   409

13.9. Lagrangian Measurements of Subsurface Currents   410

13.10. Comprehensive Study of Oceanic Circulation   415

References   416

Index   419

I first thought of writing this book back in 1982, when I was gaining working experience in Eulerian-style current measurement technologies at the Marine Instrumentation Division of CSIR—National Institute of Oceanography, Goa, India. My boss, Dr. Ehrlich Desa, supported me in this endeavor. A manuscript I prepared in 1985 was reviewed by Dr. Narayana Swamy and Dr. Desa, who encouraged me to expand the scope and contents. My well-wisher, the late Dr. Rabin Sen Gupta (a chemical oceanographer by profession), frequently used to inspire me to quickly complete the writing and get it published by Elsevier. Elsevier showed interest in this project and even sent me a reminder.

However, during this time my attention was diverted to sea-level measurement technologies, the importance of which became so obvious in the aftermath of the unfortunate December 2004 Sumatra tsunami event; therefore, destiny had it that I should first publish a book on tsunamis. Thus, my first book was published in February 2011 by Elsevier, New York.

By the time I again put my hand to the task of completing the unfinished work on current measurement technologies in March 2011, the subject had blossomed to a much bigger, and so to say unwieldy, branch of oceanography, having fundamentally and profoundly changed our capability to see the ocean in motion. After much effort I managed to categorize the whole subject matter into different disciplines, put them together as a full manuscript, and submitted to Elsevier.

Dr. Albert J. Williams 3rd (Scientist Emeritus, Woods Hole Oceanographic Institution, USA) helped me a good deal in finding other experts from which to seek suggestions. Several experts in various branches of oceanic current measurement technologies provided comments and suggestions related to the content and organization, enabling me to include all the important aspects of these technologies and properly organize them into the present book. Their suggestions in finalizing the title and subtitle of the book were also valuable.

It is a pleasure to acknowledge the contributions of several individuals in the preparation of this work and to thank them for their support. Included in this list are Dr. Brian K. Haus, Associate Professor and Chair, Rosenstiel School of Marine & Atmospheric Sciences, University of Miami; Dr. Arata Kaneko, Professor, Graduate School of Engineering, Hiroshima University; Dr. Peter Spain, Scientist, Teledyne RD Instruments and IEEE Current Meter Technology Committee; and Dr. Mal Heron, IEEE. I thank Mrs. Surekha Nagvekar for helping me in the voluminous literature search. Mr. V. Githin and Mr. H. Bharat provided support in the preparation and modification of several diagrams. Apart from providing valuable suggestions, Dr. Albert J. Williams 3rd donated some beautiful pictures from his own archives. He also kindly agreed to write the foreword for this book. Dr. Philip Woodworth and Mr. Peter Foden, National Oceanography Centre, United Kingdom, and several scientists and technologists of the National Institute of Oceanography, Goa, India; Dr. T. Pankajakshan, Dr. S. Prasanna Kumar, Dr. M.R. Ramesh Kumar, Dr. P. Vethamony, Dr. M. T. Babu, Dr. V. Sanil Kumar, and Mr. V. Fernando supported me in several ways in this endeavor. I am profoundly grateful to all of them.

Once again, I express my gratitude to the late Dr. Rabin Sen Gupta, who stimulated my interest in writing a comprehensive book on measuring ocean currents.

Antony Joseph

Antony Joseph

In this volume, Antony Joseph has comprehensively examined the history and practice of current measurements, including the mean current, wavelike motions, and turbulent velocities that constitute flow in the ocean. This text should be interesting to all oceanographers who have a need to understand ocean currents, flow in waves, and the measurement of turbulence. Joseph's treatment is accessible to anyone with a scientific or engineering background, but it will be especially useful to one who needs to know exactly what a prospective technology might deliver. If you want to know where the water goes, how it goes, or how to measure how fast it goes, this book will help you. It covers the field comprehensively from about 1910 to about 2010, with ample references to original work in nearly every case. As such it can serve as a reference as well as a tutorial.

Flow of water and its measurement are ancient and honorable concepts with which to be concerned. I myself built a boat with a sail as a child and thus became concerned with fluid flow and behavior of water as it supported the boat but also resisted the lateral force of the wind on the sail. Current was not at the forefront of my thinking until another homebuilt sailboat took a companion and me through Barnegat Inlet in New Jersey and current prevented us from landing on the shore until we had passed through to the bay inside. Lift and drag on the sail and hull presented aspects of fluids that interested me. But it was my introduction to oceanography after graduate work in upper atmospheric physics that again took me back to water and its movement. In fact, the movement was mixing, flow at the millimeter scale and molecular diffusion of heat and salt responsible for *salt fingers*, a process that permits stable stratification of the ocean with warm water above cooler water to lower its center of gravity through the differential diffusivity of heat and salt. The term *current* would hardly describe the centimeter-scale flow of cells of warm, salty water descending adjacent to cooler, ascending columns of fresher water. My shadowgraph images showed the region above the narrow horizontal fingering-interface to exhibit turbulent plumes of fresher water above and saltier water beneath. And it was this turbulent flow requiring measurement that brought me into the world of current measurement in 1973.

Flow consists of current, waves, and turbulence. Though my own odyssey started with attempts to measure turbulence, I subsequently directed my efforts to measurement of current and eventually to waves. Turbulence is zero-mean and dissipative, at least in boundary layers. Physical oceanographers may dismiss it as not really contributing to the global circulation budget, but it is responsible for all of the mixing, except for salt fingers, at the final diffusive scale. Waves are also zero-mean (except for a small Stokes drift contribution to current), but they transport energy and carry stress from wind at the wave-generating region to erosive processes and possibly destruction of property at great distances. What is left over when these zero-mean processes are subtracted or averaged is current, and current carries heat and material great distances, dispersing across ocean basins what may have been mixed by turbulent diffusion upstream.

As Antony Joseph explores topics in this volume, I too have touched on some and can report on them from my own experience. As I sought a technical solution to a measurement problem that was impeding understanding of an oceanographic process, I tried many of the solutions that the author explains and, as he has done, I found some less suitable for my purposes than those I finally selected. I have probably learned the most from blind alleys or solutions that didn't work, and my experiences may be instructive to others. So, in the spirit that these experiences may be useful, I will describe my career in current measurement.

In the laboratory where I was a postdoctoral investigator at the Woods Hole Oceanographic Institution (WHOI), there was a project to develop a new version of the Savonius rotor and vane internally recording moored current meter that became the VACM, for vector averaging current meter. It used digital electronics, the new RCA CMOS circuits, with a custom lookup table to accumulate each eighth turn of the rotor into north and east registers from the sine and cosine of the difference in angle between the vane and the internal compass. Watching this and the difficulties with bearings, magnetic damping of the compass, calibrations in a tow tank, and development of a suitable tape recorder for the data, I vowed that I would never become involved in the measurement of current, although I appreciated the instruction in digital electronics the exposure afforded me.

Not long after, I needed to measure the flow in the mixed layers adjacent to the fingering interface that I was obtaining in shadowgraph images, but clearly the Savonius rotor wasn't going to do it. A small and very sensitive flow meter was required, and in my thinking I considered laser Doppler, electromagnetic, drag-force spheres, heated thermistor, heated thin films, and acoustic differential travel-time methods. Fortuitously, a visitor from Norway traded me his design for an acoustic travel-time velocimeter for my help in duplicating the Neil Brown CTD that I was building to obtain temperature and salinity measurements on my salt-finger detector microstructure probe. Trygve Gytre was providing his acoustic shear probe to Thomas Rossby, who in turn needed the CTD, and we all benefited by the cooperative sharing of technology. Although the salt-finger detector, an optical system with film recording, was lost at sea soon after discovering salt fingers in the Mediterranean Outflow, the shear probe illustrated nicely the critical shear at density interfaces associated with Richardson-number instabilities that were potentially causing mixing through rolls on the sheared interface.

Having listened to seminars from physical oceanographers visiting WHOI and hearing their difficulties with mixing problems from a turbulence perspective, I vowed to perhaps measure current but never turbulence. Yet not long after that a student, John Tochko, asked me if my shear probe might be used to measure deep-sea currents and Reynolds stress. Reynolds stress, as Joseph explains in his boundary layer chapter, is the turbulent exchange of momentum from a flow above the bottom with the slower-moving fluid closer to the boundary. Small-scale, rapid, and sensitive 3D velocity measurements would be needed. I thought we might manage to do that with the acoustic velocity probe and by a serendipitous mistake, confusing fathoms with meters on the ship's echo sounder, managed to put the shear probe on the bottom and make such a current measurement. Reynolds stress awaited John Tochko's three-dimensional revision of the shear probe into BASS, the benthic acoustic stress sensor.

Antony Joseph describes nicely the benthic boundary layer investigations that were undertaken in the 1980s. I will summarize my own contributions here. My instrument, BASS, was a tower of six 3D velocimeters covering the benthic boundary layer from 35 cm to 5 m above the bottom with a logarithmic spacing. Indeed, over flat topography at 4,800 m depth on the lower rise off New England, the Reynolds stress was uniform over several meters most of the time, a true inertial sublayer without acceleration. The benthic storms that suspended sediment as thickly as 100 m above the bottom as determined from transmissometers on moorings during the High Energy Benthic Boundary Layer Experiment (HEBBLE) occurred a dozen times a year, whereas high-stress episodes also occurred about 10 times a year, but not at the same times as the turbidity storms, causing me to think of the events observed at the HEBBLE site as principally advective, bringing material eroded from the bottom and suspended into the nepheloid layer through the agency of storms somewhere upstream. Having learned

as much as we could from this single HEBBLE site, most of us moved on to other sediment transport or bedform studies, but I was convinced we needed to learn more about upstream processes.

The prospect of capturing the actual erosional event and understanding the forcing that increased shear stress above the limited shear strength of the sediment seemed to require fluid stress sensors that might be deployed in an array extending upstream as much as 100 km. With that diameter and with a sensor every 10 km, about 100 sensors would be required. This became an economic as well as a technical issue, since the unit cost per sensor had to be modest, comparable at least to the ship time cost of delivery over a 10-km interval between launches—say, an hour of ship time. The only way to reduce the cost was to automate the construction of the BASS type of sensor, and the National Science Foundation (NSF) awarded me a development grant for what became the modular acoustic velocity sensor (MAVS), a small and thus inexpensive but vector-resolving sensor, subsequently manufactured at first by General Oceanics and later by Nobska Development, Inc. As a sensor, the MAVS served an imagined need for adding current measurement capability to a bottom lander with its own power and data-logging capability, much like a SeaBird temperature or conductivity sensor. This presented a potential market that would support the mass production that could lower unit costs to the level that would permit me to study the origin of the benthic storms that eroded sediment to be later transported downstream. I had to learn things about business that were not in my scientific training, but I had learned much about current meters.

Almost no customer wanted a modular sensor, so a data logger and battery were immediately added to make MAVS a current meter. Shortly after adding the data logger, a compass and then a tilt meter were added so that MAVS became a self-contained current meter. Since that time, capability has been added to accommodate the needs of customers, whom I viewed as clients since this isn't a commodity but a custom instrument supporting the researcher's or monitor's specific needs. Precision temperature or temperature arrays, pressure, conductivity, optical transmission, fluorescence, pH, and other sensors have been added, making MAVS less a modular sensor than a complete benthic platform supporting other modular sensors. There have been times when my concept of measuring current almost seems to be an afterthought for my clients. But advances in capability for current measurement in MAVS have also continued. It is this progression in my own experience in current measurement and technology to fill a specific need that has interested me in Joseph's comprehensive consideration of the history of current measurement technology, including the development of understanding the current systems of the world. He also addresses the processes mediated by currents that affect the oceans and the techniques that have been applied to their understanding.

Although the experiences leading to MAVS form a synopsis of my career in instrumentation, I watched and participated in experiments involving current measurement, drew conclusions, and learned lessons from my own and others' misadventures as well as successes. I will relate some of these and the lessons I learned.

Deep-sea current measurement is easy in contrast to shelf measurements in the presence of waves. But waves are flow and can be measured with current meters too. In the Coastal Ocean Dynamics Experiment (CODE), we discovered from BASS tripods and other measurements that at depths of 90 m on the California shelf, the ubiquitous 17 s Pacific swell was responsible for suspension and reworking of sea-floor sediment. It was the combination of waves and current that did the sediment transport at that depth.

At that depth waves were important, but in shallow water waves are much more important. Measurements with either three axes of current or two axes of current plus pressure can produce directional wave spectra plots that are informative for beach processes. Getting such data in real time may even help shorefront property owners make decisions about evacuation or armoring their beach. By running a cable ashore from a MAVS jetted into the bottom off a beach in Nantucket Island off Massachusetts, we were able to understand sand movement during a winter storm affecting a small community. Jetting provided a stable platform from which to make these measurements. Furthermore, providing a rigid support to make such measurements has been important and surprisingly difficult.

Directional wave spectra were reported in real time to coordinate construction operations at a ferry terminal in Martha's Vineyard near Woods Hole. Although the Nantucket measurements were made from MAVS on a jetted pipe, the MAVS at the ferry terminal in Martha's Vineyard was strapped to a piling on the pier. Divers had to move the attachment to a new piling as old pilings (one of which supported MAVS) were removed. Cabling to an Internet-connected computer on the dock allows construction crews and even ships' captains to assess wave conditions at the dock.

The design of the Martha's Vineyard ferry itself benefited from directional wave spectra, measured where the ferry to Martha's Vineyard turns on its run from Woods Hole, since there were no buoy observations in Vineyard Sound to inform naval architecture ferry design near the turning location. The directional wave spectra for this project were obtained from a small tripod with an internally logging MAVS weighted heavily to stay on the bottom. It was to have had support from an ADCP to extend the velocity profile to the surface that was mounted on one tripod foot, but this instrument had to be removed at the last minute, unbalancing the tripod, which rolled over shortly after launch. Such field experiences are instructive. It turned out that even after the tripod rolled over, good

directional wave spectra were obtained, since the pressure sensor gave sufficient information to compute the spectrum from the vector velocities.

Obtaining wave measurements from a rigid mounting (even a rolled-over tripod) are easy compared to those beneath a surface buoy, yet that is where current meters are often mounted, either attached to the surface buoy or on the mooring line beneath the buoy. This totally violates the goal of a rigid mounting but nevertheless must be accommodated. Vector averaging flow measurements to obtain current is further complicated by the sensors that determine attitude, which are susceptible to errors from wave accelerations. Gimbaled magneto-inductive elements swing when subjected to wave accelerations, and strapped-down three-axis magneto-inductive sensors leveled by solid-state two-axis linear accelerometers for tilt sensing are also affected by wave accelerations. Most recently, inertial sensors combining three-axis rate gyros, three-axis magnetic sensing, and three-axis linear accelerometers have become available at a reasonable price. These can provide heading, pitch, and roll at a frequency of at least 5 Hz for rotating instrument frame current measurements into Earth frame coordinates. The remaining motion of the current meter that needs to be subtracted from the measured flow can be computed from the integral of linear acceleration, although this remains a research development problem.

Here I want to interject a few words about the community of those of us who measure current, use measurements of current, manufacture current meters, and sell current meters or current measurements. Under sponsorship of the IEEE Oceanic Engineering Society (OES), a technology committee, originally named the Current Measurement Technology Committee, holds workshops about every four years. The value of our techniques for measurement of waves and turbulence caused us to rename ourselves the Current, Waves, and Turbulence Measurement Technology Committee, and CWTM workshops have been held most recently in Charleston, South Carolina, and Monterey, California, and will be held in 2015 in St. Petersburg, Florida. As Antony Joseph explains and describes in this volume, many techniques for observing water movement and waves and precisely measuring these and turbulence are presented and discussed at these workshops. At some workshops the problems with existing current meters were the theme, at others new technologies were stressed. Every year, and more recently twice every year, an Oceans conference is sponsored by the Marine Technology Society (MTS) and the OES, at which there is generally a session or two at which current measurement talks are presented. HF Radar, also described by Antony Joseph, is often presented at Oceans conferences, since it is considered part of ocean remote sensing as well as current, wave, and turbulence measurement, all of which are topics in Oceans conferences. Less is presented about physical oceanographic topics in currents at Oceanic Engineering events, as presented by Antony Joseph in his first chapter, yet it has been a principal driver of measurement of ocean currents and sensors. Though I have learned something from physical oceanographers at Oceans conferences, it is really at WHOI where I have learned the most about physical oceanography from talks and from cruises on which I have been present.

Western boundary currents like the Gulf Stream, Kuroshio, and Agulhas were first observed by ship drift but later by moorings, motivating the development of the VACM for capturing variability in the Gulf Stream. An interesting experience that this instrument endured in the Agulhas (I learned from talks) was the nearly total destruction of its electronics from Strouhal oscillation of the mooring cable in the very high-velocity Agulhas current. This experience motivated more secure construction of the chassis and even the circuit boards as well as some efforts to reduce the Strouhal excitation. I learned that one possible approach to reducing Strouhal oscillation could be to wrap lines around the current meter housings in a spiral to force detachment of the vortices at defined points along the housing and thus reduce the forcing. There was even some consideration given to use of faired mooring lines, including Science Applications International Corporation's Quiet Cable, which, though developed to support hydrophones, had a sharp extruded plastic shape along the cable. The total drag is purportedly reduced by suppression of the Strouhal oscillation. I found attractive the possibility that rather than building great strength into the moorings for current meters, a lightweight solution might achieve as much with a beer-can-sized current meter and 3,000 m of 100 Kg test-strength Quiet Cable.

The discovery of rings and eddies has been well described by Antony Joseph. I learned of rings shortly after my arrival at WHOI. On occasion, a warm ring from the Gulf Stream was observed at moorings north of the mean location of the Stream. The strong current dragged buoys under, moved anchors, and sometimes caused moorings to go missing. But there were important processes going on in these rings. Inertial and near-inertial oscillations in these rings had been estimated to increase internal shear sufficiently to cause shear instabilities and consequent mixing. WRINCLE, the Warm Ring Inertial Critical Layer Experiment of 1990, investigated these layers with XCTDs (expendable CTD probes), XBT (expendable bathy thermograph) drops, my RiNo (Richardson number) float, and numerous CTD lowerings to obtain density profiles and current shear at 700 m depth. RiNo was essentially a BASS tripod covering 5 m vertically without a weighted part that measured velocity at six locations and density at the ends to characterize Richardson number, a measure of resistance to shear instabilities. It was neutrally buoyant and acoustically

commanded to occupy an intermediate depth (700 m in WRINCLE) until recalled for data recovery after a week. In the ring, it was tracked with a low-frequency beacon so that it could be found after this duration. I enjoyed my collaboration with Ray Schmitt and Eric Kunze in these measurements. Rings became recognized as important structures for bioproductivity as well as strong current anomalies in the western ocean regions.

Diapycnal mixing, the turbulent or possibly diffusive mixing between water masses separated by a horizontal interface or gradient, called for both integrative measurements with tracer releases and subsequent sampling and direct measurements of shear velocity. The RiNo float mentioned with respect to the WRINCLE project was designed to measure shear over a 5-m vertical span and simultaneously measure the temperature gradient over that range and the density difference between the two ends. RiNo was neutrally buoyant and could be acoustically commanded to float stably at a selected depth for periods as long as a year (although 10 days was the longest deployment that was ever recovered). The ratio of the density gradient to the velocity gradient squared, the Richardson (Ri) number, is a measure of stability to overturning, where $Ri = 1/4$ is the critical value below which finite perturbations grow into overturning events that are moderately ($\sim 30\%$) efficient at mixing fluid. The expected scale of this mixing is order 1 m, and the intermittency even with $Ri = 1/4$ is low, perhaps $1-5\%$ of the time. The RiNo float was deployed in WRINCLE and allowed to drift around the warm ring for about half a revolution of the ring and then recovered to obtain the data, later reported by Eric Kunze.

Two RiNo floats were deployed subsequently in the North Atlantic Tracer Release Experiment (NATRE) in 1991, west of the Cape Verde islands near Seiberling Seamount. Although one float was attached to a bobber float (a variable ballast CTD-measuring Swallow float, acoustically tracked through the SOFAR channel), this float became silent after seven months and was never recovered. The other was lost during ballasting operations, showing that not every technically challenging operation is successful. Observations with RiNo and from other probes capable of measuring Richardson number have shown that in regions where diapycnal mixing seems to be occurring, $Ri = 1/4$ is frequently observed and hovers near that value or slightly above much of the time. This gives the idea that much of the thermocline is marginally stable and susceptible to some trigger to cause active mixing, something as small as an internal wave superimposed on a steadier shear from inertial oscillations.

As a drifter near the surface, RiNo offered a glimpse of the forces affecting surface drifters since the velocity at depths spanning 5 m were logged as well as the surface float position being tracked. In an experiment Rocky Geyer performed in 1989 at WHOI, the behaviors of several surface drogues were compared to RiNo. Although it had been assumed that the surface layer in Buzzards Bay near Woods Hole was well mixed in a moderately high wind event in April, before surface heating was significant, RiNo showed a shear layer at several meters' depth that had a strong effect on the drogues, depending on the type of subsurface drag (rope or window shade) and the shape of the buoy (spherical, half sphere, or cylindrical) exposed to the wind. These instrumental validations were important since Lagrangian drift measurements have been used to discover currents acting over great distances in remote parts of the ocean. More can be read about these surface drift measurements in Antony Joseph's chapter on that topic.

In his chapter on Lagrangian-style subsurface current measurements through tracking of subsurface drifters, Joseph amply describes and explains the techniques used to study deep currents. Subsurface drifters are certainly better current-measuring probes than surface drifters and can even provide real-time tracks from SOFAR acoustic transmissions, as demonstrated by the floats employed by Douglas Webb and Thomas Rossby. If data in real time are not required, the SOFAR system that uses sound transmitted by the float and received at fixed deep listening stations can be inverted. This inverted RAFOS system developed by Rossby uses sound transmitted from moored sound sources to permit floats to listen and track themselves from the arrival times of the pulses from three or more beacons. Having done work with both Webb (I was assigned the job of repolling one of the original low-frequency SOFAR transducers) and Rossby (I participated in the cruise where his shear probe received its sea trial), I followed these developments with interest.

The arrival times logged over the deployment life of the float are relayed by satellite after several months of Swallow-float drifting when a recovery weight is dropped and the float surfaces to tell its story before becoming another dead floating glass object on the surface or sinking. I found this mode of recovering data from an instrument that might be expendable economically compelling, since ship time is so expensive. In fact, I attempted to use this strategy for an array of MAVS expendable benthic landers, XBLs, to study benthic weather. This proposal was not funded, however. There is a time to study benthic processes, and there is a time to focus on other scientific puzzles. The time to investigate benthic weather has passed.

The highly successful ARGO array (Argo is named for Jason's ship from Greek mythology) has floats that drift subsurface for about a week and then pump ballast to the surface and give a satellite fix and transmit temperature, and often salinity and pH or other measured properties, from its excursion to the surface before pumping ballast to return to the sampling layer. More than 3,000 of these floats in nearly every part of the world ocean now are providing a global picture of currents, eddies, gyres, and the water masses that are associated with these current structures. As Joseph tells in his discussion of subsurface floats, the time on the surface where the float is not at its target depth causes its drift to deviate from that desired and must be corrected with surface current estimates. The degree of autonomy in this program is most impressive and takes us to the new generation of ocean large-scale observations.

In considering large-scale ocean currents, nothing is as large scale as the tides, and these are important even on the seafloor, where they can give benthic larvae a ride to a new home. By chance I had two roles to play in such tidal flow measurements at the Endeavor site of the Juan de Fuca hydrothermal vent. Twenty MAVS current meters, some with temperature arrays to monitor flow and thermal gradient in the bottom meter, were deployed for a year by the University of Washington, and then several years later WHOI deployed three MAVS instruments to measure tidal flow. This was at a site called Easter Island and determined that heat apparently modulated by the tide was in fact only advected by the tide from steady flow through the diffuse vents there. Tidal flow, rectified along the axis of the valley, is capable of dispersing larvae and may therefore be the mechanism for colonization of new vents. This is another example of the utility of current measurements in inaccessible regions. But this had not been my first exposure to a hydrothermal vent.

In 1979 Fred Grassle at WHOI invited me to bring my prototype BASS acoustic velocity sensor on a dive by the deep submersible, Alvin, to the newly discovered hydrothermal vent, Garden of Eden near the Galapagos, where a previous attempt to measure flow with a propeller had suffered from crab infestation. With low velocities and probability of fouling by crabs, the BASS current probe measured a vertical flow of 12 cm/s and a temperature anomaly of about 8°C above ambient temperature.

At the surface, Lagrangian trajectories of drifters tracked by Argos satellite (Argos is Advanced Research and Global Observation Satellite) and processed by CLS Service Argos have given us low-cost information about eddies and gyres and, almost as an afterthought, data from the transmitters on the surface. The identification number that permits several or many PTTs (the surface transmitters) to be distinguished from one another also allows data to be transmitted about temperature, conductivity, acoustic arrival time delays, and recent profiles of these quantities so that the ARGO floats spending most of their time submerged can send not only position but profile data to the user. As Joseph explains, the certainty of a single Argos transmission being received and then captured without error is low so that repeated transmissions are required to ensure essentially total data transfer. The RAFOS floats that remain submerged for periods as long as half a year and then surface to transmit all the acoustic arrival times of beacon transmission during their submerged deployment continue to transmit the full data log for several weeks until their battery is exhausted, with about 80% of the log eventually recovered, much of it in duplicate. This capability has opened up a great window on ocean-scale processes. Service Argos also offers a relocation service for instruments that have popped to the surface unexpectedly. I have subscribed to this service, but the few times when it might have allowed me to recover a lost instrument, it has not worked, probably because there is a low probability of detecting a transmission when it first occurs. However, its utility for recovering data from an expendable instrument is excellent.

As Joseph describes, many different techniques have been used to measure current and, more particularly, turbulent flow in a boundary layer. My own contribution to this technology has been acoustic travel-time techniques (ATT), but my admiration for Doppler techniques is great; I particularly envy their freedom from flow distortion by structure of the sensor. Certain problems in making measurements from a moving surface buoy plague both techniques: motion of the current sensor. As mentioned previously, this affects both knowledge of in which direction the measurement of flow velocity has been made and the uncertainty of the velocity of the sensor itself. The new attitude sensors that have become available are inexpensive in power, volume, and cost and will be utilized in the next generation of current meters mounted on mooring or surface buoys exposed to surface waves. As configured, the inertial measurement units (IMUs) are capable of removing the rotations of the current meter housings, whereas the attitude and heading reference sensor (AHRS) units can permit direct rotations of instrument frame velocity measurements to Earth frame coordinate velocities. The second problem, determining current meter velocities in waves, remains unsolved, although using linear accelerations is clearly a path to its solution.

Measurements of current from moving vehicles are becoming more important, with gliders and AUVs observing water properties and flow. Ships have long obtained current measurements with ADCPs and can remove the ship motion with GPS measurements of ship velocities in deep water and with bottom-tracked Doppler-shifted returns in shallow water. But when the vehicle is submerged as a glider or AUV, the GPS signal is not available and in deep water a bottom-tracked Doppler signal is not available either. A Doppler profile of velocity beneath a descending vehicle has been used to extend surface measurements of velocity where GPS is available to greater depths by a bootstrap extension downward. This remains a problem, however, and measurements of velocity on a submerged moving vehicle are still tricky. Of course, the AUV doing a survey may have navigational transponders on the bottom that give precise AUV velocities from which profiles with an ADCP can be extended. One of the more difficult AUV navigational tasks is from the underside of ice, where "bottom tracking" with an upward-looking ADCP might be hoped to provide precise AUV velocities, but the ice is not as reliable a backscatter target as seafloor generally is. And whenever the reference surface is very rough, as the underside of ice may be, and hydrothermal vent chimneys are the bottom track, it is difficult to maintain. My own experiences with these problems are hearsay from my colleagues at WHOI and the National Oceanography Center (NOC) in Southampton, where the navigation problems are taken very seriously and the current profiles are almost an afterthought.

There are practical issues with current meters, such as power, sensitivity, sampling rate, calibration, linearity, and freedom from flow disturbance. The last is partly dependent on the measurement technique so that, for example, a Doppler sensor is less influenced by the structure of the sensor than a mechanical or acoustic travel-time sensor. But even when the distortion of flow by the structure of the sensor is accounted for in the design of the structure, there is often an issue of bio-fouling, and that is quite serious and difficult to prevent. My own earlier applications of the BASS current measurements were in deep water, and bio-fouling was not experienced. But in shelf depths there were barnacle infestations that didn't affect the acoustic sound transmission but surely had some effect on the flow around the transducer supports. This condition became more severe in the near-surface region where sunlight permitted seaweed to grow as well as tunicates, so that in several months there was sufficient material growing on the transducer supports to obstruct the flow through the sensor. The acoustic signal was not affected unless seaweed with flotation bladders obstructed the acoustic path. Tri-butyl tin antifouling worked to some extent, and Desitin (zinc oxide) cream also delayed the accumulation of bio-fouling, but the solution was only found in 2010 with a silicone coating, two sources being ClearSignal and PropSpeed. These coatings did not kill organisms but rather prevented them from attaching to the substrate tightly, and the organisms washed off or could be removed with a light rubbing.

Power limits deployments with stored energy (batteries), so it is desirable to keep the power consumption low. Mechanical sensors like the Aanderaa RCM4 and the VACM derive the power for the sensor of fluid movement from the flow itself, and these have lower power consumption than Doppler or ATT sensors. But even in active flow sensors, the acoustic power is often less responsible for the power budget than the digital processor controlling the process. When I encountered the problem with deployment time limitations in MAVS from microprocessor power consumption, the short-term remedy was to switch to a controller that could be put into low-power sleep between measurement bursts and where the clock frequency of the microprocessor could be reduced between samples within a burst. This extended the available deployment duration from two weeks to a year or more. Recent applications are now cable-supported for observatories, which reduces the demand somewhat for low power. But other techniques to reduce power, if the need is great enough, have not yet been explored. BASS, for example, used the COSMAC microprocessor as a controller with significantly lower power than modern microprocessors, but this system did not offer the user the friendliness we now expect of current meters.

Sensitivity is rarely a serious limit except for turbulence studies or critical deep-sea flows where the flow is very slow and there is little turbulence to overcome hysteresis or *sticktion* in the sensor. But some of these conditions do require such sensitivity. In mechanical sensors there is residual bearing friction so that no motion occurs with flows below some threshold near 2 cm/s, depending on design and condition. Since direction of this unresolved flow is sometimes more important than the actual volume of transport, it is argued that applying this sticktion velocity threshold for no motion to the vane-indicated direction of flow still yields valuable information. Something similar can occur with acoustic Doppler sensors as well, particularly in very clear (no acoustic scatterers) water. The Doppler signal may be so weak that it cannot be distinguished from the side lobe scatter from mooring hardware, and this zero Doppler-shifted signal might be interpreted as no motion and accounted for in some way that is not representative of the actual flow. ATT sensors do not have hysteresis or sticktion issues since they respond linearly right through zero flow, but the determination of the zero-flow reading is a calibration requirement in the absence of this distinctive reading.

Calibration, linearity, and sampling rate are important considerations for any current meter. Calibration might mean electronic calibration against some standard. For the Savonius rotor and vane instruments, this means readout of turns of the rotor and angle of the vane. For an ATT instrument, it means readout versus nanosecond delay in a simulated acoustic path. For each of these, the model of behavior indicates that these calibrations are a good representation of the linearity of the instrument and the limitations of hysteresis, drift, and noise of the measurement. But a secondary calibration is necessary for each technology, at least every time a major change is made, and that is a tow through still water and other actual fluid calibrations. These will include the effects of flow disturbance and vortex shedding by the structures so that readout can be related to tow speed. To obtain more comprehensive calibration confidence, there should be variation in angle of the sensor during the tow, variations in towing speed, addition of oscillatory motion during the tow, and such possible effective environmental conditions as temperature, salinity, bubbles, small-scale turbulence, and depth or pressure.

Sample rate limits how high a frequency of flow variability one can accurately capture. To first approximation, this is the number of independent samples per unit of time. In some instruments, a single measurement taken at each sample interval is sufficient as an independent sample, whereas in other instruments the measurement rate may be much greater than the sample rate, and the measurements are averaged to obtain a lower noise sample at the sample rate. Acoustic Doppler current meters generally make many measurements or pings for each recorded sample, since the noise in a single ping is fairly large, but measurement frequency is limited only by the effective range, which in turn is a function of the acoustic frequency. By contrast, the ATT instruments achieve a lower noise from a single ping, so that is the sample rate limit. High sample rates use more power but permit tracking flow variations at a higher frequency.

These practical concerns for current meters affect what problems can be addressed. Turbulence requires high sampling rate and may require high sensitivity. Creeping flow in the deep sea requires sensitivity and linearity near zero velocity; all require calibration.

Current measurement technology develops at the intersection of oceanographic problems and technological capabilities in society at large. Modern HF radars, acoustic Doppler current meters and profilers, satellite altimeters, and SAR systems all depend on electronic and cybernetic innovations. Even such basic capabilities as titanium and ceramic materials for housings were not available formerly. Meanwhile, the demand for more detailed information about high-frequency variability at finer vertical separations has driven the need for profilers and real-time cabling of sensors to the user. Antony Joseph has assembled a comprehensive and scholarly story of the technologies and their applications in this volume, and I congratulate him.

A final thought in such a volume is, what is next in current measurements? The author has restricted himself to what has been done and reported, but I feel freer to project to what might be done. The surprises from communications, electronics, remote sensing, and data handling will continue to influence current measurement technology. As significant will be the influence of problems ready for observations to illuminate. Global circulation changes in response to global climate change are at the forefront. Diffusive mixing, interleaving of water masses, dispersion in horizontal flows, and surface forcing by heating, evaporation, wind stress, and freshwater runoff are all drivers and need technology to measure and track them. The scales are small for mixing processes and large for dispersive processes, whereas the circulation scales are global and require extension of the ARGO array to very high latitudes. Gliders and other autonomous vehicles, including the Waveglider for near-surface surveys, will become more important since they provide a multiplier in human effort and investment. But they stress data handling, as do the observatories where data continues to pour in in ever-increasing streams. Processes at the boundaries are still among the most important—not just the air-sea surface but the bottom (especially at vents) and the shore. These regions are all more difficult to measure than the interior and will require new technologies such as vorticity sensors and flux probes for heat, salt, CO_2, and nutrients. Biological observations in concert with physical measurements are critical for understanding many of the most important interactions, and each practitioner, both biologist and physical oceanographer, will need to ask for help from the other. The task in the broadest sense is to understand the Earth, at least the ocean part of the Earth and its atmosphere, sea floor, shores, and the water budget that results from storms, insolation, evaporation, and precipitation. Finally, the past is a key to the future, and understanding periods of freezing and heating in the geological record, paleooceanography, may enable us better to anticipate the future.

Albert J. Williams 3rd,
Scientist Emeritus
Woods Hole Oceanographic Institution
Woods Hole, Massachusetts
May 3, 2013

Ever since climate change began to be recognized as a serious threat to living comfortably on this planet, ocean-current and mean sea-level monitoring began to attract considerably more attention from climatologists and environmentalists across the globe. Although more than 71 percent of the Earth is covered by ocean, the myriad large- and fine-scale water motions taking place within the ocean are hardly known to many. Over the past few decades, however, scientific debate on the nature and possible causes underlying climate change and the role of ocean currents in this change has intensified.

Present knowledge of ocean currents in the world oceans has accumulated from measurements made through a variety of ingenious devices, including satellite-borne, seafloor-mounted, and vessel-mounted sensors as well as moored and freely drifting instrumentation deployed at various depths in different regions of the oceans. This book summarizes the methods used for surface, subsurface, and abyssal ocean-current measurement and highlights some of the important applications of such measurements in academic and operational interests.

The methods of ocean-current measurement discussed in this book are broadly categorized under Lagrangian measurement (i.e., finding the path of particular parcels of water masses as they traverse horizontally from place to place), Eulerian measurement (i.e., water-current measurement at geographically fixed positions), vertical profiling techniques (i.e., measurement of currents at different depths along specified vertical columns of water

in the ocean), and imaging of seawater motion signatures (e.g., surf zone circulation cells, upwelling, fronts, etc.). While describing the technological aspects of oceanic current measurement, which are the major thrust of this book, adequate attention has also been paid to the history and evolution of ocean-current measurement technologies, from primitive to state-of-the-art. Thus, attempts have been made to cover all major aspects of ocean-current measurement to the present. This historical perspective allows the reader to gain awareness of the difficulties with which current measurements were made in the past and how advances in technology made ocean-current measurement relatively much simpler in the present era.

In the course of describing the technology of ocean-current measurement, the incredible acumen and perseverance exhibited by a few erudite technologists and oceanographers and the important role played by them in the invention of some of the state-of-the-art remote sensing tools are covered in some detail. Shining examples of four such inventions that deserve special mention in this context are HF Doppler radar technology for real-time time-series remote mapping of sea surface current vectors; Swallow floats and their improved variants for tracing the trajectories of subsurface and abyssal currents; acoustic Doppler profilers for measuring both directional waves and current profiles; and acoustic tomography techniques for integrated measurement of water motions across large ocean basins, straits, coastal waters, and even rivers. During the initial untiring experiments that culminated in these path-breaking inventions, thoughtful technologists and oceanographers found invaluable "signals" in what mediocre oceanographers who lacked acumen would have termed "instrument noise," "high-frequency contamination," or "deployment error." It is my hope that discussions of the step-by-step technological leaps made in the field of ocean-current measurement over the years will kindle sufficient enthusiasm in the minds of young researchers and academicians to cause them to continue these traditions.

This book consists of 13 chapters. The introductory chapter provides a brief overview of the major ocean-current circulations in the world oceans and their implications for mankind. Special emphasis has been given to the planet-spanning thermohaline circulation known as the ocean's *conveyor-belt circulation* and its crucial role in climate change, a hot topic in recent decades. Apart from this and the technological aspects of ocean-current measurement, the book provides a brief discussion of the importance of ocean-current measurement for a wide spectrum of applications, extending from the domain of scientific research to the domain of a multitude of ocean engineering applications.

When I began to write this book, there was a dearth of comprehensive and authentic books that cover various aspects of ocean-current measurement technologies and implications of ocean currents. This book addresses these issues in a detailed manner, keeping in mind the broad spectrum of oceanographic and academic communities as well as the general public. While addressing the state-of-the-art in ocean current measurement, the book emphasizes that every technology has its own merits and limitations. These aspects are addressed at appropriate places in this book. This approach is expected to enable the reader to capture a holistic view of the technologies that are available and help the oceanographic communities choose the most appropriate technologies for their specific needs and applications. I am hopeful that both the nonspecialist readers who have an interest in the broad subject of global ocean circulation and its impacts and the specialists who want to know more about the recent technologies used for ocean-current measurements will find this book a source of valuable information. The book will be of particular interest to advanced students in ocean engineering and, more generally, to oceanographers in a variety of disciplines.

Researchers at universities and oceanographic institutions need to stay up to date with the latest technological developments. Keeping this need in mind, this book endeavors to provide valuable information compiled from an extensive knowledge base scattered in a variety of journal publications and conference proceedings in the domains of technology and science.

I felt honored by the encouraging review of my first Elsevier book, *Tsunamis: Detection, Monitoring, and Early-Warning Technologies,* published in February 2011, which awoke great interest internationally. This success was a great motivation for me to work on and complete this present book, *Measuring Ocean Currents: Tools, Technologies, and Data.*

Antony Joseph

Antony Joseph

Chapter 1

Oceanic Currents and Their Implications

Chapter Outline

1.1. **Oceans' Thermohaline Conveyor Belt Circulation and Global Climate Change**   2
1.2. **Meandering Currents, Eddies, Rings, and Hydrographic Fronts**   5
  1.2.1. Meandering Currents   5
  1.2.2. Eddies   5
  1.2.3. Rings   9
  1.2.4. Hydrographic Fronts   10
1.3. **Influence of Eddies and Fronts on Fishery and Weather**   11
1.4. **Major Current Systems in the World Oceans**   12
  1.4.1. Antarctic Circumpolar Current   13
  1.4.2. Western Boundary Currents in the Atlantic Ocean   15
    1.4.2.1. The Gulf Stream   15
    1.4.2.2. The Brazil Current   17
  1.4.3. Western Boundary Current in the Pacific Ocean: The Kuroshio Current   20
  1.4.4. Western Boundary Currents in the Indian Ocean   22
    1.4.4.1. The Agulhas Current   22

    1.4.4.2. The Somali Current   24
  1.4.5. Equatorial Undercurrents   24
    1.4.5.1. Equatorial Undercurrent in the Atlantic Ocean   25
    1.4.5.2. Equatorial Undercurrent in the Pacific Ocean   27
    1.4.5.3. Equatorial Undercurrent in the Indian Ocean   28
1.5. **Currents of Different Origins**   29
  1.5.1. Wind-Driven Current   31
    1.5.1.1. Ekman Spiral   31
    1.5.1.2. Langmuir Circulation   33
  1.5.2. Inertia Current   33
  1.5.3. Tidal Currents in Open Seas, Estuaries, and Ridge Valleys   34
  1.5.4. Rip Currents   36
1.6. **Implications of Ocean Currents**   40
References   42

Climate and oceanic researchers consider that one of the most important roles played by the planet's oceans is the regulation of the Earth's climate; naturally this has become the focus of the global approach to research in recent decades. We are increasingly concerned about global climate change (i.e., long-term fluctuations in temperature, precipitation, wind, and all other aspects of the Earth's climate) and its regional impacts. The sea ice extent in the Arctic Ocean has decreased to record minimums in recent years (e.g., Serreze et al., 2003; Comiso, 2006), and such Arctic processes might amplify changes in global climate.

An examination of autonomous drifting float observations collected during the 1990s and historical shipboard measurements suggests that the mid-depth (700–1,100 m) Southern Ocean temperatures have risen since the 1950s (Gille, 2002). This warming is faster than that of the global ocean and is concentrated in the *Antarctic Circumpolar Current* (ACC), where temperature rates of change are comparable with Southern Ocean atmospheric temperature increases (Gille, 2002). It has been found that the warming is associated with a southward migration of the ACC since the 1950s of about 50 km in the Pacific (Swift, 1995) as well as the Atlantic and Indian oceans (Gille, 2002). A more recent analysis of 32 yr (1966–98) of subsurface layer (200–900 m) temperature observations in the Indian sector of the Southern Ocean similarly show a warming trend concentrated in the ACC, indicative of a southward shift of the ACC of about 50 km (Aoki et al., 2003).

These changes seem to be associated with long-term changes in the overlying atmospheric circulation (Thompson and Solomon, 2000; Fyfe, 2003; Marshall, 2003). From the global ocean modeling standpoint it is known that a change in the position of the surface wind stress over the Southern Ocean can affect a change in the position of the ACC (Hall and Visbeck, 2002; Oke and England, 2004). Transient climate-change simulations carried out by Fyfe and Saenko (2005) suggest that about half of the observed poleward shift of the ACC seen since the 1950s is the consequence of human activity. In the future the ACC is predicted to continue to shift poleward as well as to accelerate. Based on theoretical studies, these changes appear to be indicative of the oceanic response to changing surface wind stress. The potential impacts of these changes on the global climate system merit investigation.

Global climate change continues to be a hot topic among scientists, climatologists, and the general public across the globe. It has become important to focus closely on the issue of climate change because the climate change is expected to increase the frequency and intensity of weather- and climate-related hazards (Goldenberg et al., 2001; Meehls et al., 2007; Ulbrich, et al., 2009) and will deplete and stress the planet's ecosystems, upon which we all depend (Holmes, 2009). Between 1980 and 2007, more than 8,000 natural disasters killed 2 million people, and more than 70% of casualties and 75% of economic losses were caused by extreme weather events (Jarraud, 2009). It is feared that there is an increased threat of future hurricanes as a result of climate change.

Analysis of the sea-level measurements made across the globe over the past several decades indicates that sea level is currently rising at an accelerating rate of 3 mm/year as a result of global warming. Arctic sea ice cover is shrinking and high-latitude areas are warming rapidly. Extreme weather events cause loss of life and place an enormous burden on the insurance industry. Globally, 8 of the 10 warmest years since 1860, when instrumental records began, occurred in the past decade. Their impacts are in some cases beneficial (e.g., opening of Arctic shipping routes) and in others adverse (increased coastal flooding, more extreme and frequent heat waves and weather events such as severe tropical cyclones). Likewise, the response of ocean boundary currents to climate change may directly affect marine ecosystems and regional climate (e.g., Stock et al., 2011).

1.1. OCEANS' THERMOHALINE CONVEYOR BELT CIRCULATION AND GLOBAL CLIMATE CHANGE

One of the ways the sun's energy is transported from the Earth's equator toward its poles is through the globally interconnected movement of ocean waters (i.e., ocean currents). Currents and countercurrents were first noticed in the oceans by ancient mariners. In understanding the system of oceanic circulation, they made a very simple assumption that from whatever part of the ocean a current is found to run, to the same part a current of equal volume is obliged to run. The whole system of ocean currents and countercurrents is based on this principle. Dr. Smith appears to have been the first to conjecture in 1683 (vide *Philosophical Transactions*) the existence of an undercurrent in the ocean. His conjecture was based on

the finding of a high-salinity surface current in the Mediterranean Sea. This current was found to carry an immense amount of salt into the Mediterranean from the Atlantic Ocean. Because the Mediterranean is not salting up (i.e., its salinity is not increasing day by day), it was logical to infer the existence of an undercurrent through which this salt finds its way out into the Atlantic Ocean again, thus preventing a perpetual increase of saltness (i.e., salinity) in the Mediterranean Sea beyond that existing in the Atlantic. The proofs derived exclusively from reason and analogies were clearly in favor of an undercurrent from the Mediterranean to the Atlantic.

Seawater temperature and salinity together play an important role in the preservation of equilibrium in the ocean; thus there exists a special category of ocean currents known as *thermohaline circulation* (*thermo* for heat and *haline* for salt). This circulation, primarily driven by differences in heat and salt content, influences the net transport of mass in the ocean. Thermohaline circulation transports large quantities of warm water from the equator to the polar regions and cold water from the high-latitude regions to the low-latitude regions via various pathways and therefore plays an important role in distributing heat energy across the oceans and seas. The ocean currents can thus warm or cool a large region. The planet-spanning thermohaline circulation, also called *meridional over-turning circulation* (MOC), is referred to as the oceans' thermohaline *conveyor belt* circulation (see the main image on the front cover of this book).

The wonderful conveyor system of global oceanic circulation consisting of a chain of surface, subsurface, and deep-ocean circulation paths and its role in controlling the global climate have attracted considerable attention in recent years. For maintenance of an efficient conveyor belt circulation system, there should be "sinking" regions as well as "rising" regions at diverse locations in the global oceans to connect them together as a closed circuit.

Careful observations by navigators in the 19th century brought to light the existence of an efficient *rising region* in the Arctic Ocean. The movement of right whales provided a reliable indicator in identifying such a region. Examination of the log books containing the records maintained by various ships for hundreds of thousands of days for preparation of mariner wind and current charts led to the interesting discovery that the tropical regions of the oceans are to right whales as a sea of fire, through which they cannot pass and into which they never enter. Note that whereas sperm whales are warm-water mammals, right whales are a special category of whale that delights in cold water. It was also found that the right whales of the northern hemisphere are a different species from those of the southern hemisphere (Maury, 1855).

It was the custom among the ancient whale hunters to have their harpoons marked with the date and the name of their ship before they fired the harpoons at whales. The presence of a region in the Arctic Ocean that was not covered by ice sheets was identified in the 19th century based on a very short travel time made by the harpoon-stricken right whales in traveling from the Atlantic side to the Pacific side of the Arctic Ocean. The calculation of the short travel time was arrived at based on the logic that the harpoon-stricken whale could not have traveled below the ice sheet for such a long distance, stretching across the entire Arctic path, nor could they have traveled around either Cape Horn (at the southern tip of South America) or the Cape of Good Hope (at the southern tip of South Africa) because of their proven dislike for warm waters. (Right whales are a class that cannot cross the equator because their habits are averse to the warm waters of the equatorial belt.) In this way circumstantial evidence afforded the most irrefragable and irrefutable proof that there is, at times at least, the presence of open water (i.e., water free from ice cover) in the Arctic Sea.

Further, based on rapid drifting of icebergs against a strong surface current, it was inferred that there is a powerful undercurrent through Davis Strait. The most dominant meridianal overturning cross-equatorial thermo-haline circulation, which traverses northward across the ocean surface all the way up from Antarctica (the return flow of cold water underneath traversing all the way into the middle of the Pacific), is found in the Atlantic Ocean (see the main image on the front cover of this book), and the climatic effect of this arm of the global thermohaline circulation is due to its large heat transport in the North Atlantic. It has been estimated that the amount of heat transported into the northern North Atlantic (north of 24°N) warms this region by ~5°K. This is indeed roughly the difference between sea surface temperatures (SST) in the North Atlantic compared to that in the North Pacific at similar latitudes (Rahmstorf, 2006). As a result of the existence of warmer surface currents in the North Atlantic sector compared to the North Pacific, the sea ice margins in the Atlantic sector are pushed back, which in turn leads to reduced reflection of sunlight and thus warming (albedo feedback).

In particular, it is recognized that since 1860 the net thermal effect of the northbound currents has caused a decrease in the extent of the ice cover in the Nordic Seas in spring. A continuation of this trend is predicted by global circulation models. If these predictions turn out to be correct, a permanent warming of the Arctic's climate and a decrease of the ice extent of the Barents Sea and the Arctic Ocean are likely to occur. Climate change will, in turn, affect large-scale circulation volume transport, which causes changes in water masses and subsequently affects

the Arctic marine ecosystem (Scottish Association for Marine Science [SAMS] report). Part of the cross-equatorial thermohaline circulation in the Atlantic Ocean is known as the *Gulf Stream*, providing some of the heat that keeps Europe warmer in winter than regions of North America at the same latitude.

The "conveyor belt" circulation spanning the world oceans is maintained primarily by the following four factors (Rahmstorf, 2006):

1. Downwelling (sinking) of water masses from the sea surface into the deep ocean in a few localized areas (e.g., the Greenland-Norwegian Sea, the Labrador Sea, the Mediterranean Sea, the Weddell Sea, and the Ross Sea)
2. Spreading of deep waters (e.g., North Atlantic Deep Water, and Antarctic Bottom Water) mainly as deep western boundary currents
3. Upwelling of deep waters into the sea surface (mainly in the Antarctic Circumpolar Current region)
4. Near-surface currents (required to close the flow)

It is feared that if the northern surface waters somehow become less salty (and therefore less dense)—as might happen if melting of ice sheets due to global warming diluted the upper ocean layer in this region with fresh water—then there would be no scope for the North Atlantic water to sink to the ocean bottom and thus maintain the continuity of the conveyor belt loop. Climate change could thus interfere with the formation of the cold, dense water that drives the thermohaline conveyor belt circulation and thus could bring about further changes in climate. Oceanic current circulation might, therefore, be crucial for climate change, not just over geological time but more immediately (Weart, 2009).

The Southern Ocean waters also play a role in climate change. For example, the ocean waters that move toward the Antarctic sink as their temperatures drop. Once this occurs, these waters then move northward. Water at greater depths has been found to have a significantly different impact than the high-nutrient waters that flow northward at intermediate depths. The circulation in the regions around Antarctica, where water sinks to depths greater than 1.5 km, was shown to be largely responsible for controlling the air-sea balance of carbon dioxide. The circulation in the sub-Antarctic regions that feeds water to depths between 0.5 and 1.5 km controls biological productivity. Further, the sinking water masses in the Southern Ocean are part of the overturning in this region and thus play a major role in the global climate. Interestingly, scientists from the National Aeronautics and Space Administration (NASA) and Columbia University in New York have used computer modeling (the Goddard Institute for Space Studies [GISS] climate model) to successfully reproduce an abrupt climate change that took place 8,200 years ago after the end of the last Ice Age. At that time, the beginning of the current warm period, climate changes were caused by a massive flood of freshwater into the North Atlantic Ocean. Scientists believe that the massive freshwater pulse interfered with the ocean's overturning circulation (the conveyor belt), which distributes heat around the globe. The GISS climate model is also being used for simulation by the Intergovernmental Panel on Climate Change (IPCC) to simulate the Earth's present and future climate.

One of the most persistent concerns among some climatologists is that with the collapse of the MOC or even catastrophic transitions in its structure, the climate system might lurch into a new state. These climatologists go even to the extent of fearing that a new glacial period could begin in the absence of the vast drift of the colossal amount of tropical warm waters northward, near the surface of the Atlantic, as a result of the collapse of the conveyer belt circulation. Some climate models have indicated that a shutdown of the conveyer belt is especially likely with rapid increase of greenhouse gas emissions, although a dramatic shift of the ocean circulation is expected to be very unlikely within the 21st century (International panel, 2007). Thus, the oceanic current circulation—together with the general circulation of the atmosphere—plays a very important role in global climate change.

Lack of sustained observations of the atmosphere, oceans, and land has hindered the development and validation of climate models, which are required to understand and eventually make predictions for changes in both the atmosphere and the ocean. Such predictions are needed to guide international actions, to optimize governments' policies, and to shape industrial strategies. An issue that has been debated in recent times is an example that came from a recent analysis, which concluded that the currents transporting heat northward in the Atlantic and influencing the western European climate had weakened by 30% in the past decade. This result had to be based on just five research measurements spread over 40 years. Questions have been asked as to whether this change could be part of a trend that might lead to a major change in the Atlantic circulation, or due to natural variability that will reverse in the future, or an artefact of the limited observations?

The Greenland Sea (a region that links the Arctic Ocean and the North Atlantic) has a unique worldwide impact on the properties of deep-ocean waters. The surface waters of the Greenland Sea area are sufficiently cold and salty in the winter that they become denser than the water below and then sink to great depths. This creates a source of cold water that can be traced around the globe through the deep oceans, extending from the north Atlantic to the south Atlantic around the tip of Africa through the Indian Ocean and up into the north Pacific.

Oceanographic research programs sponsored by multiple nations have evolved to observe in detail the movement of cold Arctic water as it travels through the Greenland Sea into all the world's oceans and to study the interactions of the atmosphere, ice, and ocean in the Greenland Sea. The deep-water current formed by conditions in the Greenland Sea is important to the processes of ocean mixing worldwide. This cold-water mixing has an effect on not only the physical properties of the global ocean but on biological activities as well because it affects the distribution of nutrients. Some atmospheric carbon dioxide is absorbed into the ocean, and the rate of deep-water mixing has an effect on the amount of that absorption.

The Atlantic Ocean has a prominent role among all other oceans in terms of its influence on climate, because this is the only ocean that provides a direct link between the equatorial warm waters and the cool waters of both the southern and the northern polar regions. The principal current system of the Atlantic Ocean is the Gulf Stream, which is often likened to a grand river in the ocean. There is no other such majestic flow of waters in the world. Its current is more rapid than the Mississippi or the Amazon. The Gulf of Mexico is well known for strong currents associated with the Loop Current and Loop Current eddies (LCE) and mesoscale (order 100 kilometers) features in near geostrophic balance. LCE currents usually extend to 400 m or deeper. The ancient sailors who navigated in simple wooden boats dreaded storms in the Gulf Stream more than they did in any other part of the ocean. It was not the fury of the storm alone that they dreaded, but it was the "ugly sea" that those storms raised. The current of the stream running in one direction and the wind blowing in another direction creates a sea that is often frightful. Lieutenant M. F. Maury (Maury, 1855) provides a telling account of how 179 officers and soldiers of a regiment of U.S. troops on board a fine new steamship, the *San Francisco*, were washed overboard and drowned after the ship was terribly crippled by a gale of wind in the Gulf Stream in December 1853.

The net transports of mass, heat, and salt through *straits* (i.e., narrow passages between basins of oceans or marginal seas) may represent a control for interior processes or forcing and thus is of interest for the functioning of a basin or as a boundary condition for modeling studies. Because straits give integrals of the fluxes over the interior basin, they potentially provide important observation sites for various applications. For example, monitoring the variability of the low-frequency transport through the Strait of Gibraltar is important for a wide range of problems. Send et al. (2002) has provided an overview of the water-current circulation studies on the Strait of Gibraltar in the context of climate change.

On the largest scale, the Mediterranean outflow plays an important role in the circulation of the North Atlantic

Ocean. The Mediterranean salt tongue is one of the main components of the Atlantic water mass system, extending across nearly the entire basin from the outflow at the Strait and strongly influencing the salinity budget of the Atlantic Ocean. Recent studies suggest that correct inclusion of the mass flow out of the Mediterranean (in Sverdrups) is necessary for a correct representation of the salt tongue in numerical models (Gerdes et al., 1999), as opposed to the frequently used salinity source at the Strait of Gibraltar.

The Mediterranean outflow and its intensity also may play a role in the dynamics of the North Atlantic. Reid (1979) speculated that the high-salinity Mediterranean Water (MW) may influence North Atlantic Deep Water formation and thus the global thermohaline circulation (THC). In a series of idealized global ocean models, Cox (1989) showed that the salty MW enabled the North Atlantic Deep Water to penetrate deeper and farther south in the Atlantic and into the Indian and Pacific Oceans. Rahmstorf (1998) used a simple coupled model to compare the THC with and without the Mediterranean outflow. He found that the THC intensifies by $1-2$ Sverdrups (1 Sv $\equiv$ 10^6 m^3/s) and that northern Atlantic surface temperatures increase by a few tenths of a degree if the outflow is included. The outflow from the Strait of Gibraltar even appears to be relevant for some large-scale, upper-layer current systems. There is a large flow of chilled polar waters into the Atlantic Ocean, and the preservation of the equilibrium of the Atlantic Ocean needs to be accounted for with the presence of undercurrents, which play an important part in the system of oceanic circulation.

It is interesting to note that probably the first observations of the penetration of Antarctic intermediate water into the North Atlantic were made in 1886 by the British oceanographer J. Y. Buchanan during the *Challenger* Expedition (December 1872–May 1876) for investigating "the physical and biological conditions of the great ocean basins." His charts and vertical sections showed the global distribution of salinity for the first time and partially revealed the conveyor belt circulation, which was unheard of before that time.

1.2. MEANDERING CURRENTS, EDDIES, RINGS, AND HYDROGRAPHIC FRONTS

The complexity of the mesoscale ocean flow field has often been visualized as an evolving mosaic of eddies, rings, frontal systems, plumes, lenses, and filamentous jets.

1.2.1. Meandering Currents

Meandering currents (often called *meanders*) are those category of currents that "snake around," exhibiting a characteristic wavy nature in the horizontal and vertical

planes. Under some conditions, the meanders grow exponentially, close upon themselves, and pinch off a closed loop with either cyclonic or anticyclonic circulation (depending on whether the meander is convex or concave, looking northward).

Trajectories of surface floats and subsurface floats over large areas in the oceans have shed much light on the complex motions of oceanic waters such as meandering currents, eddy motions, and rings. Several subsurface floats have been launched in the Gulf Stream (off the east coast of North America) since the beginning of 1983. Rossby et al. (1986) have provided an example of the space-time evolution of the meandering current, based on the trajectory of a subsurface float. The trajectory exhibited the characteristic wavy nature of the meandering Gulf Stream. In this example, the mean speed of the float was 55 cm/s. A striking aspect of this and all other float tracks in the Gulf Stream was found to be the tendency for floats to shoal from meander trough to crest and to deepen from a meander crest to the next trough. It was observed that during the float's 2,100 km journey to the east, its lateral displacements relative to the current were less than 100 km. From the pressure record it was noticed that these motions are not random but were clearly a result of the dynamics of curvilinear motion. For example, Figure 1.1 shows an artist's rendition of a multiscale circulation of the Gulf Stream (GS) meander and ring region. Large-scale meandering of oceanic currents results in shedding of mesoscale eddies and rings.

Apart from the most stable meander found in the Gulf Stream, meanders are commonly also found in the other major western boundary current systems. For example, the Kuroshio Current south of Japan exists in one of two stable paths, a zonal path and a meander path, each lasting for a period of years. Mizuno and White (1983) reported identification of a quasi-stationary meander pattern (Kuroshio Meander) in the Kuroshio Extension in the vicinity of the Shatsky Rise (160°E). The Kuroshio Meander becomes unstable as a consequence of increased eddy activity and ring production.

1.2.2. Eddies

The term *eddy* refers to rotation (vortex motion) of water mass. The motion of an eddy can be likened to that of a spinning top, having both rotary and translatory motions. The horizontal cross-section of most eddies is essentially circular in shape, whereas others are much more elongated (both these types of eddies have been noticed in the vicinity of Sugarloaf Point off the east coast of Australia; see Tilburg et al., 2001). Twin eddies have also been noticed in some regions. Eddies usually taper toward the bottom, but cylindrical-shaped eddies are not uncommon. In the case of essentially elliptical eddies, the swiftness of the current in

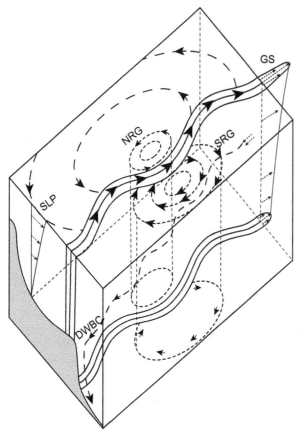

FIGURE 1.1 Artist's rendition of a multiscale circulation of the Gulf Stream (GS) meander and ring region. The large-scale meandering of the Gulf Stream results in mesoscale eddies while interacting with the subbasin-scale gyres (NRG and SRG). The deep western boundary current flows along the 3,400 m isobath and crosses under the Gulf Stream near Cape Hatteras to flow further southward. *(Source: Gangopadhyay et al., 1997. © American Meteorological Society. Reprinted with permission.)*

certain portions of the eddy can be much different from the other portions. Eddy diameters are in the tens and hundreds of kilometers. Some are warm-core eddies; others are cold-core eddies.

Cold-core eddies cause transport of nutrient-rich water from ocean depths to the sea surface, thereby providing them a low-temperature signature. A horizontal spatial gradient in nutrient-rich water relative to the vertical axis of the core results in a corresponding spatial gradient in chlorophyll concentration. Whereas the spatial gradient in temperature relative to the core allows remote detection of eddies using satellite-borne thermal radiometers, a spatial gradient in chlorophyll concentration relative to the core allows remote detection of eddies using satellite-borne visible wavelength radiometers. Whereas cold-core eddies have their centers depressed, *warm-core eddies* have elevated centers. This enables their detection by satellite-borne altimeters. *In situ* hydrographic measurements

are often employed for detailed examination of eddy structure. The core of the eddy current is defined as the region of fastest-flowing surface current (Andrews and Scully-Power, 1976).

The mechanisms that generate eddies in the open ocean include baroclinic (i.e., surface of equal hydrostatic pressure crossing surface of equal water density) instability of large-scale currents, topographic steering, and direct atmospheric forcing (Kamenkovich et al., 1986, Ikeda et al., 1989). In general, Antarctic Circumpolar Current and western boundary currents (i.e., currents flowing in the vicinity of the western boundaries of oceans) such as the Gulf Stream and the Brazil Current in the Atlantic, Kuroshio and East Australian Currents in the Pacific, and the Agulhas Current in the southern Indian Ocean have the potential for generating warm- or cold-core eddies, which are an integral part of the general circulation (see descriptions of these currents in section 1.4 (Major Current Systems in the World Oceans)).

With the deployment and monitoring of increasing numbers of subsurface floats at various depths, oceanographers have gained increased knowledge of some facets of the extremely complex oceanic circulation features in the interior of the ocean basins. An illuminating account of the complex movements of a cluster of subsurface floats at a depth of about 1,300 meters is provided by Rossby et al. (1975) and Rossby (1983). They described a variety of exciting time-dependent events such as a chain of eddy lenses, each about 600 m thick and 100 km in diameter. Some floats exhibited striking anticyclonic motions with time periods as large as eight months. In fact, Rossby and co-researchers observed repeated cyclonic and anticyclonic oscillations of a cluster of some floats in direct response to the effect of the Earth's rotation, exhibiting perhaps "the simplest and most explicit geophysical demonstration of the conservation of angular momentum to date." These interesting effects indicate that eddies not only command considerable interest in their own right but are also a key to our understanding of the general oceanic circulation (Rossby, 1983). New and unexpected findings will certainly help develop appropriate theories to support the observed facts. Hopefully, such theories will permit oceanographers to unravel, at least partially, the hidden secrets of nature.

Eddies are found in several parts of the world oceans. Davis (2005) conducted a comprehensive analysis of the voluminous dataset obtained from Argo floats, using various methods. He found that well-developed strong subtropical gyres feed western boundary currents. It was also found that tropical gyres are separated by eastward flow along the equator in both hemispheres of both the Indian Ocean and the South Pacific Ocean basins, although the Indian subcontinent splits the North Indian tropical gyre. The Indian Ocean's subtropical gyre, and perhaps part of the South Atlantic's, reaches east to a retroflection just upstream of the Campbell Plateau, south of New Zealand. It was found that encounter between two different currents gives rise to the generation of eddies. It was also noticed that several Argo floats went from speeds of a few centimeters per second north of the Agulhas (which is the strongest flow, exceeding 50 cm/s, observed off the South African coast in the Indian Ocean), passed through the boundary current in one or two submergence cycles of 25 days, and, after clearing the tip of Africa, decelerated again to a few cm/s where the Agulhas encounters the eastward-flowing South Atlantic Current and apparently breaks up into eddies. In fact, the eddy motion is so vigorous that it clouds the general ocean circulation patterns. In the face of ubiquitous eddy variability, averaging is often required to describe patterns of mean circulation.

Another piece of information obtained from analyses of Argo float datasets is that outside the equatorial zone ($\pm10°$) seawater circulation is dominated by subtropical gyres bounded by strong eastward flow in the ACC and the South Indian Current (SIC) or South Pacific Current (SPC). It was found that the Indian Ocean's subtropical gyre is at least twice as strong as the South Pacific's, thereby making the powerful intermediate-depth Agulhas Current much stronger than the East Australia Current. Remarkably, the eastern limbs of the southern subtropical gyres in all three basins occur away from land.

Based on satellite-derived data, hydrographic data, and data obtained from moored current meters, Lee et al. (1981) found that on the Georgia continental shelf the Gulf Stream frontal eddies control the residence time of the outer shelf waters, defined as the mean separation time between eddy events. Upwelling in the cold core extends into the euphotic (i.e., light penetrating) zone (~45 m) and shoreward (35 to 40 km) beneath the southward-flowing warm filament in a 20-m-thick bottom intrusion layer. The annual nitrogen input to the shelf waters by this process is estimated at 55,000 tons each year, about twice all other estimated nitrogen sources combined.

Long-term seawater temperature observations have also shed ample light on the presence and lifetimes of several eddies in the world oceans. Such measurements indicated that the Kuroshio Current region in the China Sea is well known for prevalence of cold-core eddies. Stommel and Yoshida (1971) examined the subsurface temperature arranged chronologically for a selected region (Enshunada) in the Kuroshio to clearly document the bimodal state of the Kuroshio in this area. The great Cold Eddy in the Kuroshio, often referred to as the Cold Water Mass, appears from time to time south of Enshunada. Surveys carried out for a decade from 1955 to 1964 by the Maritime Safety Agency have shown the appearance and disappearance of this cold eddy very convincingly. It was found that there exists

a pronounced bimodality between 150 m and 600 m, thus verifying the on and off nature of the eddy. Interestingly, the stream passes through intermediate states quite quickly. The Cold Eddy was present from mid-1934 until the beginning of 1941. The eddy was present also during the last half of 1941 and all of 1942. Stommel and Yoshida (1971) found that the Cold Eddy remained absent until mid-1959, when it abruptly appeared again and stayed turned on for three and a half years until early 1963, when it disappeared once more.

Depending on the local topography and proximity to islands, eddies demonstrate varying characteristics. For example, Robinson and Lobel (1985) report that the ocean eddies that form off the Island of Hawaii can remain in the lee for weeks before moving or dissolving.

Eddies are frequently found in western boundary currents, but they are also found in eastern boundary currents. In fact, eddies are considered to be a general feature of the general circulation. In the Gulf Stream, eddies can form within a week of meander generation and then persist for another one to two weeks (see Lee and Mayer, 1977; Legeckis, 1979). They appear to form all along the Gulf Stream boundary at any time of year, and at times they serve to dissipate kinetic energy from the mean flow (Lee, 1975). They also serve as an effective mechanism for exchange of shelf and Gulf Stream waters. In the Florida Straits, vortex diameters are in the order of 10–30 km, with downstream axes two to three times the cross-stream dimension (Lee, 1975; Lee and Mayer, 1977). The vortex travels northward along the shelf break at speeds less than the mean speed of the Gulf Stream. The vortex occurs on the average of one per week in the Florida Straits and results in large-amplitude, cyclonic flow reversals over the shelf that distort the temperature and salinity fields to a depth of approx. 200 m (Lee, 1975; Lee and Mayer, 1977). In the case of the Brazil Current, eddies are spun off from its northern end as it moves up the north coast of South America. These eddies generate speeds of up to three knots. Lee et al. (1981) investigated the eddy-induced sea surface temperature (SST) features in the Florida Straits and termed them "spin-off eddies" due to their cyclonic rotation, cold core, and exchange of heat and salt with adjacent shelf waters. However, that terminology has been discontinued in favor of a more general definition of the features as *frontal eddies*.

Warm-core eddies off East Australia have been observed from time to time. Wyrtki (1962) produced dynamic topographies of the southwest Pacific that suggested that these eddies are separated from the main current and drift south down the coast. In this study, eddy diameters were found to be in the range 200–250 km, whereas drift rates were estimated to lie within about 5–8 km/day. Hamon (1965) and Boland and Hamon (1970) have also noticed the dynamic height maps of this region usually showing several intense anticyclonic eddies just offshore. Hynd (1969) tracked a "pool of warm water" with an airborne radiation thermometer for a month in mid-summer, finding a drift rate of 5.5 km/day. The pool was about 50–70 km in diameter and was surrounded by marked temperature fronts. Boland (1973) found that the average time between the appearances of successive eddy structure in the Tasman Sea off the southeast coast of Australia at about 33.5°S is about 73 days and that strong currents appear and disappear within 40-day intervals. Hamon and Cresswell (1972) deduced a dominant length scale (distance between like eddies) of 500 km. Andrews and Scully-Power (1976) examined the structure of an intense, anticyclonic, warm-core winter eddy off the east coast of Australia and found that the eddy had a diameter of 250 km and a mixed layer depth extending to over 300 m in the core. Such deep mixed layer is the most pronounced feature that makes the warm-core eddies off east Australia somewhat unique.

The East Australian Current seems to consist of strong anticyclonic eddies, which tend to move irregularly southward along the coast. In particular, the eddies south of Sugarloaf Point are not separate entities but probably form by the pinching off of current loops that start near Sugarloaf Point (Godfrey et al., 1980). An excellent graphical presentation of a multitude of eddies generated off the southeast coast of Sydney during 1977–1979 has been reported by Godfrey et al. (1980). In the Agulhas Current System, eddies in the Mozambique Channel have spatial scales of approximately 300–350 km and propagate southward at speeds of approximately 3–6 km/day (Schouten et al., 2002, 2003). In the North Pacific, Bernstein and White (1981) found eddy activity in the Kuroshio Extension to be intensified over the abyssal plane between the Izu Ridge and the Shatsky Rise.

Seafloor topography exerts an important role in the generation of eddies. For example, the circulation in the coastal region of the Gulf of Maine and Georges Bank (GOMGB) is associated with features such as the buoyancy-driven Maine Coastal Current (MCC), Georges Bank anticyclonic frontal circulation system, the basin-scale cyclonic gyres, the deep inflow through northeast channel, and the shallow outflow via the great south channel (Gangopadhyay and Robinson, 2002). A schematic of the surface circulation of this coastal region with the features identified is presented in Figure 1.2a; Figure 1.2b shows various processes and the topographic pathways that the coastal features tend to follow. In addition to processes that govern the deep regional dynamics (large-scale wind-driven, barotropic and baroclinic instabilities, meandering and eddy-mean flow interaction), significant variability in the shallow region is determined from water mass formation, topography, freshwater influence, tides, winds, and heating/cooling.

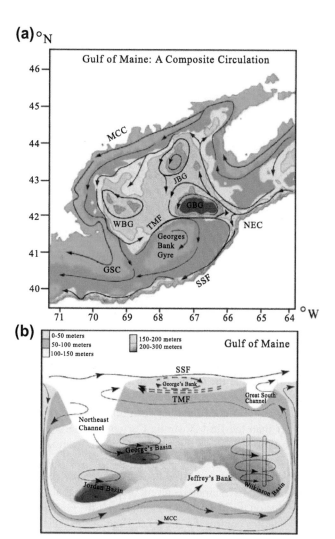

FIGURE 1.2 (a) A schematic of circulation features in the Gulf of Maine: The great south channel (GSC); shelf-slope front (SSF); northeast channel (NEC); Georges basin gyre (GBG); tidal mixing front (TMF); Wilkinson basin gyre (WBG); Maine coastal current (MCC); Jordan basin gyre (JBG). (b) A three-dimensional bathymetric perspective of the regional circulation features. The basins are the three deep regions in the interior gulf. The vertical mixing region is predominantly in the Wilkinson basin. *(Source: Gangopadhyay and Robinson, 2002.)*

Subsurface eddies that do not reach up to the sea surface have also been reported. For example, based on hydrographic measurements, Babu et al. (1991) reported identification of a cold-core subsurface eddy in the Bay of Bengal (in the Indian Ocean). The thermal structure observed across the eddy (Figure 1.3a) indicates that it was confined to a level well below the mixed layer, between 50 m and 300 m, and that it had a diameter of about 200 km. A temperature drop of 4–5°C relative to the periphery was observed at the center of the eddy. The surface dynamic topography derived from the conductivity, temperature, depth (CTD) data, relative to 500 db, showed that the eddy

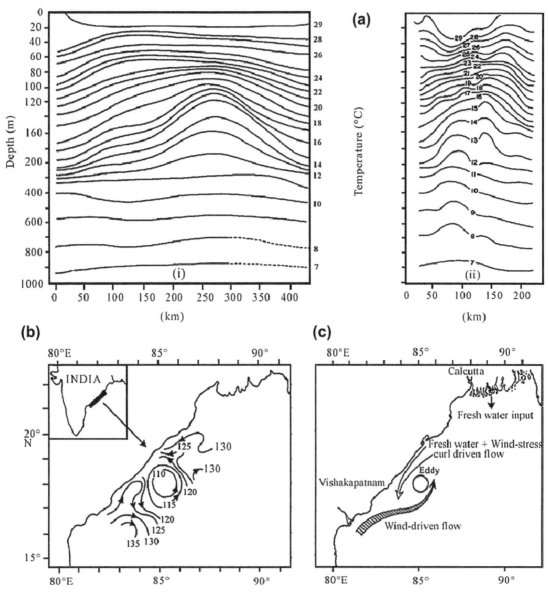

FIGURE 1.3 (a) Vertical distribution of water temperature (°C) across a subsurface eddy in the Bay of Bengal off the east coast of India along transects (i) parallel and (ii) normal to the coast (note break in scale at 200 m depth). (b) Surface dynamic topography (in dynamic cm) relative to 500 db. (c) Schematic representation of the eddy and associated currents. *(Source: Babu et al., 1991. Reproduced with kind permission from the lead author.)*

center was depressed by approximately 10 dynamic cm compared with the waters adjacent to the 200-km-wide eddy (Figure 1.3b), thus indicating cyclonic circulation. Just below the mixed layer, the eddy was elliptical in shape, with a major axis of about 400 km and a minor axis of 200 km. Figure 1.3c gives a schematic view of the eddy and the regional currents. The eddy was apparently generated at the interface of two opposing currents along the western boundary of the Bay of Bengal. High stratification caused by freshwater influx prevented the eddy from being detected at the surface.

1.2.3. Rings

Rings are that category of eddy that exhibits very definitive annulus characteristics; therefore rings can be identified by their distinctive thermal characteristics at the sea surface. Rings are usually identified by a closed contour line, the temperature amplitude size of which is distinctly different relative to the mean background temperature field. For example, Kawai (1979) defined cold rings in the North Pacific characteristically as having water colder than 16°C at a depth of 200 m in the area south of the Kuroshio

Current. The temperature difference is usually greater than 0.5°C. Ring formation usually takes two weeks to one month, as observed by Fuglister and Worthington (1951). But Kawai (1972) showed an example of ring formation that took place more slowly, over a six-month period. Ring shedding is not unusual for western boundary currents. For example, ring shedding from the Gulf Stream, the Kuroshio, the East Australian Current, and the Brazil Current are either well known or have been recognized. The Kuroshio Extension east of Japan is known as a highly variable current, with cold and warm rings generated from the season-to-season unstable growth of meander.

The generation of rings is a fascinating process. Because there is open water on both sides of the jets, the rings are free to meander and snake around. Ring formation is one of the important processes of mesoscale activity in the Kuroshio Extension. Cold/warm rings can be found over the entire Kuroshio Current System, west of 170°E. The warm rings in the Kuroshio Extension adjacent to Japan have been examined by several researchers (for example, see Kawai, 1972; Hata, 1974; Kitano, 1975; Tomosada, 1978). Kawai (1979) studied the geographical distribution of cold rings, finding them to occupy the area southeast of Japan. In the North Atlantic, Richardson (1981) found Gulf Stream meander activity and ring generation intensification by the presence of the New England Seamounts. The upstream or downstream drift of rings and their subsequent reabsorption by the mother current has been observed in the Gulf Stream, the Kuroshio, and the East Australian Current. The coalescence of cast-off rings has also been observed. In the case of the East Australia Current, closed rings are the rule and not the exception.

Lai and Richardson (1977) estimated an average life-time of 650 days for Gulf Stream Rings, which compares favorably with that estimated for East Australian Current eddies by Nilsson and Cresswell (1981). The Ring Group (1981) estimates an even longer lifetime of up to four years, suggested by the decay rate. The distribution of Gulf Stream rings in the North Atlantic has been studied in some detail by Richardson (1980) and Richardson et al. (1978) and has been shown to be extensive. The drift of eddies cast off from the East Australian Current (Nilsson and Cresswell, 1981) as well as the Brazil Current (Legeckis and Gordon, 1982) has been shown to be small. In many cases the effect of mid-ocean or other ridges on the area in which rings or eddies are to be found is significant (Dantzler, 1976; Roden et al., 1982; Bernstein and White, 1981). This is found to be true in the Agulhas Current rings as well. For example, the ring-shedding area of the Agulhas retroflection is bordered by the Atlantic-Indian mid-ocean ridge to the south, the Atlantic mid-ocean ridge to the west, and the Walvis Ridge to the north. However, it is difficult to find a clear relation between bottom bathymetry and the location of ring generation, as was shown by Richardson (1981) for the North Atlantic.

The location of maximum ring population probably coincides with the location of maximum generation. Interestingly, warm rings are generated more uniformly along the Kuroshio Extension from 140–170°E, whereas cold-ring generation is concentrated between 140–150°E. Rings of both types propagate to the west at an average speed of 1 cm/s. Whereas cold rings move consistently southward, at about 0.6 cm/s, warm rings move generally northward, at about 0.4 cm/s, but less consistently.

1.2.4. Hydrographic Fronts

Hydrographic fronts in the ocean have always attracted oceanographers' attention. *Ocean fronts* are narrow zones of enhanced gradients separating different water masses. The frontal structure of seawater motion is defined by steeply sloping isopycnals (equal density contour). Fronts, across which there are large horizontal variations of temperature and salinity, are known for the presence of prominent oceanic fine structures (down to a meter-scale wavelength) and microstructures (down to a centimeter-scale wavelength). A front operates to block exchange across it (in the absence of diapycnic mixing). Strong western boundary currents (WBCs) such as the Gulf Stream, Kuroshio, and Brazil Currents are well known for large-scale meandering frontal systems. The Kuroshio warm current and the Oyashio cold current meet in the Kuroshio Extension, east of Japan and form a high gradient frontal structure.

Along-front length scales vary from 100 m to 10,000 km; cross-front length scales from 10 m to 100 km; vertical scales, from 10 to 1000 m; and time scales as short as 1 h. The cross-frontal temperature/salinity changes sometimes exceed 10°C/1 ppt over an order 100 m distance. Ocean fronts are associated with, or accompanied by, strong mixing and stirring, current jets, water mass boundaries, bioproductivity maxima, acoustical wave guides, and atmospheric boundary layer fronts. Subduction, upwelling, and phytoplankton blooms are commonly observed features at oceanic fronts.

Fronts, where two different water masses meet, are ubiquitous features in the ocean and have long been known to be sites of elevated primary production and fertile fishing grounds. The elevated primary production in the frontal regions is supported by the nutrient supply from lower layers associated with the frontal circulation that consists of a primary circulation, the motions of which are mostly directed parallel to the front, and a secondary circulation, the motions of which are orthogonal to the front. The subduction near fronts is a major mechanism for the transmission of atmospheric and near-surface properties, such as heat and carbon dioxide, into the ocean's interior.

Therefore, a proper understanding and proper representation of the subduction process near fronts are necessary to quantify the role of the ocean as a reservoir and sink of heat and carbon in the climate system.

Eddies and rings, which have unique water masses, are considered to be circular fronts. With the availability of satellite data, the unique signature of fronts is now identifiable in the gradients of SST, sea surface height, and sea surface color. SST images derived from satellite infrared radiometers consistently show folded-wave patterns in the western boundary of the Gulf Stream from Cape Canaveral to Cape Hatteras (see Lee et al., 1981). Frontal eddies of this type manifest in the surface waters as warm, tongue-like extrusions of the Gulf Stream oriented toward the south around cold upwelled cores. They were first measured as a succession of overlapping thermal segments termed *shingles* in the pioneering mapping of the Gulf Stream cyclonic front off the southeast United States by Von Arx et al. (1955). The shingle shapes of frontal eddies are more similar to *roll vortices,* which are produced by wave-like rolling up of a shear zone (Rouse, 1963), or *wake vortices,* which are formed in the wake of islands as Karman vortex streets (Wille, 1960). Sailing captains were probably the first to be aware of the transient southward flow generated on the shoreward side of the Gulf Stream by these events. In 1590 it was reported (White, 1590) that a sailing vessel bound from Florida to Virginia had to stand far out to sea to avoid "eddy currents setting to the south and southwest." It has been suggested that atmospheric forcing can trigger a disturbance in the front that travels with the stream as an unstable wave, eventually evolving into a cyclonic edge-eddy (see Lee and Mayer, 1977; Duing et al., 1977; Lee and Brooks, 1979).

It is important to recognize the role of fronts in describing regional seawater circulation. For example, a large-scale Gulf Stream meandering frontal system also defines the boundaries of unique water masses, which in turn defines the boundary of the basin and subbasin-scale gyres in a synoptic state (see Gangopadhyay et al., 1997). A list of different kinds of fronts that are observed in the eastern boundaries of the world ocean is presented in the first table of Hill et al. (1998) and discussed by many authors in *The Sea* (Robinson and Brink, 1998). These front types include upwelling fronts, equatorial fronts, water mass fronts, plume fronts, coastal current fronts, and shelf-break fronts. It may be noted that the large-scale, wind-driven current transport and its interaction with the adjacent gyres play important roles in the dynamics of the western boundary current fronts (in addition to their role as the boundary of two distinct water masses). The constituent water masses and buoyancy forcing primarily contribute to the formation and maintenance of the coastal and water mass fronts (Hill et al., 1998; Church et al., 1998).

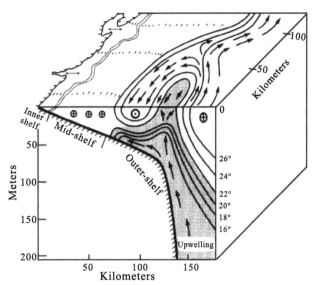

FIGURE 1.4 Schematic characterization of a Gulf Stream frontal eddy on the Georgia shelf. *(Source: Lee et al., 1981.)*

The deep Gulf Stream Meandering and Ring (GSMR) region has the large-scale meandering Gulf Stream front, which interacts with the subbasin-scale gyres and generates mesoscale eddies, the deep western boundary current, and other mesoscale and submesoscale transients. Schematic characterization of a Gulf Stream frontal eddy on the Georgia shelf is shown in Figure 1.4. Perturbations of the Gulf Stream cyclonic front are commonly observed as folded wave patterns in routine satellite-derived analyses of the western boundary of the Gulf Stream between Cape Hatteras and Miami. The features are defined as cyclonic, cold-core frontal eddies due to their flow and water mass properties (Lee et al., 1981).

1.3. INFLUENCE OF EDDIES AND FRONTS ON FISHERY AND WEATHER

Eddies have vital influence on biological productivity and fishery. Cold-core eddies are associated with *upwelling,* which is an outstanding phenomenon in the ocean whereby cold and nutrient-rich deep water is displaced toward the surface, exerting widespread influence on the ecology, regional climate, and meteorological conditions of the adjacent land (La Fond, 1980). Estimates of speeds of vertical motions in regions of upwelling are from 10−20 m/month to about 80 m/month. Because the upwelled water is rich in nutrients, it could contribute significantly to the biological productivity of the area. There are examples where eddies and associated features in the ocean enhanced plankton and primary productivity, resulting in concentration of large fish schools (Cram and Hanson, 1974; Tranter et al., 1986). Robinson and Lobel (1985) investigated the role of seasonal low-frequency oscillations of mesoscale currents (eddies) on the dispersal and recruitment of reef fishes in the early 1980s. Many coral reef fishes possess a pelagic larval phase that ranges from several days to several months in duration, depending on the species. Early discoveries of Lobel and co-researchers at Boston University have revealed mesoscale eddies as one of the physical oceanographic mechanisms that can function in favor of the local retention of the pelagic larvae from coastal marine species. Lobel and Robinson (1983, 1986, 1988) proposed that reef fish larvae could be entrained and retained near natal reefs on the basis of the structure and movement of an ocean eddy and the trajectory of current. It was found that the peak seasonality of fishes spawning in Hawaii coincided with a seasonal oceanographic regime that included frequent eddies (Lobel, 1978, 1989).

It has been hypothesized that the larvae during an eddy event would benefit from passive drift by favorable currents, which would bring them regularly near the coastline. Once near the coast, fish larvae would be able to transition (metamorphosis during the settlement process) at the youngest competent ages. Reef fish larvae that are adrift when eddies are absent may spend longer periods at sea due to the less predictable and more chaotic patterns of advection that would bring larvae within range of the shorelines. Studies conducted by Lobel (2011) indicate that mesoscale ocean currents can influence the recruitment pattern of coral reef fishes. For example, the cyclonic ocean eddy off Hawaii that remained in one general area for 70 days from July to September 1982 allowed many reef fish larvae to complete their planktonic development from an embryo to a settlement-stage larva. There are indications suggesting that the rotating stationary eddy (i.e., an eddy having no translatory motion) could be bringing larvae near reefs more regularly than at other times when such eddies are absent. It has also been suggested that ocean currents can be influential in the recruitment process. The frequency with which larval fishes could be transported past a coast is a function of an eddy's swirl speed. An eddy is predicted to act as a major entrapment and near-island retention mechanism. Lobel (2011) found that eddies can enhance local replenishment of populations as well as being a mechanism for transport away. If an eddy develops and remains in a given area adjacent to a coast and thereby retains larvae in local waters, recruitment may be enhanced. Alternatively, if an eddy develops and entrains a quantity of larvae but moves far offshore quickly, it will likely have a diminished effect on recruitment back to the natal area but may result in enhanced recruitment elsewhere.

Links between oceanic fronts in the mid-latitude North Atlantic and North Pacific and the large-scale atmospheric circulation in the Northern Hemisphere have been a subject of study over the last few decades. Efforts to study the impact of oceanic fronts on the large-scale atmospheric circulations, using various data products and atmospheric general circulation models with high spatial resolutions, have revealed that the oceanic fronts do play a major role in the large-scale atmospheric circulations. It has been found that SST anomalies in the vicinity of the fronts can generate at least local response in the atmosphere (see Kwon et al., 2010; Kelly et al., 2010; Nakamura, 2012). The variations in the Gulf Stream and the Kuroshio—Oyashio Extension exert major impacts on the storm tracks and low-frequency flow in the Northern Hemisphere.

1.4. MAJOR CURRENT SYSTEMS IN THE WORLD OCEANS

Water circulation in the oceans results from a number of primary forcing mechanisms such as thermohaline driving, surface wind stress, tide-generating forces, and so on. Other major influences include the piling-up effect occurring at the western boundaries of the ocean and the presence of ocean ridges, gorges, sandbanks, islands, and the proximity of the seabed. The just-mentioned forces and perturbations produce a spectrum of water motions, with temporal and spatial scale variations spanning from turbulent motions at millimeter scales to interbasin exchanges and large-scale boundary currents that have representative distance scales up to several hundred km and perturbations at time scales of months or longer.

Oceanic water circulation can be broadly classified into steady or irregular motions. In any turbulent flow, heat, chemical components, and dispersed materials are transported not only by the mean currents but also by the turbulent components. Surface waves and shear flows induced by wind stress are the main sources of small-scale turbulence in the upper few meters of the ocean, whereas internal waves and eddies cause turbulent motions at depths. A turbulent motion is made up of a large number of components with different frequencies and length scales. In some cases instantaneous current supplies energy to the mean current. The resolution of the large-scale turbulence problem is especially important in the oceanographic context because the fluctuating turbulent motions are normally much stronger than the velocity of the mean circulation (Zenk et al. 1988). Measurements of turbulent velocity pulsations permit evaluation of energy dissipation rates in the near-surface layer. It is, therefore, important to measure not only the mean currents but also their fluctuations with sufficient accuracy and resolution. To study the dynamics of the prevalent eddies in the ocean, precision instruments capable of measuring current fluctuations become necessary. Air-sea interaction studies of momentum, heat, and salt fluxes in the upper ocean require a detailed knowledge of the mean and fluctuating components of near-surface flows, whereas bottom turbulence measurements are required to describe the benthic boundary layer dynamics and for sediment transport studies near the ocean floor. Flow sensors used for such studies must respond fast enough to capture the signals of interest and must be small enough to resolve their spatial variations. Such measurements hold out the promise of collection of useful information that will throw some light on the mechanisms of many strong ocean currents.

The global ocean current circulation is strongly structured in terms of dynamically distinct regions. In particular, the western boundary regions of the oceans are dominated by large-scale meandering currents (Loder et al., 1998). Thus there exist major currents such as the *Antarctic Circumpolar Current* in the Antarctic Ocean, the *Gulf Stream* and the *Brazil Current* in the Atlantic Ocean, the *Kuroshio Current* in the Pacific Ocean, the *Agulhas Current* and the *Somali Current* in the Indian Ocean, and the *Equatorial Undercurrent* beneath the equator in the Atlantic Ocean, Pacific Ocean, and Indian Ocean.

With the observation that most of the kinetic energy of the ocean circulation is associated with mesoscale variabilities, namely major ocean gyres, the general circulation is broadly regarded as ocean climatology and the mesoscale variability as the ocean weather (Munk et al., 1979). Mesoscale variabilities are ocean variabilities on a 100-km scale. It is now known that, at any given time, a snapshot of the ocean circulation is dominated by the mesoscale eddies, which can be regarded as "ocean storms" (Monk, 1983). Near-monochromatic anticyclonic oscillations in the ocean have been reported, and no atmospheric counterparts to these eddies are known to exist (Rossby, 1983). The variety of oceanic motion that falls under the general title of eddy motion is enormous. Interestingly, theoretical model studies indicate (Burkov et al., 1981) that the horizontal structure of the general oceanic circulation can be represented mainly by a system of semi-enclosed gyres nongeostrophically connected to each other through their boundary fronts. In fact, the earlier notion of the ocean as a steady-state system of large-scale, wind-driven currents and a weaker thermohaline abyssal circulation, and the hitherto classical view of the ocean circulation as a climatological mean taken over several years, have been overhauled with the discovery of these mesoscale variabilities supported by theoretical model studies of global circulation patterns. It is now known that several well-known coastal currents of the world oceans are influenced by these mesoscale variabilities.

Western boundary currents generally exhibit different modes such as large meander, straight path, and short meander. It is now well recognized that western boundary coastal currents are merely the limbs of one or more gigantic whirls referred to as *gyres*. The water mass structures across the deep-sea free inertial jets can be rather complex. For the Gulf Stream, the high-transport jet core is situated at the boundary between two distinct water masses (slope and Sargasso). Additionally, the core of the stream carries several different water masses from outside this region, which is usually originated during its inertial and wind-driven regime further south (Gangopadhyay and Robinson, 2002). Similar situations must be anticipated in other regions of the world oceans.

1.4.1. Antarctic Circumpolar Current

In terms of the mean ocean circulation, the Southern Ocean is distinguished from all other oceans by the presence of a strong eastward-flowing circumpolar current, namely the *Antarctic Circumpolar Current* (ACC). The ACC is the only current that circumnavigates the globe unimpeded by continental barriers and coastlines (see Figure 1.5). The ACC connects the three major ocean basins (Atlantic, Pacific, and Indian) and redistributes active and passive oceanic tracers such as heat, salt, and nutrients. Thus, anomalies created by atmospheric forcing in one basin can be carried around the globe, affecting the global oceanic mass balance, ocean stratification, circulation, and consequently heat transport and climate.

The ACC encircles the Antarctic continent, flowing eastward through the southern portions of the Pacific, Atlantic, and Indian oceans. It is the world's largest, and arguably most influential, ocean current (Nowlin and Klinck, 1986; Rintoul et al., 2001). Although the speed of the ACC is not extraordinary (about 50 cm/s at the surface), its depth (about 4 km) and width (500–1,000 km) result in a massive transport of about 140×10^6 m^3 of water per second, equivalent to about 150 times the flow of all the world's rivers combined. Because the Pacific, Atlantic, and Indian ocean basins are almost entirely surrounded by land except at their southern boundaries, the ACC is the primary means by which water, heat, and other properties are exchanged between ocean basins. At the latitudes of the Drake Passage there are no barriers to zonal flow (i.e., current flow parallel to the equator) in the upper ocean. Water from abyssal depths in the ocean at low latitudes rises along the sloping density surfaces associated with the ACC and toward the surface in the Southern Ocean, helping to sustain the conveyor belt circulation. In fact, the characteristics of more than 50% of the world ocean volume reflect the air–sea ice interactions taking place in the Southern Ocean.

The ACC forms the northern boundary for the Southern Ocean, which represents an important component of the climate system. Antarctic bottom water, one of the coldest and densest water masses, is formed in the Southern Ocean. It is this water mass that cools and ventilates most of the volume of the deep oceans (Schmitz, 1995). There is

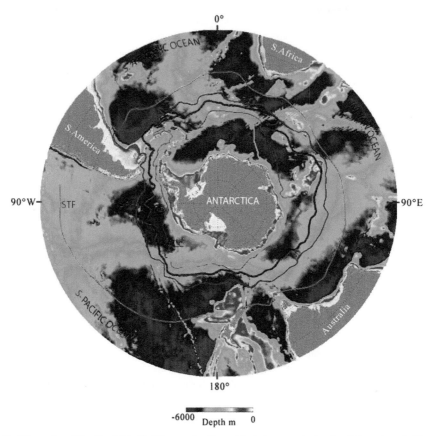

FIGURE 1.5   Trajectory of the eastward flowing Antarctic Circumpolar Current that circumnavigates the globe unimpeded by continental barriers and coastlines and connecting the three major ocean basins (Atlantic, Pacific, and Indian). *(Source: Orsi et al., 1995.)*

considerable interannual and decadal variability at high southern latitudes, and these features tend to propagate along the ACC. The anomalies circle around Antarctica in roughly 8–10 years at an average speed of 6–8 cm/s (White and Peterson, 1996). The coupling of the atmosphere and the ocean takes place via anomalous wind stresses and heat fluxes. In short, the ACC profoundly influences, and is influenced by, the regional and global climate (Fyfe and Saenko, 2005).

The absence of land barriers in the ACC results in distinct dynamical features that have no direct counterpart in the theory of mid-latitude ocean gyres. Sverdrup dynamics, the cornerstone of subtropical thermocline theory (Rhines and Young, 1982; Luyten et al., 1983), do not apply here. Both wind and buoyancy forcing play a role, as do geostrophic eddies, which appear to be crucial in determining the stratification and transport of the ACC. Recently, however, residual-mean theories have been applied (see Karsten et al., 2002; Marshall and Radko, 2003) that appear to capture the essence of the zonally averaged circulation and stratification of the Southern Ocean and fully embrace the central role of eddies. It is assumed that the Eulerian meridional circulation driven by the westerly winds (the Deacon cell), tending to overturn

isopycnals, is largely balanced by the geostrophic eddies that act in the opposite sense.

It is generally believed that wind stress is the main source of zonal momentum for the current, although thermohaline processes may also be important in driving the ACC (Olbers and Wubber, 1991). The dynamics of the ACC are complex and strongly depend on interactions among mesoscale eddies, topography, and the mean flow (Gouretski et al., 1987; Wolff et al., 1990, 1991). Munk and Palmev (1951) were the first to show that the flow can establish a pressure difference across meridional ridges, with the high pressure on the western side of the ridge. A necessary ingredient of the flow is the presence of mesoscale eddies that transfer the surface stress to the bottom, where it is dissipated by the topographic form stress. Marshall (1995) investigated the topographic steering of the ACC. He found that the ACC is partially steered by bottom topography and therefore does not exactly follow latitude circles. As a result, the surface heat flux into the ocean increases (relative to the streamline average) where the ACC meanders equatorward into warmer regions downstream of Drake Passage and decreases when the ACC drifts poleward in the Pacific sector. Wind stress also varies along the path of the ACC, being significantly larger in the Atlantic-Indian sector. Marshall (1995) found that transient eddies are important in the general maintenance of the current. Ivchenko et al. (1996) investigated the dynamics of the ACC in a near-eddy-resolving model of the Southern Ocean (FRAM) and found that the topographic form drag is the main sink of the momentum that is input by the wind.

The ACC is traditionally thought to be composed of a series of hydrographic fronts associated with sloping isopycnals and relatively strong meridional property gradients. Fronts are strongly steered by the topography. Sallee et al. (2008) found three typical frontal regimes, namely merging, shoaling, and lee meandering, depending on their position relative to the bathymetry. Thus topography influences the pathway of the fronts, as seen in several previous studies, but also influences the mean intensity of the jet.

1.4.2. Western Boundary Currents in the Atlantic Ocean

Two major currents in the Atlantic Ocean are the *Gulf Stream* and the *Brazil Current*. Of these two currents, the former commands more prominence.

1.4.2.1. The Gulf Stream

The GS originates in the Gulf of Mexico and then follows the North American continental shelf until the shelf takes a major bend at Cape Hatteras (35°N, 75°W), where it turns eastward away from the coast and flows straight out to sea, rather than following the bend (see Figure 1.6). Eventually the current breaks up in several directions south and east of the Grand Banks. The surface currents have typical maximum speeds of about 200 cm/s (4 knots). Except in summer, the current is conspicuously warmer at the surface than in the surrounding waters.

The GS System is a complex of currents in the western and northern North Atlantic Ocean. Based on the outcome of an extensive survey reported by Fuglister (1960), it is found that the GS occupies an extensive area on the western and northern edges of the relatively warm, saline, central Atlantic water mass where the main thermocline layer rises toward the sea surface. The principal part of the GS lies off the east coast of North America between Florida and Newfoundland. To the east of Newfoundland, the GS is separated from the continental shelf by the cold, southward-flowing Labrador Current. The currents of the GS System generally contain a core of water at the surface that is warmer than the surroundings, suggesting a transport from lower latitudes. Consequently, the westward flow of

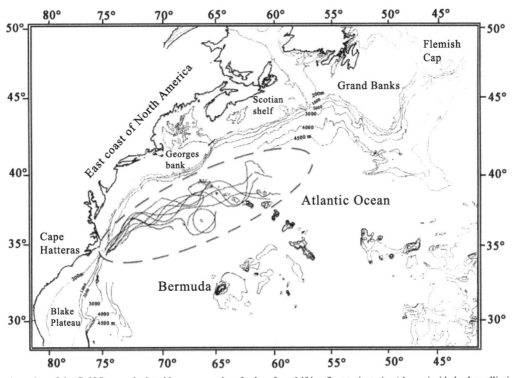

FIGURE 1.6 A portion of the Gulf Stream depicted by a composite of subsurface drifting float trajectories (shown inside broken elliptical boundary). The float trajectories clearly reveal the meandering character of the Gulf Stream and the formation of circular eddies. *(Source: In part from Rossby et al., 1985, © American Meteorological Society. Reprinted with permission.)*

FIGURE 1.7 Thermal infrared image of the Norwegian Current hugging the coast of Southern Norway. *(Source: Johannessen et al., 2000.)*

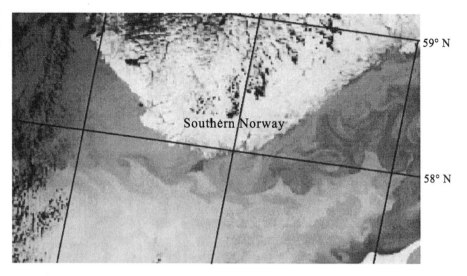

relatively warm water south of Iceland (the Irminger Current) and the northward flow off Norway (see the thermal infrared image of the Norwegian Current in Figure 1.7) are considered to be parts of the system. The Bay of Biscay and the Azores (see Figure 1.8 for the Azores Current region) are two regions into which some of the currents in the central Atlantic Ocean flow.

South of Cape Hatteras (see Figure 1.6), the GS System presses against the western boundary of the ocean basin. This boundary is not a vertical wall but consists, at the

surface, of the shore line, then a shelf roughly 60 miles wide out to the 200-m depth contour, then a broad plateau averaging 800 m in depth (the Blake Plateau), and, finally, a relatively steep slope down to the floor of the basin 5,000 m below. Flowing northward on the plateau, close to the shelf, is the strong current sometimes referred to as the Florida Current but more generally called the Gulf Stream. This current meanders, the amplitude of the meanders being about equal to the width of the stream (Webster, 1961), and it reaches to the bottom as evidenced by ripple

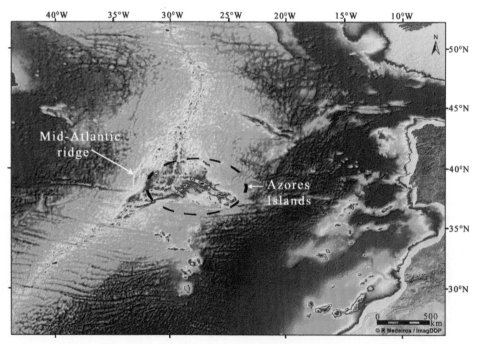

FIGURE 1.8 The Mid-Atlantic Ridge across which the Azores Current (a meandering jet 60–100 km wide with an eastward velocity of 25–50 cm/s) heads southeastward to the south of the Azores Islands chain and then flows mainly eastward at a latitude of about 35°N to the Gulf of Cadiz. *(Source: Courtesy of Dr. Ricardo S. Santos, University of Azores, Portugal.)*

marks and current observations. Stommel (1957) hypothesized a deep southward current along this boundary, and Swallow and Worthington (1961) observed a southward flow at depths near 2,800 m off Charleston, South Carolina. This flow has been referred to as a *deep countercurrent* to the Gulf Stream, although it is not actually beneath the stream in this area.

The Gulf Stream south of Hatteras is known to extend to the ocean bottom, i.e., to about 800 m. Profiles across this current show a horizontal density gradient at great depths, even near the bottom in 5,000 m of water. North of the principal current of the GS System, the main thermocline again rises abruptly toward the surface. This latter horizontal temperature gradient, or current, is not always present just north of Cape Hatteras but is a permanent and quite pronounced feature to the eastward, south of the Laurentian Channel. Based on the results of extensive surveys, Fuglister (1955) suggested that the Gulf Stream may not be a single continuous current between Cape Hatteras and the Grand Banks. It is generally agreed that the Gulf Stream forms the boundary zone between the Sargasso Sea and the slope water. At the surface it contains the warm core of the stream, which is characteristically fresher than the water at the same level in the Sargasso Sea and has less dissolved oxygen than the water to either side. The Slope Water Current and the Gulf Stream are both parts of the GS System, but the interrelationship between the two currents is not clear. Measurements have indicated that the subsurface currents in the GS are essentially in the same direction as the surface flow.

Detailed surface- and subsurface-temperature measurements made in the western area showed a banded structure parallel to the GS, which was undoubtedly associated with the streaky, "discontinuous edge" of the stream as observed from the air (Von Arx et al., 1955). The 1960 survey revealed the presence of an eddy, which was moving slowly toward the north along an anticyclonic curve. The eddy was found to have been circling in the area over a period of a month apparently before it moved again downstream.

The GS is well known for meander activity. Sudden change in the pattern of meanders is a permanent feature of the Gulf Stream. Consequently, the GS is perceived as a giant conduit wiggling about in the ocean. The meander pattern of the current exhibits a sharp line of demarcation near 65°W longitude, the longitude of Bermuda, separating the area of relatively small amplitude meanders in the west from the eastern area of much larger north-south meanders. The shapes of these large meanders may be influenced by the various geographical obstacles such as sea mounts in this area. For example, the GS path is deflected and curved around by Kelvin Sea Mount and the New England Seamounts. Ring generation is intensified by the presence of such Seamounts (Richardson, 1981). Lai and Richardson (1977) identified

cold rings in the North Atlantic characteristically as having greater than a 150-m upward displacement in the thermocline isotherms. They investigated the movement of GS rings and obtained a mean distance of 250 km during a season. The GS separates into two branches at the southeast of the Grand Banks of Newfoundland. The northern branch turns northeastward and becomes the North Atlantic Current (NAC). The southern branch, which becomes the Azores Current (AC), heads southeastward across the Mid-Atlantic Ridge to the south of the Azores (see Figure 1.8), then flows mainly eastward at a latitude of about 35°N to the Gulf of Cadiz (GoC).

Jia (2000) has provided an analysis of the Azores Current and shows that its existence and strength depends critically on the volume flow out of the Mediterranean Sea. Associated with AC is a front with significant temperature and salinity contrasts. The eastward flow of the AC extends all the way to the African coast, with southward branches in the Canary Basin as part of the subtropical gyre recirculation (Stramma 1984; Olbers et al., 1985; Klein and Siedler, 1989). The hydrographic database of Lozier et al. (1995) reveals a coherent AC that stretches across the eastern half of the basin, with divergences to the south and convergences from the north such that the downstream transport does not change much. Hydrographic surveys also indicate the eastward extension of the AC to the Moroccan continental slopes (Fernandez and Pingree, 1996; Pingree, 1997). Drifting buoys deployed in the AC are found to travel eastward and reach the western side of the GoC, then move northward or southward along the continental slopes. Based on the just-mentioned surveys, the AC is observed to be a meandering jet 60–100 km wide with an eastward velocity of 25–50 cm/s. The eastward flow is mostly in the upper few hundred meters but can reach as deep as 2,000 m. The current carries a large fraction of the water entering the eastern recirculation region of the Canary Basin. The estimates of the AC transport are in the range of 10–15 Sv (1 Sv $\equiv 10^6$ m^3/s). The surface temperature and salinity changes across the front can be as large as 2°C and 0.3 psu. The front marks the northern boundary of the 18°C Sargasso Sea water in the central North Atlantic. Both drifter data (Richardson, 1983; Krauss and Kase, 1994; Brugge, 1995) and satellite altimetry data (e.g., Le Traon et al., 1990; Wunsch and Stammer, 1995; Stammer, 1997) show a band of high eddy kinetic energy (EKE) associated with the AC. Kase and Siedler (1982) observed considerable meandering of the front southeast of the Azores with mesoscale eddies on both sides of the front.

1.4.2.2. The Brazil Current

The Brazil Current (BC) was a great bugbear to ancient mariners, principally on account of the difficulties that a few dull vessels falling to leeward of St. Roque found in beating up against it. It was said to have caused the loss of some English transports in the 18th century; they fell to leeward of the cape on a voyage to the other hemisphere, and navigators, accordingly, were advised to shun it as a danger.

The BC is a highly baroclinic western boundary current. This current has a southern segment and a northern segment. The formation region of the North Brazil Current (NBC) is generally agreed to be near 10°S, where waters flowing westward in the South Equatorial Current (SEC) first begin to concentrate into a northward boundary current. The bulk of the volume transport of the BC is concentrated between 25 and 40 Sv, in the top 500 m of the water column. The collision of the Brazil Current and the Malvinas Current (which is a swift, barotropic, and narrow branch of the Antarctic Circumpolar Current that flows north along the continental slope of Argentina up to approximately 38°S), known as the Brazil/Malvinas Confluence, occurs near the mouth of the La Plata River, where it creates a region of intense mesoscale variability (Figure 1.9). The interaction between the poleward flow of the BC and the bottom topography greatly influences the nearshore circulation, particularly in the bottom boundary layer.

Temporally growing frontal meandering and occasional eddy shedding are observed in the BC as it flows adjacent to the Brazilian Coast. Silveira et al. (2008) reported a study of the dynamics of this phenomenon in the region between

22°S and 25°S. Large frontal meanders and intense meso-scale activity are often observed off the southeastern Brazilian coast (Mascarenhas et al., 1971; Signorini, 1978; Schmid et al., 1995; Silveira et al., 2004). The vertical structure of the BC system between 20°S and 25°S presents a unique regime in terms of a subtropical western boundary current. From the surface down to intermediate depths (400−500 m), the BC flows south−southwestward. The BC has maximum surface speeds of 40−70 cm/s and a width of about 100−120 km. It transports two water masses: Tropical Water (TW) at surface levels and South Atlantic Central Water (SACW) at pycnocline levels (Silveira et al., 2008). The BC transport ranges from 5 to 10 Sv in this region. Below 500 m, there is a direction reversal associated with the Intermediate Western Boundary Current (IWBC) flow toward the north−northeast (Boebel et al., 1999). The IWBC transports 2−4 Sv of Antarctic Intermediate Water (AAIW) and has its core centered between 800 and 1,000 m, with maximum speeds of about 30 cm/s. The frontal meander growth inferred from thermal satellite images (Figure 1.10) suggests that a mechanism of geophysical instability occurs in the region.

The NBC is an intense low-latitude western boundary current in the western tropical Atlantic Ocean that transports upper-ocean waters northward across the equator (Figure 1.11). Near 6−8°N the NBC separates sharply from the South American coastline and curves back on itself (retroflects) to feed the eastward North Equatorial Countercurrent (e.g., Csanady, 1985; Ou and DeRuijter, 1986; Johns et al., 1990, 1998; Garzoli et al., 2003). The NBC retroflection is present year-round but is most intense in boreal autumn. The NBC occasionally retroflects so severely as to pinch off large, isolated warm-core rings exceeding 450 km in overall diameter. The anticyclonic rings, with azimuthal speeds approaching 100 cm/s, move northwest-ward toward the Caribbean Sea on a course parallel to the South American coastline (Goni and Johns, 2003). After translating northwestward for three to four months, the rings decompose in the vicinity of the Lesser Antilles. During their brief lifetime, and especially upon encountering the islands of the eastern Caribbean, the strong and transient velocities associated with NBC rings episodically disrupt regional circulation patterns, impact the distributions of near-surface salinity and icthyoplankton (e.g., Cowen et al., 2003), and pose a physical threat to expanding deep-water oil and gas exploration on the South American continental slope (e.g., Summerhayes and Rayner, 2002).

In the Atlantic the NBC plays a dual role, first in closing the wind-driven equatorial gyre circulation and feeding a system of zonal countercurrents and second in providing a conduit for cross-equatorial transport of South Atlantic upper-ocean waters as part of the Atlantic meridional over-turning cell (Johns et al., 1998). Seasonal variability of the NBC between 10°S and the equator appears to be quite small.

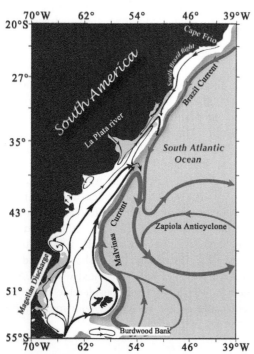

FIGURE 1.9 Schematic representation of the depth-averaged circulation including Brazil Current and Malvinas Current in the southwestern Atlantic region. The shelf (depths smaller than 200 m) is marked by a white background. *(Source: Matano et al., 2010.)*

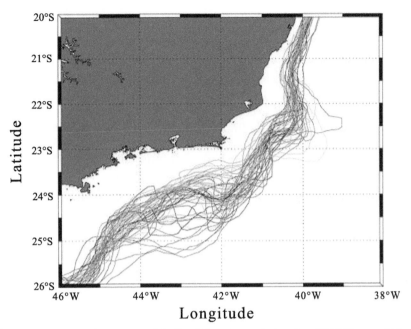

FIGURE 1.10 Thermal front patterns representing a section of the Brazil Current meander based on AVHRR images obtained at the Brazilian National Institute for Space Research, INPE. *(Source: Silveira et al., 2008.)*

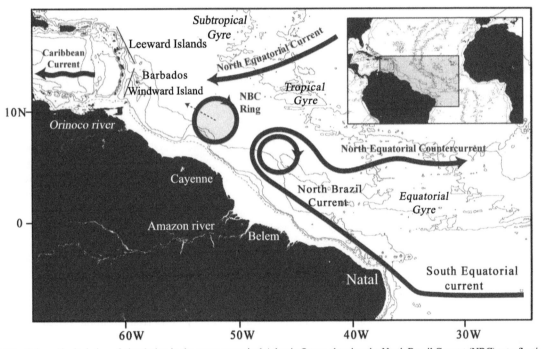

FIGURE 1.11 Schematic depiction of circulation in the western tropical Atlantic Ocean showing the North Brazil Current (NBC) retroflecting into the North Equatorial Countercurrent near 6°N. The NBC retroflection occasionally collapses upon itself, resulting in the generation of anticyclonic NBC rings that translate northwestward toward the Caribbean and the arc of the Lesser Antilles. *(Source: Fratantoni and Richardson, 2006, © American Meteorological Society. Reprinted with permission.)*

However, models (e.g., Philander and Pacanowski, 1986) indicate that the NBC has a large seasonal cycle north of the equator, related to the seasonal migration of the intertropical convergence zone (ITCZ) and associated changes in the wind stress curl across the interior of the basin.

In contrast to the BC, the NBC regularly sheds rings primarily because of the small inclination of coastline between 5° and 8°N (Zharkov and Nof, 2010). Both the NBC and its rings contribute to the dispersal of fresh, nutrient-rich outflow from the Amazon River and provide a mechanism for transport of this water northwestward toward Tobago and Barbados (e.g., Muller-Karger et al., 1988; Johns et al., 1990; Fratantoni and Glickson, 2002). As described by Fratantoni and Glickson (2002), NBC rings undergo a rapid evolution, driven primarily by interaction with topography and neighboring rings. Fratantoni and Richardson (2006) have reported a detailed account of the evolution and demise of North Brazil Current rings.

1.4.3. Western Boundary Current in the Pacific Ocean: The Kuroshio Current

The major boundary current in the Pacific Ocean is the *Kuroshio Current* flowing along and off the Japan coast. The other current in the Pacific Ocean is the southward-flowing East Australian Current (EAC). Extending from the Coral Sea to the Tasman Sea, the EAC system generates numerous eddies and has several branches including the Tasman Front, the East Auckland Current (EAUC), the East Cape Current (ECC), and the EAC extension. Unlike other western boundary currents such as the Gulf Stream and the Kuroshio, the strength of the EAC varies substantially with time. The mass transport within eddies spawned by the EAC can be much larger than its mean flow (Ridgway and Godfrey, 1997; Lilley et al., 1986). The flow patterns in the EAC are so complex and variable that it is often difficult even to decide whether a single, continuous current exists (Godfrey et al., 1980). In the light of the preceding description of the EAC, this current is not considered in this book. Instead, some general ideas on the Kuroshio Current System are given here.

In recent years, much curiosity has been generated concerning the quasi-stationary meander in the Kuroshio Current System in the North Pacific as it flows eastward past the Japanese islands. When the meander is absent, the Kuroshio flows along the continental slope near the 1000-m isobath, but when it is well formed, the Kuroshio bends away from the slope near Shikoku and enters deep water (see Figure 1.12). As just mentioned, the Kuroshio Current System consists of two limbs: the Kuroshio south of Japan and the Kuroshio Extension east of Japan. The Kuroshio and its variations may exert great influence on the climate of East Asia, particularly China, Japan, and Korea. On the other hand, it may respond to some big oceanic/climatic events such as El Nino. Great efforts have been made to describe Kuroshio and its relationship with some variations of marine environment and atmosphere. The Kuroshio south of Japan exists in one of two stable paths, a zonal path (i.e., path parallel to the equator) and a meander path, each lasting for a period of years. Taft (1972) has shown that the meander may remain in evidence for a number of years, disappear for some years, and then reappear. Past studies of the temporal and spatial variability of the Kuroshio Current System have revealed the presence of both annual and interannual variability (Taft, 1972; Shoji, 1972). Their studies can be summarized as follows:

- The amplitude of the Kuroshio meander is as large (250 km) as Kyushu Island and its adjacent shelf and slope region.
- When the meander is present, its wavelength seems directly related to the magnitude of the transport.
- The transport of the Kuroshio downstream from Kyushu is generally about 30% greater than the transport upstream from Kyushu, this being the result of recirculation of some kind. In addition, the maximum surface speed of the Kuroshio (generally in the vicinity of the Kii Peninsula) can increase by a factor of two or more from speeds measured off Kyushu, indicating that the Kuroshio downstream from Kyushu can be both faster and narrower than upstream.
- There is an inverse relationship between the transport of the Kuroshio and the presence of the meander; below the critical transport (i.e., $\sim 40 \times 10^6$ m^3/s) the meander is present, above the critical transport the meander disappears, leading to an apparent bimodal character of the Kuroshio meander.

These conclusions have been corroborated by White and McCreary (1976) in their studies.

Based on a numerical study, Chao and McCreary (1982) showed three separate paths that arise from Rossby wave resonance with the coastline topography; two of these were similar to the observed paths. According to the studies carried out by Mizuno and White (1983), the Kuroshio Extension bifurcates near the Shatsky Rise. One of the branches, represented by the 7–8°C isotherm, consistently goes northeastward along the Shatsky Rise, while the main branch extends eastward along 36°N, encountering the Emperor Seamounts. After the regional change, this bifurcation occurs much farther to the west, near 150°E.

The Kuroshio Extension east of Japan is known as a highly variable current, with cold and warm rings generated from unstable meander growth. Ring formation is usually a fast process, taking two weeks to one month as observed by Fuglister and Worthington (1951). However, Kawai (1972) showed an example of ring formation that took place more slowly over a six-month period. Kitano (1975) observed warm rings in the Kuroshio Current System to move 50–250 km during a season in the northeast direction. Hata (1974) observed one warm ring almost continuously for 21 months, moving also in the northeast direction 110 km during a season on average. The typical speed of translatory movement of rings is <2 km/day (i.e., <2.3 cm/s); and due to their relatively slow speed the detection of ring movement is much easier. According to a study of Mizuno and White (1983), both warm and cold rings in the Kuroshio Current System propagate to the west

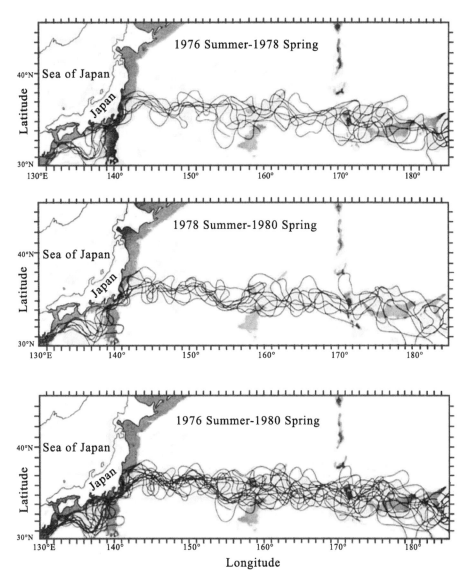

FIGURE 1.12 Path of the Kuroshio Current System during various years. *(Source: Mizuno and White, 1983, © American Meteorological Society. Reprinted with permission.)*

at an average speed of 1 cm/s. Cold rings were observed to have moved consistently southward, at about 0.6 cm/s, whereas warm rings moved generally northward, at about 0.4 cm/s, but less consistently. The fact that warm rings tend to have moved poleward while cold rings tend to have moved equatorward indicates that a net flux of heat is being conducted poleward by the meridional movement of rings in the Kuroshio Current System.

The Kuroshio Current System is characterized by the quasi-stationary existence of meanders (Kawai, 1972). Bernstein and White (1977) first observed the eastward extension of intense meander and eddy activity in the Kuroshio Extension. It extended eastward from the coast of Japan to at least 175°W, which is much farther east than had previously been realized. The quasi-stationary meander pattern in the Kuroshio Extension becomes

unstable, associated with increased eddy activity and ring production.

Using a series of multiple-ship expendable bathythermograph (XBT) surveys, Wilson and Dugan (1978) and Bernstein and White (1981) produced a series of thermal maps of the eddy activity in the Kuroshio Extension and showed westward propagation of meanders and eddies over this region. Ring formation is one of the important processes of mesoscale activity in the Kuroshio Extension. Whereas warm rings occupy the Kuroshio Extension adjacent to Japan (Kawai, 1972; Hata, 1974; Kitano, 1975; Tomosada, 1978), cold rings occupy the area southeast of Japan (Kawai, 1979).

Bathymetric perturbations such as the Izu Ridge, the Shatsky Rise, and the Emperor Seamounts and their valleys have been found to deflect the Kuroshio Current System (see Taft, 1972; Bernstein and White, 1981).

Abyssal plane also plays a role in the eddy activity. For example, Bernstein and White (1981) found eddy activity in the Kuroshio Extension to be intensified over the abyssal plane between the Izu Ridge and the Shatsky Rise. According to the studies conducted by Mizuno and White (1983), the Kuroshio Extension east of Japan can be traced as a high-gradient frontal feature.

1.4.4. Western Boundary Currents in the Indian Ocean

Two major boundary currents in the Indian Ocean are the *Agulhas Current* and the *Somali Current*, both flowing along the east coast of the African continent. In other regions in the Indian Ocean (except the equatorial under-current), the general patterns of circulation are found to be masked by eddy variability, and some form of averaging is needed to define patterns of general circulation.

1.4.4.1. The Agulhas Current

The Agulhas Current is the strongest western boundary current in the Southern Hemisphere. It transports about 65 Sverdrup (Sv, 1 Sv $\equiv 10^6$ m^3/s) of warm tropical water southward along the southeast coast of Africa (Stramma and Lutjeharms, 1997). This is also one of the strongest western boundary currents in the world's oceans (Lutjeharms, 2007). The water movement in the southwest Indian Ocean is largely dominated by the anticyclonic sweep of the Agulhas System (Barlow, 1935). Figure 1.13 shows trajectories of the Agulhas Current and eddies constructed from thermal infrared images for a period from December 1984 to December 1985. The three major sources in the south Indian Ocean for the Agulhas Current are considered to be: (1) recirculation in the south-west Indian Ocean, (2) flow through the Mozambique Channel, and (3) the East Madagascar Current. This interrelation is also subject to seasonal variation due to direction changes in the monsoon winds (Barlow, 1935; Darbyshire,

1964). The flow in the Mozambique Channel is dominated by southward-moving anticyclonic eddies (Sætre and da Silva, 1984; Biastoch and Krauss, 1999; de Ruijter et.al., 2002). As a major western boundary current, the Agulhas Current, which rushes toward the south pole in the southwest Indian Ocean, exhibits a unique and prominent turnabout near 20°E, at which the current direction is more or less reversed back eastward; this is known as the *Agulhas retroflection* (Gordon et al., 1987; Jacobs and Georgi, 1977; Lutjeharms and van Ballegooyen, 1988). It retroflects and flows back into the mid-latitude South Indian Ocean as the Agulhas return current (Lutjeharms and van Ballegooyen, 1988) before eventually becoming the South Indian Ocean current (Stramma, 1992).

Figure 1.14 shows a conceptual image based on seven years of thermal data of the Agulhas retroflection and environment. Thus, forming part of the Southwest Indian Ocean subgyre, the Agulhas Current flows poleward along the southeastern coast of Southern Africa from 27°S, eventually retroflecting and flowing eastward back into the South Indian Ocean south of Africa between 40°S and 42°S (Gordon, 1985; Stramma and Lutjeharms, 1997). Although the Mozambique Channel eddies and the East Madagascar Current do not form a continuum with the Agulhas Current, they both affect its dynamics (Lutjeharms, 2007) and contribute to the fluxes of volume, heat, and salt. For example, interaction of the Mozambique Channel eddies with the Agulhas Current has been shown to influence the timing and frequency of Agulhas ring-shedding events at the retroflection.

The terminal region of the Agulhas Current is also populated by a range of eddies. During the flow of the relatively warm waters of the retroflection and the Agulhas return current in the east-southeastward direction across the southern mid-latitudes, they release large amounts of heat to the overlying atmosphere. Each year, a number of Agulhas rings are shed off the retroflection region and track northwest into the southeast Atlantic Ocean, again with

FIGURE 1.13 Trajectories of Agulhas Current and eddies constructed from thermal infrared images for a period from December 1984 to December 1985. *(Source: Lutjeharms and Van Ballegooyen, 1988, © American Meteorological Society. Reprinted with permission.)*

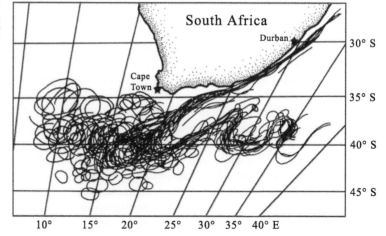

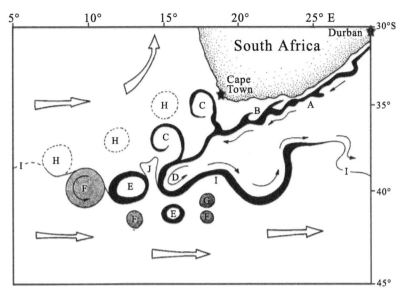

FIGURE 1.14 A conceptual image based on seven years of thermal data on the Agulhas retroflection and environment. *(Source: Lutjeharms and Van Ballegooyen, 1988, © American Meteorological Society. Reprinted with permission.)*

substantial heat fluxes to the atmosphere. It is likely that such large fluxes may have a significant impact on the regional atmospheric circulation and help to enhance the intensity of transient weather disturbances passing across southern Agulhas waters (e.g., Walker, 1990). It is also conceivable that heat transfer over the Agulhas return and South Indian Ocean currents may influence the intensity and track of mid-latitude depressions that contribute significantly to winter rainfall downstream in southern Australia (Reason, 2001). A number of studies have linked the variability in the greater Agulhas Current region with rainfall over large areas of South Africa (e.g., Walker, 1990; Jury et al., 1993; Mason, 1995).

The Agulhas retroflection is unstable, and it can coalesce and form Agulhas rings (Lutjeharms, 1981a) with an average diameter of 320 km (Lutjeharms, 1981b). The factor responsible for the singular current behavior manifested in the retroflection may stem from the principle of conservation of potential vorticity. The exchange process of water south of Africa depends to a large extent on ring shedding at the Agulhas retroflection (Lutjeharms and Gordon, 1987; Olson and Evans, 1986). A narrow filament of much colder water lies directly to the west of the retroflection, spreading laterally north of the current. The narrow, warm ribbon of the Agulhas Current (about 90 km wide, on average) exhibits a number of meanders south of Africa.

Because of retroflection, the volume exchange between the Indian Ocean and the Atlantic Ocean due to the Agulhas Current is relatively small, that is, about 10% of the Agulhas transport. Within the Agulhas retroflection region, the flow has a strong variability, and about six times every year a large ring is separated from the main current that propagates westward into the South Atlantic. Although the

exchange due to the rings is small, it may be sufficiently large to cause changes in the Atlantic thermohaline circulation because the properties of the Indian Ocean water are substantially different from those in the Atlantic (Gordon, 1985). Paleoceanographic studies have indicated that this exchange has varied significantly in the past, although a "supergyre" that fully connected the Indian and Atlantic Oceans probably never existed. The physics of the Agulhas Current, particularly its retroflection, has been examined by Dijkstra and de Ruijter (2001).

Because the Agulhas Current experiences strong topographic control, first by the continental slope of Africa, then by the Agulhas Bank, it does not exhibit the continuous meandering of the detached, eastward-flowing Gulf Stream. As a result, such a distinct steering level within the Agulhas is not expected. However, solitary meanders, often referred to as *Natal pulses* (owing to their apparent origin over an area of wide shelf known as the *Natal Bight* offshore of Port Elizabeth) propagate downstream in the Agulhas Current about six times per year (Lutjeharms and Van Ballegooyen, 1988; Lutjeharms and Roberts, 1988). The gentler continental slope and wider shelf at Natal Bight, between 29°S and 30°S, present favorable conditions for the occurrence of instabilities and subsequent growth of meanders (de Ruijter et al., 1999). Thus, mesoscale variability in the northern Agulhas Current occurs in the form of Natal pulses (intermittent cyclonic meanders). Natal pulses propagate downstream at rates of 10 km/day (Lutjeharms and Roberts, 1988).

Using current-meter time series measurements, Bryden et al. (2005) showed that solitary meanders are the dominant mode of variability in the Agulhas Current. Van Leeuwen et al. (2000) demonstrated that solitary meanders (i.e., Natal pulses) on the trajectory of the Agulhas Current are responsible for the timing of ring shedding at the Agulhas retroflection. These Natal pulses in turn are triggered by offshore eddies. Schouten et al. (2002) demonstrated a case in which an eddy from the Mozambique Channel triggered a Natal pulse that was subsequently responsible for the occlusion of an Agulhas ring at the Agulhas retroflection. It is believed that a Natal pulse event enables cross-frontal mixing within the Agulhas Current in addition to causing straightforward path variability. The shedding of nine rings per year, even if not all of them were to drift northward, has significant implications for the energy contribution of these features to the total kinetic energy balance of the South Atlantic Ocean. A detailed study by Lutjeharms and Van Ballegooyen (1988) of this ring-spawning process shows that in almost all cases the shedding of a ring is preceded by the genesis and growth of a cold wedge of Sub-Antarctic Surface Water at the subtropical convergence. It is believed that the spawning of rings may be influenced or precipitated directly by perturbations in the Agulhas Current itself.

A major feature of the Agulhas Current is a very sharp meander in the Agulhas Return Current at about 27°E. It has been identified as a cold-core eddy being spawned at the subtropical convergence. Another feature is a large eddy southwest of Cape Town, centered at 15°E. A distinctive wedge of cold sub-Antarctic water separates the ring from the retroflection loop. The ring has an elliptical configuration with evidence of some shear-edge features. According to some estimates, the Agulhas rings have diameters of ~250 km and circular velocities of 90−100 cm/s and move at an average speed of 7 cm/s. Olson and Evans (1986) have shown that these rings translate at speeds of 4.8−8.5 cm/s. Sarukhanyan (1982) identified intense and less intense mesoscale eddy-like features south of the subtropical convergence; Lutjeharms (1988) has described the budding off of warm eddy across the convergence and its subsequent drift in the Sub-Antarctic zone. Some satellite images of the area show warm-water features with undifferentiated surface temperatures, i.e., discs with more or less the same surface temperature throughout, as well as distinct rings of warmer water with higher temperatures in an annulus shape, i.e., with slightly cooler water in the center. Although exceptional in its complete retroflection, the ring shedding from the Agulhas Current is not unusual for a western boundary current. However, the rate of ring shedding by the Agulhas Current seems high.

The average diameter for the retroflection loop (~342 km) does not vary dramatically from year to year. It has been suggested that the westward protrusion of warm Agulhas Current water may extend far into the South Atlantic.

Observations by Beal et al. (2006) suggest that the water mass properties on either side of the dynamical core of the Agulhas Current are significantly different. Inshore of its velocity core are found waters of predominantly Arabian Sea, Red Sea, and equatorial Indian Ocean origin, but the offshore waters are generally from the Atlantic Ocean, the Southern Ocean, and the southeast Indian Ocean.

1.4.4.2. The Somali Current

The Somali Current system is well known to be forced significantly by the wind field along and off the coast of Somalia. This current system is made up of two gyres. There is a northern gyre slightly to the north of the equator and a southern gyre a little to the south of the equator, and the Indian monsoon is related to the movements of these gyres (Anderson and Rowlands, 1976). As the monsoon sets over India, the southern gyre begins to move northward, and eventually the two gyres coalesce. The two-gyre configuration is most prominent during good monsoon years, but in years of weak monsoons the southern gyre is hardly perceptible. The alongshore winds over the Somalia coast are generally of a jet-like structure and can be very strong, and they reverse direction seasonally with the monsoons. In rapid response to the changing winds, the Somali Current also reverses seasonally. Soon after the onset of the Indian Southwest Monsoon, an intense coastal circulation exists north of 5°N. Cold water appears all along the coast of Somalia, although it is concentrated in wedges (Brown et al., 1980). The surface poleward flow is very strong (~100−200 cm/s) and forms one or two quasi-stationary eddies (Duing et al., 1980). There is an undercurrent with an instantaneous maximum speed as large as 60 cm/s (Leetmaa et al., 1982), although the monthly average speed has a maximum of only about 20 cm/s (Schott and Quadfasel, 1982; Quadfasel and Schott, 1983). The width scale of both currents is of the order of 50−100 km. The wind at this time is in the form of an alongshore jet, which reaches a maximum speed near 10−15°N, well north of the region of the undercurrent (Luther and O'Brien, 1985; Knox and Anderson, 1985). One western-boundary upwelling region that has been studied extensively is the coast of Somalia; it is found that this current system is a classic upwelling regime (Schott, 1983; Knox and Anderson, 1985). Based on model studies, McCreary and Kundu (1985) found the existence of an undercurrent in the south, which is in agreement with observations off Somalia, where a southward undercurrent has been observed at 5°N during the Southwest Monsoon.

1.4.5. Equatorial Undercurrents

The *Equatorial Undercurrent* (EUC), which is one of the most outstanding branches in the equatorial current system, is a narrow ribbon of eastward flow (termed *jet*) centered on the equator in the upper thermocline. It is a permanent feature of the general circulation in the Atlantic and Pacific oceans and is present in the Indian Ocean in northern winter and spring during the northeast monsoon. It reaches speeds of 50–100 cm/s below the westward flow of the South Equatorial Current, and in the Pacific the jet transports as much mass on average (40×10^6 m³/s) as the Florida Current, which feeds the Gulf Stream. Intense vertical shear of the undercurrent both above and below its core produces turbulent mixing. Turbulence tends to homogenize the thermal structure near the equator and so works in concert with the field of vertical velocity to weaken the thermocline. McPhaden (1986) has provided an interesting review of the chronology of historical events surrounding the "multiple discoveries" of the Equatorial Undercurrent.

It is now generally agreed that the driving force of the Equatorial Undercurrent is primarily the horizontal pressure gradient along the equator, given by the downward slope from west to east of the sea surface and of isobaric surfaces in the upper strata of the ocean. Among other reasons, speed variations along the course of the current have to be expected as a result of slope variations in zonal direction. The eastward zonal pressure gradient along the equator is generated by prevailing easterly wind distribution (i.e., wind blowing from the east) over the tropical and subtropical regions of the ocean. In general, the trade winds on both sides of the equator produce a westward transport of water in the upper strata of the sea. This water piles up along the western boundaries of the equatorial ocean basin, and the general slope of the sea surface in a zonal direction is downward from west to east.

An equatorial zonal slope downward from west to east, comparable to that in the Atlantic Ocean, is also found in the Pacific Ocean and in the Indian Ocean during the period of existence of an Equatorial Undercurrent. If the zonal slope downward to the east is a necessary requirement for the development of the Undercurrent, this current should be missing in the region where the slope is reversed. A peculiar physical-chemical structure, typical for the Equatorial Undercurrents in all three equatorial oceans, is a pronounced weakening of the vertical temperature gradient in the thermocline and water with a high oxygen content reaching down from the surface to a depth of 200 m and more (Neumann, 1966).

1.4.5.1. Equatorial Undercurrent in the Atlantic Ocean

The first observations of the Equatorial Undercurrent were made in 1886 by the British oceanographer J. Y. Buchanan in the Gulf of Guinea in the eastern Atlantic Ocean (Figure 1.15) during the *Challenger* Expedition. While making subsurface seawater temperature measurements, Buchanan found that the surface water had a very slight westerly set, and the water below 30 fathoms (180 ft) was running so strongly to the southeast direction that it was impossible to make temperature observations, since the heavily loaded lines drifted straight out and could not be sunk by any weight of which they could bear the strain. By using a makeshift current drogue composed of a surface buoy and a weighted biological sampling tow net suspended at about 55 m depth, Buchanan estimated the speed of this "very remarkable undercurrent" at more than 50 cm/s at three equatorial stations. He observed a weakening of the thermocline within 2° of the equator, a feature now commonly associated with equatorial upwelling and enhanced vertical mixing in the undercurrent.

On a later submarine cable-laying cruise from Senegal to the island of Fernando Noronha, Buchanan again observed the undercurrent in the western Atlantic and noted that the equatorial undercurrent is a constant and important factor of the oceanic circulation. Subsequent to the first current-meter measurements of the undercurrent in the Atlantic Ocean by Soviet oceanographers from on board the *R. V. Mikhail Lomonosov*, the eastward subsurface equatorial current in the Atlantic Ocean basin has frequently been referred to as the *Lomonosov Current* (Philander, 1973).

The Equatorial Undercurrent in the Atlantic Ocean received considerable attention in the years since 1960 (Neumann, 1960; Metcalf et al., 1962). It was found that the Atlantic Equatorial Undercurrent is a thin, swift, and relatively narrow current flowing from west to east at, or in close proximity to, the geographical equator. Its thickness is between 150 and 250 meters and its lateral extent 250 to 300 kilometers. The maximum speed in the core of the current is found at depths between 50 and 100 meters. In general, this core is deeper in the western part of the ocean and rises toward the east. The speed of the Equatorial Undercurrent varies along its course and with time at a given longitude. The average speed is probably near 70 cm/s. However, maximum speeds as high at 130 cm/s have been observed by Metcalf et al. (1962). They suggested that the high-salinity core of the Undercurrent has as its source region the high-salinity shallow water south of the equator off the east coast of Brazil. Cochrane (1963) described the Undercurrent as being an equatorial extension of a narrow saline current setting east-southeast at the surface near (2°N, 42°W), combined near 38°W with a retroverse branch of the northwesterly flowing Guiana Current along the South American coast.

Vertical oscillations of the core of maximum speed at a fixed location have been noted by several investigators. Evidence for lateral oscillations and lateral current displacements to the north and south of the equator, respectively, have been reported by Metcalf et al. (1962), Stalcup and Metcalf (1966), and Sturm and Voigt (1966). There is evidence that the current may sometimes split into two cores flowing just a little north and a little south of the equator (Sturm and Voigt, 1966). The temperature/oxygen relationship indicates that most of the Undercurrent water comes from the South Atlantic by way of the North

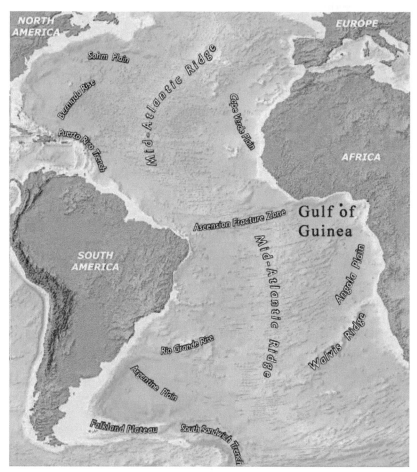

FIGURE 1.15 Map of a portion of the Atlantic Ocean that covers the equatorial belt showing the Gulf of Guinea region from where the "first discovery" of the equatorial undercurrent was made by the British oceanographer J. Y. Buchanan during the *Challenger* Expedition. *(Source: http://mccainsocialstudies.wikispaces.com/E1.+Geography.)*

Brazilian Coastal Current and that the contribution of North Atlantic water is very minor (Metcalf and Stalcup, 1967). They provided evidence on the source of the distinctive water type that makes up the Undercurrent's core and the complex relationship between the Undercurrent and other nearby currents. Figure 1.16 provides a detailed schematic drawing of the circulation in the western equatorial Atlantic as reported by Philander (1973).

Neumann (1966) examined the Equatorial Undercurrent in the Atlantic Ocean, with special emphasis given to the sources of high-salinity water that characterizes the core of the EUC along a zonal distance of about 5,000 km. He found that the beginning, or source, of the Atlantic Equatorial Undercurrent is in the vicinity of 38°W at the confluence of two currents coming from the Northern and Southern Hemisphere, respectively. Both confluent currents are characterized by a maximum salinity in the upper thermocline. However, the branch coming from the Southern Hemisphere has a considerably higher salinity maximum. These conclusions were supported by the salinity distribution in the layer of maximum salinity at depths between about 60 and 100 meters. Philander (1973) reports that from about 30°W to within 80 nautical miles (150 km) of the African coast, the Atlantic EUC is a permanent feature of the equatorial circulation. Its most outstanding feature is a subsurface core of water of very high salinity (usually in excess of 36.2‰). The undercurrent derives this core from the North Brazilian Coastal Current; there is no evidence of advection of high-salinity water into the undercurrent east of 35°W.

Cochrane (1963, 1965) found the high-salinity core of the Undercurrent in the western part of the ocean always a little south of the equator (near 1°S). Neumann (1966) also found evidence that the eastward extent of the Atlantic EUC varies with the season. Parachute drogue tracking of current indicated a north-south oscillation in the course of the Undercurrent with an approximate period of 2 days and 8 hours. However, different investigators found different periods of oscillation, ranging from ½ day (Stalcup and Metcalf, 1966) to 3 days (Federov, 1965). However, an outstanding feature of the Atlantic EUC is the core of high-salinity water that

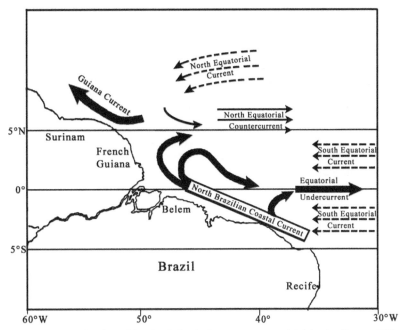

FIGURE 1.16 A schematic drawing of the circulation in the western equatorial Atlantic. *(Source: Philander, 1973.)*

coincides with, or is at least very close to, the core of maximum current speed. The high-salinity core is traceable far to the east into the Gulf of Guinea.

The Atlantic EUC plays a crucial role in the dynamics of the Atlantic oceans. For example, field measurements and model simulation studies by Giarolla et al. (2005) revealed the shallowing of the Atlantic EUC between January and May, concurrent with the reversal of the easterly trade winds to westerly and the deepening of the EUC from May to December.

1.4.5.2. Equatorial Undercurrent in the Pacific Ocean

It was the Japanese oceanographers who first measured an eastward subsurface equatorial current in the western Pacific around 1925. Subsequently, the undercurrent in the Pacific was *rediscovered* in August 1952 by Townsend Cromwell and Raymond Montgomery, who suggested the name *Equatorial Undercurrent*. Surprisingly, neither of these two groups was aware of Buchanan's earlier equatorial subsurface current measurements in the Atlantic Ocean way back in 1886 as well as the speculations around this current. The rediscovery of the EUC in the Pacific Ocean in 1952 inspired more comprehensive ocean surveys and dynamic theories of equatorial circulation. Although theories of the general circulation developed by Sverdrup (1947), Stommel (1948), and Munk (1950) had successfully accounted for the existence of major ocean currents such as the Gulf Stream and Kuroshio and explained the western intensification of wind-driven gyres in terms of the

sphericity of the Earth, the Equatorial *Undercurrent* had not been predicted by these theories. Thus, observation of the EUC was a "new" discovery.

Cromwell et al. (1954) made the first attempt to extensively study the EUC in the central and eastern parts of the Pacific Ocean. There arose some confusion as regards an appropriate name for the current. Shortly after Cromwell's death, Knauss and King (1958) proposed that the current be called the *Cromwell Current* after its first explorer.

During the International Geophysical Year (IGY, 1957–1958) and afterward into the 1960s, the United States, France, Japan, and the Soviet Union launched expeditions in the Pacific to determine the meridional, vertical, and zonal extent of the undercurrent and its relation to temperature, salinity, and chemical tracer distributions. Theories developed by Stommel (1960) and Charney (1960) established the central importance of the trade winds in setting up a baroclinic zonal pressure gradient to provide the source of eastward momentum for the undercurrent. Moreover, this pressure gradient eliminates singularities in the Ekman layer and so allows for a reconciliation of undercurrent dynamics and Sverdrup dynamics. It was found that in the Pacific Cromwell Current, the high-salinity core is absent or at least not well developed (Montgomery and Stroup, 1962; Knauss, 1966).

Perhaps the most dramatic observations of EUC variability in recent years were made during the 1982–1983 El Niño. The trade winds reversed over much of the western and central equatorial Pacific in late 1982, which led to a collapse of the zonal pressure gradient (Cane, 1983). Direct velocity measurements indicated the virtual disappearance of the undercurrent from September 1982 until January 1983 at 159°W (Firing et al., 1983) and during January and February 1983 at 110°W (Halpern, 1983). Equatorial upwelling ceased as a consequence of these radical wind, current, and pressure changes, causing major disruptions in primary productivity (Barber and Chavez, 1983) and in global cycling of carbon dioxide (Gammon et al, 1986). Cessation of upwelling also led to basin-scale equatorial sea surface temperature anomalies of ~3°C. These were associated with intense atmospheric convection that maintained the most pronounced El Niño of the century (Rasmussen and Wallace, 1983).

The EUC is a quasi-permanent feature of the equatorial Pacific. Its thinness and length are remarkable. It has a meridional width of about 200–400 km, is centered on the equator in the thermocline between about 50 and 200 m depth, and extends nearly across the whole length of the basin. The EUC reaches a speed of 1 m/s in its core and transports around 30–40 Sv (1 Sv $\equiv 10^6$ m^3/s). Thus, the Pacific Equatorial Undercurrent, or Cromwell Current, system is characterized by a large transport of water from the western Pacific toward the east, producing a large vertical shear. An important factor in understanding the development and structure of this current system is the role of turbulence, turbulent friction, and turbulent mixing (Williams and Gibson, 1974). The EUC has strong interannual variations in mass transport (Picaut and Tournier, 1991; Johnson et al., 2000) and temperature (Izumo et al., 2002) and is part of both subtropical and tropical shallow meridional overturning cells (Liu et al., 1994; McCreary and Lu, 1994; Sloyan et al., 2003). The EUC feeds equatorial upwelling (Bryden and Brady, 1985) and may have strong influences on SST in the eastern Pacific and thus on El Niño Southern Oscillation (ENSO).

The EUC, the shallow meridional overturning cells feeding it, and their role in El Niño and decadal variability in the equatorial Pacific have been studied by Izumo (2005) using both *in situ* data and an ocean general circulation model. His study has evidenced how all the branches of the shallow meridional overturning cells strongly co-vary with ENSO, slowing down and even vanishing during strong El Niño events and strengthening during La Niña events, leading to important changes in zonal and meridional heat exchanges. The study reveals that mass transport variations of the different branches of the cells are indeed all very well correlated with eastern Pacific SST.

1.4.5.3. Equatorial Undercurrent in the Indian Ocean

Results obtained during the Indian Ocean expeditions have shown that an Equatorial Undercurrent exists in the Indian Ocean at least during part of the year, when winds on both sides of the equator blow toward the West (Knauss and Taft, 1964; Swallow, 1964). Measurements in the Indian Ocean provided an important test of the theories developed to explain the EUC. The Indian Ocean is dominated by seasonally reversing monsoons and mean westerly winds along the equator. Easterlies prevail only during the northeast monsoon, which lasts from approximately December to April. Researchers therefore expected that an undercurrent and eastward pressure gradient would be present in the thermocline only during the northern winter and spring. This was confirmed independently by Soviet, British, and U.S. scientists, who found eastward subsurface flow along the equator at speeds of 50–100 cm/s during the International Indian Ocean Expedition (Knauss and Taft, 1964; Swallow, 1964). Furthermore, the zonal (i.e., parallel to the equator) pressure gradient associated with this flow reversed during the southwest monsoon, at which time the undercurrent was either absent or poorly developed.

Unlike in the Pacific and the Atlantic oceans, a distinct seasonal cycle associated with the Asian-Australian monsoon dominates the atmospheric circulation over the Indian Ocean (see Figure 1.17). Thus different oceanic structures are expected to occur in the Indian Ocean in response to such wind forcing. The undercurrent is driven by a pressure gradient and flows within the thermocline. The undercurrent in the Indian Ocean is associated with the northeast monsoon season. A subsurface salinity maximum, similar to that found in the Atlantic Ocean and associated with the Equatorial Undercurrent, is indicated also in the Indian Ocean during the Northern Hemisphere winter season. The sea level responds to both the accumulation of warm water and the wind stress. According to Wyrtki (1973), during the two months when the easterly winds blow, a slope of the sea surface of about 20 cm over a distance of 5,000 km from west to east is generated. Measurements by Luyten and Swallow (1976) in the equatorial region of the western Indian Ocean suggest a complex vertical structure in the horizontal velocity field close to the equator; this structure was found to be equatorially trapped. Based on long uninterrupted time-series measurements from a single point at (0.5°S, 73°E) in the equatorial Indian Ocean, Cane (1980) found that an undercurrent was absent in 1974.

Iskandar et al. (2009) investigated variations of subsurface zonal current in the eastern equatorial Indian Ocean by examining six-year data (December 2000 through November 2006) from an acoustic Doppler current profiler mooring at (0°S, 90°E). They found that during winter, the generation of an eastward pressure gradient, which drives an eastward flow in the thermocline, is caused primarily by upwelling equatorial Kelvin waves excited by prevailing easterly winds. The subsurface current reveals a distinct seasonal asymmetry. The maximum eastward speed of 63 cm/s is observed in April and a secondary maximum of 49 cm/s is seen in October. The subsurface current during summer undergoes significant interannual variations; it was absent in 2003, but it was anomalously strong during 2006 (Iskandar et al., 2009).

1.5. CURRENTS OF DIFFERENT ORIGINS

Currents of different origins and of various types are a fascinating permanent feature of the oceans. Thus there exist wind-driven currents that are confined to the surface Ekman layer, geostrophic currents that are maintained by horizontal pressure gradients, tidal currents that are present all over the ocean at all depths but change periodically, surf zone currents that result from offshore wave breaking, and so forth. Ocean currents are often complicated in structure because superimposed on the major currents that transport enormous volumes of water, there are gigantic whirls called *gyres* that may reach to great depths. Interestingly, currents at some regions happen to be merely the arms of such gyres.

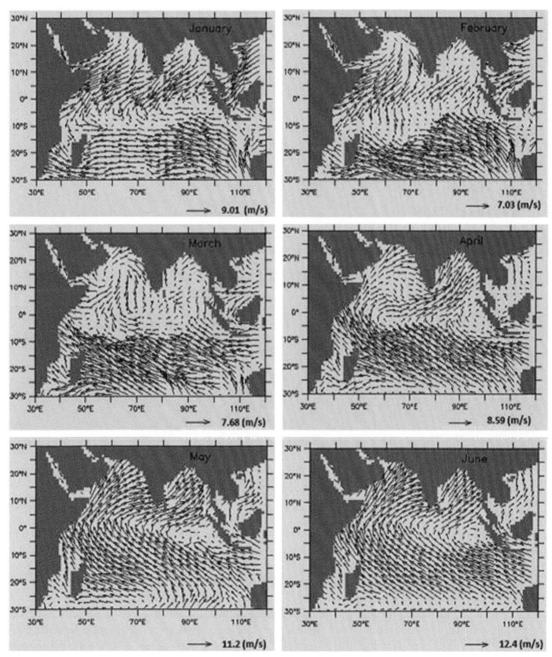

FIGURE 1.17   Pattern of European Remote Sensing (ERS) satellite-derived monthly mean sea surface wind circulation in the Indian Ocean region. *(Source: Courtesy of Dr. M. R. Ramesh Kumar, Chief Scientist, National Institute of Oceanography, Goa, India.)*

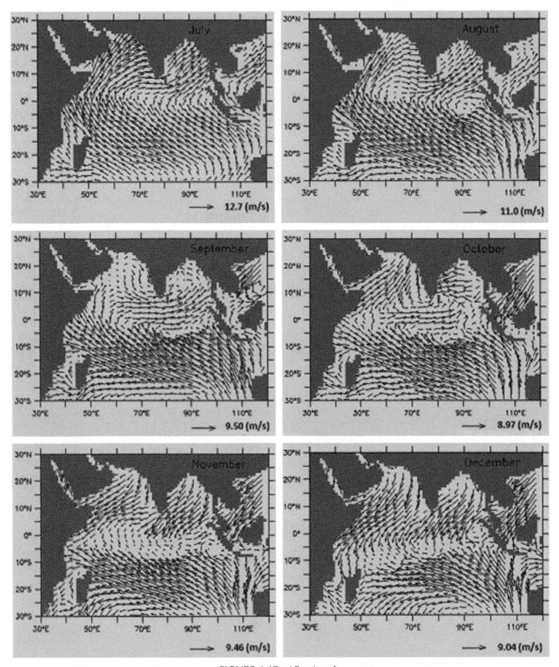

FIGURE 1.17 (*Continued*).

An outstanding characteristic of oceanic currents is the temporal and spatial variability of their speed and direction. Accurate observations and analysis of ocean currents have, for a long time, been a difficult task in oceanography. For a realistic understanding of any large-scale oceanic circulation feature, it is necessary to first obtain an overall picture of the temporal and spatial scales of the event as well as its variability. In the absence of such an overall picture, detailed time-series observations solely from a few selected locations are of limited use, and can sometimes yield misleading interpretations and even lead to erroneous conclusions, when applied to large-scale events.

The modern concept of large-scale oceanic circulation features such as meanders and gyres was achieved mostly from remote monitoring of the trajectories of surface and subsurface floats over vast areas, together with remote sensing using various kinds of satellite-borne sensors. Such measurements have guided several oceanographic communities, which were indeed groping in the dark for quite some time in the past, seeking to achieve a clearer understanding of the general features of large-scale oceanic circulation.

The 1957 joint British-American cruise in the research vessel *Aries,* sparked by Henry Stommel's ideas on deep-water movements, began to show that the deep circulation might consist of a wide spectrum of motions, some of them with velocities at least an order of magnitude faster than the mean velocities. There has since been a great volume of work on deep-water currents in open oceans.

1.5.1. Wind-Driven Current

Mixing at the surface of the oceans is important to a wide variety of concerns, including global climate and the health of marine life. Two major causes of this mixing are (1) when surface water is cooled, it becomes denser and sinks, mixing downward either to the bottom or until limited by the pre-existing stratification; and (2) the wind can mechanically stir the surface layer to some depth, usually limited by stratification. Wind also generates sea surface water motions. In the upper layers of the oceans, the currents are primarily wind-driven. When the wind blows over the ocean surface, there is transfer of momentum from air to water, part of which causes a net forward motion of the water. Generation of an Ekman spiral and Langmuir cells are two important concepts associated with wind-driven ocean currents.

1.5.1.1. Ekman Spiral

On studying the observations of wind and ice drift in the Arctic Ocean, the Norwegian oceanographer Professor Fridtjof Nansen found that the drift produced by a given wind did not follow the wind direction, as would generally have been expected, but the direction of ice drift deviated $20°$ to $40°$ to the right. Nansen hypothesized that under the influence of the Coriolis force (a force resulting from the spinning of Earth and which increases with increasing latitude of Earth), the surface water current would get progressively deflected from the wind direction. He concluded further that the water layer immediately below the surface must have a somewhat greater deviation than the latter and so on, because every water layer is put in motion by the layer immediately above it, sweeping over it like a wind. It might, therefore, be assumed *a priori* that the water current would at some depth run even in the *opposite* direction to the surface current, and there would consequently be a limit to the capability of the wind in generating currents (Kullenberg, 1954).

On Professor Nansen's suggestion, the anticipated progressive deflection (from the wind direction) of the water current in the upper layers of a large water body was given a mathematical foundation by the Swedish researcher Vagan Walfrid Ekman. To simplify the mathematical treatment of the problem, at the same time invoking the influence of Earth's rotation and the friction in the water, Ekman imagined a large ocean of uniform depth and without differences of water density affecting the motion of the water. The influences of neighboring ocean currents and continents were also left out of the account. Finally, the curvature of the ocean surface was also disregarded within this region, and the sea surface was treated as plane. These assumptions would allow the water to freely enter into or flow from the region under consideration. Considering water as an incompressible fluid and using the equations of motion under the simplified assumptions mentioned earlier, Ekman (1905) expressed the water current velocity components u and v in the X- and Y- directions, respectively, at depth z below the sea surface, as follows:

$$u = V_o \exp\left(-\frac{\pi}{D}z\right)\cos\left(45^o - \frac{\pi}{D}z\right) \qquad (1.1)$$

$$v = V_o \exp\left(-\frac{\pi}{D}z\right)\sin\left(45^o - \frac{\pi}{D}z\right)$$

In these expressions, V_o is the absolute velocity of the water at the sea surface, given by the expression:

$$V_o = \frac{T}{\sqrt{2\mu\rho\omega\sin\phi}} \qquad (1.2)$$

Here, T is the tangential pressure of the wind on the sea surface, directed along the positive axis of Y (i.e., the direction of the wind velocity relative to the water); μ is the coefficient of viscosity; ρ is the density of water at the region considered; ω is Earth's angular velocity of rotation ($=7.29 \times 10^5$ radian/s); ϕ is the geographical latitude of the region under consideration; D is the depth of frictional influence (i.e., depth of wind current), also known as the *Ekman depth,* expressed as $D = \pi\sqrt{(2\mu/\rho f)}$; and f is the *Coriolis parameter* (defined as twice the vertical component of the Earth's angular velocity at geographical latitude ϕ) and is given by $f = (2\omega \sin\phi)$. The velocity of the water is as a rule much smaller than that of the wind—a few hundredths of the latter. Equations 1.1 and 1.2 show that in the northern hemisphere, the drift current at the very surface will be directed $45°$ to the right of the velocity of the wind relative to the water. In the southern hemisphere it is directed $45°$ to the left. This angle further increases uniformly with the depth.

Equations 1.1 and 1.2 predict a spiral structure for the currents in the surface layers of the sea. The major elements of the Ekman spiral are the following:

- Deviation of $45°$ of the surface current (to the right of the wind in the northern hemisphere, and to the left of the wind in the southern hemisphere)

- Decrease of current speed with increasing depth from the sea surface
- Progressive deviation in direction of the deeper current

The just-mentioned current profile is known as the *Ekman current profile.* The direction and velocity of water current at different depths are represented by the arrows in Figure 1.18, in which the longest arrow refers to the surface, the next one to the depth $z = \pi/10a$, and the other ones to 2, 3, 4, etc. times this depth. In this, $a = +\sqrt{\dfrac{\rho\,\omega\,\sin\phi}{\mu}}$. These arrows, if conceived as drawn in position at their respective depths, would appear in the form of a spiral staircase, the breadth of the steps decreasing rapidly downward. Coming back to the meaning of the terminology *depth of wind current*, it is simply the depth down to that level where the velocity of the water is directed opposite to the velocity at the surface. At depths exceeding $z = D$, the wind-driven current velocity, and consequently the friction between the water layers, is zero. Thus, D is the depth down to which the effect of the wind is noticeable. It may be noted that at the equator, the solution embodied in Equations 1.1 and 1.2 does not hold true. It has been noted that whereas the deflection increases with increasing depth from the sea surface, the magnitude of the wind-driven current decreases with increasing depth and finally vanishes at a certain depth. Thus, the deep oceans have a more or less clearly defined upper layer, which represents the portion of the ocean that is influenced by wind stress. Below this layer, there is a much larger body of water of greater density that shows little response to the prevailing atmospheric state but undergoes motion under the influence of a pressure gradient and a host of other factors unrelated to wind stress. Further, independently of any other circumstances except the geographical latitude, a wind-driven current would become practically fully developed a very short time after the rise of the generating wind—in a day or two outside the tropics.

In the initial investigations of Ekman, the very important influence of continents, differences of density of the water, and other complicating circumstances were expressly left out of accounting. In a detailed investigation, some of these restrictions, particularly the influence of continents and of neighboring currents, were examined. The calculation showed, as might be expected, a decided tendency on the part of the surface current to follow the direction of the shorelines, though with a deviation *more or less* to a direction 45° to the *right* of the wind (in the northern hemisphere).

Following the theoretical concepts underlying the formation of the *Ekman spiral* (named after Ekman), quantitative work by several researchers ensued to test these concepts under various oceanic environments. Thus,

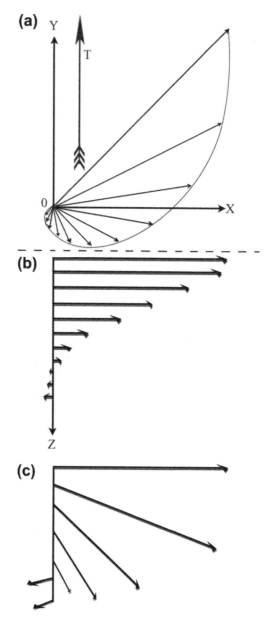

FIGURE 1.18 Schematic diagram of an Ekman spiral: (a) plan view in the horizontal plane (source: Ekman, 1905); (b) elevation view in the vertical plane; (c) isometric view.

a qualitative indication of the Ekman spiral in the deep-ocean region was observed by Katz et al. (1965). A detailed Ekman spiral structure has been measured under the polar ice by Hunkins (1966). Assaf et al. (1971) conducted field experiments in the Bermuda area to study the vertical structure of near-surface currents through aerial photography of surface-floating cards and colored plumes at different depth layers below the sea surface. The colored plumes were generated through a variety of dye injections under different wind, sea state, and thermal profile conditions. The vertical profile of horizontal currents obtained by them contained all the major elements of the Ekman spiral.

Calculations by Ekman long ago suggested that the dominant flow direction in wind-driven shallow waters near coasts should be primarily along the coast except under winds that are nearly perpendicular to the coast. Murray (1975) conducted detailed observations (using precision theodolite) of the movement of drogues within 800 m of a long, straight, sandy shoreline, and his measurements indicated that the direction of currents driven by local winds is predominantly alongshore and that there is little dependence on wind speed. It was found that under moderate conditions, current speed is strongly controlled by the wind angle to the shoreline and, to a lesser degree, by wind speed. In the moderately stratified waters off the coast of the Florida Gulf, a subtle three-layered pattern was persistent in the coastal normal component of the velocity profile such that onshore motion was present in a near-surface and a near-bottom layer, whereas intermediate depths were characterized by offshore motion. In unstratified (i.e., well-mixed) coastal waters of Lake Superior, similar drogue experiments conducted by Saylor (1966) showed a dominant coastal parallel flow despite large wind angles to the shore. The lack of stratification in the Lake Superior data apparently results in a two-layered velocity profile in which onshore components are in the upper layer and offshore components are in the lower layer.

It must be noted that incorporation of vertical structure near the equator was not straightforward, because the Coriolis force, an important component of Ekman theory, vanishes at the equator. This led to singularities in the surface Ekman layer flow.

1.5.1.2. Langmuir Circulation

When wind blows steadily over the sea surface, lines are often visible on that surface, running roughly parallel to the wind. These lines are manifestations of a special kind of water motion known as *Langmuir circulation*. Langmuir circulation consists of a series of shallow, slow, rotating vortices at the sea surface. Irving Langmuir discovered this phenomenon after observing windrows of seaweed in the Sargasso Sea in 1938. In an exemplary series of experiments, Langmuir (1938) found these to be lines of convergence along the sea surface, with downwelling below each line. He found also that maxima in the downwind surface currents occur along these lines. Thus the mixing layer is organized into rolls of alternating sign, aligned with the wind, and water parcels follow helical paths downwind (the helix forming bands of divergence and convergence at the surface).

This form of water circulation in the surface layers of water has come to be called *Langmuir circulation* after its discoverer. Historically, this term has not implied any particular mechanism of formation, and it generally has not been used in reference to other similar structures (e.g., roll vortices in the atmosphere). The form of helical vortices set up by wind is efficient in transporting momentum, energy, and matter throughout the surface layer of water. At the convergence zones, there are commonly concentrations of floating seaweed, foam, and debris along these bands. Along the divergence zones, the sea surface is typically clear of debris because diverging currents force material out of this zone and into adjacent converging zones.

Langmuir circulation in the mixing layer in the Bermuda area has been examined by Assaf et al. (1971) based on aerial photographs of floating cards and a variety of dye injections. The *Sargassum* weeds present at the experimental site provided natural surface floats to study the Langmuir circulation pattern. As expected, the photographed lines of *Sargassum* weed were parallel to the lines of convergence of the dye patches. The experimental results showed that the maximum horizontal spacing between adjacent cells is approximately the same as the depth of the mixing layer. Faller (1969) also found that the spacing between the *Sargassum* lines was the same, on average, as the mixing depth.

Langmuir circulation is often observed to form within minutes after the onset of wind. Smith (1992) reported observations from a time period including a sudden increase in wind, followed by rapid evolution of Langmuir circulation from small to larger scale. It was found that the evolution from initially small scales to a "quasi-steady state" occurred within a time span of less than an hour. Another finding obtained from Smith's observations is that the Langmuir circulation oriented with streaks parallel to the dominant wave direction but not absolutely uniform in that direction (i.e., irregular streaks). Theoretical aspects of Langmuir circulation have been reported by Craik and Leibovich (1976), Garrett (1976), Leibovich and Paolucci (1981), Leibovich (1983), and Leibovich et al. (1989). Based on numerous studies, Scott et al. (1969) and Gordon (1970) conclude that the most common and effective mechanism of vertical transport in the mixing layer is Langmuir circulation (and under moderate to strong winds this process dominates the other mechanisms).

1.5.2. Inertia Current

It has been known that under favorable conditions (e.g., if, for some reason, the horizontal pressure gradient as well as friction, that is, dissipative forces, become zero), the water particles in the ocean can move with a constant speed with a constant rate of change in direction. This circulation, described in the literature as *inertia current*, must move in a circle, termed the *circle of inertia*, with constant speed. At depths below the surface Ekman layer of frictional influence, the two conditions of zero horizontal pressure gradient and zero frictional forces will be satisfied if the oceanic region under consideration is homogenous (i.e., of uniform density). It has been shown that in the Northern Hemisphere this inertia current must move around the inertia circle in a clockwise sense and in the Southern Hemisphere in a counter-clockwise sense (Neumann, 1968). When water particles move with a velocity c on a rotating Earth of angular velocity ω, the Coriolis force comes into the picture. Because the Coriolis force always acts at a right angle to the motion, the only acceleration to balance the Coriolis acceleration, $2\omega c \sin \phi$, in the absence of friction is the centrifugal acceleration, (c^2/r), when the water moves in a circle of radius r. To maintain the inertia motion in a circle, both accelerations must be equal and must act in opposite directions, so that:

$$\frac{c^2}{r} = 2\omega c \sin \phi \qquad (1.3)$$

The radius of the inertia circle, therefore, becomes

$$r = \frac{c}{2\omega \sin \phi} \qquad (1.4)$$

In these expressions, ϕ is the local geographical latitude.

It can be seen from Equation 1.4 that, for a given flow speed (c), the radius (r) of the inertia circle becomes infinite at the equator ($\phi = 0°$) and a minimum at the poles ($\phi = 90°$). The time needed to complete a full path around the circle of inertia, known as the *inertia period T_p*, is given by the expression

$$T_p = \frac{\pi}{\omega \sin \phi} \qquad (1.5)$$

This suggests that the inertia period depends only on the geographical latitude. Thus, at the poles, T_p is approximately 12 hours; at latitude 30°, T_p is approximately 24 hours; and at the equator T_p is infinite. Existence of these circulating currents in the oceans was established by direct observations as early as 1931. Currents of elliptical as well as circular shapes have since been observed at varying depths below 1,000 meters. It has been found (Neumann, 1968) that the observed inertia period of these circulating currents closely matched with the theoretically predicted periods, and their measured radius was related to the current c and the geographical latitude predicted by the equation for r. It has been noticed that some of the observed circulating currents have exhibited a translatory motion. The trajectory shape may, perhaps, change under topography-related reasons.

1.5.3. Tidal Currents in Open Seas, Estuaries, and Ridge Valleys

The periodic rise and fall of tides, in concert with astronomically induced tide-generating forces and topographically induced interactions among various tidal constituents, are associated with a periodic horizontal flow of water, known as *tidal current*. Tidal currents are strongest in restricted areas such as funnel-shaped estuaries with wide mouths and narrow heads, shallow areas, and over the sills. A decrease of cross-sectional area of the flow channel produces faster flows. In the open sea, where the direction of flow is not restricted by any barriers, the tidal current is rotary; i.e., it flows continuously, with the direction changing through 360° during the tidal period. The tendency for the rotation in direction has its origin in the deflecting force of the Earth's rotation, known as *Corioli's force*. The current speed usually varies throughout the tidal cycle, passing through two maximums in approximately opposite directions and two minimums about halfway between the maximums in time and direction. Rotary current, depicted by a series of arrows representing the speed and direction of the current, is usually known as a *current rose*. A line joining the extremities of the radius vectors will form a curve roughly approximating an ellipse. The cycle is completed in one-half tidal day or in a whole tidal day, according to whether the tidal current is of the semidiurnal or the diurnal type. A current of the mixed type will give a curve of two unequal loops each tidal day. Because of the elliptical pattern formed by the envelope joining the ends of the arrows, it is also referred to as a *tidal current ellipse*. Figure 1.19 shows a graphic representation of a rotary current in which the velocity of the current at different hours of the tidal cycle is represented by radius vectors and vectoral angles.

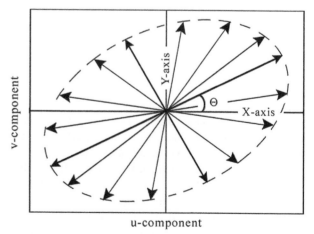

FIGURE 1.19 Graphic representation of a rotary current in which the velocity of the current at different hours of a tidal cycle is represented by radius vectors and vectoral angles. By convention, the u component is the east component and the v component is the north component.

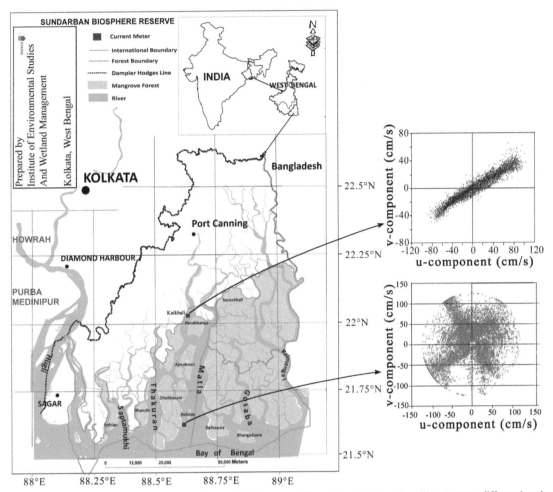

FIGURE 1.20 Graphic representations of two extreme forms of tidal current ellipses (i.e., circular and rectilinear) at two different locations in Sunderbans mangrove reserve forest delta. The tidal current circle observed at the southern station appears to have resulted from a combination of two near-rectilinear currents flowing through the two adjacent channels. *(Source: Courtesy of Kakoli Sen Sarma and Somenath Bhattacharyya, Institute of Environmental Studies & Wetland Management, Kolkata, India.)*

The characteristics of tidal ellipse are (1) semi-major axis, (2) semi-minor axis, (3) eccentricity (ratio of minor to major axes; positive for counter-clockwise rotation), (4) orientation, and (5) sense of rotation. With a one-year-long dataset it is possible to extract the characteristics of tidal ellipses for the main semidiurnal, diurnal, fortnightly, monthly, semiannual, and annual constituents. Depending on the regional peculiarities, tidal ellipses are often found to exhibit large spatial variability. The tidal ellipse can change from cyclonic (counter-clockwise in both hemispheres) to anticyclonic (clockwise in both hemispheres) and from rectilinear (major axis » minor axis) to circular (major axis = minor axis). Depending on the topographic peculiarities, the eccentricity has been found to vary considerably, both laterally and vertically; also, different tidal current constituents (e.g., M_2, K_1, etc.) can rotate in different directions (Marinone and Lavin, 2005). Tidal current is expected to be constant from sea surface to bottom, unless modified by stratification or bottom friction.

Figure 1.20 shows the presence of two extreme forms of tidal current ellipses (i.e., *circular* and *rectilinear*) at two different locations in Sunderbans, which is the largest delta in the world, with 10,200 km² of Mangrove Reserve Forest extending over two countries, namely India (~4,200 km²) and Bangladesh (~6,000 km²). In a true tidal current ellipse, the current direction must exhibit a smooth change over a full tidal circle (i.e., in the range 0–360°). Based on the concentration of the u-components (i.e., east components) and v-components (i.e., north components) of currents along two dominant directions, it can be suggested that the tidal current circle observed at the southern station results primarily from a combination of two near-rectilinear tidal currents flowing through the two adjacent channels. Plotting of tidal current ellipses of different constituents (of tidal currents) at different depths will throw more light on the character of tidal currents in a region.

There are instances in which the tidal current ellipses are not the result of the Coriolis force but are generated by the alternating cross-shore water fluxes due to the tide. Houwman and Hoekstra (1998) have reported such a case happening at a multiple-bar system of the barrier island of Terschelling in the Netherlands. At this site, the direction of rotation of the current vector has nothing to do with the Coriolis force but is determined by the position of the coast, left or right, relative to the progressive tidal wave.

Tidal currents in the open ocean are generally weak. However, tidal currents in coastwater bodies belong to the most vigorous flow phenomena in the sea. Nonlinear instabilities give rise to a large variety of "secondary currents," ranging from the smallest three-dimensional eddies in the dissipation range—with a scale of ~10^{-3} m, carried along by the tidal flow (Grant et al., 1962)—to the topographically "frozen" quasi-two-dimensional residual eddies (Zimmerman, 1978; 1980) with length scales of 10^4 m. Topographic obstacles in the path of large currents give rise to production of vortex streets in the wakes of such obstacles, and the vortices are carried along by the current. Any existing quasi-two-dimensional turbulence in tidal currents must have a pronounced influence on the physical transport processes in tidal areas and hence on the total physical environment (Veth and Zimmerman, 1981).

In estuaries or straits or where the direction of water flow is more or less restricted to certain channels, the tidal current is *reversing*, i.e., it flows alternately in approximately opposite directions with a short period of little or no current, called *slack water*, at each reversal of the current. During the flow in each direction, the speed varies from zero at the time of slack water to a maximum, called *strength*, about midway between the slacks. The water current movement from the sea toward shore or upstream is the *flood current*, and the water current movement away from shore or downstream is the *ebb current*. Tidal currents may be of the semidiurnal, diurnal, or mixed type, depending on the type of tide at the location. Offshore rotary currents that are purely semidiurnal repeat the elliptical pattern each tidal cycle of 12 h and 25 min duration. If there is considerable diurnal inequality (i.e., if there is considerable difference in the two successive tidal ranges), the arrows representing the instantaneous tidal current vector describe a set of two ellipses of different sizes during a period of 24 h and 50 min (one lunar day). The difference between the sizes of the two ellipses is dependent on the diurnal inequality. In a completely diurnal rotary current, the smaller ellipse disappears and only one ellipse is produced in duration of 24 h and 50 minutes.

The magnitude of the tidal current speed varies with the tidal range (i.e., the difference between successive high-tide and low-tide elevations). Thus, the stronger *spring* and *perigean* currents occur, respectively, near the times of new and full moon and near the times of the moon's

perigee. The weaker *neap* and *apogean* currents occur, respectively, at the times of neap and apogean tides. An important role of the fast tidal currents is to produce intense tidal mixing, which, by pumping nutrients into the surface layers, is one of the ultimate causes of a location's biological richness.

Tidal currents in ridge valleys are particularly interesting. For instance, Garcia-Berdeal et al. (2006) have reported fascinating vertical structures of time-dependent water currents in the axial valley at Endeavour Segment of the Juan de Fuca Ridge, which lies along the divergent boundary between the Pacific and Juan de Fuca plates, approximately 400—500 km off the coasts of Washington and Oregon (Figure 1.21).

Diurnal tidal current motions are trapped to the ridge and experience an amplification of their clockwise rotary component that is attributed to the generation of anticyclonic vorticity by vortex squashing over the ridge. Whereas the across-valley water currents diminish with increasing depth, the along-valley currents are accelerated and intensified with depth. Alignment of the current flow with the along-valley direction yields a velocity vector that spirals with depth.

At the central valley site, all diurnal tidal ellipses are found to be clockwise rotary. At the central valley site, both a gradual decrease in the eccentricity of the ellipses and an amplification of the major axis with increasing depth down to 15 m above the bottom has been noticed. In addition, the orientation of the major axis of the ellipses rotates predominantly counter-clockwise with increasing depth and aligns with the along-valley direction.

In contrast, in the case of the diurnal tidal current ellipses at the Easter Island, the higher ellipses tend to be clockwise rotary and the deeper ones, counter-clockwise, although they are all rather rectilinear. The major axes of the K_1 ellipses at Easter Island (2—4 cm/s) are larger than those from the central valley record (2—3 cm/s) and do not undergo a progressive trend with depth. The ellipse orientation rotates clockwise with increasing depth and becomes aligned along the valley.

1.5.4. Rip Currents

The shallow-water coastal zone encompassing the onshore and offshore sides of the wave-breaking region (known as the *surf zone*) is one of the very energetic regions of the oceans. Changes in wave breaking owing to variations in wave height and direction along curving coastlines, irregular bathymetry, and manmade structures cause complex surf zone circulation, primarily rip currents, along with relatively weak mean alongshore currents that can change direction both across the surf zone and along a depth contour (Shepard and Inman, 1950). *Rip currents* are a generic name given to approximately shore-normal,

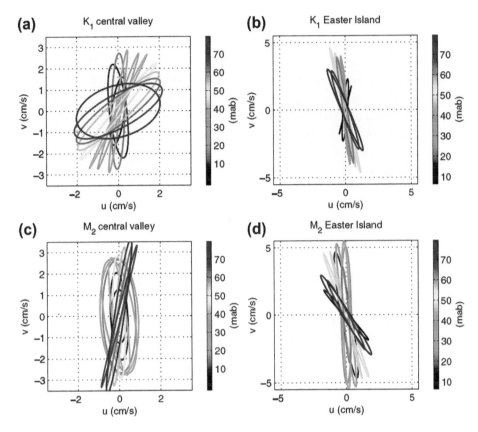

FIGURE 1.21 Observed tidal current ellipses in across-valley (u-component) and along-valley (v-component) coordinates for the diurnal (top panels) and semidiurnal (bottom panels) frequencies from the central valley (a), (c) and Easter Island (b), (d) records. Ellipses are plotted at 6.5–78.5 m above bottom (mab), in intervals of 8 m. Blue corresponds to levels closer to the bottom. Note change of velocity scale in (b) and (d). *(Source: Garcia-Berdeal et al., 2006.)*

strong, narrow, offshore-directed (i.e., seaward-flowing) current jets with a strong and constrained flow that originate within the surf zone, extend seaward, and broaden outside the breaking region, forming a head region once they pass beyond the surf zone (see Figure 1.22).

Figure 1.23 shows a photograph of a rip current in Monterey Bay, California. Once initiated, a rip current develops into a distinct current system consisting of three main features: (1) a *feeder region* where the currents are directed toward the center of the rip and provide the volume flux for the offshore flow, (2) the *rip neck*, which is the narrow offshore-flowing section of the rip that usually has the highest current velocities and may extend a significant distance outside the surf zone, and (3) the *rip head*, which is the seaward end of the rip current, wherein distinct vortex features such as spinning down eddies are often seen. The rip head can often be observed and clearly identified in photographic and satellite images as a mushroom-shaped feature at the offshore limit. According to Smith and Largier (1995), episodic instability plumes are often observed outside the breaker region. Outside the surf zone, Haas and Svendsen (2002) found the mean velocities to be near zero, although instantaneous velocities could be substantial. This is associated with the pulsing and unstable nature of the rip current (Haller and Dalrymple, 2001). They found that the jet flow is surface dominant

seaward of the breaking region. Therefore, the visual markers such as sediment plumes or bubbles associated with rip currents seen outside of the surf zone are surface dominated and their average propagation speed is minimal. As the rip current moves offshore into deeper water, the streamlines move closer together, creating a stronger and narrower current, which explains why even when a rip channel occurs over a gentle alongshore depression, the offshore jet is well confined (Peregrine, 1998; Kennedy and Thomas, 2004).

Although the rip neck is characterized by intense water current velocities in the offshore direction, the currents decelerate rapidly in the head region. The spatial gradients of the current velocities across the rip neck and the diverging region of the head are strong. In addition, the region of strong offshore flow is tightly constrained. In several instances, shore-perpendicular rip current trajectories have been found to swing to become oblique to the shore as a result of the complex hydrodynamics involved. In a complete rip current cell, a diffuse onshore flow in the feeder region supplies the rip necks. Examination of trajectories of freely drifting drogues deployed in the surf zone have indicated that quite often the rip current undergoes a large meander while still maintaining a coherent jet form before spreading in a head-type feature. This general circulation feature is often

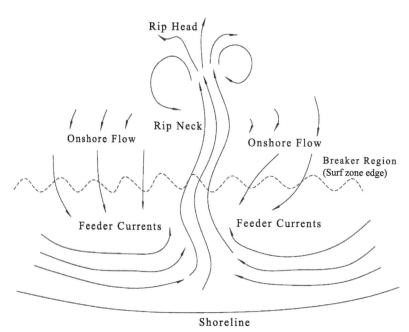

FIGURE 1.22   A schematic of a rip current. *(Source: MacMahan et al., 2006.)*

observed even though the rips themselves are transient in time and space. Most field observations have been for rip currents coupled to the underlying beach morphology. There have been a few observations of nonstationary rip currents, which are often referred to as *transient rip currents* (Tang and Dalrymple, 1989; Fowler and Dalrymple, 1990; Johnson and Pattiaratchi, 2004). Whereas the exact arrangement in a rip system varies greatly, the law of conservation of mass alone suggests that all rip current systems must possess these three basic regions in some form.

Early qualitative observations suggested that the rip currents are coupled to the morphology, the rip currents pulsate, and the velocities increase with increasing wave height (Shepard, 1936; Shepard et al., 1941; Shepard and Inman, 1950; McKenzie, 1958; Bowman et al., 1988a,b). Later studies have increased both spatial and temporal resolution of the observations and with greater accuracies, which have quantified earlier qualitative assessments.

There are weaker and broader onshore-directed flows in the region neighboring the rip current, contributing to the alongshore-directed feeder currents that converge to the rip

FIGURE 1.23 A photograph of a rip current in Monterey Bay, California. The rip current is the dark patch. There is intense wave breaking on both sides of the rip currents with little breaking within the deeper rip channel, where bubbles are advected seaward. *(Source: MacMahan et al., 2006.)*

current. Early studies obtained crude estimates of rip current velocities, which include 1 m/s (Shepard and Inman, 1950), 50 cm/s (Sonu, 1972), 30 cm/s (Huntley et al., 1988), and 70 cm/s (Short and Hogan, 1994). Mega-rips are generally observed on embayed beaches with suggested velocities exceeding 2 m/s (Short, 1999), but there are no *in situ* field measurements of mega-rips. Rip currents shape the sandy shoreline and are considered to be important for transporting sediments offshore (Cooke, 1970; Komar, 1971; Short, 1999). The seaward-directed high-speed rip current flow often transports sediments, pollutants, foam, surfers, and swimmers offshore and is dreaded by beach swimmers. Based on survey results, rip currents account for more than 80% of lifeguard rescue efforts and are the number-one natural hazard in the state of Florida (MacMahan et al., 2006). More people fall victim to rip currents in Florida than to lightning, hurricanes, and tornadoes (Luschine, 1991; Lascody, 1998).

Recently there have been a significant number of laboratory and field observations to study rip currents, which has led to advances in our understanding of these systems. An overview of rip current kinematics based on these observations and the scientific advances obtained from these efforts have been synthesized by MacMahan et al. (2006). Johnson and Pattiaratchi (2004) and MacMahan et al. (2006) have given brief surveys of rip currents based on available literature and their own findings from field measurements and model-based studies. It is generally agreed that the rip current flow consists of various temporal contributions, all of which have their own forcing mechanism. Accordingly, rip current flows can be partitioned in terms of the following contributions (MacMahan et al., 2006):

1. Infragravity band (i.e., period of oscillation in the range 25–250 seconds)
2. Very low-frequency band (i.e., period of oscillation in the range 4–30 minutes)
3. A mean flow based on the rip current system and wave conditions
4. Modulation associated with the slow variations in the water level due to changing tidal phase

Each of these contributes significantly to the total, and the combination results in significant flow speeds that have the potential for transporting sediment and catching beachgoers off-guard.

Longuet-Higgins (1970) showed that water circulation in the surf zone is driven primarily by breaking waves. All dynamical models of rip currents are forced by alongshore variations of wave height that result in alongshore variations in wave-induced momentum flux, termed *radiation stress* by Longuet-Higgins and Stewart (1964). When the waves break, the changes in radiation stresses generate slowly migrating rip currents traveling alongshore.

Bowen (1969) was the first to show that alongshore perturbations in bathymetry result in alongshore variations in wave height, which generate rip currents. Rip currents are most often observed to occur when the waves approach at near-normal incidence and where there are alongshore variations in bathymetry with the alongshore sandbar incised by rip channels. When adjacent pairs of wave groups offset in time and space enter the surf zone, the interaction of the generated vortices can lead to narrow, offshore-directed, quasi-steady flows. Over an alongshore homogeneous bathymetry, the circulation cells, in general, propagate slowly alongshore, simulating migrating rip currents. Thus, rip current flows are forced by the incoming wave energy but influenced by tidal elevation and the shape of the morphology.

Rip currents pulsate at various temporal scales, which have different forcing. The pulsations are composed of infragravity motions, modulations of wave group energy, shear instabilities, and tides. The summation of these flow contributions can lead to strong offshore rip currents that last for several minutes. The time-averaged pulsations are minimal outside the surf zone, yet when the pulsations occur, they are surface dominated. Data from the laboratory and the field suggest that rip current strength increases with increasing wave energy and decreasing water depths. Rip currents can occur under various bathymetric perturbations, even for beaches with subtle alongshore variations. The maximum mean current occurs inside the surf zone, where the maximum forcing is present owing to the dissipation of waves. Wave–current interaction may define the energy of a rip current system and feedback mechanisms.

Rip currents are tidally modulated such that decreases in tidal elevation increase rip current flows to a relative maximum (Sonu, 1972; Aagaard et al., 1997; Brander, 1999; Brander and Short, 2001; MacMahan et al., 2005), and the presence and danger of rip currents are often linked to lower tidal elevations. Rip currents are nonexistent at high tides (Shepard et al., 1941; McKenzie, 1958; Short and Hogan, 1994; Luschine, 1991; Lascody, 1998; Engle et al., 2002). During times of spring tides, the threat of rip currents to beachgoers and the potential influence of the rip currents transporting sediments increase until the water depth over the shoal becomes too shallow or dry. Rip currents are morphologically controlled, and therefore the mean velocity of a rip current varies for different beaches, wave forcing, and tidal elevation. Subtle bathymetric perturbations can induce significant three-dimensional flows. Laboratory and field data indicate that rip current velocity increases with increasing wave height and decreasing water depth.

The maximum mean current occurs inside the surf zone, where the maximum forcing is present owing to the dissipation of waves. Rip currents have been observed to pulsate in association with wave groups at the infragravity band, and such infragravity rip current pulsations increase the instantaneous rip current maximum flow to over 1 m/s. Stochastically varying wave groups can further increase the rip velocity.

Numerical modeling experiments by Svendsen et al. (2000) have indicated that the total transport through the rip channel does not depend on the spacing of the rip channels, and preliminary analysis from the Nearshore Canyon Experiment (NCEX) supports this finding. In topographic rips, the extent of the feeder region is also largely independent of rip spacing and the rip is only supplied by a small local area (Svendsen et al., 2001). However, whether this is the case in transient rips is not known at present. The rip neck may be directed perpendicular or obliquely to the shore, may have a meandering shape, and may move around. The rip head is often clearly visible as a sediment-laden patch of water at the end of the narrower neck. This patch of water may persist as a coherent feature for some time after the rip flow has ceased.

Surf zone eddies associated with rip currents have been predicted theoretically and numerically by several researchers (e.g., Peregrine, 1998; Chen et al. 1999; Svendsen et al., 2000) and noted in field and laboratory observations. There exist transient rips, which are temporary features that develop in varying locations. These have a specific lifetime and subsequently decay. Rip currents are ubiquitous in the near-shore regions of several long beaches. The world-famous Calangute beach in Goa (on the west coast of India), which finds a prominent place on tourist maps, is just such a beach; it is notorious for rip current generation, leading to human casualties almost every year.

1.6. IMPLICATIONS OF OCEAN CURRENTS

Ocean currents and atmosphere interact to produce large-scale changes in climate. For example, the baroclinic transport variability of the Kuroshio Current system is considered to have an important effect on the winter climate of the Northern Hemisphere (White, 1975). The timing and strength of the Indian monsoon depend at least partially on the behavior of the Somali Current (the only known ocean current that reverses its direction in consonance with the overlying winds) off the East African Coast (Anderson and Rowlands et al., 1976; Das, 1988), as well as to a global-scale atmospheric circulation feature called the Southern Oscillation (Shukla, 1987), the temperature anomalies of the El Niño current and of the western Indian Ocean (Rameshkumar et al.,1986; Shukla, 1987), and some as-yet unknown large-scale, low-frequency forcings (Shukla, 1987). The poleward transport of heat by the oceans and the eddies shed by them also play an important role in this activity. The poleward "counter-currents"—a characteristic feature of boundary current regions—are well known for their influence on activities such as fishing

and sound ranging as well as on coastal marine climate and small-scale, near-shore circulations for which they may be a significant driving mechanism (Wickham, 1975).

Apart from their influence on climate change, ocean currents affect us in several other ways. For example, transport of large volumes of water with anomalous temperature differences is believed to be responsible for triggering unusual weather patterns affecting entire continents. As carriers of thermal energy, ocean currents have a direct influence on fish migration and weather patterns. It is well known that El Niño, a current of warm water that appears in some years off the coast of Peru, causes a rapid rise of SST in normally cool-water regions. This leads to instabilities in the overlying atmosphere, which cause excessive rainfall and damaging floods, and erosion in regions where the normal precipitation is very low. The anomalous condition also has a disastrous effect on the fishing industry due to catastrophic destruction of plankton and fish life. Finding nothing to feed on, the surviving fish move on to newer pastures. Consequently, sea birds die of hunger in large numbers (e.g., in Peru during some El Niño events) or migrate, abandoning their young ones. Interestingly, the El Niño event, at least partially, is an indirect indicator of deficient monsoon over India and may provide very useful guidance for long-range forecasting of monsoon rainfall over India (Shukla 1987a,b).

Currents within approximately a meter of the sea surface transport floating matter and, therefore, are of great importance in coastal areas, where considerable damage can be inflicted by surface-borne pollutants and spilled oil. From pollution monitoring and control perspectives as well as for determination and forecasting of oil-spill drift in the case of oil tanker accidents, knowledge of currents is of great practical interest. Surface current measurements are also important for studies related to pollutant distribution patterns and identification of suitable areas for disposal of these pollutants.

From an environmental perspective, seasonal mapping of coastal surface currents is useful to determine the most optimal outfall sites for emerging industries, harbors, thermal and nuclear power plants, and so forth. With more and more pollutant outfalls being brought to the coastal waters, such mapping procedures have become all the more important to control the spread of effluent emissions and thereby save the coastal waters from excessive pollution levels. Because a vast majority of humans' ocean-related activities are influenced by sea surface conditions, timely information on the surface currents and their perturbations in such areas is of vital importance. The behavior of surface phenomena such as thermal plumes can be properly studied only if the coastal water current circulation patterns are known.

Radioactive emissions from nuclear power stations in Japan as a result of the damage inflicted on them by the

March 2011 Japanese tsunami and the fear of the possible spread of radioactive-contaminated water to distant locations through ocean currents were great concerns among a large section of people in Japan and the neighboring coastal nations in the days immediately after the Japanese tsunami episode. This concern is justifiable in view of the fact that tsunamis are known to generate strong water-current velocities in the coastal zones and inside partially confined water bodies such as bays and harbors (Koshimura et al., 2009), and such strong currents can quickly spread radioactive contamination offshore, facilitating further transport across the oceans. In the mid-1980s, there was considerable interest in the potential for disposal of radioactive waste either below or on the seabed. The feasibility studies included research on the interaction of radionuclides with sediment but also called for an investigation of ocean circulation around potential disposal sites. This required information on both the subthermocline mean circulation and on eddy-induced lateral mixing at these depths.

Currents and turbulence are essential factors in understanding many biological processes. Turbulence is one of the factors that regulate the supply of nutrients to the photic zone, and the nutrient supply ultimately controls the productivity of the area. The upper portion of the sea, especially the photic (light-penetrating) zone, carries zooplankton and phytoplankton, which are the dominant components at the bottom of the food chain and are responsible for the production of most of the world's oxygen. Ocean surface pollution, which destroys these vital components required for the survival and prosperity of mankind, is of serious concern. Fish eggs are borne by surface currents, which are, therefore, of concern to the fishing industry. Knowledge of surface currents is useful in understanding the distribution of fish such as salmon and tuna. The mass transport of plankton and marine invertebrates to and from the bays and shelf waters is greatly influenced by the currents of the near-shore waters. Detailed knowledge of currents is necessary to understand the transport of larval forms of oysters, crabs, and the like from the point of hatching to the point of attachment or residence.

One of the critical needs of the offshore industries is for real-time information on water currents during events such as severe storms. Near-real-time data are also highly desirable during quieter periods to aid on-site construction. Knowledge of ocean currents is essential for several ocean engineering applications such as offshore pipe laying, offshore mining (e.g., minerals, hydrocarbon deposits, polymetallic nodules), and coastal erosion control. Estimation of hydrodynamic loads required for the design of marine structures such as offshore drilling platforms, pipelines, and dock facilities also require ocean current data. With the advancement of the offshore industry to deeper waters, oil and gas operators face new environmental challenges. There

is an ever-increasing need for ocean current data for applications in the design and operation of offshore structures. Whether in the Gulf of Mexico, southeast of Brazil, offshore South Africa, or in Indonesia, the oceanic upper layer often encompasses very strong and extremely variable currents (jets, eddies, meanders, filaments, etc.). Such energetic features present major risks for marine operations and have already caused costly losses of structures over various areas of the world.

Potentially dangerous ocean current eddies have threatened several drillships. It is important to track and forecast eddies as they head toward the drilling vessel so that their effects can be minimized by careful planning of subsea operations. Eddies are fast enough to severely impact drillship operations. Tracking and forecasting the progress of eddies help reduce the delays and the harm they can cause. There is a special requirement for near-real-time observations during installation of offshore structures. For instance, drilling superintendents on exploration rigs need real-time information on ocean currents in order to assess present and upcoming stresses on marine risers, which may produce vortex-induced vibrations. In extreme cases these can cause the riser to separate from the blowout preventer stack. Other operations, such as the deployment of underwater vehicles and subsea systems, also benefit from knowledge of the ocean current regime. Real-time data measured from rigs are used not only to assist day-to-day drilling operations but also to provide valuable information for future design engineering plans.

A major component of understanding a unique system called a *hydrothermal vent field* involves characterizing near-bottom currents at mid-ocean ridges and determining the flux of heat and the dispersal of biological/chemical species associated with hydrothermal fluid. The transport of planktonic larvae by bottom currents plays a key role in the dispersal potential of vent species that have limited mobility yet are constantly faced with the challenge of colonizing new sites because of the ephemeral, unpredictable, and highly variable nature of hydrothermal vent systems. Because the highest abundances of these larvae are found close to the bottom and to their vent source communities (Mullineaux and France, 1995) and because hydrothermal venting is largely concentrated within the axial valley of the ridge, characterizing the current flow patterns in the valley is central to understanding larval dispersal mechanisms and pathways. In addition, it has been suggested that transport by near-bottom currents may be a more prevalent larval pathway than entrainment into rising plumes and dispersal within the neutrally buoyant plume, which is favored only during periods of slack tide (Kim and Mullineaux, 1998).

Abyssal currents of remarkably great widths (30 to 500 km) have been discovered and surveyed in most major ocean basins (Stommel and Arons, 1972). Bottom currents are considered major contributors that control sediment distribution in the oceans. Deep thermohaline currents have given rise to many large sediment drifts comprising silt and clay located in a variety of places, including the foot of the continental rise (Heezen et al., 1966). There are examples where deep currents were responsible for the formation of sediment ridges several hundred km long, several tens of km in width, and as much as a km in thickness (Ewing and Edgar, 1966). In other areas, strong evidence of erosion on the deep-sea floor is indicative of the important role of currents in the distribution of oceanic sediments. Large forms such as *longitudinal triangular ripples* (LTR) and furrows occur at the sea bottom under strong currents and greatly influence abyssal sediment dynamics. Variations in deep-ocean thermohaline circulation merit investigation.

At the estuarine level, data on water currents are important to hydrodynamic numerical modeling for prediction of sediment transport and deposition along shipping channels and estuaries (Biswas and Chatterjee, 1987). Such predictions aid hydrographic surveys and are useful for selective dredging along long and complex navigation channels (e.g., Hugli River in Calcutta, India, and the Waterway to Rotterdam Port, which often branch and wend their way between the port and the sea).

Coastal current circulation measurements are needed for studies related to the dynamics of coastal waters. Current measurements have important applications for coastal surveillance, rescue operations, and proper planning for effective utilization of coastal waters. Information on surface currents is vital to assess the impact of possible offshore oil recovery operations. It is the surface currents that primarily determine whether an oil spill or leak may dissipate harmlessly at sea or do irreparable damage to the coast. A vast majority of mankind's ocean-related activities take place on or near the ocean's surface, and most coastal operations are influenced by surface conditions.

It is now our understanding that each oceanic region, however unique in individual behavior, consists of a number of "common" characteristic synoptic circulation structures and water masses. These synoptic entities, or "features," when put together in the background climatology of a particular region, interact and evolve together to generate the combined circulation variability due to various factors. A regional basin may include a set of multiscale features such as large-scale meandering currents and fronts, basin-scale and subbasin-scale gyres, mesoscale eddies, and vortices. A coastal region may include circulation structure features such as mesoscale buoyancy-driven fronts, upwelling fronts, and sub-mesoscale eddies and fronts. Thus, the scales of motions in the oceans range from major ocean dimensions down to millimeters. As would be expected, the spectrum of motions has a wide range. During the last few decades,

there has been significant advancement of our accrued knowledge and understanding of oceanic circulation and variability in regional oceans. This has been possible due to technological advances in multiplatform at-sea and space-borne instrumentation capabilities, in the availability of highly efficient computing platforms and numerical modeling capabilities, and in model data assimilation and integration methodologies.

The oceanographer, with his inaccessible environment, has long sought to unravel all the characteristics of oceanic variabilities and statistics of fluid motion. Measurements of variabilities ranging from microscale to mesoscale are, therefore, important. Because it is impossible to develop a single technique to make measurements that could be used to describe all scales of motion, it becomes necessary to divide them into subranges and study them separately. Although studies of all scales of motions are important, a limited range of motions might be more significant to a specific problem and can be studied in depth if techniques are developed to look only at the range of interest. The opportunity to learn about a phenomenon in part offers the advantage that the complexity of instrumentation can be reduced to practically achievable bounds. The last few decades have witnessed a rapid evolution of new technologies for measuring ocean currents. These technologies and the associated problems and prospects are addressed in the following chapters.

REFERENCES

Aagaard, T., Greenwood, B., Nielsen, J., 1997. Mean currents and sediment transport in a rip channel. Mar. Geol. 140, 25–45.

Anderson, D.L.T., Rowlands, P.B., 1976. The Somali Current response to the southwest monsoon: the relative importance of local and remote forcing. J. Mar. Res. 34, 395–417.

Anderson, L.T., Rowlands, P.B., 1976. The Somali current response to the southwest monsoon: the relative importance of local and remote forcing. J. Mar. Res. 34 (1), 395–417.

Andrews, J.C., Scully-Power, P., 1976. The structure of an East Australian Current anticyclonic eddy. J. Phys. Oceanogr. 6, 756–765.

Aoki, S., Yoritaka, M., Masuyama, A., 2003. Multidecadal warming of subsurface temperature in the Indian sector of the Southern Ocean. J. Geophys. Res. 108, 8081.

Assaf, G., Gerard, R., Gordon, A.L., 1971. Some mechanisms of oceanic mixing revealed in aerial photographs. J. Geophys. Res. 76 (27), 6550–6572.

Babu, M.T., Kumar, S.P., Rao, D.P., 1991. A subsurface cyclonic eddy in the Bay of Bengal. J. Mar. Res. 49, 403–410.

Barber, R.T., Chavez, F.P., 1983. Biological consequences of El Niño. Science 222, 1203.

Barlow, E.W., 1935. The 1910–1935 survey of the currents of the Indian Ocean and China Seas. Mar. Observer 12, 153–163.

Beal, L.M., Chereskin, T.K., Lenn, Y.D., Elipot, S., 2006. The sources and mixing characteristics of the Agulhas Current. J. Phys. Oceanogr. 36, 2060–2074.

Bernstein, R.L., White, W.B., 1977. Zonal variability in the distribution of eddy energy in the mid-latitude North Pacific Ocean. J. Phys. Oceanogr. 7, 123–126.

Bernstein, R.L., White, W.B., 1981. Stationary and traveling mesoscale perturbations in the Kuroshio Extension Current. J. Phys. Oceanogr. 11, 692–704.

Biastoch, A., Krauss, W., 1999. The role of mesoscale eddies in the source regions of the Agulhas Current. J. Phys. Oceanogr. 29, 2303–2317.

Biswas, A.N., Chatterjee, A.K., 1987. Sedimentation at Auckland in the Hugli Estuary, Proc. Coastal and Port Engineering in Developing Countries. In: China Ocean Press, 2, 1309–1320.

Boebel, O., Davis, R.E., Ollitraut, M., Peterson, R., Richard, P., Schmid, C., Zenk, W., 1999. The intermediate depth circulation of the Western South Atlantic. Geophys. Res. Let. 26 (21), 3329–3332.

Boland, F.M., 1973. A monitoring section across the East Australia Current. CSIRO Div. Fish. Oceanogr. Tech. Pap., 34.

Boland, F.M., Hamon, B.V., 1970. The East Australian Current, 1965–1968. Deep-Sea Res. 17, 777–794.

Bowen, A., 1969. Rip currents: 1. Theoretical investigations. J. Geophys. Res. 74, 5479–5490.

Bowman, D., Arad, D., Rosen, D., Kit, E., Goldbery, R., Slavicz, A., 1988a. Flow characteristics along the rip current system under low energy conditions. Mar. Geol. 82, 149–167.

Bowman, D., Rosen, D., Kit, E., Arad, D., Slavicz, A., 1988b. Flow characteristics at the rip current neck under low energy conditions. Mar. Geol. 79, 41–54.

Brander, R., 1999. Field observations on the morphodynamic evolution of a low-energy rip current system. Mar. Geol. 157, 199–217.

Brander, R.W., Short, A.D., 2001. Flow kinematics of low-energy rip current systems. J. Coast. Res. 17 (2), 468–481.

Brown, O.B., Bruce, J.G., Evans, R.H., 1980. Evolution of sea surface temperature in the Somali Basin during the southwest monsoon of 1979. Science 209, 595–597.

Brugge, B., 1995. Near-surface mean circulation and kinetic energy in the central North Atlantic from drifter data. J. Geophys. Res. 100, 20543–20554.

Bryden, H.L., Brady, E.C., 1985. Diagnostic model of the three-dimensional circulation in the upper equatorial Pacific Ocean. J. Phys. Oceanogr. 15, 1255–1273.

Bryden, H.L., Beal, L.M., Duncan, L.M., 2005. Structure and transport of the Agulhas Current and its temporal variability. Japan. J. Oceanogr. 61, 479–492.

Burkov, V.A., Bulatov, R.P., Neyman, V.G., 1981. Large scale features of water circulation in the world ocean. Mar. Phys., 325–332.

Cane, M.A., 1980. On the dynamics of equatorial currents, with application to the Indian Ocean. Deep-Sea Res. 27A, 525–544.

Cane, M.A., 1983. Oceanographic events during El Niño. Science 222, 1189.

Chao, S., McCreary, J.P., 1982. A numerical study of the Kuroshio south of Japan. J. Phys. Oceanogr. 12, 679–693.

Charney, J.G., 1960. Non-linear theory of a wind-driven homogeneous layer near the equator. Deep Sea Res. 6, 303.

Chen, Q., Dalrymple, R., Kirby, J., Kennedy, A., Haller, M., 1999. Boussinesq modelling of a rip current system. J. Geophys. Res. 104, 20,617–20,637.

Church, J.A., J,-Bethoux, P., Theocharis, A., 1998. Semienclosed seas, islands and Australia (S). In: Robinson, A.R., Brink, K.H. (Eds.). The Sea, vol. 11. Wiley, New York, NY, USA, pp. 79–124.

Cochrane, J.D., 1963. Equatorial Undercurrent and related currents off Brazil in March and April 1963. Science. 142 (3593), 669–671.

Cochrane, J.D., 1965. Equatorial currents of the western Atlantic. ONR Progress Report, Dept. of Oceanography and Meteorology, Texas A&M University, College Station, TX, USA. June 1965.

Comiso, J., 2006. Arctic warming signals from satellite observations. Weather 61 (3), 70–76.

Cooke, D.O., 1970. The occurrence and geologic work of rip currents off southern California. Mar. Geol. 9, 173–186.

Cowen, R.K., Sponaugle, S., Paris, C.B., Lwiza, K., Fortuna, J., Dorsey, S., 2003. Impact of North Brazil Current rings on local circulation and coral reef fish recruitment to Barbados, West Indies, Interhemispheric Water Exchange in the Atlantic Ocean. In: Goni, G.J., Malanotte-Rizzoli, P. (Eds.). Elsevier Oceanographic Series, vol. 68. Elsevier, pp. 443–455.

Cox, M.D., 1989. An idealized model of the world ocean. Part I: the global scale water masses. J. Phys. Oceanogr. 19, 1730–1752.

Craik, A.D.D., Leibovich, S., 1976. A rational model for Langmuir circulation. J. Fluid Mech. 73, 401–426.

Cram, R., Hanson, K., 1974. The detection by ERTS-1 of wind-induced ocean surface features in the lee of the Antilles islands. J. Phys. Oceanogr. 4, 594–600.

Cromwell, T., Montgomery, R.B., Stroup, E.D., 1954. Equatorial undercurrent in the Pacific Ocean revealed by new methods. Science. 119, 648–649.

Csanady, G.T., 1985. A zero potential vorticity model of the North Brazilian Coastal Current. J. Mar. Res. 43, 553–579.

Dantzler, H.L., 1976. Geographic variations in intensity of the North Atlantic and North Pacific oceanic eddy fields. Deep-Sea Res. 23, 783–794.

Darbyshire, J., 1964. A hydrological investigation of the Agulhas current area. Deep-Sea Res. 11, 781–815.

Das, P.K., 1988. Why monsoons fail. Science Reporter. Council of Scientific and Industrial Research, New Delhi, India.

Davis, R.E., 2005. Intermediate-depth circulation of the Indian and South Pacific Oceans measured by autonomous floats. J. Physical Oceanogr. 35, 683–707.

de Ruijter, W.P.M., Ridderinkhof, H., Lutjeharms, J.R.E., Schouten, M.W., Veth, C., 2002. Observations of the flow in the Mozambique Channel. Geophys. Res. Lett. 29 (10), 1502–1504.

de Ruijter, W.P.M., van Leeuwen, P.J., Lutjeharms, J.R.E., 1999. Generation and evolution of Natal pulses: Solitary meanders in the Agulhas Current. J. Phys. Oceanogr. 29 (12), 3043–3055.

Dijkstra, H.A., de Ruijter, W.P.M., 2001. On the physics of the Agulhas Current: Steady retroflection regimes. J. Phys. Oceanogr. 31, 2971–2985.

Duing, W., Mooers, C.N.K., Lee, T.N., 1977. Low frequency variability in the Florida Current and relations to atmospheric forcing from 1972 to 1974. J. Mar. Res. 35, 129–161.

Duing, W., Molinari, R.L., Swallow, J.C., 1980. Somali Current: Evolution of surface current. Science 209, 588–590.

Ekman, V.W., 1905. On the influence of the earth's rotation on ocean currents. Ark. Mat. Astron. Fys. 2, 1–53.

Engle, J., MacMahan, J., Thieke, R.J., Hanes, D.M., Dean, R.G., 2002. Formulation of a rip current predictive index using rescue data. Florida Shore and Beach Preservation Association National Conference.

Ewing, J., Edgar, T., 1966. Abyssal sediment (thickness). Encyclopedia of Oceanography 1, 6–10.

Faller, A.J., 1969. The generation of Langmuir circulations by the eddy pressure of surface waves. Limnol. Oceanogr. 14, 504–513.

Federov, K.N., 1965. Equatorial seiches. Oceanology 5 (l), 37. (English translation).

Fernandez, E., Pingree, R.D., 1996. Coupling between physical and biological fields in the North Atlantic subtropical front southeast of the Azores. Deep-Sea Res. 43, 1369–1393.

Firing, E., Lukas, R., Sadler, J., Wyrtki, K., 1983. Equatorial undercurrent disappears during 1982–1983 El Nino. Science 222, 1121.

Fowler, R.E., Dalrymple, R.A., 1990. Wave group forced nearshore circulation, Proceedings of the 22nd International Conference on Coastal Engineering. Am. Soc. of Civ. Eng. Delft, the Netherlands, 729–742.

Fratantoni, D.M., Glickson, D.A., 2002. North Brazil Current Ring generation and evolution observed with SeaWiFS. J. Phys. Oceanogr. 32, 1058–1074.

Fratantoni, D.M., Richardson, P.L., 2006. The evolution and demise of north Brazil current rings. J. Phys. Oceanogr. 36, 1241–1264.

Fuglister, F.C., 1955. Alternative analyses of current surveys. Deep-Sea Res. 2 (3), 213–229.

Fuglister, F.C., 1960. Gulf Stream '60. Woods Hole Oceanographic Institution Report, 265–372.

Fuglister, F.C., Worthington, L.V., 1951. Some results of a multiple ship survey of the Gulf Stream. Tellus 3, 1–14.

Fyfe, J.C., Saenko, O.A., 2005. Human-induced change in the Antarctic Circumpolar Current. J. Climate 18, 3068–3073.

Fyfe, J.C., 2003. Extratropical Southern Hemisphere cyclones: Harbingers of climate change. J. Climate 16, 2802–2805.

Gammon, R.H., Sundquist, E.T., Fraser, P.J., 1986. History of carbon dioxide in the atmosphere, In: Report to the U.S. Congress on the CO_2 Question. U.S. Dept. of Energy, Washington, D.C., USA, 62 pp.

Gangopadhyay, A., Robinson, A.R., Arango, H.G., 1997. Circulation and dynamics of the western north Atlantic. I. Multiscale feature models. J. Atmos. Oceanic Technol. 14 (6), 1314–1332.

Gangopadhyay, A., Robinson, A.R., 2002. Feature oriented regional modeling of oceanic fronts. Dynamics of Atmospheres and Oceans 36, 201–232.

Garcia-Berdeal, I., Hautala, S.L., Thomas, L.N., Johnson, H.P., 2006. Vertical structure of time-dependent currents in a mid-ocean ridge axial valley. Deep-Sea Res. I 53, 367–386.

Garrett, C.J.R., 1976. Generation of Langmuir circulations by surface waves—A feedback mechanism. J. Mar. Res. 34, 117–130.

Garzoli, S.L., Ffield, A., Yao, Q., 2003. North Brazil Current Rings and the variability in the latitude of retroflection, Interhemispheric Water Exchange in the Atlantic Ocean. In: Goni, G.J., Malanotte-Rizzoli, P. (Eds.). Elsevier Oceanographic Series, vol. 68, pp. 357–373. Elsevier.

Gerdes, R., Koeberle, C., Beckmann, A., Herrmann, P., Willebrand, J., 1999. Mechanisms for spreading of Mediterranean Water in coarse-resolution numerical models. J. Physic. Oceanogr. 29, 1682–1700.

Giarolla, E., Nobre, P., Malagutti, M., Pezzi, L.P., 2005. The Atlantic Equatorial Undercurrent: PIRATA observations and simulations with GFDL modular ocean model at CPTEC. Geophys. Res. Lett. 32, L10617. http://dx.doi.org/10.1029/2004GL022206.

Gille, ST., 2002. Warming of the Southern Ocean since the 1950s. Science 295, 1275–1277.

Godfrey, J.S., Creswell, G.R., Golding, T.J., Pearce, A.F., Boyd, R., 1980. The separation of the East Australian Current. J. Phys. Oceanogr. 10, 430–440.

Goldenberg, S.B., Landsea, C.W., Mestas-Nunez, A.M., Gray, W.M., 2001. The recent increase in Atlantic hurricane activity: Causes and implications. Science 293, 474–479.

Goni, G.J., Johns, W.E., 2003. Synoptic study of warm rings in the North Brazil Current retroflection region using satellite altimetry,

Interhemispheric Water Exchange in the Atlantic Ocean. In: J.Goni, G., Malanotte-Rizzoli, P. (Eds.). Elsevier Oceanographic Series, vol. 68. Elsevier, pp. 335–356.

Gordon, A.L., 1970. Vertical momentum flux accomplished by Langmuir circulation. J. Geophys. Res. 75, 4177–4179.

Gordon, A.L., 1985. Indian-Atlantic transfer of thermocline water at the Agulhas Retroflection. Science 227, 1030–1033.

Gordon, A.L., Lutjeharms, J.R.E., Gründlingh, M.L., 1987. Stratification and circulation at the Agulhas retroflection. Deep-Sea Res. 34, 565–599.

Gouretski, V., Danilov, A., Ivchenko, V.O., Klepikov, A., 1987. Modelling of the Southern Ocean Circulation. Hydrometeorologisher Publ., Leningrad, U.S.S.R.

Grant, H.L., Stewart, R.W., Moillet, A., 1962. Turbulence spectra from a tidal channel. J. Fluid Mech. 12, 241–268.

Haas, K.A., Svendsen, I.A., 2002. Laboratory measurements of the vertical structure of rip currents. J. Geophys. Res., 107.

Hall, A., Visbeck, M., 2002. Synchronous variability in the Southern Hemisphere atmosphere, sea ice, and ocean resulting from the annular mode. J. Climate 15, 3043–3057.

Haller, M.C., Dalrymple, R.A., 2001. Rip current instabilities. J. Fluid Mech. 433, 161–192.

Halpern, D., 1983. Variability of the Cromwell Current before and during the 1982–83 warm event. In: Trop. Ocean Atmos. Newslett, 21. Univ. of Washington, Seattle, WA, USA.

Hamon, B.V., 1965. The East Australian Current, 1960–1964. Deep-Sea Res. 12, 899–921.

Hamon, B.V., Cresswell, G.R., 1972. Structure functions and intensities of ocean circulation off east and west Australia. Aust. J. Mar. Freshwater Res. 23, 99–103.

Hata, K., 1974. Behavior of a warm eddy detached from the Kuroshio. J. Meteor. Res. 7, 295–321.

Heezen, B.C., Hollister, C.D., Ruddiman, W.F., 1966. Shaping of the continental rise by deep geostrophic contour currents. Science 152, 502–503.

Hill, A.E., Hickey, B.M., Shillington, F.A., Strub, P.T., Brink, K.H., Barton, E.D., Thomas, A.C., 1998. Eastern ocean boundaries (E). In: Robinson, A.R., Brink, K.H. (Eds.). The Sea, vol. 11. Wiley, New York, NY, USA, pp. 29–68.

Holmes, J., 2009. Proc. Second Session of the Global Platform for Disaster Risk Reduction. Switzerland, Geneva, 16–19 June 2009, pp. 3–5.

Houwman, K.T., Hoekstra, P., 1998. Tidal ellipses in the near-shore zone (−3 to −10 m): Modeling and observations. Coastal Eng., 773–786.

Hunkins, K., 1966. Ekman drift currents in the Arctic Ocean. Deep-Sea Res. 13, 607–620.

Huntley, D.A., Hendry, M.D., Haines, J., Greenidge, B., 1988. Waves and rip currents on a Caribbean pocket beach, Jamaica. J. Coast. Res. 4, 69–79.

Hynd, J.S., 1969. Isotherm maps for tuna fisherman. Aust. Fish 28 (7), 13–22.

Ikeda, M., Johannessen, J.A., Lygre, K., Sandven, S., 1989. A process study of mesoscale meanders and eddies in the Norwegian coastal current. J. Phys. Oceanogr. 19, 20–35.

Iskandar, I., Masumoto, Y., Mizuno, K., 2009. Subsurface equatorial zonal current in the eastern Indian Ocean. J. Geophys. Res. 114, C06005.

Ivchenko, V.O., Richards, K.J., Stevens, D.P., 1996. The dynamics of the Antarctic Circumpolar Current. J. Phys. Oceanogr. 26, 753–774.

Izumo, T., 2005. The equatorial undercurrent, meridional overturning circulation, and their role in mass and heat exchanges during El Niño events in the tropical Pacific ocean. Ocean Dynamics 55, 110–123.

Izumo, T., Picaut, V., Blanke, B., 2002. Tropical pathways, equatorial undercurrent variability and the 1998 La Niña. Geophys. Res. Lett. 29 (22), 2080–2083.

Jacobs, S.S., Georgi, D.T., 1977. Observations on the southwest Indian/Antarctic Ocean. Deep-Sea Res. 24, 43–84.

Jarraud, M., 2009. Reducing risk in a changing climate, United Nations International Strategy for Disaster Reduction, Proc. Global Platform for Disaster Risk Reduction, 20–21.

Jia, Y., 2000. Formation of an Azores Current due to Mediterranean overflow in a modeling study of the North Atlantic. J. Physical Oceanogr. 30, 2342–2358.

Johns, W.E., Lee, T.N., Schott, F.A., Zantopp, R.J., Evans, R.H., 1990. The North Brazil Current retroflection: Seasonal structure and eddy variability. J. Geophys. Res. 95 (C12), 22103–22120.

Johns, W.E., Lee, T.N., Beardsley, R.C., Candela, J., Limeburner, R., Castro, B., 1998. Annual cycle and variability of the North Brazil Current. J. Phys. Oceanogr. 28, 103–128.

Johnson, D., Pattiaratchi, C., 2004. Transient rip currents and nearshore circulation on a swell-dominated beach. J. Geophys. Res. 109. (C02026).

Johnson, D., Pattiaratchi, C., 2004. Transient rip currents and nearshore circulation on a swell-dominated beach. J. Geophys. Res. 109. (C02026).

Johnson, G.C., McPhaden, M.J., Rowe, G.D., McTaggart, K.E., 2000. Upper equatorial ocean current and salinity during the 1996–1998 El Niño–La Niña cycle. J. Geophys. Res. 105, 1037–1053.

Jury, M.R., Valentine, H.R., Lutjeharms, J.R.E., 1993. Influence of the Agulhas Current on summer rainfall along the southeast coast of South Africa. J. Appl. Meteor 32, 1282–1287.

Kamenkovich, V.M., Koshlyakov, M.N., Monin, A.S., 1986. Synoptic eddies in the ocean, D. Reidel, the Netherlands.

Karsten, R., Marshall, J., 2002. Constructing the residual circulation of the ACC from the observations. J. Phys. Oceanogr. 32, 3315–3327.

Kase, R.H., Siedler, G., 1982. Meandering of the subtropical front southeast of the Azores. Nature 300, 245–246.

Katz, B., Gerard, R., Costin, M., 1965. Response of dye tracers to sea surface conditions. J. Geophys. Res. 70, 5505–5513.

Kawai, H., 1972. Hydrography of the Kuroshio Extension. In: Stommel, H., Yoshida, K. (Eds.), Kuroshio, Its Physical Aspect. University of Tokyo Press, Tokyo, Japan, pp. 235–352.

Kawai, H., 1979. Rings south of the Kuroshio and their possible roles in transport of the intermediate salinity minimum and in formation of the skipjack and albacore fishing ground, Kuroshio IV. Proc. Fourth CSK Symp. Tokyo, 250–273.

Kelly, K.A., Small, R.J., Samelson, R.M., Qiu, B., Joyce, T.M., Kwon, Y.-O., Cronin, M.F., 2010. Western boundary currents and frontal air–sea interaction: Gulf Stream and Kuroshio Extension. J. Climate 23, 5644–5667.

Kennedy, A.B., Thomas, D., 2004. Drifter measurements in a laboratory rip current. J. Geophys. Res. 109, C08005.

Kim, S.L., Mullineaux, L.S., 1998. Distribution and near-bottom transport of larvae and other plankton at hydrothermal vents. Deep-Sea Res. II 45, 423–440.

Kitano, K., 1975. Some properties of the warm eddies generated in the confluence zone of the Kuroshio and Oyashio Current. J. Phys. Oceanogr. 5, 670–683.

Klein, B., Siedler, G., 1989. On the origin of the Azores Current. J. Geophys. Res. 94, 6159–6168.

Knauss, J.A., King, J.E., 1958. Observations of Pacific Equatorial Undercurrent. Nature 182, 601–602.

Knauss, J.A., 1966. Further measurements and observations on the Cromwell Current. J. Mar. Res. 24 (2), 205–240.

Knauss, J.A., Taft, B.A., 1964. Equatorial Undercurrent of the Indian Ocean. Science 143 (3004), 354–356.

Knox, R.A., Anderson, D.L.T., 1985. Recent advances in the study of the low-latitude ocean circulation. Prog. Oceanogr. 14, 259–317.

Komar, P.D., 1971. Nearshore cell circulation and the formation of giant cusps. Geol. Soc. Amer. Bull. 82, 2643–2650.

Koshimura, S., Yanagisawa, H., Miyagi, T., 2009. Mangrove's fragility against tsunami, inferred from high-resolution satellite imagery and numerical modeling. Abstract; 24th International Tsunami Symposium (ITS-2009) and Technical Workshop on Tsunami Measurements and Real-Time Detection, held at Novosibirsk, Russia (14–16 July 2009).

Krauss, W., Kase, R.H., 1994. Mean circulation and eddy kinetic energy in the eastern North Atlantic. J. Geophys. Res. 89, 3407–3415.

Kullenberg, B., 1954. Vagn Walfrid Ekman 1874–1954: A biography by B. Kullenberg. Journal du Conseil international pour l'exploration de la mer 20 (2), 1–52.

Kwon, Y.-O., Alexander, M.A., Bond, N.A., Frankignoul, C., Nakamura, H., Qiu, B., Thompson, L., 2010. Role of the Gulf Stream and Kuroshio–Oyashio systems in large-scale atmosphere–ocean interaction: A review. J. Climate 23, 3249–3281.

La Fond, E.C., 1980. Upwelling. McGraw-Hill Encyclopedia of Ocean and Atmospheric Sciences, 523–525.

Lai, D.Y., Richardson, P.L., 1977. Distribution and movement of Gulf Stream Rings. J. Phys. Oceanogr. 7, 670–683.

Langmuir, I., 1938. Surface motion of water induced by wind. Science 87, 119–123.

Lascody, R.L., 1998. East central Florida rip current program. National Weather Service In-house Report, 10.

Le Traon, P.Y., Rouquet, M.C., Boissier, C., 1990. Spatial scales of mesoscale variability in the North Atlantic as deduced from Geosat data. J. Geophys. Res. 95, 20267–20285.

Lee, T.N., Atkinson, L.P., Legeckis, R., 1981. Observations of a Gulf Stream frontal eddy on the Georgia continental shelf, April 1977. Deep-Sea Research 28 A (4), 347–378.

Lee, T.N., 1975. Florida current spin-off eddies. Deep-Sea Res. 22, 753–765.

Lee, T.N., Brooks, D.A., 1979. Initial observations of current, temperature and coastal sea level response to atmospheric and Gulf Stream forcing on the Georgia shelf. Geophys. Res. Letts. 6, 321–324.

Lee, T.N., Mayer, D., 1977. Low-frequency current variability and spin-off eddies on the shelf off southeast Florida. J. Mar. Res. 35, 193–220.

Lee, T.N., Atkinson, L.P., Legeckis, R., 1981. Observations of a Gulf Stream frontal eddy on the Georgia continental shelf, April 1977. Deep-Sea Res. 28A (4), 347–378.

Leetmaa, A., Quadfasel, D.R., Wilson, D., 1982. Development of flow field during the onset of the Somali Current, 1979. J. Phys. Oceanogr. 12, 1325–1342.

Legeckis, R., 1979. Satellite observations of the influence of bottom topography on the seaward deflection of the Gulf Stream off Charleston, South Carolina. J. Phys. Oceanogr. 9, 483–497.

Legeckis, R., Gordon, A.L., 1982. Satellite observations of the Brazil and Falkland Current. Deep-Sea Res. 29, 375–401.

Leibovich, S., 1983. The form and dynamics of Langmuir circulations. Annu. Rev. Fluid Mech. 15, 391–427.

Leibovich, S., Paolucci, S., 1981. The instability of the ocean to Langmuir circulations. J. Fluid Mech. 102, 141–167.

Leibovich, S., Lele, S.K., Moroz, I.M., 1989. Nonlinear dynamics in Langmuir circulations and in thermosolutal convection. J. Fluid Mech. 193, 471–511.

Lilley, F.E.M., Filloux, J.H., Bindoff, N.L., Ferguson, I.J., 1986. Barotropic flow of a warm-core ring from seafloor electric measurements. J. Geophys. Res. 91, 12979–12984.

Liu, Z., Philander, S.H.G., Pacanowski, R.C., 1994. A GCM study of tropical–subtropical upper-ocean water exchange. J. Phys. Oceanogr. 24, 2606–2623.

Lobel, P.S., 1978. Diel, lunar and seasonal periodicity in the reproductive behavior of the pomacanthid fish, *Centropyge potteri* and some other reef fishes in Hawaii. Pac. Sci. 32, 193–207.

Lobel, P.S., 1989. Ocean current variability and the spawning season of Hawaiian reef fishes. Environ. Biol. Fish. 2 (3), 161–171.

Lobel, P.S., 2011. Transport of reef lizardfish larvae by an ocean eddy in Hawaiian waters. Dynamics of Atmospheres and Oceans 52, 119–130.

Lobel, P.S., Robinson, A.R., 1983. Reef fishes at sea: ocean currents and the advection of larvae. In: Reaka, M.L. (Ed.), The Ecology of Deep and Shallow Reefs. Symposia Series for Undersea Research, vol. 1. Office of Undersea Research, NOAA, Rockville, MD, USA, pp. 29–38.

Lobel, P.S., Robinson, A.R., 1986. Transport and entrapment of fish larvae by ocean mesoscale eddies and currents in Hawaiian waters. Deep-Sea Res. 33, 483–500.

Lobel, P.S., Robinson, A.R., 1988. Larval fishes and zooplankton in a cyclonic eddy in Hawaiian waters. J. Plank. Res. 10 (6), 1209–1223.

Loder, J.W., Petrie, B., Gawarkiewicz, G., 1998. The coastal ocean off northeastern North America: a large-scale view (1, W). In: Robinson, A.R., Brink, K.H. (Eds.), The Sea, Vol. 11. Wiley, New York, NY, USA, pp. 125–134.

Longuet-Higgins, M.S., 1970. Longshore currents generated by obliquely incident seawaves, 1. J. Geophys. Res. 75, 6790–6801.

Longuet-Higgins, M.S., Stewart, R.W., 1964. Radiation stress in water waves, a physical discussion with applications. Deep-Sea Res. 11 (4), 529–563.

Lozier, M.S., Owens, W.B., Curry, R.G., 1995. The climatology of the North Atlantic. In: Progress in Oceanography, 36. Pergamon, 1, 44.

Luschine, J.B., 1991. A study of rip current drownings and weather related factors. Natl. Weather Dig., 13–19.

Luschine, J.B., 1991a. A study of rip current drownings and weather related factors. Natl. Weather Dig., 13–19.

Luther, M.E., O'Brien, J.J., 1985. A model of the seasonal circulation in the Arabian Sea forced by observed winds. Prog. Oceanogr. 14, 353–385.

Lutjeharms, J.R.E., Gordon, A.L., 1987. Shedding of an Agulhas Ring observed at sea. Nature 325, 138–140.

Lutjeharms, J.R.E., 1981a. Features of the southern Agulhas Current circulation from satellite remote sensing. S. Afr. J. Sci. 77, 231–236.

Lutjeharms, J.R.E., 1981b. Spatial scales and intensities of circulation in the ocean areas adjacent to South Africa. Deep-Sea Res. 28, 1289–1302.

Lutjeharms, J.R.E., 1988. Meridional heat transport across the subtropical convergence by a warm eddy. Nature 331, 251–253.

Lutjeharms, J.R.E., 2007. Three decades of research on the greater Agulhas Current. Ocean Science 3, 129–147.

Lutjeharms, J.R.E., Roberts, H.R., 1988. The Natal pulse: An extreme transient on the Agulhas Current. J. Geophys. Res. 93, 631–645.

Lutjeharms, J.R.E., van Ballegooyen, R.C., 1988. The Agulhas Current retroflection. J. Phys. Oceanogr. 18, 1570–1583.

Lutjeharms, J.R.E., Van Ballegooyen, R.C., 1988a. The retroflection of the Agulhas Current. J. Phys. Oceanogr. 18, 1570–1583.

Luyten, J.R., Swallow, J.C., 1976. Equatorial undercurrents. Deep-Sea Res. 23, 999–1001.

Luyten, J., Pedlosky, J., Stommel, H., 1983. The ventilated thermocline. J. Phys. Oceanogr. 13, 292–309.

MacMahan, J.H., Thornton, E.B., Reniers, A.J.H.M., 2006. Rip current review. Coastal Engineering 53, 191–208.

MacMahan, J., Thornton, E.B., Stanton, T.P., Reniers, A.J.H.M., 2005. RIPEX: Rip currents on a shore-connected shoal beach. Mar. Geol. 218, 113–134.

Marinone, S.G., Lavin, M.F., 2005. Tidal current ellipses in a three-dimensional baroclinic numerical model of the Gulf of California. Estuarine Coastal Shelf Sci. 64, 519–530.

Marshall, D., 1995. Topographic steering of the Antarctic Circumpolar Current. J. Phys. Oceanogr. 25, 1636–1650.

Marshall, G.J., 2003. Trends in the southern annular mode from observations and reanalyses. J. Climate 16, 4134–4143.

Marshall, J., Radko, T., 2003. Residual-mean solutions for the Antarctic Circumpolar Current and its associated overturning circulation. J. Phys. Oceanogr. 33, 2341–2354.

Mascarenhas, A.S., Miranda, L.B., Rock, N., 1971. A study of oceanographic conditions in the region of Cabo Frio. In: Costlow, Jr., D. (Ed.), Fertility of the Sea, Vol. 1. Gordon and Breach, New York, NY, USA, pp. 285–308.

Mason, S.J., 1995. Sea-surface temperature: South African rainfall associations, 1910–1989. Int. J. Climatol 15, 119–135.

Matano, R.P., Palma, E.D., Piola, A.R., 2010. The influence of the Brazil and Malvinas Currents on the Southwestern Atlantic Shelf circulation. Ocean Sci. 6, 983–995.

Maury, M.F., 1855. The physical geography of the sea. Harper & Brothers, New York, NY, USA.

McCreary, J.P., Kundu, P.K., 1985. Western boundary circulation driven by an alongshore wind: With application to the Somali Current system. J. Mar. Res. 43 (3), 493–516.

McCreary, J.P., Lu, P., 1994. Interaction between the subtropical and equatorial ocean circulations: The subtropical cell. J. Phys. Oceanogr. 24, 455–497.

McKenzie, P., 1958. Rip-current systems. J. Geol. 66, 103–113.

McKenzie, P., 1958a. Rip-current systems. J. Geol. 66, 103–113.

McPhaden, M.J., 1986. The Equatorial Undercurrent: 100 years of discovery. Eos 67 (40), 762–765.

Meehls, G.A., Stocker, T.F., Idlingstein, W.D., Gaye, A.T., Gregory, J.M., Kitoh, A., Knutti, R., Murphy, J.M., Noda, A., Raper, S.C.B., Watterson, I.G., Weaver, A.J., Zhao, Z.-C., 2007. Climate change: The physical science basis. Cambridge University Press, New York, NY, USA. Chap. Global Climate Projections.

Metcalf, W.G., Voorhis, A.D., Stalcup, M.C., 1962. The Atlantic Equatorial Undercurrent. J. Geophys. Res. 67, 2499.

Metcalf, W.G., Stalcup, M.C., 1967. Origin of the Atlantic Equatorial Undercurrent. J. Geophys. Res. 72 (20), 4959–4975.

Mizuno, K., White, W.B., 1983. Annual and inter-annual variability in the Kuroshio Current System. J. Phys. Oceanogr. 13, 1847–1867.

Montgomery, R.B., Stroup, E.D., 1962. Equatorial waters and currents at 150°W in July–August 1952. Johns Hopkins Oceanogr. Studies (No. 1), 68.

Muller-Karger, F.E., McClain, C.R., Richardson, P.L., 1988. The dispersal of the Amazon's water. Nature 333, 56–59.

Mullineaux, L.S., France, S.C., 1995. Dispersal mechanisms of deep-sea hydrothermal vent fauna. In: Humphris, S.E., Zierenberg, R.A., Mullineaux, L.S., Thomson, R.E. (Eds.), Seafloor Hydrothermal

Systems: Physical, Chemical, Biological, and Geological Interactions. American Geophysical Union, Washington, DC, USA, pp. 408–424.

Munk, J.W., 1983. Acoustic and ocean dynamics. In: Brewer, P.G. (Ed.), Oceanography: The present and future. Springer-Verlag, New York, Heidelberg, Berlin, pp. 109–126.

Munk, J.W., Wunsch, C., 1979. Ocean acoustic tomography: A scheme for large scale monitoring. Deep-Sea Res. 26A, 123–161.

Munk, W., 1950. On the wind-driven ocean circulation. J. Meteorol. 7, 79–93.

Munk, W.H., Palmen, E., 1951. Note on the dynamics of the Antarctic Circumpolar Current. Tellus 3, 53–55.

Murray, S.P., 1975. Trajectories and speeds of wind-driven currents near the coast. J. Phys. Oceanogr. 5, 347–360.

Nakamura, M., 2012. Impacts of SST anomalies in the Agulhas Current System on the regional climate variability. J. Climate 25, 1213–1229.

Neumann, G., 1960. Evidence for an equatorial undercurrent in the Atlantic Ocean. Deep-Sea Res. 6, 328–334.

Neumann, G., 1966. The Equatorial Undercurrent in the Atlantic Ocean. Proc. Symposium on Oceanography and Fisheries Resources of the Tropical Atlantic, 1–17.

Neumann, G., 1968. Ocean Currents. Elsevier, Amsterdam- London- New York.

Nilsson, C.S., Cresswell, G.R., 1981. The formation and evolution of East Australian Current warm-core eddies. Progress in Oceanography vol. 9, 133–183. Pergamon.

Nowlin Jr., W.D., Klinck, J.M., 1986. The physics of the Antarctic Circumpolar Current. Rev. Geophys. 24, 469–491.

Oke, P.R., England, M.H., 2004. Oceanic response to changes in the latitude of the Southern Hemisphere subpolar westerly winds. J. Climate 17, 1040–1054.

Olbers, D.J., Wenzel, M., Willebrand, J., 1985. The inference of North Atlantic circulation patterns from climatological hydrographic data. Rev. Geophys. 23, 313–356.

Olbers, D., Wubber, C., 1991. The role of wind and buoyancy forcing of the Antarctic Circumpolar Current, Strategies for Future Climate Research. In: Latif, M. (Ed.), Max-Planck Institute for Meteorology, pp. 161–192.

Olson, D.B., Evans, R.H., 1986. Rings of the Agulhas. Deep-Sea Res. 33, 27–42.

Ou, H.W., DeRuijter, W.P., 1986. Separation of an inertial boundary current from a curved coastline. J. Phys. Oceanogr. 16, 280–289.

Peregrine, D., 1998. Surf zone currents. Theor. Comput. Fluid Dyn. 10, 295–309.

Peregrine, D., 1998a. Surf zone currents. Theor. Comput. Fluid Dyn. 10, 295–309.

Philander, S.G.H., 1973. Equatorial Undercurrent: Measurement and theories. Rev. Geophys. Space Phys. 11 (3), 513–570.

Philander, S.G.H., Pacanowski, R.C., 1986. A model of the season cycle in the tropical Atlantic Ocean. J. Geophys. Res. 91, 14192–14206.

Picaut, J., Tournier, R., 1991. Monitoring the 1979–1985 equatorial Pacific current transports with bathythermograph data. J. Geophys. Res. 96, 3263–3277.

Pingree, R.D., 1997. The eastern subtropical gyre (North Atlantic): Flow rings recirculations structure and subduction. J. Mar. Biol. Assoc. United Kingdom 77, 573–624.

Quadfasel, D.R., Schott, F., 1983. Southward subsurface flow below the Somali Current. J. Geophys. Res. 88, 5973–5979.

Rahmstorf, S., 1998. Influence of Mediterranean outflow on climate. Nature 79 (24), 281–282.

Rahmstorf, S., 2006. Thermohaline ocean circulation. In: Elias, S.A. (Ed.), Encyclopedia of Quaternary Sciences. Elsevier, Amsterdam, the Netherlands.

Rameshkumar, M.R., Sadhuram, Y., Rao, L.V.G., 1986. Pre-monsoon sea surface temperature anomalies in western Indian Ocean. Proc. Symposium on Long range forecasting of monsoon rainfall, held at New Delhi, India (16–18 April 1986).

Rasmussen, E.M., Wallace, J.M., 1983. Meteorological aspects of the El Niño/Southern Oscillation. Science 222, 1195.

Reason, C.J.C., 2001. Evidence for the influence of the Agulhas Current on regional atmospheric circulation patterns. J. Climate 14, 2769–2778.

Reid, J.L., 1979. On the contribution of the Mediterranean Sea outflow to the Norwegian-Greenland Sea. Deep-Sea Res. 26, 1199–1223.

Rhines, P.B., Young, W.R., 1982. A theory of the wind-driven circulation I: Mid-ocean gyres. J. Mar. Res. 40 (Suppl.), 559–596.

Richardson, P.L., 1981. Gulf Stream trajectories measured with free-drifting buoys. J. Phys. Oceanogr. 11, 999–1010.

Richardson, P.L., 1983. Eddy kinetic energy in the North Atlantic from surface drifters. J. Geophys. Res. 88, 4355–4367.

Richardson, P.L., 1980. Gulf Stream ring trajectories. J. Phys. Oceanogr. 10, 90–104.

Richardson, P.L., Cheney, R.E., Worthington, L.V., 1978. A census of Gulf Stream rings. J. Geophys. Res. 83, 6136–6144.

Ridgway, K.R., Godfrey, J.S., 1997. Seasonal cycle of the East Australian Current. J. Geophys. Res. 102, 22921–22936.

Rintoul, S.C., Hughes, C., Olbers, D., 2001. The Antarctic Circumpolar Current system. *Ocean Circulation and Climate*. In: Siedler, J.C.G., Gould, J. (Eds.), Academic Press, pp. 271–302.

Robinson, A.R., Brink, K.H. (Eds.), 1998. The Sea, vol. 11. Wiley, New York, NY, USA, p. 1062.

Robinson, A.R., Lobel, P.S., 1985. The impact of ocean eddies on coastal currents. In: Magaard, L., et al. (Eds.), The Hawaiian Ocean Experiment Proceedings. Hawaii Inst. Geophysics Spec. Publ, pp. 325–334.

Robinson, A.R., Lobel, P.S., 1985. The impact of ocean eddies on coastal currents. In: Magaard, L., et al. (Eds.), The Hawaiian Ocean Experiment Proceedings. Hawaii Inst. Geophysics Spec. Pub, pp. 325–334.

Roden, G.I., Taft, B.A., Ebbesmeyer, C.C., 1982. Oceanographic aspects of the Emperor seamounts region. J. Geophys. Res. 87, 9537–9552.

Rossby, H.T., 1983. Eddies and the general circulation. In: Brewer, P.G. (Ed.), Oceanography — the present and future. Springer-Verlag, New York, Heidelberg, Berlin, pp. 137–161.

Rossby, H.T., 1983a. Eddies and the general circulation. In: Brewer, P.G. (Ed.), Oceanography: The present and future. Springer-Verlag, New York, Heidelberg, Berlin, pp. 137–161.

Rossby, H.T., Voorhis, A.D., Webb, D., 1975. A Quasi-Lagrangian study of mid-ocean variability using long range SOFAR floats. J. Mar. Res. 33, 355–382.

Rossby, T., Dorson, D., Fontaine, J., 1986. The RAFOS system. J. Atmos. Oceanic Technol. 3, 672–679.

Rouse, H., 1963. On the role of eddies in fluid motion. American Scientist 51, 285–314.

Sætre, R., da Silva, A.J., 1984. The circulation of the Mozambique Channel. Deep-Sea Res. 31, 485–508.

Sallee, J.B., Speer, K., Morrow, R., 2008. Response of the Antarctic Circumpolar Current to atmospheric variability. J. Climate 21, 3020–3039.

Sarukhanyan, E.I., 1982. The three-dimensional structure of the west wind drift in the region between Africa and Antarctica. Dokl. Akad. nauk SSSR 250, 234–237.

Saylor, J.H., 1966. Currents at Little Lake Harbor. U.S. Lake Survey Res. Rept. No. 1–1. Lake Survey District, Corps of Eng, Detroit, MI, USA.

Schmid, C.H., Schafer, H., Podesta, G., Zenk, W., 1995. The Vitória eddy and its relation to the Brazil Current. J. Phys. Oceanogr. 25 (11), 2532–2546.

Schmitz Jr., W.J., 1995. On the interbasin scale thermohaline circulation. Rev. Geophys. 33, 151–173.

Schott, F., 1983. Monsoon response of the Somali Current and associated upwelling. Prog. Oceanogr. 12, 357–381.

Schott, F., Quadfasel, D.R., 1982. Variability of the Somali Current system during the onset of the southwest monsoon, 1979. J. Phys. Oceanogr. 12, 1343–1357.

Schouten, M.W., de Ruijter, W.P.M., van Leeuwen, P.J., 2002. Upstream control of Agulhas Ring shedding. J. Geophys. Res. 107 (C8), 3109–3120.

Schouten, M.W., de Ruijter, W.P.M., van Leeuwen, P.J., 2002a. Upstream control of the Agulhas ring shedding. J. Geophys. Res. 107, 3109.

Schouten, M.W., de Ruijter, W.P.M., van Leeuwen, P.J., Rifferinkhof, H., 2003. Eddies and variability in the Mozambique Channel. Deep-Sea Res. 50, 1987–2003.

Scott, J.T., Myer, G.E., Stewart, R., Walther, E.G., 1969. On the mechanism of Langmuir circulations and their role in epilimnion mixing. Limnol. Oceanogr. 14, 493–503.

Send, U., Worcester, P.F., Cornuelle, B.D., Tiemann, C.O., Baschek, B., 2002. Integral measurements of mass transport and heat content in the Strait of Gibraltar from acoustic transmissions. Deep-Sea Res. II 49, 4069–4095.

Serreze, M.C., Maslanik, J.A., Scambos, T.A., Fetterer, F., Stroeve, J., Knowles, K., Fowler, C., Drobot, S., Barry, R.G., Haran, T.M., 2003. A record minimum Arctic sea ice extent and area in 2002. Geophys. Res. Lett. 30 (3), 1110.

Shepard, F.P., 1936. Undertow, riptide, or "rip current". Science, 84.

Shepard, F.P., Inman, D.I., 1950. Nearshore water circulation related to bottom topography and wave refraction. Trans. Amer. Geophys. Union 31, 196–212.

Shepard, F.P., Inman, D.I., 1950. Nearshore water circulation related to bottom topography and wave refraction. Trans. Amer. Geophys. Union 31, 196–212.

Shepard, F., Emery, K., LaFond, E., 1941. Rip currents: A process of geological importance. J. Geol. 49, 337–369.

Shoji, D., 1972. Time variation of the Kuroshio south of Japan. In: Stommel, H., Yoshida, K. (Eds.), Kuroshio, Its Physical Aspect. University of Tokyo Press, Tokyo, Japan, pp. 217–234.

Short, A.D., 1999. Handbook of Beach and Shoreface Morphodynamics. Wiley. p. 379.

Short, A.D., Hogan, C.L., 1994. Rip currents and beach hazards, their impact on public safety and implications for coastal management. In: Finkl, C.W. (Ed.), Coastal Hazards. J. Coastal Res., Special Issue, 12, pp. 197–209.

Shukla, J., 1987a. Inter-annual variability of monsoons. In: Fein, J.S., Stephens, P.L. (Eds.), Monsoons. Wiley Interscience, New York, NY, USA.

Shukla, J., 1987b. Long-range forecasting of monsoons. In: Fein, J.S., Stephens, P.L. (Eds.), Monsoons. Wiley Interscience, New York, NY, USA.

Signorini, S.R., 1978. On the circulation and the volume transport of the Brazil Current between the Cape of São Tomé and Guanabara Bay. Deep-Sea Res. 25, 481–490.

Silveira, I.C.A., Lima, J.A.M., Schmidt, A.C.K., Ceccopieri, W., Sartori, A., Francisco, C.P.F., Fontes, R.F.C., 2008. Is the meander growth in the Brazil Current system off Southeast Brazil due to baroclinic instability? Dyn. Atmos. Oceans 45, 187–207.

Silveira, I.C.A., Calado, L., Castro, B.M., Cirano, M., Lima, J.A.M., Mascarenhas, A.S., 2004. On the baroclinic structure of the Brazil Current-Intermediate Western Boundary Current system at 22°–23°S. Geophys. Res. Lett. 31.

Sloyan, B.M., Johnson, G.C., Kessler, W.S., 2003. The Pacific cold tongue: an indicator of hemispheric exchange. J. Phys. Oceanogr. 33 (5), 1027–1043.

Smith, J.A., 1992. Observed growth of Langmuir circulation. J. Geophys. Res. 97 (C4), 5651–5664.

Smith, J., Largier, J., 1995. Observations of nearshore circulation: Rip currents. J. Geophys. Res. 100, 10,967–10,975.

Sonu, C., 1972. Field observations of nearshore circulation and meandering currents. J. Geophys. Res. 77, 3232–3247.

Stalcup, M.C., Metcalf, W.G., 1966. Direct measurements of the Atlantic Equatorial Undercurrent. J. Mar. Res. 24 (1), 44–55.

Stammer, D., 1997. Global characteristics of ocean variability estimated from regional TOPEX/Poseidon altimeter measurements. J. Phys. Oceanogr. 27, 1743–1769.

Stock, C.A., Alexander, M.A., Bond, N.A., Brander, K.M., Cheung, W.W.L., Curchitser, E.N., Delworth, T.L., Dunne, J.P., Griffies, S.M., Haltuch, M.A., Hare, J.A., Hollowed, A.B., Lehodey, P., Levin, S.A., Link, J.S., Rose, K.A., Rykaczewski, R.R., Sarmiento, J.L., Stouffer, R.J., Schwing, F.B., Vecchi, G.A., Werner, F.E., 2011. On the use of IPCC-class models to assess the impact of climate on living marine resources. Prog. Oceanogr. 88, 1–27.

Stommel, H., 1948. The western intensification of wind-driven ocean currents. Eos Trans. AGU 29, 202.

Stommel, H., 1960. Wind drift near the equator. Deep Sea Res. 6, 298.

Stommel, H.M., 1957. A survey of ocean current theory. Deep-Sea Res. 4 (3), 149–184.

Stommel, H., Arons, A.B., 1972. On the abyssal circulation of the World Ocean, V. The influence of bottom slope on the broadening of inertial boundary currents. Deep-Sea Res. 19, 707–718.

Stommel, H., Yoshida, K., 1971. Some thoughts on the Cold Eddy south of Enshunada. J. Oceanographical Society of Japan 27 (5), 213–217.

Stramma, L., 1984. Geostrophic transport in the Warm Water Sphere of the eastern subtropical North Atlantic. J. Mar. Res. 42, 537–558.

Stramma, L., 1992. The South Indian Ocean Current. J. Phys. Oceanogr. 22, 421–430.

Stramma, L., Lutjeharms, J.R.E., 1997. The flow field of the subtropical gyre in the South Indian Ocean. J. Geophys. Res. 99, 14053–14070.

Sturm, M., Voigt, K., 1966. Observations on the structure of the Equatorial Undercurrent in the Gulf of Guinea in 1964. J. Geophys. Res. 71 (12), 3105–3108.

Summerhayes, C.P., Rayner, R., 2002. Operational oceanography, Oceans 2020: Science, Trends, and the Challenge of Sustainability. In: Field, J., Hempel, G., Summerhayes, C. (Eds.), Island Press, pp. 187–207.

Svendsen, I.A., Haas, K.A., Zhoa, Q., 2000. Analysis of rip current systems. Proc. Coastal Eng. 2000, Sydney, Australia. Amer. Soc. Civil Engineers, pp. 1127−1140.

Svendsen, I., Haas, K., Zhao, Q., 2001. Analysis of rip current systems. Am. Soc. of Civ. Eng, Sydney, Australia. paper presented at *27th International Conference on Coastal Engineering*.

Sverdrup, H.U., 1947. Wind-driven currents in a baroclinic ocean, with application to the equatorial currents of the eastern Pacific. Proc. Natl. Acad. Sci. 33, 318.

Swallow, J.C., 1964. Equatorial Undercurrent in the western Indian Ocean. Nature 204, 436−437.

Swallow, J.C., Worthington, L.V., 1961. An observation of a deep counter-current in the Western North Atlantic. Deep-Sea Res. 8 (1), 1−19.

Swift, J.H., 1995. Comparing WOCE and historical temperatures in the deep southeast Pacific. International WOCE Newsletter, No. 18, WOCE International Project Office, Southampton, U.K, 15−26.

Taft, B.A., 1972. Characteristics of the flow of the Kuroshio south of Japan. In: Stommel, H., Yoshida, K. (Eds.), Kuroshio, Its Physical Aspect. University of Tokyo Press, Tokyo, Japan, pp. 165−216.

Tang, E.-S., Dalrymple, R., 1989. Nearshore Circulation: Rip Currents and Wave Groups. In: Seymour, R.J. (Ed.), Plenum, New York, NY, USA.

The Ring Group, 1981. Gulf Stream cold-core rings: Their physics, chemistry, and biology. Science 212, 1091−1100.

Thompson, D.W.J., Solomon, S., 2002. Interpretation of recent Southern Hemisphere climate change. Science 296, 895−899.

Tilburg, C.E., Hurlburt, H.E., O'Brien, J.J., Shriver, J.F., 2001. The dynamics of the East Australian Current System: The Tasman Front, the East Auckland Current, and the East Cape Current. J. Phys. Oceanogr. 31, 2917−2943.

Tomosada, A., 1978. A large warm eddy detached from Kuroshio east of Japan. Bull. Tokai Reg. Lab. 94, 59−103.

Tranter, D.J., Carpenter, D.J., Leech, G.S., 1986. The coastal enrichment effect of the East Australian current eddy field. Deep-Sea Research 33 (11/12 A), 1705−1728.

Ulbrich, U., Leckebusch, G.C., Pinto, J.G., 2009. Cyclones in the present and future climate: a review. Theor. Appl. Climatol.. published online, doi: 10.1 0071s00704-008-0083-8.

van Leeuwen, P.J., de Ruijter, W.P.M., Lutjeharms, J.R.E., 2000. Natal pulses and the formation of Agulhas rings. J. Geophys. Res. 105, 6425−6436.

Veth, C., Zimmerman, J.T.F., 1981. Observations of quasi-two-dimensional turbulence in tidal currents. J. Physical Oceanogr. 11, 1425−1430.

Von Arx, W.S., Bumpus, D.F., Richardson, W.S., 1955. On the fine structure of the Gulf Stream front. Deep-Sea Res. 3, 46−65.

Walker, N.D., 1990. Links between South African summer rainfall and temperature variability of the Agulhas and Benguela Current systems. J. Geophys. Res. 95, 3297−3319.

Weart, S., 2009. Ocean currents and climate. In: The discovery of global warming. Harvard University Press.

Webster, T.F., 1961. A description of Gulf Stream meanders off Onslow Bay. Deep-Sea Res. 8 (2), 130−143.

White, J., 1590. Narrative of the 1590 voyage to Virginia. In: Quinn, D. (Ed.), The Roanoke voyages, vol. 2. Hakluyt Society, London, England.

White, W.B., 1975. Secular variability in the large-scale baroclinic transport of the North Pacific from 1950−1970. J. Mar. Res. 33 (1), 141−155.

White, W.B., McCreary, J.P., 1976. On the formation of the Kuroshio meander and its relationship to the large-scale ocean circulation. Deep-Sea Res. 23, 33−47.

White, W.B., Peterson, R.G., 1996. An Antarctic Circumpolar Wave in surface pressure, wind, temperature, and sea-ice extent. Nature 380, 699−702.

Wickham, J.B., 1975. Observations of the California counter current. J. Mar. Res. 33 (3), 325−340.

Wille, R., 1960. Karman vortex streets. Advances in Applied Mechanics 4, 185−196.

Williams, R.B., Gibson, C.H., 1974. Direct measurements of turbulence in the Pacific Equatorial Undercurrent. J. Phys. Oceanogr. 4 (1), 104−108.

Wilson, W.S., Dugan, J.P., 1978. Mesoscale thermal variability in the vicinity of the Kuroshio Extension. J. Phys. Oceanogr. 8, 537−540.

Wolff, J.O., Olbers, D., Maier-Reimer, E., 1991. Wind-driven flow over topography in a zonal β-plane channel: A quasigeostrophic model of the Antarctic Circumpolar Current. J. Phys. Oceanogr. 21, 236−264.

Wolff, J.O., Ivchenko, V.O., Klepikov, A., Olbers, D., 1990. The topographic influence on the dynamics of zonal fluxes in the ocean. Dokl. Acad. Nauk SSSR 313, 970−974.

Wunsch, C., Stammer, D., 1995. The global frequency-wave-number spectrum of oceanic variability estimated from TOPEX/Poseidon altimetric measurements. J. Geophys. Res. 100, 24895−24910.

Wyrtki, K., 1973. An equatorial jet in the Indian Ocean. Science 181, 262−264.

Wyrtki, K., 1962. Geopotential topographies and associated circulation in the western south Pacific ocean. Aust. J. Marine Freshwater Res. 13, 89−105.

Zenk, W., Muller, T.J., 1988. Seven-year current meter record in the eastern North Atlantic. Deep-Sea Res. 35 (8), 1259−1268.

Zharkov, V., Nof, D., 2010. Why does the North Brazil Current regularly shed rings but the Brazil Current does not? J. Phys. Oceanogr. 40, 354−367.

Zimmerman, J.T.F., 1978. Topographic generation of residual circulation by oscillatory (tidal) currents. Geophys. Astrophys. Fluid Dyn. 11, 35−47.

Zimmerman, J.T.F., 1980. Vorticity transfer by tidal currents over an irregular topography. J. Mar. Res. 38, 601−630.

Chapter 2

The History of Measuring Ocean Currents

Chapter Outline

2.1. Surface Current Measurements   **52**
  2.1.1. Measurements Based on Motion of Drifting
        Surface Bodies   52
  2.1.2. Imaging of Surface Water Motion Trajectories
        and Patterns   57
        2.1.2.1. Aerial Photography   57
        2.1.2.2. Radiometry   60
        2.1.2.3. Active Microwave Radar Imaging   61
  2.1.3. Vector Mapping Based on Current-Driven Sea
        Surface Wave Transport   62
2.2. Subsurface and Abyssal Current Measurements   **64**
  2.2.1. Early Mariners' Contributions   64
  2.2.2. Spatially Integrated Measurements Based on
        Earth's Magnetism and Oceanic Sound Speed   65
        2.2.2.1. Electromagnetic Method   65
        2.2.2.2. Acoustic Tomography   66
  2.2.3. Measurements Based on Motion of Drifting
        Subsurface Floats   69

  2.2.4. Measurements from Fixed Locations at
        Predetermined Depths   70
        2.2.4.1. Suspended Drag   70
        2.2.4.2. Propeller Revolution Registration
               by Mechanical Counters   71
        2.2.4.3. Unidirectional Impeller Current Meters   71
        2.2.4.4. Savonius Rotor Current Meters   72
        2.2.4.5. Ultrasonic Acoustic Methods   74
        2.2.4.6. Thermal Sensors for Measurements
               of Turbulent Motions   77
        2.2.4.7. Laser Doppler Sensor   78
        2.2.4.8. Acoustic Doppler Current Meter   81
2.3. Seafloor Boundary Layer Current Measurements   **82**
  2.3.1. Mechanical Devices   83
  2.3.2. Nonmechanical Devices   83
        2.3.2.1. BASS and MAVS   84
2.4. Vertical Profiling of Horizontal Currents   **86**
References   **88**
Bibliography   **91**

The world came to know of the existence of ocean currents from the early mariners, who painstakingly crossed the world seas on their long and tiring trading voyages through the pathless ocean. As a matter of necessity in terms of navigational safety and in an attempt to substantially reduce the travel time to the destination ports, it began to be a routine practice among the early mariners to collect the experiences of every navigator about the ocean currents and winds he encountered, and then discuss their observations during leisure times. They realized that by systematically recording on a chart the tracks of several ships on the same voyage but at different times and in different years, and during all seasons, and by graphically indicating the currents and winds daily encountered by them on each track, these charts would eventually turn out to be fine expressions of the results of the combined experiences of all whose tracks were thus drawn.

The initiative for methodical documentation of all observations of all the navigators, as just mentioned, came from a veteran navigator named Lieutenant M. F. Maury of

the U.S. Navy, and the charts so prepared came to be known as *Maury's Wind and Current Charts*. Eventually the charts describing the currents and winds in different seas and oceans became dependable guides to the early navigators. Maury's charts subsequently began to be considered as time-honored treasures by the maritime communities of the whole civilized world to guide them across the vast oceans, to save themselves from the wrath of the sweeping currents and winds of the oceans, in the best possible manner, for the improvement of commerce and maritime navigation.

A serious attempt at the systematic study of ocean currents began in the 18th century, primarily as a need to decrease travel time for cargo between England and America. In this attempt the first map was created of the notorious currents in the Gulf Stream in the Atlantic Ocean. This map showed the routes that ships should sail in order to decrease their travel time. According to Maury (1855), the National Observatory at Washington, D.C., showed immense interest in converting every ship that navigated the high seas into a floating observatory and a temple of

Measuring Ocean Currents. http://dx.doi.org/10.1016/B978-0-12-415990-7.00002-8
Copyright © 2014 Elsevier Inc. All rights reserved.

science. Accordingly, more than a thousand navigators began to engage day and night, and in all parts of the ocean, in making and recording observations according to a uniform plan and in furthering this attempt to increase our knowledge about the currents and winds of the seas, and several other phenomena that manifest themselves at sea and that might have implications for safe and proper ocean navigation. This attempt proved a giant stride in the advancement of knowledge and a great step toward its spread upon the waters.

In the present era, ocean currents are monitored for several applications, such as port operations, coastal environmental monitoring, offshore explorations, and other commercial purposes; studies of climate change; and out of purely academic interest because, with proper knowledge of the ocean currents, it is possible to predict a variety of useful as well as damaging events in the oceans and effectively deal with the consequences.

With advances in oceanographic studies, application of measurements, together with modeling techniques, became more and more common in helping humans understand several oceanographic processes. Thus, in the context of ocean current circulation studies, there are now two kinds of approaches to understanding the state of the ocean: (1) the direct or indirect observation of current, and (2) model simulation based on the available knowledge about the governing laws. The difficulty in the former approach is that the number of observation points is usually insufficient to get adequate spatial coverage of the parameters of interest due to instrumentation cost and environmental regulation. Insufficient observation points lead to data being badly interpolated in space. Model simulation can represent only an approximated aspect of the particular problem under consideration and must therefore be validated appropriately by the field observation data. Fortunately, several methods of data assimilation have been developed to strengthen the performance of observation and model by combining these two approaches optimally.

2.1. SURFACE CURRENT MEASUREMENTS

As noted in Chapter 1, knowledge of sea surface currents is important for a variety of applications. Lagrangian method (i.e., a method that provides a quantitative description of both track and speed of water flow) is often employed to obtain an overall picture of water motions over large oceanic regions. In a Lagrangian sense, ocean currents are the channels through which the waters of the oceans flow. The Lagrangian method originally meant measurement of water movements by tracing the path of water parcels over sufficiently long time intervals. The combination of many such paths, termed *trajectories*, mapped on a chart (called

a *current chart*) was then used to describe the overall pattern and speed of horizontal currents in a given oceanic region.

The study of ocean currents was important to mariners for several reasons. First, gaining insight into the weather pattern at sea involves knowledge of its currents, both cold and warm. Second, oceanic currents are the chief agents for distributing heat to the various parts of the oceans and thus play an important role in regulating the global climate. With advances in offshore exploration of marine resources, knowledge of ocean currents at various depth layers became important. Crude oil exploration and its maritime transport (via submarine pipelines or ships) resulted in occasional oil slicks. Timely mitigation of hazards arising from such unfortunate episodes required knowledge of surface currents and their trajectories over large areas. There are numerous other reasons that justify gathering information on sea surface currents. Consequently, several methods were attempted for the measurement of sea surface currents and their trajectories. Such methods threw much light on the nature of sea surface currents at different regions of the world oceans. It is interesting to examine the gradual growth achieved in this subject over the years with the use of more and more improved technologies. To appreciate these achievements, we first glance through the primitive methods and then skim through more and more sophisticated methods developed over the years. Detailed discussions on the sophisticated and the state-of-the-art technologies in each category of measurements are reserved for the subsequent chapters of this book.

2.1.1. Measurements Based on Motion of Drifting Surface Bodies

In the previous years of oceanographic studies, the trajectories traced by freely drifting seaweeds and driftwoods provided an indication of the approximate sea surface water circulation paths. An excellent example of the application of this method came from the Sargasso Sea, which is located in the Atlantic Ocean, off the west coast of North Africa. In past centuries, the Sargasso Sea was well known as a general receptacle of the driftwood and seaweeds of the Atlantic Ocean. Likewise, to the west from California a pool existed into which driftwoods and seaweeds generally gathered.

The natives of the Aleutian Islands (which are notorious for fogs and mists, as are the Grand Banks of Newfoundland), where trees hardly grow, depended on the driftwoods cast ashore for all the timber they used in the construction of their fishing boats, fishing tackle, and household gear. Among this timber, the camphor tree and other woods of China and Japan were said to have been often recognized (Maury, 1855). The drift, which brought the timber, was the result of the surface current of the China Stream.

Thus, driftwoods and seaweeds provided some of the earliest evidence of oceanic surface currents across the oceans.

The role of driftwoods in identifying surface current paths, although rudimentary in a practical sense, led to the notion that this idea can be effectively used for measuring surface currents and identifying surface current paths cost-effectively, although the precision in such measurements might be rather poor. Accordingly, one of the oldest methods of determining the speed and direction of oceanic surface currents was to trace the paths of floating objects that freely drifted with the currents.

Mariners have routinely noticed currents on the sea surface through ship drifts. Initially, observations of surface currents had been derived from ships' navigation logs (Sverdrup et al., 1942). In the ship drifts method, the ship itself is used as a tracer of surface currents. The vector difference between a ship's *dead reckoning* and its true position had long been used to provide a measure of sea surface current. In the early decades of maritime navigation, surface currents were determined from the difference between a ship's position by fixes from astronomical observations at least 24 hours apart and dead reckoning. This method was first used by Maury, about 1853, to examine surface currents of the Gulf Stream region, and the method was subsequently extended worldwide. *Dead reckoning*, i.e., deduced reckoning, is a method of estimating a ship's position from knowledge of the course steered and the speed maintained. The accumulating discrepancy between the dead-reckoning position and the true position is an indication of the average displacement of the surface layer of water through which the ship traversed. Over the years a large data bank of surface current statistics has been built up (e.g., Duncan and Schadlow, 1981).

A drawback of the ship-drifts method is that speed and direction are uncertain if the displacement is small, because an astronomical fix is usually not accurate to better than 1 or 2 nautical miles, and the position accuracy by dead reckoning is, as a rule, limited. Further, the ship's drifts are influenced partly by direct wind effects and partly by surface currents. In addition, wave motion can make it very difficult to maintain the ship's heading on the intended course. Even if wind speed, wind direction, and surface waves are recorded, it is difficult to fully assess their contributions to the drift of the ship. These inherent errors and limitations, together with those resulting from inaccuracies in fixing the true position of the ship, mean that observations made at intervals of 24 hours are insufficient. In olden days, the now-ubiquitous global positioning system (GPS) was unheard of, and therefore, for realistic assessment of sea surface currents using the ship-drifts method, the absolute position of the ship relative to the ground had to be very accurately determined at close time intervals. However, technology was not available for such measurements.

In coastal waters it was possible to obtain fairly accurate position fixes at fairly close intervals using navigational aids such as Loran, Shoran, Decca, Mini-Ranger, and so forth. However, further offshore, the navigator needed to rely on Omega and satellite navigation by the Navy Navigation Satellite System (see Joseph, 2000). Unfortunately, the data on sea surface currents obtained solely from ships' logs are often inadequate to construct detailed synoptic charts of surface currents over large areas because such data are concentrated solely along maritime trade routes. Furthermore, the average conditions are not represented because the number of observations is far too small. Nevertheless, several central agencies (for example, hydrographic offices of various countries) have accumulated large sets of such data from commercial shipping, thereby enabling charts to be constructed of surface currents, at least along navigation routes. Interestingly, a notable feature of the wind-driven surface layer circulation (namely, clockwise deflection in the northern hemisphere and counter-clockwise deflection in the southern hemisphere) was first observed from synoptic charts of currents prepared from navigational records.

A rather primitive method for measurement of sea surface currents utilized the drift of free bodies such as drift bottles (Sverdrup et al., 1942; Merritt et al., 1969; Chew and Berberian, 1970; Wyatt et al., 1972). These methods were generally employed near coastal regions. The drift bottles were weighted down so that they would be nearly immersed, offering only a very small surface for the wind to act on, and they were carefully sealed. They contained cards giving the number of the bottle, which established the locality and time of release and requesting the finder (often fishermen and coastal dwellers) to fill in information as to the place and time of finding and to send this information to a central office in return for a small reward. Surface current speed could be determined from the range covered by the bottle and from its travel time. The straight-line distance between the positions of release and finding of a bottle were taken as the *range*, and the difference between the times of release of a bottle and its finding were taken as the *travel time*. Being relatively cheap, drift bottles were used in great numbers. Together with ship-drift observations, this simple method provided a practical means to describe many features of the general surface circulation in coastal regions. In certain studies of ocean currents, drift bottles could provide information that was unobtainable otherwise. This was especially true in shallow coastal regions and in areas abounding in many shoals and islands or those with an irregular coastline.

A disadvantage of drift bottles is that they are fragile and may break, either in handling aboard fishing boats and ships or when washed against rocky cliffs or beaches. To use them effectively, several bottles should be deployed on a single cruise. When used on the South African coast, their recovery rate was less than 1%, presumably because of the high breakage rate on the rocky coast. There are, however, instances in which nearly all the drift bottles were recovered after having been tossed about in a severe hurricane.

A substitute for drift bottle was the drift card, first used by F. C. W. Olson in the early 1950s. The *Olson drift card*, named after its inventor, consisted of a printed card rolled hermetically in a polyethylene envelope. The Olson card thus formed an inexpensive, compact, and convenient means for studying surface currents. Because drift cards are cheap and easy to handle, several hundred of them were usually released at a given point. The method of multiple releases provided more dependable data than would be obtained if the cards were released singly. In a series of trials conducted by Olson, the drifts indicated by these cards were found to be consistent with flow patterns determined by other methods such as water mass analysis and drift bottle studies (Olson, 1951). The results also indicated that drift cards were not at the complete mercy of the wind.

However, the Olson cards had some disadvantages. It was realized that the cards recovered from sandy beaches showed definite signs of sand abrasion, which in some instances acted through the polythene envelope, thereby scouring all identification marks. Because these cards were so light, they had also been, on some occasions, whipped along by strong winds. There were also instances in which the cards displayed marks that suggested attacks by fish and sea birds that were in search of food (Stander et al., 1969). This sometimes resulted in sinking of cards or the writing on them becoming illegible.

In an attempt to retain the advantages and reduce the disadvantages of the Olson cards, a solid polythene drift card (3 mm thick) was designed by Duncan (1965). In the *Duncan drift card*, the messages in several languages (English, Afrikaans, Portuguese, and German) were heat-embossed. The Duncan cards proved to be durable and withstood rough sea action that would normally break bottles and puncture Olson-style envelopes. The heat-embossed lettering withstood sand abrasion. Being heavier, the Duncan cards were not as susceptible to wind action as the Olson cards. Further, their conspicuous red color was advantageous in their recovery. In deployments along the South African coast, Duncan cards proved both durable and convenient.

Like drift bottles, the drift cards could be released from ships and aircraft. Usually hundreds of cards were released at each of the selected stations at both high and low tides. The adjacent beaches were then monitored at regular intervals immediately following each release. This approach considerably improved the card return rate. Duncan (1965) has graphically reported an example of drift card recoveries from the southern coastal waters of South Africa. Drift card studies by Dornhelm (1977) indicated that there was a significant correlation between drift card trajectories and the local winds. He observed that the role of the tides was also

significant in the drift card motions, as evidenced by distinctive return patterns for the low- and high-tide deployments. Comparison of current velocities determined by a recording instrument, deployed at a depth of about 7 meters below the surface, with the current velocities inferred from drift card motions revealed reasonably good agreement for direction and a consistently low instrument-recorded current flow speed. This was natural because the drift cards responded to the flow in the upper few centimeters of the water column, which, in turn, was highly responsive to the surface wind stress. However, in situations where a major portion of the drifting body is exposed to wind, considerable care had to be exercised in interpreting the data, because often the wind rather than the surface current had carried the body through the water.

Several drift cards released by Stander et al. (1969) in the vicinity of Cape Town, South Africa, and at various locations in the Atlantic Ocean were recovered on the coast of North and South America and on some islands in the South Atlantic as well as in England, France, Nigeria, and Australia. Some cards released east of Cape Agulhas (on the southern tip of the African continent) traveled across the Indian Ocean and were recovered from the Australian coast after 23 to 31 months. The mean surface current velocities derived from the travel times of these cards were in reasonably good agreement with those obtained from drift bottles that were deployed four years before the drift card experiment. Stander et al. (1969) have reported the trajectories of the drift cards and the drift bottles in these experiments.

An obvious disadvantage in the use of drift bottles and cards was that, often, they remained undiscovered on a shore for some time, and in those cases it became impossible even to determine the magnitude of the average drift velocity between the start and the end points. Fortunately, in coastal waters off heavily populated localities, this problem would probably not be so severe. Interestingly, the data derived from drift bottle and drift card measurements often proved valuable in several unexpected ways. For example, drift bottle experiments could provide useful information such as the presence of eddies (Sverdrup et al., 1942). The drift bottles were also used extensively for the estimation of surface current circulation on the continental shelf off eastern North America and for observation of surface currents off Oregon state as well as those along South African coasts.

Interpretation of data obtained from drift bottle and drift card experiments presented some difficulties because the drifters, in most situations, were unlikely to follow a straight course from the place of release to the place of finding. Because only the end points of the trajectory were known, the drift bottle/drift card experiments provided only information on the average trajectories. The reconstruction of the paths taken by the bottles or cards could be aided by knowledge of the temperature and salinity distribution in

FIGURE 2.1 Application of satellite-tracked drift-bottles in the Pacific Ocean to determine the movement (trajectory and speed) of marine debris in the aftermath of the March 11, 2011 Japan tsunami. *(Source: Courtesy of CLS [France] and Dr. Koki Nishizawa, the Sustainability Research Institute, Tottori University of Environmental Studies.)*

Studies, in Japan, perhaps the most primitive sea surface current-monitoring device has ultimately been resurrected as a modern, yet the cheapest, device for Lagrangian measurement of sea surface currents.

To determine surface currents close to the landmark, *drift poles* have also been used (Pickard and Emery, 1982). The drift pole, which was a wooden pole of a few meters in height and weighted to float, with only ½ to 1 meter emergent above the water surface, functioned as a simple Lagrangian surface current indicator with a minimum of surface exposed to the wind. The pole was simply allowed to drift with the water, and its position was determined at regular intervals, either from the shore or by approaching it in a small boat and fixing its position relative to the shore. The surface current speed was determined by timing the travel time (t) of the pole between its successive positions. If the distance between these two positions is D, then current flow speed is given by D/t in appropriate units. The trajectory of the current could be mapped by joining those successive positions at regular intervals. A concept diagram of this scheme is depicted in Figure 2.2.

Although drift poles may appear to be primitive devices, their modified version, incorporating an antenna, are frequently deployed today. For example, Lobel (2011) deployed such current-drogues offshore of the Kona coastline (Island of Hawaii) and over reefs as Lagrangian tracers of surface currents during cyclonic eddy events. The drift pole drogues used by Lobel (2011) were made of polyvinylchloride (PVC) pipe with a radio transmitter attached to the emergent section (see Figure 2.3). Cross vanes of plastic sheets (0.7×1 m) were used as sea anchors below the surface (see Figure 2.4a). The overall length was about 6 m. The drogues were tracked from the nearby coast (tracking distance was about 50 km) using trucks equipped with directional 12-element yagi antennas (see Figure 2.4b). Drogue positions were determined by a running-fix calculation. Figure 2.5 indicates the date-wise total time that each drift pole drogue remained within transmitter signal range of the shoreline. The points indicate the total duration (days) that a drogue, deployed on a given date, remained within the Kona coast domain.

the surface layers. In the absence of alternative methods, synoptic drift bottle/card experiments that were of immense value to fisheries were conducted especially in coastal waters.

A major drawback associated with the Lagrangian method (i.e., a method that provides a quantitative description of both track and speed of flow) just discussed was the time-consuming process involved in tracking given water mass over a large area, requiring many days and sometimes even months of painstaking effort. Further, tracking of water parcels during rough weather conditions and in remote areas was often impractical. This lacuna was removed in 2011 with Dr. Koki Nishizawa's judicious application of satellite-tracked drift bottles in the Pacific Ocean to determine the movement (trajectory and speed) of marine debris in the aftermath of the March 11, 2011, Japan tsunami (see Figure 2.1). Thanks to Collecte Localisation Satellites (CLS) in France and Dr. Koki Nishizawa of the Sustainability Research Institute, Tottori University of Environmental

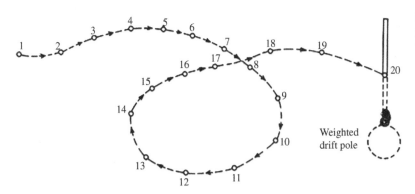

FIGURE 2.2 Concept diagram of the use of a drift-pole as a simple Lagrangian surface current indicator with a minimum of surface exposed to the wind.

FIGURE 2.3   Drift pole drogue fitted with antenna. *(Source: In part from Lobel, 2011.)*

Retention was greatest during the June to September 1982 period, when the eddy field formed and remained resident. One drogue released on July 25, 1982, in the core area of the cyclonic eddy, circulated within the eddy as the eddy remained near the Kona coast from July 25 until September 13, 1982, at which time the eddy drifted away from the Island of Hawaii. Satellite infrared photographs and bathythermograph measurements of the eddy field confirmed these patterns of the eddy (see Lobel and Robinson, 1986, 1988). Information gathered from the track and motion of the drift pole drogues demonstrated the transport from the offshore eddy to very near the coastline coral reefs and provided valuable information on the role played by passive advection in larval fish recruitment patterns. Examination of drogue drift tracks provided additional evidence that oceanic zooplankton were being transported near-shore by the cyclonic eddy currents.

Despite the vital importance of closely monitoring the sea surface circulation in polar regions in terms of climate change studies, such measurements have been one of the most difficult tasks encountered by oceanographers. Over the years, application of the Lagrangian approach with the aid of drifting ice floes as tracers of water movement proved helpful in determining surface circulation in Arctic areas. Evaluation of transport of water, salt, and heat in polar oceans, as well as study of the response of sea ice to climate changes, all require an understanding of the movement of sea ice over long periods of time. The observational basis for our present understanding of sea ice motion has been accumulated over the past several decades by tracking icebergs marked by manned research stations or, more recently, by unmanned automatic satellite transmitting stations for long periods. The trajectories of these ice floes were then used to define a mean circulation pattern, which was often considered as the response of the ice packs to the long time-average circulation of the atmosphere and ocean. Motion of an ice island in the Arctic Ocean is said to have been tracked for more than 40 years. Knowledge of the position and movement of icebergs is needed daily for safe and efficient operation of drill rigs and oil tankers in the Arctic Ocean and the Greenland Sea. In situations in which data on sea ice movement over long periods were not available, observations of sea ice displacement over short periods of time were used to deduce some statistics of the ice motion over long periods of time.

With rapid advancements in technology, the concept of Lagrangian measurements took on new dimensions and underwent significant leaps. For example, pairs of pulse-radar transponders located at fixed reference points, a few km apart on the coast or on fixed platforms in the sea, have been employed to track radio buoys and thereby to determine surface currents (see Section 3.1.1). Application of the Doppler effect brought about a radical improvement in ocean current measurements. Incorporation of a drifter-borne Doppler transponder for surface current measurements from coastal waters, regardless of weather conditions (i.e., in fog, high winds, rough seas, and at night; see Section 3.1.2), is just one example. Combined use of polar-orbiting satellite technology and the Doppler effect gave birth to an advanced method of determining surface currents and their trajectories over long distances in the

FIGURE 2.4   (a) The underwater section of the drift pole drogue with antenna; (b) Tracking using a directional radio antenna. *(Source: Lobel, 2011.)*

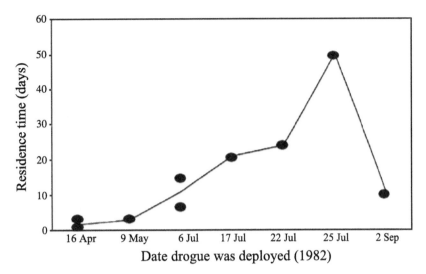

FIGURE 2.5   Residence time of drift pole drogues (i.e., the number of days that a drogue was tracked) along the Kona coast, Hawaii, in 1982, covering the times of summer eddies. *(Source: Lobel, 2011.)*

open ocean regions for long periods of time (see Section 3.1.3). There are several other examples where the Doppler effect is widely used for measurements of surface and subsurface oceanic currents, and discussions of these are spread over various chapters in this book. Nondifferential GPS, which was initially developed for military purposes and released in May 2000 for civilian application after removal of Selective Availability (SA), proved to be a valuable device for position-fixing applications for the purpose of sea surface/subsurface current measurements. Availability of small and low-cost standard GPS receivers with the capability to fix their position to an accuracy of within a few meters, and anywhere in the world, led to the development of GPS-tracked sea surface drifters for surface current measurements (see Section 3.1.4).

Introduction of such cost-effective Lagrangian drifters stimulated renewed interest in coastal sea surface current measurements, and these came as a great relief to ocean-ographers, ocean engineers, and environmentalists in tracking surf zone currents, for which a suitable device was hitherto unavailable. Advancements in cellular phone technology in the 1990s and the rapid spread of cellular towers almost all over the world began to exert considerable influence in the realm of sea surface current measurements from coastal and land-locked regions as well by way of availability of telephonically tracked drifters (see Section 3.1.5).

2.1.2. Imaging of Surface Water Motion Trajectories and Patterns

Imaging provides an important means of deriving a visually comprehensible picture of oceanic surface circulation trajectories and their shapes. Apart from academic interest, such images are vital for several operational applications such as oil slick monitoring and commercial applications such

as fishing. Generation of such images involves photography and microwave imaging.

2.1.2.1. Aerial Photography

Aerial photography was one of the earliest methods used for identifying seawater motion in coastal water bodies. For example, Inman et al. (1971) obtained excellent aerial photographs of surf zone current patterns, clearly indicating the dangerous and evenly spaced jet-like rip current paths carrying material from the surf zone and producing vortex pairs. Aerial photography has also been successful in identifying strong tidal currents from straits and tidal vortex pairs generated in tidal channels (e.g., Yamoaka et al., 2002). Images such as aerial photographs are used in the identification and study of sea surface currents because the human brain is especially good at pattern recognition. Images of seawater motion signatures (e.g., tidal vortex, gyres) are constructed based on the characteristics of electromagnetic radiations that are reflected or scattered from or emitted by the sea surface.

Sea surface roughness and color contrasts are the two major types of signatures that are often utilized in the detection of sea surface current patterns with the aid of aerial photography in the visible and infrared bands of the electromagnetic spectrum. These contrasts provide visual indications of the surface manifestations of seawater motion. Sea surface roughness often changes from the combined effect of varying wind stress in the marine atmospheric boundary layer (induced by sea surface temperature front) and surface current convergence and divergence (induced by the meandering current). In particular, the changes in the stratification of the marine atmospheric boundary layer (MABL) and subsequent wind stress changes produce a step-like drop of the sea surface roughness at the downwind side of the front (Johannessen et al., 2005). The color signatures are usually those of chlorophyll, algae, and suspended-sediment plumes.

Whereas chlorophyll and algal blooms are often indicative of upwelling motions, the suspended sediment plumes often arise from river discharges into the bluer, clearer seawater. The plume of turbid brown water is easily detectable in the visible wavelength photographs. In the initial years of oceanographic studies, time-lapse aerial photography from overhead was employed for obtaining such images. Photographic cameras produce black-and-white or color images by a simple and relatively inexpensive process that, like the human eye, is limited in wavelength response. In the past, the most commonly used aerial photographic camera was the multiband camera, which contained four to nine lenses and carried many rolls of photographic film. Different films and filter combinations permitted simultaneous photography in different optical spectral regions.

Remote sensing of circulation features in coastal and estuarine regions is a valuable complement to *in situ* observations. In the study of fluid motion, flow visualization techniques have often been used to estimate intuitively the spatial dimensions of the flow field. In fact, visible wavelength photography was historically the first remote-sensing technique developed to study the ocean from above the sea surface. This is also one of the very few techniques that "sees" directly from the surface layers of the sea down to a depth of several meters. Visible wavelength photography is particularly useful in studies of various kinds of sea surface manifestations of seawater motion such as upwelling, intense tidal currents, vortices, and surf zone circulation cells. Such photographs provide an overview to help preliminary evaluation and interpretation of these dynamic processes, which occur in a given region, before any attempt is made at field studies incorporating *in situ* measurements. In some cases, shapes or patterns of seawater motion are equally, or sometimes more, important than their speed and direction values. A typical example is the surf zone circulation cells, the dimensions and locations of which are valuable to give warning to swimmers at a beach that is characterized by this dangerous circulation.

In the initial years of oceanographic studies, aerial time-lapse photography was used for imaging of sea surface circulation patterns employing two different concepts. One was the just-mentioned truly remote-sensing technique whereby the surface circulation pattern is detected and identified from the color or roughness contrasts of the sea surface. The other was a quasi-remote-sensing technique whereby the mean flow velocities at small scales were estimated from the travel times of surface drifters such as floats, drogues (i.e., weighted floats), and fluorescent dye patches. Travel times were determined from a series of time-lapse aerial photographs taken from overhead.

Because turbulent currents play an important role in mixing processes, it is essential to measure such currents. One method of mapping turbulent surface currents is aerial photography using fluorescent dye. The dye method has been used successfully by several investigators for flow visualization and measurement of current speeds in the surf zone. Richardson et al. (1972) and Flynn and Cook (1978) mapped sea surface currents by aerial photography of surface current markers, such as fluorescent dye, using aircraft outfitted with position-fixing equipments. Radioactive or dye tracers indicate surface current as well as turbulence. In this method, dye markers or phosphorescent tracers are dropped from an aircraft or a vessel at selected locations. Currents transport the dye/tracer patches while horizontal diffusion due to the turbulence spreads them. Current velocities and diffusion coefficients are determined, respectively, from the temporal changes in position and size of the dye/tracer patches between successive photographic flights over the measurement site.

For economy in such experiments, flights can be replaced by balloons. Calibration of the photographs can be carried out using simultaneous measurements of the dye concentration by a fluorometer. High contrasts in aerial photographs can be obtained by filtering the camera, either to the band of maximum light absorption of the dye or to the band of maximum fluorescence of the dye. For the first case, on a positive black-and-white print, the seawater would appear light in tone, whereas the dye patch would appear dark. For the second case, the dye patch would be light in tone, whereas the seawater would appear dark. The dyes usually used for these investigations are Rhodamine B (red color) and Rhodamine 5GDN (yellow-red color). One of the primary problems encountered in processing the photographs obtained from vertical photography is the direct sunlight reflection from the sea surface. To avoid this problem, aerial photographs are usually taken with an oblique camera mounting.

Several investigators have observed that during the early stages of diffusion of the dye, the patch shape is very close to a circle. At later stages, the patch becomes elongated. The elongation of the dye patches occurs in the direction of the current in the absence of wind. If the directions of the current and of the wind are different, the axis of the patch can be aligned with yet another direction. Depending on the state of the sea and the associated circulation features, striation, curvature, elongation, and stripe formation of the dye patches are common. As the sea becomes rough, the finely mottled tone changes to a coarsely mottled pattern. When the wind velocity exceeds 25 knots, streaks parallel to the direction of the wind begin to appear. The elongation of a dye patch can be influenced by horizontal and vertical velocity shears. If Langmuir circulation is predominant, the dye patches will converge into parallel stripes. For photographic studies of dye plumes, the aircraft is generally flown at altitudes of 500 to 5,000 feet above the sea surface.

Sea surface current measurements using dye-patch photography method has merits and limitations. An important advantage, over other methods of circulation studies, is that this method enables measurements of turbulent fluctuations in the flow velocity field. The growth, elongation, change of shape, and breakup of dye patches are related to the nature of turbulent surface currents. Measurement of turbulence in the upper layers of the oceans is important in understanding the mechanism of surface phenomena such as slicks, streaks, and foam lines as well as such other phenomena such as heat exchange between the atmosphere and the ocean, the flushing and disposal of polluted water, mixing of different water masses, and sedimentation.

The study of turbulence has been handicapped by the lack of suitable measuring devices. The float method was inadequate in obtaining information on detailed structures of small-scale eddies owing to difficulties in simultaneously tracking a large number of floats. In contrast, studies of dye patches in the ocean were found to be much simpler in practice and yet gave more detailed pictures of turbulence and small-scale eddies. Dye patch photography has been found to be an excellent technique for detection and quantification of rare circulation phenomena such as Langmuir cells and Ekman spirals. From an operational point of view, the dye method is routinely applied in studies related to pollution monitoring and control. In designing an outfall for discharge of waste, one has to select such an outfall length that harmful concentrations do not reach the coast. The prediction of concentration near the shore for different outfall lengths is often carried out by observing the spread of radioactive tracers or fluorescent dyes such as Rhodamine released at a chosen point in the proposed disposal area. The concentration of pollutants at the shore arising out of a continuous source is determined by superposition of results from different successive releases. The fluorescent dye Rhodamine is also traced by *in situ* measurements of fluorescent concentration with the aid of a fluorometer.

Although aerial photography of dye patches in the sea is a comparatively simple remote-sensing technique, the quality of the result depends on various environmental and flight-related factors. For example, when the sea is rough, suspended sand and silts in the coastal waters can make identification of a dye patch difficult. Shadows from scattered clouds cause uneven lighting on the sea surface, and this interferes with quantitative data processing. Whitecaps and spray generated from strong winds can also interfere with the effectiveness of photographic techniques. From a flight-related point of view, side winds and turbulence make it difficult to maintain straight flight lines. These factors may also cause the camera axis to deviate from the vertical. Camera axis tilt results in distortion and variations

in scale across the photograph. To obtain better-quality aerial photographs, a variety of expensive equipment, such as gyro-stabilized camera mounts, auxiliary view-finders, radar profile recorders, and inertial guidance systems, must be used. The feasibility of large-scale current measurements in the open ocean using aerial photography is often limited because the quantities of tracers required to "tag" open-ocean water masses would be impractically large. Furthermore, both advection and diffusion act three-dimensionally to spread the tracers; therefore, the results cannot be interpreted solely in terms of advection by currents (Pickard et al., 1982).

Aerial photography of floats for circulation studies has some advantages as well as some limitations. Whereas turbulent characteristics of turbid waters are difficult to investigate using conventional dye techniques, aerial photography of clusters of floating objects gives reliable information about turbulent dispersion. It is assumed that motions of a sufficiently large number of floats adequately represent the turbulent motions in the flow field. The float method, although simple in analyzing the results, may, however, be inadequate in obtaining information on the detailed structure of small-scale eddies, owing to the limited number of floats deployable in practice. In contrast, the release of dye in water is much simpler in practice and yet gives a more detailed picture of small-scale eddies in the ocean. Another disadvantage in the use of float clusters is that they cannot be used to assess vertical mixing and the vertical distribution of horizontal turbulent components because the positive buoyancy of the float does not allow its vertical motion. For these reasons, photography of dye patches is often employed, especially in clear water, in studies of turbulent motions.

Aerial photography has been used to distinguish water masses, delineate circulation patterns, identify upwelled waters along the coast, and trace thermal plumes in the ocean. Circulation patterns in the coastal regions have been obtained from photographs taken from aircraft that are usually used by the Coast Guard and the Air Force. Aerial photographs of the sea surface, taken from Earth-orbiting satellites and manned spacecraft, have also provided useful information. For example, color photographs taken by *Skylab* astronauts using handheld cameras revealed oceanic features such as algae blooms, which are often indicative of upwelling. Photographic surveillance has played a major role in pollution monitoring of estuaries and coastal waters.

High absorption of light energy in the infrared band by clear water makes infrared black-and-white photographs of clear water surfaces normally appear black. In contrast, muddy water shows up in lighter tones than clear water. This is because suspended sediments or algae in the muddy water have high reflectance in the infrared band. This property of muddy water is often used for the investigation of transport of suspended sediments in estuaries. Sediment patterns reveal residual transport, tidal streams, and small gyres (Robinson, 1985). It may be noted that sediment transport in a tidal flow takes place as a result of residual tidal flows. Data on sediment movements have potential applications in civil engineering studies related to harbor management.

Monitoring algae blooms with the aid of infrared photography helps in the detection of a special kind of circulation signature, known as *upwelling*. Furthermore, the property of high reflection in the infrared region by muddy waters might be applied in clearly delineating patches of floating plants in turbid waters. Because floating plants are excellent Lagrangian tracers of surface circulation in coastal waters and estuaries, infrared photography may be a promising remote-sensing tool for measuring surface circulation in these regions. Once a feature has been identified, the speed and direction of its movement over time may be studied from a sequence of photographs taken in the area.

The purpose of most infrared photography is that of mapping a pattern. The technique is especially useful for detection of circulation patterns involving some kind of signature that can be detected using infrared photography. An excellent example is the detection of chlorophyll in the ocean. The property of fluorescence of chlorophyll in the infrared region of the electromagnetic spectrum permits infrared photography of these patches. The importance of chlorophyll in the detection of certain circulation features (e.g., cold-core eddy; upwelling) is briefly mentioned in Chapter 5.

A great advantage of infrared photography over conventional photography is that atmospheric dust and vapor can be penetrated to a great distance by infrared rays. These rays also have the ability to penetrate in the dark. Infrared photography requires special films and lenses. Infrared-sensing lenses use materials such as magnesium fluoride and zinc sulfide. These lenses, however, generally suffer from the problems of chromatic aberration. Infrared lenses have special coatings designed for transmission of maximum infrared radiation. To avoid the problem of aberration, lens systems that use reflective surfaces only are sometimes used. Usually a combination of reflective and refractive components is used in infrared photographic work. These lenses do not suffer from chromatic aberration and do not absorb infrared radiation.

2.1.2.2. Radiometry

In the past, radiometers of varying resolution and spectral bands were used from various satellites. They are the scanning radiometer (SR), visible and thermal infrared radiometer (VTIR), visible and infrared radiometer (VIRR), visible and infrared spin-scan radiometer (VISSR), medium-resolution infrared radiometer (MRIR), high-resolution infrared radiometer (HRIR), very high-resolution radiometer (VHRR), and advanced very high-resolution radiometer (AVHRR). The SR was flown on the National Oceanic and Atmospheric Administration (NOAA) series of Improved TROS Operational Satellites (ITOS), which began in 1970. The SR was a dual-channel, line-scanning radiometer that measured radiation in the range 0.5−0.7 μm (visible band) and in the range 10.5−12.5 μm (thermal infrared band). The ground resolution of the data at nadir was about 4 km for the visible channel and 8 km for the thermal infrared channel.

On board Seasat-1, the VIRR was a scanner for collection of digital data in the visible and thermal infrared portions of the electromagnetic spectrum over broad swaths of the Earth's surface. Scanning was accomplished by means of a rotating mirror mounted at 45° to the optical axis of the collecting telescope. The mirror rotated continuously, creating a line scan perpendicular to the motion of the spacecraft. The motion of the spacecraft provided the second dimension of scan. A relay optical system transmitted the radiation to a dichroic beam splitter, which separated the visible and infrared wavelengths. The visible energy was then focused onto a silicon photodiode and the infrared onto a thermistor bolometer. On board the GOES satellite, the VISSR also provided visible and thermal infrared images of the sea surface. The VISSR data were transmitted from the satellite in digital form. The best ground resolution at nadir was 1 km for the visible channel and 8 km for the thermal channel data.

The VHRR, a second type of radiometer on the ITOS, was also a two-channel line-scanning radiometer. The VHRR was sensitive to energy in the 0.6−0.7 μm and 10.5−12.5 μm bands. The VHRR had many similarities to the SR. The VHRR, however, had substantially better ground resolution than the SR, about 1 km at nadir for both channels. The limited capacity of the VHRR's on-board tape recorder permitted a maximum of only 10 minutes of stored data per pass to be acquired when the spacecraft was remote from the two Command and Data Acquisition Stations. The AVHRR on board the NOAA satellite scanned at a rate of 360 swaths per minute with an effective ground resolution of 1.1×1.1 km at nadir (Smith et al., 1987). With the sensor cooled to 105°K, the AVHRR achieved sensor noise levels of around 0.05°K in the very short integration time of less than 25 μs of the scanning process (Robinson, 1985). See Section 5.2 for details.

The AVHRR images have been of great value in exploring the mighty western boundary currents (WBCs) such as the Gulf Stream and the Somalia currents and in detecting large-scale oceanic circulation features such as rings, thermal jets, and frontal currents. The WBC in the Bay of Bengal (Indian Ocean) and its unstable nature were first explored using AVHRR images (Legeckis, 1987). These images played an important role in monitoring drifting gyres associated with the Somalia current system and in studying their relationship to the Indian southwest monsoon.

2.1.2.3. Active Microwave Radar Imaging

It has been noted that sea surface roughness contrast is a good indicator of sea surface circulation patterns. Some of the merits and limitations of photography and radiometry were indicated in the preceding sections. An alternative microwave device that was born out of sheer necessity is synthetic aperture radar (SAR). This device was initially developed to aid Lunar missions but was subsequently released for terrestrial imaging. SAR has been successfully used for several oceanographic studies. For example, the surface signatures of meandering fronts and eddies have been regularly observed and documented in SAR images. Wave-current interactions, suppression of short wind waves by natural films, and the varying wind field resulting from atmospheric boundary layer changes across oceanic temperature fronts all contribute to the radar image manifestation of such mesoscale (order 100 kilometers) features.

A new radar imaging model developed by Johannessen et al. (2005) provides promising capabilities for advancing the quantitative interpretation of oceanic water motion features manifested in SAR images. An example of such a SAR image expression of a 10-km-in-diameter cyclonic eddy in the Norwegian Coastal Current that generally depicts the eddy boundary has been reported by them. Due to the convergence, the passive surface-floating material accumulates to delineate the eddies. In addition, the enhanced wave breaking of intermediate waves in the vicinity of the zones of surface current convergence is the dominant source for the radar cross-section modulation and subsequent SAR image manifestation. Studies by Johannessen et al. (2005) emphasize the crucial role of current convergence and divergence that occur along meandering fronts and eddies as well as the wind direction versus the SAR look direction. In the case of convergence, the sea surface roughness modulation comes from the direct and indirect effects of the breaking of intermediate scale waves that takes place within the converging zone due to wave current interaction. This, in turn, produces the sharp delta-like intensity changes in the radar cross-section.

Topographic maps have a long history, and the advent of stereo aerial photography in the first part of the 20th century made possible photogrammetric measurements of surface elevations over substantial areas; many regions have been mapped at various scales by these means. Stereo images or photographs taken by satellites are also used to construct topographic maps. Such topographic maps have the ability to reveal sea surface roughness contrasts and, therefore, sea surface circulation signatures. A major problem for optical imagery in tropical areas is cloud cover that prevents imaging of the ground surface from space. The Interferometric

Synthetic Aperture Radar (InSAR) technique has emerged as the state-of-the-art technique of measuring dense points in an area accurately, economically, conveniently, efficiently, and without any effect from cloud cover (Gabriel et al., 1989; Massonnet et al., 1993). Studies have demonstrated the potential of the interferometric technique to produce high-resolution topographic maps, as found by Zebker (1994) in tests with three-day repeat-pass ERS-1 imagery.

The use of space-borne SARs as interferometers became popular only recently, although the basic principle dates back to the early 1970s (Graham, 1974; Richman 1971). Today it is generally appreciated that SAR interferometry is an extremely powerful tool for mapping the sea surface topography. The so-called differential InSAR method (D-InSAR) represents a unique method for detection and mapping surface displacements over large temporal and spatial scales with precision in the centimeter and even millimeter range.

Since the early 1990s, InSAR has developed from theoretical concept to a technique that is being utilized at an increasing rate for a wide range of Earth science fields. Despite the fact that none of the currently deployed imaging radar satellite platforms was designed with interferometric applications, the high quality and vast quantity of exciting results obtained from the InSAR technique have demonstrated their potential as powerful ground deformation measurement tools (for millimeter-scale ground deformation characterization). Their capability has been further considerably improved by using large stacks of SAR images acquired over the same area (multi-image InSAR technique) instead of the classical two images used in the standard configurations (Ferretti et al., 1999, 2000, 2001). With these advances the InSAR techniques are becoming more and more quantitative geodetic tools for deformation monitoring rather than simple qualitative tools.

In judging the progress made so far in SAR interferometry, it should be kept in mind that neither of today's space-borne SARs have been designed explicitly for interferometry. The first Earth observation satellite to provide SAR data suitable for interferometry was the Seafaring Satellite (SEASAT). Launched in 1978, it was operated for 100 days, whereas SAR data collection was limited to a period of 70 days. The real breakthrough in SAR interferometry was achieved through the European ERS-1 satellite and its follow-on, ERS-2. The ERS-1 mission was officially terminated in 1996. The ERS-2 SAR, identical to the ERS-1, was launched in 1995 and has the same orbit parameters as ERS-1. It continues the ERS-program with the 35-day repeat period. Most important, from the SAR interferometry point of view, was the TANDEM mission (Duchossois and Martin, 1995), during which ERS-1 and ERS-2 were operated in parallel. ERS-2 followed ERS-1 on the same orbit at a 35-min delay. Together with the Earth's rotation, this orbit scenario assured that ERS-1 and ERS-2 imaged the same areas at the same look angle at a one-day time lag. The orbits were deliberately tuned slightly out of phase such that a baseline of some 100 m allowed for cross-track interferometry. This virtual baseline between ERS-1 and ERS-2 could be kept very stable because the two satellites were affected by similar disturbing forces. The first of several TANDEM missions was executed in May 1996. The instrument was designed for oceanographic imaging and, hence, uses a very steep incidence angle of only 23°. (See Section 5.4 for details.)

2.1.3. Vector Mapping Based on Current-Driven Sea Surface Wave Transport

Imagery basically provides patterns. Although sea surface current patterns have great value in themselves, time-series images are needed to yield quantitative estimates of sea surface currents. Thus, oceanographers were badly in need of some devices that would provide time-series quantitative information directly on sea surface currents. An exciting discovery by D. D. Crombie in the early 1950s ultimately paved the way for remotely acquiring high-resolution, real-time time-series vector maps of sea surface current fields over large areas by processing the backscatter returns from the sea surface. The device used for this purpose is a Doppler radar system, which operates in the high-frequency/very high-frequency (HF/VHF) band of the electromagnetic spectrum. The functioning of HF/VHF Doppler radar systems is based on the principle of the Doppler shift of the backscattered electromagnetic radiation, caused by Bragg-scattering sea surface waves that are transported by the underlying water currents (details given in Chapter 4). In this method, radio-wave propagation is mainly accomplished by ground-wave propagation scheme. Looking at the usefulness of these systems, these devices can be legitimately termed as enduring resources for generations of oceanographic researchers to come.

Although the landmark discovery of the concept embodied in the application of the backscattered Doppler radar signal from the ocean surface for remote mapping of sea surface currents and waves was published by Crombie in 1955, his *Nature* publication (Crombie, 1955) did not elicit citation by the oceanographic researchers of his time. It is surprising that Crombie's pioneering discovery remained in the dark until the theoretical investigations on the retrieval of wave information from HF radar backscatter were reported by Hasselmann in 1971 (see Hasselmann, 1971) and Crombie himself reported the practical application of his discovery to physical oceanography at the IEEE Oceans '72 Conference (see Crombie, 1972) after a long gap of approximately 17 years, based on his "self-cited" publication!

With the reporting of the practical value of Crombie's discovery, his *Nature* publication and his conference paper began to attract considerable attention in the oceanographic community. The realization that HF Doppler radar estimates, derived based on Crombie's "Bragg scattering" principle, are capable of providing quasi-real-time information about several oceanographic parameters such as sea surface currents, sea states, and wind direction led to vigorous research and developmental activities in subsequent years. For example, Barrick and co-researchers at NOAA in the United States developed the coastal ocean dynamics applications radar (CODAR) system, which proved to be the first commercial application for remotely mapping the sea surface current fields. Since then the technology of HF Doppler radar systems has become a rapidly expanding field. Decades of technological developments have resulted in a variety of HF Doppler radar systems capable of producing current vector maps of the coastal zone over spatial scales ranging from hundreds of meters to about 200 kilometers and on time intervals from tens of minutes to days. Within a relatively short interval, the technology emerged as a viable commercial product, and radar systems of different constructions to suit installation at different coastal locations (Figure 2.6) became available. Based on the same basic technology as that used in CODAR, other systems have been developed, such as the SeaSonde, which employs broad-beam antennas and direction finding to produce maps of radial current velocity vectors. Each vector produced is the average over a radial cell, the area of which is primarily determined by the carrier frequency of the radar systems. Figure 2.7 shows a Long-Range SeaSonde surface current vector map spanning 200 km offshore, overlain on SeaWIFS imagery, capturing the complex Kuroshio Current as it meanders off the coast of Japan.

The ocean surface current radar (OSCR) developed in the United Kingdom (Prandle et al., 1993) was another milestone in the rapid expansion of the HF/VHF Doppler radar technology. The University of Hamburg, Germany, started its work on HF radars in 1980 by modifying the CODAR technology (Gurgel et al., 1986). By 1985, both hardware and software had been modified to reduce internal noise, increase sensitivity, and optimize processing algorithms, and they brought out a new system called Wellen radar (WERA) (Gurgel et al., 1999). At present, these commercial products adorn a prime position in coastal observing systems in many parts of the world. Independent of CODAR systems, other developments have been made, such as coastal ocean surface radar (COSRAD) at James Cook University in Australia (Heron et al., 1985), VHF Courants de Surface MEsures par Radar (COSMER) at the University of Toulon in France (Broche et al., 1987), PISCES at the University of Birmingham, United Kingdom (Shearman and Moorhead, 1988), high-frequency surface wave radar (HF-SWR) at C-CORE and Northern Radar Systems in Canada (Hickey et al., 1995), high-frequency ocean surface radar (HFOSR) at Okinawa Radio Observatory/Communications Research

FIGURE 2.6 HF Doppler radar systems of different constructions to suit installations at different coastal and offshore locations. *(Source: CODAR Ocean Sensors brochure, May 2003, Volume 7, Number 4; and CODAR CURRENTS newsletter, Spring/Summer 2011.)*

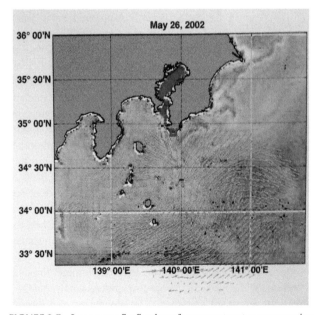

FIGURE 2.7 Long-range SeaSonde surface current vector map spanning 200 km offshore, overlain on SeaWIFS imagery, capturing the complex Kuroshio Current as it meanders off the coast of Japan. *(Source: Hydro International; CODAR Ocean Sensors, May 2003, Vol. 7, Number 4; SeaSonde data courtesy of Tomotaka Ito, Japan Coast Guard; SeaWIFS image courtesy of Dr. Robert Arnone, Naval Research Laboratory.)*

Laboratory (ORO/CRL) in Japan (Takeoka et al., 1995), multifrequency coastal radar (MCR) at the University of Michigan at Ann Arbor (Teague et al., 2001), Ocean States Measuring and Analyzing Radar (OSMAR2000) at the Wuhan University in China (Huang et al., 2002), and PortMap at James Cook University in Australia (Heron et al., 2005).

The concept of HF/VHF Doppler radar-based measurements received considerable attention in coastal oceanographic measurements to extract ocean surface parameters such as surface currents, waves, and winds. The reason for the considerable importance attached to this technology stems primarily from its ability to map coastal ocean surface currents in real time over relatively large areas (covering a radial distance of ~200 km from the radar) with high temporal and spatial resolution, which offers the possibility to track a variety of oceanographic phenomena (e.g., tidal vortex, gyres). HF/VHF Doppler radar systems do not provide a global coverage like that obtainable from the well-known satellite-based systems; however, they have the advantage of providing continuous observations in limited areas with high spatial resolution (a few kilometers down to a few hundred meters) and with high temporal resolution (30 min down to 10 min) in contrast with a repeat cycle of several days for the satellite systems (Cochin et al., 2006). Consequently, HF/VHF Doppler radar systems have now become an important technology for studying coastal circulation processes. Good spatial and temporal resolution make the HF/VHF Doppler radar technology particularly suitable for aiding search-and-rescue operations, studying pollutant and larval dispersal (e.g., Bjorkstedt and Rough-garden, 1997; Graber and Limouzy-Paris, 1997), and analyzing the physical forcing of coastal flows (e.g., Shay et al., 1998).

The applicability of HF/VHF Doppler radar systems for detecting the arrival of tsunami-induced amplified currents from as far away as the continental edge and subsequent monitoring of tsunami propagation over the continental shelf and toward the coast is expected to be a landmark achievement in the technology of tsunami early warning in the coming years. This ability comes from the fact that in case of an open-ocean tsunami entering the shelf edge and heading toward a coast, the strong ocean currents generated as a result of the considerable decrease in water depth of the continental shelf can be observed by the HF/VHF Doppler radar systems. Such systems, programmed to make measurements at fine temporal resolution, could therefore contribute immensely to the development and improvement of tsunami early warning systems.

HF/VHF Doppler radar systems are primarily operated from the seacoast (of course, there are instances in which a few deployments have been made from offshore platforms and buoys), and therefore the sea surface current maps derived from them are generally limited to the coast (extending up to a maximum distance of ∼200 km from the coast). There was a clear need for a device that would provide quantitative estimates of sea surface currents from offshore and polar regions. Satellite altimetry paved the way for such measurements (see Chapter 12 for details).

2.2. SUBSURFACE AND ABYSSAL CURRENT MEASUREMENTS

In ancient times, the oceans were ruled exclusively by fishing folk and mariners. They considered the oceans vast ponds of blue water that is turbulent at the surface and stagnant in the deeps. However, purely accidental observations by the early mariners brought to light the folly of previous notions about the stagnancy of the deep oceans.

2.2.1. Early Mariners' Contributions

Ancient mariners began to realize the existence of subsurface and abyssal currents in the ocean through rare incidents during their seafaring careers. For example, Maury (1855) mentions a shipwreck incident that took place in the year 1712, which a Dr. Hudson communicated to the Philosophical Society in 1724. This wreck pertains to a Dutch ship, with her cargo of brandy and oil, which sunk in the deep water at Ceuta Point, Spain. Surprisingly, the sunken ship arose on the shore near Tangier directly *against* the strength of the *surface current* in that region. This incident clearly demonstrated the existence of a strong *undercurrent* in the same region, in the opposite direction of the surface current, which dragged the sunken ship to the shore.

Maury (1855) mentions another incident on the discovery of oceanic subsurface currents. According to this story, an able seaman who was navigating in the midstream of the Baltic Sound from one of the king's frigates observed that the boat was carried violently by the surface current. He sunk a bucket with a heavy cannonball to a certain depth of water to give a check to the boat's violent motion. Surprisingly, sinking the weighted bucket still lower and lower, the boat was driven ahead to the windward *against* the upper current. The lower the bucket was let fall, he found the undercurrent was the stronger.

Further evidence for the presence of oceanic subsurface currents emerged when Lieutenant J. C. Walsh and Lieutenant S. P. Lee of the United States were carrying out a system of observations in connection with the *Wind and Current Charts*. They made some interesting experiments on the subject of submarine currents. In their experiments, a block of wood was loaded to sinking and, by means of a fishing line, they let the weighted wood down first to the depth of 600 feet and then to 3,000 feet. A small float, just sufficient to keep the block from sinking deeper, was then tied to the fishing line, and the combination was let down from the boat. It was wonderful indeed to see the subsurface float moving off *against the surface currents*. For the men in the boat, it really appeared as if some monster of the deep had hold of the weight below and was walking off with it, although what they witnessed was an interesting manifestation of an under-current.

Much light had been thrown on the subject of subsurface currents by the ancient mariners, based on several of their experiments in deep-sea depth measurements. It may be noted that in the early eras of navigation, ancient navigators conducted water depth measurements using a plummet fastened at the end of a roll of twine. They reckoned that there was reason to believe that undercurrents existed in almost all parts of the deep sea because during deep-sea depth measurement operations, the plummet line never ceased to run out, even after the plummet had reached the bottom. Lieutenant J. P. Parker's deep-sea depth measurement experiments off the coast of South America in 1852 also provided reliable indications that the plummet line was swept out by the force of one or more undercurrents; but their directions remained unknown. The navigators also discovered the existence of an undercurrent in the Arctic Ocean; they noticed immense icebergs drifting rapidly against a strong surface current. These icebergs protruded above the sea surface, and their depth below was approximately seven times greater than their height above the sea surface.

2.2.2. Spatially Integrated Measurements Based on Earth's Magnetism and Oceanic Sound Speed

A class of oceanic subsurface current measurements that have been found to be of immense practical value is spatially integrated measurements. Such measurements became particularly useful for investigating the transport of water through channels, straits, and ocean basins.

2.2.2.1. Electromagnetic Method

Although the earliest sign of the existence of subsurface currents in the ocean was communicated to the world as far back as 1724, an attempt to measure such currents started in 1832 with von Arx's measurement of water-flow-induced electric current flowing over a short range (see von Arx, 1950). However, little was accomplished until 1918, when F. B. Young and his associates used both moored and drifting electrodes to measure electric currents generated by tidal motions.

The duplication of Young's experiments in the autumn of 1946 led to the development of the geomagnetic electro-kinetograph, or GEK. The electromagnetic principle on which the GEK operates is the following: Because every point on the Earth's surface that is located away from the equator is permeated by a vertical component of the geomagnetic field, moving seawater mass generates a velocity-dependent potential gradient on a pair of electrodes inserted into the ocean, obeying Faraday's principle of electromagnetic induction (EMI). Note that seawater, being a solution of electrolytes, is a good electrical conductor. Although the theory is attractive in principle, there were several hurdles to be surmounted before the theory could be successfully applied for water current measurements in the ocean. The hurdles encountered were the following: Because the potential gradient developed across the pair of electrodes is of the microvolt order, a necessary requirement for the application of the electro-magnetic method is the use of stable electrodes so that electrochemical effects do not complicate the measure-ments. The biggest problem in measuring electromagnetic potentials in the ocean by means of metallic electrodes arose from the fact that, unlike metals that conduct electricity by means of free electrons, seawater is an electrolyte solution that conducts by means of "assorted" ions (i.e., both positive and negative ions). Any metallic electrode inserted into the ocean as a probe can only function with the occurrence of "electrode reactions" involving both ions and electrons. Such electrode reactions are sensitive to the ambient temperature, pressure, ionic concentration (i.e., salinity), and contamination (i.e., oxygen tension). In effect, changing environmental conditions would give rise to changing electrode potentials, thereby resulting in "base-line drifts" and possibly irregular potentials that vary with changes in the chemical environment of the seawater surrounding the electrodes. A sensor based on the EMI principle was investigated by Michael Faraday as early as 1832. However, due to several practical difficulties just mentioned and the resulting unsteady response of the copper electrodes with which Faraday experimented, he could not succeed in applying his ideas to accurately measure oceanic water currents.

As a result of untiring experimentation by several researchers, it was found that electrode noise can be reduced by the use of Ag-AgCl electrodes that are reasonably reversible to chloride ions in seawater. The Ag-AgCl electrode, which is a thin layer of silver chloride electro-chemically deposited on the silver electrode surface, can also pass on electric current without changing the chemical environment in the vicinity of the electrode and provides high porosity to the electrode surface (Offner, 1967). With the use of suitable electrodes, Faraday's EMI principle was successfully applied by von Arx in the development of a practical device that he named the geomagnetic electro-kinetograph (GEK) for measuring motionally induced potentials in the ocean.

In the GEK method, the seawater is the conductor, and when it flows horizontally across the vertical component of the lines of force of the Earth's magnetic field, a potential gradient given by $E = (\bar{V} \times \bar{B})L$ is generated across a pair of electrodes. In this, $\bar{V}$ is the horizontal component of the water current flowing perpendicular to the line joining the electrodes, L is the electrode separation, and $\bar{B}$ is the vertical component of the Earth's magnetic field. When the GEK method is utilized for seawater current measurements, the electrodes are placed on either side of the current flow path. Through measurements of the potential differences on two courses at right angles, the velocity components in these two directions are determined, and the vector sum or resultant of these velocities is the surface water current vector for that locality.

An advantage of the GEK method is that it permits measurements of comparatively large-scale water currents across straits, rivers, and channels where logistics of deploying and operating moored current meter arrays may be difficult. The GEK method has successfully been used for measurements of the Florida Current as well as tidal currents in the English Channel (Gonsalves et al., 1981). In practice, the silver-silver chloride electrodes are spaced about 100 meters apart and are streamed astern, away from the magnetic and electrical influences of the ship. The shipboard end of the cable is connected to an industrial-grade recording potentiometer.

A practical consideration in the application of the GEK method for seawater current measurements is the small magnitude of the Earth's magnetic field, together with ambient electrical noise present in nature. These hurdles limit the applicability of the GEK method to situations where electrode separations of tens of meters or more are feasible. Furthermore, because this method relies on Earth's magnetic field, the electrodes tend to be sensitive to magnetic storms as well as the electrical fields produced by passing ships. Such effects have been reported by Mangelsdorf (1961). Sensitivity to interference due to ships limits the application of the GEK method for water current measurements to areas away from harbors.

For shore stations, the GEK may be adopted for use in channel, pass, or open arms of any estuary where turbulence is slight and where upwelling and sinking are negligible. Because the electrodes may be mounted on each side of the channel and connected by a cable, the horizontal velocity of the water passing through the channel (i.e., between the electrodes) may be measured. Because it is awkward to reverse the electrodes to calibrate the instrument when making measurements at a fixed station, it is necessary to consider the measurements of electrical potential recorded during times of zero tidal flow. This is computed using the measurements made during slack water. This comparison is then used to determine the cross-sectionally averaged water current flow through the channel.

The use of the GEK instrument is limited to areas located beyond 20° latitude from the Earth's magnetic equator because the vertical component of the Earth's magnetic field is involved in the computation of water current flow based on the measured electric potential difference across the pair of electrodes. In addition, highly industrialized areas are to be avoided for installation of the GEK instrument because stray grounds and intermittently operating electrical machinery may mask the signals from natural sources.

2.2.2.2. Acoustic Tomography

Ever since humans realized that the oceans exhibit mesoscale (order 100 kilometers) water current circulation features such as cold-core/warm-core gyre systems, just as the atmosphere exhibits storms and tornados, a great deal of effort has been expended in inventing some means to track and measure such large-scale ocean circulation features. Ocean acoustic tomography (OAT) has been identified as a probable vital tool for such measurements. Beginning in 1975, Walter Munk and Carl Wunsch of the Massachusetts Institute of Technology (MIT) pioneered the development of acoustic tomography of the ocean (see Munk and Wunsch, 1979). With Peter Worcester, Munk developed the use of sound propagation, particularly sound arrival patterns and travel times, to infer important information about the ocean's large-scale temperature and current. Since the concept of acoustic tomography for mapping ocean mesoscale variability was first articulated by Munk and Wunsch in 1979, there has been a continually growing research program aimed at understanding the capabilities and limits of acoustic tomographic techniques in the ocean and an

evolving technology program to develop instruments for tomographic measurements. A significant aspect of the development of OAT has been to exploit the integrating features of acoustic measurements (see Chapter 7 for details). Acoustic transmissions inherently average observations and as such they constitute a unique ocean measurement. There is no comparable way to extract integral or large-area average information about fields of fundamental importance to large-scale ocean physics such as temperature, heat flux, horizontal velocity, vorticity, and open-ocean upwelling.

Ocean tomography, as conceived by Munk and Wunsch and as currently practiced, consists of measuring the travel time of acoustic signals transmitted between multiple points. Usually appropriate source-receiver geometry can be found that produces multipaths with good vertical distribution. Upon reception, the multipaths are distinguished by their unique arrival times and angles, thus identifying them with the path they traversed. Acoustic travel time data provide an indication of the average sound speed in the water body bounded between the source and receiver along the acoustic ray. Inversion of acoustic travel time data reveals the intervening sound speed structure, which in turn provides water temperature structure, thus shedding light on mesoscale ocean current circulation features such as cold-core/warm-core gyre systems.

One of the major early accomplishments of the tomography program was to settle some ocean acoustic issues regarding the stability of multipaths. A 1978 experiment in which pulse-like signals were transmitted over a 900-kilometer distance in the Atlantic, south of Bermuda, for a 48-day period showed a stable arrival structure. Furthermore, most of the arrivals corresponded almost exactly to those predicted by ray theory. Munk's investigations, together with the work of Spiesberger and Metzger (1991), eventually motivated the 1991 Heard Island Feasibility Test (HIFT) to determine whether manmade acoustic signals could be transmitted over antipodal (point on the Earth's surface that is diametrically opposite to another point) distances to measure the ocean's climate (water circulation and temperature). The experiment, perhaps the most ambitious program at least in terms of spatial scales and designed to see if it would be possible to measure changes in acoustic travel times over paths as long as 16,000 kilometers to estimate the magnitude of global ocean warming or cooling, came to be called "the sound heard around the world."

During the six days of the Heard Island Experiment in January 1991, acoustic signals were transmitted by sound sources lowered from the M/V Cory Chouest near Heard Island, an uninhabited island in the southern Indian Ocean located approximately 1,630 km north of Antarctica. These signals traveled halfway around the globe to be received on the east and west coasts of the United States as well as at many other stations around the world (Munk and Baggeroer, 1994). This idea of using acoustic signals as the basis for an ocean thermometer arose early in the tomography program. Unfortunately, the follow-up 1996—2006 Acoustic Thermometry of Ocean Climate (ATOC) project in the North Pacific Ocean (ATOC Consortium, 1998; Dushaw, et al., 2009) engendered considerable public controversy concerning the effects of manmade sounds on marine mammals.

Several demonstrations of mesoscale tomography followed the 1978 propagation experiment. An early, rather crude demonstration using limited bandwidth acoustic transmitters was conducted south of Bermuda in 1981. Acoustic transmissions were in one direction only, thus allowing inversion for the sound-speed field but not the absolute water current field. Maps of the changes in the sound-speed field were produced both by initializing the inversion with sound speed data acquired by conductivity-temperature-depth (CTD) casts and by using only a single historic sound-speed profile. Two experiments in 1983, one by the University of Miami in the Straits of Florida and the other by Woods Hole and Scripps institutes of oceanography, respectively, showed that subsurface ocean currents could be measured on 20- to 50-kilometer scales and 300-kilometer scales. A long-range, reciprocal transmission experiment—that is, simultaneous transmissions between two spatially distant sites in the ocean—conducted north of Hawaii in 1987 also demonstrated that differential travel times computed from reciprocal acoustic transmissions can be inverted to obtain ocean currents. In these experiments, three moored tomography instruments formed a triangle roughly 1,000 kilometers on each side. Tidal currents computed from acoustic transmissions agreed well with values derived from a numerical model and from current meter data.

Data storage, which was a limiting factor in early receivers, was subsequently handled by high-capacity, small hard disks. Mooring with satellite telemetry links was developed by Woods Hole Oceanographic Institution to provide near-real-time tomography data transmission to shore. For shorter-range experiments in the open ocean, single hydrophone instruments that transmit less intense signals at higher frequency (centered at 400 Hz) became available commercially.

For ocean tomography, the 1990s were a decade of learning and development. The primary focus had been on demonstrating that ocean acoustic tomography really works, that it can be used to map the mesoscale sound-speed and water current fields with the desired accuracy and precision, and that it is an efficient measurement tool. Moreover, it has been shown that acoustic transmissions can be used for integrating measurements. A great deal of effort has gone into technology development, refinement of techniques, and exploration of experimental limits. A steady progression of

field experiments validated most of the assumptions that are basic to the method, and a rapidly advancing technology development program provided the necessary instrumentation. Inversion models and methods resulted in efficient techniques for extracting maximum resolution sound-speed maps and current fields.

Open ocean acoustic tomographic programs have been conducted at a variety of institutions, starting with MIT, Scripps Institute of Oceanography, the University of Michigan, and Woods Hole Oceanographic Institution. These organizations have been joined by others, including the University of Miami, the University of California at Santa Cruz, the University of Washington, the Naval Research Laboratory, the Naval Postgraduate School, NOAA laboratories, IFREMER in France, JAMSTEC in Japan, the University of Kiel in Germany, and several other agencies.

Reciprocal transmissions permit measurement of water currents. Measuring ocean current with the aid of reciprocal acoustic transmissions relies on the principle that the *difference* in travel time between signals propagating in opposite directions is proportional to the average along-track component of water current velocity. The spatial distribution of transmitter-receiver pairs (transceivers) in the X-Y plane determines the horizontal resolution. Resolution in the vertical (i.e., the Z-axis) arises naturally as a consequence of the ocean's vertical sound-speed profile, which causes multipath propagation. Longer-range tomographic transmissions have hinted at the possibility of measuring ocean basin-scale variability.

Technology had been the primary factor limiting the application of ocean acoustic tomography. First, a tomographic system requires devices capable of emitting and receiving acoustic pulses that are narrow enough to resolve separate multipath arrivals; second, they must be sufficiently above the background noise for precise arrival time estimation; third, acoustic pulses must be transmitted and received with coordinated time bases. Furthermore, unless the instruments are fixed rigidly in place, provision must be made to account for travel time changes that result from instrument motion, so as not to confuse them with changes due to variations in the ocean sound-speed field. Ideally, there should be means for telemetering arrival time data to shore for inversion. In practical terms, these requirements translate into pulses of several milliseconds' duration, timed to within about 1 millisecond/year, at the requisite sound pressure levels. Instrument motion must be monitored to within about 1 meter. A major part of tomography programs has been to develop equipment that meets these specifications.

Technology development resulted in low-frequency, wideband transmitting and receiving devices having the required millisecond timing accuracy; the development of new navigation techniques for precise positioning at sea; and the development of wide time-bandwidth signal-processing schemes for fixed and moving tomographic sensors. Intense research in the field of inverse theory resulted in development of techniques for extracting maximum resolution sound-speed maps from tomographic data. Acoustic tomography experiments in the Atlantic and Pacific oceans and the Greenland Sea demonstrated the capability of acoustic tomography methods to map mesoscale sound-speed variability and to measure ocean currents. Thus, OAT proved to be a powerful tool to measure mesoscale phenomena in the ocean. OAT has special advantages for capturing snapshots of oceanic sound speed and current velocity fields by measuring the travel time of sound (Munk and Wunsch, 1979; Munk et al., 1995).

The initial decades of OAT experiments were largely confined to the open seas. However, OAT grew at a rather fast pace from its early infant stages to adulthood—so much so that the technique, with appropriate improvements, began to be used for water current measurements from straits, inland seas, and rivers. Probably the first reported tomographic experiment in a strait was carried out by Elisseeff et al. (1999) in the Haro Strait. They assimilated the acoustic tomography data and point measurements from their field experiments in the Haro Strait into an ocean model.

Most of the semi-enclosed coastal seas facing industrial areas and urban regions are damaged by various environmental problems such as water pollution and red tides. Environmental management and protection require monitoring of water and material circulation inside the semi-enclosed coastal sea and water exchange to the offshore sea. However, the well-designed measurement of water circulation that covers the entire region of the coastal sea is quite difficult to perform in the coastal sea because of strong fishing activity and heavy shipping traffic. Coastal acoustic tomography (CAT) has thus been proposed as an innovative oceanographic technology that can yield snapshots of coastal sea circulation and relax the observational limitations of conventional methods (Zheng et al., 1998; Park and Kaneko, 2000; Yamaoka et al., 2002; Yamaguchi et al., 2005; Kaneko et al., 2005; Lin et al., 2005). It should be noted that the technology and science of deep-sea acoustic tomography, initiated from the United States, are accumulated in the CAT (Munk et al., 1995).

Taking cues from the results of OAT studies in the open oceans and straits, the application of this technology began slowly expanding to the arena of the coastal water bodies as well. Application of OAT techniques to the coastal seas has been found to be a very attractive theme. Japanese researchers have made excellent progress in the development of CAT techniques. The Hiroshima University acoustic tomography group's continued efforts since the 1990s has resulted in considerable progress in CAT technology for remote measurement of coastal current velocity fields (Zheng et al., 1997, 1998; Park and Kaneko, 2000, 2001; Yamaoka et al., 2002).

A small number of experiments were carried out in the coastal seas in the 1990s for remote measurements of temperature (Chiu et al., 1994) and current field (Zheng et al., 1997, 1998). During these two experiments, which were performed using a pair of acoustic stations, attention was focused on the vertical distribution of temperature and current. Subsequently, for estimation of horizontal distribution, a tomography experiment was carried out using four moored acoustic stations (Elisseeff et al., 1999). This experiment was aimed at estimating the horizontal structure of current and temperature, not by tomography data alone but by their assimilation into an ocean model. However, for quite some time there have been no attempts to perform the coastal tomography experiment where the observation domain was surrounded by multiple acoustic stations.

There were several reasons for focusing attention on CAT research. For example, monitoring water current fields in the coastal seas can produce basic information to help predict disasters caused by oil spill, red tide, water pollution, and so on. Although point measurements of water currents with the aid of Eulerian current meters and vertical profile measurements using acoustic Doppler current profilers (ADPs) possess the ability to measure flow fields from coastal water bodies, horizontally integrated measurements require the deployment of a sufficiently large number of such devices distributed inside the observation region. Apart from the financial burden imposed by the deployment of a large network of current measuring devices, such deployments are sometimes inhibited by heavy ship traffic and fishing activities. Furthermore, current meters deployed inside the observation region are difficult to moor or fix on the bottom under strong tidal currents. In principle, difficulties in monitoring the structure of coastal currents may be drastically resolved by the judicious application of CAT technology in which velocity fields are measured by multiple acoustic stations arrayed in the periphery of the observation region, away from navigation channels and regions of vigorous fishing activities.

Whereas mapping of coastal surface currents can be performed by HF Doppler radar systems, as indicated earlier, the acoustic tomography measures depth-averaged currents with future extension to the vertical profile measurement of currents. There are regions where logistic constraints prevent deployments of HF Doppler radar systems. For example, in the Seto-Inland Sea region in Japan, because of the development of residential and industrial areas it is often difficult to find sufficient shore space for locating the array of antennae needed by some of the HF Doppler radar systems. (Note that CODAR is a comparatively compact HF Doppler radar system devoid of arrays.) In contrast, CAT, operated by multiple sets of compact mooring stations and placed near the shore, can be a practically feasible system with more flexibility and potential ability than HF Doppler radar. In the Neko-Seto

Channel, the mooring observation has been strictly prohibited by the Japan Maritime Safety Agency, except for the near-shore region, because of the risk of shipping accidents.

Practical applications of CAT technology proved to be numerous. For example, the rapid processes of growth, transition, and decay of the tidal vortices were first measured using this technology (see Yamoaka, et al., 2002). It is hoped that, in the near future, the tomography system may make further progress with new instrumentation to measure the temporal and spatial variability of flow patterns to aid navigational safety in coastal water bodies and to understand phenomena such as strong tidal mixing and dissipation in logistically difficult coastal and estuarine regions and around islands without disturbing marine traffic. When this happens, appropriate tidal models will become more useful for operational oceanography. At that stage, real-time data telemetry via satellite or cellular modems will become all the more useful for predicting water current variability in combination with the ocean model. Perhaps long-term operation at shorter intervals can be carried out by putting solar panels on a surface buoy.

Application of acoustic tomography crossed the boundaries of water flow measurements in the open ocean, straits, and coastal water bodies and spilled over to flow and discharge measurements in estuaries and rivers using an advanced in-water acoustic remote sensing technology known as river acoustic tomography (RAT). For example, a RAT system that utilizes a GPS clock and 10th order M-sequence modulation was developed by Kawanisi et al. (2009) and applied to a shallow tidal river with a complex flow field. The RAT system, composed of a couple of transducers that are installed diagonally across the tidal channel, was able to measure the cross-sectional mean velocity in the channel. Advanced signal-processing technology used here yielded a sufficiently high signal-to-noise ratio owing to the 10th order M-sequence modulation. Thus, it was found that the RAT system works well even throughout flood events in which turbidity and acoustic noise are very high. In addition to measuring water discharge, mean water temperature and salinity could also be deduced from processing the sound-speed data collected by the system. Consequently, the RAT, operated at the shallow tidal channel with large changes of water depth and salinity, successfully measured the cross-sectional mean velocity over a long duration.

The water discharge deduced from the RAT system was compared with that measured by ADP. The agreement between RAT and ADP on water discharge was satisfactory. In addition, the mean temperature/salinity obtained from RAT fulfilled an acceptable compliance with CTD data. Thus, the RAT system is found to be a promising technology for continuous measurement of water discharge and mean temperature/salinity at estuaries. It is surprising that

RAT technology could be successfully used for water current measurements, even from the highly dreaded tidal-bore-ravaged Qiantang River in China. Based on the success of this experiment in 2009, it has been suggested that the RAT method provides a prosperous way for continuous long-term monitoring of river discharge in large tidal-bore-infested rivers with heavy shipping traffic, such as the Qiantang River.

2.2.3. Measurements Based on Motion of Drifting Subsurface Floats

Apart from the GEK and acoustic tomography methods, subsurface drogues have also been tried for current measurement, initially in the mixed layer and subsequently in the abyssal depths. Freely drifting drogues, constructed of two mutually orthogonal PVC sheets, weighted to assure vertical attitude, and suspended from a float, have been used to indicate the trajectory of subsurface currents at specified depths. Color-coded flags on buoyant poles attached to the floats identified individual drogues set at different depths for optical tracking. Deployment consisted simply of dropping the drogues into the ambient current off the stern of a boat, which was used as the tracking control station. Precision theodolites were employed to track drogues (Murray, 1975).

The early major investigations into subsurface current measurements included the experiments of Stommel (1949) and Swallow (1955). The revolution in oceanic subsurface and abyssal current measurements began after John Swallow returned to Cambridge University after a wartime break, maintaining radar systems for the Royal Navy. Gould (2005) has given an absorbing account of John Swallow's pioneering initiatives in the development of a branch of subsurface current measurement technology. The prelude to Swallow's invention has been described thus: In the summer of 1954, John Swallow, a 30-year-old Cambridge Ph.D. student, made his first visit to the U.K. National Institute of Oceanography (NIO). The visit was to discuss with NIO oceanographers the possibility of making direct measurements of the vertical profile of water currents in the deep ocean by tracking a slowly sinking acoustic source and hence to derive the current profile.

Following the visit, Swallow accepted an invitation to join the staff of NIO. He left his initial interest in profiler development and, subsequently, his untiring experimentation with developing neutrally buoyant floats for Lagrangian measurements of deep ocean currents (which later came to be known as the "Swallow float," after his name) began at a time in the 1950s when little was known about the circulation of the deep ocean from direct measurements. Swallow's efforts focused on developing floats that would remain neutrally buoyant at an approximately constant specified depth (in practice, at a level experiencing constant water pressure) so that in the presence of water motion at that depth, the float will freely drift with the water motion over the specified isobaric surface (i.e., depth of constant water pressure). Tracking the float (i.e., determining the geographical position of the float) at frequent time intervals (e.g., by tracking it by ship-borne hydrophones or through acoustic receivers moored in the Sound Fixing and Ranging [SOFAR] channel) would then yield a Lagrangian description of the water motion at that depth layer.

Swallow's work at Cambridge had made him familiar with the compressibility of materials and of seawater. He had experience of the measurement of acoustic travel times and knew about the difficulty of making deep-sea pressure seals. This experience led him to consider trying to stabilize a float at some depth level within the ocean so that while it freely drifts with the ocean current at that depth, the neutrally buoyant float could be tracked acoustically and its drift integrated over an extended period. Construction of the first neutrally buoyant floats started at the beginning of 1955. Materials and money were in short supply, and so the dictum "necessity is the mother of invention" became truly applicable. Aluminum had the correct mechanical properties in terms of strength and density; the most readily available source of aluminum tubing was the scaffolding used in the construction industry. However, standard tubing had too great a wall thickness, so it had to be thinned by immersing the tubes in a bath of caustic soda. Six meters of tubing were needed to provide sufficient neutral buoyancy at the intended depth, and for ease of handling this was cut into two 3-m lengths laid side by side.

According to Gould (2005), over the next few years Swallow made further exploratory measurements in the Atlantic, gaining confidence in float ballasting, improving tracking techniques, and using floats to depths as great as 2,900 m but with tracking still lasting no more than two and half days. Subsurface current measurements were made in April and May 1956 west of Gibraltar and in the Norwegian Sea in October and November 1956. Apart from acquiring new knowledge of the variability of currents at depth and the comparison of these measurements with geostrophic shear calculations, Swallow was surprised to find that a float close to the Mediterranean water core did not move westward, as it was expected to. Gould (2005) further states that Swallow and Henry Stommel had by then started to correspond. They met for the first time in 1955 when John Swallow went to New England for a meeting to discuss radioactive waste disposal and was able to visit Woods Hole Oceanographic Institution (WHOI). Swallow's long-term collaborations with U.S. researchers, including Stommel, led to major advances in the study of deep-ocean dynamics.

The Swallow floats (see Sections 6.3 and 6.4) provided a Lagrangian view of the ocean interior and thus started the mesoscale revolution. Indeed, Swallow floats are given credit for revealing the existence of some exciting oceanic subsurface current paths, such as the strong countercurrents under the Gulf Stream. Although several differing designs of subsurface floats (drifters) emerged in the subsequent period, they were all some form of improvements of the Swallow floats, such as those listening to moored acoustic sources making use of SOFAR channels (see Section 6.5) and horizontally displaced and vertically cycling types that transmit the recorded data via satellite systems such as ARGOS (see Sections 6.6 and 6.7).

2.2.4. Measurements from Fixed Locations at Predetermined Depths

Although the rather crude Lagrangian methods employed in the olden times of ocean current measurements provided a rough indication of the trajectories of surface currents, no information was available on the temporal variability of currents at a given location of interest (i.e., Eulerian measurements). In addition, currents at depths below the sea surface needed to be known for several operational applications. These requirements led to the invention of devices known as *current meters*. Primitive methods of Eulerian-style ocean current measurements relied primarily on purely mechanical devices and sensors. Measurements of currents using such devices and sensors were based on the drag or physical rotation experienced by the sensors in response to the force exerted on them by the moving water. Some of these devices used in the early stages of oceanographic studies are no longer in use today. However, in view of their historical importance in the evolution of ocean current measurement technologies, some aspects of such measurements are briefly mentioned here, with the hope of kindling appreciation in the minds of the next-generation oceanographic researchers and academicians as to the difficulties with which current measurements were made in the past.

2.2.4.1. Suspended Drag

The simplest current meter is a *drag*, described in the literature as the *Chesapeake Bay Institute drag*, consisting of two crossed rectangular plates weighted and suspended by a thin wire (Pickard and Emery, 1982). When the drag is immersed, the force exerted by the water current (proportional to the square of the current speed) pulls the wire to an angle from the vertical. The current speed is computed using a formula relating the size of the drag, its weight in water, and the angle of the wire from the vertical. This wire angle is measured to determine the current. The device is simple and quick to use from an anchored ship. It is, however, limited to depths of a few tens of meters because the current drag on the wire increases with length and complicates the interpretation of the wire angle at greater lengths. Furthermore, the drag method is useful only for short-term *in situ* measurements of currents, not for continuous long-term measurements.

2.2.4.2. Propeller Revolution Registration by Mechanical Counters

Perhaps the most widely used current meter for some time in the past was the *Ekman current meter*. This device consisted of a propeller mounted within a frame, a vane that oriented the instrument so that the propeller faced the current, and a magnetic compass for sensing the orientation of the propeller relative to Earth's magnetic north and, therefore, the direction of flow. The instrument, attached to the end of a wire, was lowered to the desired depth. A metal weight (messenger) dropped down along the wire disengaged a lever to free the propeller to rotate with a speed directly proportional to the water current. A second messenger, dropped after a known interval of time, arrested the rotations of the propeller. The number of revolutions made by the propeller during the measurement interval was recorded by a mechanical counter. The water current speed could then be related to the number of revolutions per minute.

The current direction was recorded by an ingenious method of dropping minutely sized metal balls through a set of tubes into the indentations of a disk, at intervals, along a magnetic compass into one of 36 compartments with 10° sectors and graduated in degrees relative to Earth's magnetic north. The compass box was rigidly connected to the vane of the current meter, but its magnets could adjust themselves along the local magnetic meridian. The graduated sector of the compartment into which the ball dropped indicated the direction of flow at the moment the ball fell. The design of the current meter permitted one ball to drop for each 33 revolutions of the propeller. The average direction of water current during one single observation was obtained by computing the weighted mean according to the distribution of the balls.

A major disadvantage of the Ekman current meter was that it had to be lowered and raised for each measurement—a tedious job, often performed from small boats such as country crafts in hostile marine conditions. This drawback has been circumvented, to some extent, in the Ekman repeating current meter and the Carruthers residual current meter. However, these were also purely mechanical in design and direct reading in style and therefore unsuitable for deployments in continuous, long-term current-monitoring projects. These current meters have been used successfully by biologists in bays and near-shore waters to determine the current flow in relation to fish migration and marine life dispersal.

Recordings of current measurements were purely mechanical in nature. In such current meters, raised numbers indicating mechanically derived current speed and direction measurements were recorded by application of pressure on strips of tinfoil advanced by mechanical clockwork, or markings representing the data were recorded by a fine stylus on a paper chart run by mechanical clockwork. Two other instruments designed much later for current measurements are the Schaufelrad (paddlewheel) current meter, as reported on by Boehnecke (1949), and the Roberts radio current meter (Roberts, 1950). These were also mechanical instruments.

The Schaufelrad (paddlewheel) current meter is a self-contained, continuous recording instrument. The current meter is moored into position with an anchor and held off the bottom with a buoy. The buoy also aids in the recovery of the instrument. The size of the current meter, 170 cm long and 58 cm wide, makes it impossible to be deployed closer than 100 cm to the bottom. In this current meter, a paddlewheel mounted at the head of a bomb-shaped shell is turned by the prevailing currents. An intermittent light source records on a photographic film strip the turns of the paddlewheel and the direction as indicated by a compass. After the desired period of operation, up to four weeks, the current meter is raised and the film strip removed and developed. The traces of light are scaled to determine the water current's speed and direction adjacent to the current meter during the period of immersion.

Later current meters recorded speed and direction measurements on chronographs, magnetic tapes, and photographic plates. Modern versions of current meters record data in semiconductor recording media. Data acquisition techniques as well as recording and retrieval formats gradually improved with advances in technology. Modern recording current meters permit direct transfer of the recorded time-series vector-averaged current measurements to a computing device. This permits rapid analysis and documentation of results in the desired format. Ocean current measurement technologies have undergone rapid evolution both in terms of modernizing the basic sensors and adapting to the fast-changing hardware and embedded-software technologies incorporated in data acquisition and data recording/displaying devices. It is only recently that oceanographers have been able to make the continuous, long-term recordings required for achieving a realistic picture of ocean currents and to apply theories that take reasonable account of both winds and density gradients.

2.2.4.3. Unidirectional Impeller Current Meters

Probably the first electronic counterpart to the purely mechanical current meter was a modification of the Ekman current meter, popularly known as the *Roberts radio current meter* (*Manual of Current Observations*, revised (1950) edition; supplement to C&GS special publication No. 215, November 1961). The Roberts radio current meter consists of three parts: the meter, the buoy, and the transmitter. In this current meter a rotating impeller, which is actuated by the current, is connected through a magnetic drive to an enclosed interior mechanism that opens and closes an electric circuit by means of two contacting devices. One lever mounted on the compass is always oriented toward Earth's magnetic north. The other lever, fixed to the meter, is always oriented in the direction of the water current flow. A post, carried on a gear, revolves past the switch-actuating levers and causes the lever mounted on the compass shaft to close its switch contacts once for each revolution of the gear (five revolutions of the impeller). Each contact produces a radio signal. The frequency (i.e., number of cycles per second) of these signals is determined by the velocity of the impeller and therefore can be translated into the water current velocity through the use of a dedicated rating table. In the Roberts meter, a speed signal derived from an impeller and a direction signal derived from a magnetic compass as described are electrically transmitted (telegraphed) to the surface and recorded on a shipboard deck unit. The Roberts current meter was an important step in the direction of current meter design—in fact, a forerunner of the modern electronic current meters incorporating mechanical sensors.

In 1958 the Roberts radio current meter was modified by the addition of larger impellers and tailfins made out of 1/16" fiberglass. Known as the Model 1V low-velocity meter (Figure 2.8), to distinguish it from the older conventional Model III high-velocity meter, it had been modified to enable accurate measurement of current velocities as low as 0.1 knot (i.e., 5 cm/s).

In operation, a maximum of three current meters can be suspended below a buoy (of ship-like design) anchored at a particular station (Figure 2.9). A sequence switch, carried in the buoy, selects in proper order one current meter at a time. The buoy, which houses a radio transmitter, also houses batteries that supply power to the current meter. An antenna mounted on the buoy is used for transmission of the acquired data as a radio signal. At a radio receiving station, either ashore or afloat, the signals are received, amplified, and recorded by means of a chronograph. Observers tune the individual radio buoy frequencies, record the signals, scale the tape, and tabulate the values for each current station. Amplitude modulation (AM) radio gear has been used in the transmission of current signals from the buoy to the receiving station since the system was first developed. In 1958 frequency modulation (FM) radio gear was introduced to obtain separate velocity and direction radio signals, which are recorded by a three-stylus chronograph.

A disadvantage of this type of current meter is that the rolling of the ship (from where measurements are made) or the up-and-down movement of the mooring line (on which the current meter is mounted) may cause the impeller to turn and cause inaccuracies in the speed measurement. A hollow cylinder mounted around the impeller minimizes this effect. This is, in fact, implemented in the Toho Dentan

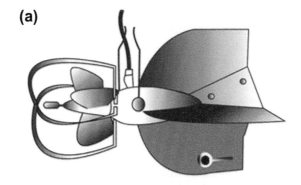

(a)

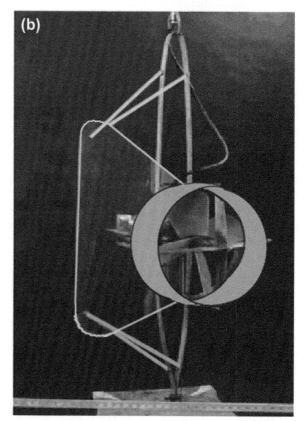

(b)

FIGURE 2.8 Model IV Roberts radio current meter: (a) side view; (b) front view. *(Source: NOAA Central Library and National Climate Data Center (NCDC), www.reference@nodc.noaa.gov.)*

direct-reading current meter (Figure 2.10). In the Toho Dentan current meter, the impeller diameter is 10 cm, and its time constant is 2 seconds. The current speed is continuously recorded on an analog tape recorder. The tape is subsequently processed on a signal analyzer.

2.2.4.4. Savonius Rotor Current Meters

Although the history of ocean current measurements is long, the advent of modern current meter development dates to the 1960s and early 1970s. A well-known current meter that, so to say, ruled the realm of Eulerian-style ocean current

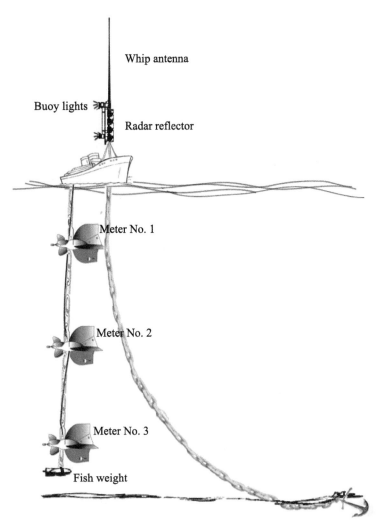

Whip antenna

Buoy lights

Radar reflector

Meter No. 1

Meter No. 2

Meter No. 3

Fish weight

FIGURE 2.9 Multisuspension of Roberts radio current meters and data transmission from radio buoy. *(Source: NOAA Central Library and National Climate Data Center (NCDC), www.reference@nodc.noaa.gov.)*

measurements for some decades after the 1960s is the Savonius rotor current meter, popularly known as the *Aanderaa current meter* (see Chapter 8 for details). It is worthwhile to ponder the birth of this current meter and the success stories that followed. In 1960, the Christian Michelson Institute (CMI) was awarded the contract from the North Atlantic Treaty Organization (NATO) Subcommittee on Oceanographic Research to develop a recording oceanographic current meter. The project coordinator was Helmer Dahl, although the daily management of the project was handled by Ivar Aanderaa. After the project's end in 1966, Ivar Aanderaa left CMI and founded Aanderaa Instruments A/S on the basis of this current meter. The company still exists in Bergen, Norway, as well as in the United States and Japan. The Aanderaa current meter has been the single most popular oceanographic current meter worldwide for decades.

The story of the success of Aanderaa current meter is simply the keen interest Ivar Aanderaa showed in the field

performance of this current meter. Several problems were observed in the initial years of Aanderaa current meter deployments. These included errors in the measured flow direction at large depths and over-speed registration when currents were weak. Such problems, reported by researchers and published in oceanographic journals, were taken seriously by Aanderaa, and corrective measures were implemented as early as possible. These measures included changing the material of the pressure housing and redesigning the basic sensor. My own personal interaction with Aanderaa in 1995 at his laboratory in Bergen on the latter issue was awe inspiring. The foundation of Aanderaa Instruments is a very early example of a commercial spin-off from the oceanographic research activity at CMI.

A highlight in the history of the Aanderaa current meter is the successful recovery, in the Weddell Sea (Antarctica), of one of the first units manufactured by Aanderaa Instruments. The instrument was deployed in 1968 and successfully

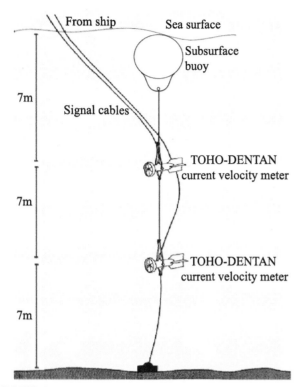

FIGURE 2.10 Toho Dentan direct reading current meter, in which a hollow cylinder mounted round the impeller minimizes inaccuracies in the speed measurement arising from turning of the impeller in response to the rolling of the ship (from where measurements are made) or the up-and-down movement of the mooring line (on which the current meter is mounted). *(Source: Veth and Zimmerman, 1981. © American Meteorological Society. Reprinted with permission.)*

recovered five years later, in 1973. This was the first long-term current meter data series ever to be collected from Antarctica. By January 1982, this company had built 7,000 instruments for recording ocean currents all over the world, and to commemorate this joyous occasion Mr. Aanderaa printed an exact copy of the first edition of the book *The Physical Geography of the Sea*, by M. F. Maury, an early pioneer who laid the foundation for modern oceanography by carrying out accurate observations of phenomena at sea during his many voyages. By recording his findings in his book, Maury created what became the first textbook on oceanography.

2.2.4.5. Ultrasonic Acoustic Methods

Although current meters incorporating mechanical sensors were widely used for Eulerian-style measurements of currents, they are incapable of measuring weak currents and turbulence. This led to the development of current meters incorporating nonmechanical sensors. Ultrasonic acoustic sensor is one among such nonmechanical sensors.

The principle used in the design of acoustic current meters is that differential travel times computed from reciprocal acoustic transmissions between two acoustic transceivers can be inverted to obtain water current speed parallel to the acoustic path (see Section 9.2 for details). Such measurements along two mutually orthogonal acoustic paths enable estimation of the water current velocity vector within the volume bounded by the four transceivers. In practice, the straight-line distance between a pair of transceivers in an acoustic current meter is about 15 cm.

An advantage of the acoustic travel time (ATT) difference method is that it permits simple multiplexing, which allows current measurements along many axes using one set of circuits. In some designs (e.g., Lester, 1961), "zero calibration" can be accomplished by reversing the direction of the acoustic path of one of the probes. With the signals in the same direction, the water path time delays of both signals are identical, and therefore the electrical delays of the system can be adjusted to "zero."

A disadvantage, arising from the small distance (~ 15 cm) between the two transceivers, is that it requires the arrival time differences of acoustic pulses to be measured with sufficient speed to resolve currents less than 1 cm/s. For a practical transducer pair of this dimension, the travel time difference will be 1.3 nanoseconds for current flow speed of 1 cm/s along the acoustic axis. Time resolution as short as this requires ultra-high-speed electronic circuitry. Nevertheless, flow resolution of 1 mm/s has been achieved in a commercial instrument using this principle (Simrad ultrasonic current meter [UCM] manual).

It has been shown theoretically (Gytre, 1976) that the differential travel time type current sensors exhibit ideal cosine response in the azimuth. That is, if the acoustic path between a given transceiver makes an angle θ with the flow direction, the transit time difference is proportional to $v \cos \theta$. To achieve the theoretically ideal response of the sensor, it is required to optimize its hydrodynamic performance at all current speeds and directions. Good design demands that the sensor not disturb the current flow that is being measured. To obtain a true estimate of the flow field, the dimensions of the transceivers should be minimal. Practical designs employ piezoelectric crystals of about 1 cm diameter mounted in metallic tubes. The distance between these two piezoelectric crystals in a transducer pair is typically 15 cm.

If an orthogonal two-axis sensor possessing ideal cosine response in the azimuth is rotated in a horizontal plane while being towed at various constant speeds, the V_x and V_y current signals should follow perfect sine and cosine responses, and the measured mean currents at different speeds would ideally appear as perfect concentric circles. Laboratory experiments have shown that the circles are slightly squared (Gytre, 1980), probably due to the effect of mirror holders of the current meter. In a recent design (Williams, 1985), the use of a folding mirror is avoided by supporting four pairs of acoustic transducers on two vertically separated rings. Each acoustic path is inclined 45° to the horizontal and spaced 90° in azimuth. Thus, disturbances to the flow within the measurement volume are minimal for near-horizontal flows. The horizontal and vertical responses have been reported to be cosine with a maximum error of 5% of the speed along either axis until the vertical angle caused the flow to come within 20° of an acoustic axis. Near-ideal responses could be achieved because of the small size of the transducers (approximately 1 cm³) and their thin (0.3 cm diameter) supporting structures. Thus, the wakes from the structures are reasonably low, and these wakes cross the measurement path only over a small percentage of its length.

In an alternate design, two low-frequency signals are derived from the original high-frequency signals (by a process known as *heterodyning*), which are transmitted between the two acoustic transceivers. The phase difference between these two low-frequency signals is then related to the speed of water current. This relation depends on the particular heterodyning technique used for deriving the two low-frequency signals.

In one method, reported by Lawson et al. (1976), ultrasonic signals of frequency f_1 are continuously transmitted from transducer-1 toward transducer-2 of a given pair, while similar signals of frequency f_2 are simultaneously sent from transducer-2 to transducer-1. Thus, each transducer simultaneously transmits one ultrasonic frequency and receives another frequency from the other transducer. Application of appropriate signal-handling and heterodyning (beating) techniques provides signals with a phase difference related to the acoustic path length, the square of the velocity of sound in water at rest, the transmission frequencies, and the component of the water current speed parallel to the acoustic path.

In a slightly different phase difference method (Robbins et al., 1981), continuous wave bursts of a single frequency are simultaneously transmitted by both the transducers of a given pair. Before the acoustic signals arrive, the transducers are simultaneously connected to the receiving circuits. The received bursts resynchronize "slave" oscillators, which maintain phase information between bursts. The continuous output of the slave oscillators is heterodyned with a local oscillator, resulting in low-frequency outputs. Phase difference between these low-frequency outputs is a linear function of the water current speed, v, governed by the relation:

$$\phi = \frac{2vl\omega}{c^2} \qquad (2.1)$$

where ϕ is the phase difference, ω is the acoustic frequency in radians/sec, c is the local speed of sound in water at rest, and l is the transducer spacing.

An advantage of this method is that because the phase difference arising from the differential travel times is measured from low-frequency signals, timing requirements in the electronics circuitry are substantially less stringent than if performed at the acoustic frequency. In addition,

because all the critical signal processing is implemented at the low-frequency beat signal, the circuitry is simplified. Furthermore, sensitivity and stability better than 1 cm/s become achievable at the low-frequency level because at a low frequency, it is not necessary to correct for changes in internal circuit delays due to temperature and other factors.

A disadvantage with this method, as with the travel time difference measuring method, is the dependence of the current speed measurements to the square of the velocity of sound in water—a parameter whose variability in different environments cannot be ignored.

Ocean current measurements using the travel time difference as well as the phase difference methods greatly depend on the sound velocity in seawater—a parameter that has large variability in space and time. A technique that has less dependence on the sound velocity in seawater is the swing-around method. A swing-around acoustic current meter consists essentially of two swing-around velocity-of-sound meters configured in such a way that the transmission paths in the water are side by side and of equal length and the directions of acoustic pulse travel are opposite in the two meters. The instrument is oriented by a large tailfin so that the directions of acoustic pulse travel are parallel to the current flow direction. This orientation ensures that the time of pulse travel is greater in one meter than in the other. One sensor measures the sum of the speeds of sound (in seawater) and the current flow; the other sensor measures the difference between the two speeds. The difference in swing-around frequencies of the two sensors is then proportional to the speed of the water current. It has been shown (Suellentrop et al., 1961) that the output frequency (f) of such a swing-around acoustic current meter is related to the speed (v) of the water current by the expression

$$f = \left(\frac{2v}{L}\right)\left(\frac{1 - 2tc}{L}\right) \qquad (2.2)$$

where L is the separation between the transmitter and the receiver; c is the velocity of sound propagation in seawater at rest; and t is the time delay in the electronic circuitry of each sensor. For typical values of L = 0.15 m and t = 0.6 µs, the percentage error in f caused by a change in c is only 0.012 times the percentage change in c. This makes the error practically negligible.

An alternate swing-around method reported by Hardies (1975) provides an automatic correction for the variations of the speed of sound in water, without its separate measurement, by the use of an ingenious closed-loop technique. In this method, two swing-around loops are configured for the transducer pair such that they transmit in opposite directions along the same fluid path. Bursts of ultrasonic signals are first transmitted from transducer 1 to transducer 2. Reception of a leading edge at transducer 2 causes the transmission of a new burst of signals from transducer 1 to transducer 2, and a closed-loop A is established. After an initial displacement of a small time period, similar transmissions are initiated from transducer 2 to transducer 1, and another closed-loop, B, is established. Once started, these two loops continue to swing around, and the number of times the loops swing around during a fixed time period (T) is registered. After the elapse of time duration T, the swing-around loop is continued for an "additional" swing around. If there is a variation in c from a preset value, there occurs a corresponding variation in the duration of the swing-around loop such that the sound velocity term vanishes from the expression for the speed of current flow. Half the duration of the swing-around loop is equal to the magnitude of the speed of water current.

An advantage of this technique is that because the dependence of sound velocity on water current measurements is eliminated, it offers a promising solution to current measurements in estuaries and other water bodies where sound velocity may considerably vary in time and with depth.

A drawback of all types of ultrasonic acoustic current meters, however, is the response of acoustic signals to air bubbles in a wave field. Tiny, almost invisible air bubbles (micro bubbles) can have a resonance frequency the same as that of the acoustic signal in water. When this happens, the air bubble is a very strong scatterer and absorber of sound waves (Clay et al., 1977). Thus, aeration of water occurring in a wave field may limit the minimum depth at which ultrasonic acoustic sensors can operate successfully. In situations where these current meters are mounted on subsurface buoys or on bottom-mounted tripods, this drawback is not of any serious consequence. Because air bubbles are virtually absent in such areas, measurement errors will be minimal.

High accuracy current measurements are often needed by both the scientific and the commercial community. Acoustic current measurement technique employed by Falmouth Scientific Inc. (FSI) is based on measuring and comparing direct path acoustic phase shifts along multiple paths. Because the speed of sound in water is very large compared to the water current velocity in natural water bodies such as seas, inlets, rivers, and the like, historically it was very difficult to accurately measure the differential travel time of acoustic pulses propagated between pairs of acoustic transducers mounted at fixed, known locations in space. To circumvent this difficulty, FSI acoustic current meter (ACM) uses the *phase-difference* between upstream and downstream acoustic signals to accurately determine the water current speed.

FSI ACM has four "fingers" (Figure 2.11). Each finger houses two acoustic transceivers. The transceivers are used to create four acoustic paths. The flow velocity is measured by comparing phase shifts of sound pulses traveling along three of the four acoustic paths. One path that is contaminated by the wake from the center support strut is always disregarded (A. L. Kun and A. J. Fougere, Falmouth Scientific, Inc.).

FIGURE 2.11 Falmouth Scientific, Inc. 3-D Acoustic Current Meter, incorporating acoustic phase-shift measurement principle, in protective frame. *(Courtesy of Falmouth Scientific Inc.; www.falmouth.com/sensors/currentmeters.html.)*

The acoustic current meter transmits a 1-MHz continuous wave signal for a period of 1 ms, first in one direction where the total phase shift including the receiver phase shift is measured, and then in the opposite direction where the total phase shift is measured again, using the same receiver. The current velocity is proportional to the difference in phase for the two directions. The advantage of measuring phase shift is that it can be accomplished with slower circuits than measuring time of travel. Slower circuits in turn require less power to operate.

The first acoustic current meter developed by FSI is the three-dimensional acoustic current meter (3D-ACM). The sample volume of the 3D-ACM is small; it is a sphere with a radius of roughly 13 cm. Individual measurements take a very short time (about 32 ms), and the single ping uncertainty is only 0.01 cm/s. The instrument includes a *3-axis* fluxgate compass, which measures the Earth's magnetic field, and a *2-axis* electrolytic tilt sensor, which measures the instrument's angle to the vertical. Using the internal compass and the tilt sensor it is possible to determine the heading of the instrument, and consequently estimate the water flow direction relative to the horizontal surface and magnetic north (in Earth-coordinates), without the need for specific orientation of the current meter during deployment. Vector averaging of the acoustic current sensor data also yields mean flow speed and direction.

In April 2012, FSI released a new generation ACM, the *ACM-PLUS* family, which allows the user to select between 2D and 3D measurement along with several other feature enhancements (*ACM-Plus* Compact, Vector Averaged Current Speed and Direction Meter User Manual, April 2012, P/N 8000-ACM-PLUS, Rev. 2). According to FSI, the ACM-PLUS uses the most accurate and stable current measurement techniques available today and is configured with features such as extended on-board data memory, fast download capability, high-accuracy real-time clock, and high speed data sampling. The unit's compact size and light weight make the ACM-PLUS well suited for multiple meter arrays. Windows-based software for meter setup, data collection and data visualization make the FSI ACM-PLUS very user-friendly. The ACM-PLUS is available in either shallow-water or deep-water housings. The device may also be equipped with an optional conductivity-temperature-depth (CTD) module and can be configured to log up to two analog inputs from external sensors (e.g., dissolved oxygen sensor, optical backscatter sensor, fluorometer, transmissometer).

2.2.4.6. Thermal Sensors for Measurements of Turbulent Motions

The requirement to measure microscale fluctuations in the oceanic velocity field and physical (laboratory) models and to investigate water current flows in areas that are difficult to access (e.g., rock crevices, micro-habitats) led to the development of thermal sensors such as thin films, thermistor probes, and temperature-sensitive quartz crystals. These sensors have no moving parts and are smaller than other water current-measuring sensors we have addressed so far.

Thermal sensors operate on the mechanism that the heat exchange between a heated solid body and the surrounding fluid medium is a function of the flow rate of the surrounding fluid. For use as water current meters, thermoresistive devices are electrically heated to a temperature higher than that of the surrounding water medium and are placed in the flow field.

Use of miniature metal-film probes for measurements of oceanic turbulence arises from their superior high-frequency response, fine spatial resolution, and high sensitivity. The thin-film probe consists of a thin metal coating such as platinum, with resistance of 5 to 20 Ohms, deposited on a substrate of high electrical and thermal insulating properties. A very thin layer of insulating coating, such as silicon dioxide or quartz, over the metallic film prevents its electrical shorting to the conductive saltwater yet provides good heat transfer. Platinum film is particularly suited to measurements of oceanic turbulence because it is chemically inert and offers low thermal time constants and nearly linear R-T (i.e., resistance versus temperature) characteristics. Experiments have shown (Brech et al., 1971) that the measurements tend to be unreliable when the angle of attack is near 90° to the axis of the probe, probably because of secondary flows caused by the probe. To make it insensitive to small variations in the direction of the current flows, the thin-film probes are usually provided with a conical shape. Additionally, this shape provides favorable hydrodynamic characteristics to the probe. The thin-film probe, as a result of its conical shape, is sensitive to the direction of flow velocity and exhibits a near-cosine response.

In operation, the thermal probe is usually connected in one of the arms of a DC bridge and used in a constant-temperature mode. The probe is electrically overheated, in the range of tens of degrees above the ambient temperature, and a feedback bridge maintains this constant temperature during the measurements. The bridge voltage required to maintain this constant temperature bears a relationship to the heat transfer from the probe and, therefore, to the velocity of the immediately surrounding flow field.

Although platinum thin-film probes have many ideal properties suitable for measurements of turbulence, their mechanical brittleness is a major hindering factor to its long-term use in the marine environment. They often develop pinhole leaks, and the insulating coating over the metallic thin-film flakes off (Irish and Nodland, 1978). Long-term accurate measurements of turbulent oceanic current velocity fluctuations are, therefore, difficult with these sensors. In fact, it has been observed (Dingwell and Weiskopf, 1981) that there is a strong correlation between the operating temperature and the probe's lifetime. The average operating lifetime of a thin-film probe is estimated to be approximately 200 hours.

The drawbacks of the thin-film probes necessitated the design and development of alternative thermoresistive elements such as thermistors for long-term measurements of the turbulent motions in the oceanic velocity field. As in the case of all thermal sensors addressed so far, the thermistor probe also operates on the mechanism that, when placed in a flow stream, the heated thermistor dissipates an amount of heat proportional to the flow. The amount of electrical power required to restore the lost heat is directly related to the flow speed. At low velocities the output of the thermistor-based water current meter is independent of the direction of flows, but above about 50 cm/s the thermistor probes are sensitive not only to the magnitude but also to the direction of flow (Barbera and Vogel, 1976). At high water current speeds the probes must, therefore, be oriented into the direction of the current flow.

Basically, there are two types of probe construction. In one construction the thermistor is directly heated. This construction is used where small size of the probe is an important requirement. With this construction, useful working ranges of 0.02−25 cm/s have been reported (Forstner and Rutzler, 1969). In the other construction, the thermistor is heated by a wire, which is wound over the thermistor. This construction is employed in situations where stronger currents are to be measured. Its useful working range is 1−100 cm/s. Probes of the second type have a much slower response than the directly heated types. The time constant is typically of the order of a few seconds. The essential criterion in any construction is that the thermistor be correctly mounted to achieve the best possible heat conduction to the flow medium. This is normally achieved by mounting it in a thin-walled, miniature silver cone filled with silicone grease for better heat conduction.

There are two classical bias conditions for heated thermistor probes used as flow meters—namely, constant current and constant temperature. The latter configuration is more suitable for compensation of the fluid temperature dependence (Catellani et al., 1982). One method used for compensation of the dependence on water temperature is the use of two thermistors (Forstner et al., 1969). In this, one thermistor is heated to approximately 5°C above the ambient temperature of the flow medium. The two thermistors are connected in the parallel arms of a Wheatstone bridge and the resistance changes in the thermistors are measured. The flow velocity is directly related to the differential temperature of the two thermistors. The usefulness of thermistor flow meters is limited to the measurement of currents and turbulences of low to medium speeds.

Heated temperature-sensitive quartz crystal-based sensors used for measurement of water current operates on the mechanism that, at low velocities, it is cooled by forced convection (Resch and Irish, 1972). A quartz crystal transducer utilizes two temperature-sensitive quartz crystals that are mounted in two separate pressure cases and maintained a small distance apart, having a common axis. One crystal is maintained at the temperature of the surrounding water; the other is electrically heated. The former measures the ambient temperature and the latter measures the increase above the ambient temperature. The difference in temperature is a function of the magnitude of the current flow. The transducer detects the currents perpendicular to the common axis of the two sensors. It measures low speeds that the mechanical devices cannot. Because deep-sea currents are typically 0.1 to 10 cm/s, this sensor is particularly well suited for deep-sea current measurements. However, because the time constant of the quartz transducer is approximately 10 to 20 seconds, it is not suitable for measurements of turbulence.

2.2.4.7. Laser Doppler Sensor

The Doppler effect, which was discovered in 1842 by the Austrian scientist Christian Doppler, is an effect in physics according to which the frequency of any harmonic wave motion at a receiver differs from the frequency at its source whenever the receiver or the source or both are in motion relative to one another. Subsequently, this effect began to be used for diverse applications in different areas of practical interest. In the areas of oceanographic research and operational oceanography, the Doppler effect is extensively used for remote measurement of ocean currents.

Laser Doppler techniques were applied to fluid-flow measurements as early as 1964. However, its marine application began as late as 1980. The laser Doppler current meter (LDCM) operates on the mechanism that the waterborne suspended particles in a flow field, illuminated by a monochromatic light radiation, introduce frequency shifts in the scattered light radiation by virtue of the Doppler effect. The scattering particles are assumed to be passive tracers of the water velocity. The Doppler shift in frequency thus introduced is linearly related to the water flow velocity, given entirely by the system geometry. The Doppler frequency shift is measured from the heterodyned signal produced as a result of photo mixing the received scattered radiation with the direct radiation or with another scattered radiation.

Two types of laser Doppler systems are in common use: the forward-scatter system and the backward-scatter system. The forward-scatter system requires two watertight enclosures, one containing the source of the laser radiation and the other containing the receiver. These enclosures are to be maintained in strict optical alignment. Scattering of light from natural particles is much stronger in the forward direction. With strong forward-scattered optical signals, continuous as well as periodic samples of water flow velocities may be obtained. A disadvantage of the forward-scatter system is that the struts joining the laser source and the receiver could disturb the flow that is intended to be measured. The backscatter configuration permits a "monostatic" design (i.e., source and receiver at the same location), enabling flow measurements to be made remotely without disturbing the flow. A backscatter system would thus require only one enclosure but a more powerful laser. This is because the signal-to-noise ratio for a backscattering system is generally poorer than for the forward-scattering system.

In a forward-scatter system reported by Fowlis et al. (1974), the original laser beam is split into two parallel beams by a beam splitter. To detect the magnitude as well as the direction of flow, it is necessary to introduce a known frequency offset, say f_R, in one of the laser beams. One method employed to introduce such a shift in frequency is to rotate a radial diffraction grating in one of the beams. Based on the rate of rotation and the total number of lines on the grating, the frequency of the diffracted light is shifted by a constant multiple of the original frequency, depending on which diffraction order is observed.

Another method employed to generate a known frequency offset is the introduction of a Bragg cell into one of the two parallel beams. A lens then focuses the two parallel beams, one of original frequency and the other of an artificially introduced frequency offset, to a common point within the flow field. The point of intersection (known as the *crossover point*) of these two beams defines the location of the scattering volume. This location is the position of water flow measurement. This geometry produces interference fringes at the common volume of intersection of the two beams. The volume of intersection of the beams, often termed the *probe volume* or *fringe region*, is related to the spatial resolution. The particles in the flow field, passing through the crossover volume, scatter light from both the incident beams. When a single scattering particle passes through a sinusoidal optical interference pattern, the intensity of the scattered light varies sinusoidally in time with a frequency proportional to the velocity of the particle. When more than one particle passes the interference pattern at the same time, phase shifts will occur between scattered signals corresponding to each particle. When the signal is large, the multiparticle phase shifts are of no concern (Stachnik and Mayo, 1977). Because there is an angle between the two incident beams, the scattered light from each beam is shifted in frequency (Doppler shift) by a different amount. The two forward-scattered light beams are then directed to a second collecting lens, which focuses the scattered light onto a photo detector. This arrangement is known as the *dual-scatter* or *fringe system*. The light beams scattered by the waterborne particles are mixed (i.e., heterodyned) on a photo-detector surface, resulting in the generation of beat frequencies. Because the voltage output from the photo detector is the

superposition of a large number of pulses scattered from particles of different sizes and spatial separation, this signal will have a randomly varying amplitude and a randomly fluctuating phase (Greated and Durrani, 1971), the front-end electronic circuitry separates out, usually by a "frequency-domain" method, the difference frequency ($f_R \pm f_D$), where f_D is the water flow-induced Doppler shift. The sign of f_D depends on the flow direction. In the dual-scatter geometry, the flow velocity is related to f_D by the formula

$$f_D = \frac{2\mu v \sin\theta}{\lambda} \qquad (2.3)$$

where μ is the refractive index of the fluid, λ is the wavelength of the laser radiation in vacuum, θ is half the angle between the two converging incident beams, and v is the component of the flow perpendicular to the plane containing the optic axis and the two beams of the system. In practical designs (Fowlis et al., 1974), this angle is less than $6°$. Note that if an additional frequency offset f_R were not introduced into one of the incident beams, it would have been impossible to determine the sign of f_D, and then the system would have been sensitive only to the flow speed and not its direction.

A single set of incident beams can sense only one component of the water flow. Rotation of the beam splitter by $90°$ about the optic axis, or having a second beam-splitter mounted with its optic axis parallel to the original system but rotated by $90°$ with respect to it, would make the combined system sensitive to the two mutually orthogonal flow components. The former method would involve temporal separation of the measurement of the two flow components, whereas the latter method would involve spatial separation. For most ocean flow measurements such minor separations would not contribute to any significant errors. Simultaneous measurements of the two orthogonal flow components at the same position in space would require a multicomponent laser Doppler system.

An ingenious approach employed by Agrawal and Belting (1988) to obtain two separate velocity components using a forward-scatter system is the use of three beams from a single laser, but with widely differing offsets (Figure 2.12). The photocurrent from the detector is then amplified and mixed with local oscillators, the frequencies of which are very near the laser offset frequencies and bandpass-filtered to separate the two velocity axes.

In the backscatter systems, laser beams of different frequency offsets derived from the same source are focused to the point of measurement through a glass window. The backscattered radiation from the crossover volume is then steered to a photo detector, followed by signal processing. An advantage of the backscattering system is that, by a process of optical zooming, water current velocity profiles can be remotely measured (Agrawal and Belting,

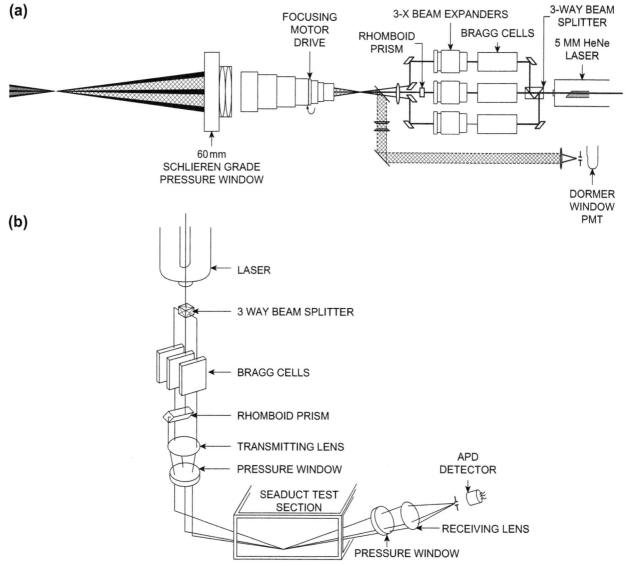

FIGURE 2.12   (a) Forward and (b) backward scatter system of Laser Doppler Velocimeter (LDV). *(Source: Agrawal and Belting, 1988.)*

1988). Optical zooming of the point of measurement along the optic axis is achieved by turning the incident beam-focusing lens with a stepping motor. This arrangement used in backscatter geometry produces range-dependent probe-volume geometry.

A noteworthy merit of laser Doppler velocimeters is that they can achieve extremely small sensing volumes (often less than one cubic millimeter) and can remotely (of the order of a meter) measure flow velocities without themselves disturbing the flow field. They also possess inherently fine accuracy and stability and do not require recalibration. They have extremely low flow speed thresholds and their responses are intrinsically linear. Flow speeds of 0.1 cm/s or less can be accurately measured. Their spatial and temporal resolutions are good and are capable of high sampling rates. The capability of the laser

Doppler technique to detect slow flows suggests its suitability in the designs of vertical-flow-measuring current meters. These characteristics arise from the fact that the short wavelength of coherent laser radiation allows it to be focused to very small sensing volumes and the high frequency-stability of laser sources allow optically interfering beams to form stable interference patterns in space and in time (Stachnik, 1977).

A disadvantage of the LDCMs is that, although they have no moving parts external to them, many practical designs need to have internal moving parts. Rotations of the beam splitter and the radial grating, optical zooming, and so forth are some of the areas where the use of a motor becomes inevitable. Furthermore, a relatively high power is required to operate the laser, Bragg cell, photo multiplier, and the like. As a result of these logistic limitations, laser Doppler flow velocimeters are now being used only for specialized applications such as study of sediment transport in the deep sea and turbulent velocity fields closer to the sea bottom, where other types of current meters may not be suitable.

2.2.4.8. Acoustic Doppler Current Meter

In the initial designs of Eulerian style acoustic Doppler current meters, a bistatic configuration (i.e., the transmitter and the receiver are located at different points) has been employed. These transducers were inclined to each other so that the volume of intersection of the transmitted and the received beams defined the scattering volume. Depending on the spatial separation and the angle between the transducers, the Doppler-shifted echoes were received from a reverberation volume located at a known distance from the transceivers (Koczy et al., 1962). An advantage of the acoustic Doppler measurement technique over conventional ocean current measuring devices was that the geometry could be set to receive the backscattered signals from a region sufficiently away from the body of the meter so that the velocity of flow could be nonintrusively detected from the undisturbed flow field. The basic accuracy of measurements is determined by the stability of transmission frequency, precision of mechanical alignment of the transducers, and knowledge of the ambient velocity of sound in water. The frequency used for transmission was in the range of 1–10 MHz.

In situations in which a vane oriented the instrument into the direction of the flow (Figure 2.13), only a single transceiver set has been used (Kronengold and Vlasak, 1965). In this current meter, a 10-MHz transmitter projected a narrow beam of acoustic energy into the seawater. A similar transducer, which is inclined to the transmitter, received the backscattered signal from a location 10 inches away, where the transmitter and the receiver beam pattern intersected, thus providing measurements from a 1 cm^3 volume in the flow field. Because a large vane fixed to the instrument oriented the transceivers into the flow, the current meter measured flows away from its wake field.

A geometry that does not require a vane to detect the direction of flow is the use of two pairs of transceivers. In a design developed by EDO Western Corporation, two sets of mutually orthogonal transceiver systems (i.e.,

transducers mounted 90° apart in the azimuth) transmitted narrow-beam, high-frequency continuous tones. Two other transducers below each transmitter received the back-scattered signals from a region 0.5 to 1 meter ahead of the respective transmitter. The transmitting and the receiving narrow-beam patterns of each pair of the transceivers were inclined on a vertical plane at an angle to each other so that the "volume" of flow measurement was defined by the intersection of the transmitting and the receiving beams. An advantage of this system is that it can measure two orthogonal components of the flow field, which is assumed to be uniform in the vicinity of the instrument. Because two orthogonal components of the mean horizontal flow field are measured, there is no necessity for a direction-orienting tailfin. A disadvantage of this configuration, however, is that the common axis of each transceiver pair lies in the same horizontal plane, and there exists a probability for any one of the transceiver pairs to sense distorted flows from the wake field of the current meter.

The drawback associated with contamination of current measurements by water current flows from the wake field of the current meter was circumvented in later designs. In a bistatic Doppler sonar system of a tripod shape reported by Wiseman et al. (1972), an upward-looking transmitter positioned at the axis of symmetry of the sonar platform transmitted a narrow-beam ultrasonic sinusoidal signal (approximately 10 MHz). Three receiving transducers, located in a horizontal plane, were symmetrical about the transmitter. Each of these receiving transducers, making an angle of 45° to the transmitter beam, detected the back-scattered acoustic waves. Because the sensing volume was located sufficiently above the instrument, the measurements were free of contamination from the disturbances due to the flow-induced wakes, irrespective of the direction of flow. Because the conventional north and east components of the flow field are derived from measurements at the same sensing volume, errors normally introduced in other configurations as a result of the assumption of the uniformity of the flow field over a sufficiently large area in the vicinity of the current meter are nonexistent. Additionally, this configuration permitted remote measurements of true vector averages of the three-dimensional velocity.

For the geometries mentioned, continuous transmission is possible because the transmitter and the receiver are separate. In principle, precision of the current flow velocity estimates is higher for continuous transmission than pulse transmission, but the range resolution is very poor (Pinkel, 1980). For an Eulerian-style current meter, poor range resolution is not a constraint because measurements are made from a given fixed range. A disadvantage with continuous transmission is electrical and acoustical crosstalk between the transmitter and the receiver. This problem can be reduced by ensuring that the receiving transducer does not lie in the path of the side lobes of the transmitter.

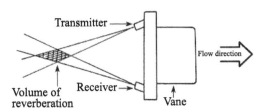

FIGURE 2.13 Schematic of an acoustic Doppler current meter where a vane oriented the instrument into the direction of the flow. *(Source: In part from Kronengold and Vlasak, 1965.)*

Another solution to this problem is the use of parametric transducers, which have suppressed side lobes and possess high directivity from a relatively small aperture (Konrad and Moffett, 1977). Insertion of an acoustically opaque medium between the transmitter and the receiver may be another method to prevent crosstalk between the transceivers.

In earlier designs, Doppler shift of the received signals was estimated using conventional electronic hardware-based frequency measurement methods such as detection of zero crossings of received pulse streams or using the frequency-locked-loop method. Although the Doppler method permits flow measurements in a nonintrusive manner, for a long time Doppler current meters suffered from severe signal dropout problems arising from the inadequacy of Doppler shift estimation techniques. As just mentioned, a common difficulty associated with most of the Doppler systems in the past was with regard to efficient extraction of the Doppler-shift information. Because the volume reverberation effect is statistical in nature with Gaussian distributions, extraction of Doppler-shift information using analytical methods introduced errors. However, these problems are now adequately handled using advanced signal-processing techniques, thereby enhancing the accuracy of measurements. Such modern acoustic Doppler current meters are now commercially available from Aanderaa Instruments.

2.3. SEAFLOOR BOUNDARY LAYER CURRENT MEASUREMENTS

Ocean currents and waves exert frictional forces on the seafloor by transport of momentum through a boundary layer. The flow in this region scales with the velocity shear, seafloor roughness, unsteadiness in the flow, and stratification of the fluid. It is turbulent, three-dimensional, and varies principally with distance above the seafloor (e.g., see Nowell, 1983; Grant and Madsen, 1986). Measurement of stress and turbulent-kinetic-energy (TKE) generation and dissipation in turbulent boundary layers is fundamental to understanding how the ocean works. Bottom boundary-layer stress is very important in sediment erosion, transport, and deposition. Understanding TKE production and dissipation is important in modeling the mixing of oxygen, heat, waste, and nutrients in the ocean. Precise vector flow measurements in the inner region of the bottom boundary layer are required to determine stress and to describe the momentum field during erosional and depositional periods. A vertically distributed array of current meters is necessary to estimate the stress profile to reveal the inner boundary layer structure. Because the water current measured by any sensor is a weighted average over its volume, miniature sensors are required to make observations of turbulent current fields closer to the deep seafloor, where the flow structure tends to be spatially inhomogeneous.

The water flow speed within the boundary layer, under certain conditions, increases logarithmically with distance from the bottom. The slope and intercept of the logarithmic profile provides two parameters, friction velocity and roughness length, which are needed to characterize the flow. More generally, the flow can be characterized by the turbulent kinetic energy and the Reynolds stress as a function of distance from the seafloor. *Reynolds stress* is the turbulent exchange of momentum in a frictional flow that carries stress from the interior flow to the boundary. It is determined by correlating the fluctuations in vertical velocity with the fluctuations in downstream velocity. It is the need to resolve small advected eddies contributing to the turbulent kinetic energy and the Reynolds stress that determines the sensor size and defines the sampling rate.

The mean velocity, $\overline{U}$, at 1 meter above the seafloor, is a valuable parameter because of its requirement in the computation of shear stress, τ, which is expressed as:

$$\tau = C_d \times \rho \times \overline{U}^2 \qquad (2.4)$$

where C_d is a drag coefficient and ρ is the fluid density. The drag coefficient, when waves are not present, is typically between $2 - 4 * 10^{-3}$ (Sternberg, 1968). In a tidally driven shallow flow, ignoring wind stress, one expects the stress to vary linearly from the seafloor to zero at the top surface, as given by the expression (Thwaites and Williams, 2001):

$$\tau(z) = \tau_b \left(1 - \frac{z}{h}\right) \qquad (2.5)$$

In this expression, $\tau(z)$ is the shear stress at a height z above the seafloor, τ_b is the seafloor shear stress, and h is the total water depth. The log-fit stress estimate is made by fitting mean velocities at the different sensor heights above the bottom to a logarithmic layer model, expressed as (Monin and Yaglom, 1971):

$$\frac{\overline{u}}{u_*} = \frac{1}{k} \ln\left(\frac{z-d}{h_o}\right) + B \qquad (2.6)$$

In this expression, $\overline{u}$ is the mean velocity at height z; u_* is the friction velocity, defined as the square root of the shear stress divided by density; k is von Karman's constant, taken to be 0.4; d is a displacement height; h_o is the roughness scale, and B is an empirical function of the roughness Reynolds number. Log-fit stress estimates require that equilibrium exists and the boundary layer flow is fully developed over the sensors. The log-fit stress estimate further requires currents from at least three heights to compute the stress and displacement height.

To study seafloor boundary layer flows, vertical arrays of current meters have been used either as mean current sensors (Weatherly and Wimbush, 1980) or as stress sensors

(Heathershaw, 1979). For some measurements where the flow directions cannot be known in advance, the sensor must be omnidirectional and its supports must not shed appreciable wakes into the measurement volume (Wyngaard et al., 1982). At the same time, the velocity fluctuations responsible for the turbulence signal may be small, so the structure supporting the sensor must be stiff and not vibrate, lest it contaminate the measurements. A tripod-type platform provides such stiffness with minimum flow disturbance. Instrument pressure cases and buoyancy modules on the measurement platform are severe flow disturbers, but these can be confined to a certain height level where measurements can be sacrificed.

2.3.1. Mechanical Devices

Attempts to measure currents near the seafloor in the open ocean have employed variations of three basic techniques: (1) current meter lowered to the bottom with wire from a ship (Pratt, 1963); (2) current meter suspended from an anchored buoy (Richardson et al., 1963); and (3) current meter attached to submersible bodies (LaFond, 1962). Currents have also been inferred from bottom photographs of ripple marks on the bottom (Hurley and Fink, 1963).

Knauss (1965) described a device that, when launched, sinks to the bottom, remains there for a predetermined length of time, and then surfaces. The instrument system consists of five components: current meter, release mechanism, weights, floatation device, and the transmitter. In air, the system weighs ~ 217 kg. The current meter is a self-recording instrument, recording on photographic film. The release mechanism can be set to go off in any 20-minute interval up to 140 days. At a predetermined time, an explosive device releases the line attached to the anchor and the gear floats to the surface.

A purely mechanical current meter used exclusively for bottom-current measurements is the deflecting wand current meter (Niskin, 1965). This device operates on the principle that a positively buoyant wand tethered at one end will assume an angle of inclination and a direction that are functions of the water current speed and direction, respectively. To make seafloor current measurements, the device, with an anchored weight, is dropped down to descend to the seafloor, where it preserves a single recording of the current. Upon dissolution of a soluble link, the instrument is automatically released from the anchor to float to the sea surface, where it is recovered and its recording noted. If further measurements are desired, another soluble link is inserted, an anchor weight attached, and the operation is repeated. This current meter could record even abyssal currents.

Another purely mechanical device for measuring very weak currents near the seafloor and designed to mount on a tripod is the Nansen pendulum current meter. In this current meter, a light pendulum freely swung above a slightly concave disk, which was carried by a magnet and covered by graduated waxed paper. At short intervals of time, a clock-work lowered the pendulum and a fine stylus attached to its bottom marked the paper, indicating the speed and direction of the current.

Since the early 1970s, bottom boundary layer and sediment transport experiments have been conducted with rigid bottom-mounted frames, which are arrayed with a variety of instruments. Experiments have obtained mean water current flow velocity data with Savonius rotors (Sternberg et al., 1973; Butman and Folger, 1979; Weatherly and Wimbush, 1980). Although the rotors have low power requirements and are durable, their usefulness is limited by their inability to measure the smaller time and length scales of the fluctuating turbulence components.

Direct measurement of turbulent velocity fluctuations in sand-transporting flows were made with the rugged impellor clusters of Smith and McLean (1977). Three small ducted rotors allowed measurements of the three components of the velocity vector from which direct measures of the fluctuating Reynolds stress, $u'w'$ (where u' is the fluctuating part of the along current velocity and w' is the fluctuating part of the vertical velocity) and three-dimensional spectra could be obtained. Thus, no dependence on the velocity profile being logarithmic was required. However, these sensors had to be oriented into the flows and were thus unsuitable for long deployments in reversing flows.

2.3.2. Nonmechanical Devices

Nonmechanical sensors used for water current flow measurements close to the seafloor include electromagnetic (EM), acoustic travel time difference (ATT), acoustic Doppler (AD), laser Doppler (LD), thermistors, thin-film probes, and quartz crystals. In addition to the sensors just mentioned, photographic techniques have also been employed for detection of currents at the deep-sea boundary layer (Sternberg, 1969) and for measurement of currents at the tip of a core (Ewing et al., 1967). Heated thermistor probes have been successfully employed for studies of boundary layer turbulence at small heights above the deep ocean bottom (Chriss et al., 1982). Their main advantages are small size and sturdiness. Nevertheless, they are not intrinsically fast devices and have nonlinear R-T characteristics. Doppler current meters were also used for study of sediment transport in the deep sea and turbulent velocity fields closer to the seafloor, where conventional current meters may not be suitable.

Thorpe et al. (1973), Heathershaw (1979), and Swift et al. (1979) used two-axis electromagnetic current meters to measure Reynolds stress, but these instruments had to be oriented into the flow via a fin on the supporting frame. All of these instruments have difficulty measuring water current flows in which the mean velocities are low relative to the fluctuations occurring in boundary layers. Cacchione and Drake (1979) added four electromagnetic current meters in a vertical array to measure two dimensions of the turbulence velocity vector. The finer spatial resolution, absence of speed threshold, and omnidirectional response allowed average current profiles to be measured even under wave reversals on the shelf. Most of these instruments depend on the velocity profile technique to estimate stress, and this requires a logarithmic velocity profile over the vertical region covered by the sensors.

2.3.2.1. BASS and MAVS

Seafloor boundary layer current measurement technology saw great advances under the able laboratory and field work carried out by Sandy Williams and co-researchers at Woods Hole. He showed keen interest in this topic because optical microstructure observations in 1972 and 1973 showed close-spaced vertical bands on thermo-haline interfaces. These bands are the signature of salt fingers. But there were also jumbles of curved lines in other regions that he presumed were the shadowgraphs of a stirring event, mechanical folding, or rolling of an interface and the cascade to smaller scale of the turbulence, as is expected in 3-D turbulence. He wanted to see the velocity structure at these irregular shadowgraph images, but there were no current meters suitable for this task. Trygve Gytre from Bergen, Norway, was visiting Dr. Williams' lab at Woods Hole for several months, wanting to add the new Neil Brown CTD to an acoustic shear meter he had developed and that Dr. Williams was building for his own use, and they agreed to help each other—Dr. Williams to help Gytre with the CTD and Gytre to help Dr. Williams with the acoustic shear meter. This was in 1973.

The shear meter worked fine for the intended purposes and obtained an excellent shear profile. But a student of Williams, John Tochko, queried whether they could adapt the acoustic shear meter to benthic boundary layer studies where geologist Charley Hollister had seen photographic evidence that there was active sediment transport in the deep sea. For this purpose, the two horizontal acoustic paths of the shear meter had to be augmented with additional paths to obtain the vertical component of flow so that the Reynolds stress could be measured.

Because Reynolds stress is determined by correlating the fluctuations in vertical velocity with the fluctuations in downstream velocity, a 3-D current sensor was required. The acoustic shear meter had a sensitivity of 1 mm/s in velocity fluctuations, so it seemed to have the capability. And they built the first Benthic Acoustic Stress Sensor (BASS). The BASS is a novel deep-sea tripod-supported ATT current meter array instrument that was specifically designed to study bottom turbulence. The BASS was expected to meet the necessary performance criteria and at the same time be deployable in a wide range of oceanic and near-shore environments. The purpose of this vertical array of instruments was to measure acoustically the average water flow along each of four axes in 15 cm diameter volumes at six heights above the bottom (Williams and Tochko, 1977). This array was expected to avoid the problems of flow disturbance, flow alignment, limited sensitivity, and unknown measurement volume common to other boundary layer flow sensors.

It is interesting to ponder the development cycle of BASS, in which teamwork was the key. Sandy Williams had electronic help from a Woods Hole engineer, Richard Koehler. He addressed the now-critical problem of zero point drift. Early in his experiments to improve the shear meter to BASS, he became tired of getting a 300-V shock when he inadvertently touched the part of the circuit with the high voltage. First he marked this part of the circuit with red and yellow striped tape, but next he revolted entirely against using high voltage and trying to excite a spike to achieve high-amplitude acoustic pulse transmission. Replacing the single switched voltage with a burst of low-voltage square waves tuned to the resonant frequency of the piezoceramic transducers produced the same steep voltage excursion at the receiver transducer as the single spike had in the shear meter. The steep slope was on the 14th negative going zero crossing. Because the excitation waveform was a simple low-voltage square wave at 1.8 MHz, standard digital circuits could be used and the danger of shock was a thing of the past. Williams was satisfied that the added complexity was worth it. But drift of the "zero" was still present. The electronic drift was eliminated with further modifications in the circuitry. This was confirmed in tests done in a bucket of still water and later in cast gelatin without convection currents.

Williams was fortunate to have eager geologists wanting to know what the deep-sea currents might be, and a few fluid dynamics people also showed interest in examining whether a one-dimensional boundary layer model was appropriate for flow in the deep sea over simple flat topography. Because of this interest and a concern of U.S. Navy about acoustic detection of Soviet submarines, Williams received support from the National Science Foundation (NSF) for the first and Office of Naval Research (ONR) for the second concern, and he was able to continue design and testing and eventually deployed his BASS instrument (see Section 9.2 for details). About that time, a valuable colleague, William Grant, joined the Ocean Engineering Department at Woods Hole from MIT, where he had studied the problem of wave-current interaction. He wanted to get his hands on the BASS to test his models in the coastal zone. In the early history of the BASS, it suffered from "zero-point" uncertainty, and this needed to be solved. The issue was that the BASS transducers were connected to the housing with the electronics by 6-meter coaxial cables, and pressure caused the capacitance of the cables to change. With appropriate correction to this problem, the BASS turned out to be the gold standard for boundary layer current measurements against which other current meters might be compared.

On the continental shelf and in estuaries, the BASS must meet additional requirements. Here the logarithmic velocity profile method often fails because the wave boundary layer is a major influence, whereas at other times stratification or internal wave excitation is dominant. In these situations, the ability to measure turbulent kinetic energy at a series of sensor heights and to compare the stress estimates derived from them with the velocity profile is valuable. BASS is unique in permitting turbulent quantities to be measured in reversing flows of a wave field with low enough disturbance to estimate stress. Here, fast sampling (5 Hz) is necessary. The large datasets are accommodated in shallow deployments by telemetering the data by radio from a moored transmitter near the tripod.

BASS has been proven to be a valuable instrument for turbulent flow measurements. It has been used in near-shore studies of large swell and wind-wave-current interaction off the coast of California in the CODE experiment (Grant et al., 1984). Measurements of velocity profiles, turbulent energy, and velocity spectra resolving the inertial range and bottom wave influence were made. Adaptations to deploy the BASS for long-term boundary layer monitoring in the HEBBLE site enabled quantitative evaluation of the effect of benthic storms (Gross et al., 1986; Grant et al., 1985). These two- and six-month deployments allowed changes in bottom roughness over timescales of hours to weeks to be observed using logarithmic velocity profiles and direct measurements of Reynolds stress and kinetic energy. The BASS's unique ability to measure across timescales from half a second to several months with velocity resolution of 0.03 cm/s has made possible studies of the benthic boundary layer, which was unattainable by any other instrument in use.

It may be worthwhile to note that the ultimate success came after several difficulties suffered by the instrument designers and the technologists. According to Williams et al. (1987), transducer failures were common in the early days, sometimes affecting as many as half the acoustic axes. Initially the problems were broken wires, then intermittent connectors, misaligned transducers, and delaminated ceramic elements. These faults were eliminated by changing materials, changing construction techniques, modifying designs, and inspecting subassemblies in the molding process. In 1984 over 500,000 transducer hours were logged and there were no failures (Dunn, 1984). An additional 280,000 transducer hours were logged in 1985–86 on these same sensors, still with no failures. As of

1987, total immersion was over one year for 80 of these transducers.

The full value of ultrasonic acoustic current meters for measurements of weak currents and turbulence in the bottom boundary layer can be achieved only if flow obstruction by the body of the current meter is arrested. This is easier said than done. As we have seen already, in the history of acoustic current meter development much effort has indeed gone in this direction. Unfortunately, although the novel ATT current meter known as the BASS has been designed to achieve improved performance to meet the needs of near-bed benthic boundary layer studies, this device required the use of external electrical cables, which are thicker than the structure cage rods.

In continuation of intense efforts toward achieving better performance, Thwaites and Williams III (1996) started developing the derivative of the BASS, known as the Modular Acoustic Velocity Sensor (MAVS) to achieve better performance by removing some of the obstacles found in the BASS. Featured was a new sensor that had no exposed coaxial cables to compress and change the "zero point"; it also had faired rings supporting the acoustic transducers, to reduce the vertical cosine response error. This sensor is shown in Figure 2.14. (By chance, the sensor shown in this figure, a photograph taken on February 21, 2013, was coated with a silicone coating called ClearSignal and had spent six months about 1 meter beneath the surface in a local salt water pond over the summer. Organisms were easily washed off under the tap with mild rubbing between thumb and forefinger.) At the end of the MAVS development, it was offered to a company called General Oceanics for manufacture and marketing. The issues that arose during development were the shape of the sensor rings, their molding (injection molding of PVC and later ABS

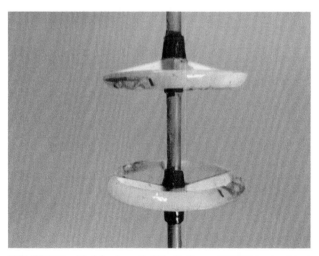

FIGURE 2.14   Modular Acoustic Velocity Sensor (MAVS) having faired rings to reduce the wake along the acoustic axes (*Source: By permission of Dr. Albert J. Williams 3rd.*)

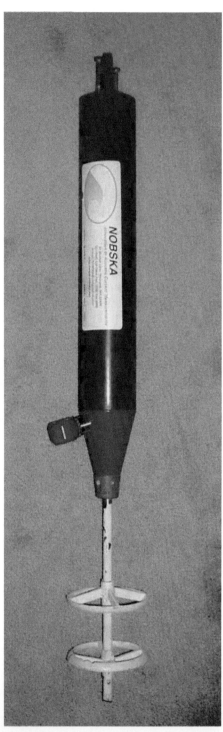

FIGURE 2.15 MAVS marketed by Nobska Development, Inc. *(Source: Kobe Tutorial; MAVS Current Measurement, Nobska Development, Inc., Albert J. Williams 3rd, and Archie Todd Morrison III. By permission of Dr. Albert J. Williams 3rd.)*

plastic), the mounting and electrical wiring to the transducers, the overcoating and sealing of the transducers and wire, and finally the pressure compensation of the support tube with its pressure-resisting through-hull conductors.

MAVS is also a three-axis ATT current meter that measures differential-acoustic-travel time in a small measurement volume. Each acoustic axis on MAVS is oblique, with the axes 45° to the plane of the rings and passing from one ring to the other. The paths are spaced 90° in azimuth around the rings, one path going obliquely up, the next obliquely down, the third obliquely up, and the fourth obliquely down. Manufacturing passed from General Oceanics to Nobska Development, Inc., and it went well. (A MAVS marketed by Nobska is shown in Figure 2.15.) MAVS was deployed at a 2,300-m depth at the Juan de Fuca hydrothermal vent field and successfully measured the flow characteristics there (a color picture of this deployment is available at WHOI archives). It is also deployed in shallow waters including coral reef sites. Figure 2.16 is the picture of MAVS mounted on a lander amidst the gorgeous coral reefs in the Florida Keys.

2.4. VERTICAL PROFILING OF HORIZONTAL CURRENTS

In the early years of oceanographic studies, linear arrays of multiple current meters deployed at different depths, using mooring lines, were used for measurements of vertical profiles of horizontal currents. However, such measurements became unfeasible for high-resolution measurements over large depths, primarily because of the prohibitively large expenditure involved in terms of current meter cost.

The just mentioned difficulty forced oceanographic technologists to develop alternative methods for such measurements. Taking a cue from the successful use of meteorological balloons for wind-profile measurements, oceanographers developed dropsondes for current profile measurements in the ocean. Freely falling dropsondes have been extensively used to determine vertical profiles of horizontal currents in the ocean. The basic concept underlying the dropsonde technique is that any submerged body will be accelerated by the horizontal drag forces arising from differences between the velocity of the local fluid and that of the body. If the equilibration to the flow is rapid enough, successive determinations of the position of the body can be used to estimate the velocity of the field. Accurate navigation is the key to the system. The technologies used in the early years for determination of vertical profiles of horizontal currents in the ocean were essentially variants of those used in meteorology for tracking meteorological balloons, which provided a large fraction of the basic wind profile data.

In oceanography, the technique of determination of vertical profiles of horizontal currents using freely falling/rising dropsondes was pioneered by Richardson and Schmitz (1965) in the Florida Straits. They used Decca

FIGURE 2.16 Picture of MAVS current meter mounted on a lander amidst the gorgeous coral reefs on a Florida Key. The picture is taken by an Olympus Digital Camera and creation date is 3/5/2012 *(Source: Courtesy of Prof. Wade R. McGillis.)*

Hi-Fix navigation with land-based stations. The limited range of Hi-Fix navigation and insufficient accuracy of the Loran-C or Omega navigation have restricted the Richardson-Schmitz technique to coastal situations. Largely because of such limitations, a few researchers were motivated to use acoustic navigation techniques to track freely falling or rising probes. Rossby (1974) and Rossby and Sanford (1976) developed acoustic dropsondes using underwater acoustic technique. Both instruments are based on pulsed navigation, in which the travel time is determined between a fixed transmitter-receiver (transceiver) and the freely falling instrument (see Chapter 10 for details). Subsequently, other techniques such as continuous wave phase tracking (Spindel et al., 1976) and correlation sonar became available. These more advanced technologies could provide more accurate navigation and could be utilized for dropsonde systems.

The marvelous success achieved over the years in the development of various kinds of ADPs brought about a drastic change in the history of current profile measurements from diverse types of ocean water bodies (see Chapter 11 for details). The most important feature of the ADP is its ability to measure current profiles, which are divided into uniform segments called *depth cells* or *bins*. ADPs significantly altered the face of current profile measurements, since a single instrument is now available to remotely profile the speed and direction of water currents throughout the water column (horizontal or vertical), as opposed to capturing only an isolated point measurement.

In essence, a single ADP can now acquire water current measurements that are obtainable from a full string of single-point current meters. Bottom-mounted ADPs as part of "light" or "heavy" trawl-resistant deployment configurations are suitable for measurements of currents in the coastal regions of vigorous fishing activity, where conventional

current meter moorings have difficulty surviving. ADPs have been designed for deployments on stationary platforms (e.g., offshore structures and moorings) and mobile platforms (e.g., ships) of various kinds, merging trends in the field of sensor and platform development. It has been found that lowered ADCPs (L-ADCPs) can be used as deep as 5,000 m, to make vertical profile measurements at very low marginal cost in conjunction with CTD profiles, and can provide scientifically useful measurements of currents such as inertial oscillations and the equatorial deep jets with vertical wavelengths less than about 1,000 m.

In the early 1980s, RD Instruments (RDI) commercialized the ADCP for remotely measuring 3-D water current vectors throughout the water column. Subsequently, several other agencies began to manufacture ADPs, each manufacturer vying to outshine the other in terms of miniaturization, improved quality of data, user-friendly features, and so on. With the passage of time, ADPs began to be deployed by several users across the world. The present scenario is that in coastal and continental shelf waters, the ADPs have largely replaced mooring strings of current meters; moving boat measurement of river discharge using the ADP is replacing the traditional method; and using ADPs from coastal crafts has provided a previously unavailable survey capability for measuring circulation patterns in inshore and coastal waters.

Closer examination of high-resolution, high-frequency ADP current measurements in wave-laden waters led to new discoveries. It was noticed that current profile measurements from ADPs deployed in shallow wave-laden waters or located closer to the surface (e.g., on an offshore pier) in deep waters differed in some ways from conventional measurements. To those pessimistic users who are wedded to the old technologies, who lacked acumen, and who have no nose for exploring the unknown, the observed peculiarities in ADP measurements would have been "instrument noise" or a result of "deployment error." There have been instances in which voluminous high-resolution shallow-water ADP measurements have been thrown to the winds, blaming them as having been contaminated by "instrument noise"! However, to those erudite technologists and devoted researchers who have a nose for mining the hidden treasures from the deep, the "noise" was indeed an invaluable signal.

Research initiated in the 1980s (e.g., Krogstad et al., 1988; Terray et al., 1990) and bin-wise and beam-wise analysis of ADP current measurements from flow-cum-wave tank upward-looking deployments by Appell et al. (1991) as well as continuation of research beyond the 1990s (e.g., Herbers et al., 1991; Zedel, 1994; Visbeck and Fischer, 1995; Terray et al., 1999; Pedersen and Lohrmann, 2004) showed that clearly, sea surface wave orbital velocities were also being measured, and the shallow-water ADP measurements are indeed a combination of currents, wave orbital velocities, and

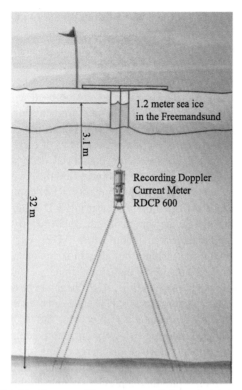

1.2 meter sea ice
in the Freemandsund

3.1 m

32 m

Recording Doppler
Current Meter
RDCP 600

FIGURE 2.17 Current profile measurements below polar ice sheets using an ADCP. *(Source: Aanderaa News Letter, B137, pp- 1–4, May 2004).*

turbulence. This realization ultimately culminated in the invention of the now-famous seminal piece of oceanographic remote-sensing device known as the directional wave-measuring ADP. Thus, apart from the well-established capability of ADPs for remote measurement of horizontal and vertical water current profiles, they can also be used for measuring sea surface wave height and direction from a single upward-looking multibeam ADP package deployed in shallow depths relative to the sea surface. With inclusion of the additional capability of ADPs for directional surface wave measurements since the late 1990s, the survey capability of ADPs has been doubled. Consequently, ADPs can now be used to measure both directional wave spectra and current profiles at the same time. The abilities to sample the flow field without disturbing it and, most important, to simultaneously measure the three components of the fluid velocity are important features of the multitransceiver acoustic Doppler current meters. Figure 2.17 demonstrates the capability of ADP to make current profile measurements below polar ice sheets.

REFERENCES

Agrawal, Y.C., Belting, C.J., 1988. Laser velocimetry for benthic sediment transport. Deep-Sea Research 35 (6), 1047–1067.

Appell, G.F., Bass, P.D., Metcalf, M.A., 1991. Acoustic Doppler Current Profiler performance in near surface and bottom boundaries. IEEE J. Ocean Eng. 16 (4), 390–396.

Barbera, M.L., Vogel, S., 1976. An inexpensive thermistor flow meter for aquatic biology. Limnology & Oceanography 21, 750–756.

Bjorkstedt, E., Roughgarden, J., 1997. Larval transport and coastal upwelling: An application of HF radar in ecological research. Oceanography 10, 64–67.

Boehnecke, G., 1949. German oceanographic work, *Transactions.* American Geophysical Union 29 (1), 59–68.

Brech, R., Bellhouse, B.J., Bellhouse, F.H., 1971. A directionally sensitive thin film velocity probe. J. Phys. E: Scientific Instrumen. 4, 464–465.

Broche, P., Crochet, J.C., de Maistre, J.C., Forget, P., 1987. VHF radar for ocean surface current and sea state remote sensing. Radio Sci. 22, 69–75.

Butman, B., Folger, D.W., 1979. An instrument system for long-term sediment transport studies on the continental shelf. J. Geophys. Res. 84, 1215–1220.

Cacchione, D.A., Drake, D.E., 1979. A new instrument to investigate sediment dynamics on continental shelves. Mar. Geol. 30, 299–312.

Catellani, A., Stacchicitti, R., Taroni, A., Canali, C., 1982. Performance and temperature stability of an air mass flow meter based on a self-heated thermistor. Sensors and Actuators 3 (1), 23–30.

Chew, F., Berberian, G.A., 1970. Some measurements of current by shallow drogues in the Florida current. Limnol. Oceanogr 15, 88–99.

Chiu, C.S., Miller, J.H., Lynch, J.F., 1994. Inverse technique for coastal acoustic tomography. In: Lee, D., Schultz, M.H. (Eds.), Theoretical and Computational Acoustics. World Scientific, 2, Singapore, pp. 917–931.

Chriss, T.M., Caldwell, D.R., 1982. Evidence for the influence of form drag on bottom boundary layer flow. J. Geophys. Res. 87, 4148–4154.

Clay, C.S., Medwin, H., 1977. Acoustical oceanography: Principles and applications. Wiley-Interscience.

Cochin, V., Mariette, V., Broche, P., Garello, R., 2006. Tidal current measurements using VHF radar and ADCP in the Normand Breton Gulf: Comparison of observations and numerical model. IEEE J. Ocean. Eng. 31 (4), 885–893.

Crombie, D.D., 1955. Doppler spectrum of sea echo at 13.56 Mc/s. Nature 175 (4459), 681–682.

Crombie, D.D., 1972. Resonant backscatter from the sea and its application to physical oceanography. Proc. IEEE Oceans '72 Conference, 172–179.

Dingwell, R.E., Weiskopf, F.B., 1981. A hot film anemometer for ocean turbulence measurements. Proc. Oceans '81 Conferene. 1, 468–472.

Dornhelm, R.B., 1977. Current measurements offshore Humboldt bay, California for proposed ocean outfall. Proc. Oceans '77 Conference. pp. 46C-1 to 46C-7.

Duchossois, G., Martin, P., 1995. ERS-1 and ERS-2 Tandem Operations. ESA Bull. 83, 54–60.

Duncan, C.P., 1965. Disadvantages of the Olson drift card and description of a newly designed card. J. Marine Res. 23 (3), 233–236.

Duncan, C.P., Schadlow, S.G., 1981. World surface currents from ships' drift observations. International Hydrographic Review LVIII, 101–111.

Dunn, C.V.R., 1984. A miniature urethane-molded acoustic transducer. Woods Hole Oceanogr. Inst. Tech. Rep. WHOI-84-30.

Dushaw, B.D., et al., 2009. A decade of acoustic thermometry in the North Pacific Ocean. J. Geophys. Res. 114, C07021.

Elisseeff, P., Schmidt, H., Johnson, M., Herold, D., Chapman, N.R., McDonald, M.M., 1999. Acoustic tomography of a coastal front in Haro Strait, British Columbia. J. Acoust. Soc. Amer. 106, 169–184.

Ewing, M., Hayes, D.E., Thorndike, E.M., 1967. Corehead camera for measurement of currents and core orientation. Deep-Sea Research 14, 233–253.

Ferretti, A., Prati, C., Rocca, F., 1999. Permanent scatterers in SAR interferometry. In: International GeoScience and Remote Sensing Symposium. Germany, Hamburg, 28 June–2 July, 19991–3.

Ferretti, A., Prati, C., Rocca, F., 2000. Nonlinear subsidence rate estimation using permanent scatterers in differential SAR interferometry. IEEE Trans. GeoSci. Remote Sensing 38 (5), 2202–2212.

Ferretti, A., Prati, C., Rocca, F., 2001. Permanent scatterers in SAR interferometry. IEEE Trans. GeoSci. Remote Sensing 39 (1), 8–20.

Flynn, T.L., Cook, D.C., 1978. Charting of outer continental shelf surface currents by aerial tracking of tracers. Proc. Oceans '78 IEEE Conference, 315–320.

Forstner, H., Rutzler, K., 1969. Two temperature-compensated thermistor current meters for use in marine ecology. Journal of Marine Research 27 (2), 263–271.

Fowlis, W.W., Thompson, J.D., Terry, W.E., 1974. A laser-Doppler velocimeter with ocean applications,. J. Marine Res. 32 (1), 93–102.

Gabriel, A.K., Goldstein, R.M., Zebaker, H.A., 1989. Mapping small elevation changes over large areas: differential radar interferometry. J. Geophys. Res. 94 (B7), 83–91.

Gonsalves, W.D., Brainard II, E.C., 1981. Geomagnetic Electrokinetograph (GEK): A technological advancement in long-term continuous monitoring of fluid flow velocity. Proc. Oceans '81 Conference 1, 233–238.

Gould, W.J., 2005. From Swallow floats to Argo: The development of neutrally buoyant floats. Deep-Sea Res. II 52, 529–543.

Graber, H.C., Limouzy-Paris, C.B., 1997. Transport patterns of tropical reef fish larvae by spin-off eddies in the Straits of Florida. Oceanography 10, 68–71.

Graham, L.C., 1974. Synthetic interferometer radar for topographic mapping. Proc. IEEE 62 (6), 763–768.

Grant, W.D., Williams III, A.J., Glenn, S.M., 1984. Bottom stress estimates and their prediction on the northern California Continental Shelf during CODE-I: the importance of wave-current interaction. J. Phys. Oceanogr. 14, 506–527.

Grant, W.D., Williams III, A.J., Gross, T.F., 1985. A description of the bottom boundary layer at the HEBBLE site: low-frequency forcing, bottom stress and temperature structure. Mar. Geol. 66, 219–241.

Grant, W.D., Madsen, O.S., 1986. The continental-shelf bottom boundary layer. Ann. Rev. Fluid Mech. 18, 265–305.

Greated, C., Durrani, T.S., 1971. Signal analysis for laser velocimeter measurements. J. Phys.E: Instrum. 4, 24–26.

Gross, T.F., Williams III, A.J., Grant, W.D., 1986. Long-term in situ calculations of kinetic energy and Reynolds stress in a deep-sea boundary layer. J. Geophys. Res. 91, 8461–8469.

Gurgel, K.W., Antonischki, G., Essen, H.H., Schlick, T., 1999. Wellen radar (WERA), a new ground-wave based HF radar for ocean remote sensing. Coastal Eng. 37 (3-4), 219–234.

Gurgel, K.W., Essen, H.H., Schirmer, F., 1986. CODAR in Germany: A status report valid November 1985. IEEE J. Ocean. Eng. 11 (2), 251–257.

Gytre, T., 1976. The use of a high sensitivity ultrasonic current meter in an oceanographic data acquisition system. The Radio and Electronic Engineer 46 (12), 617–623.

Gytre, T., 1980. Acoustic travel time current meters. In: Dobson, F., Hasse, L., Davis, R. (Eds.), Air-sea interaction; Instruments and methods, pp. 155–170.

Hardies, C.E., 1975. An advanced two-axis acoustic current meter. Proc. Offshore Technology Conference, II, 465–476.

Hasselmann, K., 1971. Determination of ocean wave spectra from Doppler radio return from the sea surface. Nature Physical Science 229, 16–17.

Heathershaw, A.D., 1979. The turbulent structure of the bottom boundary layer in a tidal channel. Geophys. J.R. Astron. Soc. 58, 395–430.

Herbers, T.H.C., Lowe, R.L., Guza, R.T., 1991. Field verification of acoustic Doppler surface gravity wave measurements. J. Geophys. Res. 96 (9), 17023–17035.

Heron, M.L., Dexter, P.E., McGann, B.T., 1985. Parameters of the air-sea interface by high-frequency ground-wave HF Doppler radar. Aust. J. Mar. Freshwater Res. 36, 655–670.

Heron, M.L., Helzel, T., Prytz, A., Kniephoff, M., Skirving, W.J., 2005. PortMap: A VHF ocean surface radar for high spatial resolution. Proc. IEEE OCEANS–Europe 1, 511–515.

Hickey, K., Khan, R.H., Walsh, J., 1995. Parametric estimation of ocean surface currents with HF radar. IEEE J. Ocean. Eng. 20 (2), 139–144.

Huang, W., Wu, S., Gill, E., Wen, B., Hou, J., 2002. HF radar wave and wind measurement over the Eastern China Sea. IEEE Trans. Geosci. Remote Sens. 40 (9), 1950–1955.

Hurley, R.J., Fink, L.K., 1963. Ripple marks show that countercurrent exists in Florida Straits. Science 139 (3555), 603–605.

Inman, D., Tait, R., Nordstrom, C., 1971. Mixing in the surf zone. J. Geophys. Res. 76, 3493–3514.

Irish, J.D., Nodland, W.E., 1978. Evaluation of metal film temperature and velocity sensors and the stability of a self-propelled research vehicle for making measurements of ocean turbulence. Proc. Oceans '78 Conference, 180–187.

Johannessen, J.A., Kudryavtsev, V., Akimov, D., Eldevik, T., Winther, N., Chapron, B., 2005. On radar imaging of current features: 2. Mesoscale eddy and current front detection. J. Geophys. Res. 110, C07017.

Johannessen, J.A., Kudryavtsev, V., Akimov, D., Eldevik, T., Winther, N., Chapron, B., 2005. On radar imaging of current features: 2. Mesoscale eddy and current front detection. J. Geophys. Res. 110, C07017.

Joseph, A., 2000. Applications of Doppler effect in navigation and oceanography. In: Encyclopedia of Microcomputers, vol. 25. Marcel Dekker, New York, NY, USA, 17–45.

Kaneko, A., Yamaguchi, K., Yamamoto, T., Gohda, N., Zheng, H., 2005. A coastal acoustic tomography experiment in Tokyo Bay. Acta Oceanologica Sinica 24 (1), 86–94.

Kawanisi, K., Watanabe, S., Kaneko, A., Abe, T., 2009. River acoustic tomography for continuous measurement of water discharge. 3rd International Conference and Exhibition on Underwater Acoustic Measurements: Technologies & Results, 613–620.

Knauss, J.A., 1965. A technique for measuring deep ocean currents close to the bottom with an unattached current meter, and some preliminary results,. J. Marine Res. 23 (3), 237–245.

Koczy, F.F., Kronengold, M., Loewenstein, J.M., 1962. A Doppler current meter. In: Marine Sciences Instrumentation, 2. Proc. Symposium on transducers for oceanic research, pp. 127–134.

Konrad, W.L., Moffett, M.B., 1977. Specialized drive waveforms for the parametric acoustic source. Proc. MTS-IEEE Oceans '77 Conference. pp. 10B-1 to 10B-4.

Krogstad, H.E., Gordon, R.L., Miller, M.C., 1988. High resolution directional wave spectra from horizontally mounted Acoustic Doppler Current Meters. J. Atmos. Oceanic. Technol. 5, 340–352.

Kronengold, M., Vlasak, W., 1965. A Doppler current meter. In: Marine Sciences Instrumentation, 3. Proc. Third National Marine Sciences Symposium, pp. 237–250.

LaFond, E.C., 1962. Deep current measurements with the bathyscaphe Trieste. Deep-Sea Res. 9 (2), 115–116.

Lawson Jr., K.D., Brown, N.L., Johnson, D.H., Mattey, R.A., 1976. A three-axis acoustic current meter for small scale turbulence. Instrument Society of America. pp. 501–508.

Legeckis, R., 1987. Satellite observations of a western boundary current in the Bay of Bengal. J. Geophys. Res. 92 (C1 2), 12974–12978.

Lester, R.A., 1961. High-accuracy, self-calibrating acoustic flow meters. In: Gaul, R.D., Ketchum, D.D., Shaw, T.T., Snodgrass, J.M. (Eds.), Marine Sciences Instrumentation, 1, pp. 200–204.

Lin, J., Kaneko, A., Gohda, N., Yamaguchi, K., 2005. Accurate imaging and prediction of Kanmon Strait tidal current structures by the coastal acoustic tomography data. Geophy. Res. Lett. 32, L14607.

Lobel, P.S., 2011. Transport of reef lizardfish larvae by an ocean eddy in Hawaiian waters. Dynamics of Atmospheres and Oceans 52, 11977 vol. 25, 130.

Lobel, P.S., Robinson, A.R., 1986. Transport and entrapment of fish larvae by ocean mesoscale eddies and currents in Hawaiian waters. Deep-Sea Res. 33, 483–500.

Lobel, P.S., Robinson, A.R., 1988. Larval fishes and zooplankton in a cyclonic eddy in Hawaiian waters. J. Plank. Res. 10 (6), 1209–1223.

Mangelsdorf Jr., P.C., 1961. The world's longest salt bridge. In: Gaul, R.D., Ketchum, D.D., Shaw, T.T., Snodgrass, J.M. (Eds.), Marine Sciences Instrumentation, Vol. 1. A publication of Instrument Society of America, pp. 173–185.

Massonnet, D., et al., 1993. The displacement field of the Landers Earthquake mapped by radar interferometry. Nature 364, 138–142.

Maury, M.F., 1855. The physical geography of the sea. Harper & Brothers, New York, NY, USA.

Merritt, S.R., Pattullo, J.G., Wyatt, B., 1969. Subsurface currents off the Oregon coast as measured by parachute drogues. Deep-Sea Res. 16, 449–461.

Monin, A.S., Yaglom, A.M., 1971. Statistical Fluid Mechanics. MIT Press, Cambridge, MA, USA. pp. 257–364.

Munk, W.H., Wunsch, C.A., 1979. Ocean acoustic tomography: A scheme for large scale monitoring. Deep-Sea Res. 123–161.

Munk, W., Baggeroer, A., 1994. The Heard Island Papers: A contribution to global acoustics. J. Acoustical Soc. of America 96 (4), 2327–2329.

Munk, W., Worcestor, P.F., Wunsch, C., 1995. Ocean acoustic tomography. Cambridge University Press, Cambridge, U.K. pp. 433–433.

Murray, S.P., 1975. Trajectories and speeds of wind-driven currents near the coast. J. Phys. Oceanogr. 5, 347–360.

Niskin, S.J., 1965. A low-cost bottom current velocity and direction recorder. Marine Sciences Instrumentation 3, 123–131.

Nowell, A.R.M., 1983. The benthic boundary layer and sediment transport. Rev. Geophys. Space Phys. 21, 1181–1192.

Offner, F.F., 1967. Electrodes. In: Electronics for biologists. McGraw-Hill, pp. 141–147.

Olson, F.C.W., 1951. A plastic envelope substitute for drift bottles,. J. Marine Res. 10 (2), 190–193.

Park, J.-H., Kaneko, A., 2000. Assimilation of coastal acoustic tomography data into a brotropic ocean model. Geophys. Res. Lett. 27 (20), 3373–3376.

Park, J.-H., Kaneko, A., 2001. Computer simulation of the coastal acoustic tomography by a two-dimensional vortex model. J. Oceangr. 57, 593–602.

Pedersen, T., Lohrmann, A., 2004. Possibilities and limitations of acoustic surface tracking. Proc. Oceans, 2004, Kobe, Japan.

Pickard, G.L., Emery, W.J., 1982. Instruments and Methods. In: Descriptive physical oceanography: An introduction, 77, vol. 25. Pergamon Press, 124.

Pinkel, R., 1980. Acoustic Doppler techniques. In: Dobson, F., Hasse, L., Davis, R. (Eds.), Air-sea interaction: Instruments and methods, pp. 171–199.

Prandle, D., Loch, S.G., Player, R.J., 1993. Tidal flow through the Straits of Dover. J. Phys. Oceanogr. 23 (1), 23–37.

Pratt, R.M., 1963. Bottom currents on the Blake Plateau. Deep-Sea Res. 10, 245–249.

Resch, F.J., Irish, J.D., 1972. Quartz crystals as multipurpose oceanographic sensors — II speed. Deep-Sea Research 19, 171–178.

Richardson, W.D., Stimson, P.B., Wilkins, C.H., 1963. Current measurements from moored buoys. Deep-Sea Res. 10, 369–388.

Richardson, W.S., Schmitz Jr., W.J., 1965. A technique for the direct measurement of transport with application to the Straits of Florida,. J. Marine Res. 16, 172–185.

Richardson, W.S., White Jr., H.J., Nemeth, L., 1972. A technique for the direct measurement of ocean current from aircraft. J. Marine Res. 30 (1), 259–268.

Richman, D., 1971. Three-dimensional azimuth-correcting mapping radar. United Technologies Corporation, USA.

Robbins, R.J., Morrison, G.K., 1981. Acoustic direct reading current meter. Proc. IEEE Oceans '81 Conference, 506–511.

Roberts, E.B., 1950. Roberts radio current meter mod. II operating manual. U.S. Coast and Geodetic Survey, Revised.

Robinson, I.S., 1985. Satellite Oceanography, An introduction for oceanographers and remote sensing scientists. Ellis Horwood Limited (a division of Wiley).

Rossby, H.T., 1974. Studies of the vertical structure of horizontal currents near Bermuda. J. Geophys. Res. 79, 1781–1791.

Rossby, H.T., Sanford, T.B., 1976. A study of velocity profiles through the main thermocline. J. Phys. Oceanogr. 6, 766–774.

Shay, L.K., Lentz, S.J., Graber, H.C., Haus, B.K., 1998. Current structure variations detected by high-frequency radar and vector-measuring current meters. J. Atmos. Oceanic Technol. 15, 237–256.

Shearman, E.D., Moorhead, M.D., 1988. PISCES: A coastal ground-wave radar for current, wind and wave mapping to 200 km ranges. Proc. Int. Geosci. Remote Sens. Symp. (IGARSS), 773–776.

Smith, J.D., McLean, S.R., 1977. Spatially averaged flow over a wavy surface. J. Geophys. Res. 82, 1735–1746.

Smith, R.C., Brown, O.B., Hoge, F.E., Baker, K.S., Evanas, R.H., Swift, R.N., Esaias, W.E., 1987. Multiplatform sampling (ship, aircraft, and satellite) of a Gulf Stream warm core ring. Applied Optics 26 (11), 2068–2081.

Spiesberger, J.L., Metzger, K., 1991. Basin-scale tomography: A new tool for studying weather and climate. J. Geophys. Res. 96, 4869–4889.

Spindel, R.C., Porter, R.P., Marquet, W.M., Durham, J.L., 1976. A high resolution pulse 2— Doppler underwater acoustic navigation system. IEEE J. Ocean Eng. 1, 6–13.

Stachnik, W.J., Mayo Jr., W.T., 1977. Optical velocimeters for use in sea water. Proc. Oceans '77 Conference. pp. 18A-1 to 18A-5.

Stander, G.H., Shannon, L.V., Campbell, J.A., 1969. Average velocities of some ocean currents as deduced from the recovery of plastic drift cards. J. Marine Res. 27 (3), 293–300.

Sternberg, R.W., 1968. Friction factors in tidal channels with differing bed roughness. Mar. Geology 6, 243–260.

Sternberg, R.W., 1969. Camera and dye-pulser system to measure bottom boundary-layer flow in the deep sea. Deep-Sea Research 16, 549–554.

Sternberg, R.W., Morrison, D.R., Trimble, J.A., 1973. An instrument system to measure near bottom conditions on the continental shelf. Mar. Geol. 15, 181–189.

Stommel, H., 1949. Horizontal diffusion due to oceanic turbulence. J. Mar. Res. 8, 199–225.

Suellentrop, F.J., Brown, A.E., Rule, E., 1961. An acoustic ocean-current meter. Marine Sciences Instrumentation 1, 190–204.

Sverdrup, H.U., Johnson, M.W., Fleming, R.H., 1942. The oceans: Their physics, chemistry, and general biology. Prentice Hall, New York, NY, USA.

Swallow, J.C., 1955. A neutral-buoyancy float for measuring deep current,. Deep-Sea Res. 3 (1), 74–81.

Swift, M.R., Reichard, R., Celikkol, B., 1979. Stress and tidal current in a well-mixed estuary. J. Hydraul. Div. Am. Soc. Civ. Eng. 105, 785–799.

Takeoka, H., Tanaka, Y., Ohno, Y., Hisaki, Y., Nadai, A., Kuroiwa, H., 1995. Observation of the Kyucho in the Bungo Channel by HF radar. J. Oceanogr. 51 (6), 699–711.

Teague, C.C., Vesecky, J.F., Hallock, Z.R., 2001. A comparison of multifrequency HF radar and ADCP measurements of near-surface currents during COPE-3. IEEE J. Ocean. Eng. 26 (3), 399–405.

Terray, E.A., Brumley, B.H., Strong, B., 1999. Measuring waves and currents with an upward-looking ADCP. In: Proc. IEEE Sixth Working Conference on Current Measurement. Institute of Electrical and Electronic Engineers, pp. 66–71.

Terray, E.A., Krogstad, H.E., Cabrera, R., Gordon, R.L., Lohrmann, A., 1990. Measuring wave direction using upward-looking Doppler sonar. In: Appell, G.F., Curtin, T.B. (Eds.), Proc. of the IEEE 4th Working Conf. on Current Measurement. IEEE Press (IEEE Catalog No. 90CH2861-3), New York, NY, USA, pp. 252–257.

The ATOC Consortium, August 28, 1998. Ocean climate change: Comparison of acoustic tomography, satellite altimetry, and modeling. Science Magazine, 1327–1332.

Thorpe, S.A., Collins, E.P., Gaunt, D.I., 1973. An electromagnetic current meter for measuring turbulent flow near the ocean floor. Deep-Sea Research 20, 933–938.

Thwaites, F.T., Williams 3rd, A.J., 1996. Development of a modular acoustic velocity sensor. Proc. Oceans '96, 607–612.

Thwaites, F.T., Williams III, A.J., 2001. BASS measurements of currents, waves, stress, and turbulence in the North Sea bottom-boundary layer. IEEE J. Oceanic Eng. 26 (2), 161–170.

Veth, C., Zimmerman, J.T.F., 1981. Observations of quasi-two-dimensional turbulence in tidal currents. J. Physical Oceanogr. 11, 1425–1430.

Visbeck, M., Fischer, J., 1995. Sea surface conditions remotely sensed by upward-looking ADCPs. J. Atmos. and Oceanic Technol. 12, 141–149.

von Arx, W.S., 1950. An electromagnetic method for measuring the velocities of ocean currents from a ship under way, Papers in Physical Oceanography and Meteorology. Massachusetts Institute of Technology and Woods Hole Oceanographic Institute 3 (3).

Weatherly, G.L., Wimbush, M., 1980. Near bottom speed and temperature observations on the Blake-Bahama Outer Ridge. J. Geophys. Res. 85, 3971–3981.

Williams III, A.J., 1985. BASS, an acoustic current meter array for benthic flow-field measurements. Marine Geology 66, 345–355.

Williams III, A.J., Tochko, J.S., 1977. An acoustic sensor of velocity or benthic boundary layer studies. In: Nihoul, J.C.J. (Ed.), Bottom turbulence. Elsevier Oceanography Series, 19, pp. 83–98. Amsterdam, the Netherlands.

Williams III, A.J., Tochko, J.S., Koehler, R.L., Grant, W.D., Gross, T.F., Dunn, C.V.R., 1987. Measurement of turbulence in the oceanic bottom boundary layer with an acoustic current meter array. J. Atmos. Oceanic Technol. 4, 312–327.

WisemanCrossby Jr., R.M., Pritchard, D.W., 1972. A three-dimensional current meter for estuarine applications,. J. Marine Res. 30 (1), 153–158.

Wyatt, B., Burt, W.V., Pattullo, J.G., 1972. Surface currents off Oregon as determined from drift bottle returns. J. Phys. Oceanogr. 2, 286–293.

Wyngaard, J.C., Businger, J.A., Kaimal, J.C., Larsen, S.E., 1982. Comments on 'A reevaluation of the Kansas mast influence on measurements of stress and cup anemometer overspeeding,'. Bound. Layer Meteor. 22, 245–250.

Yamaguchi, K., Lin, J., Kaneko, A., Yamamoto, T., Gohda, N., 2005. A continuous mapping of tidal current structures in the Kanmon Strait. J. Oceanogr. 61, 283–294.

Yamaoka, H., Kaneko, A., Park, J.-H., Zheng, H., Gohda, N., Takano, T., Zhu, X.-H., Takasugi, Y., 2002. Coastal acoustic tomography system and its field application. IEEE J. Ocean. Eng. 27 (2), 283–295.

Yamaoka, H., Kaneko, A., Park, Jae-Hun, Zheng, H., Gohda, N., Takano, T., Zhu, Xiao-Hua, Takasugi, Y., 2002. Coastal acoustic tomography system and its field application. IEEE J. Oceanic Eng. 27 (2), 283–295.

Zebker, H., Werner, C., Rosen, P., Hensley, S., 1994. Accuracy of topographic maps derived from ERS-1 interferometric radar. IEEE Trans. Geosci. Remote Sens. 32, 823–836.

Zedel, L., 1994. Deep ocean wave measurements using a vertically oriented sonar. J. Atmos. and Oceanic Technol. 11 (1), 182–191.

Zheng, H., Yamaoka, H., Gohda, N., Noguchi, H., Kaneko, A., 1998. Design of the acoustic tomography system for velocity measurement with an application to the coastal sea. J. Acoust. Soc. Jpn. (E) 19, 199–210.

Zheng, H., Gohda, N., Noguchi, H., Ito, T., Yamaoka, H., Tamura, T., Takasugi, Y., Kaneko, A., 1997. Reciprocal sound transmission experiment for current measurement in the Seto Inland Sea, Japan. J. Oceanogr. 53, 117–127.

BIBLIOGRAPHY

Breivik, Ø, Ø. Sætra, 2001. Real-time assimilation of HF radar currents into a coastal oceanmodel. J. Mar. Syst. 28, 161–182.

Cornuelle, B., Wunsch, C., Behringer, D., Birdsall, T., Brown, M., Heinmiller, R., Knox, R., Metzger, K., Munk, W., Spiesberger, J., Spindel, R., Webb, D., Worcester, P., 1985. Tomographic maps of the ocean mesoscale; Part I: Pure acoustics. J. Phys. Oceanogr. 15, 133–152.

Davis, R.E., 1985. Drifter observation of coastal surface currents during CODE: The statistical and dynamical views. J. Geophys. Res. 90, 4756–4772.

Dever, E.P., Hendershott, M.C., Winant, C.D., 1998. Statistical aspects of surface drifter observations of circulation in the Santa Barbara Channel. J. Geophys. Res. 103, 24781–24797.

Gill, A., 1982. Atmosphere-ocean dynamics. Academic Press, New York, NY, USA, 662–662.

Gould, W.J., 2001. Direct measurement of subsurface ocean currents: a success story. In: Deacon, M., Rice, T., Summerhayes, C. (Eds.), Understanding the ocean. UCL Press, pp. 193–211.

Howarth, 1999. Wave measurements with an ADCP. IEEE 1999, 41–44.

Howe, B.M., Mercer, J.A., Spindel, R.C., Worcester, P.F., 1989. Accurate positioning for moving ship tomography. Proc. Oceans'89, Seattle, WA, USA, 880–886.

Krause, G., 1986. *In situ* instruments and measuring techniques. In: Sündermann, J. (Ed.), Landolt–Bornstein Numerical Data and Functional Relationships in Science and Technology, n.s. Group V, vol. 3a. Oceanography, Springer-Verlag, Berlin, Germany, pp. 134–232.

Lane, A., 1997. Currents and SPM measurements, Holderness, East Coast, England, November–December 1993, October 1994–February 1995, and October 1995–January 1996. Proudman Oceanographic Laboratory Report No. 45.

Lane, A., Prandle, D., Harrison, A.J., Jones, P.D., Jarvis, C.J., 1997. Measuring fluxes in tidal estuaries: sensitivity to instrumentation and associated data analyses. Estuarine Coastal and Shelf Sci. 45, 433–451.

Lewis, J.K., Shulman, I., Blumberg, A.F., 1998. Assimilation of Doppler radar current data into numerical ocean models. Cont. Shelf Res. 18 (5), 541–559.

Longuet-Higgins, M.S., Stem, M.E., Stommel, H., 1954. The electrical field induced by ocean currents and waves, with application to the method of towed electrodes,. Papers in Physical Oceanography and Meteorology 13, 1–37.

Medwin, H., 1975. Speed of sound in water: A simple equation for realistic parameters. J. Acoust. Soc. Amer. 58, 1318–1319.

Oke, P.R., Allen, J.S., Miller, R.N., Egbert, G.D., Kosro, P.M., 2002. Assimilation of surface velocity data into a primitive equation coastal ocean model. J. Geophys. Res. 107 (C9), 3122.

Paduan, J., Niiler, P., 1993. Structure of velocity and temperature in the northeast Pacific as measured with Lagrangian drifters in fall 1987. J. Phys. Oceanogr. 23, 585–600.

Schmidt, W.E., Woodward, B.T., Millikan, K.S., Guza, R.T., 2003. A GPS-tracked surf zone drifter. J. Atmos. Ocean. Technol. 20, 1069–1075.

Smith, S.R., Jacobs, G.A., 2005. Seasonal circulation fields in the northern Gulf of Mexico calculated by assimilating current meter, shipboard ADCP, and drifter data simultaneously with the shallow water equations. Cont. Shelf Res. 25 (2), 157–183.

Spiesberger, J.L., Spindel, R.C., Metzger, K., 1980. Stability and identification of ocean acoustic multipath. J. Acous. Soc. Am. 67, 2011–2017.

Spindel, R.C., Worcester, P.F., 1991. Ocean acoustic tomography: A decade of development. Sea Technol. 32 (7), 47–52.

Spydell, M., Feddersen, F., Guza, R.T., Schmidt, W.E., 2007. Observing surf-zone dispersion with drifters. J. Phys. Oceanogr. 37, 2920–2939.

Stommel, H., 1954. Exploratory measurements of electrical potential differences between widely spaced points in the North Atlantic Ocean, Archiv fur Meteorologie. Geophysik und Bioklimatologie, A 7, 292–304.

Stommel, H., von Arx, W.S., Parson, D., Richardson, W.S., 1953. Rapid aerial survey of the Gulf Stream with camera and radiation thermometer. Science 117, 639–640.

Sundermeyer, M.A., Price, J.F., 1998. Lateral mixing and the North Atlantic tracer release experiment: observations and numerical simulations of Lagrangian particles and a passive tracer. J. Geophys. Res. 103 (C10), 21481–21497.

Swallow, J.C., 1954. Seismic investigations at sea. Ph.D. thesis. University of Cambridge, Cambridge, U.K.

Vachon, W.A., 1977. Current measurement by Lagrangian drifting buoys—Problems and potential. Proc. Oceans '77. pp. 46B-1 to 46B-7.

Williams III, A.J., Terray, E.A., 2000. Measurement of directional wave spectrum with a Modular Acoustic Velocity Sensor. Oceans 2000, 1175–1180.

Zhang, C., Zhu, X.-H., Kaneko, A., Wu, Q., Fan, X., Li, B., Liao, G., Zhang, T., 2010. Reciprocal sound transmission experiments for current measurement in a tidal river, 2010 IEEE.

Zhurbas, V., Oh, I.S., 2003. Lateral diffusivity and Lagrangian scales in the Pacific Ocean as derived from drifter data. J. Geophys. Res. 108, 3141.

Chapter 3

Lagrangian-Style Surface Current Measurements Through Tracking of Surface Drifters

Chapter Outline

3.1. Radio Buoys   94
   3.1.1. Drifter-Following Radar Transponder   95
   3.1.2. Drifter-Borne Doppler Transponder   96
   3.1.3. Radio Buoys Tracked by Polar-Orbiting
       Satellites   97

   3.1.4. GPS-Tracked Drifters   100
   3.1.5. Telephonically Tracked Drifters   104
3.2. Limitations of Surface Drifters   105
References   105
Bibliography   107

The small-scale water current structures found near fluid boundaries, such as the coast and strong hydrographic fronts, are poorly described owing to a lack of adequate measurements. Likewise, away from such boundaries, the fine-scale structure of horizontal dispersion is poorly described. In response to this void, several researchers have attempted and succeeded in developing drifter systems that would resolve small-scale, high-frequency coastal and estuarine flows. Trajectories of freely drifting bodies in coastal regions have been utilized for several other applications as well, such as understanding the pattern of coastal circulation and its role on the dispersion or retention of larvae (Edwards et al., 2006), determining the advection of discharged ballast ship water (Larson et al., 2003), and forecasting and containment of oil spills and other pollutants (Abascal et al., 2009). Furthermore, search-and-rescue operations require prediction of the path of drifting targets and of an optimal search region based on the initial location and on the coastal current field (Ullman et al., 2006).

In oceanography, the so-called *Lagrangian method*, a method that provides a quantitative description of both track and speed of flow, is often employed to obtain an overall picture of oceanic water currents over large areas. The Lagrangian method originally involved measurement of water movements by tracing the path of water parcels over sufficiently long time intervals. The combination of many such paths, termed *trajectories*, mapped on a chart (called a *current chart*), was then used to describe the overall pattern and speed of horizontal currents in a given region. Thus, the Lagrangian method is useful in establishing large as well as small water circulation routes and in detecting and identifying major ocean gyres, which contain almost 99% of the kinetic energy of the ocean circulation (Monk and Wunsch, 1979). In fact, much of the general knowledge gained on large-scale oceanic circulation has come from Lagrangian methods of measurement. Unfortunately, Lagrangian data of surf zone flows are relatively scarce, but they are important for understanding the spatial and temporal variability of water currents in the surf zone domain.

Lagrangian techniques have been used widely in the study of water currents in oceans and large lakes, both for fundamental understanding of the associated fluid dynamics as well as for solving environmental problems. The data provided by water current-following *drifters* are particularly valuable in observing the spatial and temporal structure of the flow field and providing a different insight into the flow dynamics than that which is obtainable from Eulerian data. Lagrangian data also allow diffusion coefficients to be estimated more realistically than with fixed current meters (Pal et al., 1998). This is important for ecological investigations of, for example, the fate of pollutants, algal blooms, and artificial fertilization.

Drifters have often been used to complement fixed current meters in the deep ocean (e.g., McPhaden et al., 1991) and on the continental shelf (e.g., Davis, 1985). A comprehensive overview of the development of oceangoing Lagrangian

Copyright © 2014 Elsevier Inc. All rights reserved.

drifters is given by Davis (1991). Early examples include the experiments of Stommel (1949) and Swallow (1955). Recent drifters make use of SOFAR channels for subsurface applications (Rossby and Webb, 1970) and satellite systems such as Advanced Research and Global Observation Satellite (ARGOS) for near-surface applications. In the past, the use of Lagrangian drifters had been largely limited to the deep ocean and large lakes. Some work has been done over smaller scales, such as, for instance, in coastal regions (Davis, 1983; List et al., 1990; George and Largier, 1996).

Lagrangian field data of water current systems in the surf zone are extremely rare but are valuable for understanding their detailed structure, confirming model predictions of transient features, and making estimates of dispersion. The few existing Lagrangian measurements in the surf and near-shore zone were obtained with the application of a variety of techniques. Surface floats and drogued drifters were used by Shepard et al. (1941), Shepard and Inman (1950), and Sonu (1972) to investigate rip currents. Positions were obtained by compass fixes from boats and the shore. Floats were tracked using sequential aerial photographs taken from a balloon in the experiments of Sasaki and Horikawa (1975, 1978). The method of Short and Hogan (1994) used "live" floats, whereby swimmers floating in rip currents were tracked by theodolite fixes. This technique was used by Brander and Short (2000) to investigate the dynamics of a large rip system. Dye has also been used for flow visualization and measurement of current speeds (e.g., Sonu, 1972; Brander, 1999).

With rapid advancements in technology, the concept of Lagrangian measurements underwent significant leaps, particularly in the realm of surface circulation. Lagrangian paths can now be directly measured *in situ* by tracking neutrally buoyant phosphorescent tracers, although this technique is limited in scope through restricted spatial and time scales (Gaskin et al., 2002) and by satellite-tracked drifters (Gawarkiewicz et al., 2007). Today remote sensing using satellite-borne sensors as well as fixed platform-based HF/VHF Doppler radar systems permits collection of near-real-time "snapshot pictures" of oceanic surface circulation over large areas. In terms of monitoring the ever-increasing menace of oil spills from tankers and oil-drilling platforms, as well as conducting successful rescue operations, the capability of modern Lagrangian techniques for all-weather, near-real-time mapping of oceanic surface circulation is of immense practical utility.

Ocean surface currents being frequently decoupled from those at the depths, data from moored current meters cannot be readily extrapolated to the surface. One of the best viable methods for quantitative inference of water currents in the upper few meters of a wave-laden ocean is the Lagrangian technique. Basically, this technique has provided an enormous amount of data that have served to construct mean sea surface current charts as they are found today in most geographical atlases. A bird's-eye view of the primitive technologies of Lagrangian style surface current measurements and their timely upgradation in tune with technological leaps was provided in Chapter 2. As indicated, with the rapid advancement in technology, Lagrangian-style surface current measurements took on new dimensions. Several advanced methods are currently in use and are addressed here.

3.1. RADIO BUOYS

Advancements achieved in the technologies of sensor design, signal detection, data logging and communication electronics, and satellite technologies in the last few decades had a significant positive impact on detection, monitoring, and telemetric reporting of oceanic surface current measurements. Application of radio buoys is just one example. In this method, a buoy freely drifting on the sea surface is tracked by radio signals. Its position is periodically determined as it drifts along under the influence of the drag force exerted on it by the sea surface currents and winds. Radio buoys are usually released from ships. Speed and direction of the surface currents are determined by observing the distance and direction the buoy drifts in a given time interval. These buoys, meant to drift freely with the water mass, are equipped with colored flags, flashing lights, radio location beacons, or satellite transmitters.

Flow speed errors in the drifter measurements greatly depend on the design of the drogue system; the drogue is to be designed to minimize the effects of wind drag on the surface buoy. A crossed vane rigidly mounted to the bottom end of the buoy or canvas "window-blind" drogue located below the air-water interface ensures that the buoy is less influenced by wind. In bygone days of surface current mapping, the buoys were usually tracked using boats or aircrafts with the aid of HF direction-finding systems (Whelan et al., 1975). In the 1940s, measurements of tidal current structures generated at the Hayatomono-Seto of the Kanmon Strait (a strait famous not only as an important shipping traffic route to China and Korea but also as a dangerous passage with quite strong tidal currents exceeding 5 m/s at the narrowest point), located in the Sea of Japan, were traditionally made with the use of drifting floats tracked by many small boats (Fukunishi, 1948a, b). An overall feature of current structures is understood by this kind of measurement.

Many buoys can be tracked at the same time if means are provided to distinguish them individually. One method of accomplishing this is to provide individual buoys with some sort of identification number and to transmit these numbers either periodically or when interrogated from shore-based transponders or from a satellite. These buoys provide both track and speed of flow, i.e., a Lagrangian description, and have revealed many new details of eddies associated with ocean currents (note that the term *eddy* refers to rotation of water mass).

A more sophisticated drifting buoy technique is to provide the buoy with a radio transponder, which replies when interrogated from a satellite in orbit (Pickard et al., 1982). By this method the buoy's position can be determined more accurately than with the lower-frequency radio direction-finding techniques. The position of such a satellite-tracked radio buoy is calculated on board the satellite using the Doppler shift of the buoy's VHF signal, then transmitted to a ground station for recording (Royer et al., 1979; Briscoe et al., 1987).

3.1.1. Drifter-Following Radar Transponder

In situations in which surface current monitoring was desired only for short periods, the buoys were accompanied by a boat. The locations of the buoys were periodically determined by a microwave tracking system. The boat, which was used to deploy the buoy, carried the master station, which continuously interrogated two pulse-radar transponders located at fixed reference points, a few km apart on the coast or on fixed platforms in the sea. The elapsed time between the interrogation pulse (transmitted from the master station on the boat) and each of the two reply pulses (from the reference stations) was used to determine the distance (from the boat) to each fixed reference station. This information, together with the known locations of the reference station, was trilaterated to obtain a position fix on the boat. This process is shown schematically in Figure 3.1.

The tracking system accurately located the position of the boat relative to two known geographical locations (reference points). The position of the buoy at a given instant in time was measured by periodically approaching the buoy and carefully maneuvering the boat to within a few meters of the buoy, then noting the distance (i.e., range) of

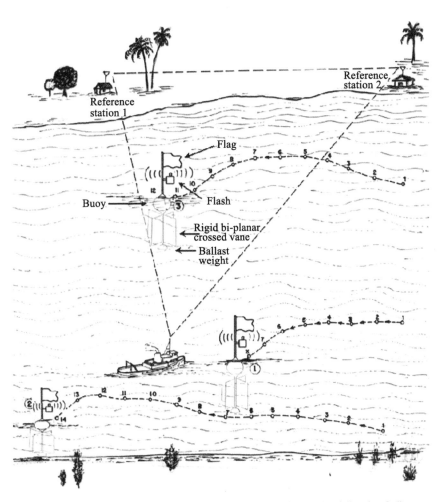

FIGURE 3.1 Illustration of radio buoys being tracked using a boat with the aid of direction-finding systems.

the boat to each of the two geographically fixed transponders. Because the boat approached a buoy only when measurements were desired, the surface current-driven free motion of the buoy remained largely unaffected by the movements of the boat. In this case, the buoy was not strictly a "radio" buoy, but the master station on the accompanying boat effectively made it a radio buoy when measurements were taken.

The buoy-position measurements were repeated at time intervals of 1 hour or longer, and sea surface current velocities were calculated from the differences in position at successive time intervals. This methodology had been employed for surface current measurements in the year 1978 during the Joint Air-Sea Interaction (JASIN) international experiment conducted in the North Atlantic off the coast of Scotland. During the development stages of the HF Doppler radar system, drifting buoys provided data for comparison of surface currents derived from the HF Doppler radar system (Teague, 1986).

3.1.2. Drifter-Borne Doppler Transponder

As noted earlier, floating objects that drift freely with the moving sea surface water parcels are considered to be Lagrangian tracers of sea surface currents. An improved radio-buoy system, for ranges less than 100 km, is a drifter-borne Doppler transponder that is tracked by a pair of HF Doppler radar systems located on shore or on offshore platforms. Utilization of the radio buoy technique for sea surface current measurements could be enhanced by the use of radar tracking, because such tracking operations can be performed regardless of weather conditions, i.e., in fog, high winds, rough seas, and at night.

Although several methods exist for measurement of sea surface current trajectories, the Doppler effect plays a key role in the technology addressed in this section. As the name suggests, the Doppler transponder system works on the Doppler effect principle, according to which the frequency of any harmonic wave motion at a receiver differs from the frequency at its source whenever the receiver or the source are in motion relative to one another. Thus, in the present case, the relative motion between a stationary radar and a radio buoy (transponder) that is freely drifting under the influence of sea surface current causes a variation in the received radio frequency. With the introduction of microcomputer-based technology, HF Doppler transponder systems have been operationally introduced to obtain a Lagrangian description of surface currents in coastal water bodies.

CODAR Ocean Sensors Ltd. (Los Altos, California) developed a system which makes radial velocity measurement of a drifting Doppler transponder. In operation, HF Doppler radar transmits a high-frequency pulsed signal from a stationary interrogating site, which is located on the

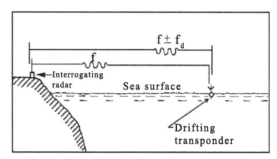

FIGURE 3.2 Diagram illustrating the principle of radial velocity measurement of a drifting Doppler transponder. *(Source: CODAR brochure, CODAR Ocean Sensors Ltd., Los Altos, CA, reproduced with kind permission of CODAR Ocean Sensors Ltd.)*

coast (see Figure 3.2) or an offshore platform. As a consequence of the relative motion between the drifting radio buoy (transponder) and the stationary radar, the constant-frequency radio transmission from the radar system is received by the transponder with an apparent variation in frequency (Doppler effect). The Doppler-shifted frequency received by the transponder is immediately retransmitted to the stationary radar. The frequency received by the stationary radar suffers an additional Doppler shift in the return travel. Measurement of this Doppler shift yields an accurate estimate of the radial velocity of the transponder relative to the radar.

The Doppler shift, Δf, in the frequency received by the stationary radar is given by:

$$\Delta f = \frac{2fV_R}{c} \tag{3.1}$$

In this expression, f is the transmission frequency of the stationary radar, V_R is the radial velocity of the drifting transponder relative to the radar, and c is the velocity of propagation of the electromagnetic wave in air. Following conventional practice, some forms of statistical methods are applied online to extract the Doppler shift from the returned signal. This is usually achieved using a microcomputer-controlled digital signal processor (DSP) in the receiver section. The range R of the transponder from the stationary radar is computed from the round-trip travel time *(t)* using the relation:

$$c = \frac{2R}{t} \tag{3.2}$$

Measurement of velocity and position of the transponder in the Earth coordinate system requires two stationary interrogating sites separated by a known straight-line distance, usually a few tens of kilometers (see Figure 3.3). Under normal operating conditions, the radial velocity is claimed to be accurate to within ± 0.1 cm/s. Position accuracy is estimated to be between ± 50 and ± 500 m.

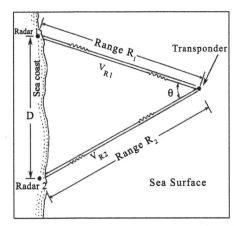

FIGURE 3.3 Diagram illustrating the scheme used for velocity and position determination of a drifting transponder with the aid of two known spatially separated stationary interrogating-sites. *(Source: CODAR brochure, CODAR Ocean Sensors Ltd., Los Altos, CA, reproduced with kind permission of CODAR Ocean Sensors Ltd.)*

In the case of CODAR transponder, the two interrogating sites need to be separated by approximately 30 km to obtain the best accuracy and resolution. With a dual-site system, the absolute velocity and position of the transponder are determined by the usual triangulation methods commonly employed by marine surveyors. The transponder established on each freely drifting radio buoy is activated by a unique address code, which allows sequential interrogation of several transponders. The CODAR systems are capable of simultaneously measuring the positions and instantaneous velocities of as many as 126 different transponders drifting at range up to 100 km with velocity accuracies typically 0.2 cm/s and position accuracies between ± 50 and ± 500 meters. The maximum range at which a transponder can be interrogated and the accuracy of the surface current velocity measurements are both governed by the strength of the transponder signal relative to the atmospheric and manmade noises at the operating frequency as well as the dielectric properties of the medium over which the signals travel.

3.1.3. Radio Buoys Tracked by Polar-Orbiting Satellites

In situations in which trajectories of surface currents are to be tracked over long distances in the open ocean regions for long periods of time, the radio buoys are tracked by polar orbiting satellites. Satellite-tracked drifting radio buoys, with operating lifetimes of a year or more, are attractive tools for measuring the spatial structure of sea surface currents. Deployed in large arrays, they offer an unmatched capability of mapping the two-dimensional field of surface/near-surface currents in the open oceans over long periods of time (D'Asaro, 1992). Freely drifting radio buoys

deployed in the open ocean are tracked by polar orbiting satellites. The satellite technology meets the majority of drifting buoy-tracking needs and has the virtues of continuity over time, remote monitoring, true global coverage, all-weather operation, relatively good immunity to natural and manmade interferences, frequent availability of buoy position fixes, and good location accuracy. Such features also make satellite tracking far more affordable than more conventional techniques. In fact, satellite-tracked radio buoy experiments have helped in locating several large-scale eddies. When the existence of large current paths or current loops with large radii are to be studied, comprehensive ship surveys prove very costly and often ineffective. Alternatively, satellite-tracked buoys, with an estimated life of about 300 days, provide an ideal tool to study the movement of water bodies. Such studies, initially supported by NASA satellite NIMBUS and later by NOAA satellites, have produced direct Lagrangian current measurements on large time and spatial scales in many regions. The presence of cyclonic and anticyclonic drifting eddies could be readily identified from the spiraling motion of the buoy track. Today NOAA is a leading organization that provides such satellite facilities.

The radio buoys are equipped with transmitters (platform transmit terminals) to facilitate their tracking by polar orbiting satellites. Since 1978, the Service Argos, France, offers capabilities for satellite-based position fixing of radio buoys. The footprint of the polar orbiting satellites on the surface of the Earth is ~5,000 km in diameter (see Figure 3.4). A satellite can receive signals from any radio buoy located at any point within its footprint. Further, the visibility time of a satellite is only 10−13 min. Special

FIGURE 3.4 Footprint of a polar orbiting satellite centered on its ground track on the Earth's surface. *(Source: ©CLS 2012, reproduced with kind permission of CLS Service Argos, Toulouse Cedex, France.)*

precautions are therefore needed to identify each drifting buoy and to determine its location. The buoy identification is achieved from its assigned unique *identification number,* which is transmitted by the radio buoy. Determination of the position of a radio buoy within the 5,000-km-diameter footprint of the satellite, without ambiguity, requires the measurement of the Doppler shift in the frequency received by the satellite.

The platform transmit terminal (PTT) onboard the radio buoy, featuring microcircuitry, transmits at a nominal frequency of $\sim$400 MHz. The whole PTT message is transmitted in less than 1 second and includes 160 ms of unmodulated carrier to allow the satellite's receiver to lock onto the carrier. To maximize the probability of accurate message reception by the satellite, the PTT message is transmitted several times. The probability of message reception is claimed to be 0.9920 for PTT messages repeated three times and 0.9999 for messages repeated six times.

The satellite-borne data collection and location system (DCLS) receives and records all transmissions from the radio buoys, which are located in the visibility zone of the orbiting satellite. The relative motion between the polar orbiting satellite and the radio buoy causes a Doppler shift in the PTT signal received by the satellite. As the satellite passes through its point of closest approach to the PTT, there is a Doppler shift in the carrier frequency received by the satellite. As the satellite approaches the PTT, the received frequency f_r is higher than the transmitted frequency f_t (i.e., the Doppler shift is positive); at the point of closest approach, f_r is equal to f_t (i.e., the Doppler shift is zero); and as the satellite goes away, f_r is less than f_t (i.e., the Doppler shift is negative). The Doppler shift is a function of the relative velocity between the satellite and the PTT. If the PTT is stationed below the orbital path of the polar orbiting satellite (which is along a longitude over the Earth), the Doppler shift would remain constant at all times, the only change being a sudden jump from positive to negative as the satellite passed over the PTT. However, if the PTT is a little farther away from the orbital path of the satellite, the closest approach range is larger, and therefore the slope of the Doppler curve will be less steep (Figure 3.5). There is thus a direct correlation between the change of slant range and the shape of the Doppler curve.

Because the frequency transmitted by the PTT and the latitude and longitude of the satellite at every instant are known, measurement of the Doppler shift defines the field of possible positions for a given PTT (a radio buoy in the present case). The field is in the form of a half-cone, with the satellite at its apex, and the satellite velocity vector (**V**) as the axis of symmetry (Figure 3.6). The Doppler shift is related to the apex half-angle *(A)* of the cone, the satellite velocity *(v)* relative to the PTT, and the velocity *(c)* of the

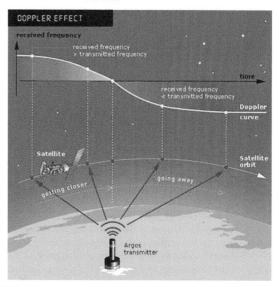

FIGURE 3.5 Doppler curve corresponding to differing distances of the buoy-borne platform transmit terminal (PTT) from the satellite's orbital path. *(Source: ©CLS 2012, reproduced with kind permission of CLS Service Argos, Toulouse Cedex, France.)*

electromagnetic wave by the expression (Carpiniello and Buell, 1972):

$$\cos A = \frac{(f_r - f_t)c}{v f_t} \tag{3.3}$$

In this expression, f_t and f_r are the transmitted and the received frequencies, respectively. Different location cones, obtained from successive Doppler measurements in a given satellite pass, intersect the sea surface to yield the two possible positions of the radio buoy. Such positions are symmetrical with respect to the trajectory (longitude) of the satellite ground track. Additional information such as previous position of the buoy, range of possible speeds, and so forth are used to determine which of the two possible positions are realistic. The processor in the satellite calculates the location of that radio buoy, from which at

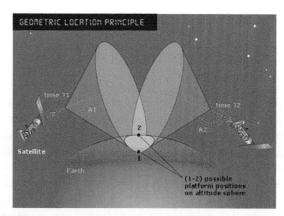

FIGURE 3.6 Half-cone that defines the field of many possible positions for a PTT. *(Source: ©CLS 2012, reproduced with kind permission of CLS Service Argos, Toulouse Cedex, France.)*

least four messages were collected, and of which the first and the last messages were separated by at least 4 min. To minimize the ambiguity in the estimation of the buoy location, we use certain criteria, such as lowest least-squares value, absurdity test on distance covered by the PTT since last estimated location, and so forth. The estimated buoy location is rejected by the location estimation algorithm if the number of PTT messages were lower than a preset value or if the PTT was not within the required range of the satellite. The location estimation algorithm assumes the PTT to be stationary. Any movement of the PTT, which occurs in the case of a drifting radio buoy, therefore causes an error in the location estimation unless it is factored into the location estimation as uniform motion from one satellite pass to the next.

The real-time or recorded location information for each radio buoy is accessible to the user from the processing centers via international telex network, switched telephone network, data transmission network, or the Global Tele-communication System (GTS), which is reserved for meteorological and oceanographic data.

Two NOAA satellites are simultaneously in circular mutually orthogonal polar orbits. The orbit altitudes are different (~ 830 km and ~ 870 km), producing a ~ 1-min difference in the orbital period. Each satellite makes ~ 14 revolutions in a day. As the satellite orbits the Earth, the visibility zone sweeps a swath of $\sim 5,000$ km in width. As a result of the Earth's rotation to the east, this swath shifts $25°$ west about the polar axis on each revolution (Figure 3.7), corresponding to a distance of $\sim 2,800$ km at the equator. As a result, the satellite orbits provide complete coverage of the Earth's surface. In polar regions, the PTTs can be deployed on icebergs, the drift of which provides a description of the surface circulation in the region.

Different PTTs (distinguished by their unique identification numbers) have different repetition periods and transmission frequencies. Furthermore, transmissions by different PTTs are asynchronous. Theoretically, these schemes would enable the onboard DCLS to pick up and sort messages from all the PTTs in the visibility zone. However, the PTT is ignorant of the three possible states of its message (Sherman, 1992): (1) there is no satellite in reception range, (2) due to bad signal quality, the satellite has rejected the message completely, and (3) the message has been received, but with some undetermined number of bit errors. Further, the PTT receives no acknowledgment from the satellite and does not know whether the message has been successfully received. An analysis by Sherman (1992) revealed that only 6% of the transmitted messages are received by the *Argos* satellite, with 17% rejected while a satellite was in view and 9% of received messages containing at least one error. Inclusion of the satellite's orbital information in the PTT's memory and transmission only when the satellite is in its visibility zone are expected to improve the overall efficiency.

A source of error in the sea surface currents, estimated from radio buoys, is the random error in the position fixes. The root-mean-squared error in position fixes is quoted to be less than 350 m. For moving drifters, additional nearly random errors will be contributed by unresolved high-frequency motions induced by surface and internal gravity waves. Messages received at the ARGOS center in Toulouse, France, are sorted according to the user identification number and coded with the time of reception at the satellite-borne ARGOS DCLS (Bellamy and Rigler, 1986).

Despite many limitations, satellite tracking of radio buoys has revealed many new details of eddies associated with ocean currents. In fact, the existence of gigantic gyres of more than 100 km in diameter, described in the literature as *mesoscale eddies* (and sometimes as *ocean storms*), has been confirmed after an accidental discovery during remote monitoring of some radio buoys that were trapped for several weeks within the periphery of a drifting gyre. Further observations have indicated that such gyres are occasionally present in many regions of the oceans. A conceptual impression of the satellite tracking of a gyre based on messages received from a conglomeration of drifting buoys is given in Figure 3.8. Synoptic surface current measurements by satellite tracking of freely drifting buoys in offshore areas have helped in the past in identifying ocean eddies (gyres) as large as 200 km in diameters and in monitoring their movement (Cresswell, 1977; Grundlingh, 1977).

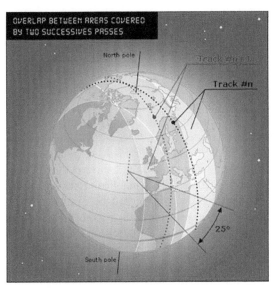

FIGURE 3.7 Schematic picture illustrating the mechanism of satellite orbit providing complete coverage of the Earth's surface as a result of the satellite's visibility zone (comprising a swath of $\sim 5,000$ km in width) shifting $25°$ west about the polar axis of the Earth on each of its revolution to the east. *(Source: ©CLS 2012, reproduced with kind permission of CLS Service Argos, Toulouse Cedex, France.)*

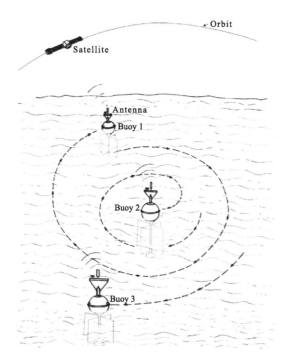

FIGURE 3.8 A conceptual impression of the satellite tracking of a gyre based on messages from a conglomeration of drifting buoys.

Figure 3.9 shows a plot of the trajectory of a satellite-tracked drifting buoy deployed off Goa, India.

Starting on 15 March 2011, the Argos user communities were provided the option to choose between two location processing algorithms:

- The algorithm based on the classical least-squares method that has been employed since Argos processing began in 1986
- The algorithm based on Kalman filtering

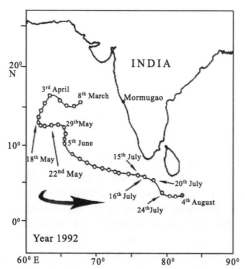

FIGURE 3.9 Plot of the trajectory of a satellite-tracked drifting buoy deployed off Goa, India. (*Source: Nayak et al., 1993, Reproduced with kind permission of Osservatorio geofisico sperimentale [OGS].*)

The trajectory of an Argo float in the Pacific calculated with the algorithm based on Kalman filtering and least-squares analysis is given in Figure 3.10.

3.1.4. GPS-Tracked Drifters

Resolving spatially complex water current circulations, such as the surf zone circulation, would require deployment of a large number of current meters. The surf zone is a challenging environment in which to make hydrodynamic measurements through deployment of drifters as well. Instruments must be very robust to withstand wave breaking, and there are significant difficulties in deployment and retrieval. Although Lagrangian measurements are valuable in revealing the horizontal current structure in the surf zone, few such measurements have been made in the past because of the practical difficulties involved. The main problem to overcome in the deployment of drifters is the tendency of any floating object to surf shoreward when caught in a breaking wave; its velocity is then the phase speed rather than water particle motion. The less an instrument penetrates into the water below the breaking region of the wave, the worse the surfing effect. A second problem is that the depth varies greatly from deep water offshore to the swash zone, where there is water only part of the time. Any floating instrument, therefore, has to have the apparently incompatible requirements of significant drag in the deeper section of a wave and the ability to move into shallow water. Finally, the only way to deploy and recover drifters within the surf zone is to physically carry them; they need to be small and light enough to accomplish this easily and safely.

Lagrangian drifters equipped with GPS came as a great relief to the hitherto difficult situation faced by oceanographers, ocean engineers, and environmentalists in the tracking of surf zone currents (Muzzi and McCormick, 1994; George and Largier, 1996). GPS is a worldwide radio-navigation system that employs a constellation of 24 satellites; up to eight are used at any time to determine the position of a receiver. Until May 2000, Selective Availability (SA) deliberately degraded the publicly available signal for military purposes and limited the accuracy to approximately 100 m. This practice effectively restricted the scales of motions that could be resolved. Improved position fixing was possible with differential correction, but this required a fixed base station and additional signal processing. Since the removal of SA, the nondifferential GPS proved to be a valuable device for position-fixing applications. Standard GPS receivers are small and low-cost and can fix their position within a few meters anywhere in the world.

With the availability of GPS coverage, surface drifters began to be tracked using GPS techniques. The small spatial (of the order of 5 m) and short temporal (of the order

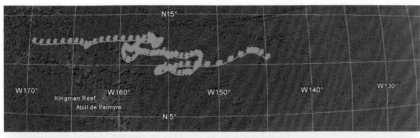

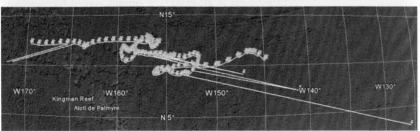

FIGURE 3.10 Trajectory of an Argo float in the Pacific Ocean calculated with the algorithm based on Kalman filtering (top image) and least-squares analysis (bottom image). *(Source: ARGOS SYSTEM newsletter, FLASH #20, February 2011, ©CLS 2012, reproduced with kind permission of CLS Service Argos, Toulouse Cedex, France.)*

of 1 min) scales of interest in surf zone circulation are re-solved with GPS techniques used in coastal and near-ship drifters (e.g., real-time DGPS, 3–5-m accuracy [George and Largier, 1996] and carrier phase post-processing, 1-m accuracy [Doutt et al., 1998]). Drifters that autonomously record their position enable high-frequency Lagrangian data to be collected.

Near-shore drifters are intended for use in confined or near-shore environments over timescales of up to several days and are a low-cost alternative for applications that do not require drifters with full ocean-going capabilities. According to George and Largier (1996), near-shore drifters must be inexpensive so that large numbers can be deployed simultaneously to provide both statistical reli-ability and spatial resolution over the domain of interest. The system must be self-contained, reliable, rugged, and easily transportable. The drifters are to be retrievable. The drifter package should be modular, allowing for the possi-bility of using different drogue, float, and visibility device designs. The drifter should be easily adjusted for drogue at depths from 1 to 15 m for use in vertically sheared flow in a wide variety of applications. Positioning error must be small (less than 5 m) in order to resolve the small-scale structure of the flows of interest. Sample interval must be short (less than 10 m) and not dependent on the number of drifters deployed at one time. Position accuracy and sample rate must be maintained at distances of up to 20 km from the base of operations. In the presence of moderate wind and wave conditions, the errors in the water-following performance of the drifter must be known and the wind-induced drift-to-wind-speed ratio must not exceed 0.005.

A near-shore GPS drifter constructed of canvas moun-ted on a fiberglass frame has been described by George and Largier (1996). The projected area of this drogue is approximately 1 m², and the electronics package is carried

in a central PVC pressure case. This drifter has known surface-following characteristics in a wave field, has low mass, and can be inexpensively constructed. However, high impacts from breaking waves and the tendency to "surf" shoreward often precluded the use of this device.

Schmidt et al. (2003) described a surf zone drifter and results from a deployment near a rip current. In this system, an impact-resistant body of tubular PVC is ballasted for nearly complete submergence (Figure 3.11). They cir-cumvented the problem of surfing by attaching a PVC disk to the base of their 50-cm-long main casing. This disc strongly dampens the vertical response of the drifter, allowing broken and near-breaking waves to pass over without rapidly pushing the drifter ashore. This mechanism was found to be effective in resisting surfing. Their longer receiver casing also penetrates deeper into the water below, breaking wave rollers, and thereby precludes some of the need for an additional drogue. The heave and roll of the drifter, which can degrade GPS position estimates by causing large, rapid deviations of the antenna orientation from vertical, are reduced by the damping plate and by the vertical separation of the centers of buoyancy and mass. A shoreward mass flux is associated with breaking waves and bores, but by design, bores and breaking waves pass over the drifters.

In the design of Schmidt et al. (2003), each drifter records GPS pseudo-range and carrier phase data at 1-Hz for post-processing and transmits this information to shore at 10-s intervals for real-time differential GPS (DGPS) tracking and partial data backup (Figure 3.12). The telemetry range is about 5 km for a shore antenna elevation of 10 m. Battery storage limits the deployment duration to 24 h. Figure 3.13 shows a conglomeration of drifter trajectories observed in a surf zone. Field intercomparison measurements have indicated that the mean alongshore

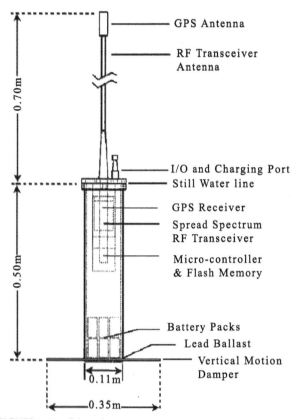

FIGURE 3.11 Schematic of a surf zone drifter. Surface-piercing antennae for receiving GPS signals and for radio-frequency (RF) communication with shore are molded permanently to the drifter top cap. *(Source: Schmidt, W. E., B. T. Woodward, K. S. Millikan, R. T. Guza, B. Raubenheimer, and S. Elgar: A GPS- tracked surf zone drifter, J. Atmos. Ocean. Technol., 2003 (20) 1069−1075. ©American Meteorological Society. Reprinted with permission.)*

currents estimated from trajectories of the 0.5-m-draft drifters in 1−2-m water depth and nearby fixed current meters within the surf zone agreed well with measurements obtained from nearby bottom-mounted acoustic current meters (correlation 0.95 and rms differences of 10 cm/s). Drifters deployed near a rip current often followed eddy-like trajectories before being advected seaward of the surf zone (Figure 3.14).

In Johnson et al. (2003), a simple, robust receiver unit suitable for use in environments such as the near-shore zone, lakes, and estuaries was described, and specialized drogue arrangement for use in the surf zone was briefly discussed.

Johnson and Pattiaratchi (2004a) reported a simple, low-cost drifter design for surf zone applications based on nondifferential GPS position fixing and having the capability to collect high-frequency, accurate, Lagrangian data. They also addressed the main issues of the dynamic response of the drifters, a robust design capable of autonomous position fixing, position-fixing accuracy, and field validation.

The solution found by Johnson and Pattiaratchi (2004a) to overcome the challenging environment of the surf zone was to use a series of "soft" drogues attached to a small, compact receiver unit. The drifter arrangement, shown in Figure 3.15, is a cylindrical receiver unit connected to a series of soft parachute drogue elements. This type of drogue opens and dramatically increases its drag when there is a differential velocity between the upper and lower part of the water column, as is the case in wave breaking. The parachute drogue also stabilizes the drifter and

FIGURE 3.12 Drifter system schematic. The base station performs DGPS and data-logging functions and can serve a fleet of 10 drifters. A mobile tracking station monitors drifter DGPS positions in real time. *(Source: Schmidt, W. E., B. T. Woodward, K. S. Millikan, R. T. Guza, B. Raubenheimer, and S. Elgar: A GPS- tracked surf zone drifter, J. Atmos. Ocean. Technol., 2003 (20) 1069−1075. ©American Meteorological Society. Reprinted with permission.)*

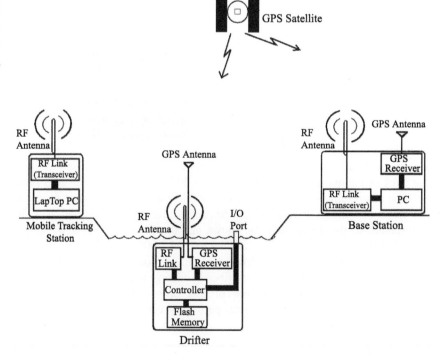

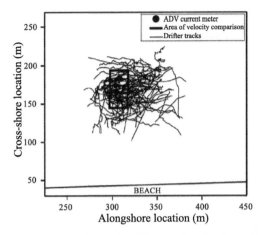

FIGURE 3.13 A conglomeration of drifter trajectories observed in a surf zone. Thin curves and filled circles represent drifter trajectories and ADV current meter locations, respectively. *(Source: Schmidt, W. E., B. T. Woodward, K. S. Millikan, R. T. Guza, B. Raubenheimer, and S. Elgar: A GPS- tracked surf zone drifter, J. Atmos. Ocean. Technol., 2003 (20) 1069–1075. ©American Meteorological Society. Reprinted with permission.)*

prevents it from rolling excessively. The drifter floats with only 2 cm of the receiver casing above the water, so the effect of windage (i.e., stress exerted by the wind directly on the float) is expected to be very small.

The receiver units are 32 cm long and 10 cm in diameter and obtain and record a GPS position fix at 1-second interval. These units consist of an integrated GPS antenna/receiver wired to a data logger and a power source in a highly robust waterproof housing. Details of their construction and the internal components can be found in Johnson and Pattiaratchi (2004b). They are deployed and recovered manually and can easily be used for repeated

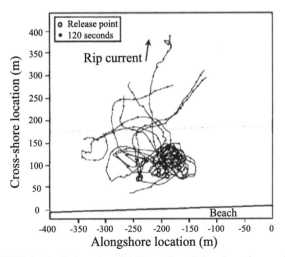

FIGURE 3.14 Trajectories of drifters released (locations shown with large circles) near the base of a rip current. The small filled circles on each trajectory indicate 120-s intervals. *(Source: Schmidt, W. E., B. T. Woodward, K. S. Millikan, R. T. Guza, B. Raubenheimer, and S. Elgar: A GPS-tracked surf zone drifter, J. Atmos. Ocean. Technol., 2003 (20) 1069–1075. ©American Meteorological Society. Reprinted with permission.)*

FIGURE 3.15 Drifter specifically designed for the surf zone, with drogue attached. *(Source: Johnson and Pattiaratchi, 2004a.)*

runs over a short period of time; their ease of deployment makes them effective for measuring rapidly developing transient features. To minimize windage and inertial effects, the drifter units are designed for nearly neutral buoyancy so that only the upper surface covering the internal GPS antenna projects above the water. In calm water, only 2 cm of the instrument projects above the surface; therefore, it is reasonable to assume that windage is negligible. A wire with a small ribbon is extended above the main unit to serve as a visual aid to enhance visibility.

The drogue is a series of parachute-shaped elements that hang below the GPS drifter casing. The parachutes are made of a cone of Dacron sailcloth with webbing attachments and are almost neutrally buoyant. A small weight is attached at the end of the parachutes to keep them hanging below the receiver unit. When the receiver unit is pulled shoreward by the breaking section, the parachutes open and anchor the drifter to the orbital velocities below the breaking region. In nonbreaking waves, the parachutes are closed and hang almost vertical and therefore present only their cross-sectional area. The drogue has also been found to be very effective in stabilizing the receiver unit by strongly damping the oscillatory motions that an "undrogued" receiver experiences. In water depths less than the length of the complete drifter arrangement, the drogue may touch the seabed, which inevitably causes some measurement error. This means that the minimum operating depth is the length of the receiver unit (32 cm), though the drogue still provides sufficient resistance to surfing in large waves.

Careful visual observation by Johnson and Pattiaratchi (2004a) of the drifter in breaking waves indicates that the parachutes are extremely effective in resisting surfing. When in the overturning section of a plunging breaker, the parachutes prevent the receiver from accelerating up the face and forward with the plunging lip. In spilling breakers, the receiver unit ducks underneath the foaming roller section and reappears at the surface a couple of seconds later as the wave passes over. The only situation in which the drifters do not perform well is when they're caught at the plunge point of a strongly plunging wave. However, due to the strong downward velocities, the whole assembly may be completely disrupted and rolled by the wave, the parachutes no longer correctly oriented, and the drifter then tends to surf in the developing roller section. When caught in strong breaking events, the drogue is ineffective at anchoring the drifter to the wave orbital motion and the whole unit travels at close to the wave phase speed. Surfing events are very easy to identify in the data, since the phase speed is much greater than orbital velocities of water particles and typical wave-averaged current speeds. Data contaminated by surfing can then easily be excluded from any analysis.

There are errors inherent in both the position fixing of the GPS receiver and in the calculation of velocities and accelerations from raw position data. Although the removal of SA has greatly improved the performance of non-differential GPS, there are still errors in the reported positions due to various factors (Hofmann-Wellenhof et al., 1997). The magnitude of errors can be greatly reduced and decimeter accuracy obtained in a differential mode using either code phase or carrier phase data. However, this does introduce an additional level of complexity and significant additional cost into the drifter design.

?>In the surf zone, there are motions over a wide range of frequencies, and it is necessary to determine how much different frequency ranges are affected by the positioning errors. Error analysis by Johnson and Pattiaratchi (2004a) indicates that nondifferential GPS position fixing is sufficient for accurately measuring motions with frequencies below about 0.05 Hz. The relative RMS error can be greatly reduced using differential methods and effectively allows accurate positioning for motions at frequencies of 1 Hz (Schmidt et al., 2003).

Intercomparison studies (drifter versus vertical profile measurements using a bottom-mounted acoustic Doppler current profiler) have suggested that the drifters slightly underestimate the depth-averaged velocity in both directions. It must, however, be borne in mind that direct validation of the drifters in the surf zone is somewhat problematic. For example, whereas the drifter measures near-surface velocities, the profile instrument near the bed measures near-bed velocities. This highlights the difficulties of comparing different types of surf zone measurements and the care required in assessing exactly what drifters are measuring. The type of drifter design, in terms of receiver and drogue arrangement, is also clearly important in determining the response in the cross-shore direction. Unfortunately, in the absence of any "true" Lagrangian velocity information, it requires comparison of wave-averaged Lagrangian drifter velocities with fixed Eulerian depth- and wave-averaged data. Therefore, the two instruments are not actually measuring the same parameter. This is particularly the case in the surf zone, because the speed and short length scales of typical surf zone currents mean that the two instruments are quickly separated and measure different parts of the water current field.

The cost, size, and weight of the GPS-based surf-zone drifter (based on the design of Schmidt et al., 2003) reported by MacMahan et al. (2009) allow for a large number of independent observations to be obtained for various meteorological and oceanographic applications. In a field experiment they conducted in May 2007 at Monterey Bay, California (a natural beach with persistent rip currents), a cluster of 10 drifters was simultaneously released with dye. Fluorescent dye represents a "true" Lagrangian measurement of water movement. The bulk of the dye patch and the drifter cluster remained co-located for one rip-current circulation, suggesting that the drifters provide accurate Lagrangian estimates. Both the dye and the drifters were spreading slightly, but they remained together in a similar patch/cluster for the initial circulation. The dye followed the average drifter velocity estimates and completed a revolution in 5 min, similar to the drifters. The experimental results enabled them to arrive at a confident conclusion that the drifters represent Lagrangian estimates.

The ability to obtain Lagrangian velocity observations in a rip-current system is a good example of the usefulness of the inexpensive handheld GPS. The flow field of rip currents has a large spatial variability that is difficult to measure with *in situ* instruments owing to expense and deployment complexities. The inexpensive system reported by MacMahan et al. (2009) has the ability to fill in the voids between *in situ* instruments, thereby advancing our understanding of the hydrodynamics of a rip-current system.

3.1.5. Telephonically Tracked Drifters

A coastal environment is usually characterized by more stratified conditions, thinner layers of water, and higher vertical shears than the open ocean; thus the requirements are for wider and shorter drogues. Several drifter designs seek to satisfy such criteria. The one reported by Zervakis et al. (2005) is such a drifter, which addresses the need to perform coastal studies.

The great expansion of Global System for Mobile (GSM) communication technologies that took place in the late 1990s permitted the exploitation of a low-cost, readily available and widely expanded technology in the design of surface/subsurface drifters for use in coastal and land-locked seas, archipelagos, estuaries, and lakes. The idea of developing a GSM/GPS drifter suitable for coastal studies but fully satisfying the needs for small seas and archipelagos was born in the Hellenic Centre for Marine Research (HCMR). The technology was subsequently adopted by MARAC Electronics.

The novelty of the design lies in the use of a hollow PVC cylinder as a spine to which the drogue is attached. The surface module, equipped with the electronics (GSM, GPS, and ultra high-frequency, or UHF, modules; micro-processor; and batteries) has a cylindrical body of slightly smaller diameter than the drogue spine. In the configuration for measuring surface currents, the surface module is placed inside the drogue spine and the four donut-shaped floats are attached on the four top corners of the drogue.

The power source comprises four alkaline D-cell batteries. There are no cables and connectors for program-ming and data transfer. The drifter is provided with a GSM and a short-range UHF communication module. As soon as the drifter is turned on, it calls the base station (using both modules) and receives the program of its new mission. The GSM sampling rate as well as the frequency of data trans-mission to the base station are fully controlled by the user-friendly software at the base station provided with the drifters. All communication, programming and data exchange are performed through either the GSM or the UHF modules. The GSM technology offers the capability to monitor the drifter fleet anywhere, provided there is a GSM signal both at the measurement site and at the base station site. The base station can also be located on a vessel that follows the drifters. The software can foresee whether the condition of the GSM signal is too low for communication. In that case, the drifter position values are stored in the drifter's memory and are transmitted to the base station when the GSM or UHF signals allow such communication. Thus, there is no loss of data regardless of the GSM coverage.

The base station software provides the ability not only to program and monitor the drifter fleet, but also to analyze the data and facilitate the following and recovering of the drifters. A special module of the software enables the connection to a GPS onboard the vessel for the recovery of the floats. As soon as a drifter is selected for recovery, the software provides the necessary information (heading and distance) for finding and recovering the instrument. This module, along with the two-way communication, changes the nature of the drifter from an expendable instrument to a recoverable one that can be reused. This change in the use of the instrument significantly lowers the actual purchase cost due to the added value through its repeated use.

Another software module enables monitoring of the status of each drifter (power of each battery, memory

capacity, and level of GSM and UHF signal) before and during the measurements. If the energy levels of certain drifters are running dangerously low, it is possible through the two-way communication capability to reprogram the sampling and reporting strategy of that particular drifter in order to continue the measurements uninterrupted.

The telephonically tracked drifters are aimed not only for oceanographic research applications but also to support scientists dealing with coastal constructions, pollution, coastal managements, and so on. The main application of the telephonically tracked drifters is real-time monitoring and analysis of coastal (as well as lake and reservoir) advection and dispersion. The spatial scale of the phenomenon is limited only by the GPS position error on one hand (a few meters) and, on the other hand, the scale of GSM coverage of the order of 30 kilometers from the coast. An example of use in a small aquatic basin is their deployment in the rowing facilities for the 2004 Olympic Games, a reservoir about 1,500 by 120 meters.

3.2. LIMITATIONS OF SURFACE DRIFTERS

We have seen that the devices used for Lagrangian-style current measurements are of several functional types. Trajectories of the surface water masses of the ocean can be determined by following the drift of floating bodies that are carried by the currents. It is necessary, however, to exercise considerable care in interpreting the data derived from such bodies, because often the wind has carried them through the water. In practice, a Lagrangian drifter of any design is only a quasi-Lagrangian device because it never "perfectly" locks to a particular water mass (Vachon, 1977, 1980). This is particularly true in the surf zone current measurements wherein the drifter performance can be degraded by both the rectification of oscillatory wave motions and by windage. To ensure that the buoys do move with the water and to mini-mize the effect of wind, they are frequently fitted with a subsurface drogue to provide additional water drag and more effective coupling with the water motions. This drogue may be in the form of either a parachute or a window shade. Despite these precautions, rectification can occur because the drifter behaves as a damped, nonlinear oscillator forced by buoyancy, flow drag forces, and pressure gradients (Davis, 1985). For example, if the area submerged or the tilt of the drifter from vertical (and hence drag) depends on the wave phase, the mean drifter velocity may be nonzero, even for a zero-mean orbital velocity.

REFERENCES

Abascal, A.J., Castanedo, S., Medina, R., Losada, I.J., Fanjul, E.A., 2009. Application of HF radar currents to oil spill modeling. Mar. Poll. Bull. 58, 238–248.

Bellamy, I., Rigler, J., 1986. Utilization of "Service ARGOS" for the remote monitoring of oceanographic data acquisition systems. Oceanology (Society for Underwater Technology published by Graham & Trotman) 6, 69–76.

Brander, R., Short, A., 2000. Morphodynamics of a large-scale rip current system at Muriwai Beach. New Zealand, Mar. Geol. 165, 27–39.

Brander, R., 1999. Field observations on the morphodynamic evolution of a low-energy rip current system. Mar. Geol. 157, 199–217.

Briscoe, M.G., Frye, D.E., 1987. Motivations and methods for ocean data telemetry. Marine Technol. Society J. 21 (2), 42–57.

Carpiniello, F., Buell, H., 1972. Doppler systems applied to area navigation. Navigation 9 (3), 260–265.

Cresswell, G.R., 1977. The trapping of two drifting buoys by an ocean eddy. Deep-Sea Res. 24, 1203–1209.

D'Asaro, E., 1992. Estimation of velocity from Argos-tracked surface drifters during ocean storms. J. Atmos. Ocean. Technol. 9, 680–686.

Davis, R., 1983. Oceanic property transport, Lagrangian particle statistics, and their prediction. J. Marine Res. 41, 163–194.

Davis, R.E., 1985. Drifter observations of coastal surface currents during CODE: The method and descriptive view. J. Geophys. Res. 90, 4741–4755.

Davis, R., 1991. Lagrangian ocean studies. An. Rev. Fluid Mech. 23, 43–64.

Doutt, J.D., Frisk, G.V., Martell, H., 1998. Using GPS at sea to determine the range between a moving ship and a drifting buoy to centimeter-level accuracy. Proc. Oceans '98, 1344–1347. (Nice, France, Institute of Electrical and Electronic Engineering).

Edwards, K.P., Hare, J.A., Werner, F.E., Blanton, B.O., 2006. Lagrangian circulation on the southeast U.S. continental shelf: Implications for larval dispersion and retention. Cont. Shelf Res. 26, 1375–1394.

Fukunishi, M., 1948a. The tidal current in Kanmon Strait (I). J. Japan Soc. Civil Engineers 33 (2), 10–13 (in Japanese).

Fukunishi, M., 1948b. The tidal current in Kanmon Strait (II). J. Japan Soc. Civil Engineers 33 (2), 16–19 (in Japanese).

Gaskin, S., Kemp, L., Nicell, J., 2002. Lagrangian tracking of specified flow parcels in an open channel embayment using phosphorescent particles. ASCE Conf. Proc. 113, 19.

Gawarkiewicz, G., Monismith, S., Largier, J., 2007. Observing larval transport processes affecting population connectivity progress and challenges. Oceanography 20 (3), 40–53.

George, R., Largier, J.L., 1996. Description and performance of finescale drifters for coastal and estuarine studies. J. Atmos. Oceanic Technol. 13, 1322–1326.

Grundling, M.L., 1977. Drift observations from Nimbus VI satellite-tracked buoys in the southwestern Indian Ocean. Deep-Sea Res. 24, 903–913.

Hofmann-Wellenhof, B., Lichtenegger, H., Collin, J., 1997. Global positioning system: Theory and practice, 4th ed. Springer-Verlag, Wien, Austria.

Johnson, D., Pattiaratchi, C., 2004a. Application, modelling and validation of surfzone drifters. Coastal Engineering 51, 455–471.

Johnson, D., Stocker, R., Head, R., Imberger, J., Pattiaratchi, C., 2003. A compact, low-cost GPS drifter for use in the oceanic nearshore zone, lakes and estuaries. J. Atmos. Ocean. Technol. 18, 1880–1884.

Johnson, D., Pattiaratchi, C., 2004b. Transient rip currents and nearshore circulation on a swell dominated beach. J. Geophys. Res. 109, C02026.

Larson, M.R., Foreman, M.G.G., Levings, C.D., Tarbotton, M.R., 2003. Dispersion of discharged ship ballast water in Vancouver Harbour, Juan De Fuca Strait, and offshore of the Washington Coast. J. Environ. Eng. Sci. 2, 163–176.

List, E., Gartrell, G., Winant, C., 1990. Diffusion and dispersion in coastal waters. J. Hydraul. Eng. 116, 1158–1179.

MacMahan, J., Brown, J., Thornton, E., 2009. Low-cost handheld global positioning system for measuring surf-zone currents. J. Coastal Res. 25 (3), 744–754.

McPhaden, M.J., Hansen, D.V., Richardson, P.L., 1991. A comparison of ship drift, drifting buoy, and current meter mooring velocities in the Pacific South Equatorial Current. J. Geophys. Res. 96, 775–781.

Munk, J.W., Wunsch, C., 1979. Ocean acoustic tomography: a scheme for large scale monitoring. Deep-Sea Res. 26A, 123–161.

Muzzi, R., McCormick, M., 1994. A new global positioning system drifter buoy. J. Great Lakes Res. 3, 1–4.

Nayak, M.R., Peshwe, V.B., Tengali, S.B., 1993. Satellite tracked drifting buoys for ocean circulation studies. Bull. Oceanol. Teor. Applic. X1 (2), 125–130.

Pal, K., Murthy, R., Thomson, R., 1998. Lagrangian measurements in Lake Ontario. J. Great Lakes Res. 24 (3), 681–697.

Pickard, G.L., Emery, W.J., 1982. Instruments and Methods. In: Descriptive physical oceanography: An introduction. Pergamon Press, New York, pp. 77–124.

Rossby, T., Webb, D., 1970. Observing abyssal motions by tracking swallow floats in the SOFAR channel. Deep Sea Res. 17, 359–365.

Royer, T.C., Hansen, D.V., Pashinski, D.J., 1979. Coastal flow in the northern Gulf of Alaska as observed by dynamic topography and satellite-tracked drogue drift buoys. J. Phys. Oceanogr. 9, 785–801.

Sasaki, T., Horikawa, K., 1975. Nearshore current system on a gently sloping bottom. Coast. Eng. Jpn. 18, 123–142.

Sasaki, T., Horikawa, K., 1978. Observation of nearshore current and edge waves. Proc. 16th International Conference on Coastal Engineering, ASCE, 791–809.

Schmidt, W.E., Woodward, B.T., Millikan, K.S., Guza, R.T., Raubenheimer, B., Elgar, S., 2003. A GPS-tracked surf zone drifter. J. Atmos. Ocean. Technol. 20, 1069–1075.

Shepard, F., Emery, K., LaFond, E., 1941. Rip currents: a process of geological importance. J. Geol. 49, 337–369.

Shepard, F.P., Inman, D.I., 1950. Nearshore water circulation related to bottom topography and wave refraction. Trans. Amer. Geophys. Union 31, 196–212.

Sherman, J., 1992. Observations of Argos performance. J. Atmos. Ocean. Technol. 9 (3), 323–328.

Short, A., Hogan, C., 1994. Rip currents and beach hazards: their impact on public safety and implications for coastal management. J. Coast. Res., SI 12, 197–209.

Sonu, C., 1972. Field observations of nearshore circulation and meandering currents. J. Geophys. Res. 77, 3232–3247.

Stommel, H., 1949. Horizontal diffusion due to oceanic turbulence. J. Mar. Res. 8, 199–225.

Swallow, J., 1955. A neutral-buoyancy float for measuring deep currents. Deep-Sea Res. 3, 74–81.

Teague, C.C., 1986. Multi-frequency HF radar observations of currents and current shears. IEEE Journal of Oceanic Engineering, OE-11(2), 258–269.

Ullman, D.S., O'Donnell, J., Kohut, J., Fake, T., Allen, A., 2006. Trajectory prediction using HF radar surface currents: Monte Carlo simulations of prediction uncertainties. J. Geophys. Res. 111, C12005.

Vachon, W.A., 1977. Current measurement by Lagrangian drifting buoys —Problems and potential. Proc. Oceans '77, 46B-1 to 46B-7.

Vachon, W.A., 1980. Drifters. In: Dobson, F., Hasse, L., Davis, R. (Eds.), Air-Sea Interaction, Instruments and Methods. Plenum Press, London, UK, pp. 201–218.

Whelan, E.T., Tornatore, H.G., Murray, S.P., Roberts, H.H., Wiseman, W.J., 1975. An over-the-horizon radio direction-finding system for tracking coastal and shelf currents. Geophys. Res. Letters 2, 211–213.

Zervakis, V., Ktistakis, M., Georgopoulos, D., 2005. TELEFOS: A New Design for Coastal Drifters. Sea Technol. 46 (2), 25–30.

BIBLIOGRAPHY

Brander, R., Short, A., 2000. Morphodynamics of a large-scale rip current system at Muriwai Beach, New Zealand, Mar. Geol. 165, 27–39.

Cresswell, G.R., 1976. Adrifting buoy tracked by satellite in the Tasmanian Sea. Australian J. Marine and Freshwater Res. 27, 251–256.

Kirwan Jr., A.D., McNally, G., Change, M.S., Molinari, R., 1975. The effect of wind and surface currents on drifters. J. Phys. Oceanogr. 5, 361–368.

Kirwan Jr., A.D., McNally, G., Pazan, S., 1978. Wind drag and separations on undrogued drifters. J. Phys. Oceanogr. 8, 1146–1150.

Krauss, W., Dengg, J., Hinrichsen, H.H., 1989. The response of drifting buoys to currents and wind. J. Geophys. Res. 94, 3201–3210.

Longuet-Higgins, M.S., 1970. Longshore currents generated by obliquely incident seawaves, 1. J. Geophys. Res. 75, 6790–6801.

MacMahan, J.H., Thornton, E.B., Reniers, A.J.H.M., 2006. Rip current review. Coastal Engineering 53, 191–208.

Montiero, L.S., Moore, T., Hill, C., 2005. What is the accuracy of DGPS? J. Navigation 58, 207–255.

Murray, S.P., 1975. Trajectories and speeds of wind-driven currents near the coast. J. Phys. Oceanogr. 5, 347–360.

Murthy, C., 1975. Dispersion of floatables in lake currents. J. Phys. Oceanogr 5 (1), 193–195.

Niiler, P., Davis, R., White, H., 1987. Water-following characteristics of a mixed-layer drifter. Deep Sea Res. 34, 1867–1881.

Niiler, P.P., Sybrandy, A.S., Bi, K., Poulain, P.M., Bitterman, D., 1995. Measurements of the water-following capability of holey-sock and TRISTAR drifters. Deep Sea Res. 42, 1951–1964.

Rao, R.K., Sarma, A.D., Kumar, R.Y., 2006. Technique to reduce multipath GPS signals. Current Sci. 90, 207–211.

Saeki, M., Hori, M., 2006. Development of an accurate positioning system using low-cost L1 GPS receivers. Computer-Aided Civil & Infrastructure Engineering 21, 258–267.

Spydell, M., Feddersen, F., Guza, R.T., Schmidt, W.E., 2007. Observing surfzone dispersion with drifters. J. Phys. Oceanogr. 37 (12), 2920–2939.

Witte, T.H., Wilson, A.M., 2005. Accuracy of WAAS-enabled GPS for the determination of position and speed over ground. J. Biomechanics 38, 1717–1722.

Chapter 4

Remote Mapping of Sea Surface Currents Using HF Doppler Radar Networks

Chapter Outline

4.1. Crombie's Discovery	110
4.2. Peculiarities of Pulse Doppler-Radar Echo Spectra	112
4.3. Estimation of Sea Surface Current Using the Bragg Resonance Principle	113
4.4. Depth Extent of Doppler Radar-Based Sea Surface Current Measurements	113
4.5. Technological Aspects of Doppler Radar-Based Surface Current Mapping	114
4.6. Experimental Developments	115
4.7. Instrumentation Aspects	118
4.8. Radial and Total Vector Currents	120
4.9. Developments on Operational Scales	120
4.9.1. CODAR	121
4.9.2. OSCR	121
4.9.3. SeaSonde	123
4.9.4. WERA	124
4.9.5. Systems for Special Applications	126
4.10. Intercomparison Considerations	126
4.11. Advantages of Radio-Wave Doppler Radar Measurements	129
4.12. Round-the-Clock Coast- and Shelf-Observing Role of Doppler Radar	131
4.13. Detection and Monitoring of Tsunami-Induced Sea Surface-Current Jets at Continental Shelves	132
References	133
Bibliography	135

Real-time sea surface current measurements are needed for applications such as port operations, surveys, and oil-slick dissipation. Operational oceanography has not yet reached meteorology's operational level, primarily because progress in numerical modeling still has to be made and routine data assimilation is still to be developed. Until the arrival of remote sensing technologies and their routine application on a real-time basis, surveying companies and similar agencies had to rely heavily on *in situ* real-time current measurements for assisting offshore operations. Although providing valuable information, this approach is costly (human and ship costs for at-sea operations, instrument deployment, and so on), time consuming (often moorings have to be deployed for lengthy periods to gather enough data for reliable statistics to be established), and somewhat incomplete in terms of budget considerations in the sense that only scattered point measurements can be made (Vigan, 2002).

Forecasts cannot be established with *in situ* data. To predict local water circulation, even at relatively short time scales, a large area must be monitored synoptically with a high spatial and temporal coverage. Based on these considerations, and thanks to the technological leap achieved in the last few decades, the present trend in sea surface current measurements is to use remote sensing technologies.

Remote sensing in the oceanographic context is the science of making measurements from a distance without placing measuring instruments into physical contact with the sea surface. Applications of remote sensing technologies are especially valuable, and sometimes crucial, in coastal waters when the survival of conventional instrumentation becomes rather difficult because of rough weather conditions or other local factors.

The ideal tool for measurement of surface currents in coastal water bodies should be a land-based system that remotely scans the sea surface day and night at a rapid rate in all weather conditions and provides real-time vector maps of the surface current patterns at close spatial and temporal intervals. Although the backscattering mechanism underlying the radio wave Doppler radar technique for sea surface current mapping (Figure 4.1) was discovered by D. D. Crombie in the early 1950s (Crombie, 1955), nonavailability of miniature computers required for online digital signal processing hampered the development of an operational sea surface current mapping system for more than two decades after the discovery was made.

The basic mechanism underlying the application of radio wave (HF/VHF) Doppler radar systems for sea surface current mapping is that the phase velocity of sea

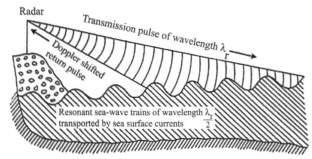

FIGURE 4.1 Schematic of the Bragg resonant backscattering of radio wave pulses from a Doppler radar impinging on the sea surface. *(Source: In part from Barrick et al., 1977.)*

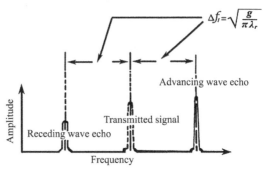

FIGURE 4.2 Schematic illustration of the Doppler shift Δf_1 of the ideal first-order backscattered sea echo spectrum of HF/VHF Doppler radar signal in the presence of sea surface gravity waves propagating on still water. λ_r is the radar wavelength. *(Source: Barrick et al., 1977.)*

surface waves is altered by the underlying water currents. The basic interaction between electromagnetic (EM) and ocean waves is *backscattering*. When EM waves transmitted from radar impinge upon a rough sea surface, a certain amount of incident energy is reflected, or *scattered*, back to the radar. This return signal contains information about the sea surface. The two basic processes or mechanisms responsible for backscatter of EM energy from the sea surface are specular reflection and Bragg scattering. If there exists some relative motion between the radar and the target (sea surface in the present case), the return signal exhibits a shift in frequency (i.e., the frequency of the received signal is different from the frequency of the transmitted signal). This shift in frequency, known as *Doppler shift*, is directly related to the relative velocity and inversely proportional to the radar wavelength. The return signal, displayed as a Doppler frequency spectrum, is analyzed to provide information on the motion of the target.

4.1. CROMBIE'S DISCOVERY

In the 1950s, while monitoring the HF radio wave signals backscattered from the sea surface, D. D. Crombie, a communications engineer, observed two peculiar and unique dominant peaks in its spectrum. The observed spectrum was unique because whereas the familiar Doppler spectrum of the radio wave echoes from a discrete target, such as an aircraft, shows only one peak, the spectrum of backscattered radar signals from the sea surface revealed the following spectacular peculiarities:

- Two well-defined peaks were present, and these were placed on either side of the radar transmission frequency, representing positive and negative Doppler peaks (Figure 4.2).
- The displacement of these peaks from the transmission frequency varied as a function of the square root of the transmission frequency rather than in direct proportion to the transmission frequency, as in the case of the Doppler echo from a discrete target.

- The strength of the positive and the negative Doppler peaks was proportional to the height of the advancing and the receding sea waves but was much larger than that expected from a discrete wave (target).

From conventional knowledge, we know that one target can generate only one echo, and therefore a "casual" observer would have attributed the second echo to the presence of "electrical noise" or "instrument noise" in the radar system. However, Crombie's "observing" mind did not allow him to neglect the observed double peaks as mere "instrument noise." Rather, he believed that he could be on the verge of a great discovery.

Based on the hitherto unreported unique peculiarities that he observed in the spectrum of the backscattered signals from the sea surface, Crombie sought to establish his discovery of the scattering mechanism as follows:

- Because the Doppler echoes consist of two discrete peaks placed on either side of the radar transmission frequency, these echoes must be originating from two targets (located on the sea surface) that move apart at a constant velocity.
- These two targets must be the advancing and the receding sea surface waves.
- The considerably enhanced strength of the two Doppler peaks compared to those of discrete targets of the same size (in the present case, sea surface waves) must be attributed to a resonant backscattering mechanism.

Based on this sound reasoning, Crombie postulated that the strong backscatter of HF radio waves emanating from the sea surface is a form of Bragg scattering, resulting from resonant sea surface wave trains lining up to form a diffraction grating for the radio waves. Conceptually, Bragg resonance can be described as the constructive reinforcement of radio waves reflected back selectively from those sea surface waves having half the incident radio wavelength, in a manner analogous to X-ray diffraction from the arrays of atoms in a crystal lattice. Although the heights of ocean waves are generally small compared to the HF/VHF radar wavelength, the backscattered echo is surprisingly large and readily interpretable in terms of its Doppler features.

Based on the arguments Crombie put forward, he postulated that, in the absence of sea surface currents, the sea-echo spectrum consists, to first order, of two sharp lines with positive and negative Doppler shifts. These Doppler shifts correspond to the phase velocity of radially advancing and receding sea surface waves (i.e., those waves of which the wavefronts are oriented in the direction of the radar) of one-half the radar wavelength. A steady sea surface current transports the sea surface waves so that their apparent phase velocity, as measured by a stationary observer, is the sum of their phase velocity in still water (i.e., nonmoving water) and the component of near-surface current in the direction of wave travel. Any shift in the measured phase velocity of the sea surface wave from its theoretical still-water value in deep water can, therefore, be related directly to the vector component of the sea surface current moving toward or away from the radar site.

It may be noted that phase velocity of deep-water waves is considered here for convenience, because of its mathematical simplicity relative to that of shallow-water waves. HF/VHF radar can therefore infer radial surface current from the amount of displacement of the "Bragg lines" of the echo spectrum from symmetry about zero Doppler (see Figure 4.3). The HF/VHF part of the electromagnetic spectrum is important because one-half the incident radio wavelength, corresponding to the HF/VHF frequencies, is comparable to the wavelengths of typical sea surface gravity waves.

In principle, Crombie's hypothesis provided a means for estimation of radial components of water currents at different cells on the sea surface. Measurements of range-Doppler spectra from different cells on the sea surface using two widely separated radars would thus enable mapping of two-dimensional sea surface current vectors at every individual cell.

The theory behind the application of the Doppler radar principle for remote measurement of sea surface currents is as follows: In the absence of underlying water current, sea surface wave trains of a given wavelength L propagate in deep water with a phase velocity U_o given by

$$U_o = \sqrt{\frac{g}{k}}, \qquad (4.1)$$

where $k = \frac{2\pi}{L}$ and g is the acceleration due to gravity. Thus, from Equation 4.1,

$$U_o = \sqrt{\frac{gL}{2\pi}} \qquad (4.2)$$

Assuming that the radar "sees" the sea surface gravity wave as a "target," the velocity V_1 of the sea surface gravity wave trains in the direction of the radar can be calculated from the well-known Doppler shift Δf_1 for target speed, namely:

$$\Delta f_1 = \frac{2V_1}{\lambda_r} \qquad (4.3)$$

where λ_r is the radar wavelength. Alternatively, substituting the relation $c = f\lambda_r$ in Equation 4.3, we get

$$\Delta f_1 = \frac{2fV_1}{c}, \qquad (4.4)$$

where f and c are the frequency and propagation speed, respectively, of the radar transmission signal.

From Equation 4.3,

$$V_1 = \lambda_r \frac{\Delta f_1}{2}. \qquad (4.5)$$

Equating the "target" velocity to the phase velocity of deep-water gravity waves in the absence of an underlying water current, we get

$$\lambda_r \frac{\Delta f_1}{2} = \sqrt{\frac{gL}{2\pi}}, \qquad (4.6)$$

where L is the wavelength of the sea surface gravity wave. According to Crombie's hypothesis, Bragg resonance occurs when $L = \frac{\lambda_r}{2}$. By substitution of $L = \frac{\lambda_r}{2}$ in Equation 4.6, we get

$$\Delta f_1 = \sqrt{\frac{g}{\pi\lambda_r}}. \qquad (4.7)$$

This frequency depends only on the radar transmission wavelength and is termed the *Bragg frequency*.

Substituting $\frac{c}{f}$ for λ_r in Equation 4.7, we get

$$\Delta f_1 = \sqrt{\frac{gf}{\pi c}}, \qquad (4.8)$$

This explains the observed square-root relationship between the Doppler shift and the radar transmission frequency. Any Fourier decomposition of a random but

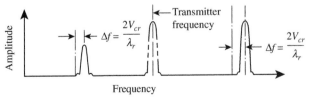

FIGURE 4.3 Schematic illustration of the displacement Δf of the first-order backscattered sea-echo spectrum of HF/VHF Doppler radar signal in the presence of sea surface gravity waves propagating over an underlying radial current V_{cr}. Note that λ_r is the radar wavelength. *(Source: Barrick et al., 1977.)*

finite patch of sea surface gravity waves always contains wave spectral energy at or near the required wavelength and direction to satisfy the Bragg resonance condition. Thus, discrete resonant Doppler peaks are always observed in the spectrum of the backscattered echoes from the sea under all sea states (except totally calm sea).

The presence of underlying water current displaces the Doppler spectra unidirectionally from their symmetrical positions, relative to the radar transmission frequency (Figure 4.3). It is from this displacement that an estimate of the radial current component is made, making the assumption that sea surface gravity waves and the underlying surface currents are linearly superimposed. The first-order scattering mechanism is highly selective in terms of the direction of wave propagation; that is, only those resonant sea surface gravity waves that are traveling radially toward or away from the radar contribute appreciable energy to the first-order echo. A second-order spectrum is also generated, but this is of no interest in the remote measurement of sea surface current.

4.2. PECULIARITIES OF PULSE DOPPLER-RADAR ECHO SPECTRA

An examination of the Doppler frequency spectrum of the return echo shows two distinct regimes. The first-order regime, or echo, is characterized by a sharp and distinct peak at a frequency corresponding to the sea surface wave satisfying the Bragg resonant condition. The second-order regime is the continuum portion of the spectrum and contains more information on the sea surface properties. This portion is caused by scatter from all sea surface waves and provides, after analysis, the directional wave-height spectrum. Thus, because of the presence of "first-order" and "higher-order" components, the observed Doppler spectra of the backscattered sea echo are more complicated than the ideal situation depicted in Figures 4.2 and 4.3.

A real situation is illustrated in Figure 4.4. The dominant spectral features explained by the simple, lowest-order terms of the perturbation analysis (Barrick, 1972) are referred to as a *first-order* sea echo, and the remaining less dominant features are termed *higher-order* because they arise from the smaller (i.e., second-order, third-order, etc.) terms. The large first-order resonant peaks are usually more dominant than the higher-order peaks but are insensitive to the sea state. The first-order lines are usually very sharp, often only a few millihertz wide. The second-order features, which result from wave-wave interactions, are highly sensitive to the sea state and manifest themselves as sidebands of the first-order echo (Weber and Barrick, 1977). This results in a natural broadening of the first-order peaks. However, the presence of the second-order side bands does not alter the position of the

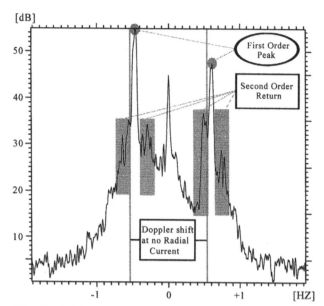

FIGURE 4.4 Measured HF Doppler spectrum of the backscattered radio-wave signal from the sea surface. First-order lines and second-order side bands are indicated. *(Source: Gurgel et al., 1999.)*

first-order Bragg lines. As long as its position can be accurately determined, the second-order continuum does not affect the current measurement capability of the Doppler radar system.

For all practical purposes, the backscattered sea echo-signal amplitude is a Gaussian random variable. This applies to both the first- and the second-order portions of the echo. The echo is random because of the statistical nature of the scattering sea surface. Furthermore, the sea surface current flow field immediately beneath the waves is not laminar but turbulent. This also contributes to the randomness of the received backscattered echo. The observed spectra are, therefore, often complicated in structure and of peculiar definitions, depending on the nature of the sea surface, currents, and the shallowness of the water body. Accordingly, the observed spectra can be categorized into different types (Eccles, 1989), such as:

- Classical (i.e., two narrow Bragg lines)
- Typical (two Bragg lines with a higher-order continuum surrounding them)
- Split lines (two close sets of Bragg lines)
- Turbulence (broad spiky Bragg lines)
- Shallow water (definite Bragg lines at less than normal Bragg separation)
- Stormy sea (large second-order component)
- Complex (several large peaks and confusing features)

In regions of strong spatial gradients in the surface circulation, two sets of Bragg lines are frequently observed, indicating the presence of two current regions. In such cases, to derive a representative radial current value, the two current values are weighted according to the signal strength of the Bragg lines. In shallow waters, the phase velocity of the sea surface gravity waves decreases compared to that of the deep-water waves. This results in the Bragg lines of the Doppler spectrum to be at less than normal separation. If the sea surface is devoid of any disturbances (i.e., gravity waves are absent), the frequency of the backscattered signal received by the radar will be equal to the frequency of the transmitted signal, and consequently the Bragg resonance will be absent. The resulting spectrum is called a *single-line spectrum* because the Bragg lines are absent. Some of the complications of the received signals and various aspects of signal detection schemes have been discussed by Leise (1984). Shay et al. (1995) have observed that even in a clearly wave-laden sea, normally only one Bragg peak is available in the spectra of the return signal originating from extreme ranges (i.e., close to the theoretical limit of the radar range). Measurements by Skop et al. (1995) indicated that propagation across land (during current measurements from bays and estuaries) decreases the signal strength, as expected, but does not result in extraneous Doppler peaks. The reduction in power at the Doppler peaks can be related to the amount of land mass swept by the signal. Certain anomalous behavior in the Doppler spectra appears in shoal regions during periods of low tide. The anomalous returns and the complexities in very shallow-water spectra are yet to be fully understood.

4.3. ESTIMATION OF SEA SURFACE CURRENT USING THE BRAGG RESONANCE PRINCIPLE

The mathematical expression for sea surface current measurements using Doppler radar systems is derived as follows: It is assumed that sea surface waves are accompanied by an underlying water current, at least the Stokes' drift generated by the waves. Let the sea surface wave ride on current with a radial velocity V_2. The Doppler radar recognizes this wave as a target, which moves with a velocity V_2. The Doppler shift Δf_2 detected by the radar is given by the Doppler equation:

$$\Delta f_2 = \frac{2fV_2}{c} \qquad (4.9)$$

Current-driven displacement of the observed Bragg resonance peak from the expected still-water Bragg resonance peak is given by

$$\Delta f = \Delta f_2 - \Delta f_1 \qquad (4.10)$$

From Equations 4.8–4.10,

$$V_2 = \left(\frac{c}{2f}\right)\left[\Delta f + \left(\sqrt{\frac{gf}{\pi c}}\right)\right] \qquad (4.11)$$

However, V_2 is the vector sum of the radial velocity of the surface current V_{cr} and the phase velocity U_o of the resonant sea surface gravity waves propagating toward and away from the radar in still water. Thus,

$$V_{cr} + U_o = V_2 \qquad (4.12)$$

However, for $k = \dfrac{2\pi}{L}$ (from wave theory) and $L = \dfrac{\lambda_r}{2}$ (Bragg resonance condition), the phase velocity of the resonant gravity waves can be expressed as:

$$U_o = \sqrt{\frac{gL}{2\pi}} = \sqrt{\frac{g\lambda_r}{4\pi}} = \sqrt{\frac{gc}{4\pi f}} \qquad (4.13)$$

From Equations 4.11–4.13, we get:

$$V_{cr} = \frac{c\Delta f}{2f} \qquad (4.14)$$

Thus, the displacement Δf by the Doppler peak of the radar echoes (see Figure 4.3), resulting from a radial current V_{cr}, is given by the expression:

$$\Delta f = \frac{2fV_{cr}}{c} = \frac{2V_{cr}}{\lambda_r} \qquad (4.15)$$

Thus,

$$V_{cr} = \frac{\Delta f \lambda_r}{2} \qquad (4.16)$$

The radar transmission frequency f is usually in MHz and the displacement Δf is usually in Hz. Therefore, for convenience, if f is expressed in MHz and the displacement Δf is expressed in Hz, the radial surface current V_{cr} (in m/s) can be denoted by the expression

$$V_{cr} = 150\left(\frac{\Delta f}{f}\right) \qquad (4.17)$$

4.4. DEPTH EXTENT OF DOPPLER RADAR-BASED SEA SURFACE CURRENT MEASUREMENTS

The depth extent of sea surface current measurements using the HF/VHF Doppler radar-based Bragg resonance principle is not related to the penetration depth of the incident electromagnetic radiation. The penetration depth δ of the incident electromagnetic wave is given by the expression

$$\delta = \sqrt{\frac{2}{\omega\mu\sigma}} \qquad (4.18)$$

where ω is the angular frequency (radian/s) of the radar transmission wave, μ is the permeability of the medium ($\mu = 4\pi \times 10^{-7}$), and σ is the conductivity ($\sigma = 4.3$ Siemens/m). Based on the above considerations, the penetration depth of the incident electromagnetic radiation of ~27 MHz is approximately 4.7 cm.

The depth-extent of Doppler radar-based sea surface current measurements is a complicated function of the depth extent of those sea surface waves that are responsible for Bragg scattering. Thus, the radar-derived sea surface current measurements represent a depth-integrated value in the uppermost water column. The phase velocity U of a sea surface wave superimposed on an underlying water current is a function of the wave number k of the sea surface wave, the vertical current profile $V(z)$ over depth z, and $\exp(-2kz)$ as given by the expression:

$$U = U_o \pm 2k \int_0^\infty V(z)\exp(-2kz)dz \qquad (4.19)$$

where U_o is the phase velocity of the sea surface wave in still water. Thus, the unidirectional displacement Δf of the Bragg spectra is the result of a depth-averaged current, with $\exp(-2kz)$ as a weighting function. However, the vertical current profile $V(z)$ is sheared and depends on various factors such as the hydrodynamic characteristics of the locality, wind characteristics, and so forth. Accordingly, the vertical current may have (1) an exponential profile, (2) a linear profile, or (3) a logarithmic profile. The effective radar sampling depth thus depends on the near-surface water-current structure. The sampling depths for linear, exponential, and logarithmic profiles are given by $\left(\frac{\lambda_r}{8\pi}\right)$, $0.7\left(\frac{\lambda_r}{8\pi}\right)$, and $0.011\lambda_r$, respectively (Collar, 1993). The sampling depth is typically 2–3 m, which is extremely close to the sea surface compared with other more familiar current measurements. The dependence of the mean sampling depth on the wavelength of the incident electromagnetic wave (i.e., radar wavelength, λ_r) can provide a means of investigating the vertical structure of water current very close to the sea surface if multifrequency transmissions are made. Because longer sea surface waves are affected by deeper currents than are shorter sea surface waves, the water-current shear close to the ocean surface can be estimated by varying the radar transmission frequency.

4.5. TECHNOLOGICAL ASPECTS OF DOPPLER RADAR-BASED SURFACE CURRENT MAPPING

Vertically polarized electromagnetic signals are used in the HF/VHF Doppler radar technology of sea surface current measurement because seawater is an excellent electric conductor at these frequencies. Thus, in addition to the radar wavelength, the radar polarization is another important specification of the system. *Polarization* is the orientation (i.e., direction) of the electric vector in the electromagnetic radiation related to the terrain. A sketch of a vertically polarized electromagnetic wave is shown in Figure 4.5. All electromagnetic waves propagate along straight lines in free space and, therefore, cannot generally propagate to other surface points below the horizon. However, in *ground-wave* mode, the vertically polarized propagating fields at these frequencies follow the curvature of the Earth and continue well into the shadow region beyond the horizon, even in the absence of atmospheric and ionospheric refractive-index anomalies (Barrick, 1972; Barrick et al., 1974; Barrick et al., 1977). Thus, using the ground-wave mode of propagation (Ponsford and Srivastava, 1990), targets can be detected at a range far greater than conventional sky-wave microwave radar. It may be noted that horizontally polarized propagating fields are never used in ground-wave propagation because a horizontally polarized sea echo is quite sensitive to the incidence angle (Barrick et al., 1974). High-frequency (HF) over-the-horizon (OTH) radars have been used for mapping sea surface currents over very large ocean areas. In this case, although the measurements are not affected by clouds or precipitation, they are subject to ionospheric distortion. In the case of OTH radars, ionospheric motions shift the sea-echo spectrum in the same way that currents do, thereby making it difficult to separate the two effects. One method to reduce the effects of ionospheric motions in OTH radars is to employ the nearby land echo as a zero-Doppler reference (Maresca and Carlson, 1980). Another possibility is to use the less stable ionospheric reflections by averaging the measurements taken at different times (i.e., low-pass filtering), taking advantage of the shorter

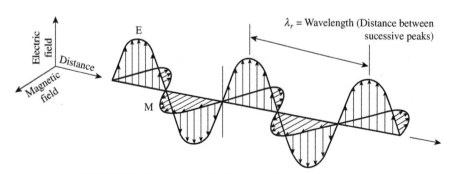

FIGURE 4.5 Sketch of a vertically polarized electromagnetic wave.

time variability of the ionosphere compared with that of ocean currents (Georges and Harlan, 1995).

It may be noted that the overall displacement of the entire Doppler spectrum, as a result of the transport of the sea surface waves by the underlying water current, is considerably smaller compared to the Doppler shift in the spectrum solely due to the phase velocity of the sea surface waves. Nevertheless, this small overall displacement is measurable. The Doppler spectra illustrated in Figures 4.2 and 4.3 are idealized representations. The observed spectra are often buried in noise. Discrimination of the signal from noise is a problem of digital signal processing. Online signal processing and estimation of sea surface currents are accomplished by special-purpose algorithms using micro-computers or minicomputers, which form an integral part of the electronic hardware of the Doppler radar system.

4.6. EXPERIMENTAL DEVELOPMENTS

Crombie employed vertically polarized HF radio waves propagating at grazing incidence to make practical use of his concept for remote mapping of sea surface currents using Doppler radar (Crombie, 1972). In his experiments, the backscattered echoes from the sea surface were measured from two known narrow azimuthal sectors on the sea surface. As with all pulse-Doppler radar systems, time gating of the received echo signal referenced to the trans-mitted pulse determines the range to the measurement location (i.e., the radar footprint, or illuminated sea surface area, from where the sea-echo originated). The footprint size is determined by range and radar beam width, which is a function of antenna size. For a real-aperture radar system, radar beam width and aperture are related by the expression:

$$\phi = 1.2\left(\frac{\lambda_r}{D}\right) \tag{4.20}$$

where ϕ and D are the radar beam width and the antenna aperture, respectively. Thus, the narrow beam width required for fine spatial resolution can be achieved only with a large antenna.

At any given instant in time, the backscattered signals are received from an angular section of the sea surface located at a given distance (say, R) from the antenna, given by the following equation:

$$R = \frac{ct}{2} \tag{4.21}$$

where c is the velocity of the radar transmission signal (i.e., 3×10^8 m/s) and t is the elapsed time between pulse transmission and return. The radial width β of the annulus (Figure 4.6) depends on the transmitted pulse width τ given by the expression:

$$\beta = \frac{c\tau}{2} \tag{4.22}$$

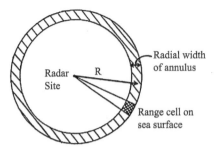

FIGURE 4.6 Backscatter from a range cell on a circular annulus of radius R.

whereas the angular width (of the range cells) corresponds to the receiving-antenna beam width. Thus, for pulse-Doppler radar, the radial range resolution is determined by the width of the transmitted radar pulse. For example, if the transmission pulse width τ is 20 μs, the signal sample from each range (time) gate would represent the echo from an annulus of the sea surface 3 km in width, concentric with the radar location. A noteworthy feature in this arrange-ment is that the system processes data from all azimuths in the same spectral analysis integration time. For a total distance of 75 km from the radar, time-gated echoes from a total of 25 consecutive concentric cells $\left(\text{i.e.,}\frac{75}{3}\right)$ are received and digitized. A pulse-Doppler radar system typically produces several velocity values in a given radar cell, which is a segment of a range cell defined by the azimuth angle from the radar and azimuth increment (see illustration in Lipa et al., 2006). These values are averaged to obtain the final output value for that location, and the standard deviation is calculated (Lipa et al., 2006). It is evident that their standard deviation increases with range as more vectors are included in the larger radar cells.

HF radar systems use different methods of spatial resolution in both range and azimuth. Range resolution is conventionally achieved by means of short pulses. Azi-muthal resolution is achieved by means of direction finding (phase comparison) and beam forming. In both cases it is assumed that the transmit antenna is omnidirectional or slightly directed toward the sea, and azimuthal resolution is performed by means of an array of receive antennas. Each single antenna receives echoes from different azimuthal directions, which are superimposed.

Historically, sea surface current-mapping, direction-finding (DF) antennas and algorithms have taken several forms. Barrick et al. (1977) and Gurgel (1997) used antenna elements separated by short distances and relied on phase-path differences to extract bearing, which is an ingenious technique originally conceived by Crombie. The direction-finding method is based on the Fourier decomposition of the time series received by three or four antennas. Each Fourier line is attributed to a Doppler shift and in turn to a radial speed. It is assumed that the different radial speeds arrive from different directions, which are determined by comparing the amplitudes and phases at the receive antennas. Figure 4.7 illustrates the principle of direction finding for a single plane wave, i.e., one Fourier component. To avoid ambiguities, the diagonal of the antenna array must be equal to or less than $\frac{\lambda_r}{2}$. Some further information on direction-finding algorithms is presented by Gurgel and Essen (1998) and Gurgel et al. (1999).

In the DF technique, the phase difference between the echo signals received at any two of the receiving antenna elements is used to obtain the angular direction of arrival of the echoes with reference to the line joining the receiving antenna elements. This is shown in Figure 4.8. The phase difference (ϕ degrees) of the plane wave front received at the two antenna elements is related to the corresponding path difference l. Because a phase difference of 360° corresponds to a path difference of λ_r, the measured phase

difference (ϕ degrees) corresponds to a path difference l, given by the expression:

$$l = \frac{\lambda_r \phi}{360} \qquad (4.23)$$

where l is expressed in the same unit as λ_r. Referring to Figure 4.8,

$$l = d(\cos\theta) \qquad (4.24)$$

From Equations 4.23 and 4.24, we get

$$\theta = \cos^{-1}\left(\frac{\lambda_r \phi}{360 d}\right) \qquad (4.25)$$

Because λ_r and d are known for a given arrangement of the receiving antenna elements and ϕ is measured, the direction of arrival θ of the echo with reference to the receiving antenna elements can be estimated. Thus, by comparing the phases between two noninteracting antennas separated by a known distance (usually less than $\lambda_r/2$), it is possible to uniquely determine the direction of arrival of the signal over a sector of 180° or less. In the experimental stages, Barrick et al. (1977) employed three collinear independent receiving antennas, each separated by distance $\lambda_r/4$ and aligned parallel to a straight coastline. The three-element system was later changed to a four-antenna configuration (arranged in a square) to resolve two signals from 360°. This permitted the operation of the radar on peninsula or island. A detailed description of the four-element receiving antenna system is given by Jeans and Donnelly (1986).

The main advantage of direction finding is the small extend of the receive antennas array (i.e., $\lambda_r/2$). The direction-finding method of achieving azimuthal resolution has been applied successfully for surface current measurements. The main disadvantage of the direction-finding method is that it is not appropriate for extracting information on the sea state. This is because the second-order signal (which contains the sea surface wave information) from a certain direction is masked by first-order signals from deviating directions (Gurgel et al., 1999). Another disadvantage is the assumption that the different radial velocities all come from different directions, which might be invalid in a very inhomogeneous current field. Subdividing the time series in overlapping parts and using sophisticated algorithms help overcome this limitation. However, the beamforming method of achieving azimuthal resolution does not depend on such an assumption.

An alternative way to gain directional information on currents, and additionally on waves, is to form a set of real narrow beams. As mentioned earlier, large antenna arrays are needed to do this (see Figure 4.9), and they can be both expensive and cause siting problems. The output of a linear array of several receiving antennas (typically, 16) can be combined, in a process known as *beam forming*, to make narrow beams that are generally limited to a minimum

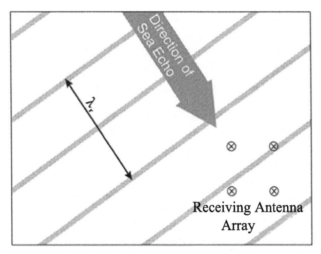

FIGURE 4.7 Principle of direction finding for a single plane wave, i.e., one Fourier component. The diagonal of the square antenna array is equal to or less than $\frac{\lambda_r}{2}$. *(Source: Gurgel et al., 1999.)*

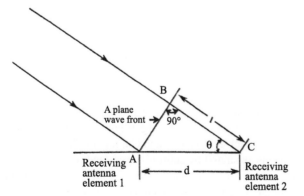

FIGURE 4.8 Diagram showing the principle of estimation of the direction of arrival (θ) of an echo with reference to the line joining the receiving antennas. l is the path difference of a plane wave front at the two receiving antenna elements, and d is the separation between the receiving antenna elements.

FIGURE 4.9 An example of installation of a linear array of HF Doppler radar receive antenna close to the seacoast. *(Source: Hammond et al., 1987.)*

width of about λ_r/D radians, where λ_r is the radar wavelength and D the overall length of the array. The optimum distance of adjacent antenna elements of the array is $\lambda_r/2$.

The advantage of beam forming is that the beam can be steered to achieve particular area coverage. Beam forming generally increases the antenna gain, which in turn increases the signal-to-noise ratio of the echoes received. Modern digital beam formers can also be configured to form nulls in the antenna pattern in the direction of any interfering radio systems.

The oldest types of beam formers make use of time delays in cables. If all the cables from the antennas to the receiver are of the same length, then a beam is formed along the boresight perpendicular to the array. By introducing a progressive increase in cable length from one end of the array to the other, the beam can be steered to either side of the boresight as required, up to a practical maximum of perhaps 45°. The just mentioned time-delay beam formers are slow and unreliable but have the great advantage of having a very wide instantaneous bandwidth.

The effect of a time delay at each antenna can be reproduced by electronically inserting the equivalent phase change. It is much quicker to steer a beam using such phase beam steering. Furthermore, the unreliability, expense, and weight of cables and switches can be avoided. There are many ways of electronically changing the phase of a radio signal, and the only real disadvantage of these systems, the limited instantaneous bandwidth, is not usually a problem for HF radar sea sensing (Gurgel et al., 1999).

The most advanced type of forming antenna patterns is digital beam forming. In this scheme, each antenna element in the array has its own receiver and analog-to-digital converter, and the beams are formed by digitally processing all the outputs. The system is very flexible, and beams can be recalculated to any shape and even formed after the experiment. Different weighting (window) functions can be

applied to control the antenna side lobes, and correction factors can be built in to account for the end elements of the array behaving in a different way from those at the center. The WERA radar of Gurgel et al. (1999) uses this type of beam forming, as shown in Figure 4.10. A typical idealized beam formed from a 16-element array parallel to the coast is shown in Figure 4.11.

A practical difficulty affecting most long linear arrays is the problem of erecting them on a straight coastline (see Figure 4.9). When two radar systems are used to map the sea surface currents from an area of water, the two arrays must be installed at an angle to the coast; otherwise their beams will not intersect (which is a necessary criterion to generate surface current vector maps of the region under consideration). When the two arrays are installed at an angle to the coast, antenna elements at one end of the array must necessarily be erected closer to the sea than those at the other. The differential radio propagation along land/sea paths is so great that the beams formed have much larger side lobes than

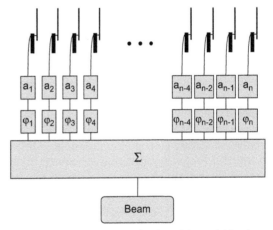

FIGURE 4.10 Principle of beam forming. After weighing by a_i and phase shifting by φ_i, the signals of n receive antenna signals are added. *(Source: Gurgel et al., 1999.)*

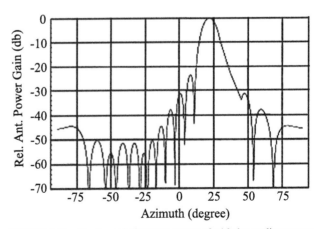

FIGURE 4.11 The theoretical antenna pattern of a 16-element linear array when the beam is steered 22° off boresight. *(Source: Gurgel et al., 1999.)*

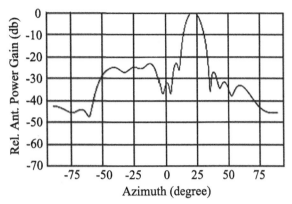

FIGURE 4.12 Same as Figure 4.11 but the pattern when the antenna is installed on real ground at an angle to the coastline. *(Source: Gurgel et al., 1999.)*

desired. Figure 4.12 shows the pattern of beams formed from a 16-element array when the array is set at an angle to a real coastline. Deviation of the pattern of beam from the ideal does not usually create problems in measuring currents, but it can cause corruption of some wave measurements. There is really no easy way around this problem, but the more compact the antenna array can be made, the smaller the effect, and it is now generally recognized that there is a need to develop such arrays (Gurgel et al., 1999).

It may be noted that only at HF (with a radar wavelength between 6 and 60 meters) can the sea echo be neatly decomposed. Microwave radars see an echo that is so complex as to be indecipherable in terms of the important underlying sea surface current information. In addition, HF radars with their long wavelengths see well beyond the visible horizon that limits all microwave radars. This is by virtue of the well-known rule that the lower the frequency, the farther the signal penetrates. If conventional radar wisdom is followed in forming and scanning a beam to derive bearing angle, the long wavelengths demand a huge coastal antenna facility. This has been the single major impediment blocking the acceptance of this tool by the ocean community, since it imposes large initial and operating costs, besides restricting the most obviously desirable site locations.

4.7. INSTRUMENTATION ASPECTS

Sea surface current mapping HF Doppler radar systems can be roughly divided into two types based on the method used to determine bearing to a sector of the ocean surface: (1) beam forming, or (2) direction finding. As mentioned earlier, beam-forming radars electronically steer a linear phased array of receive antennas toward a sector of sea surface; direction-finding radars exploit the directional properties of loop antennas to determine bearing. Digital signal processing and computing play a vital role in the operation of radio-wave pulse-Doppler radar systems used for mapping sea surface currents. The Doppler radar system consists of a transmit

antenna, an array of receiving antennae, an antenna control and multiplexing unit, digital control, digital beam-forming and data-acquisition units, front-end data processing and archival modules, and online sea surface current estimation and data communication algorithms run on a computer. Data processing involves several steps, including Fourier processing of the radar echo signal, formation of narrow receiving antenna beams in several different directions by digital beam-forming techniques, and estimation of Doppler shifts of the backscattered signal from which radial sea surface current velocities can be estimated.

The transmitting antenna system radiates electromagnetic energy to the sea surface, usually using a wide-beam omnidirectional transmitting antenna, as a stream of continuous wave (CW) short bursts (i.e., pulses) of width τ (usually less than 20 μs) at every few milliseconds, to floodlight the survey area. Timing signals in the hardware, controlled by a computer, ensure coherence between the transmitter and receiver internal signals. A steerable narrow receiving beam is usually employed for the selection of the desired azimuth and directivity to achieve gain and for rejection of noise, interference, and sea echo from other directions (Shearman, 1986). The receiving beam can be steered in several directions by varying the relative complex weights applied to the signals emanating from each receiving element of the linear receiving antenna array. This permits velocity measurements in several directions. The beam formation is done in software in the frequency domain (Teague, 1986). Electronic steering of the receive beam becomes necessary because, in contrast to microwave radars, mechanically rotating an HF/VHF antenna is impractical due to its prohibitively larger size (see Figure 4.9).

In the receiver section, under computer control, the receiver gain is increased at an interval of τ to compensate for the decrease of echo strength with increasing range. This procedure, known as sensitivity time control (STC), is similar to the automatic gain control (AGC) commonly used in echo sounders and acoustic Doppler current profilers (ADCPs). In the receiving circuitry, after amplification, the echo signal is coherently mixed down in a mixer to produce an in-phase and quadrature (I and Q) intermediate frequency (IF) centered at 0 Hz (i.e., zero intermediate frequency). The mixing process demodulates the Doppler frequency-shifted echo signal (usually a few tens of megahertz) and provides an output signal that contains only the Doppler shift component (a few Hertz). The use of I and Q signals allows the separation of positive and negative Doppler shifts so that approaching waves can be distinguished from receding waves (Teague, 1986). After suitable amplification, the I and Q signals are digitized with a high-resolution analog-to-digital (A/D) converter every τ s (i.e., corresponding to each range gate). Subsequently, these signals are digitally filtered over $1/n$ seconds to reduce the signal bandwidth to n Hz. This process filters out noise beyond this bandwidth, which originated from various natural and manmade sources, near-DC components, higher-order scattering components, spurious spikes due to interference, and so forth. The value of n is typically 2 to 4.

Because the radio signals backscattered by the sea surface consist of a pair of very narrow side bands displaced above and below the carrier frequency by a few Hz or a fraction of a Hertz at HF/VHF, this value of n easily accommodates the cumulative Doppler shifts in the received echo arising from the phase velocity of the sea surface waves and sea surface currents. A number of such digitally filtered complex samples (i.e., real part and imaginary part) for each range gate and each receiving antenna are then collected for N s and added together in an accumulator and averaged to improve the S/N ratio. These time-averaged signals are then input to a complex fast Fourier transform (FFT), thereby converting the time-domain signal into frequency-domain information, which reveals the Doppler shift of the backscattered echo frequency with reference to the transmitted frequency. The first-order echoes are extracted from the second-order continuum by first searching for the peak of the power spectrum in a narrow region centered on the theoretical position of the Bragg line. Then the nulls between the first-order peak and the second-order continuum on either side of the Bragg line are found by searching for the points at which the local slope of the smoothed frequency spectrum changed sign. The nulls thus define the limits of the first-order energy band. This technique has been successfully used by Teague (1986).

The displayed spectrum will then have a Doppler resolution of $1/N$ Hz over a window from $-n$ to $+n$ Hz, where N is the integration time. The $1/N$-Hz Doppler resolution yields a radial current velocity resolution of $\lambda_r/2N$ cm/s. For $N = 256$, a 25-MHz (i.e., $\lambda_r = 12$ m) Doppler radar provides a radial current velocity resolution of $\sim$2.5 cm/s.

The essence of data processing is to detect the centroid of the first-order spectrum (the first-order moment). The centroid is found by integrating the spectrum over the first-order energy band. The frequency corresponding to the centroid of the spectrum is taken as the frequency of the Bragg line. The surface-current-induced displacement (Δf) in the Doppler shift (see Figure 4.3) is determined by subtracting the theoretical Bragg frequency from the measured frequency. The difference (Δf) is converted to current velocity using the relation $V_{cr} = \dfrac{\lambda \Delta f}{2}$. The processing software also determines the range and bearing corresponding to each of the unique radial velocity estimates.

Although the backscattered energy from the Bragg scattering process is much larger than that from other reflections, in practice many observations must be averaged to obtain a statistically reliable estimate of the locations of the spectral peaks. Because of the requirement of a large

spatial and temporal averaging, a single radar observation bin is usually of the order of 1–3 km wide and approximately 1 h in duration (Paduan, 1995).

A particular transceiver system (i.e., a pair consisting of a transmitter and a receiver) can measure only the radial components of currents with respect to an antenna from various range cells defined by the geometry of the transceiver system. To obtain an unambiguous map of two-dimensional sea surface current flow pattern at maximum resolution, it is necessary to obtain measurements from at least two radar systems located in different spatially distant locations. The use of two systems to determine radial currents at a given coordinate from two directions allows calculation of the resultant current speed and direction (i.e., total vector current).

To accomplish this, two separate radar systems (one master and one slave) are to be installed and positioned at a sufficiently large distance apart (typically 20–30 km). When the two radars are properly sited, their beams overlap over the survey area of the sea surface at sufficiently large angles to permit accurate estimation of sea surface current vectors at different surface bins (defined by a distinct range cell and angle section). Because each individual radar system measures the component of current along its line of sight, two radars together give unambiguous estimates of sea surface currents.

In practice, two sites at least 5 km apart are chosen. This is illustrated in Figure 4.13. During the sampling interval, the master radar system transmits, receives, and processes the backscattered energy and interpolates the information from each of the numerous grid points. This entire process is repeated several times over a few minutes. During this period, the slave radar system also transmits, receives, and processes its signals, followed by a few minutes of data reduction. Subsequently, slave data are sent to the master radar system's computer via a UHF radio link, to combine the radial currents gathered by the two radar systems into two-dimensional current vectors (speed and direction), store the time-series datastream, and display the sea surface current map on a video monitor. To accomplish this, the latitudes and longitudes of the two radar sites are entered into the master site computer, along with the azimuthal bearings of the two receiving antenna arrays. The software then converts the radar-oriented polar coordinates of each bin into rectangular coordinates. Details of this aspect may be found in Lipa and Barrick (1983) and Leise (1984).

If the signal from a given bin is too weak, the sea surface current velocity at that bin is usually calculated by interpolation from the adjacent range azimuth cells. Averaging hundreds of current maps enables the removal of random errors. The angular sectors very near the shore (i.e., baseline) are sometimes excluded because the nearly parallel radial velocities seen from each site at these grid points give rise to large vector errors. Baseline instability arises from the fact that the transverse velocity component cannot be sensed. Consequently, only one degree of freedom is measured and

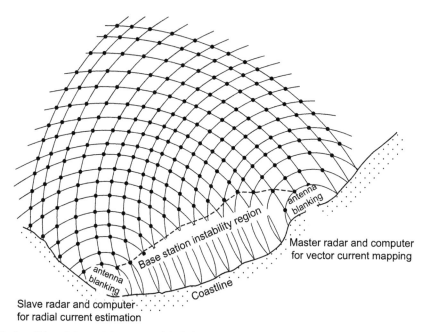

FIGURE 4.13 Conceptual rendition of the spatial coverage of dual-site HF/VHF Doppler radar systems using two wide-beam transmit antennas. The dots represent the locations on the sea surface from which radial water-current vectors are derived using echoes received at both the radars.

the vector inversion breaks down. Similar instability exists at large distances from the radar sites as well (Lease, 1984).

4.8. RADIAL AND TOTAL VECTOR CURRENTS

Two remotely located stations are required to measure the sea surface current vectors. By positioning the two stations so that beams from each of the two radar systems intersect near-orthogonally (Figure 4.14), sea surface current vectors can be accurately determined from the separately measured

surface current velocity component directed toward or away from the radar, called the *radial velocity component*. The radial components of currents are monitored along various orientations from the radar. Along each beam, surface currents are measured in range bins of a certain radial and azimuthal dimensions (see Figure 4.8). Therefore, the two-dimensional (2D) surface current velocity vector magnitude and direction (v, θ) can be extracted by adding the two radial velocity components (v_1, θ_1) and (v_2, θ_2) obtained at each radar station as follows (Mantovanelli et al., 2010):

$$v^2 = v_x^2 + v_2^y \tag{4.26}$$

$$\theta = \tan^{-1}\left(\frac{t_y}{t_x}\right), \tag{4.27}$$

where v_x and v_y correspond to the components of surface current vector with velocity (v) along the x- and y-axis, respectively, and t_x and t_y are parameters given by:

$$t_x = v_1 \sin\theta_2 - v_2 \sin\theta_1 \tag{4.28}$$

$$t_y = v_2 \cos\theta_1 - v_1 \cos\theta_2 \tag{4.29}$$

$$v_x = t_x/\sin(\theta_2 - \theta_1) \tag{4.30}$$

$$v_y = t_y/\sin(\theta_2 - \theta_1) \tag{4.31}$$

4.9. DEVELOPMENTS ON OPERATIONAL SCALES

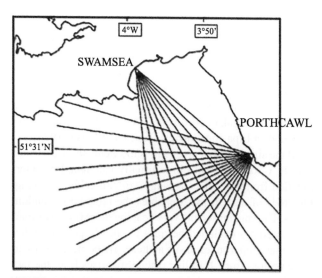

FIGURE 4.14 Positioning two radar stations so that beams from each radar intersect near-orthogonally. *(Source: Hammond et al., 1987.)*

Subsequent to the pioneering theoretical and experimental accomplishments achieved by Crombie (1955, 1972), Wait

(1966), Barrick (1972), and Stewart and Joy (1974) on remote mapping of sea surface currents using the HF Pulse-Doppler radar system, the technique rose from an experimental to an operational stage when Barrick et al. (1977) designed and developed an operational HF radar system. This system could measure and map near-surface current vectors to ranges about 70 km from the shore. However, these ranges are usually affected by the attenuation of radio waves propagating along the sea surface, the scattering properties of ocean surface waves, and atmospheric radio noise. Both the attenuation and noise are strongly frequency-dependent.

On one hand, the attenuation of radio ground waves increases with frequency (Hill and Wait, 1980). On the other hand, the atmospheric (manmade) noise decreases above a certain frequency limit because of the absence of ionospheric reflection. At night, the S/N ratio is considerably larger than during daylight hours. However, during daylight hours, radio-wave interference reduces the possible range for current measurements to 25 km at 25 MHz (Essen et al., 1983).

The first commercial HF Doppler radar remote sensing system for measurement of sea surface current was the coastal ocean dynamics applications radar (CODAR). However, developments on HF radar, independent of CODAR, have been made in Canada at C-CORE (Hickey et al., 1995); in the United Kingdom at the University of Birmingham (Shearman and Moorhead, 1988); in France at the University of Toulon (Broche et al., 1987); in Australia (Heron et al., 1985); and in Japan (Takeoka et al., 1995). Based on CODAR (Barrick et al. 1977), further developments have been made in the United Kingdom by Marconi with the ocean surface current radar (OSCR) (Prandle et al., 1992). In the United States, CODAR was upgraded to SeaSonde (Paduan and Rosenfeld, 1996). Recently, the University of Hamburg developed a new system known as Wellen Radar (WERA) using frequency-modulated continuous wave (FMCW) techniques. The most popular systems among these are discussed in the following sections.

4.9.1. CODAR

The earliest system in operational use was the 25.6-MHz CODAR developed by Barrick et al. (1977). The original version of the CODAR system developed at NOAA transmitted short CW pulses within a wide beam and performed range resolution by means of the pulse length τ. In order not to disturb the receive signal, successive transmit pulses are separated by a period of the order of 100 τ. Because of the relative short transmit time, a high peak power is needed. The main advantage of this kind of range resolution is the simple technical design. The CODAR used a four-element square array (see Figure 4.7) with a direction-finding

method for azimuthal resolution. The receiver processor used FFT routines to resolve equirange echoes within the beam, and the signals from the individual aerials (i.e., receiving antenna elements) were compared to estimate direction (Leise, 1984). Subsequently, estimates of radial current velocity versus range and bearing were constructed. A sample of sea surface currents mapped by CODAR is shown in Figure 4.15. Arrows represent current speed and direction; magnitude of current speed is indicated also by grey shading.

Researchers sought to reduce some drawbacks of a wide-beam system of the original CODAR by the use of a modified antenna system. This compact antenna system consists of three electrically separate elements, so excited that a beam of "cardioid" pattern on the sea surface is electronically steered to any angle in the horizontal (Figure 4.16). A single common steerable antenna system with the capability of transmission and reception, and meeting the stringent requirement of the compactness needed for deployment on offshore platforms, has been described as the most noticeable feature of the CODAR system. Paduan et al. (1995) reported an example of the complex nature of surface currents within and offshore of Monterey Bay as mapped by a multiple-site HF radar (CODAR) network.

While operating from off-shore oil rigs, the metals below the antenna affect the performance in two ways with respect to coastal installations (CODAR brochure). First, antenna patterns become slightly distorted by the metal directly below the antenna. For instance, Gurgel et al. (1986) observed that the ship-borne Doppler radar system's performance deteriorates as a result of the reflections of sea echo by the ship's metal. Second, the emitted radar signals getting into the metal structure of the rig can cause interference with other radio and TV equipment onboard if care is not taken during installation. The first problem is usually circumvented by measuring the actual antenna patterns by a radio source placed on a boat. The measured antenna pattern is then included in the inversion software as a lookup table. Barrick and Lipa (1986) discussed correction for distorted antenna pattern. The interference problem is usually handled by high-power bandpass filters at the transmitter output.

4.9.2. OSCR

An essential requirement in the use of radio-wave pulse-Doppler radar systems for measurement of sea surface currents is that the first-order peak in the echo spectrum must be identifiable from the higher-order continuum surrounding it. This requirement is always satisfied for a narrow receiving beam radar because the radar cell size on the sea surface is sufficiently small that the radial current variation across it is minimal. An important advantage of

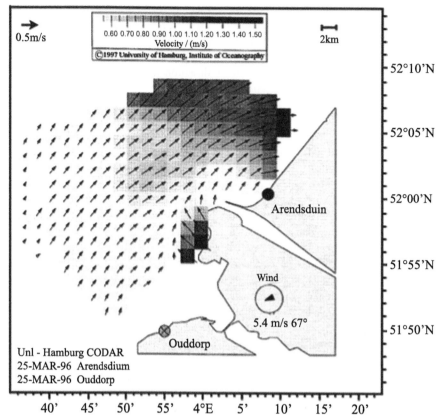

FIGURE 4.15 Sea surface currents mapped by CODAR. Arrows represent current speed and direction; magnitude of current speed is indicated also by grey shading. *(Source: Gurgel et al., 1999.)*

using a narrow-beam antenna instead of a wide-beam antenna is that the direction and range of the cells can be accurately determined (Eccles, 1989). Reception of a narrow beam of a pulsed, coherent radio wave in a known direction ensures that the direction of the radial sea surface flow component is precisely known. Based on these considerations, a number of systems are operational in various countries. These include the ocean surface current radar (OSCR), which originated at the Rutherford-Appleton Laboratory in the United Kingdom and was further developed by MAREX Ltd.

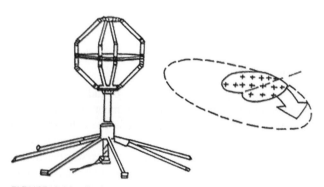

FIGURE 4.16 Compact antenna system used in CODAR. *(Source: CODAR Ocean Sensors brochure.)*

The system was designed to operate in one of two frequency ranges: 27 MHz (HF) for ranges approaching 35 km, and 51 MHz (VHF) for high-resolution measurements for ranges approaching 15 km. In the 51-MHz mode, the range cell size is a factor of 8 smaller than that used at 27 MHz, thereby improving the spatial resolution by this factor. The OSCR also transmits a wide beam but differs from the CODAR approach in using an 85-m-long, 16-element (for HF) or 32-element (for VHF) switched delay line matrix to form a phased-array antenna to achieve a narrow receive beam, which is electronically steered over the illuminated ocean surface area. In such pulse-Doppler radar, the angular resolution is controlled by the length (in wavelengths) of the phased array that is used for signal reception. An antenna configuration employing a long phased array and digital beam-forming techniques can "point" the radar (Figure 4.17). The beam width is a function of the radar wavelength divided by the length of the phased array, which is approximately 8° for the HF mode of OSCR. With a beam dwell time of approximately 2 min, the sea surface current resolution is approximately ±1 cm/s within cells of typically 1.2 $R\Theta$ km, where R is the range and Θ is approximately ±0.1 rad. The OSCR system could achieve these improvements in terms of accuracy (2 cm/s for radial current and 4 cm/s for total vector current) and

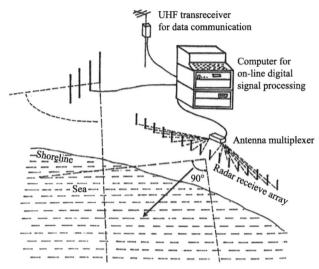

FIGURE 4.17 A typical configuration of OSCR transceiver system. *(Source: In part from an OSCR brochure.)*

spatial resolution (1 km) due to the use of a narrow-beam receive array by enhanced digital beam-forming techniques. The accuracy is a function of the spectral resolution, angle of intersection between the master and slave radial beams, and the positioning accuracy of the antennas. Other effects such as atmospheric noise and sea echo may induce additional uncertainty in the velocity field or limit the range performance (Barrick, 1980).

Figure 4.18 shows a map of sea surface current distribution in the Dover Strait produced by an OSCR system.

The OSCR system has no ambiguities and offers better resolution than does the CODAR, although the shore station requires greater space and is less portable.

To achieve the best possible result from radio-wave pulse-Doppler radar systems, a lot of care needs to be taken in their installation. The radar location needs to be chosen so that the antenna has an unobstructed view without being shadowed either by buildings or hills. The antenna should be located as close to the water as possible, preferably within 50 m. Barrick et al. (1977) found that the optimal location for the antennas is at sea level on the beach, as close to the water as possible, with the grounding system beneath the antennas making electrical contact with seawater. Land of greater extent than 100 m severely attenuates the signal. This observation was supported by Lane et al. (1999), who noticed in their field measurements in the U.K. waters that analysis of the OSCR data gave good results for the cells nearest the radar sites, but the data return and associated quality decreased with increasing range, especially close to the Northern Irish coast where the water is shallow and there are small islands present. There appeared to have been localized interference from a nearby broadcast transmitter.

4.9.3. SeaSonde

The SeaSonde, manufactured by Codar Ocean Sensors, Ltd., Los Altos, California, evolved out of the original CODAR developed by NOAA's Wave Propagation Laboratory (now

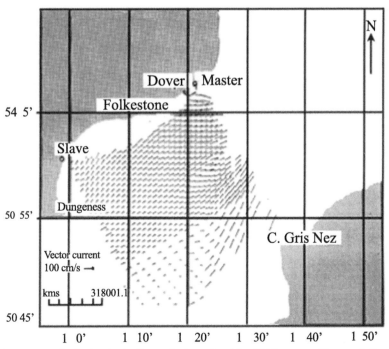

FIGURE 4.18 Map of surface current distribution in the Dover Strait produced by OSCR system. *(Source: Collar, 1993, reproduced with kind permission from National Oceanography Centre, Southampton SO14 3ZH, United Kingdom.)*

the NOAA Environmental Technology Laboratory). The CODAR SeaSonde system uses a three-element crossed-loop/monopole receive antenna along with a variant of the multiple signal classification (MUSIC) algorithm (Schmidt, 1986) to determine bearing. This configuration allows a deployment in a smaller area compared to a phased array antenna system. However, the system may be more sensitive to antenna response patterns, and the effect of this on surface current data is not well documented.

Historically, sea surface current-mapping direction-finding (DF) antennas and algorithms have taken several forms. Barrick et al. (1977) and Gurgel (1997) used linear array antenna elements separated by short distances and relied on phase-path differences to extract bearings. On the other hand, as old as radio itself, the simplest DF system is a loop antenna rotated until the incoming signal vanishes. Knowledge of this null direction and the angular response function of the antenna provides information about the direction from which the radio signal is originating. The most compact realization of DF techniques for this purpose has been the CODAR-type HF radars, which employ two crossed loops mounted around a whip (a vertical monopole). Such an antenna configuration is shown in Figure 4.19, along with stylized plots of the ideal angular amplitude patterns for each of the three elements. In addition to the theoretical (and achieved) shape of the antenna patterns, the accuracy of all DF algorithms also depends on the S/N ratio of the measured backscatter.

SeaSonde is an enhanced version of CODAR. Like CODAR, the SeaSonde overcomes the obstacle of requiring a huge coastal antenna facility with the use of compact antennas that rely on direction finding rather than beam forming. The receive antenna system is mounted up on a post, out of reach, or on the roof of a building rather than sprawling over hundreds of meters of valuable seafront property at the shoreline. This, combined with a novel, highly efficient

FIGURE 4.19 Stylized view of a coastal SeaSonde crossed-loop/monopole receive antenna. Indicated at the antenna base are idealized patterns of the two loops (yellow and pink) and the vertical whip/monopole (white). *(Source: Barrick and Lipa, 1997.)*

patented waveform that works well at HF, allows the SeaSonde to be very compact and portable. This system uses the frequency-modulated interrupted continuous wave (FMICW) technique for range resolution. Azimuthal resolution is achieved by a very small loop antenna combined with a special direction-finding algorithm. Controlled by unattended desktop PCs, real-time files are sent to the desired user office. There they are combined with similar data from other SeaSondes viewing the same area from tens of kilometers away to produce total vector fields. Typically they generate hourly current maps, and often data are automatically uploaded to Web sites for immediate access by the public.

The standard SeaSonde operates between 12-27 MHz to ranges of 30−60 km with typical resolution of 1 to 3 km. The Hi-Res SeaSonde provides 200−500-meter range resolution over bays with smaller areas such as 10−15 km. The latest addition is the Long-Range SeaSonde, achieving average daytime ranges of 170−200 km. The Long-Range SeaSonde makes it possible for a country to map currents along large sections of its coastline, continuously, with low initial and operating costs.

Rutgers University operates four Long-Range Sea-Sonde radars. They monitor the entire continental shelf off New Jersey. The current vectors are averaged and overlain on satellite-derived AVHRR sea surface temperatures. The SeaSonde can provide near-real-time data updates, even when satellite data are unavailable due to cloudy weather or other reasons.

At present, SeaSondes are used worldwide to monitor coastal sea surface currents to as far as 220 kilometers offshore. This has proven to be one of the technologies for emerging coastal ocean observing systems. HF Doppler radar is also making an impressive contribution to ocean forecasting programs.

4.9.4. WERA

Wellen Radar (WERA) is a new HF radar developed at the University of Hamburg, Germany. WERA uses frequency-modulated continuous wave (FMCW) chirps (without transmit/receive switching) instead of CW pulses. Chirp waveform and some of its specialties are given in chapter 12. One main advantage of this system is the possibility of connecting different configurations of receive antennas. WERA can be operated with up to 16 receive antennas. When operated with such a linear array, information on the sea state can also be obtained via second-order spectral bands. The capability of measuring surface wave spectra by means of WERA has been discussed by Wyatt et al. (1999). A further advantage is the flexibility in range resolution between 0.3 and 1.2 km instead of a fixed resolution of about 2 km for CODAR. In addition, because both transmitter and receiver operate continuously, the "chirp" technique avoids the blind range of about 3 km in front of the CODAR. Range resolution by means of FMCW chirps requires a more advanced technique than the resolution by means of CW pulses. However, the chirp method is advantageous in several aspects. It allows more flexibility in altering transmit frequency and range resolution. In particular, the S/N ratio can be improved by avoiding aliasing problems, and it should be possible to increase the currently realized working range.

HF radars transmit electromagnetic waves to the sea surface and record the backscattered signal, which contains information on surface current and waves. To locate the scattering area, spatial resolution has to be achieved in range and azimuth. Unlike all other HF Doppler radar systems described before, WERA transmits linear-frequency chirps. The range cell extent is related to the bandwidth of the chirp. The transmit signal is:

$$s(t) = \sin\left[2\pi\left(v_o + \frac{b}{2T}t\right)t\right], \qquad (4.32)$$

During the chirp period T, the frequency (i.e., the derivative of the phase with respect to time) linearly increases from v_o to $(v_o + b)$. After reaching the maximum frequency, the chirp is repeated. The received signal is a superposition of HF waves, which have been backscattered at different distances from the radar. Range resolution is performed by Fourier transforming each single chirp. The resolution of the frequency $v = b\tau/T$ is determined by the length T of the chirp such that $\Delta v = 1/T$. Thus, the resolution of the propagation time $\Delta\tau$ and, in turn, range Δr becomes (Gurgel et al., 1999b):

$$\Delta\tau = \frac{T\Delta v}{b} = \frac{1}{b}, \quad \Delta r = \frac{c}{2}\Delta\tau = \frac{c}{2b}, \qquad (4.33)$$

where c is the speed of light. It should be mentioned that the Fourier transform requires some caution. Due to the leakage problem in spectral analysis, near-range high-energy spectral lines may mask far-range lines of low energy. This problem is accounted for by applying a window filter to the chirps prior to the Fourier transform.

Advantages of range resolution chirp technique are the possibility of altering the range cell extent by simply modifying the width of the chirp, omitting the blind range in front of the radar, and a lower data rate to be processed in the receiver. However, transmitter and receiver must be designed for extreme dynamic range and linearity, since the superimposed near-range high-energy and far-range low-energy signals have to be separated.

The most crucial component of FMCW radar is the chirp generator. Advantages of the direct digital synthesizer (DDS) WERA uses are high flexibility and optimum realization of a linear chirp at low phase noise. The chirp is used by both transmitter and receiver. A block diagram of the hardware is presented in Figure 4.20. The DDS is controlled by a high-speed counter. Start and stop frequency can be

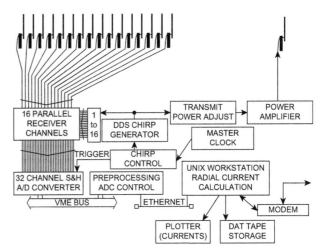

FIGURE 4.20   Block diagram of WERA system. *(Source: Gurgel et al., 1999.)*

changed easily, which permits modification of mean carrier frequency and range resolution. The system master clock is a stable synthesizer. Its long- and short-term stability and low phase noise are essential for good performance.

In WERA the generated chirp is split into 17 channels. One controls the transmitter. The others are needed for the phase-coherent I/Q demodulation of the backscattered signals. This is performed by 16 independent direct-conversion receivers connected to the receive antennas. Before A/D conversion, a low-pass filter suppresses the frequencies above the Nyquist frequency, and a high-pass filter attenuates the strong signals received directly from the transmit antenna. The measured characteristics of the filters are compensated for by software during data analysis. To extend the dynamic range of the 16-bit A/D converter, the signal is oversampled 13 times. The next steps of signal processing are implemented in software.

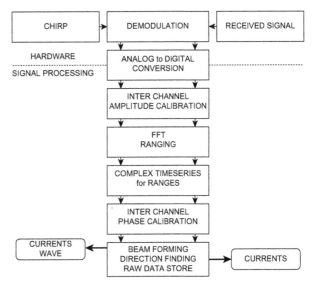

FIGURE 4.21   Signal processing implemented in WERA. *(Source: Gurgel et al., 1999.)*

Figure 4.21 illustrates the signal processing implemented in WERA. After A/D conversion of the complex demodulated signal, an interchannel amplitude calibration is applied to compensate for gain variations of the 16 receivers. Range-resolving FFT is then applied to all channels. A windowing function, as described by Gurgel et al. (1999a), has to be used in this context. After sorting the data into range cells, the complex time series are available. An interchannel calibration of the phase is performed and the data are stored for further processing. Amplitude and phase calibration are crucial for the accurate performance of azimuthal resolution techniques.

4.9.5. Systems for Special Applications

In addition to the systems described thus far, a few others have been developed for special applications (Gurgel et al., 1999b). The Pisces radar is one among them. Pisces is based on the FMICW technique for range resolution. Beam forming is provided in hardware by switchable cables used as phase shifters. FMICW introduces transmit/receive switching to overcome dynamic range limitations.

Another system is the C-CORE Cape Race. This system provides a high azimuthal resolution by a 40-element linear array, which is some 1 km long. This is a very large permanent installation for research, namely, tracking icebergs.

The Australian coastal ocean surface radar (COSRAD), developed at James Cook University, is operated at 30 MHz and uses pulses of 20 μs duration, yielding 3-km range resolution. In contrast to the other systems mentioned, which perform illumination of the measurement area by a wide-angle transmit antenna, a common antenna array for transmit and receive is used to perform beam forming. This has the advantage of squared side-lobe directivity performing a narrower beam, and this scheme increases the S/N ratio. However, because the different directions are scanned step by step, the azimuthal surveillance is slow, and technical problems in switching the antenna between transmitter and receiver arise.

Another pulsed Doppler radar system is the Courants de Surface MEsures par Radar (COSMER). This VHF system is a radio-oceanographic instrument developed by Laboratoire de Sondages Electromagnetiques de l'Environnement Terrestre (LSEET), University of Sud-Toulon-Var, France (Broche et al., 1987). It includes two Doppler radars, set on the coast, operating respectively at 45 and 47.8 MHz. The radars are operated with a linear phased array of eight receive whip antennas parallel to the coast (total length: 25 m or 50 m for 16 antennas) and with a transmitting network of four whip antennas perpendicular to the coast (total length: 12 m). The radars operate in pulsed mode, with a pulse width of 4 μs and a pulse repetition rate of 200 μs. The maximum range is 30 km. Both radars measure the radial components of the current in the range direction, at all the sampled distances, with an azimuth resolution of 14° (with eight antennas, beam-forming processing). The radar derives vertically averaged quantities over 25 cm (at 50 MHz) and temporally averaged over 9 minutes (Cochin et al., 2006).

4.10. INTERCOMPARISON CONSIDERATIONS

No measurement is complete without its error bars; therefore, uncertainty estimates are indispensable in comparing different datasets. Understanding the data involves clear insight into uncertainties as well. In attempting intercomparison among different instruments that measure a given parameter, it is necessary to clearly understand the way each instrument measures that parameter.

Several studies of sea surface current measurements made remotely by HF Doppler radar systems have validated the fundamental schemes embodied in them through comparisons with *in situ* measurements. Seminal studies by Stewart and Joy (1974), Barrick et al. (1977), and Frisch and Weber (1980) verified the underlying physics of the HF Doppler radar-based integrated surface current measurements by comparisons with Lagrangian drifters. These were soon followed by Holbrook and Frisch (1981), who compared the output of the CODAR system with records from a vector averaging current meter (VACM) moored at 4, 10, and 20 m depth; Janopaul et al. (1982), who intercompared three different radar systems with instruments at 7 and 10 m depth; and Schott et al. (1986), who included comparisons between direction-finding HF Doppler radar systems and moored current meters. Collar and Howarth (1987) compared data from the OSCR Doppler radar system with corresponding observations made by surface current buoys measuring at 1 m depth. Such comparisons have limitations, in part because the radar and current meters are responsive to different horizontal scales of motion, although use of a small current meter array distributed through a radar cell can improve matters (Collar, 1993).

More fundamentally, the radar senses a depth-integrated current, whereas the current meter samples at a point. Furthermore, in virtually all experiments reported, the moored instruments have sampled well below the effective mean depth of the radar and cannot therefore be expected to record the same current when the current structure is strongly sheared with depth under the influence of wind stress. The weight of evidence from studies conducted by the Institute of Oceanographic Sciences (IOS) in the United Kingdom has confirmed that the mean depth of response for radars operating at 27 MHz lies within 1 m of the sea surface. There is also the view that the radar includes a contribution from Stokes Drift in its measurement.

Finally, the point should be made that in making comparisons in shallow coastal waters, it may be necessary to make allowance for the current shear in the bottom boundary layer (Collar, 1993).

Several simultaneous observations have been made also using Doppler radar and surface drifters (Barrick et al., 1977; Ha, 1979; Collar and Howarth, 1987). Drifters can be drogued so as to follow the local water mass at depths commensurate with the radar averaging depth, though care is required in drogue design. Agreement between the two techniques to within a few cm/s has been reported in several of the studies, although mostly in relatively calm conditions, in which variability within the radar footprint almost certainly provides the ultimate limitation.

Hammond et al. (1987) reported results of intercomparison of current speed and direction outputs between OSCR and rig-mounted Aanderaa current meters. The study was carried out within an area of high tidal currents and complex bottom topography, located along the northern coastline of the Bristol Channel, U.K. Compared with simultaneous data obtained from conventional self-recording current meters, the radar measurements of surface currents were to within 5–10% in speed and 10° in direction. Thus, it was found that in the absence of wind influence, excellent agreement between the various techniques for measuring current speed and direction is the characteristic of the majority of the outputs (Figures 4.22 and 4.23). The OSCR data, which represent currents at the air-sea interface, were the highest of the set; they were in reasonable agreement with the speeds recorded by the upper meters on each of the rigs. However, substantial differences between the two datasets (i.e., OSCR and current meters) were found during short-term changes in the superimposed wind field.

Further studies such as Shay et al. (1995), Graber et al. (1997), and Chapman et al. (1997) built on earlier works by comparing OSCR measurements with a variety of moored and shipboard measurements. These studies increased the understanding of errors and limitations present in all HF

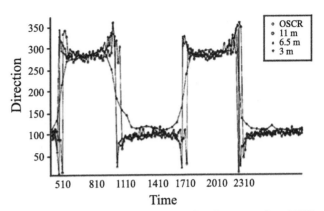

FIGURE 4.23 Intercomparison of current direction outputs from OSCR and self-recording current meters moored at 3, 6.5, and 11 m above the seabed. *(Source: Hammond et al., 1987.)*

Doppler radar current measurements, but fewer studies of CODAR SeaSonde employing MUSIC algorithm for direction finding were available. Exceptions include Paduan and Rosenfeld (1996), who reported comparisons of total vector currents that combined data from newer SeaSondes and an older CODAR HF radar; Hodgins (1994), who reported comparisons with drifters as well as modeled currents; and subsequently Kohut et al. (1999) and Paduan et al. (2001). Studies by Emery et al. (2004) sought to add to the understanding of the CODAR SeaSonde and its MUSIC direction-finding algorithm based on comparisons with an array of nine moorings in the Santa Barbara Channel and Santa Maria basin deployed between June 1997 and November 1999. Eight of the moorings carried vector-measuring current meters (VMCMs), and the ninth had an upward-looking acoustic Doppler current profiler (ADCP). Coverage areas of a network of five CODAR SeaSonde HF Doppler radar systems (broadcasting near 13 MHz and using the MUSIC algorithm for direction finding) and moorings included diverse flow and sea-state regimes. Measurement depths were ~1 m for the radar systems, 5 m for the VMCMs, and 3.2 m for the ADCP bin nearest the surface. Comparison of radial current components from 18 radar-mooring pairs yielded rms speed differences of 7–19 cm/s and correlation coefficients squared (R^2) in the range of 0.39–0.77. Noise levels corresponding to 6 cm/s rms were evident in the radar data. Errors in the radar-bearing determination were found in 10 out of 18 comparisons, with a typical magnitude of 5–10° and a maximum of 19°. The effects of bearing errors on total vector currents were evaluated using a simple flow field and measured bearing errors, showing up to 15% errors in computed flow speeds and up to ~9° errors in flow directions.

Lipa (2003) carried out field intercomparison studies between SeaSonde radar and ADCPs, which are two different kinds of modern instruments for remote (i.e., noncontact) measurements of ocean currents. These two

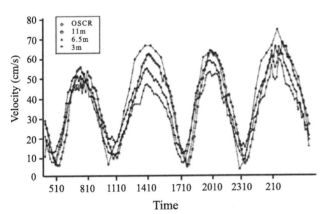

FIGURE 4.22 Intercomparison of current speed outputs from OSCR and self-recording current meters moored at 3, 6.5, and 11 m above the seabed. *(Source: Hammond et al., 1987.)*

instruments function in different ways, although they measure the same oceanographic parameter. Many sources of uncertainty in SeaSonde velocities are familiar, such as statistical variation, nonoptimal analysis parameters, etc. In addition, each radar cell contains different current velocities due to horizontal velocity shear on the ocean surface. Such instances are often encountered in the oceanic upper layer because in a variety of oceanic regions, this layer often encompasses very strong and extremely variable currents (e.g., jets, eddies, meanders, filaments, etc.). Such energetic features have presented major risks for marine operations and have caused costly losses of structures over various areas of the world. The HF radar averages over these variable current velocities, and therefore any of these values may apply at a single point within the radar cell. The standard deviation of these individual velocities may be referred to as *spatial uncertainty* (Lipa, 2003). Thus, when HF radar (spatial average) and point measurements are compared, discrepancies are bound to exist in the presence of current shear.

Spatial uncertainty in the SeaSonde HF radar average can be estimated, however. During data processing, the radial velocity value is defined by the signal frequency, and analysis of the antenna voltage signals yields the corresponding direction of arrival. Usually, several velocities fall within a radar cell, and their standard deviation is a measure of the spatial uncertainty, which tends to increase with range from the radar along with the size of the radar cell. Even if the spatial uncertainty is high, the uncertainty in the average can be low. Thus, SeaSonde HF radar measurements appear stable from time to time, and two SeaSonde HF radars operated side by side will produce similar results. However, it can be a different matter when a SeaSonde "area measurement" is compared with an ADCP "point measurement." In the presence of sizable velocity shear, good agreement between the two measurements can be expected only if the radar cell size is small. Poor agreement does not necessarily indicate inconsistency if the differences are less than the error bars (Lipa, 2003). Lipa suggests that in comparing HF radar (SeaSonde) and ADCP data, it is best to compare radial velocities, resolving the ADCP velocities into components radial to the radar site. Then, differences between measurements can be compared with the radar uncertainties, which are different for each of the two radar systems. Individual uncertainties are lost when radial velocities from two sites are combined to form resultant velocities, and therefore comparing resultant velocities is not as informative.

Lipa (2003) provides the so-called *baseline test* as another example of SeaSonde velocity comparisons. A baseline test is a consistency check on radial velocities from two SeaSondes at points on the baseline joining the two sites. Because the SeaSondes "see" the same radial current along the baseline, the two systems will ideally give the same estimates of radial velocity. However, at all baseline points except the midpoint, the sizes of the radar cells differ, and spatial uncertainty may play a role.

Lipa (2003) has reported baseline tests performed on SeaSondes at Montauk and Misquamicut on the Long Island Sound. It was found that the Montauk and Misquamicut rms radial speeds differed by as much as 40 percent on the baseline near Montauk. This discrepancy was found to be due to the different radar cell sizes and the large velocity shear close to Montauk, where the current swirls around Montauk Point. Because the radar cell size is proportional to the distance from the radar, the cell sizes for Montauk in this region are much smaller than for Misquamicut. A smaller cell size results in less averaging-down of the velocity. Thus, the Montauk site produces current velocities with higher values and lower spatial uncertainties. The large baseline deviations are completely accounted for by the large spatial uncertainties in the Misquamicut data.

Based on this discussion, it may be concluded that in comparing sea surface current velocity measurements, it is not sufficient to simply compare the values; the uncertainties also need to be considered. For the area-averaged measurements produced by HF radar systems, uncertainty estimates should include uncertainties due to horizontal velocity shear on the sea surface. Surface current velocity measurements from two instruments can be considered to be consistent if the differences are within the uncertainty limits.

Intercomparison experiments of land-based HF Doppler radar systems against *in situ* and remote measuring subsurface devices have been addressed so far. It is worth examining the results of intercomparison measurements from different makes of HF Doppler radar systems. Gurgel et al. (1999b) reported such studies between CODAR (which achieves azimuthal resolution using direction finding in Fourier domain) and WERA (which achieves azimuthal resolution using a beamforming method). Both CODAR and WERA systems were installed at two sites north and south of the Rhine estuary mouth. The CODAR was equipped with the four-antenna square array; the 16 receive antennas of WERA were separated into a 12-antenna linear array and a four-antenna square array.

Figure 4.24 shows radial speeds measured by CODAR (triangles) and WERA (crosses) at two different positions, A and B, in the Rhine mouth. The time series extended over two M2 tidal cycles with ½ h sampling. At position A, both measurements agree well with an rms difference of less than 5 cm/s. At position B, the CODAR measurements revealed some spikes within the almost continuous time series, whereas the WERA time series was stable. Position B is situated within the main shipping route toward Rotterdam harbor. For this reason, it was concluded that

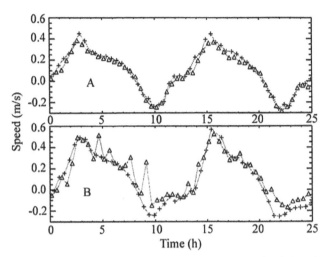

FIGURE 4.24 Radial speeds measured by CODAR (triangles) and WERA (crosses) at two different positions A and B in the Rhine mouth. *(Source: Gurgel et al., 1999.)*

the CODAR direction-finding algorithm is less robust than beam forming with respect to perturbations by ship traffic. This finding was confirmed by investigations of the S/N ratio of both measurements (Gurgel et al., 1997).

With regard to HF Doppler radar systems, it is worth noting that the maps of ocean surface currents are incomplete in space and time for the following reasons (Kim et al., 2007): (1) the algorithm on the measured Doppler spectrum does not provide a solution for all bearing angles; (2) the estimate of current vectors along the baseline between two radars where the measurements of radial velocities are nearly aligned suffers from poor geometric dilution of precision (GDOP) and frequently results in spurious total vector currents; the region with radial velocities crossing at angles less than 15–20° between two radars is commonly considered to produce unusable current vectors; and (3) hardware or software problems can lead to the temporary shutdown of individual radar sites. For example, Lane et al. (1999) reported that operational difficulties were experienced at the slave site of an OSCR system at Crammag Head (U.K.), where power failures occurred repeatedly and the system failed to reboot at times. Consequently, the data return over the 15 months of recordings was 62 percent.

In comparing surface current measurements using different technologies, the merits and limitations of each technology need to be taken into account. Major differences between surface current velocities from HF radar and other platforms are the following (Pandian et al., 2010):

- Measurements from HF radar, drifters, and current meters are all inexact. The frequency resolution of computed radar cross-spectra, which depends on FFT length, limits radial velocity resolution to ~5 cm/s and 2.5 cm/s for 12 MHz and 25 MHz systems, respectively.

Drifters can slip at ~1 to 2 cm/s from the seawater parcel they follow (Ohlmann et al., 2005).

- Vertical scales of measurement differ. The HF radar gives vertically integrated values from the surface; drifters give integrated values over their drag elements; and current meters give values for specific depths (in the case of Eulerian current meters) or depth bins (in the case of ADCP).

- Horizontal scales of measurement differ. Typically, HF radars average over extensive horizontal areas (up to several km^2), whereas other platforms give point measurements or limited spatial measurements following motion.

- Measurements are not necessarily coincident in time, and Stokes drift may not be reconciled consistently among platforms (Ohlmann et al., 2007).

The plus point about HF Doppler radar networks is that despite several limitations, the two-dimensional maps of sea surface currents they provide represent a useful and unique resource for the improvement of coastal ocean circulation models, particularly in the critical depth range encompassing the euphotic zone (Paduan and Shulman, 2004).

4.11. ADVANTAGES OF RADIO-WAVE DOPPLER RADAR MEASUREMENTS

The main attraction of the coastally located Doppler radar technology lies in its ability to rapidly produce real-time two-dimensional maps of sea surface current distribution over a wide area in a way that would be difficult and prohibitively expensive using conventional instrumentation. It is thus especially useful in mapping flow patterns and their characteristics in waters around headlands, islands, or in estuaries. Experiments by several researchers have demonstrated the feasibility and practicality of mapping bay and estuarine flow patterns using land-based Doppler radar systems. Being a land-based system, sea operations are avoided, thereby enormously reducing weather dependence and other limiting factors. Sea surface current measurements using Doppler radar systems have some unique advantages over conventional methods of measurement. Probably the most important among them is its capability for remote measurements in time as well as space domains. This enables detection and mapping of time-varying as well as space-varying events such as gyres and complicated current structures, as demonstrated by Barrick et al. (1985) and Gurgel et al. (1986). It could also reveal fine structures of near-shore circulation (Prandle, 1987). Using Doppler radar, Shay et al. (1998) could obtain a unique view of the development and dissipation of complex features and spatial variabilities in the sea-surface circulation, such as mesoscale eddies and convergence zones on a shelf, associated with a meandering western boundary current.

Another noteworthy feature of the Doppler radar technology is its inherent ability to measure currents without discriminating against Stokes drift (Barrick, 1986). The surface drift current has two components: (1) the shear current generated by the wind stress and (2) the Stokes current, which is related to wave statistics such as the phase velocity and height and length of the waves. Because the Stokes transport varies between 5 and 13 percent of the total surface drift (Wu, 1975), its inclusion in surface-current observation systems is clearly desired. Because most of the conventional sea surface current measuring devices, such as near-surface moored current meters, exclude Stokes surface transport, the inherent capability of the radio-wave Doppler radar systems to measure the total surface drift current is an added advantage. Results of sea surface current surveys in coastal waters and estuaries can be used to develop computer models that would assist in antipollution measures. Thus, Doppler radar systems can play a leading role in major investigations designed to help protect the marine environment.

A remarkable feature of the Doppler radar system is its ability to measure sea surface current shear at the topmost layer of the sea surface. This capability stems from the fact that the phenomenon of Bragg resonance, used for sea surface current measurements, occurs when a constituent of the predominant sea surface waves has a wavelength equal to one-half the radio wavelength of the radar. The sea surface wave trains contain numerous linear waves of differing wavelengths. Thus, with the use of multifre-quency radars (3–50 MHz), it is possible to measure vertical current shear within the uppermost few meters of the ocean surface. This is a measurement that is very difficult to make by any other means.

Since their introduction in the 1970s, Doppler radar systems have been used extensively to deduce information on synoptic sea surface currents to ranges about 70 km from the shore (Barrick et al., 1977; Stewart and Joy, 1974; Frish and Weber, 1980; Georges, 1981; Prandle and Ryder, 1985; Prandle, 1987; Mathews et al., 1988; Parker, 1989). The radar system has the capability to cover vast areas of high shipping and fishing activities, where other methods would be impractical for long-term monitoring of sea surface currents. For example, under an Anglo-French collaborative research program with the leadership of the Proudman Oceanographic Laboratory (U.K.) to gauge the Dover Straits' capacity to transport waterborne pollutants into the North Sea, a master-slave pair of Doppler radar systems on the British coast was effectively used in tandem with a second pair on the French coast. Between them, the two radars could monitor sea surface current flows across the entire width of the strait at a spatial resolution of 1 km. This monitoring program exemplifies the advantages of Doppler

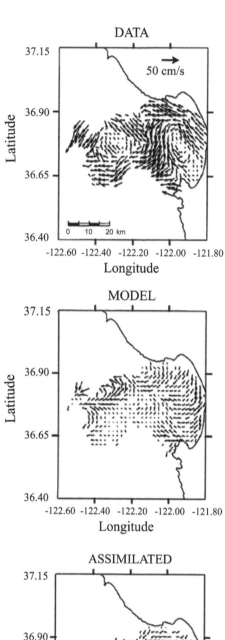

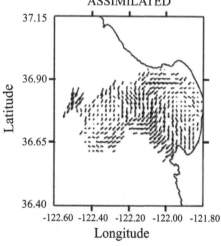

FIGURE 4.25 Pattern of surface currents as determined by HF Doppler currents (top), the numerical ocean model without any assimilation of the Doppler current (middle), and the model with the optimized assimilation of the Doppler currents (bottom). *(Source: Lewis et al., 1998.)*

radar systems to provide continuous year-long sea surface-current observations in a hostile environment where conventional instruments are unlikely to survive. A fairly wide area coverage (exceeding 500 km²), high temporal resolution (entire surveillance area surveyed in a few minutes), immunity to weather conditions and suspended sediments, and land-based long-term monitoring are some of the key features of the Doppler radar systems. It is expected that these systems will play an important role in a wide variety of activities such as outfall design, effluent transport, pollution management, vessel traffic management in ports and harbors, and search and rescue of persons and crafts in distress as well as a host of other oceanographic and fisheries research programs. Doppler radar systems can gather a large amount of information. No other technique would enable such a comprehensive picture to be built up in a short period of time. The Doppler radar systems will definitely be of significant benefit to applied oceanography.

HF Doppler radar systems have begun to be a convenient tool for ocean current modelers as well. For example, Lewis et al. (1998) presented a technique for the assimilation of ocean surface currents determined from Doppler radar systems into numerical ocean models. They demonstrated an approach in which the Doppler radar current data act as though there were an additional layer of water overlying the ocean surface. A pseudo-shearing stress resulting from the difference between the model-predicted velocity in the Monterey Bay, California, region and the Doppler radar velocity was added to that of the wind in order to force a model. The numerical ocean model, with the optimized assimilation of the Doppler currents, provided a more realistic representation of the Monterey Bay currents relative to one without any assimilation of the Doppler currents (Figure 4.25).

4.12. ROUND-THE-CLOCK COAST- AND SHELF-OBSERVING ROLE OF DOPPLER RADAR

Many climatologists believe that in the coming decades, coastal oceans throughout the world are likely to undergo changes associated with climate change; increased levels of radioactive and other waste disposal into the seas; and increased levels of plumes, slicks, and so forth. Therefore, understanding the transport of sediment and the associated material from the harbor to the coastal ocean is a fundamental problem for water-quality managers. The plumes, which are usually dynamic in space and time, are modified by bottom topography, shoreline geometry, atmospheric forcing, tides, and river outflow. Fundamentally limiting our understanding of the seas around us is the inability to sample the ocean on relevant scales in both space and time.

Our lack of understanding fuels the motivation to build regional integrated coastal ocean-observing networks. Such observing stations should be equipped to post the data to the Internet in real or near-real time through a series of Websites for distribution to people.

Coastally located land-based and offshore-located platform-mounted Doppler radars are well suited to indirectly examine several meteorological and environmental related activities happening on a large area of the sea surface through their capability for all-weather, round-the-clock monitoring of sea surface current and wave spectra distribution maps. Such a network would allow a person to sit at home during storms, tsunamis, oil slicks, or other oceanogenic hazards and monitor the sea surface motions in real time. This remote interaction is one of the hallmarks of coastal ocean-observing networks, and with several other parameters included in the observation scheme, it would be a transformational step for oceanographic research, material transport, routine environmental monitoring, search-and-rescue operations, and hazard mitigation efforts. For example, the New Jersey Shelf Observing System (NJSOS) uses hourly data from the sea surface current radar to indirectly monitor the Hudson River plume. Ocean color imagery and sea surface temperature from satellite-borne sensors provide maps that help define the spatial extent of this plume. These daily composites are then advected through time using the hourly data obtained from the radar. Through the combined application of the HF Doppler radar network and satellite imagery, environmental managers benefit from a real-time picture of the plume, allowing adaptive sampling and increased understanding of potential deposit centers of pollutants and other materials flowing out of the harbor (Schofield et al., 2005).

Continental shelf circulation is well known for its spatial variability. Until recently, the lack of data forced scientists, the U.S. Coast Guard, the U.S. Navy, and hazardous materials response groups to rely on the climatology of circulation patterns to conduct operations. Climatological approaches are not efficient enough in capturing the true variability of the circulation. Fortunately, an HF Doppler radar system, which is capable of remotely measuring sea surface currents as far offshore as 200 km, generates hourly maps indicating the speed and direction of the current at a number of spatially dense locations. These maps have great potential for search-and-rescue operations. Using the high-frequency Doppler radar network off the coast of New Jersey and near the mouth of Long Island Sound, demonstration projects have been conducted to provide proof of concept for using these surface current maps in Coast Guard search-and-rescue operations. Doppler radar data were found to be of immense practical use in defining the search areas. It was found that continuous spatial circulation data generated by the HF Doppler radar network allows a robust means to understand advection of material on the continental shelf (thus benefiting science) while simultaneously providing the potential to greatly improve Coast Guard search-and-rescue operations (Schofield et al., 2005). Thus, HF Doppler radar networks, which are deployed and sustained on an operational basis, would serve as important ingredients for providing timely support for commercial and environmental communities. Many such networks are being deployed globally to complement other multisensor coastal-observing networks. Some are focused exclusively on research problems, but others have an applied focus. Success in the future will be measured by how the observatories simultaneously serve both of these needs.

4.13. DETECTION AND MONITORING OF TSUNAMI-INDUCED SEA SURFACE-CURRENT JETS AT CONTINENTAL SHELVES

When tsunami waves encounter steep gradients at the edges of continental shelves and at the coast, the waves become nonlinear, and conservation of momentum in the water column produces squirts (i.e., jets) of sea surface currents at areas of depth discontinuities and shallow regions. As discussed by Barrick (1979), a sinusoidal tsunami wave appears as a periodic surface current. Its wave orbital velocity at the surface transports the much shorter waves seen by the radar, adding to the ambient current field and producing a clear signature detectable by the radar. The tsunami, which is assumed to propagate perpendicular to the depth contours, produces sea surface current velocities that superimpose on the slowly varying ambient current velocity background. To some degree, there is an *a priori* pattern of large surface currents that occur when a tsunami encounters steep benthic gradients at the edge of a continental shelf. Tsunami currents have a characteristic signature due to their coherence over large distances, thereby allowing them to be detected when they arrive in the radar coverage area (Lipa et al., 2006).

Barrick (1979) originally proposed the use of shore-based HF radar systems for tsunami warning. HF radar systems currently operate continuously from many coastal locations around the globe, monitoring ocean surface currents and waves to distances of up to 200 km. For each HF radar location, it is possible to calculate a tsunami response pattern by numerical modeling methods (Lipa et al., 2006; Heron et al., 2008). To a first-order approximation, the response of the sea surface currents to the tsunami approaching the continental shelf is assumed to be independent of the direction of the source of the tsunami. This is because, as mentioned earlier, tsunami wave fronts are refracted in deep water and will approach the shelf edge within a small range of angles around orthogonal. This

assumption can be tested for each site by numerical calculations provided by Greenslade et al. (2007). To simulate the signals seen by HF radar in case of a tsunami traveling toward the coast, Dzvonkovskaya et al. (2009) calculated the tsunami-induced sea surface current velocity using the oceanographic HAMburg Shelf Ocean Model (HAMSOM), then converted it into modulating signals and superposed to the measured radar backscatter signals. HAMSOM involves the friction and Coriolis terms and thus can simulate wave propagation from the deep ocean to shelf areas where nonlinear processes play an important role. After applying conventional signal processing techniques, the sea surface current maps contain the rapidly changing tsunami-induced current features, which can be compared to the HAMSOM data. The specific radial tsunami current signatures can clearly be observed in these maps, if appropriate spatial and temporal resolution is used. Gurgel et al. (2011) described a tsunami detection algorithm that can be used to issue an automated tsunami warning message. The sea surface current map based on these spectra has a pattern that changes very quickly in the shelf area before the tsunami wave reaches the beach. Specific radial tsunami current signatures are clearly observed in these maps. If the shelf edge is sufficiently far off the coast, the first appearance of such signatures can be monitored by an HF radar system early enough to issue a warning message about an approaching tsunami. The sea surface current response therefore becomes a signature that can be looked for in the data analysis process. Heron et al. (2008) have provided model calculations of sea surface current vectors when the first wave of the December 26, 2004, Indian Ocean tsunami encountered a section of the edge of the continental shelf of the Seychelles Island.

Lipa et al. (2006) demonstrated that HF Doppler radar systems in operation today are capable of detecting tsunami currents and providing vital information well before impact, when the adjacent continental shelf is wide. Heron et al. (2008) found that HF Doppler radar is well conditioned to observe the surface current bursts at the edge of the continental shelf and give a warning of 40 minutes to 2 hours when the shelf is 50 to 200 km wide. However, in the use of HF radar technology, there is a trade-off between the precision of surface current speed measurements and time resolution. An advantage in S/N ratio can be obtained from the prior knowledge of the spatial pattern of the squirts at the edge of the continental shelf. It was shown by Heron et al. (2008) that the phased array HF Doppler radar deployed in the Great Barrier Reef in Australia (where the shelf depth is about 50 m) and operating in a routine way for mapping sea surface currents can resolve surface current squirts from tsunamis in the wave period range 5–30 minutes and in the wavelength range greater than about 6 km. This network is found to be well conditioned for use as a monitor of small as well as larger tsunamis and has the potential to contribute to the understanding of tsunami genesis.

When Doppler radar operates in its routine sea surface current-mapping mode, each station records a time-averaged (a few minutes) time series at a convenient sampling interval (say, 10 minutes). In this mode the radar could detect only tsunamis with wave periods greater than twice the sampling interval (Nyquist's sampling criterion). In other words, if the sampling interval is 10 minutes, the radar would detect only tsunamis with a wave period greater than 20 minutes. However, if the HF Doppler radar is to be used for detection of tsunami-induced magnified sea surface current jets (generated at depth discontinuities) for warning purposes, the radar would need to be switched to an "alert mode" of operation, presumably following a seismic alert. Lipa et al. (2006) suggested that in the event of a tsunami threat, tsunami watch software (producing current velocities and local wave information at the many HF radars in operation around the coastlines of the world) could run in parallel (in the background), activating a tsunami warning. This information would be available to local authorities and would be invaluable if international communications fail or are too general in their predictions. Global models may be inadequate for localized areas for which the available bathymetry may not be of adequate resolution. In addition, when a quake epicenter is close to shore, there may be insufficient time for the international communication chain to be activated. In such cases local systems would provide the only advance warning. Such a system may also alleviate the false-alarm problems that plague existing tsunami watch systems. Computer prediction models and early-warning schemes apply only to tsunamis generated by earthquakes; HF radar networks would also be able to detect tsunamis generated by underwater rockslides and tidal bores.

According to Heron et al. (2008), the most effective utility of an "alert mode" of operation would be in assisting the warning network by filling the gap between deep-ocean sensors and coastal sea-level gauges and, in particular, in avoiding false alarms because of its high sensitivity compared with other sensors. However, it is to be expected that a tsunami will be more difficult to detect if it is small or if the background current velocities due to tides, winds, or density gradients in the area monitored are large and rapidly varying. For the task of tsunami detection, the background current velocities can be considered a kind of "background noise" that needs to be removed to obtain the tsunami-induced currents more clearly.

The best way to handle this difficulty is to use an oceanographic model to simulate this "background noise." To keep the model result close to the actually measured ocean currents, it can be "guided" by applying a data assimilation technique (Gurgel et al., 2011). In a field application, a feasibility study would be required for each location, based on radar transmit frequency and taking into account the typical current regimes for the location, in addition to the bathymetry. It needs to be emphasized that if oceanographic radars are used for tsunami detection, they have to be operated in a high temporal (2 min) and spatial (1.5−2.0 km) resolution mode in order to have the best sensitivity and be able to resolve the quickly changing tsunami signatures. Gurgel et al. (2011) have found that a tsunami-induced sea surface current jet signature disappears completely at integration times larger than 25 min. They described a proposal for a new algorithm for automatic detection of tsunamis using a constant false alarm rate (CFAR) approach.

REFERENCES

Barrick, D., 1979. A coastal radar system for tsunami warning. Remote Sens. Environ. 8, 353−358.

Barrick, D.E., Headrick, J.M., Bogle, R.W., Crombie, D.D., 1974. Sea backscatter at HF: Interpretation and utilization of the echo. Proc. the Institute of Electrical & Electronic Engineers 62 (6), 673−680.

Barrick, D.E., Evans, M.W., Weber, B.L., 1977. Ocean surface currents mapped by radar. Science 198 (4313), 138−144.

Barrick, D.E., 1972. First-order theory and analysis of MH/HF/VHF scatter from the sea. IEEE Trans. Antennas Propagation AP-20, 2−10.

Barrick, D.E., 1980. Accuracy of parameter extraction from sample averaged sea echo Doppler spectra. IEEE Trans. Antennas Propagation 28, 1−11.

Barrick, D.E., 1986. The role of the gravity-wave dispersion relation in HF radar measurements of the sea surface. IEEE J. Ocean. Eng. OE-11 (2), 286−292.

Barrick, D.E., Lipa, B.J., 1986. Correcting for distorted antenna patterns in CODAR ocean surface measurements. IEEE J. Ocean. Eng. OE-11 (2), 304−309.

Barrick, D.E., Lipa, B.J., Crissman, R.D., 1985. Mapping surface currents with CODAR. Sea Technol. 26, 43−48.

Broche, P., Crochet, J.C., de Maistre, J.L., Forget, P., 1987. VHF radar for ocean surface current and sea state remote sensing. Radio Sci. 22, 69−75.

Broche, P., Crochet, J.C., de Maistre, J.C., Forget, P., 1987. VHF radar for ocean surface current and sea state remote sensing. Radio Sci. 22, 69−75.

Chapman, R.D., Shay, L.K., Graber, H.C., Edson, J.B., Karachintsev, A., Trump, C.L., Ross, D.B., 1997. On the accuracy of HF radar surface current measurements: Intercomparison with ship-based sensors. J. Geophys. Res. 102 (C8), 18,737−18,748.

Cochin, V., Mariette, V., Broche, P., Garello, R., 2006. Tidal current measurements using VHF radar and ADCP in the Normand Breton Gulf: Comparison of observations and numerical model. IEEE J. Ocean. Eng. 31 (4), 885−893.

Collar, P.G., 1993. A review of observational techniques and instruments for current measurements from the open sea. Report No. 304, Institute of Oceanographic Sciences, Deacon Laboratory, Brook road, Wormley, Godalming, Surrey, UK.

Collar, P.G., Howarth, M.J., 1987. A comparison of three methods of measuring surface currents in the sea: Radar, current meters and surface drifters. Department of Energy Offshore Technology Report No. OTH 87272.

Crombie, D.D., 1955. Doppler spectrum of sea echo at 13.56 Mc/s. Nature 175 (4459), 681–682.

Crombie, D.D., 1972. Resonant backscatter from the sea and its application to physical oceanography. Proc. IEEE Oceans '72 Conference, 172–179.

Dzvonkovskaya, A., Gurgel, K.W., Pohlmann, T., Schlick, T., Xu, J., 2009. Simulation of tsunami signatures in ocean surface current maps measured by HF radar. Proc. IEEE.

Eccles, D., 1989. The Ocean Surface Current Radars. Rutherford Appleton Laboratory Report.

Emery, B.M., Washburn, L., Harlan, J.A., 2004. Evaluating radial current measurements from CODAR high-frequency radars with moored current meters. J. Atmos. Oceanic Technol. 21, 1259–1271.

Essen, H.H., Gurgel, K.W., Schirmer, F., 1983. Tidal and wind-driven parts of surface currents as measured by radar. Dt. Hydrogr. Z 36, 81–96.

Frisch, A.S., Weber, B.L., 1980. A new technique for measuring tidal currents by using a two-site HF Doppler radar system. J. Geophys. Res. 85, 485–493.

Georges, T.M., 1981. Measuring ocean currents with HF radar. Proc. Oceans '81 IEEE Conference record 1, 228–232.

Georges, T.M., Harlan, J.A., 1995. Mapping surface currents near the Gulf Stream using the Air Force over-the-horizon radar. Proc. IEEE Fifth Working Conference on Current Measurements, 115–120.

Graber, H.C., Haus, B.K., Chapman, R.D., Shay, L.K., 1997a. HF radar comparisons with moored estimates of current speed and direction: Expected differences and implications. J. Geophys. Res. 102, 18,749–18,766.

Greenslade, D.J.M., Simanjuntak, M.A., Burbidge, D., Chittleborough, J., 2007. A first-generation real-time tsunami forecasting system for the Australian Region. BMRC Report No. 126. Bureau of Meteorology, Australia.

Gurgel, K.W., Dzvonkovskaya, A., Pohlmann, T., Schlick, T., Gill, E., 2011. Simulation and detection of tsunami signatures in ocean surface currents measured by HF radar. Ocean Dynamics.

Gurgel, K.W., Antonischki, G., Schlick, T., 1997. A comparison of surface current fields derived by beam-forming and direction-finding techniques as applied to the HF radar WERA. IGARSS Proceedings Singapore.

Gurgel, K.W., Antonischki, G., Schlick, T., 1997. A comparison of surface current fields derived by beam-forming and direction-finding techniques as applied to the HF radar WERA. IGARSS Proceedings.

Gurgel, K.W., Antonischki, G., Essen, H.H., Schlick, T., 1999b. Wellen Radar (WERA): A new ground-wave HF radar for ocean remote sensing. Coastal Engineering 37, 219–234.

Gurgel, K.W., Antonishki, G., Essen, H.H., Schlick, T., 1999a. Wellen Radar (WERA), a new ground-wave based HF radar for ocean remote sensing. Coast. Eng. 37, 219–234.

Gurgel, K.W., Essen, H.H., Schirmer, F., 1986. CODAR in Germany: A status report valid November 1985. IEEE Journal of Oceanic Engineering OE-11 (2), 251–257.

Gurgel, K.W., Essen, H.-H., Kingsley, S.P., 1999a. High-frequency radars: physical limitations and recent developments. Coastal Engineering 37, 201–218.

Ha, E.C., 1979. Remote sensing of ocean surface current and current shear by HF backscatter radar. Ph.D. dissertation. Stanford University, Stanford, CA, USA.

Hammond, T.M., Pattiaratchi, C.B., Eccles, D., Osborne, M.J., Nash, L.A., Collins, M.B., 1987. Ocean surface current radar (OSCR) vector measurements on the inner continental shelf. Contin. Shelf Res. 7, 411–431.

Heron, M.L., Prytz, A., Heron, S.F., Helzel, T., Schlick, T., Greenslade, D.J.M., Schulz, E., Skirving, W.J., 2008. Tsunami observations by coastal ocean radar. International Journal of Remote sensing 29 (21), 6347–6359.

Heron, M.L., Dexter, P.E., McGann, B.T., 1985. Parameters of the air–sea interface by high-frequency ground-wave HF Doppler radar. Aust. J. Mar. Freshwater Res. 36, 655–670.

Hickey, K., Khan, R.H., Walsh, J., 1995. Parametric estimation of ocean surface currents with HF radar. IEEE J. Oceanic Eng. 20, 139–144.

Hill, D.A., Wait, J.R., 1980. Ground-wave attenuation function for a spherical earth with arbitrary surface impedance. Radio Sci. 15, 637–643.

Holbrook, J.R., Frisch, A.S., 1981. A comparison of near-surface CODAR and VACM measurements in the Strait of Juan de Fuca, August 1978. J. Geophys. Res. 86, 10,908–10,912.

Janopaul, M.M., Broche, P., De Maistre, J.C., Essen, H.H., Blanchet, C., Grau, G., Mittelstaedt, E., 1982. Comparison of measurements of sea currents by HF radar and by conventional means. International Journal of Remote Sensing 3, 409–422.

Jeans, P.K., Donnelly, R., 1986. Four-element CODAR beam forming. IEEE J. Oceanic Eng. OE-11 (2), 296–303.

Kim, S.Y., Terrill, E., Cornuelle, B., 2007. Objectively mapping HF radar-derived surface current data using measured and idealized data covariance matrices. J. Geophys. Res. 112 (C06021), 1–16.

Kohut, J., Glenn, S., Barrick, D., 1999. SeaSonde is integral to coastal flow model development. Hydro Int. 3, 32–35.

Lane, A., Knight, P.J., Player, R.J., 1999. Current measurement technology for near-shore waters. Coastal Engineering 37, 343–368.

Leise, J.A., 1984. The analysis and digital signal processing of NOAA's surface current mapping system. IEEE Journal of Oceanic Engineering OE-9 (2), 106–113.

Lewis, J.K., Shulman, I., Blumberg, A.F., 1998. Assimilation of Doppler radar current data into numerical ocean models. Cont. Shelf Res. 18, 541–559.

Lillesand, T.M., Kiefer, R.W., 1979. Remote sensing and image interpretation, second ed. Wiley.

Lipa, B., 2003. HF radar current velocity uncertainties. Sea Technol. 44 (9), 81–81.

Lipa, B.J., Barrick, D.E., 1983. Least-squares methods for the extraction of surface currents from CODAR crossed-loop data: Application at ARSLOE. IEEE J. Ocean. Eng. OE-8 (4), 226–253.

Lipa, B.J., Barrick, D.E., Bourg, J., Nyden, B.B., 2006. HF Radar detection of tsunamis. Journal of Oceanography 62, 705–716.

Lipa, B., Nyden, B., Ullman, D.S., Terrill, E., 2006. SeaSonde radial velocities: Derivation and internal consistency. IEEE J. Oceanic. Eng. 31 (4), 850–861.

Mantovanelli, A., Heron, M.L., Prytz, A., 2010. The use of HF radar surface currents for computing Lagrangian trajectories: Benefits and issues. IEEE 2010, Washington, DC, U.S.A.

Maresca Jr., J., Carlson, C., 1980. Comment on "Longshore currents on the fringe of hurricane Anita," by N. P. Smith. J. Geophys. Res. 85 (C3), 1640–1641.

Matthews, J.P., Simpson, J.H., Brown, J., 1988. Remote sensing of shelf sea currents using a high-frequency ocean surface current radar system. J. Geophys. Res. 93 (C3), 2303–2310.

Ohlmann, C., White, P., Washburn, L., Terrill, E., Emery, B., Otero, M., 2007. Interpretation of coastal HF Radar derived surface currents with high-resolution drifter data. J. Atmos. Oceanic Technol. 24, 666−680.

Ohlmann, J.C., White, P.F., Sybrandy, A.L., Niller, P.P., 2005. GPS-cellular drifter technology for coastal ocean observing systems. J. Atmos. Oceanic Technol. 22, 1381−1388.

Paduan, J.D., Shulman, I., 2004. HF radar data assimilation in the Monterey Bay area. J. Geophys. Res. 109 (C07 S09), 1−17.

Paduan, J.D., Rosenfeld, L.K., 1996. Remotely sensed surface currents in Monterey Bay from shore-based HF radar (Coastal Ocean Dynamics Application Radar). J. Geophys. Res. 101, 20,669−20,686.

Paduan, J.D., Barrick, D., Fernandez, D., Hallock, Z., Teague, C., 2001. Improving the accuracy of coastal HF radar current mapping. Hydro Int. 5, 26−29.

Paduan, J.D., Petruncio, E.T., Barrick, D.E., Lipa, B.J., 1995. Surface currents within and offshore of Monterey Bay as mapped by a multiple-site HF radar (CODAR) network. Proc. 1EEE Fifth Working Conference on Current Measurements, 137−142.

Pandian, P.K., Emmanuel, O., Ruscoe, J.P., Side, J.C., Harris, R.E., Kerr, S.A., Bullen, C.R., 2010. An overview of recent technologies on wave and current measurement in coastal and marine applications. Journal of Oceanography and Marine Sciences 1 (1), 1−10.

Parker, K., 1989. Surface current radar: Boon to coastal engineers. Sea Technol. 30 (2), 60−65.

Ponsford, T., Srivastava, S., 1990. Groundwave over-the-horizon radar development at NORDCO. NORDCO Limited.

Prandle, D., 1987. The fine structure of near-shore tidal and residual circulations revealed by HF radar surface current measurements. J. Phys. Oceanogr. 17, 231−245.

Prandle, D., Ryder, D.K., 1985. Measurements of surface currents in Liverpool Bay by high-frequency radar. Nature 315, 128−131.

Prandle, D., Loch, S.G., Player, R., 1992. Tidal flow through the straits of Dover. J. Phys. Oceanogr. 23, 23−37.

Schmidt, R.O., 1986. Multiple emitter location and signal parameter estimation. IEEE Trans. Antennas Propag. AP-34 (3), 276−280.

Schofield, O., Kohut, J., Glenn, S., 2005. The New Jersey shelf observing system. Sea Technol. 46 (9), 15−21.

Schott, F.A., Frisch, S.A., Larsen, J.C., 1986. Comparison of surface currents measured by HF Doppler Radar in the Western Florida straits during November 1983 to January 1984 and Florida current transports. J. Geophys. Res. 91 (C7), 8451−8460.

Shay, L.K., Graber, H.C., Ross, D.B., Chapman, R.D., 1995. Mesoscale ocean surface current structure detected by high-frequency radar. J. Atmos. Ocean. Technol. 12, 881−900.

Shay, L.K., Graber, H.C., Ross, D.B., Chapman, R.D., 1995. Mesoscale ocean surface current structure detected by high-frequency radar. J. Atmos. Oceanic Technol. 12, 881−900.

Shay, L.K., Lentz, S.J., Graber, H.C., Haus, B.K., 1998. Current structure variations detected by high-frequency radar and vector-measuring current meters. J. Atmos. Oceanic Technol. 15, 237−256.

Shearman, E.D.R., 1986. A review of methods of remote sensing of sea-surface conditions by HF radar and design considerations for narrow-beam systems. IEEE J. Ocean. Eng. OE-11 (2), 150−157.

Shearman, E.D.R., Moorhead, M.D., 1988. PISCES: A coastal groundwave HF radar for current, wind, and wave mapping to 200-km ranges. Proceedings IGARSS '88, 773−776.

Skop, R.A., Graber, H.C., Ross, D.B., 1995. VHF radar measurements of flow patterns in bays and estuaries. Proc. IEEE Fifth Working Conference on Current Measurements, 143−147.

Stewart, R.H., Joy, J.W., 1974. HF radio measurements of surface currents. Deep-Sea Res. 21, 1039−1049.

Takeoka, H., Tanaka, Y., Ohno, Y., Hisaki, Y., Nadai, A., Kuroiwa, H., 1995. Observation of the Kyucho in the Bungo Channel by HF radar. J. Oceanogr. 51, 699−711.

Teague, C.C., 1986. Multifrequency HF radar observations of currents and current shears. IEEE Journal of Oceanic Engineering OE-11 (2), 258−269.

Vigan, X., 2002. Upper-layer ocean current forecasts. Sea Technol. 43 (10), 15−19.

Wait, J.R., 1966. Theory of HF ground wave backscatter from sea waves. J. Geophys. Res. 71 (20), 4839−4842.

Weber, B.L., Barrick, D.E., 1977. On the nonlinear theory for gravity waves on the ocean's surface, Part I: Derivations. Journal of Physical Oceanography 7, 3−10.

Wu, J., 1975. Sea surface drift-currents. Proc. Offshore Technology Conference 2, 477−484.

Wyatt, L.C., Thompson, S.P., Burton, R.R., 1999. Evaluation of HF radar wave measurements. Coastal Engineering 37, 259−282.

BIBLIOGRAPHY

Abascal, A.J., Castanedo, S., Medina, R., Losada, I.J., Fanjul, E.A., 2009. Application of HF radar currents to oil spill modeling. Mar. Poll. Bull. 58, 238−248.

AHQ Survey Coy, 1942. AHQ Survey Coy by Authority of Director of Survey, 1942.

Anderson, S., 2011. HF skywave radar performance in the tsunami detection and measurement role. In: Czerwinski, F. (Ed.), InTechOpen (Open Access to Knowledge). The tsunami threat: Research and technology Chapter 31, pp. 641−666.

Backhaus, J.O., 1985. A three-dimensional model for the simulation of shelf sea dynamics. Dt. Hydrogr. Z. 38, 165−187.

Barrick, D., 2003. Proliferation of SeaSonde coastal current-mapping radars. Hydro Int. 7 (1), 30−33.

Barrick, D.E., 1971. Theory of HF and VHF propagation across the rough sea: 2. Application to HF and VHF propagation above the sea. Radio Sci 6, 527−533.

Barrick, D.E., 1977. Extraction of wave parameters from measured HF radar sea-echo Doppler spectra. Radio Science 12, 415−424.

Barrick, D.E., 1978. H. F. radio oceanography: A review. Bound Layer Meteor 13, 23−43.

Barrick, D.E., 1998. Grazing behaviour of scatter and propagation above any rough surface, IEEE Trans. Antennas. Propag. 46, 73−83.

Barrick, D.E., Lipa, B.J., 1996. Comparison of direction-finding and beam-forming in HF Radar Ocean surface current mapping. Phase 1 SBIR Final Report. Contract No. 50-DKNA-5-00092. NOAA, Rockville, MD, USA.

Barrick, D.E., Lipa, B.J., 1997. Evolution of bearing determination in HF current mapping radars. Oceanography 10, 72−75.

Barrick, D.E., Lipa, B.J., 1986. Correcting for distorted antenna patterns in CODAR ocean surface measurements. IEEE J. Ocean. Engr. OE-11, 304−309.

Barrick, D.E., Snyder, J.A., 1977. The statistics of HF sea-echo Doppler spectra. IEEE Trans. Antennas Propag. AP-25 (1), 19−28.

Bendat, J.S., Piersol, A.G., 2000. Random data analysis and measurement procedures. Wiley Interscience.

Bilham, R., Engdahl, R., Feldl, N., Satyabala, S.P., 2005. Partial and complete rupture of the Indo-Andaman plate boundary, 1847−2004. Seismol. Res. Lett. 76 (3), 299−311.

Bjorkstedt, E., Roughgarden, J., 1997. Larval transport and coastal upwelling: An application of HF radar in ecological research. Oceanography 10, 64−67.

Broche, P., Salomon, J.P., Demaistre, J.S., Devenon, J.L., 1986. Tidal currents in Baie de Seine: Comparison of numerical modeling and high-frequency radar measurements. Estuarine Coastal Shelf Sci. 23, 465−476.

Bryant, E., 2001. Tsunami: The underrated hazard. Cambridge University Press, Cambridge, UK.

Collar, P.G., Howarth, M.J., 1987. A comparison of three methods of measuring currents in the sea: RADAR, current meters and surface drifters. Department of Energy, Offshore Technology Report, OTH 87 272 London, HMSO.

Collar, P.G., Howarth, J.M., Millard, N.W., Eccles, D., 1985. An intercomparison of HF radar observations of surface currents with moored current meter data and displacement rates of acoustically tracked drogued floats: Evaluation, comparison and calibration of oceanographic instruments, Proceedings of an international conference. Advances in Underwater Technology and Offshore Engineering. Graham & Trotman, 163−182.

Dankert, H., Rosenthal, W., 2004. Ocean surface determination from X-band radar-image sequences. J. Geophys. Res. 109, C04016.

Dzvonkovskaya, A., Gurgel, K.W., Rohling, H., Schlick, T., 2008. Low power, high frequency surface wave radar application for ship detection and tracking. Proc. Radar 2008 Conf. Adelaide, Australia, 654−659.

Dzvonkovskaya, A., Gurgel, K.W., Pohlmann, T., Schlick, T., Xu, J., 2009a. Simulation of tsunami signatures in ocean surface current maps measured by HF radar. Proc. Oceans 2009 Conf. Bremen, Germany.

Dzvonkovskaya, A., Gurgel, K.W., Pohlmann, T., Schlick, T., Xu, J., 2009b. Tsunami detection using HF radar WERA: A simulation approach. Proc. Radar 2009 Conf. Bordeaux, France.

Essen, H.H., Gurgel, K.W., Schlick, T., 2000. On the accuracy of current measurements by means of HF radar. IEEE J. Oceanic. Eng. 25 (4), 472−480.

Essen, H.H., Gurgel, K.W., Schirmer, F., Sirkes, Z., 1995. Horizontal variability of surface currents in the Dead Sea. Oceanol. Acta. 18, 455−467.

Essen, H.-H., Gurgel, K.W., Schirmer, F., 1983. Tidal and wind-driven parts of surface currents as measured by radar. Dtsch. Hydrogr. Z. 36, 81−96.

Essen, H.-H., Gurgel, K.W., Schirmer, F., 1989. Surface currents in the Norwegian Channel measured by radar in March 1985. Tellus 41A, 162−174.

Fernandez, D.M., Graber, H.C., Paduan, J.D., Barrick, D.E., 1997. Mapping wind direction with HF radar. Oceanography 10, 93−95.

Fernandez, D.M., Vesecky, J.F., Teague, C.C., 1995. Measurement of upper-ocean surface currents with high-frequency radar. Proc. IEEE Fifth Working Conference on Current Measurements, 109−114.

Frisch, A.S., Weber, B.L., 1980. A new technique for measuring tidal currents by using a two-site HF Doppler Radar System. J. Geophys. Res. 85 (C1), 485−493.

Frisch, A.S., Weber, B.L., 1982. Application of dual-Doppler HF radar measurements of ocean surface currents. Remote Sensing of Environment 12, 273−282.

Gill, E.W., Walsh, J., 2001. High-frequency bistatic cross-sections of the ocean surface. Radio Sci. 36 (6), 1459−1475.

Gill, E.W., 1999. The scattering of high frequency electromagnetic radiation from the ocean surface: An analysis based on bistatic ground wave radar configuration. Ph.D. thesis. Memorial University of Newfoundland, St. John's, Canada.

Gonella, J., 1972. A rotary-component method for analysing meteorological and oceanographic vector time series. Deep-Sea Res. 19, 833−846.

Gower, J., 2005. Jason 1 detects the 26 December 2004 tsunami. American Geophysical Union EOS 86, 37−38.

Graber, H.C., Limouzy-Paris, C.B., 1997b. Transport patterns of tropical reef fish larvae by spin-off eddies in the Straits of Florida. Oceanography 10, 68−71.

Griffiths, C.R., MacDougall, N., 1998. The Tiree Passage mooring 1981−1997. Dunstaffnage Marine Laboratory, Marine Technology Report No. 145.

Gurgel, K.W., 1994. Shipborne measurement of surface current fields by HF radar. L'Onde Electrique 74, 54−59.

Gurgel, K.W., 1997. Experience with ship-borne measurements of surface current fields by radar. Oceanography 10, 82−84.

Gurgel, K.W., Barbin, Y., Schlick, T., 2007. Radio frequency interference suppression techniques in FMCW modulated HF radars. Proc. IEEE/OES Oceans 2007 Europe, Aberdeen, Scotland, UK.

Hasselmann, K., 1971. Determination of ocean wave spectra from Doppler radio return from the sea surface. Nature Physical Science 229, 16−17.

Heron, M.L., Prytz, A., Heron, S.F., Helzel, T., Schlick, T., Greenslade, D.J.M., Schulz, E., Skirving, W.J., 2008. Tsunami observations by coastal ocean radar. Int. J. Remote Sens. 29 (21), 6347−6359.

Hodgins, D.O., 1994. Remote sensing of ocean surface currents with the SeaSonde HF radar. Spill Sci. Technol. Bull. 1, 109−129.

Holbrook, J.R., Frisch, A.S., 1981. A Comparison of near-surface CODAR and CACAM measurements in the Strait of Juan De Fuca, August 1978. J. Geophys. Res. 86 (C11), 10,908−10,912.

Howarth, M.J., Harrison, A.J., Knight, P.J., Player, R.J., 1995. Measurement of net flow through a channel. IEEE Fifth Working Conference on Current Measurement, 121−126.

Janopaul, M.M., Broche, P., de Maistre, J.C., Essen, H.H., Blanchet, C., Grau, G., Mittelstaedt, E., 1982. Comparison of measurements of sea currents by HF radar and by conventional means. Int. J. Remote Sens. 3, 409−422.

Kaplan, D.M., Largier, J., Botsford, L.W., 2005. HF radar observations of surface circulation off Bodega Bay (northern California, USA). J. Geophys. Res. 110, C10020.

Khan, R.H., 1991. Ocean-clutter model for high-frequency radar. IEEE J. Oceanic Eng. 16, 181−188.

King, J.W., co-authors, 1984. OSCR (Ocean Surface Current Radar) observations of currents off the coasts of Northern Ireland, England, Wales and Scotland, Current measurements off shore. Soc. Underwater Technol.. London, UK.

Kinsman, B., 1965. Wind waves. Prentice-Hall, Englewood Cliffs, NJ, USA.

Klinck, J.M., 1985. EOF analysis of central Drake Passage currents from DRAKE 79. J. Phys. Oceanogr. 15, 288−298.

Knight, P.J., Howarth, M.J., 1998. The flow through the North Channel of the Irish Sea. Continental Shelf Res. 19, 693–716.

Kohut, J.T., Glenn, S.M., 2003. Improving HF radar surface current measurements with measured antenna beam patterns. J. Atmos. Oceanic Technol. 20, 1303–1316.

Kosro, P.M., Barth, J.A., Strub, T.P., 1997. The coastal jet: Observations of surface currents over the Oregon continental shelf from HF radar. Oceanography 10 (2), 53–57.

Laws, K.E., Fernandez, D.M., Paduan, J.D., 2000. Simulation-based evaluations of HF radar ocean current algorithms. IEEE J. Oceanic Eng. 25, 481–491.

Ledgard, L.J., Webster, S., Wyatt, L.R., 1996. The measurement of ocean waves along the Holderness Coast using OSCR. UK Oceanography '96 Programme and abstracts, IEEE Conference publication.

Leise, J.A., 1981. The analysis and digital signal processing of NOAA's surface current mapping radar. IEEE J. Oceanic Eng. 9, 106–113.

Leise, J.A., 1984. The analysis and digital signal processing of NOAA's surface current mapping system. IEEE J. Oceanic Eng. OE-9, 106–113.

Lipa, B.J., Barrick, D.E., 1983. Least-squares methods for the extraction of surface currents from CODAR crossed-loop data: Application at ARSLOE. IEEE J. Ocean. Eng. OE-8 (4), 226–253.

Lipa, B., Barrick, D., Bourg, J., Nyden, B., 2006b. HF radar detection of tsunami. J. Oceanogr. 62, 705–716.

Lyzenga, D., Nwogu, O., Trizna, D., 2009. Ocean wave field measurements using coherent and non-coherent radars at low grazing angles. IGARSS 2010. Honolulu, HI, USA, 26–30.

Martin, R.-J., Kearney, M.J., 1997. Remote sea current sensing using HF radar: An autoregressive approach. IEEE J. Oceanic Eng. 22, 151–155.

Matthews, J.P., Fox, A.D., Prandle, D., 1993. Radar observation of an along-front jet and transverse flow convergence associated with a North Sea front. Continental Shelf Res. 13, 109–130.

Melton, D.C., 1995. Remote sensing and validation of surface currents from HF radar. M.S. thesis. Naval Postgraduate School.

Orr, W.I., 1978. Radio Handbook, twentyfirst ed. Editors and Engineers, Indianapolis, IN, USA.

Osborne, M.J., 1991. OSCR and Interocean S4 current measurements in Poole Bay. Underwater Technology 17, 10–18.

Paduan, J.D., Graber, H.C., 1997. Introduction to high frequency radar: Reality and myth. Oceanography 10, 36–39.

Paduan, J.D., Kim, K.C., Cook, M.S., Chavez, F.P., 2006. Calibration and validation of high-frequency radar ocean surface current observations. IEEE J. Ocean. Eng. 31 (4), 862–875.

Peters, N.J., Skop, R.A., 1997. Measurements of ocean surface currents from a moving ship using VHF radar. J. Atmos. Oceanic Technol. 14, 676–694.

Pierson, W., Moskowitz, L., 1964. A proposed spectral form for fully developed seas based upon the similarity theory of S.A. Kitaigorodskii. J Geophys. Res. 69 (24), 5181–5190.

Player, R.J., Radar, H.F., 1995. (OSCR) surface current measurements in the North Channel. July 1993–August 1994, Proudman Oceanographic Laboratory, Joseph Proudman Building, 6 Brownlow street, Liverpool LB 5DA, United Kingdom. Report No. 40.

Prandle, D., 1985b. Measuring Currents at the sea surface by H.F. Radar (OSCR). J. Soc. Underwater Technol. 11 (2), 25–27.

Prandle, D., 1987. The fine structure of near-shore tidal and residual circulations revealed by HF radar surface current measurements. J. Phys. Oceanography 17, 231–245.

Prandle, D., 1991. A new view of near-shore dynamics based on observations from H.F. radar. Progress in Oceanography 27, 403–438.

Prandle, D., Ryder, D.K., 1985. Measurement of surface current in Liverpool Bay by high-frequency radar. Nature 315 (6015), 128–131.

Prandle, D., Howarth, J., 1986. The use of HF radar measurements of surface currents for coastal engineers. In: Int. Conf. on Measuring Techniques of Hydraulics Phenomena in Offshore Coastal and Inland Waters. London, BHRA, Cranfield, UK.

Richman, J.G., de Szoeke, R.A., Davis, R.E., 1987. Measurements of near-surface shear in the ocean. J. Geophys. Res. 92 (C3), 2851–2858.

Rohling, H., 1983. Radar CFAR thresholding in clutter and multiple target situations. IEEE Trans. Aerosp. Electron. Syst. AES 19, 608–621.

Shearman, E.D.R., 1983. Propagation and scattering in mf/hf groundwave radar. IEEE Proc. (F) 130, 579–590.

Sixt, M., Parent, J., Bourdillon, A., Delloue, J., 1996. A new multibeam receiving equipment for the Valensole Skywave HF radar: description and applications. IEEE Trans. Geosci. Remote Sens. 34, 708–719.

Smith, W.H.F., Scharroo, R., Titov, V.V., Arcas, D., Arbic, B.K., 2005. Satellite altimeters measure tsunami. Oceanography 18, 11–13.

Stewart, R.H., Joy, J.W., 1974. H.F. radio measurements of surface currents. Deep-Sea Res. 21, 1039–1049.

Trizna, D.B., 2001. Errors in bathymetric retrievals using linear dispersion in 3D FFT analysis of marine radar ocean wave imagery. IEEE Trans. Geosciences and Remote Sensing 39, 2465–2469.

Trizna, D.B., 2009. A coherent marine radar for decameter-scale current mapping and direct measurements of directional ocean wave spectra. OCEANS 2009. Biloxi MS, USA, 1–6.

Trizna, D.B., 2010. Coherent marine radar measurements of properties of ocean waves and currents. Proceedings of IGARSS 2010. Honolulu, HI, USA, 23–27.

U.S. Government Accounting Office, 2010. U.S. tsunami preparedness. GAO Report 10–490, April 2010.

Ullman, D.S., O'Donnell, J., Kohut, J., Fake, T., Allen, A., 2006. Trajectory prediction using HF radar surface currents: Monte Carlo simulations of prediction uncertainties. J. Geophys. Res. 111, C12005.

Wait, J.R., 1962. Electromagnetic waves in stratified media. Pergamon Press, New York, NY, USA.

Wyatt, L.R., Ledgard, L.J., 1996. OSCR Wave Measurements—some preliminary results. IEEE J. Oceanic Eng. 21 (1), 64–76.

Yoshikawa, Y., Masuda, A., Marubayashi, K., Ishibashi, M., Okuno, A., 2006. On the accuracy of HF radar measurement in the Tsushima Strait. J. Geophys. Res. 111 (C04009), 1–10.

Young, I.R., Rosenthal, W., Ziemer, F., 1985. A three-dimensional analysis of marine radar images for the determination of ocean wave directionality and surface currents. J. Geophys. Res. 90, 1049–1059.

Chapter 5

Imaging of Seawater Motion Signatures Using Remote Sensors

Chapter Outline

5.1. Aerial Photography in the Visible and Infrared Bands 140
5.2. Remote Detection by Radiometers 141
  5.2.1. Passive Radiometry in the Visible-Wavelength Band 142
  5.2.2. Active Radiometry in the Visible-Wavelength Band 145
  5.2.3. Passive Radiometry in the Thermal Infrared Band 146
    5.2.3.1. Detectors 147
    5.2.3.2. Atmospheric Effects and Correction 148
    5.2.3.3. SST Imaging 148
  5.2.4. Microwave Radiometers 149
    5.2.4.1. Principle of Passive Microwave Radiometry 150
    5.2.4.2. Passive Microwave Radiometer Instrumentation 150
    5.2.4.3. Scanning Multichannel Radiometers 151
5.3. Active Microwave Radar Imaging of Sea Surface Current Signatures 153
5.4. Active Microwave Radar Imaging Technologies 154
  5.4.1. Active Microwave Imaging by the RAR Systems 155
  5.4.2. Active Microwave Imaging by SAR Systems 158
    5.4.2.1. Synthesis of Long Antenna from Small Antenna 159
    5.4.2.2. Enhancement of Azimuth Resolution 160
    5.4.2.3. Sources of Errors 163
    5.4.2.4. Interferometric SAR (InSAR) System 165
5.5. Advances in the Development of SAR Technology 166
  5.5.1. Interpretation of Image Data 167
5.6. Detection of Seawater Circulation Features Using RAR and SAR 168
5.7. Measurement of Sea Surface Currents Using Imaging of Ice Floes 170
References 171
Bibliography 174

The human brain is especially good at pattern recognition; therefore, an advantage of obtaining imagery is that it helps us achieve clear understanding of several visually observable phenomena (e.g., signatures of various kinds of seawater motions) through images of them. Images are obtained through a wide variety of remote sensing technologies (i.e., technologies for making measurements from a distance without having direct physical contact). Images of seawater motion signatures are constructed based on the characteristics of electromagnetic radiations that are reflected or scattered from or emitted by the sea surface. Modern remote sensing devices include passive and active systems; they employ different bands in the visible, infrared, microwave, very high-frequency (VHF), ultra high-frequency (UHF), and high-frequency (HF) regions in the electromagnetic spectrum.

Fortunately, the Earth's atmosphere is almost completely transparent to all these radiation bands when cloud-free and nearly so to radio waves, even when clouds are present. This is one of the reasons for the wide application of space technologies to remotely sense several of the sea surface signatures, such as those pertaining to ocean currents and their manifestations. A remarkable feature of remote sensing technologies is their ability to obtain synoptic and repetitive coverage of large and inaccessible areas. Technical and economic feasibilities of remote sensing technologies for regional and global monitoring of the changing oceanic environmental conditions have been demonstrated through several experimental, semi-operational, and operational studies the world over. Apart from academic interest, remote sensing through space-based systems offers unique possibilities for systematic and timely acquisition of information on sea surface features, with short turnaround times for use by environmental and fisheries scientists as well as coast guard and tactical personnel. Ocean surveillance by remote sensors has already grown to a very useful operational level.

Sensors that detect natural radiation, either emitted or reflected from the sea surface, are known as *passive*

Measuring Ocean Currents. http://dx.doi.org/10.1016/B978-0-12-415990-7.00005-3
Copyright © 2014 Elsevier Inc. All rights reserved.

sensors. In the oceanographic context, sensors that produce and then transmit their own electromagnetic radiation onto the sea surface and subsequently detect the return radiations are called *active sensors*. Sensors can be either *imaging* types or *nonimaging* types. The focus of this chapter is the imaging types of sensor for remotely detecting the sea surface manifestations of various types of seawater motions. Satellite remote sensing observations (sea surface temperature, color, and roughness) are utilized to characterize the circulation processes and patterns in the upper layer of the oceans.

5.1. AERIAL PHOTOGRAPHY IN THE VISIBLE AND INFRARED BANDS

Sea surface roughness and color contrasts are the two major types of signatures that are often utilized in the detection of sea surface current patterns with the aid of aerial photography in the visible and infrared bands. These contrasts provide visual indication of the surface manifestations of seawater motions. Sea surface roughness changes from the combined effect of varying wind stress in the marine atmospheric boundary layer, or MABL (induced by the sea surface temperature front), and surface current convergence and divergence (induced by the meandering current). In particular, the changes in the stratification of the MABL and subsequent wind stress changes produce a step-like drop of the sea surface roughness at the downwind side of the front (Johannessen et al., 2005). The color signatures are usually those of chlorophyll, algae, and suspended-sediment plumes. Whereas chlorophyll and algal blooms are often indicative of upwelling motions, the suspended sediment plumes often arise from river discharges into the bluer, clearer seawater. The plume of turbid brown water is easily detectable in the visible wavelength photographs. In the initial years of oceanographic studies, time-lapsed aerial photography from overhead was employed to obtain information on water motion from such images.

The turbid brown water often assumes a lobate or relatively smooth convex front that maintains its integrity as it moves in response to the sea surface currents. The size and history of movement of the plumes can be investigated from a series of time-lapse aerial photographs. The photographic gray tone and, in the case of colored imagery, the color tone are very important in the interpretation of sea surface water circulation features. Relative ages of multiple plume fronts representing discrete water bodies in gulfs could also be determined from aerial photographs by examining the overlapping relations of plumes with differing sediment concentrations; the older, less turbid plumes probably have lower sediment concentrations. This means that aerial photography can provide a synoptic view of discrete water masses that can be differentiated by suspended sediment

load. If clearer water is absent, the prevailing circulation pattern can be deduced from density patterns of turbid water. There were instances in which small-scale eddies could be detected using photo-density analysis of aerial photographs of suspended-sediment plumes (Verstappen, 1977). For photographic investigation of plumes, the aircraft is generally flown at altitudes of 10,000 feet or less.

Aerial photography has been used as an effective tool for identifying water-motion patterns in gulfs, estuaries, and frontal regions. Another remote sensing technology effectively used for detection of sea surface circulation patterns is the sensing of sea surface roughness. Oceanic regions that are dominated by intense currents are often marked by a roughness contrast of the sea surface compared to those of the adjacent regions with no significant currents. For example, mariners knew for centuries that the Gulf Stream often exhibited atypical waves. When waves oppose currents, a much rougher surface is produced. When currents are present away from shallow depths, the initial waves are transformed into higher-amplitude waves with shorter wavelengths and more whitecaps. Although the wave period remains essentially unchanged, the roughness (root-mean-square wave slope) is greatly increased, i.e., the wavelength decreases and amplitude increases. The waves ultimately break when their steepness increases to a point where they become unstable.

The amplitude and maximum slope of a progressive wave just starting to break lie near the theoretical values of the wave's height-to-length ratio of 0.14 and 30°, respectively. In some instances, wave-current interaction, particularly wave blocking, is thought to be the mechanism responsible for increased roughness of the sea surface. Witting (1984) observed that wave blocking, which can occur when a wave field propagates into a spatially increasing current, is especially effective in altering the properties of the wave field. Wave blocking by a current, u, occurs when the component of the group velocity of the wave aligned with the current matches the magnitude of u. Wave blocking, breaking, and refraction may occur even when the waves do not advance directly into the current. Similarly, if a train of surface gravity waves runs in the direction of a surface current, their absolute progression is enhanced, with a resulting decrease in wave amplitude. The wave-current interaction resulting in wave-amplitude modification becomes more noticeable for short waves because their phase velocity is comparable to the prevailing sea surface currents. Swell waves for which the phase velocities are appreciably larger than the prevailing surface circulation velocities do not contribute to surface roughness contrasts arising from wave-current interactions.

If there is a convergence or divergence of the current, the roughness "contrast" along the line of convergence or divergence, arising from wave amplitude contrast, will be very predominant compared to that in the surrounding regions. It is known that there is often a surface convergence associated with fronts due to secondary circulation in the vertical plane at right angles to the geostrophic flow of the front.

As a result of wave-current interaction, the water flow path will be associated with a sharp contrast in the appearance of the surface wave pattern. These contrasts can be seen in the sea surface images. Therefore these images permit identification of the sea surface current circulation path. This is one of the principles behind remote detection of sea surface circulation using visible-wavelength photography.

A classical example of a current circulation pattern that can be clearly seen in visible-wavelength photography is the surf zone circulation cells consisting of longshore and rip currents. Inman et al. (1971) provided an excellent aerial photograph of the surf zone current showing an evenly spaced jet-like rip current carrying material from the surf zone and producing vortex pairs. The spacing between the rip currents determines the longshore dimensions of the nearshore circulation cell. In the wave zone, sea surface currents and waves are each influenced by the other in different ways. For example, Strong et al. (2000, 2003) examined the modifications undergone by waves propagating through regions with a significant mean current.

An important photographically detectable circulation feature is isolated vortices in the flow field. Such vortices are often produced by the shear between an intense current and its adjacent stagnant water mass. It is known that the shear region between the Kuroshio current in the East China Sea and the Pacific coast is populated with several isolated vortices of various scales. The motions of these vortices can be surveyed from time-series aerial photographs. The large roughness signature of these vortices in contrast to the surrounding water masses makes them clearly visible in the aerial photographs. Very fast tidal currents in straits are also associated with isolated vortices. For example, the intense tidal current (approximately 10 knots) in the Naruto Strait in Japan has many vortices for which the diameter usually exceeds 20 meters and water surface depression exceeds 2 meters. Yamoaka et al. (2002) reported an aerial photo of the tidal vortex pair generated in the Neko-Seto Channel. Inman et al. (1971) reported an excellent aerial photograph of evenly spaced rip currents in a surf zone.

Another factor that causes surface roughness contrast is the accumulation along a line of surface debris, seaweeds, surface pollutants, oil slicks, and the like, which are drawn into the convergence zone as a result of horizontal convergence of surface water. If there is a surface convergence in the circulation pattern, the near-surface water is pulled into the line of convergence and is then drawn down into deeper depths. However, the up-thrust on the floats prevents them from sinking down with the current. This causes the floating materials at the surface to be concentrated at the sea surface along the line of convergence. Conversely, a divergence zone results in a weakening or removal altogether of any floating material on the surface. A combination of convergence and divergence cells, such as those associated with Langmuir circulation, causes floating materials to be distributed on the sea surface in coherent rib patterns. Trapping of Sargassum weeds along the lines of convergence of Langmuir circulation cells and frontal regions is a familiar phenomenon. Furthermore, turbulence generated at the convergence zones adds to the surface roughness, which can be photographed. Thus, aerial photography of sea surface permits imaging of patterns resulting from some of the dynamic processes in the upper ocean. It is evident that aerial photography plays an important role as a seawater flow pattern visualization technique. Aerial photographs have also been valuable in detection of the boundaries of discrete water masses.

5.2. REMOTE DETECTION BY RADIOMETERS

Radiometers have been used extensively in the remote sensing of seawater current circulation features such as gyres, meanders, fronts, rings, upwelling, jets, and intrusions. The underlying principle in the use of radiometers for such studies is the fact that these features exhibit certain sea surface signatures such as color, temperature, or roughness that can be detected by radiometers.

In the oceanographic context, the radiometer is a sensor capable of measuring the flux of electromagnetic energy that impinges on it from the sea surface. Radiometers are employed to measure fluxes in the visible, infrared, and microwave bands of the electromagnetic spectrum and can be broadly classified into passive and active sensors.

Passive radiometers operate based on the principle that all materials that have temperatures greater than absolute zero (i.e., $-273°C$) emit electromagnetic radiation, the intensity and spectral compositions of which are functions of the physical properties of the emitting surface. An active radiometer illuminates the sea surface with its own energy. The energy scattered back from the sea surface in the direction of the sensor is processed to generate images and thereby elucidate the surface manifestations of seawater motion signatures. Various types of radiometric technologies that are employed for studies of such signatures are briefly discussed here.

Frontal and ring-dominated regions are associated with upwelling. In the upwelling process, the cooler and nutrient-rich water brought to the photic zone leads to greatly enhanced productivity of phytoplankton (known as a *phytoplankton bloom*) relative to that in the adjacent water masses. Phytoplankton (unicellular plants living in the sunlit sea surface layers) contains a variety of light-absorbing photosynthetic pigments. The most abundant pigment is chlorophyll-a, which absorbs most strongly in the blue part of the optical spectrum (near 443 nm), with a secondary absorption peak at red wavelength (near 670 nm), whereas accessory pigments have absorption maxima at intermediate wavelengths. This absorption causes the solar radiation backscattered out of the sea at 443 nm to decrease rapidly with increasing chlorophyll-a concentration. The chlorophyll-a absorption is much weaker at 520 to 550 nm (the green part of the spectrum). Thus, waters that are poor in chlorophyll-a will appear as deep blue in sunlight, whereas waters rich in chlorophyll-a will appear green (Gordon et al., 1980).

The influence of phytoplankton pigments on the color of the seawater under natural light has been discussed by Yentsch (1960). Thus, seawater color contrast over a spatial domain on the sea surface is one of the water-motion signatures (upwelling) detected by a visible-wavelength radiometer. Sea surface color is expressed as the spectral (i.e., wavelength) composition of radiance exiting the sea surface in the visible waveband (approximately 400–700 nanometer). The sea surface color is dependent on the absorption and scattering properties of the water as well as dissolved and particulate constituents such as organic matter, phytoplankton cells, and suspended mineral particles.

Phytoplankton can also be excited by laser radiation to induce fluorescence in the red region of the spectrum. Thus, phytoplankton bloom, often representative of dynamic subsurface seawater motion extending up to the surface layers, has a color signature under natural as well as induced radiations. Time-series images of the ocean color contrasts permit an experienced oceanographer to study the evolution, growth, movement, and ultimate decay of these dynamic features. This is the basis underlying the use of visible wavelength radiometers for remote sensing of seawater motion signatures.

Passive as well as active sensing technologies have been used in visible-wavelength radiometers. The satellite-borne Coastal Zone Color Scanner (CZCS) and aircraft-borne oceanographic LiDAR (AOL) are, respectively, examples of passive and active radiometers.

5.2.1. Passive Radiometry in the Visible-Wavelength Band

The visible-wavelength passive radiometer measures the color of the sea surface under natural light. As mentioned earlier, this color is recognized as one of the manifestations of sea surface motion. The term *color* is used here to mean the spectrum of upwelling radiance just beneath the sea surface.

The possibility of remote sensing of the ocean color was demonstrated first by aircraft-borne radiometers (Clarke

et al., 1970). Subsequently, in October 1978 the CZCS was flown on the Nimbus-7 satellite on a "proof-of-concept" mission. The CZCS was the only sensor in orbit specifically designed to detect the upwelling solar radiance reflected from within the ocean.

The CZCS was a spatially imaging multispectral scanner (MSS). The MSS systems operate on the principle of selective sensing in multiple spectral bands. Accordingly, they are designed to sense radiation energy in a number of narrow spectral bands simultaneously. By analyzing a scene (feature) in several spectral bands, it becomes feasible to distinguish the color identity of the scene.

A commonly used scanning device is a rotating or oscillating mirror which moves the field of view along a scan line perpendicular to the direction of flight. At any instant in time, the scanner has an instantaneous field of view (IFOV). The IFOV is usually expressed as the cone angle (β) within which the incident radiation is focused onto the detector and is given by $\beta = \dfrac{D}{H}$ radians, where D is the diameter of the ground footprint and H is the flying height (see Figure 5.1). The rotating or oscillating mirror causes the IFOV to move from one side of the satellite flight path to the other. This allows the radiometer to scan a strip of the sea surface perpendicular to the flight path. The width of this strip below the radar is equal to the diameter, D, of the instantaneous footprint (see Figure 5.2). A consequence of increased swath angle is continuous increase of the footprint size as the IFOV moves outward from the satellite's nadir. This happens because as the swath angle increases, the range between the satellite and the radiometer footprint increases. Consequently, the width of the strip (on the sea surface), perpendicular to the flight path, gradually increases as the distance from the satellite's nadir increases. The forward motion of the satellite advances the viewed strip between successive scans,

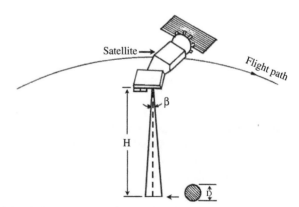

FIGURE 5.1 Schematic diagram illustrating the instantaneous field of view of a satellite-borne radiometer in relation to flying height, H, and beam width, β.

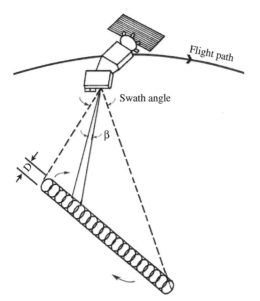

FIGURE 5.2 Schematic diagram illustrating a strip on the sea surface viewed in a single scan of a satellite-borne scanning radiometer.

causing a two-dimensional image dataset to be recorded (see Figure 5.3). For simplicity, the strip is shown to be of the same width.

The CZCS multispectral scanner had a geometric IFOV of approximately 0.05°. This corresponds to an area of a circle of approximately 830 m diameter on the sea surface from an altitude of 955 km. The scan was ±39.32° about the nadir, corresponding to a swath width of 1,659 km. Two

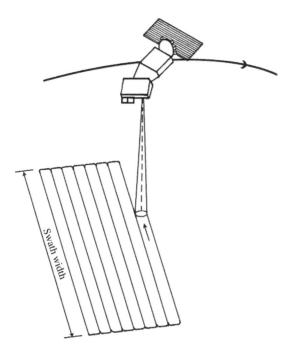

FIGURE 5.3 Schematic diagram illustrating the method of obtaining a two-dimensional image dataset from the sea surface using a satellite-borne scanning radiometer.

minutes of alongtrack scanning was packaged as an image for the purpose of data storage. With an orbit period of 104.15 minutes, this corresponded to a distance along the subsatellite track of 767 km. Thus, each two-minute image corresponded to an area of 767 × 1,659 (roughly a million) square kilometers. The image was slightly oversampled, and there were 1,968 pixels per scan line and 970 scan lines per image. The power available for sensor operation was limited; hence the CZCS was in use for only two hours per day (Austin, 1979). The CZCS incorporated a novel mechanism whereby its scan plane could be tilted up to 20° from nadir ahead of or behind the direction of travel of the satellite in 2° increments. This mechanism avoided masking a large area of the satellite field of view by sunglint (i.e., specular reflection of sunlight from the sea surface). The gain of the sensor and its tilt angle were controlled from the ground and, in normal operation, were determined by the solar elevation at the center of the scene (Gagliardini et al., 1984).

In an MSS system, the total incoming radiation is separated into several spectral components and is then sensed independently. A dichroic grating is used to separate the nonthermal wavelength from the thermal wavelength in the incoming radiation. The nonthermal wavelength component is directed from the grating through a prism or diffraction grating that splits the energy into a continuum of UV, visible, and IR wavelengths. At the same time, the dichroic grating disperses the thermal components of the incoming signal into its constituent wavelengths. By placing an array of detectors at the proper geometric positions behind the "dispersing" devices, each spectral band can be measured independently.

Silicon photodiodes and charge-coupled devices (CCDs) are the commonly used detectors in the visible region of the spectrum. Photodiodes are very sensitive light detectors whose output voltage varies linearly with the incident light intensity. Integrated photo-detector arrays with many photodiode elements are usually employed. A CCD is a microelectronic silicon chip that detects light. When light strikes the CCD's silicon surface, electric charges are produced. The magnitude of the charge thus produced is proportional to the light intensity and exposure time. CCD arrays containing thousands of CCD elements are being employed in a number of different types of remote sensing systems because of their small size and weight, durability, and light sensitivity. The CCD receivers are designed for photo-electric scanning. Scanning is performed via a simple shift operation, reading sequentially through the array, because the sensor elements are coupled electrically as a shift register. CCDs have the advantage of wide dynamic range combined with rugged construction.

The CZCS scanning radiometer viewed the ocean in several spectral bands. These included bands in the visible, near-infrared, and thermal regions of the spectrum. The visible bands were optimized for ocean color observations with narrow bandwidths placed in high, medium, and low regions of the chlorophyll absorption spectrum. The four visible bands were 443 ± 10 nm (blue), 520 ± 10 nm (green), 550 ± 10 nm (yellow), and 670 ± 10 nm (red). These very narrow bands (20 nm wide) were meant for discrimination of very subtle water reflectance differences. High signal-to-noise (S/N) ratio and high sensitivity enabled differentiation of slight changes in water-leaving radiances. The near-infrared channel (750 ± 50 nm) was meant to aid in locating cloud and water-land boundary prior to processing the data in the other bands (Lillesand et al., 1987). Because near-infrared radiation is not influenced by plant pigments but is strongly reflected from suspended sediments (Stumpf and Tyler, 1988), visible-band passive radiometry is also used for investigation of transport of suspended sediments in estuaries. Sediment patterns reveal residual transport, tidal streams, and small gyres (Robinson, 1985). The thermal infrared channel ($10.5-12.5$ μm) permitted measurements of sea surface temperature.

There are many factors that influence the quality of the data generated by an MSS system. Because these systems are designed to sense energy over a very small IFOV (to optimize spatial resolution) and in narrow wavelength bands (to optimize spectral resolution), a very limited amount of energy is incident on each of the individual detectors of the MSS system. This means that the detectors must be very sensitive to output a signal, which is significantly stronger than the level of system noise. The S/N ratio of an airborne MSS system is expressed as follows (Lillesand et al., 1987): $(S/N)_\lambda \propto D_\lambda \beta^2 (H/V)^{1/2} \Delta\lambda L_\lambda$. In this expression, D_λ is detectivity, β is instantaneous field of view, H is flying height, V is satellite velocity, $\Delta\lambda$ is spectral bandwidth of the channel, and L_λ is spectral radiance of ground feature.

Of these parameters, D_λ is a measure of the performance quality of the detector. The parameters β, H, and V together determine the spatial resolution, and $\Delta\lambda$ is a measure of spectral resolution. It is seen that the S/N value is partially based on wavelength-dependent parameters. Accordingly, different S/N values are applicable to each channel of an MSS system. Other factors that influence the S/N performance of any given scanner system are atmospheric attenuation, design of the system's optical components, and the noise characteristics of the system electronics.

In the CZCS scanner, the continuous electrical output from the detectors were digitized at such a rate that successive samples, or picture elements, represented contiguous areas on the sea surface. The angular size of the aperture was such that the forward motion of the spacecraft allowed successive adjacent scan lines of data to be added (Holligan et al., 1989). The CZCS radiance data was 8-bit digitized onboard. The Nimbus-7 contained three onboard tape recorders. With these, it was possible to acquire imagery for areas out of range of the land-based receiving stations.

Data received from the Nimbus-7 CZCS visible bands were processed to a certain level at NASA (Hovis et al., 1980). The first level of processing converted the digital counts to radiances. Through NOAA, the timestamped and geocoded digital data are available to users on computer-compatible tapes (CCTs).

The depth to which the CZCS could "see" into the ocean was typically $15-30$ meters in the open ocean. For relatively high pigment or turbid waters, it was generally less than 15 m at the most penetrating wavelengths. In certain coastal regions, large concentrations of sediments and dissolved organic substances from land runoff rendered the penetration depth even less than a meter. Thus, the remotely sensed signal represented a depth-weighted average of surface properties (Smith et al., 1987; Holligan et al., 1989).

During its travel to the visible-band passive radiometry sensor, the water-leaving radiance was considerably absorbed and scattered by the ozone layer. Furthermore, the atmosphere scattered light into the field of view of the sensor. Thus, only a small portion of the radiance received at the sensor aperture represented the radiance actually originating from the sea surface. In fact, the phytoplankton spectral signature represented less than 20% of the overall upwelling signal that reached the satellite sensor. This made it necessary to correct the data for atmospheric influences. Extraction of water-leaving radiance from the contaminated total radiance is a problem of remote sensing. Details on this topic may be found in Gordon et al. (1980), Sturm (1981), Gordon et al. (1982), Smith and Baker (1982), Robinson (1985), and Holligan et al. (1989).

After suitable atmospheric corrections of reflectances from various visible-band channels of the sensor, appropriate algorithms are applied to derive phytoplankton pigment concentration. Subsequently, color composite images are produced on a color video monitor of the image-processing computer system. Different colors are then attributed to different pigment concentration ranges. This permits the viewer to obtain a clear overall view of phytoplankton distribution. In many cases, the differences in ocean color may be used as a biological tracer or indicator of water mass origin when physical parameters give diffuse or conflicting signals (Tyler and Stumpf, 1989).

The initial imagery from Nimbus-7 satellite confirmed that the radiometer data could be processed to a level that reveals subtle variations in the concentration of phytoplankton pigments. The potential of visible-band passive radiometry imagery for the study of water mass boundaries and mesoscale seawater circulation patterns had been predicted by Hovis et al. (1980).

Today, sea surface color imaging is considered one of the potential techniques for remotely sensing oceanic water motion patterns. Through analysis of cloud-free radiometric images, Caraux and Austin et al. (1983) demonstrated that satellite remote sensing of chlorophyll could disclose characteristic seasonal phytoplankton patterns associated with hydrodynamic features such as coastal upwelling, river plumes, and offshore cyclonic activity. The passive radiometry image contains the most salient features of an eddy. For example, the chlorophyll concentration in the center portion of a warm-core ring (eddy) is depressed relative to the high-velocity boundary (Smith et al., 1987). The congruence between the physical and biological structure of the ring was remarkable. The ocean color imagery is indeed an interesting source of information for investigating upwelling motions.

Smith et al. (1987) have provided excellent CZCS images of two warm-core rings developed in the Gulf Stream/Sargasso Sea off the east coast of the United States. Analysis of successive images permits investigation of seasonal changes of oceanic boundaries. Once an image of a dynamic feature is obtained on a geographical map, it is possible to make quantitative measurements of the spatial features, which are apparent in imagery. The dominant length scales of isolated gyres are sometimes assessed by measuring with a ruler on a hard copy of the image. Looking at the whole scene, the human brain tends to automatically filter out or ignore irrelevant patterns. The speed and direction of movement of the identified feature, as well as its spatial growth or decay over time, may be studied from a sequence of images in the vicinity of the given area. Analysis of radiometer-derived images could reveal Gulf Stream meander and the rings shed by it.

5.2.2. Active Radiometry in the Visible-Wavelength Band

Interpretation of ocean color differences, especially for quantitative information, is often difficult because of interfering effects from the atmosphere. The atmosphere also produces spectrally dependent attenuation of the sea-scatter return and an overlapping upwelling radiance of its own. The effects of interfering signals can be reduced by using the pulsed nature and narrow spectral bandwidth radiation of a laser transmitter. A light detection and ranging (LiDAR) transmitter is an ideal pulsed illuminator for fluorescence studies because much of the optical background can be eliminated by time gating the receiver at the proper platform-to-surface range. Oceanographic LiDAR systems are used to make rapid, wide-area measurements from airborne platforms.

With excitation at an appropriate wavelength, natural waters are found to exhibit fluorescence from a number of naturally occurring substances, including chlorophyll,

algae, dye tracers, and pollutants, and the fluorescence emission spectra can be distinctive (Philpot and Vodacek, 1989). *Fluorescence* in this context refers to the characteristic of various objects on the sea surface to absorb energy at one wavelength and then emit a lower energy at different wavelengths shortly after excitation by the original energy source. Because different materials tend to fluoresce at different wavelengths, laser-induced fluorescence (LIF) can be used to discriminate among material types. Laser fluorescence makes use of this property. Because fluorescence intensity is usually proportional to the concentration of the fluorescing material, the LIF technique is potentially used to detect and measure the concentration of various fluorescing substances. Material particles in the ocean waters are passive tracers of water motions. Time-series spatial images of these substances permit identification and quantification of seawater motion. This is the basis underlying the use of a visible-wavelength active radiometer for detection of ocean dynamic features. LIF is routinely used for airborne monitoring of chlorophyll and is a potential technique for remote sensing of coastal ocean dynamics.

The active sensor used in the LIF remote sensing technique is a LiDAR system. In this technique, a pulsed laser is used to illuminate the top layers of the sea surface, and a multichannel fluorometer (spectroradiometer) records the spectral characteristics of the radiation emitted by the fluorescent objects in the water. The fluoresced radiation is often focused by a lens onto the entrance slit of a spectrograph (an optical grating). The spectrally dispersed light from the spectrograph is detected with a multichannel photodiode array and the associated active elements. The spectrum is usually corrected for instrument sensitivity and configuration.

Aircraft-borne oceanographic LiDAR (AOL) systems have been used operationally to map chlorophyll concentration and, therefore, circulation features of coastal seawater bodies. A conceptual rendering of an AOL system is given in Figure 5.4. Smith et al. (1987) used an AOL for detection of a Gulf Stream warm-core ring. In this system a pulsed laser with a wavelength of 532.1 nm (green) was used to stimulate fluorescence from the chlorophyll and phycoerythrin photopigments in phytoplankton from a subsurface water column within a footprint of ~ 0.5 m diameter. The LIF from these photopigments, centered at 685 nm (red) and 580 nm (yellow), respectively, were received and recorded. Along with the laser-induced fluorescence, the LiDAR recorded the short-range distance between the aircraft and the sea surface. The aircraft position was determined from an inertial navigation system.

The distinct and important characteristic of an active spectrometer is its specificity. The laser's spectral bandwidth is sufficiently narrow (typically subnanometer) to

FIGURE 5.4 Conceptual rendering of an aircraft-borne LiDAR system.

allow identification of spectral lines, which are easily resolved and unambiguously assigned to chlorophyll and phycoerythrin pigments in the phytoplankton.

5.2.3. Passive Radiometry in the Thermal Infrared Band

The use of thermal infrared radiometry for detection of seawater circulation features relies on the fact that some of these features often have a thermal signature. For example, western boundary currents (WBCs) are associated with recognizable surface temperature gradients. Well-known WBCs, such as the Gulf Stream and the Somali Current, are known to have distinct surface temperature gradients. As the name suggests, well-known circulation features such as warm-core and cold-core eddies, warm-water and cold-water jets and intrusions, fronts, and the like provide distinct spatial sea surface temperature (SST) gradients. For coastal waters, the spatial temperature gradients are often indicative of advective motions of estuarine waters. The estuarine water advection gives rise to relatively sharp horizontal gradients at the frontal boundary of warm stratified water and cool tidally mixed waters. Similarly, the upwelled waters are cooler than the surrounding surface waters. The boundary between the cooler upwelled water and the surrounding warmer surface water often takes the form of a sharp front, which can be

readily detected from sharp variations in spatial SST. In the case of water currents and intrusions, frequent monitoring of the leading edge of the SST boundary allows one to estimate the variability of their speed and direction. In the case of meanders and detached eddies, such monitoring permits determination of their spatial scales and shape as well as speed and direction of their drifts.

The sea surface temperature is remotely measured from ships, aircraft, or satellites by observing the intensity of the infrared radiation (within the wavelength range 8–14 μm) emitted from the sea surface. The radiated energy is an external manifestation of the thermal energy state of the water molecules that constitute the sea surface. The radiated energy is a function of its "radiant" temperature. Because the absorption in seawater at the infrared wavelength is rather high, only the radiation from a very thin surface layer (less than one-tenth of a millimeter) is radiated into the atmosphere.

Remote measurement of SST relies on the Stefan-Boltzmann law, which states that the total radiant exitance, M, from the surface of a black body varies as the fourth power of its absolute temperature. The *black body* is a hypothetical radiator that totally absorbs (i.e., no reflection) and reemits the entire energy incident upon it. The Stefan-Boltzmann law (applicable for a black body) is expressed in mathematical form as $M = \sigma T^4$, where M is the total radiant exitance in W/m^2; σ is the Stefan-Boltzmann constant; and T is the temperature of the black body ($^\circ$K).

The concept of a black body is only a convenient theoretical vehicle to describe radiation principles. Real bodies do not behave as black bodies; they emit only a fraction of the energy emitted from a black body at the equivalent temperature. However, blackbody radiation principles can be extended to real bodies by reducing the radiant exitance, M, by a factor termed *emissivity*, ε, such that for real bodies $M = \varepsilon \sigma T^4$. The emissivity of a material is its emitting ability relative to that of a black body.

By definition, ε is the ratio of the radiant exitance from an object at a given temperature to that from a black body at the same temperature. The emissivity can have values between 0 and 1. It can vary with wavelength and viewing angle. Depending on the material, emissivity can also vary somewhat with temperature. At thermal infrared wavelengths, the emissivity of water varies from 0.96 to 1.00; therefore, its behavior is very close to that of a blackbody radiator. The temperature T of a radiating surface can be inferred by remote measurements of the radiant exitance, M, from the surface. This indirect approach is used for remote sensing of sea surface temperature. M is measured over a discrete wavelength range. The 8–14 μm region of the spectral radiant exitance curves is of particular interest because it contains the peak energy emissions for most ocean surface features.

5.2.3.1. Detectors

Infrared radiance detectors can be classified into two types: heat-sensing and photon-sensing devices. Heating of a material or the absorption of photons by a semiconductor can both produce changes that can be detected by appropriate electrical methods. Thermal detectors respond to the total energy of the radiation impinging on them, irrespective of its spectral distribution, whereas photon detectors respond to the rate at which quanta of radiation are absorbed (Katsaros, 1980). Thermal detectors such as thermistor flakes were used in shipborne radiometers. A major disadvantage of thermal detectors is their comparatively long response time. For this reason they are rarely used from fast-moving vehicles such as aircraft and satellites. Contrary to thermal detectors, photon detectors are capable of very rapid (<1 μs) response. With their rapid-response characteristics, photon detectors are in widespread use in satellite-borne remote sensing systems. The photoconductive effect is commonly used in photon detectors. Increase in conductivity results when the energy of a photon induces free flow of electrons, or "holes," or electron-hole pairs within the crystal lattice. Detection of infrared radiation is enhanced if the background thermal excitation is minimized. For this reason semiconductor photon detectors are usually operated at temperatures approaching absolute zero. Normally, the detector is surrounded by a Dewar containing liquid helium or nitrogen ($-196°$C). Electrical (Peltier) cooling can also be used. The commonly used photon detectors are mercury-doped germanium, Indium antimonide, and mercury cadmium telluride (Lillesand et al., 1987).

In remote sensing, images of sea surface temperature contrasts are interpreted in terms of sea surface current patterns (Harris et al., 1978, Legeckis, 1987). Two-dimensional thermal images can be produced either by an array of detectors or by a scanning radiometer. Airborne radiometers are usually configured as scanners. In this configuration, only one detector is used. The target area is scanned either by a rotating mirror or by the spin-and-advance-motion of the satellite. In the spin-stabilized TIROS-series satellites, the detector views the Earth from horizon to horizon in lines perpendicular to the path of advance. In the Nimbus series, the same side of the satellite always faces the Earth, and the scan is provided by a rotating mirror. The radiation emitted from the scene is intercepted by the scan mirror, which diverts the radiation to a collecting telescope. The telescope focuses the radiation to a detector. In a normal case, the scan mirror is inclined $45°$ to the optical axis. When the scan mirror is rotated about the optical axis, the field of view of the telescope sweeps out a circle in a plane, which is normal to the rotation axis of the scan mirror. Thus, by the scan mirror rotation, radiation is received and measured from a continuous line of length corresponding to the total scan angle. When such an instrument is mounted on a moving platform (spacecraft) with the optical axis parallel to the platform motion, the motion of the platform produces successive scan lines, giving contiguous imagery. In the case of a multispectral scanner, the energy collected by the telescope is channeled to a spectral dispersing system (spectrometer) to be registered in different spectral bands.

In operation, the thermal energy radiated from a footprint on the sea surface is focused onto the detector, which converts the thermal radiation into an equivalent electrical signal. Because glass lenses absorb thermal infrared radiation, either a germanium lens or a set of gold-coated mirrors of suitable shapes is used to focus the radiant energy onto the detector. Gold is a good reflector of infrared radiation. The entering radiation comes alternately from the sea surface footprint and from an onboard temperature reference source (blackbody cavity) via a rotating toothed chopper wheel in front of the focusing system (see Figure 5.5). The radiation from the target reaches the detector through the openings between the chopper teeth. The chopper's gold coating acts as a mirror when the chopper covers the focusing system. This allows reflection of the radiation from the blackbody cavity onto the detector. Chopping of the radiation produces an alternating signal. The blackbody cavity is an electrically heated, calibrated internal temperature reference source. The electrical signal from the detector is processed so that its amplitude is related to the difference in the intensities of

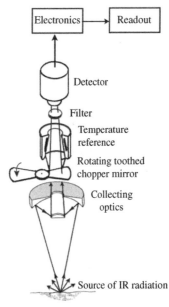

FIGURE 5.5 Schematic diagram illustrating satellite remote sensing of thermal infrared radiation from the terrestrial surface with reference to an onboard temperature reference source. *(Source: In part from Lillesand and Kiefer, 1979.)*

thermal radiation from the terrestrial (sea surface in the present case) footprint and the blackbody reference source.

5.2.3.2. Atmospheric Effects and Correction

Several environmental factors tend to degrade the accuracy of the perceived SST. The atmosphere has a significant effect on the intensity and spectral composition of the energy recorded by a thermal system. The intervening atmosphere between a thermal sensor and the sea surface can increase or decrease the apparent level of radiation coming from the sea surface. The atmospheric absorption and scattering of the signals from the sea surface tend to make the sea surface appear cooler than it is, and atmospheric emission tends to make the sea surface appear warmer than it is. Depending on atmospheric conditions during imaging, one of these effects will outweigh the other, thereby resulting in a biased sensor output (Lillesand et al., 1987). Thus one of the initial problems in making SST observations from space used to be the obscurity of the sea surface by clouds. The presence of noise levels of various satellite instruments has been another problem. However, techniques have been developed to correct for these effects (La Violette et al., 1969; Shenk et al., 1972; Hillger et al., 1988).

Several methods exist for atmospheric correction. A multispectral technique is one among them. The effects of varying water vapor concentration on the integrated atmospheric transmittance differ over different infrared spectral channels. Consequently, the differences of the perceived SST between different channels may be used to implement correction for errors induced by water vapor. Another method is to use different algorithms for processing of data from night and day. This method relies on the fact that atmospheric contributions are different at night from daytime ones. There is no reflected solar radiation at night. The daytime corrections should strictly include a dependence on the sun's zenith angle. Another major improvement in accuracy is promised from the use of a multilook sensor, which views the same piece of sea surface through different atmospheric path lengths (Robinson, 1985). The presence of clouds was a serious problem in frontal detection. However, the fact that the clouds are almost always colder than water and that they move much more rapidly than sea surface water motion features have become useful in resolving this problem. The effects of clouds are now being reduced by overlaying several daily pictures and compositing them by selecting the warmest temperatures at each point. Using this method, cloud effects have been eliminated entirely or reduced substantially in many pictures. When the clouds are extremely persistent, the process of overlaying can be extended to several days. This latter process may degrade the results because corresponding points can represent temperatures from different

days, but it has produced usable images from very poor data (Gerson et al., 1982).

5.2.3.3. SST Imaging

Beginning in 1978, passive radiometers onboard NOAA environmental satellites provided synoptic views of the sea surface temperature several times a day, with a spatial resolution of 1 to 4 km. For example, Vigan et al. (2000) reported SST images over the Gulf of Mexico and the Agulhas Current waters off South Africa. Examining SST image sequences makes it possible to track the evolution of fine-scale oceanic patterns possibly associated with strong currents. Satellite-based surveying methods benefit from high space/time resolution of radiometer data (1–4 km by 1–4 km every four to six hours) and from their synoptic spatial coverage (1,000 by 1,000 km), which sets up the possibility for short-term forecasting (48–72 hours) of ocean features evolution. This, together with cost effectiveness, contributes to making satellite-based methods superior to onsite approaches.

Vigan et al. (2000a, b) presented a variable finite-element model for the purpose of computing absolute upper-layer oceanic currents from sequences of advanced very high-resolution radiometer (AVHRR) sea surface temperature images. Satellite images have been processed daily and current-calculation algorithms and feature-tracking tools have been applied, yielding a detailed report of oceanic conditions. Comparison with simultaneously collected current measurements from an ADCP demonstrated the outstanding accuracy of the satellite-based approach. The oceanic current calculations were found to be accurate to within 10% in magnitude and 10° to 15° in direction after a comparison to simultaneous onsite current measurements.

The model is not only diagnostic but has predictive skill as well, making it a powerful tool regardless of atmospheric conditions. These researchers have also developed feature detection and tracking tools (fronts, eddies, filaments, plumes, etc.) based on satellite imagery. Tracking an oceanic front from an image series allows, for example, the monitoring of typical quantities such as frontal-wave propagation velocity or its surging or shedding, which provide forecasting clues for a 48- to 72-hour prediction of the upper-layer currents.

It was found that the calculation of currents from time-series satellite-derived SST imaging is applicable to any oceanic province featuring relatively pronounced temperature variations over space and time, typically in the Gulf of Mexico, off the U.S. east and west coasts, north and southeast Brazil, Argentina, South Africa, Namibia, or Angola, but also in the China Sea, in Indonesia, or east of Australia. As typical examples, Figures 5.6 and 5.7 show, respectively, a satellite AVHRR infrared image of a Gulf Stream frontal eddy in April 1977 and a NOAA/satellite AVHRR thermal

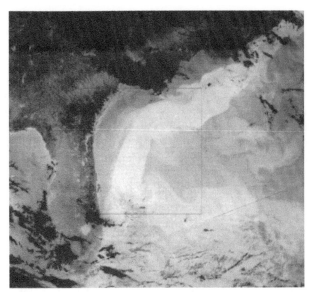

FIGURE 5.6 Satellite AVHRR infrared image of a Gulf Stream frontal eddy on 16 April 1977 at 0200 GMT. *(Source: Lee et al., 1981.)*

5.2.4. Microwave Radiometers

Remote sensing of the ocean surface by satellite-borne radiometers operating at visible and infrared wavelengths has provided valuable details of sea surface circulation features. However, the sea surface cannot be adequately viewed by visible-wavelength radiometers in darkness. Likewise, SST-derived information is obscured by cloud cover. Operational utility of these sensors is thus limited because much of the sea surface either is obscured by clouds or is in darkness. In contrast, frequencies in the microwave band have the unique capability of propagating through clouds. Also, they do not depend on the sun's illumination and therefore can operate during day and night (Calla, 1984). The ability of radiation in the microwave band to propagate through clouds renders microwave sensors effective all-weather devices, although water in the form of precipitation scatters radiation and can render the atmosphere opaque at the microwave band (Robinson, 1985).

The microwave radiometer is the passive counterpart to microwave radar. Being passive, the microwave radiometer does not supply its own radiation; rather, it senses the naturally available microwave energy within its field of view. Microwave sensing principles and sensing instrumentation are similar to those of thermal infrared sensing in many respects. As with thermal infrared sensing, blackbody radiation theory is the principle used in the operation of microwave radiometers, and both kinds of radiometer sense the sea surface temperature for detection of sea surface manifestations of seawater circulation features. The main difference between thermal infrared radiometers and microwave radiometers is that the former uses photon detection elements whereas the latter uses microwave antennae for detection of radiation emanating from their field of view.

infrared image depicting the structure of Norwegian coastal current in September 1995. Vigan (2002) provided AVHRR images of Loop Current warm waters and cold coastal water masses in the Gulf of Mexico during March 1996. Vigan (2002) also provided a July 2000 AVHRR image of the Agulhas Current waters that separate from the South African coast and flow toward the southwest describing meanders. Upper-layer current was calculated from two successive AVHRR images. Based on this calculation, the current magnitude in the Agulhas jet core was found to be about 4 knots. A METEOSAT infrared image of the meandering Agulas Current reported by Irvine and Tilley (1988) also exhibited comparable features.

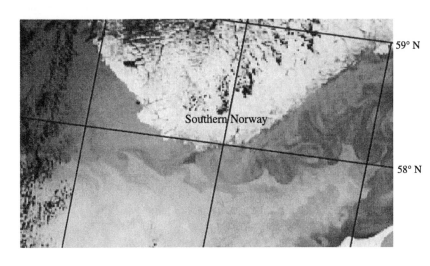

Southern Norway

59° N

58° N

FIGURE 5.7 NOAA/Satellite AVHRR thermal infrared image depicting the structure of Norwegian coastal current on 20 September 1995. *(Source: Johannessen et al., 2000.)*

5.2.4.1. Principle of Passive Microwave Radiometry

Passive microwave radiometry applied to investigation of SST involves the detection of thermally generated radiation from the sea surface at microwave frequencies. The spectral band in which passive microwave radiometry applies is approximately 1 GHz to 200 GHz, i.e., a wavelength of approximately 30 cm to 0.15 cm. Remote sensing of SST from air- and space-borne passive microwave radiometers relies on Planck's radiation law. If the temperatures to be measured lie in the vicinity of 300°K (typical of Earth's surface temperatures), and the radiation of interest is in the microwave region of the electromagnetic spectrum, the spectral radiance or brightness of the radiating body is often expressed by the Rayleigh-Jeans approximation of Planck's blackbody radiation law. This approximation is given by $B_f(\theta, \phi) = \frac{2kT}{\lambda^2}$, where $B_f(\theta, \phi)$ is the spectral brightness of the radiating surface per unit frequency of radiation in the conventional direction (θ, ϕ); k is Boltmann's constant; T is the temperature of the radiating surface in °K; and λ is the wavelength of the radiating electromagnetic wave.

It is seen that for a given wavelength, the brightness is directly proportional to the temperature of the radiating source. The approximation given above applies to an ideal black body. Because an ideal black body does not exist in nature, measurements are usually given in terms of the "brightness," T_B. In passive microwave radiometry terminology, the brightness temperature is also called *apparent antenna temperature*. The brightness temperature of the radiation is equal to the temperature T_B of the equivalent black body. This leads to the relationship of brightness temperature to physical temperature T in terms of emissivity $\varepsilon(\theta, \phi)$ as $T_B = \varepsilon(\theta, \phi)T$.

Normally, the radiometer output is expressed in terms of T_B; i.e., the system is calibrated in terms of the temperature that a black body located at the antenna must reach to radiate the same energy as was actually collected from the ground (Lillesand et al., 1987), which is sea surface in the present case. The emissivity term renders the just-mentioned relation applicable to real bodies. In the case of sea surface, the emissivity is in general a function of sea surface texture (i.e., sea state), temperature, chemical composition, and the radiating frequency. Below 5 GHz the emissivity, ε, of the sea surface at wavelength λ is expressed in terms of the molecular temperature of the sea surface, T_S, and chemical composition, S (salinity), of the surface water. Details on these aspects may be found in Blume et al. (1977).

The emissivity of the sea surface at the lower end of microwave frequencies (between 1.5 and 5 GHz) is approximately 0.3 instead of near-unity in the infrared region of radiation. This difference places stringent requirements on the absolute accuracy of measurements by the microwave radiometer system. Because passive microwave sensors operate in the low-energy tail of the blackbody radiation curve for temperatures in the region of approximately 300°K (typifying SST features), the signal received at the radiometer antenna is extremely weak. This means that a comparatively large antenna beam width is required to collect enough energy to yield a detectable signal. Consequently, passive microwave radiometers are characterized by low spatial resolution.

5.2.4.2. Passive Microwave Radiometer Instrumentation

The first microwave radiometer was reported by R. H. Dicke in 1946. Most microwave radiometers in use today are similar in concept to the one designed by Dicke. The Dicke radiometer in its simplest form consists essentially of a high-gain microwave antenna with low side lobes, a microwave switching circuit (a Dicke switch), and the required detection and recording electronics. The microwave radiation of targets is very weak. The noise temperature of a receiver can reach several thousand times the radiation from the target. For this reason microwave radiometers must be designed to detect a weak signal in a large noise.

The equation for sensitivity of a Dicke-type radiometer is given by $\Delta T_{rms} = \frac{2T_{eff}}{\sqrt{\Delta f \tau}}$, where T_{rms} is the root-mean-square output of a radiometer defined as the sensitivity; Δf is the RF bandwidth; τ is the integration time; and T_{eff} is the effective noise temperature, including the temperature received by the antenna from the target and the temperature of the receiver converted to the antenna.

It can be seen from this relation that a low effective noise temperature and a wide RF bandwidth improve the sensitivity. In conventional microwave radiometers, the antenna picks up the very low-level microwave signal. The switching circuitry permits rapid, alternate sampling between the thermal radiation power received by the antenna and an internal calibration temperature reference signal. The reference load can be either a heated or cooled source or space. Space has a brightness temperature of approximately 3°K (Thomas, 1981). The differential signal is amplified and phase-detected synchronously with the switching circuit (Tomiyasu, 1974). Through suitable calibration procedures, the difference between the antenna signal and the reference signal can be related to the input antenna temperature, which is a measure of the radiation intensity received by the antenna.

A passive microwave radiometer used for measurement of SST detects the radiation emitted from the sea surface and the intervening atmosphere. Quantitative measurement of thermal emission from the sea surface by a remote sensing radiometer, operating from high altitude, requires that correction be applied for the radiative and transmission properties of the intervening atmosphere. At microwave frequencies there are four principal categories of interference that can be classified as atmospheric effects. They are (Blume et al., 1977):

1. Extra-terrestrial background radiation
2. Radiation from discrete stellar radio sources
3. Attenuation from oxygen and water vapor in a cloudless atmosphere
4. Attenuation resulting from rain clouds

For clear atmosphere, water vapor and molecular oxygen are the major sources of attenuation in the microwave band. Precision measurement of SST using a microwave radiometer requires that all unwanted side effects such as sky background radiation, atmospheric emission, antenna beam pattern, and instrument instabilities must be accounted for. Given the very weak signal that has to be measured by a microwave radiometer, thermal noise from the antenna and the other parts of the electrical signal pathway to the preamplifier is also a major problem to be circumvented. A conventional method of correcting the errors resulting from atmospheric effects is reception of thermal radiation at different wavelengths and polarization. The problem of high-frequency noise is overcome by integrating over many samples. Thermal noise is reduced by packaging the critical parts into a single integrated unit for isothermal operation.

Initial experiments using an airborne passive microwave radiometer demonstrated its feasibility for determining SST values. Blume et al. (1977) used an aircraft-borne, 2.65 GHz radiometer for the Chesapeake Bay area. The result was compared in detail with accurately obtained sea truth data. For a calm sea, the observed temperature agreed well with that calculated from the known sea surface and atmospheric properties. For cases in which the surface wind speeds were of the order of 7 to 15 knots, an excess temperature was observed. This was attributed to emissivity variations due to surface roughness and microscale surface disturbances. The passive microwave radiometer is now recognized as a useful tool for detection of ice floes. Time-lapse images of ice floes provide a means for quantitative estimation of seawater motions in ice-infested regions such as the Greenland Sea.

5.2.4.3. Scanning Multichannel Radiometers

A scanning microwave radiometer has the advantage of providing an SST map over a large area. Conceptually, a scanning microwave radiometer operates such that its antenna's field of view is transverse to the direction of flight. Transverse scanning can be performed either mechanically or electronically. Two major types of antenna used in passive microwave radiometry are dishes and phased arrays. A typical mechanically scanning dish antenna is a parabolic dish with a suitable collecting horn. In the case of a multi-channel radiometer, the collecting horn may consist of a cluster of single-frequency horns or a single horn that receives the various frequencies coaxially (Thomas, 1981). In a phased array, multiple arrays of waveguide sections are employed. Using appropriate phase shifters, the antenna beam is electrically steered perpendicular to the direction of flight. Such an electrically scanning microwave radiometer (ESMR) was carried on the Nimbus-5 satellite, launched in 1972, and the Nimbus-6, launched in 1975. The former made use of a phased array antenna with linear polarization and operated at 19 GHz, whereas the latter measured both vertically and horizontally polarized components at a frequency of 37 GHz.

Retrieval of SST information from a passive microwave radiometer requires accurate known brightness temperatures at the antenna's collecting aperture at each frequency and polarization. The antenna receives radiation from all directions as weighted by the antenna patterns. The radiation received through the side lobes must be corrected in order to derive the true brightness temperatures emanating from the direction of the antenna boresight. Antenna pattern correction methods may be found in Njoku et al. (1980).

Passive microwave measurements from polar orbiting satellites provide global coverage of SST variability and thereby several sea surface circulation features. The first multichannel radiometer for remote sensing of terrestrial surface temperature was orbited onboard Soviet Cosmos-243 in 1968. Subsequently, a similar one was flown on Cosmos-384 in 1970. The Cosmos experiments included measurements at 8.5, 3.4, 1.31, and 0.8 cm wavelengths. The scanning multichannel microwave radiometer (SMMR) on board Nimbus-G and Seasat-A satellites measured thermal microwave emissions from the Earth at five dual linearly polarized frequencies (6.6, 10.69, 18, 21, and 37 GHz), from which SST values were derived. The SMMR instrument thus operated with 10 channels and vertical and horizontal polarizations at each of these five frequencies. Measurements were made over a swath 822 km wide below the Nimbus-G and 595 km wide below the Seasat spacecrafts. For a satellite orbiting at an altitude h, the ground resolution cell size d (i.e., the footprint on the sea surface) is given by the expression (Robinson, 1985) $d = (\lambda h)/D$, where λ is the wavelength of the received radiation and D is the antenna aperture diameter.

In practice, for a given altitude of the satellite orbit, a compromise is usually reached among footprint size, antenna size, and minimum frequency. Nimbus and Seasat SMMR had an antenna diameter of 70 cm and a highest frequency of 37 GHz (i.e., a lowest wavelength of 0.8 cm). The smallest spatial resolution cell was about 20 km at a wavelength of 0.8 cm and was proportionately larger at other wavelengths (Gloersen and Barath, 1977). For example, the spatial resolution of the SMMR was about 100 km at 6.6 GHz. On Nimbus-G, the SMMR scan pattern was forward-viewing and scanned equally to either side of the orbital track so that the swath was centered on that track. On Seasat-A the SMMR scan pattern was aft-viewing and biased toward the right of the flight path so that the center of the swath was 22° from the orbital track. Data from the SMMR were gathered from footprints of different sizes for each of the frequency channels. The SMMR instrument consisted essentially of five hardware elements:

1. An antenna assembly consisting of a reflector and a multifrequency feedhorn
2. A power supply module
3. A scan mechanism
4. An RF module containing the input and reference switching networks consisting of Dicke-type switches and detection electronics
5. A post-detection electronics module

The mechanically scanning antenna system consisted of an offset parabolic reflector with a 70-cm diameter collecting aperture and a multifrequency feed assembly. The antenna reflector was scanned with a sinusoidally varying velocity over a 50° swath angle and with a 4-s time period (Njoku et al., 1980). The antenna beams produced contiguous footprints on the terrestrial surface. During most of the scan, the radiometric signal pertaining to SST consisted of data obtained from the sea surface. At one end of the antenna scan a "hot" reference source was observed; at the opposite end a "cold" source was observed. The hot source was a microwave guide termination at the instrument's ambient temperature (approximately 300°K) and the cold source was provided by a special horn antenna viewing deep space (approximately 2.7°K). A modulator switch assembly alternated at approximately a 1-kHz rate between a reference temperature signal and an input signal.

The output of each channel—a voltage proportional to the difference between the radiometric signal and an internal reference source—was integrated over appropriate time periods, digitized, encoded, and sent to the spacecraft's data system for transmission to ground. Measurements from a number of very accurate (± 0.05°C) platinum resistance sensors, distributed at various critical points in the SMMR microwave circuitry, were combined with the telemetry datastream and transmitted to the ground. This combined dataset consisting of signal, hot calibration, cold calibration, and instrument temperature values was used to compute radiometric temperature values. The effective radiometric temperature is often referred to as *antenna temperature*. The antenna temperature was then corrected for antenna pattern effects to yield *brightness temperature*. These aspects have been discussed by Swanson et al. (1980). The SMMR retrieval algorithms have been discussed by Francis et al. (1983).

The SMMR was conceived to provide an instrument capable of obtaining (among other parameters such as surface wind speed) SST information, which is an important parameter required by oceanographers for developing and testing global ocean circulation models and other aspects of ocean dynamics. The goal of SMMR was accuracy better than 1.5°K in obtaining SST. The SMMR has indeed demonstrated sensitivities sufficient for determination of SST with an accuracy of 1°K or better, excluding conditions at significant rainfall. Because in studies of oceanic circulation features it is the difference in SST that is more important than absolute temperature, this accuracy is often sufficient for such studies. For example, owing to sharp temperature contrast between ice sheet and the surrounding water, the edge of the ice sheet is readily detectable on microwave imagery. In fact, SST studies using satellite-borne SMMR have been very useful in delineating sea ice in the Arctic and Antarctic waters (Robinson, 1985). The SMMR flown on the Nimbus satellite was able to record the movement and change in character of sea ice continuously over several years, despite cloud cover and the long polar night. Because icebergs are excellent Lagrangian tracers of water motion in the polar regions, SMMR is an excellent tool to monitor sea surface water circulation in these environmentally hostile regions. Even very slow motions can be detected using time-sequence SMMR imaging over long periods of time.

In the past, passive microwave technique has been used effectively to monitor circulation features in remote areas such as the Greenland Sea, day and night. SMMR images have been used successfully for investigation of the dynamics of the East Greenland Current, based on the creation of a mysterious large-scale ice cover phenomenon called *Odden*, in the Greenland Sea. Time-series of ice concentrations derived from the SMMR onboard the Nimbus-7 satellite have been used for this investigation. Odden is a large tongue of ice that typically appears once every year in the Greenland Sea at any time of the winter season as a result of a rather sudden decay of ice concentration solely along the shelf slope of Greenland, leaving the rest of the ice cover intact. The average size of this tongue may be 3–400 km from north to south and 1–200 km from east to west. The reason for the mysterious appearances of Odden remained largely unknown for many decades.

Close monitoring of the ice concentration in this region using data from the Nimbus-7 satellite in March 1979 indicated that Odden grew from 3–500 km in length at least within 48 hours. Based on meteorological and other data, it was argued that the process of the creation of Odden is neither a result of a current coming from the north nor a result of ice melting at the ice/air interface. These arguments led to the postulation of a vertical cellular circulation of warm water in the Greenland Sea, the driving force of which is the winds from north and northeast causing an

Ekman surface transport toward the east coast of Greenland. The proposed cellular circulation is believed to be the driving mechanism behind the formation of Odden in the Greenland Sea. It is interesting that a system inherently capable of observing solely the surface phenomena became instrumental in providing some interesting additional information about subsurface processes, which hopefully explains the mechanism behind the creation of the mysterious Odden. No wonder space agencies of various countries are including passive microwave radiometers in their ocean remote sensing programs.

5.3. ACTIVE MICROWAVE RADAR IMAGING OF SEA SURFACE CURRENT SIGNATURES

Apart from passive microwave radar imaging technologies discussed in the preceding sections, active microwave radar imaging technologies are also used in identifying sea surface water motion features. Active microwave imaging radar systems detect sea surface circulation signatures by utilizing (1) sea surface roughness contrast and (2) the sidelighted character of icebergs, which move under the influence of seawater current. The former is applicable anywhere in the ocean, but the latter is applicable only in high-latitude regions of the globe. In cases where no damping film material is present, the oceanic eddy features alter their expression into a bright radar modulation along the converging front. The indirect contribution from intermediate wave breaking via their influence on short-wind waves dominates the surface roughness modulation and thus Bragg-like scattering from the sea surface and the microwave radar image manifestation.

Radar image of the sea surface is simply a pictorial representation of the backscattering properties of the sea surface at the radar wavelength. Typically, the image is simply a photographic presentation of the backscattered signal intensity that is recorded as a function of antenna scan and slant range. The backscattered signal intensity is converted to a video signal and is usually recorded on photographic film as shades of gray ranging from black to white. These shades are known as *image tone*. The distribution of tone changes on an image is known as *image texture*. The texture is often classified as fine, smooth, coarse, grainy, speckled, mottled, irregular, and so forth. Image texture permits identification of features on an image. Because image texture depends on the distribution of tones, not on the absolute value of tones, texture is less affected by the lack of image calibration than is tone (MacDonald, 1980). For this reason, precise signal amplitude measurement is not usually necessary in radar imaging applications.

In photographic as well as radar imaging, interpretation of the picture is based on shape, texture, and context.

Production of images with good geometric fidelity and good contrast is more important than accurate calibration (Ulaby et al., 1982). Because the information content in an image is not dependent on absolute values of tones, the image texture of a given region at a given time remains relatively constant from one image to another, regardless of the radar system's gain settings.

The amount of power received by the radar from the sea surface (i.e., radar return) depends on the radar system properties and sea surface roughness properties. The former includes transmitted power, antenna gain, transmission wavelength, and slant range distance between radar and the "target" on the sea surface. The sea surface property is revealed by the scattering coefficient, also called the *normalized radar cross-section*, σ_0. When radar system properties are held constant, the gray tone on the radar image is proportional to σ_0, which, in turn, is determined by both the roughness and the electrical properties of seawater. The electrical properties of seawater do not vary appreciably from region to region. For this reason, dependence of radar return signal strength on electrical properties may be generally ignored in sea surface roughness measurement.

Surface roughness, as applied to microwave radar imaging, is a geometric property that determines the strength of the backscattered signal. The sea surface is often described statistically rather than deterministically. Scattering from statistically rough surfaces falls into two classes, namely a Bragg-resonance-like mechanism, valid for slightly rough surfaces, and a tangent plane (or facet) mechanism, valid for a gently undulating surface (Vesecky and Stewart, 1982). In the case of a slightly rough surface, it is assumed that the surface irregularity only slightly distorts the incident radar wave so that perturbation theory can be applied. Roughness in this context is not an absolute measurement but is expressed in terms of the wavelength of the radar's transmission signal. The Peake-Oliver "roughness criterion" (Sabins, 1978) considers a surface rough if $h > \lambda / (4.4 \sin \alpha)$, where h is the height of surface irregularities, λ is the radar wavelength, and α is the grazing angle between the sea surface and the direction of propagation of incident radar wave.

It can be seen from these criteria that a given surface will appear rougher at a shorter wavelength. An increased surface roughness results in the return signal being more diffuse. In the case of a rough surface, the rays of the scattered energy may be thought of as enclosed within a hemisphere, the centre of which is located at the point where the incident wave impinged on the surface. A rough surface acts as a diffuse (i.e., isotropic) scatterer where the scattering coefficient, σ_0 is essentially independent of the angle of incidence (MacDonald, 1980). When the root-mean-square (rms) surface roughness is considerably less than a wavelength, the surface appears "smooth" to the imaging radar. When the surface is smooth, scattering is almost absent and reflection becomes predominant so that the reflected signal is contained in a small angular region about the angle of reflection, obeying Snell's laws of reflection. A surface of intermediate roughness reflects a portion of the incident energy and diffusely scatters the rest. In the case of a wave-laden sea surface, a mechanism known as *Bragg resonance* enhances the strength of the return signal from those waves that have a wavelength λ_s equal to $\lambda_r/(2 \cos \alpha)$, where λ_r is the wavelength of the radar transmission, and α is the depression angle of transmission. (Some aspects of the Bragg resonance mechanism are discussed in Chapter 4.) In radar imaging, near-saturated backscatter associated with maximum roughness characterizes one expression, whereas the opposite, almost entire damping of the Bragg scattering waves, often characterizes the surface slick expression.

In most situations, the basic quantity that is measured by a microwave imaging radar is the resonant wave. Various features of ocean surface dynamics modulate the energy density of these waves. These resonant waves are often modified by wind, ocean current and its gradients, surface films, and so forth. At high sea states, nonlinear effects and scattering processes other than those just mentioned may become important. Wave breaking and the associated scattering by resonant air bubbles that exist near the surface of waves under heavy sea conditions is an example. Aside from the wind- and wave-related features, most of the features observed in microwave images are presumably related to surface currents and their variations (i.e., shear). Surface currents can amplify waves by the influence of local convergence of velocity and through changes in the direction of wave propagation, leading to the focusing of wave energy. Vesecky and Stewart (1982) showed that if the resonant sea wave train—i.e., sea waves whose length is equal to $\lambda_r/(2 \cos \alpha)$—travels with the current, its wavelength increases and the wave path tends to turn away from the current. Similarly, if the resonant sea wave train travels against the current, its wavelength decreases and the waves turn directly into the current. In the latter situation, the group velocity of the resonant wave train decreases and the resonant waves tend to pile up as they are blocked by the adverse current, reaching saturation if the blocking is strong enough. If the waves are propagating directly opposite the current, the energy density, E, of the resonant wave train is related to the energy density, E_o, of the resonant wave train in a no-current situation by this relation (Vesecky and Stewart, 1982):

$$\frac{E}{E_o} = \frac{c^2_o}{c(c + 2u)}$$

where:

u = near-surface current velocity

c = phase velocity of resonant wave train in presence of current u

c_o = phase velocity of resonant wave train in the absence of a current

This local increase in E for an adverse current (i.e., $u < 0$), and vice versa (for $u > 0$), in turn modulates σ_0 as observed by the microwave imaging radar. This variation in σ_0 of the current flow region, in comparison to the σ_0 value of the no-current region, alters the image intensity of the region where there is a surface current.

5.4. ACTIVE MICROWAVE RADAR IMAGING TECHNOLOGIES

Over the past few decades active microwave radar imaging systems have evolved into an important tool for monitoring sea surface water motion and circulation features, particularly in those parts of the globe where cloud cover presents a serious problem to optical sensors. Active microwave radar imaging systems can be broadly classified into three categories: rotating antenna radar, real aperture side-looking radar, and synthetic aperture side-looking radar. By convention, the microwave spectrum extends from 0.3—300 GHz (i.e., 1 meter to 1 millimeter in wavelength). Microwave radar was initially developed during World War II to facilitate navigation and target location using the familiar rotating antenna and circular cathode ray tube display.

It is possible to produce a good picture of the surroundings from a fixed point by rotating an antenna with a narrow beam pointing in the horizontal direction. In this type of plan position indicator (PPI) radar system, the positions of the radar "echoes" are indicated by the radial sweep on the circular display screen. A PPI radar system essentially images a continuously updated plan view map of the targets in the field of view of its rotating antenna. PPI systems are commonly employed in weather forecasting, air traffic control, and maritime navigational applications. Their resolution is rather poor for use in remote sensing for oceanographic investigations. If a rotating antenna is used on a fast-moving vehicle such as an aircraft or a spacecraft, the image will be distorted because of the forward motion of the vehicle during the antenna rotation. However, this problem is not serious on a rotating antenna system used on a relatively slow-moving vehicle such as a ship. In fact, such systems, although meant for maritime navigational applications, have been used for detection of some seawater circulation features such as plumes.

Turbulence and debris concentrated at the leading edge of the plume create sufficient relief to make the front of the plume visible on a ship's radar more than a mile away. To achieve reasonably good horizontal resolution, the antenna beam must be narrow in the horizontal (i.e., a very long antenna needs to be used). Rotating a very long antenna on a fast-moving vehicle suffers from many practical problems. Distortion in the image resulting from the motion of the antenna-bearing vehicle is another serious handicap. For these reasons, rotating antenna systems are rarely used for remote sensing of sea surface water-motion signatures.

5.4.1. Active Microwave Imaging by the RAR Systems

Real aperture radar (RAR) systems were initially developed in the 1950s for military applications and subsequently further improved during the years of the pathbreaking Apollo missions, which culminated in the successful landing of man on the moon in 1969. After military declassification, the RAR system was developed to a high degree of sophistication to generate images of the ocean surface to monitor wave climate. Although the primary application of RAR systems for oceanographic studies was to detect ocean waves from space, the images obtained from these systems were also found to be of practical value for identification of water-motion signatures in the open seas. A side-looking microwave radar system such as RAR (Figure 5.8) produces continuous strips of imagery of the terrestrial surface areas located on one or both sides of the aircraft's flight path.

Unlike a conventional PPI system, the antenna of a side-looking radar system is not rotated to achieve scanning. Nevertheless, generation of a two-dimensional image of a large sea surface area using radar requires relative motion between the sea surface and the antenna beam. In a side-looking radar system, scanning is achieved by a fixed beam pointed to the side of a moving aircraft. The aircraft's forward motion in a straight line permits the antenna beam to scan the sea surface continuously along a straight-line path that lies on the lateral side(s) of the aircraft.

In the operation of a RAR system, a transmitter sends out a short pulse of microwave energy through an aircraft-borne long and thin antenna with a small beam width, β_a, in the along-track direction and a large beam width, β_c, in the vertical plane normal to the flight direction (see Figure 5.9).

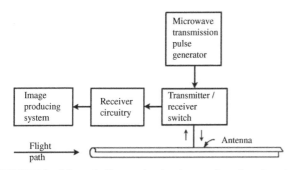

FIGURE 5.8 Schematic diagram showing the operation of a real aperture side-looking microwave radar system.

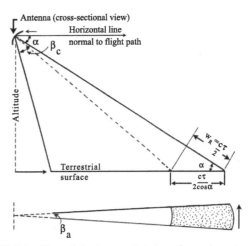

FIGURE 5.9 View of the beam orientation of a real aperture radar system. (Top) Side view in a vertical plane normal to the flight direction. Relationship between slant-range resolution, W_R, and ground-range resolution, W_C, in the cross-track direction is shown. (Bottom) Plan view as projected on the sea surface.

The beam is pointed to the side of the flight track through a large elevation angle. The beam width, β_a, determines the resolution in the flight direction. The pulse width, τ, determines the resolution normal to the flight path (i.e., range resolution). As the microwave energy impinges on the terrain surface, it gets scattered in different directions depending on the nature of the sea surface. The portion of the radiation backscattered in the direction of the radar is intercepted by the same antenna. The electrical signal so induced at the antenna is sent out to a sensitive radio receiver. The signal is amplified by the receiver in such a way that amplification increases with time so as to compensate for the decreasing echo strength at successively greater ranges. Introduction of such a time-varying gain (TVG) during signal conditioning ensures that, irrespective of range differences of various "targets" from the antenna, a signal is created for which the amplitude depends on the magnitude of the original radiation backscattered at any instant from a strip of terrain normal to the flight direction. As the forward motion of the aircraft carries the antenna forward relative to the sea surface, the footprint of the microwave beam on the sea surface automatically moves to new positions so that succeeding pulses sense the microwave reflectivity of the adjacent strips of the sea surface, which is oriented perpendicular to the flight direction. The signals received from adjacent terrain strips are then used to generate a two-dimensional image of the terrain parallel to the flight direction.

Various techniques have been used for the generation of images of the sea surface from the received signal. In the conventional method, the detected signal is made to control the brightness of a linearly moving spot of light on the face of a cathode ray tube (CRT). The speed of the moving spot of light so generated on the CRT covers the entire ground range normal to the flight path, embraced by the antenna beam. This intensity-modulated signal, displayed in a single line in the same position on the face of the CRT for each pulse transmitted, is focused by a lens system onto a single line on a photographic film that moves in front of the CRT in a direction transverse to the single-line display. Brightness of various parts of the line depends on the strength of the echo received from various points along the strip scanned on the sea surface. The forward movement of the aircraft causes a different strip of sea surface to be illuminated by the radar all the time. The motion of the film is in synchrony with the motion of the aircraft. This permits recording of the image of the adjacent sea surface strips on adjacent filmstrips. The length of the sea surface strip illuminated by the radar beam is usually 100 km or more. The film, when chemically processed, provides a map of the texture corresponding to the microwave reflectivity of the terrestrial surface swept by the antenna beam. The texture features are then interpreted to evolve useful information about the sea surface features. The interpreter must, however, take into account the possible distortions in the image, introduced by various factors, before attributing a physical meaning to the observed features in the image. The possible causes for distortions in the image include attitude or lateral changes of the aircraft, which is caused by turbulence in the atmosphere; yaw, pitch, or roll motions of the aircraft; or lack of synchrony of the drive systems in the imaging electronics with the speed of the aircraft.

In some modern systems, the detected output from the receiver is fed to a CRT or digital scan converter. When the scan converter is full, the signal is displayed in TV format on a video display monitor. Compensation for aircraft speed may be achieved in the scan converter by adjusting the scan rate (Ulaby et al., 1982). Images corresponding to each successive screenful of data are either photographed by a camera that is synchronized with the display or recorded on an onboard videotape that can be later played back on the ground. The latter method helps overcome some of the difficulties associated with setting up the contrast and brightness controls for the TV monitor during flight. Some RAR systems use telemetry for transmission of the detected signal to a ground station.

The quality of an imaging radar system is judged by its resolving power (also termed *resolution*) and is taken as its ability to distinguish between closely spaced targets. In practice, the resolution of an imaging radar system is the size of the cell area as projected on the ground, resulting from the impinging of a single radar pulse. This ground resolution cell area is conventionally called *pulse rectangle of ground-resolvable area*, although it is not a true rectangle (see Figure 5.10). The integrated reflectance from all the targets within the ground resolution cell area at any given instant in time is sensed by the radar receiver as

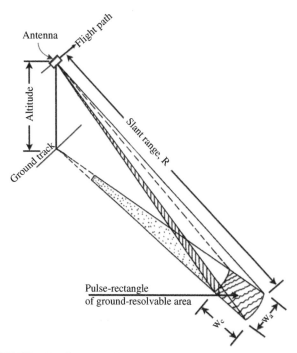

FIGURE 5.10   Schematic diagram showing pulse-rectangle of ground-resolvable area of an imaging real aperture radar system.

though they were all located at the center of the rectangle. This means that two or more targets located within this ground resolution cell will be seen by the radar as a single target. Thus, an important task of the system designer is to minimize the ground resolution cell area so that sea surface features can be better resolved. When the composite reflectance value from a given ground resolution cell differs from those in adjacent cells, the radar receiver is said to be able to *discriminate between adjacent targets*.

The dimensions of the ground resolution cell of a RAR system are determined by a combination of resolutions in the along-track and cross-track directions. These dimensions are different in the two orthogonal coordinates of the image. Spatial resolution, W_a, in the azimuth direction (i.e., in the flight direction) is determined by the angular width of the sea surface strip, which is illuminated by the radar beam at the range of interest. W_a is usually taken to be the arc length of the antenna at the range of interest; i.e., $W_a = \beta_a R$, where R is the slant distance from the antenna aperture to a given scattering target, and β_a is the 3-dB beam width, in radians, in the far field of the antenna in the along-track direction. According to the well-known Rayleigh criterion, $\beta = (\lambda / L)$ radians, where λ is the transmission wavelength, and L is the antenna length. Usually it is assumed that the antenna is focused to infinity (i.e., aperture with uniform phase), and the beam width is defined in the far zone. With very large values of L, due consideration must be given to the far-zone condition of the antenna. The far-zone distance R_{fz} is given by (Tomiyasu, 1978) $R_{fz} = (2L^2) / \lambda$.

The antenna beam width, β_a, is given by $\beta_a = (\lambda / L_a)$, where L_a is the length of the antenna in the along-track direction. By taking the relationship $W_a = \beta_a R$ into account, $W_a = (\lambda \times R) / L_a$. It can be seen from the preceding expression that, for a given transmission wavelength, resolution in the along-track direction can be enhanced by increasing the antenna length in the along-track direction. For this reason, a long antenna is used for imaging by the RAR system. It is also seen that, for a given antenna length L_a and a given transmission wavelength λ, the along-track resolution deteriorates as the antenna beam "fans out" with increasing range from the aircraft.

Spatial resolution, W_R, in the slant range direction (i.e., in the direction of propagation of the transmitted pulse from the antenna to a given target area) is determined by the transmitted pulse duration, τ, and is given by the expression $W_R = (c\tau)/2$, where c is the speed of electromagnetic radiation in the medium of travel. If the receiver bandwidth, B, of the radar system is matched to τ, then $B = (1/\tau)$ so that $W_R = c/(2B)$. The slant-range resolution $(c\tau/2)$ is the difference in range of two points of which the reflected signals differ in arrival time by τ. Thus, the cross-track resolution (i.e., range resolution in terms of the horizontal cross-track distance on the sea surface) is given by $W_c = \dfrac{c\tau}{2\cos\alpha}$, where α is the depression angle (i.e., the angle between the horizontal plane on the sea surface and the straight line connecting the antenna aperture to a given target area). Thus, as the depression angle approaches 90°, the cross-track resolution approaches infinity. For this reason, vertically downward-looking radars give poor cross-track resolution.

Cross-track resolution, however, improves as depression angle is decreased. For this reason, imaging radars are oriented to "look" to the side of the flight path rather than vertically downward. The significance of W_c is that only those targets separated by a distance more than W_c in the cross-track direction can be distinguished unambiguously by the radar as separate targets. Those targets separated by a distance less than W_c will be recognized by the radar as a single target. Better cross-track resolution means that the value of W_c is minimal. It can be seen from the expression $W_c = \dfrac{c\tau}{2\cos\alpha}$ that W_c can be enhanced (i.e., minimized) either by decreasing the depression angle α or by shortening the pulse length τ. Shortening τ reduces the amount of energy in each pulse transmitted from the antenna, which, in turn, results in reduced strength of the return signal. To meet the S/N ratio and target detection requirements, a high average radiated power can be obtained by a high transmit pulse duty cycle. These requirements on the transmitter are satisfied by transmission of a linearly swept frequency-modulated (FM) long pulse known as a *chirp pulse* (see chapter 12 for *chirp waveform*). The chirp pulse is usually generated by

transforming short wideband pulses into long pulses with the same swept bandwidth utilizing a frequency-dispersive delay line. These long pulses are then amplified by a high-power amplifier, passed through a circulator (switch), and then radiated by the radar antenna. The received signal passes through the same circulator, is amplified, and then is pulse-compressed. The pulse-compression circuit can use a delay line and a summing circuit, which converts the wideband chirp signal into a short pulse signal with the same bandwidth (see chapter 12 for details). The drawback of shortening τ is thus resolved by transmission of a long chirp pulse and then shortening the apparent pulse length of the return signal by an ingenious signal-conditioning technique, at the same time achieving enhanced signal strength.

It has been observed that decreasing the value of α results in an enhancement of W_c. However, this would also mean an increased range from the radar to a given target area. At large ranges, the spatial resolution in the along-track direction is rather poor (see Figure 5.11). This calls for a compromise in the design of an imaging radar system. Improved sea surface area coverage and good spatial resolution are achieved by careful design of the radar system. It can be seen from the relation $W_c = \dfrac{c\tau}{2\cos\alpha}$ that the cross-track resolution does not depend on beam width. However, a large beam width, β_v, in a vertical plane in the cross-track direction will provide a large coverage by the beam in the cross-track direction. This is achieved by configuring the antenna dimension in the vertical direction, L_v, to be small. In this case, $\beta_v = (\lambda/L_v)$.

The requirements of a narrow beam in the along-track direction (for better azimuth resolution) and a wide beam in the cross-track direction (for greater area coverage) are together met by the use of a long semicylindrical reflector antenna of small diameter fed by a dipole array. The antennae in some instances were as long as 15 meters and were fixed parallel to the fuselage of the aircraft (Ulaby et al., 1981). Such an antenna looks sideways with a fan beam that is wide in a vertical plane and narrow in

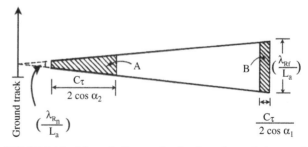

FIGURE 5.11 Schematic diagram showing dependence of along-track and cross-track resolution cell dimensions on horizontal distance from ground track. A indicates spatial resolution cell area at short slant range R_n and large depression angle α_2. Likewise, B indicates spatial resolution cell area at far slant range R_f and small depression angle α_1.

a horizontal plane. Because the beam is wide vertically, its projection on the ground contains several slices, in a linear array, the length of which in the cross-track direction is given by $\dfrac{c\tau}{2\text{Cos}\alpha}$ and in the along-track direction by $\dfrac{\lambda R}{L_a}$.

Generation of good imagery requires careful data acquisition and processing procedures. Despite transmission of a narrow pulse, the received signal is extended in time because of the long beam length projected on the sea surface in a direction normal to the flight path. The first part of the received echo has come from the *nearest* part of the illuminated area at range R_n at a time t_n after transmission. Reception continues until the echo from the *farthest* illuminated point at range R_f has arrived at time t_f after transmission. The number of pulse length τ that can be fitted between t_n and t_f is the number of independent measures of W_c that can be obtained across the swath. The depression angle α_1 corresponding to the farthest illuminated point is a minimum, and the depression angle α_n corresponding to the nearest illuminated point is a maximum. This results in an increasing degradation of W_c from the farthest to the nearest illuminated points.

If imagery obtained from radar is to be realistic, it is necessary that the pulse repetition rate vary inversely with the range of operation, because all return signals from an outgoing pulse must be received before sending out another pulse. Such a scheme avoids simultaneous reception of returns from multiple scene elements (Lowe, 1980).

Another problem inherent in an imaging radar system is geometrical distortion of image in the cross-track direction. This arises because the radar measures time delays, and therefore areas nearer to the flight path appear compressed relative to the areas farther to the flight path, unless the distortion is corrected by digital processing techniques (Elachi, 1980). All these factors require that the designer make some compromise among performance parameters such as range of operation, depression angle, swath width, and spatial resolution.

5.4.2. Active Microwave Imaging by SAR Systems

It has been indicated in the discussion on RAR that, for a given antenna length and a given transmission wavelength, the along-track resolution deteriorates as the antenna beam fans out with increasing range from the radar. The optimum RAR configuration, to achieve good range and azimuth resolutions, is that of high-frequency radar carried on a low-flying aircraft. At long ranges, as in the case of satellite-borne radar, an antenna of a few kilometers long will be necessary to achieve along-track resolution of a few tens of meters. Such an antenna is impractical for a spacecraft. The impracticability of attaining good azimuth resolution from conventional radar flying at high altitudes has been resolved with the application of an ingenious technology known as *synthetic aperture radar* (SAR).

SAR is a coherent active microwave imaging device (i.e., one in which both transmitter and receiver are phase-referenced to a stable master oscillator). In remote sensing it is used for mapping the scattering properties of a large surface area in the respective wavelength domain. Many physical and geometric parameters of the imaged scene contribute to the gray value of a SAR image pixel. Scene inversion suffers from this high ambiguity and therefore requires SAR data taken at a different wavelength, polarization, time, incidence angle, and so on. SAR was initially developed in support of various lunar mission programs in order to obtain a microwave photograph of the lunar surface with fine resolution. Microwave imaging systems were necessary to obtain images during both day and night. To realize an along-track resolution that is significantly finer than that dictated by the angular beam width of the radar antenna, the SAR system employed coherent radar in conjunction with data storage and processing of the radar signal Doppler spectrum. The data collected during the observation time were processed to synthesize a large-antenna aperture.

The radar data from the lunar mission were recorded on photographic film and were returned to Earth for processing. A combination of optical and digital processing technologies was applied to the scientific interpretation of the data. Because the return signal level was strongly influenced by the degree of flatness of the scattering surface, excellent imagery of the lunar surface could be obtained. The success of the SAR system designed for the lunar mission programs showed considerable promise for its application to imaging of terrestrial (both land and ocean) surfaces. Subsequently, aircraft-mounted SAR systems were designed and field-tested. Airborne SAR systems proved their worth in imaging some oceanic features such as waves and water-motion trajectories.

A space-borne or airborne SAR illuminates the Earth's surface in a side-looking fashion. Whereas the sensor is moving along its assumed straight path at an altitude H above some reference plane, it transmits microwave pulses into the antenna's illumination footprint at the rate of the *pulse repetition frequency* (PRF) and receives the echoes of each pulse scattered back from Earth. The SAR receiver detects the stream of echoes coherently and separates it into individual echoes, each corresponding to a transmitted pulse. For processing, the echoes are preferably arranged side by side as a 2D matrix, with coordinates of *two-way signal delay time* and *pulse number*. The pulse number relates to the satellite position along its flight path, and the delay time relates to slant range. Typical pulse-carrier wavelengths are approximately 3 cm (X-band), 6 cm (C-band), 9 cm (S-band), and 24 cm (L-band). In addition, 64 cm (P-band) might be used in the future. PRFs are in the range of 1–10 kHz. The ensemble of scatterers is assumed to be temporarily stationary and to reside in the far field of the SAR antenna. The antenna look direction will be perpendicular to the flight path, although this is never strictly true in real systems.

Thus, in SAR, a coherent phase history of the pulsed return signal is generated, and through a signal processing technique, extremely high resolution in the azimuth direction is attained without the use of a physically large antenna. In effect, a large aperture antenna is thus synthesized. A SAR achieves its characteristically fine along-track resolution by coherently summing the signals received during a period of time on the order of a second and correcting the phase of these signals so as to simulate the signal that would have been received by a single antenna having a length equal to the distance traveled by the SAR during the integration period. For stationary surfaces, this aperture synthesis technique is capable of producing an along-track resolution equal to half of the actual SAR antenna length (see Figure 5.12). However, for moving surfaces such as the ocean, the phase of the received signal fluctuates randomly during the integration time, thereby severely limiting the theoretically achievable along-track resolution.

Spatial resolution obtainable from a SAR in the cross-track direction is the same as that obtainable from a RAR in the cross-track direction. Because SAR generally operates from a very large height above the sea surface, it is necessary that the transmitted pulse be sufficiently strong. To achieve fine cross-track resolution, it is also necessary that the effective pulse width of the transmission signal be very small. However, higher power cannot be pumped into a short pulse. These two conflicting requirements of narrow pulse width and high power are met by transmission of a chirp pulse and then by ingeniously processing the received signal to make the pulse length of the return signal effectively short.

The SAR technology originated with an observation by Carl Wiley in 1951 that a radar beam oriented obliquely to the radar platform velocity vector will receive signals that have frequencies different from the radar's carrier frequency as a consequence of the Doppler effect. He noted that the Doppler frequency spread was related to the width of the antenna beams and that the desired narrow beam can be synthesized by appropriate signal processing. Wiley also noted that the narrowest angular beam would occur "broadside" to the platform velocity. Taking a cue from these observations, the ensuing years witnessed vigorous research activities in the realm of hardware and signal-processing technologies, thus culminating in the development of operational SAR systems. The SAR system is a very complex device and therefore only a bird's-eye view of the basic principles is provided here.

SAR is a phase-coherent sensor that repetitively transmits high-power pulses and detects the backscattered return signals. The received return signals are processed coherently to produce high resolution two-dimensional images of the mapped areas. The raw data acquired by coherent radar resemble a hologram rather than an image and hence require a considerable amount of signal processing for image formation (or "focusing"). The SAR is essentially composed of a high-power signal transmitter, a side-looking antenna moving at the same altitude with a constant velocity, a circulator, and a phase-coherent receiver. As in the case of RAR, the antenna of a SAR system is time-shared between the transmitter and the receiver by utilizing a circulator (switch). The typical SAR transmitter is designed to overcome the limitations of peak power in components and to satisfy stringent azimuth and range resolution requirements. High spatial resolution in the along-track direction (azimuth) is achieved by an effectively very long antenna synthesized from a small antenna. In fact, the heart of SAR technology lies in the synthesis of a very long antenna, thereby effectively generating a beam that has a very small width β_a in the far field of the antenna in the along-track direction. In achieving this goal, the really big difference between SAR systems and other radar systems lies in the signal processor. In short, in the oceanographic context, SAR maps the sea surface to a fine resolution with a fixed side-looking antenna using the artifice of simulating a large antenna by moving a comparatively smaller one. Range resolution is achieved in the standard manner with short pulses, but azimuthal resolution relies on mapping Doppler shifts into positions on the sea surface. This technique is briefly addressed in the following section.

5.4.2.1. Synthesis of Long Antenna from Small Antenna

In the synthesized antenna array concept, a very long antenna is constructed in principle essentially by an array of numerous small elemental radiators placed sufficiently

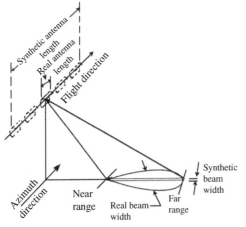

FIGURE 5.12 Schematic diagram showing the method of deriving narrow synthetic beam width for SAR from a broader real beam width.

close together to prevent grating lobes in the angular directions of interest and properly phased to focus the aperture at the midarray range, R_o to the target. Instead of exciting all elements of this hypothetical array simultaneously, each element of the long linear array is excited in sequence so that an orderly, coherent phase relationship is maintained.

The array mentioned here is only a concept, implemented in practice, using small moving radar. The ingenious scheme used in the SAR imaging system is that a hypothetical long linear array of numerous radiating elements is constructed effectively by a simple physical radiator (antenna) moving in a straight line at a constant velocity, V, with the radiator transmitting pulses periodically as it is laterally displaced by a proper amount. In this manner, the equivalent of a very long linear antenna array with synthetic aperture length, L_{sa}, is synthesized from a number of small elements equal in number to the pulses transmitted and integrated coherently. This synthesized antenna array is called a *synthetic aperture radar*. For simplicity we shall consider only a few positions of the radar antenna on the airborne vehicle with respect to a target, T, on the sea surface. In actuality there are numerous hypothetical radiators between the extreme positions, radiating at spatial intervals satisfying certain conditions discussed later. The total possible length of the synthetic aperture, L_{sa} for a given target position is the distance between points where the forward edge of the beam first intercepts the target (say, position A) and the point where the aft edge of the beam is just leaving the target (say, position C). According to the "close-target separability criterion," the azimuth resolution, W_a, obtainable from a focused long array, is given by (Tomiyasu, 1978):

$$W_a = \frac{\lambda R_o}{2L_{sa}} \tag{5.1}$$

In this expression, R_o is the midarray range to the target, and λ is the radar transmission wavelength. The radar's radian viewing angle, θ_R, relative to the target is given by:

$$\theta_R = \frac{L_{sa}}{R_o} \tag{5.2}$$

θ_R is also the one-way radar antenna beam width that can provide a dwell time T_d. The real aperture radar antenna length L_R, which can produce a beam width θ_R, is given by:

$$L_R = \frac{\lambda}{\theta_R} \tag{5.3}$$

From relations (5.2) and (5.3),

$$\lambda R_o = L_R \times L_{sa}$$

Substituting for λR_o in (5.1) we get $W_a = \dfrac{L_R}{2}$.

Thus, if the antenna is fixed to a moving vehicle, the finest possible azimuth resolution obtainable from SAR is just half the length of the real aperture. This azimuth resolution is independent of range and also independent of wavelength. The synthetic beam is narrower and also of equal width, irrespective of range. The reason that the synthetic beam is independent of range is that L_{sa} is itself directly proportional to the range, because of which a larger range results in a larger aperture. The reason it is independent of wavelength is that for longer wavelength, L_{sa} is also proportionately longer. The net result is that the azimuth resolution remains the same because the equivalent synthetic beam width remains the same (Ulaby et al., 1982).

The result also shows that very fine resolution images can, in principle, be produced with a small radar antenna. This welcome result is contradictory to that obtained for real aperture radar, where a large antenna is required to obtain fine azimuth resolution. Although it is desirable to use a very small radar antenna so as to attain the theoretically achievable fine azimuth resolution in SAR imagery, the radar antennas used in practice are not very small. For example, the Seasat-A SAR antenna system had dimensions of $10.74\,\text{m} \times 2.16\,\text{m}$ (Jordan, 1980). The deployed antenna was configured to fly with the long dimension along the spacecraft velocity vector and boresighted at an angle of $20.5°$ from the nadir direction in elevation and $90°$ from the nominal spacecraft velocity vector. The reason for the use of a relatively large antenna in SAR systems is that the reduction in size of the antenna length, L_R, is limited by the requirement to generate a sufficiently powerful signal. Resolution of SAR can be made finer than $(L_R/2)$ with the use of a scanning antenna. The scanning motion, however, will result in gaps in coverage along the flight path (Ulaby et al., 1982). This is undesirable for a radar system, which is meant for imaging a large area along its flight path. For this reason fixed antennas are used with operational SAR systems.

In the case of Seasat-A SAR, the data were collected at a rate of approximately 10^6 resolution cells per sec. Such a vast amount of information could not be stored onboard the satellite and was, therefore, telemetered to five ground stations, located in the United States (California, Florida, and Alaska), the United Kingdom, and Canada. SAR images could be obtained only when the satellite was in the horizon of one of these stations.

5.4.2.2. Enhancement of Azimuth Resolution

It has been observed that fine-resolution images can, in principle, be produced by the small radar antenna of a SAR system. In practice, high spatial resolution in the azimuth is achieved by signal processing the amplitude and total phase history of the signals reflected by the targets and collected by the SAR receiver for the dwell time period T_D. The interval during which the target area is illuminated is called the *integration time* or *dwell time*, T_D. It may be noted that the distance traversed by the radar platform while illuminating the target T during the dwell time period T_D is the synthetic aperture length, and this dimension is exactly the linear beam width of the radar antenna beam at range R_0 (R_0 is the midarray range to the target). For an isolated target, the phase history during the dwell time follows a quadratic phase function given by (Tomiyasu, 1978):

$$\phi(x) = \frac{(x - x_o)^2}{\lambda R_o} \text{ wave-lengths} \qquad (5.4)$$

In this expression, the terms have the following meanings: $\phi(x)$ is the two-way phase change as a function of the straight distance of the radar relative to the midpoint of the synthetic aperture array; x is the position of the radar platform relative to the midpoint of the synthetic aperture array; R_o is the nearest range of the radar to a given target; x_o is the radar position in the synthetic aperture array corresponding to R_o; and λ is the radar transmission wave length. The two-way phase change as a function of time, $\phi(t)$, is obtained by converting the position parameter x to time t. Accordingly,

$$\phi(t) = \frac{v^2 (t - t_o)^2}{\lambda R_o} \text{ wavelengths} \qquad (5.5)$$

where v is the constant velocity of radar platform; t is the time variable corresponding to position x. The two-way Doppler frequency, f_D, is the time derivative of $\phi(t)$. Accordingly,

$$f_D = \frac{d\phi}{dt} = \frac{2v^2 (t - t_o)}{\lambda R_o} \qquad (5.6)$$

This relation indicates that the Doppler frequency changes linearly with time. In a SAR system, the phase history is recorded over a time interval from $(-T_D/2)$ to $(+T_D/2)$ in which T_D is the coherent integration time for a "single-look" SAR system. T_D is also equal to the target illumination time, popularly known as the *dwell time*. The SAR processor expects the quadratic phase history given by the relation in Equation 5.5, and by matched filtering it positions the facet at zero Doppler. In practice, the signal amplitude and phase history are recorded on film or memory after phase-coherent addition of the received signal with a reference signal. When two signals are in phase, a higher combined level is obtained than when out of phase. Targets that cause zero Doppler shift lie on a plane strip that is oriented at right angles to the radar platform velocity vector. This plane must lie at the center of the broadside beam. Thus, measurement of Doppler shift permits processing the return signal amplitude, for sea surface imaging purposes, from only those targets that lie on a thin strip of

the sea surface and that lie at the center of the wide area illuminated by the radar beam at a given instant. This process permits us to obtain fine spatial resolution in the azimuth. From measurement of total Doppler shift during the dwell time, it is possible to calculate the azimuth resolution achieved by a SAR system. Referring to the relation in Equation 5.6, in the limiting case, $(t - t_o) = T_D / 2$ so that:

$$\pm f_D = \frac{v^2 T_D}{\lambda R_o} \qquad (5.7)$$

The total Doppler frequency change during the target illumination period is $2f_D$, so that:

$$f_{D(total)} = \frac{2v^2 T_D}{\lambda R_o} \qquad (5.8)$$

The time that the target is illuminated by the radar beam, i.e., dwell time, is given by $T_D = \dfrac{L_{sa}}{v}$, so that $L_{sa} = vT_D$. Substituting this value of L_{sa} in Equation 5.1, we get $W_a = \dfrac{\lambda R_o}{2vT_D}$, so that:

$$\lambda R_o = 2 W_a v T_D \qquad (5.9)$$

Substituting the value of λR_o in Equation 5.8, we get $f_{D(total)} = \dfrac{2v^2 T_D}{2W_a v T_D} = \dfrac{v}{W_a}$.

Thus, azimuth resolution:

$$W_a = \frac{v}{f_{D(total)}} \qquad (5.10)$$

For Seasat SAR, $v \approx 7.2$ km/s, and $T_D \approx 2.3$ s. This yielded a synthesized aperture length $L_{sa} \approx 17$ km. The single-look azimuth resolution obtainable from Seasat SAR was approximately 6.6 m. In practice, four azimuth resolution cells, or "looks," are incoherently averaged with a resulting "four−look" azimuth resolution of approximately 25 m, which matches its cross-track resolution of ≈ 25 m.

It has been seen that the theoretically achievable azimuth resolution of a SAR system is equal to one-half the antenna length in the flight direction. However, to attain this resolution in practice and to maintain continuity of the image in the azimuth direction, the radar platform displacement must not exceed one-half the antenna length L_R in the azimuth direction between successive transmit pulses. To avoid azimuth ambiguity in the SAR operation, this geometric constraint on the pulse-to-pulse displacement of the antenna beam is a particularly important feature. This means that the pulse repetition frequency (PRF) must be at least $\dfrac{v}{(L_R/2)}$. Thus, $PRF_{(Low)} = \dfrac{2v}{L_R}$. The upper limit of the PRF is governed by the requirement that the return signals arising from two successive transmitted pulses must not arrive at the receiver simultaneously. If the PRF is so high that the pulses overlap and simultaneous reception takes place, there will be ambiguity in the response. The upper limit of PRF is given by $PRF_{(upper)} = \dfrac{1}{2\tau + 2(R_f - R_n)/c}$, in which τ is the transmission pulse length; R_f is the far range; R_n is the near range; and c is the velocity of electromagnetic radiation in the medium of travel. These criteria for lower and upper limits of PRF ignore, for simplicity, the effect of the curvatures of the Earth and the satellite orbit. However, these lower and upper limits are severe constraints on the design of practical SARs.

To obtain the theoretically achievable resolution of SAR, the method employed is to transmit pulses at a suitable repetition frequency. Because the radar is in constant motion along its flight path while transmission takes place, any given target point in the field of view of the radar is illuminated several times during the dwell time. The number of times a given target point is illuminated in a given pass depends on the PRF. Consequently, in a given pass, return signals from any given target point are received at the radar receiver many times during the dwell time. The number of times the return signals from any given target point are received in a given pass again depends on the PRF. The returns from the entire target points at all ranges in the field of view of the radar are stored in memory.

Signal processing commences after collection of signals for specified time duration, often a specified number of dwell times. Because any given target point is illuminated by the successive transmission pulses at progressively changing phases with respect to their phase at the center of the radar aperture, the phases of the successive return signals from any given target point progressively vary from one extreme to the other, with a phase distribution symmetrical about the center of the radar aperture. If each of the received signals from a given target point is correlated with a reference signal appropriate for its phase history and then added, only those signals that arrived from the point of closest approach to the radar will add constructively. All the other signals received from this target point add destructively, depending on the phase geometry. This governs the effective beam width and, therefore, the azimuth resolution. Thus, by appropriate data processing of the successive return signals from a given target point, the effect of using a very wide real aperture antenna is achieved. Depending on the PRF, some side lobes may be generated in the addition process. These side lobes are generally much smaller in size than the central peak of the overall summation of the signals across the aperture.

In the preceding discussion, returns from only a single target point (a resolution cell) have been considered for simplicity in understanding the basic operation. In actuality, returns from all target points (resolution cells) that are swept by the radar beams are received over a dwell time and recorded in the receiver memory as the radar passes over the target points. Subsequently, data processing as applied in the case of a single target point is successively applied to all target points so that a high-resolution image of the sea surface strips is generated without any discontinuities. To perform this data processing, the location of each target point and its phase history must be known to the processor. Location is determined from the range, which in turn is determined from the pulse travel time. The phase history of a given target point is determined from the Doppler history of the particular target position for which the back-scattering cross-section is being evaluated. The steady motion of the radar in the azimuth direction produces a Doppler shift in the frequency of the returned signal. As the radar approaches a target point, the contribution of that point to the reflected signal is first at a higher frequency than the transmitted signal and then gradually reduces in a linear way until it is less than the transmitted signal when the radar has passed by. At any given time for a given range, the contributions from each target point along the azimuth at that range can be identified uniquely by its Doppler-shifted frequency. This enables the phase history of each target point to be determined and the summation of different contributions from different target points to be performed so as to achieve fine azimuth resolution.

For high-precision SAR processing, determination of the absolute Doppler centroid is essential. The Doppler frequency associated with a target located at the azimuth beam center line is called the *Doppler centroid*. Due to the pulsed nature of the SAR system, sampling of the target spectrum takes place discretely at the PRF. This means that the interval of unambiguous Doppler spectrum is limited to the PRF. These aspects have been addressed by Li et al. (1983). Doppler centroid error leads to significant degradation of the S/N ratio and to an azimuth shift of the pixel location. Because the antenna ephemeris data and antenna altitude data are not sufficiently precise to provide accurate Doppler centroid estimation, the Doppler information has to be derived by analysis of the coherent radar return signal. An essential requirement to be satisfied by such a Doppler centroid tracker algorithm is that it should be able to follow the Doppler frequency variation, even across the PRF band boundaries. In some cases, an accurate Doppler centroid estimate can be extracted from the SAR data through a clutter-lock process. However, most of the clutter-lock methods have been optimized for the case wherein the target is strictly homogeneous. The majority of real targets are quasi-homogeneous, wherein the backscatter coefficient varies as a function of position. Improved algorithms continue to be developed for further enhanced estimation of Doppler centroid. Three such popular methods are the ΔE method, the correlation Doppler estimator, and the sign Doppler estimator.

Because the space-borne SAR systems operate in the microwave (centimeter-to-decimeter wavelength) regime of the electromagnetic spectrum and provide their own illumination, they can acquire information globally and almost independently of meteorological conditions and sun illumination. They are, therefore, most suitable for operational monitoring tasks. There is little attenuation of the radar signal by the intervening atmosphere. Once the SAR's radiation reaches the ocean surface, the electromagnetic properties of which are quite homogeneous, penetration is limited to about a tenth of the wavelength of the radar. Thus, SAR backscatter from the ocean results from sea surface roughness elements having a wavelength on the order of that of the radar. The side-looking imaging geometry, pulse compression techniques, and the synthetic aperture concept are employed to achieve geometric resolutions in the order of some meters to tens of meters with physical antennas of modest size. The price to be paid for such favorable performance is high transmit power, a considerable amount of signal processing, and—compared to optical imagery—"unconventional" imaging geometry.

The commonly used SAR imaging geometry is known as (continuous) *stripmap mode*. Two other SAR mapping modes are of interest: ScanSAR and spotlight mode (Bamler and Hartl, 1998). The SAR integration time, i.e., the duration a scatterer is illuminated by the radar, determines the azimuth (x-) resolution of the final image. In the stripmap mode configuration, the integration time is given by the azimuth extent of the antenna pattern. In ScanSAR mode (Ahmed et al., 1990; Bamler and Eineder, 1996; Cumming et al., 1997; Monti Guarnieri and Prati, 1996; Monti Guarnieri et al., 1994; Moore et al., 1981; Moreira et al., 1996; and Tomiyasu, 1981) the integration time is deliberately shortened by operating the SAR in a bursted fashion, where it periodically transmits bunches of pulses (bursts). In the time between bursts, the look angle of the antenna beam is changed to illuminate a swath parallel to the previous one. Following this routine, the SAR sweeps its beam in a stepped manner from swath to swath before it returns to the first-look direction. Hence, a ScanSAR system images several swaths temporarily interleaved at about the same time. During processing these swaths can be stitched together to give a total swath of up to 500 km width. The consequence of the enormous coverage is the reduced resolution due to the burst-mode operation.

The complementary approach is adopted by spotlight SAR (Carrara et al., 1995; Di Cenco, 1988; Gough and Hawkins, 1997; Munson et al., 1983; and Walker, 1980).

Here the antenna is continuously steered toward a certain patch on the ground in order to keep it in view over a longer time. The increased integration time results in a higher azimuth resolution at the expense of coverage: A spotlight SAR can only image selected and isolated patches, whereas stripmap mode and ScanSAR map strips of theoretically unlimited length.

5.4.2.3. Sources of Errors

Because SAR is a phase-coherent system, generation of high-quality radar images requires that it generate accurate phase histories. The principal sources of phase errors are imperfections such as unsteady radar platform velocity, motion of target points, electromagnetic path-length fluctuations, Earth rotation effects, and instabilities in the electronic circuitry. Another cause of degradation in image quality arises from speckle. Because measurement of phase histories of returned signals plays a very important role in signal processing, phase stability is an exceedingly important factor in a SAR system. The primary oscillator that provides the signal for the transmitter as well as the reference for the receiver must be very stable. The timing of the transmit pulses must be very precise with respect to the primary oscillator.

In high-resolution SAR systems, it is important to know the precise position of the antenna during the integration time. To achieve coherent integration of the successive received pulses, which is the essence of a SAR system, it is necessary that phase errors resulting from spurious antenna-motion error (i.e., the error between the actual flight path and the nominal one) are compensated. For a SAR system mounted on aircraft, motion errors are considerably large due to atmospheric turbulence and aircraft properties. Typically the position of the antenna is calculated from the navigation data of the inertial navigation system on board the airplane.

A number of factors can cause errors in the calculated position. For example, an error can occur due to vibration of the antenna lever arm. Any such residual motion exceeding a small fraction of the carrier wavelength will cause considerable phase corruption, thereby rendering the resulting image worthless. By optimal mechanical design, the residual motions may be reduced to a certain extent. Another method is to actually measure the high-frequency residual motions and then compensate for the errors. In addition to these mechanical means, signal processing methods can also be used to estimate the residual motion and subsequently to remove its phase corruption from the radar signal.

Other motion errors arise from yaw, pitch, and roll motions of the antenna platform. If motion errors of the antenna are known, motion compensation can be realized by adjusting the PRF, applying a range-dependent phase shift to each received pulse, and delaying it. By adjusting the PRF, motion errors arising from the forward velocity variations of the antenna can be compensated.

By adjusting the range and phase delay, it is possible to compensate for the antenna displacement in the line-of-sight direction. Moreira (1990) has reported a method to extract the displacement in the line-of-sight direction, antenna velocity, and the yaw-and-drift angle from the SAR raw data. This method is based on the analysis of the azimuth spectrum of the raw data. The primary condition to implement this method is the use of a wide-azimuth antenna beam. This method for real-time motion compensation seeks to extract all the necessary motion of the aircraft from the radar backscatter signal. Hence, an inertial navigation system becomes unnecessary for several applications. The motion-compensation parameters for real-time motion-error correction are range delay, range-dependent phase shift, and PRF. The motions of the aircraft extracted by this method are displacement in the line-of-sight direction, antenna yaw-and-drift angles, and forward velocity. Extraction of the motion errors of the aircraft is based on two schemes. The first scheme analyzes only the ground reflectivity part of the azimuth spectrum and is known as the *reflectivity displacement method* (RDM). The second scheme analyzes only the antenna pattern part of the azimuth spectrum and is known as the *spectrum centroid method* (SCM).

For satellite-borne SAR, the satellite orbit planes are referenced to space, and sea surface targets have an apparent motion at Earth rotational rates. Such apparent motions cause various errors such as image shifts and azimuth defocus. These errors are greatest when the satellite is crossing the equator and least at the poles. Phase errors due to Earth rotation can be compensated using deterministic approaches.

Phase errors are also caused by targets moving on the Earth's surface. This is especially true in the case of an ever-vibrant ocean surface. The target motion can be resolved into an along-track component and a cross-track component. A steady target velocity component U_R in the range direction produces four possible effects known as *azimuth image shift*, *azimuth image smear*, *range walk*, and *amplitude reduction*. If U_R is constant, the quadratic phase history of the return signal is not affected. However, the time at which zero Doppler is encountered is changed relative to the stationary target (facet) case.

As a consequence, the SAR processor expects stationary facets and therefore erroneously allocates the azimuth position x of the moving facet to the image position x^1 given by $x^1 = (x + \Delta x)$, where Δx is the azimuth image shift. If the direction of U_R is toward the radar, the location of the moving target is shifted in the $+ve$ azimuth direction (i.e., in the direction of motion of the radar platform). Similarly, a target moving away from the radar in the range direction appears to be shifted in the negative

azimuth direction. For an airborne SAR for which the ground speed V may be considered the same as the radar platform speed, the shift Δx is given by (Robinson, 1985) $\Delta x = -(R/V)U_R$, and for a satellite with ground speed V and radar platform speed V_P, the shift Δx is given by (Robinson, 1985) $\Delta x = -\left(\dfrac{R \times V_p}{V^2}\right)U_R$, where R is the slant range. It can be seen that the facet displacement is independent of radar transmission wavelength and integration time.

For azimuth image shift to be negligible, the magnitude of Δx must be much less than the azimuth resolution W_a. This means that U_R must be less than $\dfrac{V^2 \times W_a}{R \times V_p}$. An interesting example of azimuthal image shift in an oceanographic context is the ship displacement from its wake. A ship with velocity that has a component in the range direction appears to be shifted to one side of the wake. In fact, such displacement has been utilized to calculate the velocity of a ship (Vesecky and Stewart, 1982).

If the facet undergoes acceleration in a slant-range direction, a processor that is tuned for stationary facets is mismatched for this particular facet. The result is a spreading of the image in the azimuthal direction, known in the SAR literature as *azimuthal image smear*. This makes the outline of the image obscure—an effect similar to defocusing in photography. According to Alpers (1983), for a given slant range and a given radar platform velocity, the azimuthal image smear depends linearly on $a_r\tau$, where a_r is the facet acceleration in the slant-range direction and τ is the coherent single-look integration time. Accordingly, if there is no component of the facet acceleration in the slant-range direction, there will be no azimuthal image smear. Alper's reasoning for the defocusing effect has been contradicted by Ouchi (1983), who argued that if the reference signal of the SAR processor is matched to a stationary point target, then the images of the moving facets are always defocused irrespective of their propagation direction. He further argued that defocusing cannot be due to the acceleration in the slant-range direction, nor can it be because of the change in the relative radar platform velocity. Defocusing can, however, be compensated by introducing a focal adjustment parameter to the reference signal of the SAR processor.

If the target velocity component U_R in the range direction is sufficiently large to cause the target to pass through one or more range-resolution cells within the time taken for the radar beam to pass over it (i.e., within the time the target is contributing to aperture synthesis), a phenomenon known as "range walk" occurs. The result of range walk is smearing in range as well as azimuth directions.

Another drawback that is characteristic of SAR systems is *speckle*. SAR speckle is the grainy, granular, noisy phenomenon that is observed in SAR images. This is considered noise because the limit between two different homogeneous regions is faded. This noise results from interference between signals from random scatterers on a randomly moving sea surface. A mathematical treatment of SAR speckle has been given by Barber (1983). Speckle is a natural consequence of all coherent imaging systems. Speckle in radar images is often likened to the noise in images of a rough surface illuminated with laser light (Elachi, 1988). The speckle in optical images is a result of interference of the coherent light reflected by different neighboring points. It degrades the quality of the imagery. In SAR, however, speckle seems more of a nuisance than in optical images.

Speckle is characterized by a standard deviation that is as large as its average value. When viewed, the image seems dominated by multiplicative noise. The presence of speckle in SAR images reduces detectability of facets in the images as well as reducing the capability to separate and classify distributed targets. Elimination of speckle is, therefore, necessary for reliable detection of features in an image. Reduction of speckle in SAR imagery needs to be performed in such a way that the features to be detected are retained or enhanced while the random noise is removed or reduced.

The most frequently used technique for speckle reduction is "multi-look" processing, in which the synthetic aperture length is divided into N sections. The sections are synthesized separately and then averaged incoherently. In practice, multi-looks are obtained by partitioning the available signal bandwidth and processing each look independently. The final image is produced by adding the looks incoherently, pixel by pixel. In effect, several images of the same scene are produced from different sets of data in the same pass. These N images are then added to produce a single image in which the speckle noise is reduced by a factor of $\sqrt{N}$. The SAR images resulting from independent looks of the same scene are truly independent, because the phase fluctuations that cause the speckle change rapidly with time and look angle and so give totally different results when viewed from different parts of the synthetic aperture. A drawback of multi-look processing is that the azimuth resolution is degraded by a factor N, where N is the number of looks used for processing.

Several algorithms have been developed in the last few decades to smooth speckles for retaining edge sharpness and subtle details. Moreira (1990) reported an improved algorithm for efficient multi-look processing. This technique, however, requires approximately 50% more computational time than the traditional multi-look approach.

Image enhancement by multi-look processing is just one aspect. Developments in polarimetric SAR created another dimension toward reduction of speckle, thereby increasing the interpretability of SAR images. Polarimetric SAR data with four elements (HH, HV, VH, and VV) of the scattering matrix make it possible to reduce the speckle effect by utilizing the correlations among co-polarized (HH, VV) and cross-polarized (HV, VH) images (Lee et al., 1990). The four elements have the following meanings:

HH = horizontal transmit, horizontal polarization received
VV = vertical transmit, vertical polarization received
HV = horizontal transmit, vertical polarization received
VH = vertical transmit, horizontal polarization received

The ideal situation is that the speckle statistics of these four terms are statistically uncorrelated, whereas the underlying reflectance is totally correlated. It is becoming common for SAR images to have multiple bands corresponding to different incident angles, different polarizations, and different frequencies, or a combination of these. Numerous filters have been designed for speckle reduction. These are based on various models for speckle, including multiplicative noise and non-Gaussian statistics. However, the image bands were filtered separately and independently. The correlation between bands was not utilized. Some recent methods of speckle filtering have been reported by Quelle et al. (1990), Nasr et al. (1990), and Samadani and Vesecky (1990).

5.4.2.4. Interferometric SAR (InSAR) System

Interferometric Synthetic Aperture Radar (InSAR) obtains images of sea surface motion signatures based on spatial contrast of sea surface roughness manifested as spatial variations in sea surface topography. InSARs can be operated from aircraft or satellite. Due to their comparably low flight velocity, airborne SARs with their single-pass capability are useful for ocean current monitoring by along-track interferometry (ATI). Airborne ATI has been used successfully to measure ocean currents (Bao et al., 1997; Carande, 1994; Goldstein et al., 1989; Goldstein and Zebker, 1987).

InSAR is a technique whereby at least two SAR images acquired with a nearly identical incidence angle (one usually regarded as master and the other slave) are combined to produce a phase interference image called an *interferogram* (Dixon, 1994; Massonnet, 1997; Zebker and Goldstein, 1986). SAR images consist of both magnitude (brightness) and phase values. If the phase information is retained, the SAR image is described as being complex (Henderson and Lewis, 1998). The phase in a complex SAR image is a coherent signal containing information about the distance between a resolution cell on the ground and the radar antenna, as well as information about the texture of terrain within a resolution cell. Using the phase information in the interferogram, it is possible to extract topographic height information using digital elevation models (DEM), height change information, and fine-scale temporal change measurements (Burgmann et al., 2000; Okeke, 2005). Thus, InSAR exploits the phase differences of at least two complex-valued SAR images acquired from different orbit positions and/or at different times. The time gap between two passes of satellite may not be kept large, because there may be some changes in the scene that lead to temporal decorrelation. The information derived from these interferometric datasets can be used to measure several geophysical quantities, such as roughness, topography, deformations (e.g., ice fields), and ocean currents.

During the past few years, a multitude of scientific applications of InSAR techniques have evolved, and DEMs are successfully used in topographic mapping. It has been shown that InSAR is an established technique for generating high-quality DEMs from space-borne and airborne data and that it has advantages over other methods for the generation of large-area DEMs.

Processing of InSAR data is still a challenging task. For each selected image pair, several processing steps have to be performed. One of the current challenges of the InSAR application is to bring the techniques to a level where DEM generation can be executed on an operational basis. Classical InSAR processing is known to be computationally laborious (Marinkovic et al., 2004). The amount of data to be handled is enormous, the input data often expensive, the data quality *a priori* unknown, and the algorithms require fast computers. For each selected image pair, several pre-processing steps (co-registration, interferogram generation, and so on) have to be performed. A typical scene of ERS data, for instance, occupies about 650 Mbytes of computer storage. Furthermore, the quality requirements for these pre-processing steps are generally high. These call for well-articulated processing steps and organizational workflow similar to that in photogrammetry and optical remote sensing and to the level where, for instance, InSAR topographic maps are generated on an operational basis.

Okeke (2006) has described InSAR operational steps and processing chains for DEM generation from single-look complex (SLC) SAR data. The operational steps are performed in three major stages: data search, data processing, and product validation. The data processing stage is further subdivided into five steps: data pre-processing, co-registration, interferogram generation, phase unwrapping, and geocoding. Delft Object-Oriented Interferometric Software (DORIS) InSAR processing software is an efficient tool for processing of InSAR data (Kampes and Usai, 1999; Kampes et al., 2003; Kampes, 2005a). DORIS is chosen because it is fully functional interferometric processing software in the public domain. DORIS follows the classic Unix philosophy that each tool should perform a single, well-defined function, and complex functions should be built by connecting a series of simple tools into a pipeline. DORIS consists of a series of programs (modules) that perform different interferometric tasks.

5.5. ADVANCES IN THE DEVELOPMENT OF SAR TECHNOLOGY

SAR flown on the Seasat-A satellite was the first imaging radar system dedicated to imaging the ocean surface. The main objective of the Seasat-A SAR was to detect ocean waves. This system operated at 1.275 GHz (23 cm wavelength) at HH polarization at an incidence angle of $23 \pm 3°$ across the swath. The choice of wavelength was based on the imagery obtained from aircraft-borne SAR systems during 1972–1974. The Seasat-A SAR system was turned on in an 800 km altitude orbit in July 1978 and gathered images until the spacecraft failed in orbit in October 1978. The imagery it gathered had a resolution of 25 m × 25 m for four looks over a swath of 100 km width, centered about 300 km to the right of the spacecraft's ground track. The SAR's limited memory capacity and high data rate limited its operation to times when it was in view of a few ground stations equipped with dedicated SAR signal receivers.

Subsequent to the failure of Seasat-A, two SAR systems, developed by the NASA Jet Propulsion Laboratory, were flown on the space shuttle *Challenger*. The first in this series was the Shuttle Imaging Radar (SIR) known as SIR-A, launched in 1981; the second was SIR-B, launched in 1984. The imaging radars used in these missions operated in the L-band at a frequency of 1.25 GHz with HH polarization. The SAR systems in these missions achieved resolution in the range 20–40 m and swath width in the range 30–50 km. The explosion of the *Challenger*, killing everybody on board, rendered the planned SIR-C and SIR-D missions for the years 1990 and 1991 impossible.

The imaging radar experiments on the Seasat satellite and on the space shuttle *Challenger* resulted in wide interest in the use of space-borne imaging radars. The radar sensors provided unique and complementary all-weather information on sea surface dynamic phenomena similar to what was acquired with visible and infrared imagers.

Following U.S. efforts in the field of developing and launching SAR systems for surveillance of ocean surface dynamic phenomena, the U.S.S.R. launched its Almaz satellite-borne SAR in January 1991. This satellite flew at an altitude of 300 km. The SAR transmitted 9.6 cm wavelength signal at a pulse length of 33.4 s and a PRF of 3,000 Hz at HH polarization at an incident angle of $33\pm3°$ across the swath of 40 km width, achieving sea surface resolution of 30 m (range) × 15 m (azimuth) for single look. The Almaz SAR demonstrated its ability to generate excellent images of the sea surface features.

Based on a strong conviction among oceanographers and marine engineers about the importance of obtaining long-term data on sea surface currents and waves, a sophisticated SAR system was deployed on board the first European Remote Sensing Satellite, ERS-1, flown in July

1991. Measurements leading to the understanding of ocean circulation are one of the wide ranges of primary environmental problems addressed by the ERS-1 mission, the main thrust of which is oceanographic studies.

SAR was the major payload element of the ERS-1. A major challenge in the design of its antenna was to achieve a reasonable compromise among the requirements toward maximum system performance, the constraints imposed by the in-orbit and launcher environments, and the resulting mechanical complexity. High gain, very narrow beam width in the azimuth, and light weight are the special features of the ERS-1 SAR. The first two features necessitated the use of a long, narrow antenna. To achieve low losses, good beam-shaping capability, and good mechanical compactness, a slotted waveguide antenna concept was selected over the conventional microstrip principle. Unlike conventional designs, the antenna was made considerably light in weight, employing a novel design in which all the waveguides in the resonant array of the antenna were made from metallized carbon fiber-reinforced plastics (CFRP). The manufacturing technology for the CFRP waveguides was developed in a dedicated pre-development program that ran over the course of several years. The antenna consisted of a 10 m × 1 m planar array. To permit trouble-free launching, the complete antenna was divided into 10 electrical subarrays of 1 m × 1 m. Every two subarrays were combined into one mechanical panel.

The SAR on ERS-1 operates in two modes: image mode and wave mode. Of these, the one useful for imaging circulation patterns is image mode. In this mode, the SAR onboard ERS-1 obtains high-resolution (30 m) imagery 100 km in width to the right of the satellite track. Its number of looks is 8. This SAR operates at a frequency of 5.3 GHz (C-band) with a bandwidth of 8.5 MHz and VV polarization and an incidence angle of 23° at midswath in normal operation.

Another, more sophisticated SAR is the Canadian Radarsat, with a repeat cycle of 16 days. Its unique quality permits the antenna beam to be steered electronically through a 20° to 45° angle of incidence over a 500 km track swath. Onboard, state-of-the-art tape recorders store Radarsat data, permitting worldwide information to be obtained. Sea ice and iceberg conditions detected by its SAR are transmitted from Radarsat to the ice information center at Mission Control in Ottawa, Canada. After processing, the ice information is relayed via communication satellites to ships in Arctic waters and to drilling rigs off the east coast of Canada.

There are two agencies operating SAR satellites in the civilian sector, the Canadian Space Agency (CSA) and the European Space Agency (ESA). CSA has had one SAR satellite, RADARSAT-1, in orbit since 1996. ESA launched its third SAR satellite, ENVISAT, in 2002. Its predecessors, ERS-1 and ERS-2, collected SAR images from 1992 to 2000 and from 1995 onward, respectively. Both CSA and ESA satellites have collected a large database of archival SAR images. In addition, the National Space Development Agency of Japan (NASDA) operated a SAR satellite from 1993–1998. The SAR images collected by this satellite are also available. Subsequent missions include Canadian RADARSAT 2 (C-Band), Japan ALOS (L-Band), and German TerraSAR (X-Band).

5.5.1. Interpretation of Image Data

Four important pattern elements used in human interpretation of image data are spectral, textural, contextual, and temporal features. *Spectral* features describe the overall band-to-band tonal variations in a multiband image set. *Texture* is defined as the frequency of tonal change within a given limited area. *Contextual* features contain information about the relative arrangement of large segments belonging to different categories; and *temporal* features describe changes in image attributes as a function of time.

When the range of tones in an area of interest is comparable to the range of tones in the entire range, human interpretation draws heavily on textural appearance. SAR imagery of sea ice is a good example of this phenomenon. Texture is an important spatial characteristic that is useful for identifying features or regions of interest in an image. Speckle reduction is just one aspect of SAR image enhancement. Texture enhancement is an equally significant aspect in SAR image processing. Retrieval of texture information from images buried in noise is usually performed using various kinds of digital signal processing techniques known as *texture analysis*.

In texture analysis, the most important task is to extract texture features that can then be used for the description or classification of different texture images using any one of a multitude of pattern-recognition techniques. Despite its clear-cut semantic meaning, texture is a rather vaguely defined concept. For this reason, various analysts have used different concepts for texture analysis.

Texture is an important aspect in image analysis because it involves a measure of both the spectral and spatial variations in the scene. The primitive of the image texture is a collection of pixels that share a common property and are geometrically connected. These pixels are related to the texture with a relationship that may be structural, probabilistic, or both. The properties of image texture can be described as fineness, coarseness, randomness, and regularity. Probably the most successful texture discrimination techniques are based on a spatial-statistical approach in which pixels in a selected neighborhood having special spatial relations are chosen as the data samples. A variety of numerical measures are then employed to extract the useful information necessary for differentiation. Some statistical approaches to texture analysis may be found in Haralick (1979) and Rignot and Kwok (1990).

In a texture segmentation procedure, texture discrimination techniques are used to separate and identify regions of different textures in images. The most commonly used method is to classify each pixel individually. Pixel classification requires a determination of the texture in the pixel's immediate neighborhood. The size of the neighborhood window chosen to process the texture features has to be large enough to provide a statistically meaningful sample. A faster computational approach to texture segmentation has been provided by Du (1990) based on localized spatial filtering. In this approach, filters are chosen for each class with parameters to match its frequency, bandwidth, and spatial orientation. Segmentation is then accomplished by designating each pixel as belonging to the class with the filter that produced the strongest response at that location. It is a process whereby image regions of different textures are separated by sensing the localized changes of spatial frequency and its orientation. Texture analysis is a topic unto itself, and a discussion on this aspect is beyond the scope of this work.

5.6. DETECTION OF SEAWATER CIRCULATION FEATURES USING RAR AND SAR

Aircraft- and satellite-borne real and synthetic aperture imaging radar systems are widely used to obtain a description of sea surface manifestations of ocean circulation features in two ways. One method is imaging circulation features such as large-scale oceanic convections, eddies, fronts, Langmuir circulation, tidal currents, water flow over shallow topography, and surf zone circulation cells by measuring sea surface roughness signatures of these phenomena. The other method obtains a quantitative description of polar circulation through monitoring ice movements.

In the case of phenomena such as short-lived drifting eddies, a large part of the oceanographic, commercial, and naval interest is simply locating the circulation trajectory and its features rather than quantifying the flow vectors. The success of microwave imaging radar detection of mesoscale circulation features depends on several factors, such as enhanced wave breaking at the Bragg scale, slope and refraction changes of the dominant surface gravity waves, wind discontinuity, direct interaction of Bragg waves with currents, redistribution of surfactant materials (e.g., as in the presence of Langmuir circulation), and atmospheric instability as related to sea surface temperature variations. A single mechanism does not generally dominate the observed roughness signature but rather depends on a diverse set of oceanographic and meteorological conditions.

Microwave imaging radars respond only to surface phenomena. However, these phenomena are often indicative of ocean internal dynamics, such as subsurface eddies and internal waves. The surface expressions of these internal dynamics become visible in active microwave radar images, primarily due to the modulation of backscattering cross-section strength caused by the interaction of these dynamic phenomena with Bragg-resonant sea surface waves and the resulting redistribution of short sea-wave patterns. In such a situation, the visibility of the feature is not so much a general image intensity enhancement but rather an organized pattern of small features (Vesecky et al., 1982).

Ocean fronts (i.e., regions of current shear at the interface of different water masses) are also ocean features of great interest that can be observed from active microwave imaging radars. Because gradients often exist in a large current velocity field, the motion of microwave scatterers within the flow velocity field exhibit local perturbations and therefore express local sea surface roughness differences. These differences appear as variations in image brightness. Thus, shear patterns associated with the boundary of intense current systems as well as striations associated with local variations in the flow velocity field may also be seen in the radar images. Lyzenga (1998) has reported SAR images of ocean fronts and internal waves.

Johannessen et al. (1990) have reported imaging of a variety of mesoscale sea surface roughness patterns by airborne synthetic aperture radar. These surface roughness patterns depicted a variety of surface expressions of upper-ocean circulation features such as current shear, offshore meandering jet, surface slicks with spiraling structure, long wave-current refraction, and internal waves. Comparison with "ground truth" data has shown (Johannessen et al., 1990) that the SAR frontal expression is characterized by a narrow, bright line of increased backscatter with equal darker zones of less intense backscatter on both sides.

Ocean fronts are potentially observable in radar images due to a change in surface roughness and a corresponding change in radar backscatter across the front. Changes in sea surface roughness can be caused by interaction of surface waves, with the current shear and convergence associated with the front, or damping of short waves by surfactant materials accumulated along the front. These surface roughness changes cause corresponding variations in radar backscatter due to Bragg scattering, as modified by the tilting effects of longer waves and possibly due to other mechanisms associated with small-scale wave breaking. The thin linear features often observed in the SAR images of intense current shear regions would correspond to the local saturation of SAR resonant waves as they interact with the current shear. These filamentary structures probably correspond both to the local saturation taking place as resonant waves reach their maximum penetration into the current and to the shear itself tending to transform an initially compact region of high back-scattering cross-section into multiple linear features (Vesecky et al., 1982).

Surprisingly, the SAR system could also reveal warm-water rings, although it is essentially a roughness-sensing device. It is believed that the mechanism of detection of thermal signatures in some fronts is the change in the short gravity wave field related to wind-stress changes across the front, which is a result of differences in boundary layer stability on each side of the thermal gradient associated with the front. The ring is marked by a series of straight and curved filamentary structures ∼1 km in width and as long as 150 km in length. It is believed that these filamentary features are generated due to current shear within the eddy. These rings are often likened to a midsection slice through an onion. The SAR image inside a ring is generally brighter and has a mottling distinct from the surrounding water mass. During the Norwegian Continental Shelf Experiment, simultaneous C- and X-band SAR data were examined, supported by *in situ* measurements, using a ship-mounted acoustic Doppler current profiler, a CTD, and cup anemometers, and detected a complicated mesoscale circulation pattern associated with meander growth, eddy formation, and intrusion of Atlantic water toward the Norwegian coast. SAR features imaged due to ocean circulation patterns such as current shears, frequently present along ocean fronts and eddies, cause narrow backscatter changes that appear as bright lines.

In addition to the open-ocean phenomena just discussed, SAR systems can also image tidal flows and other features of coastal inlets. When tidal currents and their gradients are large, the SAR resonant waves interacting with the current variations enhance the relative backscattering cross-section, thereby providing a bright image. Another mechanism that enhances the SAR image intensity is the steepening of long gravity waves (i.e., waves with a wavelength much larger than the resonant wavelength) by interaction with tidal currents. These tilted waves would then enhance the radar return as they become steeper. At shallow depths, tidal current parallel streaks, supposed to be responsible for the formation of longitudinal tidal bed-forms such as sand ribbons and erosional furrows, have also been detected using SAR images (Kenyon, 1983). Detection of current parallel streaks not only helps explain the formation of longitudinal bedforms but also gives a more accurate measurement of the direction of current flow.

Another circulation phenomenon often observed in SAR images is the internal wave-induced surface current. Surface manifestations of internal waves apparently arise because the currents associated with these waves modify the capillary-ultragravity wave spectrum overlying the oscillations. However, the exact mechanisms producing the

modifications are not fully understood. The radar resonant sea waves can interact strongly with the internal waves if the internal and surface waves are traveling in the same direction and the group velocity (C_g) of the surface waves is near resonant with the phase velocity (C_i) of the internal waves. In shallow water, internal wave amplitudes can be large and the induced surface current velocities also quite large. In such a situation, significant modulation of radar resonant waves can occur over a wider range of ($C_i - C_g$), and radar detection of the resulting scattering cross-section modulation is more likely. Probably this is the reason that internal wave features are so frequently seen in SAR imagery of continental shelf waters.

Another interesting circulation phenomenon that can clearly be seen on radar images is the wind-driven Langmuir circulations associated with helical roll vortices. Debris such as surface films or floating seaweeds trapped in the Langmuir circulation cell regions modulate the radar return, thereby causing image contrast enhancement. Surface films concentrated along the convergence regions of these cells damp the local resonant waves and therefore reduce radar image intensity. In current-divergence regions the dispersal of surface films would cause the energy density of resonant surface waves to be larger than that in the adjacent regions, thereby permitting enhancement of image contrast.

Strong estuarine tidal currents and surf zone circulation are yet other classes of circulation phenomena that are detectable in active microwave radar images. Tidal currents in estuaries often possess strong current gradient at certain phases of the tide. At these phases the Bragg resonant waves are strongly refracted and strained. Such modulations cause increased roughness contrast. Breaking of the shoaling waves is vigorous in the high-velocity regions of the surf zone. Because these high-velocity regions have strong backscattering cross-sections, one must expect that these regions will be seen as brighter in the radar image. On the contrary, longshore and rip-current regions in the image are darker compared to the neighboring areas. This "blacking out" in the SAR image occurs when the orbital velocities of the shoaling waves are of the order of their phase velocity (Alpers et al., 1981). In this case, the azimuthal displacement of the moving facets can become so large that they are shifted completely out of the image if the azimuthal bandwidth of the radar is limited. It is surprising that image suppression arising from fast-moving targets, which is an inherent defect of the SAR imaging process, has become a useful feature for detecting circulation patterns in the surf zone that generally has a bright background because of strong backscattering cross-section of the breaking waves.

There have been instances where the incursion of the Gulf Stream onto the shelf injected multiple water masses on the continental shelf near Cape Hatteras, North

Carolina; these were accompanied by surface-intensified fronts with strong surface shear and convergence (Marmorino and Trump, 1994). A glimpse into the frontal and water mass structure occurring in this region has been provided by microwave imaging radars (Marmorino et al., 1998). The X-band radar measurements used in this study were collected at horizontal polarization and at near-grazing angles. This type of measurement has been shown to be particularly sensitive to wave breaking and to have a large dynamic range across ocean signatures such as fronts and slicks (Askari et al., 1996; Trizna and Carlson, 1996). The observed radar frontal signatures are about a factor of 30 above background levels. Radar image of a 2- by 5-km area of the sea surface showed narrow bands of high radar backscatter (indicating locally rough water) intersecting in the shape of a narrow wedge connecting to a single band and thus having the appearance of a Y. Similar radar features appear in the works by Fu and Holt (1982), Hayes (1981), Graber et al. (1996), and Lyzenga (1998) and so are not uncommon on this part of the shelf.

Marmorino et al. (1998) proposed that features like these represent intersecting ocean fronts, which become visible in the imagery through the refraction and steepening of surface waves by across-front current convergence and shear. According to them, a Y forms because the relatively buoyant water behind one front will always overwash the water ahead of it, erasing that part of the frontal expression from the surface. Such a configuration can be likened to an atmospheric occluded front. The frontal structure was revealed in the ship's radar imagery as well as in airborne radar imagery. The width of a frontal radar signature provides an approximate estimate of the width of the velocity front because the increase in surface roughness measured by the radar is associated with both the local surface velocity gradients and the integrated velocity change across the front (e.g., Lyzenga, 1991). The widths of the frontal signatures are in the range of 50–75 m.

5.7. MEASUREMENT OF SEA SURFACE CURRENTS USING IMAGING OF ICE FLOES

With the advent of satellite-borne microwave imaging technology, continuous mapping of the motion of a large number of ice floes simultaneously has become possible. The Canadian Radarsat's SAR frequently covers the entire Arctic region, regardless of weather or darkness. Within a few hours of the satellite's passage overhead, the processed SAR image can be made available to a rig or ship.

As already noted, two techniques are currently in use for detection of icebergs using active microwave imaging technology: detection of "sidelighted character" and roughness contrast. Whereas the former method was used for the last few decades, the latter method could be operationalized only with the advent of multipolarized SAR and advanced texture analysis techniques. When the signature arising from the sidelighted character of ice floes is utilized, salinity differences of the iceberg and the surrounding seawater play no role in the imaging process. For this reason, newborn icebergs that have electrical properties that are the same as those of the surrounding water, as well as refrozen aged icebergs that have reduced salt content and, therefore, have electrical properties that are different from those of the surrounding water, can be detected by a microwave radar with the same efficiency, provided that the icebergs are large enough to protrude well above the sea surface to appear as a topographic relief.

The sidelighted character of topographic relief arises through local variations in the slope of the relief, which result in varying angles of signal incidence. This, in turn, results in relatively stronger backscatter from slopes facing the radar and weak or no backscatter from areas blocked from sufficient illumination by the radar wave. The "radar shadows" from these areas will appear in the imagery as completely black and sharply defined, unlike shadows in photography that are weakly illuminated by energy scattered by the atmosphere. The oblique illumination of the side-looking aircraft-borne microwave radar produces strong backscattered returns from the side of a large iceberg-facing antenna, in comparison to the returns from the surrounding seawater surface. The portion of the iceberg that obstructs the radar beam creates a shadow. The radar illumination becomes more oblique in the far-range direction, and therefore the shadows are proportionately longer. The contrast between the highlighted and the shadow regions of the iceberg makes the image more pronounced in the far-range direction.

The potential of microwave radar systems to observe iceberg movement and, therefore, water current circulation in the polar regions is very promising. Leberl (1979) utilized repetitive aircraft-borne synthetic aperture radar coverage of a test site in the Arctic region to measure the regional sea ice drift. *Ice drift* is the change of position of individual ice features over time. Drift is measured by comparing sequential maps of the distribution of a selected set of ice features. To extract quantitative information on ice drift from microwave radar imagery, individual image strips are usually *rectified*. In radar imaging terminology, image rectification is a process by which image measurements are transformed into geographical coordinates. Rectification requires accurate radar platform navigation data. In the study conducted by Leberl (1979) in the Arctic, radar images were taken at different seasons through clouds and fog and, in some cases, during the night. This study demonstrated that the all-time, all-weather capability makes microwave radar a remote-sensing tool well suited to providing data on sea ice movement.

A major problem with iceberg monitoring has been the difficulty in distinguishing between icebergs and ships. Large icebergs may be distinguished from ships because of the differences in their shapes. With smaller ones, distinction purely on the basis of imagery of the target is rather difficult. There is, however, an ingenious mechanism that permits the image interpreter to distinguish between icebergs moving under the influence of advection. The several waves generated by a moving ship form a V-shaped Kelvin wake behind the ship. These waves may be classified into three general categories: (1) surface waves generated by the ship, (2) vortex wakes, and (3) internal waves generated by the ship. Theoretically, the angle that the limiting shape of this Kelvin wake subtends in deep water is approximately 39°. Vesecky et al. (1982) reported ship wakes, as imaged by Seasat SAR, of lengths varying from a few kilometers to approximately 20 km and limiting angles varying from 18° to 23°. The overall V-shaped envelope of the wake shows up as a region of greater surface roughness relative to the fairly smooth region outside this envelope. Some attempts have recently been made to understand the mechanism of backscatter modulations within the ship wake field. The present view is that the vortices shed by the ship give rise to a current at the surface. This current interacts with the ocean surface waves, thereby causing backscatter modulations.

The advent of satellite-borne SARs has brought considerable interest in the use of microwave imagery to study movement of arctic ice floes. A common approach to the automated tracking of arctic ice had been to select a patch of ice floes from an early image (the source image) and to cross-correlate it with a later image (the target image) at each position that could plausibly correspond to the same patch of ice. The position that maximizes the computed correlation coefficient is deemed likely to contain the corresponding patch of ice. Sometimes consistency checks between several matches are used to identify false matches. This method is known as *area correlation*. A disadvantage of this method is its computational expense, especially in cases where ice floes rotate. To accommodate rotation, the patch must be rotated and correlated several times at each potential match position in the target image. The search space thus becomes very large, thereby increasing enormously the computational burden. As a result, there has been increasing attention to algorithms for tracking ice floes by matching feature shapes. Rather than correlating raw pixel values, these algorithms first extract features from the images. The shapes of the extracted features are then compared using shape descriptors. One method of tracking ice floes using shape comparison is based on normalized correlation. The second method is based on *dynamic time warping*. These methods effectively tackle rotation of ice floes.

The ERS-1 satellite, launched in July 1991, was a good tool for surveillance of sea-ice motion. Maps of ice motion and ice type will be routinely produced at the Alaska SAR facility using radar imagery. Detection, classification, and motion analysis of thin ice floes in open water bodies became reality with the development of advanced texture analysis techniques and ice-motion algorithms. The use of texture analysis considerably improved the contrast between ice and open water. An experienced image interpreter can distinguish textures corresponding to different types of targets.

REFERENCES

Ahmed, S., Warren, H.R.; Symonds, D., Cox, R.P., 1990. The Radarsat system. IEEE Trans. Geosci. Remote Sens. 28, 598–602.

Alpers, W., 1983. Imaging ocean surface waves by synthetic aperture radar: A review. In: Allen, T.D. (Ed.), Satellite microwave remote sensing. Ellis Horwood Limited, Chichester, England, pp. 107–119.

Alpers, W.R., Ross, D.B., Rufenach, C.L., 1981. On the detectability of ocean surface waves by real and synthetic aperture radar. J. Geophys. Res. 86 (C7), 6481–6498.

Askari, F., Donato, T.F., Morrison, J.M., 1996. Detection of oceanic fronts at low grazing angles using an X-band real aperture radar. J. Geophys. Res. 101, 20,883–20,898.

Austin, R.W., 1979. Coastal Zone Colour Scanner radiometry. SPIE, 208—*Ocean Optics*, 170–177.

Bamler, R., Eineder, M., 1996. ScanSAR processing using standard high precision SAR algorithms. IEEE Trans. Geosci. Remote Sens. 34, 212–218.

Bamler, R., Hartl, P., 1998. Synthetic aperture radar interferometry. Inverse Problems 14, R1–R54.

Bao, M., Bruning, C., Alpers, W., 1997. Simulation of ocean waves imaging by along-track interferometric synthetic aperture radar. IEEE Trans. Geosci. Remote Sens. 35, 618–631.

Barber, B.C., 1983. Some properties of SAR speckle. In: Allen, T.D. (Ed.), Satellite microwave remote sensing. Ellis Horwood Limited, Chichester, West Sussex, England, pp. 129–145.

Blume, H.C., Love, A.W., Van Melle, M.J., Ho, W.W., 1977. Radiometer observations of sea temperature at 2.65 GHz over the Chesapeake Bay. IEEE J. Oceanic Eng. OE-2 (1), 121–128.

Burgmann, R., Rosen, P.A., Fielding, E.J., 2000. Synthetic Aperture Radar interferometry to measure Earth's surface topography and its deformation. Annual Review Earth Planetic Sciences 28, 169–209.

Calla, O.P., 1984. Microwave sensors (present and future). In: Deekshatulu, B.L., Rajan, Y.S. (Eds.), Remote Sensing. Indian Academy of Science, pp. 15–25.

Carande, R.E., 1994. Estimating ocean coherence time using dual-baseline interferometric synthetic aperture radar. IEEE Trans. Geosci. Remote Sens. 32, 846–854.

Caraux, D., Austin, R.W., 1983. Delineation of seasonal changes of chlorophyll frontal boundaries in Mediterranean coastal waters with

Nimbus-7 Coastal Zone Colour Scanner data. Remote Sensing of Environment 13, 239–249.

Carrara, W.G., Goodman, R.S., Majewski, R.M., 1995. Spotlight synthetic aperture radar: Signal processing algorithms. Artech House, Boston, MA, USA.

Clarke, G.L., Ewing, G.C., Lorenzen, C.J., 1970. Spectra of backscattered light from the sea obtained from aircraft as a measure of chlorophyll concentration. Science 167, 1119–1121.

Cumming, I.G., Guo, Y., Wong, F., 1997. A comparison of phase-preserving algorithms for burst-mode SAR data processing. Int. Geoscience and Remote Sensing Symp. IGARSS '97, 731–733. (Singapore).

Di Cenco, A., 1988. Strip mode processing of spotlight synthetic aperture radar data. IEEE Trans. Aerosp. Electron. Syst. 24, 225–230.

Dixon, T.H., 1994. SAR interferometry and surface change detection. Report of Workshop. Boulder, Colorado, USA.

Du, L.J., 1990. Texture segmentation of SAR images using localized spatial filtering. Proc. 10th Annual International Geoscience & Remote Sensing Symposium vol. III, 1983–1986.

Elachi, C., 1980. Space-borne imaging radars: Geologic and oceanographic applications. Science 209, 1073–1082.

Elachi, C., 1988. Space-borne radar remote sensing: Applications and techniques. IEEE, New York, NY, USA.

Francis, C.R., Thomas, D.B., Windsor, E.P.L., 1983. The evaluation of SMMR retrieval algorithms. In: Allan, T.D. (Ed.), Satellite Microwave Remote Sensing. Ellis Horwood Limited, Chichester, England, pp. 481–498.

Fu, L.-L., Holt, B., 1982. Seasat views oceans and sea ice with synthetic aperture radar. JPL Publ., 81–120.

Gagliardini, D.A., Karszenbaum, H., 1984. Application of Landsat MSS, NOAA/TIROS AVHRR, and Nimbus CZCS to study the La Plata River and its interaction with the ocean. Remote Sensing of Environment 15, 21–36.

Gerson, D., Khedouri, E., Gaborski, P., 1982. Detecting the Gulf Stream from digital infrared data pattern recognition. In: Processes in Marine Remote Sensing, pp. 19–39.

Gloersen, P., Barath, F.T., 1977. A scanning multichannel microwave radiometer for Nimbus-G and Seasat-A. IEEE J. Oceanic Eng. OE-2 (2), 172–178.

Goldstein, R.M., Zebker, H.A., 1987. Interferometric radar measurement of ocean surface current. Nature 328, 707–709.

Goldstein, R.M., Barnett, T.P., Zebker, H.A., 1989. Remote sensing of ocean current. Science 246, 1282–1285.

Gordon, H.R., Clark, D.K., Mueller, J.L., Hovis, W.A., 1980. Phytoplankton pigments from the Nimbus-7 Costal Zone Colour Scanner: Comparisons with surface measurements. Science 210, 63–66.

Gordon, H.R., Clark, D.K., Brown, J.W., Brown, O.B., Evans, R.H., 1982. Satellite measurement of the phytoplankton pigment concentration in the surface waters of a warm core Gulf Stream ring. J. Marine Res. 40 (2), 491–502.

Gordon, H.R., Mueller, J.L., Wrigley, R.C., 1980. Atmospheric correction of Nimbus-7 Coastal Zone Colour Scanner imagery. In: Deepak, A. (Ed.), Remote Sensing of Atmospheres and Oceans. Academic Press, New York, NY, USA, pp. 457–483.

Gough, P.T., Hawkins, D.W., 1997. Unified framework for modern synthetic aperture imaging algorithms. Int. J. Imaging Syst. Technol. 8, 343–358.

Graber, H.C., Thompson, D.R., Carande, R.E., 1996. Ocean surface features and currents measured with SAR interferometry and HF radar. J. Geophys. Res. 101, 25,831–25,832.

Haralick, R.M., 1979. Statistical and structural approaches to texture. Proc. IEEE 67, 786–804.

Harris, T.F.W., Legeckis, R., Forest, D.V., 1978. Satellite infrared images in the Agulhas Current System. Deep-Sea Res. 25, 543–548.

Hayes, R.M., 1981. Detection of the Gulf Stream. In: Beal, R.C., DeLeonibus, P.S., Katz, I. (Eds.), Spaceborne synthetic aperture radar for oceanography, Johns Hopkins Oceanogr. Stud. 7. Johns Hopkins University Press, Baltimore, MD, USA.

Henderson, F.M., Lewis, A.J. (Eds.), 1998. Principles and Application of Imaging Radar, Manual of Remote Sensing, 2. Wiley, New York, NY, USA.

Hillger, D.W., Haar, T.H.V., 1988. Estimating noise levels of remotely sensed measurements from satellite using spatial structure analysis. J. Atmos. Oceanic Technol. 5, 206–214.

Holligan, P.M., Aarup, T., Groom, S.B., 1989. The North Sea: Satellite Colour Atlas,. Cont. Shelf Res. 9 (8), 667–765.

Hovis, W.A., Clark, D.K., Anderson, F., Austin, R.W., Wilson, W.H., Baker, E.T., Ball, D., Gordon, H.R., Mueller, J.L., El-Sayed, S.Z., Sturm, B., Wrigley, R.C., Yentsch, C.S., 1980. Nimbus-7 Coastal Zone Colour Scanner: System description and initial imagery. Science 210 (4465), 60–63.

Inman, D., Tait, R., Nordstrom, C., 1971. Mixing in the surf zone. J. Geophys. Res. 76, 3493–3514.

Irvine, D.E., Tilley, D.G., 1988. Ocean wave directional spectra and wave-current interaction in the Agulhas from the Shuttle Imaging Radar-B synthetic aperture radar. J. Geophys. Res. 93 (C12), 15,389–15,401.

Johannessen, J.A., Shuchman, R.A., Johannessen, O.M., 1990. NORCSEX '88: A pre-launch ERS-1 satellite study of SAR imaging capabilities of upper ocean circulation features and wind fronts. Proc. 10th Annual International Geoscience & Remote Sensing Symposium vol. 1, 707–710.

Jordan, R.L., 1980. The SEASAT-A synthetic aperture radar system. IEEE J. Oceanic Eng. OE-5 (2), 154–164.

Kampes, B.M., 2005a. Delft Object-oriented Interferometric Software (DORIS). Delft, the Netherlands, Bert Kampes, TU.

Kampes, B.M., Usai, S., 1999. DORIS: The Delft Object-oriented Radar Interferometric Software. 2nd International Symposium on Operationalization of Remote Sensing, Enschede, the Netherlands.

Kampes, B.M., Hanssen, R.F., Perski, Z., 2003. Radar interferometry with public domain tools. FRINGE 2003 Workshop, ESA/ESRIN, Frascati, Italy.

Katsaros, K., 1980. Radiative sensing of sea surface temperature. In: Dobson, F., Hasse, L., Davis, R. (Eds.), Air-Sea Interaction, Instruments and methods. Plenum Press, Newyork, pp. 293–317.

Kenyon, N.H., 1983. Tidal current bedforms investigated by Seasat. In: Allen, T.D. (Ed.), Satellite microwave remote sensing. Ellis Horwood Limited, pp. 261–270.

La Violette, P.L., Chabot, P.L., 1969. A method of eliminating cloud interference in satellite studies of sea surface temperatures. Deep-Sea Res. 16, 539–547.

Leberl, F., Bryan, M.L., Elachi, C., Farr, T., Campbell, W., 1979. Mapping of sea ice and measurement of its drift using aircraft synthetic aperture radar images. J. Geophys. Res. 84, 1827–1835.

Lee, J.S., Grunes, M.R., Mango, S.A., 1990. Speckle reduction in polarimetric SAR imagery. Proc. 10th Annual International Geoscience & Remote Sensing Symposium vol. III, 2431−2434.

Lee, T.N., Atkinson, L.P., Legeckis, R., 1981. Observations of a Gulf Stream frontal eddy on the Georgia continental shelf, April 1977. Deep-Sea Res. 28A (4), 347−378.

Legeckis, R., 1987. Satellite observations of a western boundary current in the Bay of Bengal. J. Geophys. Res. 92 (C1 2), 12,974−12,978.

Li, F.K., Johnson, W.T.K., 1983. Ambiguities in space-borne synthetic aperture radar systems. IEEE Trans. Aerosp. Electron. Syst. AES-19, 389−397.

Lillesand, T.M., Kiefer, R.W., 1979. Remote Sensing and Image Interpretation. Wiley, Newyork.

Lowe, D., 1980. Acquisition of remotely sensed data. In: Siegal, B.S., Gillespie, A.R. (Eds.), Remote sensing in geology. Wiley, Newyork, pp. 47−90.

Lyzenga, D.R., 1991. Interaction of short surface and electromagnetic waves with ocean fronts. J. Geophys. Res. 96, 10,765−10,772.

Lyzenga, D.R., 1998. Effects of intermediate-scale waves on radar signatures of ocean fronts and internal waves. J. Geophys. Res. 103, 18,759−18,768.

MacDonald, 1980. Techniques and applications of imaging radars. In: Siegal, B.S., Gillespie, A.R. (Eds.), Remote sensing in geology. Wiley, pp. 297−336.

Marinkovic, P.S., Hanssen, R.F., Kampes, B.M., 2004. Utilization of parallelization algorithms in InSAR/Ps-InSAR processing. 2004 Envisat & ERS Symposium, ESA SP-572, Salzburg, Austria.

Marmorino, G.O., Trump, C.L., 1994. A salinity front and current rip near Cape Hatteras, North Carolina. J. Geophys. Res. 99, 7627−7637.

Marmorino, G.O., Shen, C.Y., Allan, N., Askari, F., Trizna, D.B., Trump, C.L., Shay, L.K., 1998. An occluded coastal oceanic front. J. Geophys. Res. 103 (C10), 21,587−21,600.

Massonnet, D., 1997. Satellite radar interferometry. Scientific American 276 (2), 46−53.

Monti Guarnieri, A., Prati, C., 1996. ScanSAR focusing and interferometry. IEEE Trans. Geosci. Remote Sens. 34, 1029−1038.

Monti Guarnieri, A., Prati, C., Rocca, F., 1994. Interferometry with ScanSAR. EARSeL Newsletter. December 20.

Moore, R.K., Claassen, J.P., Lin, Y.H., 1981. Scanning spaceborne synthetic aperture radar with integrated radiometer. IEEE Trans. Aerosp. Electron. Syst. 17, 410−420.

Moreira, A., 1990. Improved multi-look techniques applied to SAR and SCANSAR imagery. Proc. 10th Annual International Geoscience & Remote Sensing Symposium vol. 1, 321−324.

Moreira, A., Mittermayer, J., Scheiber, R., 1996. Extended Chirp Scaling algorithm for air- and space-borne SAR data processing in stripmap and ScanSAR image modes. IEEE Trans. Geosci. Remote Sens. 34, 1123−1136.

Munson, D.C., O'Brien, J.D., Jenkins, W.K., 1983. A tomographic formulation of spotlight-mode synthetic aperture radar,. Proc. IEEE 71, 917−925.

Nasr, J.M., Fernin, P., 1990. A comparison of the detection performances obtained with a fully polarized SAR and a dual polarized SAR (HH, VV). Proc. 10th Annual International Geoscience & Remote Sensing Symposium Vol. 1, 317−320.

Njoku, E.G., Christensen, E.J., Cofield, R.E., 1980. The Seasat scanning multichannel microwave radiometer (SMMR): Antenna pattern correction — Development and implementation. IEEE J. Oceanic Eng. OE-5 (2), 125−137.

Okeke, F.I., 2005. Integrating InSAR, GRACE mission data and traditional measurements for topographic mapping and earth surface deformation monitoring for Nigeria. 1st International Conference on Geodesy and Geodynamics, Toro, Nigeria.

Okeke, F.I., 2006. InSAR operational and processing steps for DEM generation, Promoting Land Administration and Good Governance. 5th FIG Regional Conference, Accra, Ghana. March 8−11, 2006.

Ouchi, K., 1983. Effect of defocusing on the images of ocean waves. In: Allen, T.D. (Ed.), Satellite microwave remote sensing. Ellis Horwood Limited, Newyork, pp. 209−222.

Philpot, W.D., Vodacek, A., 1989. Laser-induced fluorescence: Limits to the remote detection of hydrogen, ion, aluminium, and dissolved organic matter. Remote Sensing of Environment 29 (1), 51−65.

Quelle, H., Boucher, J.M., 1990. Combined use of parametric spectrum estimation and Frost-Algorithm for radar speckle filtering. Proc. 10th Annual International Geoscience & Remote Sensing Symposium vol. 1, 295−298.

Rignot, E., Kwok, R., 1990. Extraction of textural features in SAR images: Statistical model and sensitivity. Proc. 10th Annual International Geoscience & Remote Sensing Symposium vol. III, 1979−1982.

Robinson, I.S., 1985. Satellite oceanography: An introduction for oceanographers and remote sensing scientists. Ellis Horwood (a division of Wiley), Chichester, England.

Sabins Jr., F.F., 1978. Radar imaging. In: Remote sensing principles and interpretation. W. H. Freeman, Newyork, USA.

Samadani, R., Vesecky, J.F., 1990. Using optimally pruned decision trees to find objects in speckled images. Proc. 10th Annual International Geoscience & Remote Sensing Symposium vol. III, 1975−1978.

Shenk, W.E., Salomonson, V.V., 1972. A multispectral technique to determine sea surface temperature using Nimbus 2 data. J. Phys. Oceanogr. 2, 157−167.

Smith, R.C., Baker, K.S., 1982. Oceanic chlorophyll concentrations as determined by satellite (Nimbus-7 Coastal Zone Colour Scanner). Marine Biology 66, 269−279.

Smith, R.C., Brown, O.B., Hoge, F.E., Baker, K.S., Evanas, R.H., Swift, R.N., Esaias, W.E., 1987. Multiplatform sampling (ship, aircraft, and satellite) of a Gulf Stream warm core ring. Applied Optics 26 (11), 2068−2081.

Strong, B., Brumley, B., Terray, E.A., Kraus, N.C., 2000. Validation of the Doppler shifted dispersion relation for waves in the presence of strong tidal currents, using ADCP wave directional spectra and comparison data. Proc. 6th Intl. Workshop on Wave Hindcasting and Forecasting, Monterey, California, USA. November 6−10, 2000.

Strong, B., Brumley, B., Stone, G.W., Zhang, X., 2003. The application of the Doppler shifted dispersion relationship to hurricane wave data from an ADCP directional wave gauge and co-located pressure sensor. Proc. of the IEEE/OES Seventh Working Conference on Current Measurement Technology, 119−124.

Stumpf, R.P., Tyler, M.A., 1988. Satellite detection of bloom and pigment distributions in estuaries. Remote Sensing of Environment 24 (3), 385−404.

Sturm, B., 1981. The atmospheric correction of remotely sensed data and the quantitative determination of suspended matter in marine water surface layers. In: Cracknell, A.P. (Ed.), Remote sensing in meteorology, oceanography and hydrology. Ellis Hordwood Limited, Chichester, UK, pp. 163−197.

Swanson, P.N., Riley, A.L., 1980. The Seasat scanning mutichannel microwave radiometer (SMMR): Radiometric calibration algorithm development and Performance. IEEE J. Oceanic Eng. OE-5 (2), 116–124.

Thomas, D.P., 1981. Microwave radiometry and applications, In: Remote Sensing in Meteorology, Oceanography and Hydrology. In: Cracknell, A.P. (Ed.). Ellis Horwood Series in Environment Science, pp. 357–369.

Tomiyasu, K., 1974. Remote sensing of the earth by microwaves. IEEE Proceedings 62 (1), 86–92.

Tomiyasu, K., 1978. Tutorial review of synthetic aperture radar (SAR) with applications to imaging of the ocean surface. IEEE Proceedings 66 (5), 563–583.

Tomiyasu, K., 1981. Conceptual performance of a satellite borne, wide swath synthetic aperture radar. IEEE Trans. Geosci. Remote Sens. GE-19, 108–116.

Trizna, D.B., Carlson, D.J., 1996. Studies of dual polarized low grazing angle radar sea scatter in nearshore regions. IEEE Trans. Geosci. Remote Sens. 34, 747–757.

Tyler, M.A., Stumpf, R.P., 1989. Feasibility of using satellites for detection of kinetics of small phytoplankton blooms in estuaries: Tidal and migrational effects. Remote Sensing of Environment 27 (3), 233–250.

Ulaby, F.T., Moore, R.K., Fung, A.K., 1982. Radar Remote Sensing and Surface Scattering and Emission Theory, Addison-Wesley Advanced Book Program/World Science Division. Microwave Remote Sensing (Active and Passive) vol. II, 1064. Wesley publishing company, Reading, MA.

Verstappen, H.T., 1977. Remote sensing in geomorphology. Elsevier Scientific, Amsterdam–Oxford–New York–Tokyo.

Vesecky, J.F., Stewart, R.H., 1982. The observation of ocean surface phenomena using imagery from the SEASAT synthetic aperture radar. J. Geophys. Res. 87 (C5), 3397–3430.

Vigan, X., Provost, C., Podesta, G., 2000b. Sea-surface velocities from sea-surface temperature image sequences, application to the Brazil-Malvinas Confluence area. J. Geophys. Res. 105 (C8), 19,515–19,531.

Vigan, X., 2002. Upper-layer ocean current forecasts. Sea Technol. 43 (10), 15–19.

Vigan, X., Provost, C., Bleck, R., Courtier, P., 2000a. Sea-surface velocities from sea-surface temperature image sequences, method and validation using primitive equation model output. J. Geophys. Res. 105 (C8), 19,499–19,514.

Walker, 1980. Range-Doppler imaging of rotating objects. IEEE Trans. Aerosp. Electron. Syst. 16, 23–52.

Witting, J.M., 1984. Wave-current interaction: A powerful mechanism for an alteration of the waves on the sea surface by subsurface bathymetry. In: Nihoul, J.C.J. (Ed.), Remote sensing of shelf sea hydrodynamics. Elsevier Oceanography Series, 38. Elsevier, Amsterdam, pp. 187–203.

Yamoaka, H., Kaneko, A., Park, Jae-Hun, Zheng, H., Gohda, N., Takano, T., Zhu, Xiao-Hua, Takasugi, Y., 2002. Coastal acoustic tomography system and its field application. IEEE J. Oceanic Eng. 27 (2), 283–295.

Yentsch, C.S., 1960. The influence of phytoplankton pigments on the colour of sea water. Deep-Sea Res. 7 (1), 1–9.

Zebker, H.A., Goldstein, R.M., 1986. Topographic mapping from interferometric SAR observations. J. Geophys. Res. 91 (B5), 4993–4999.

BIBLIOGRAPHY

Arons, A.B., Stommel, H., 1967. On the abyssal circulation of the World Ocean: III, An advection-lateral mixing model of the distribution of a tracer property in an ocean basin. Deep-Sea Res. 14, 441–457.

Barrick, D.E., Swift, C.T., 1980. The seasat microwave instruments in historical perspective. IEEE J. Oceanic Eng. OE-5 (2), 75–79.

Bogler, P.L., 1987. Motion-compensated SAR image ISLR. IEEE Trans. Geosci. Remote Sensing GE-25 (6), 871–878.

Born, G.H., Dunne, J.A., Lame, D.B., 1979. Seasat mission overview. Science 204, 1405–1406.

Brander, R., 1999. Field observations on the morphodynamic evolution of a low-energy rip current system. Mar. Geol. 157, 199–217.

Brown, W.J., Porcello, L., 1968. An introduction to synthetic aperture radar. IEEE Spectrum 6, 52–66.

Brown, O.B., Cheney, R.E., 1983. Advances in satellite oceanography: Reviews of Geophysics and Space Physics 21 (5), 1216–1230.

Davis, R., 1985. Drifter observations of coastal surface currents during code: The statistical and dynamical views. J. Geophys. Res. 90, 4756–4772.

Davis, R.E., 1987. Modeling eddy transport of passive tracers. J. Mar. Res. 45, 635–665.

Essen, H.-H., 1995. Geostrophic surface currents as derived from satellite SST images and measured by a land-based HF radar. Int. J. Remote Sensing 16, 239–256.

Evans, R.H., Brown, O.B., 1981. Propagation of thermal fronts in the Somali Current system. Deep-Sea Res. 28 A (5), 521–527.

Graber, H.C., Thompson, D.R., Carande, R.E., 1996. Ocean surface features and currents measured with synthetic aperture radar interferometry and HF radar. J. Geophys. Res. 101, 25,813–25,832.

Graham, L.C., 1974. Synthetic interferometer radar for topographic mapping. Proc. IEEE 62, 763–768.

Johannessen, O.M., Sandven, S., Jenkins, A.D., Durand, D., Pettersson, L.H., Espedal, H., Evensen, G., Hamre, T., 2000. Satellite earth observation in operational oceanography. Coastal Eng. 41, 155–176.

Johnson, D., Pattiaratchi, C., 2004. Transient rip currents and nearshore circulation on a swell-dominated beach. J. Geophys. Res. 109 (C02026).

Johannessen, J.A., Kudryavtsev, V., Akimov, D., Eldevik, T., Winther, N., Chapron, B., 2005. On radar imaging of current features: 2. Mesoscale eddy and current front detection. J. Geophys. Res. 110 (C07017).

Keyte, G.E., Pearson, M.J., 1983. An assessment of SEASAT-SAR image quality. In: Allen, T.D. (Ed.), Satellite microwave remote sensing. Ellis Horwood Limited, pp. 187–208.

Lyzenga, D.R., 1991. Interaction of short surface and electromagnetic waves with ocean fronts. J. Geophys. Res. 96, 10,765–10,772.

Marmorino, G.O., Thompson, D.R., Graber, H.C., Trump, C.L., 1997. Correlation of oceanographic signatures appearing in synthetic aperture radar and interferometric synthetic aperture radar imagery with in situ measurements. J. Geophys. Res. 102, 18,723–18,736.

McClam, P., Marks, R., Cunningham, G., Mc Culloch, A., 1980. Visible and infrared radiometer on Seasat-1. IEEE J.Oceanic Eng. OE-5 (2), 164–168.

Njoku, E.G., Stacey, J.M., Barath, F.T., 1980. The Seasat scanning multichannel microwave radiometer (SMMR): Instrument description and Performance. IEEE J. Oceanic Eng. OE-5 (2), 100–115.

Schwicsow, R.L., 1982. Marine LiDAR for subsurface sensing. In: Vernberg, F.J., Diemer, F.P. (Eds.), Processes in Marine Remote

Sensing, pp. 133–149. The Belle W. Baruch Library in Marine Science, No. 12.

Shay, L.K., Graber, H.C., Ross, D.B., Chapman, R.D., 1995. Mesoscale ocean surface current structure detected by high-frequency radar. J. Atmos. Oceanic Technol. 12, 881–900.

Strong, A.E., Derycke, R.J., 1973. Ocean current monitoring employing a new satellite sensing technique. Science 182 (411), 482–484.

Sturm, B., 1981. The atmospheric correction of remotely sensed data and the quantitative determination of suspended matter in marine water surface layers. In: Cracknell, A.P. (Ed.), Remote sensing in meteorology, oceanography, and hydrology. Ellis Horwood Limited, Chichester, pp. 163–197.

Teague, C.C., Tyler, G.L., Joy, J.W., Stewart, R.H., 1973. Synthetic aperture observations of directional height spectra for 7s ocean waves. Nature 244, 98–100.

Teague, C.C., Ross, D., 1979. The offshore environment: A perspective from seasat-1 SAR data. Proc. 11th annual OTC, 215–220.

Tomiyasu, K., 1978. Tutorial review of synthetic aperture radar (SAR) with applications to imaging of the ocean surface. Proc. IEEE 66 (5), 563–583.

Chapter 6

Lagrangian-Style Subsurface Current Measurements Through Tracking of Subsurface Drifters

Chapter Outline

6.1. Surface-Trackable Subsurface Drifters 177
6.2. Satellite-Recovered Pop-Up Drifters 178
6.3. Swallow Floats Tracked by Ship-Borne Hydrophones 180
6.4. Subsurface Floats Transmitting to Moored Acoustic Receivers 183
6.5. Subsurface Floats Listening to Moored Acoustic Sources 188
6.6. ALACE: Horizontally Displaced and Vertically Cycling Subsurface Float 192
6.7. Drifting Profiling Floats (Argo Floats) 195
　　6.7.1. Profiling Observations from Polar Regions 197
References 198
Bibliography 199

Because the oceans constitute a large and complex structure, surface currents provide only a portion of the database required to advance the science of oceanic water circulation dynamics. It is, therefore, important to measure currents beneath the surface Ekman layer, which are themselves complex. Subsurface current measurements are needed for the study of the general characteristics of oceanic water motion and to determine transport of energy, materials, and organisms such as plankton. Dense arrays of continuously tracked floats are desirable to describe the structure of synoptic circulation features.

But to efficiently establish mean flow, long-term observations from a float array of low but statistically uniform density are even more important. The extensive and remarkable developments in the field of underwater sound made during World War II placed in oceanographers' hands several efficient means of making subsurface current measurements by a variety of underwater acoustic methods. These have become powerful tools for measuring ocean circulation, mainly because of acoustic signals' ability to travel long distances in water and the inherently noninvasive nature of measurement. The following sections of this chapter describe various methods employed for measurements of oceanic subsurface currents and gyre systems.

6.1. SURFACE-TRACKABLE SUBSURFACE DRIFTERS

As noted in Chapter 2, the oldest method of Lagrangian-style subsurface current observation was the use of drogues. These drogues, maintained at the desired depth of measurements, were of various designs, such as wooden or metal crosses (popularly known as *biplanar crossed vanes*), sail drogues, or parachutes that offer the largest possible drag at the level of water current measurement. All these devices had to be adequately weighted to reduce the wire angle of the suspension wire and to keep the drogue at the desired depth. The drogues were connected by fine wires to small surface floats or buoys, equipped with identification signs such as numbered signaling flags for daytime visual observation, flashing lights for nighttime visual observation, or radar reflectors for remote detection through electromagnetic techniques. The data derived from each drogue consisted of a series of successive positions of the surface (visible) part of the drogue, determined at known times. The positions were determined by Long Range Navigation (LORAN) or radar. These types of subsurface Lagrangian drifters have historically seen the widest application in oceanographic studies, primarily because of their simplicity of construction and their cost effectiveness. Usually many drogues launched at different depths are tracked at nearly the same time. Drogue measurements have revealed some interesting results (Merritt et al., 1969).

The most commonly used subsurface drifters are parachutes. They are designed to fully open at a prescribed depth. The drogue depth is determined by the length of the wire connected to the supporting surface raft/buoy and the ballast weight attached to the parachute. Performance evaluation studies conducted under steady water-flow conditions have revealed that velocity-dependent shapes such as those of parachute drogues are unsuitable for measurement of small flows (Vachon, 1977). Most parachutes are made of nylon, which is approximately 12 percent denser than seawater. As a result, a nylon chute tends to hang down in closed conditions. In practice, these drogues have to be held open by devices such as rings or spreaders and must be made as neutrally buoyant as possible. Results of tow-tank tests have indicated that a weighted sail drogue, when tethered in line with the center of pressure, will align itself perpendicularly to the prevailing flow due to its hydrodynamic characteristics. Some drifters are fitted with canvas "window-blind" drogues to study currents at a depth given by the length of the drogue line. Graphical illustration of such a drogue is given in *Guide to ARGOS Systems,* Chapter 3, page 4. The telephonically tracked floats described in Chapter 3 can also be configured to measure subsurface currents. In this configuration, the two modules are deployed separately, connected with a rope line, and the four donut-shaped floats are attached on the surface module. The design is quite compact, modular, and flexible. The relatively small height of the drogue allows studies in highly stratified environments, with good vertical resolution of the velocity field, unlike the World Ocean Circulation Experiment (WOCE) open-ocean drifter design.

There are several problems associated with subsurface drogues. For realistic measurement of subsurface currents using the drogue method, it is important that the drogue move with the same velocity as the surrounding water. This requires that the drag exerted at the drogue be substantially greater than that exerted at the surface float, suspension cable, and other parts of the system. In practice, however, the surface float, the position of which is considered as that of the subsurface drogue, is constantly subject to the drag forces of surface-shear current, tension on the connecting tether, and the dynamics of waves and swells. In addition, the elasticity of the suspension cable with variable drags at the drogue will cause some up-and-down motion of the drogue, which traces the flow supposedly at a constant depth. In a dynamic environment, all these forces combine to introduce large errors in measurements, even in a system where drag characteristics of the drogue under steady-flow conditions are known. Furthermore, the surface-float-induced dynamic loading on a drogue in severe environments can shorten the life of the whole system.

In the midst of numerous environment-related problems, the data collected from subsurface drogue measurements tend to suffer from inaccuracies of various kinds. In the absence of alternate convenient techniques, the drogues were widely used in the past, and the data derived from such measurements provided valuable information on the rather complex water circulation features at depths a few hundred meters below the ocean surface (Merritt et al., 1969).

Interestingly, the validity of HF Doppler radar techniques for remote measurement of currents in the upper layers of the ocean surface (describe in Chapter 4) was first tested and convincingly proved by a series of intercomparison measurements using subsurface drogues deployed in the upper layers of the ocean (Stewart and Joy, 1974). In this experiment, the current was measured by tracking the positions of parachute drogues placed near the surface of the ocean at varying depths in the range of 2−4 meters below the surface. The drogues had a surface float that was tracked with accurate microwave radar. Their positions were measured approximately every 10−20 min, with a relative accuracy of ±2 m. The mean direction of the current was obtained from a straight line drawn through the positions, and the mean speed was determined from the distance the drogue moved in 1 hour. The close agreement, within a few cm/s, between the drogue measurements and the nonintrusive measurements made by the HF Doppler radar, in fact, started the "radio scatter measurements revolution" that is witnessed today.

6.2. SATELLITE-RECOVERED POP-UP DRIFTERS

It has been observed that the surface-trackable subsurface drogue system described in the preceding section do not permit the desired true measurement of subsurface currents, unaffected by extraneous influences, because the current–tracing subsurface drogue is "tied" to a float that is sliding on a wind–affected and wave–laden ocean surface. Because a "chained" drogue cannot freely move with the surrounding water mass, the trajectories traced by the drogue are only approximations to the true trajectories of the water parcels at the depths of the drogue. The ideal subsurface current tracing drifter is the one that is free to drift at the prescribed depth, with no attachments to the sea surface and carried along with the flow trajectory at that depth.

With the availability of satellite-based position-fixing technology, deep drifters with ARGOS radio transmitters have been developed by Rossby and Dorson (1983) to synoptically measure subsurface and abyssal currents. The drifter uses an ARGOS radio transmitter to reveal its pop-up position at the end of a programmed period of submergence, which may last from days to months. When used in clusters, these drifters yield, after prolonged submergence, an ensemble of displacement vectors, each a time-integral of Lagrangian motion. The design sought to have the following characteristics: (1) ability to withstand high hydrostatic pressure; (2) excellent corrosion resistance for long submerged periods; (3) sufficient payload (5 kg); and (4) carry a 400-MHz radio antenna at depth without damage, yet be able to extend out of the water for satisfactory transmission to the ARGOS satellite. This requires (5) adequate surface riding characteristics and (6) low cost in volume production. Glass housings were considered to be superior, primarily for corrosion resistance. Glass spheres were the initial choice, but this led to concerns about sufficient antenna exposure. Thus, the final choice was to use a standard borosilicate glass pipe of the kind widely used in the chemical and pharmaceutical industries.

Although the pipes are only slightly less expensive than spheres for the same payload, the use of penetrators is greatly facilitated by routing them through a metal endplate instead of drilling through glass. Furthermore, glass pipes chosen by Rossby and Dorson (1983) had the following features: a simple, clean, and structurally stable housing with one end rounded, the other end open, through which the electronics package is inserted. A single metal disk sealed with a Teflon gasket completes the enclosure. The metal endplate had a pressure gauge port and one penetrator for the ballast release wire. Layout of the fully assembled subsurface drifter is shown in Figure 6.1i. It weighed 9.9 kg.; 1.7-kg ballast was required for neutral buoyancy. At the surface, the drifter exposed about 0.4 m of its 1.6-m length. Thus the 0.25-m-long antenna inside the glass pipe was fully exposed.

Rossby and Dorson (1983) chose borosilicate glass for the following reasons: (1) There must be no loss of material by corrosion, which would alter the buoyancy over long periods of time; (2) borosilicate glass shows no plastic flow or deformation under stress; (3) glass (especially borosilicates, with their low metal content) is very transparent to electromagnetic radiation; thus, the antenna can be mounted inside, thereby greatly simplifying the integration of the ARGOS transmitter and the drifter; and (4) glass housings have extremely low coefficients of thermal expansion, allowing deployment of the floats for isopycnal operation (Rossby et al., 1982).

Two disadvantages of glass pipes have been identified. First, they are delicate, requiring care in handling. However, the use of robust packaging can compensate for this issue. The other problem is their end closure. The open end of the glass pipe has a conical flange with a rounded end-surface (Figure 6.1ii) with a concentric groove in the middle. In normal use, pipe sections are held together with clamps pulling from the conical flange. In this case, Teflon gaskets provide the seal, and the end surfaces are never subject to high axial loading. In the use of closed pipes as

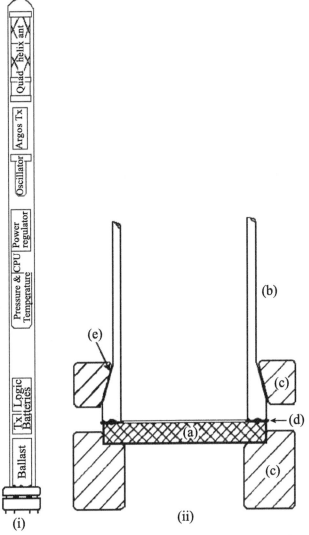

FIGURE 6.1 (i) Layout of the fully assembled subsurface drifter. (ii) Cross-section of glass pipe and endplate assembly. The aluminum endplate (a) is bolted to the glass pipe (b) by means of a polypropylene clamp assembly (c). A Teflon gasket (d) conforms to the shape of the glass end surface to provide a watertight seal. A thin vinyl or rubber pad (e) is fitted to the glass flange. *(Source: Rossby and Dorson, 1983.)*

subsurface drifters, they are subject to large external pressures. Ordinarily this should not matter, because glass is very resistant to compression. The difficulty is that the end surface is not truly flat. Thus, the axial loading from an endplate under high hydrostatic pressure will be nonuniform, both radially and along the perimeter. In light of extensive laboratory studies, it was found that 3-in. glass pipes cannot be reused reliably at pressures exceeding 1,000 psi.

The most serious failure mode of the glass in the flange area is spalling due to tensile stresses in the glass, caused by friction between the glass and the endplate as the glass is compressed. Thus, the choice of gasket material is very important. Rossby and Dorson (1983) found that Teflon acts as an excellent lubricant and, compared to the earlier methods, greatly reduces spalling. In their deployments, the glass flange was prestressed into compression by means of an external clamp or collar (Figure 6.1ii). The arrangement has been reported to be very reliable, rendering it possible to be deployed at greater depths, possibly to 2,000 m. As an additional precaution against leaks, the outside of the gasket was sprayed with an automotive gasket sealant, which seemed to flow readily into all tight spots and to form a sticky, continuous film around the pipe. However, pipes had been sealed many times without the sealant and used at depth with no problems.

At the end of its submergence mission, the microprocessor turns on the ballast circuit. This circuit consists of a transistor that causes a thin wire to dissolve electrolytically and to drop a 2.9-kg ballast weight. The process takes 2 minutes. Thereafter, the instrument ascends at a speed of approximately 2 m/s, so the time to surface ranges up to 30 min, depending on its depth.

As the deep drifter pops up to the surface at the end of a programmed period of submergence, it broadcasts the collected information to the ARGOS satellite. The float starts transmitting 30 min after release. The format for the radio transmission is structured to conform to the RF requirements of the ARGOS system. Approximately every 43 seconds, a message of 32 bytes is transmitted. Thirty of the bytes contain pressure and temperature data; the other two are used for identification and error detection. Forty-eight different messages and a total of 35 min are required to transmit the entire memory. Since a single satellite pass rarely lasts more than 10 to 12 min, a number of passes is required for the entire dataset to be obtained.

The pop-up position of the drifter is determined using the ARGOS satellite-borne relocation system by processing the Doppler shift in the signal transmitted by the ARGOS radio transmitter on the drifter (Briscoe et al., 1987). Subsequent developments in ARGOS-tracked drifter designs permit accurate measurements of mean Lagrangian displacements of water parcels in the mixed layer as well (Niiler et al., 1987).

To sum up, functionally the deep drifter has two operational modes: submerged and surface. While submerged, the instrument is drifting under the influence of the ambient water current velocity, thus providing a Lagrangian view of the oceanic water circulation in its neighborhood. It also makes seawater pressure and temperature measurements on a regular schedule to identify regions of upwelling, fronts, and the like. On the surface, it broadcasts the collected information to the ARGOS system.

Although a satellite-recovered pop-up subsurface drifter has the advantage of not being tied to a surface float, its numerous geographical positions during its current-driven motion between the two successive pop-up positions are unknown. This is a drawback of this method.

6.3. SWALLOW FLOATS TRACKED BY SHIP-BORNE HYDROPHONES

The first successful long-term operating and ideal subsurface water current tracer, the geographical positions of which can be determined at regular intervals—the neutrally buoyancy float—was developed and used by Dr. John Swallow in 1950s; this float began to be known as *Swallow float* after his name. Conceptually, a Swallow float is a neutrally buoyant device with an acoustic transducer that transmits CW pulses according to a precisely timed schedule. The float functions as an acoustic beacon for tremendous distances. In practice, however, a considerable engineering effort is required to create a float with sufficient acoustic range and with an expected life length of 9—12 months. These requirements imposed special demands on reliability, corrosion resistance, efficiency of operation, and conservation of power. Fortunately, the standard Swallow floats were used with great success and operated for as long as seven weeks (Swallow and Hamon, 1960).

In classic isobaric operation, the compressibility of the float must be significantly less than that of the seawater. Thus, the philosophy behind the operation of a Swallow float is that a body that is less compressible than seawater will gain buoyancy as it sinks; and if its excess weight at the surface is small, it may at some depth gain enough to become neutrally buoyant, at which time no further sinking will occur (Swallow, 1955). In isobaric operation, if a float is displaced upward from its equilibrium surface, it will expand less than the water around it. The resulting density difference will force it back to its equilibrium depth. In other words, the Swallow float can be adjusted so that, at a certain depth, the weight of the float equals the weight of the water displaced by it. At this depth, the float remains neutrally buoyant so that it is perfectly free to move with a prevailing horizontal flow of water (Weidemann, 1966; Sturges, 1980; Pickard and Emery, 1982).

The operation of a Swallow float is as follows: After being released from a ship, a float sinks to the selected depth and then drifts horizontally with the water around it. The water-flow trajectory is traced by periodically determining the position of the subsurface float for a sufficiently long period of time (Swallow, 1955). Following the movement of such a float would give a direct measurement of the current at that depth, free from the uncertainties involved in using a conventional current meter from an anchored ship. The possibility of using this method for measuring deep-drift currents over a long period was first suggested by Stommel (1955).

Recovery of the subsurface float after a specified mission requires the use of an expendable ballast weight attached to the bottom end of the float. Ballasting a float means adjusting its weight to neutral buoyancy to a certain depth or for a certain density (σ_t) surface (i.e., isobaric or isopycnal operation). Isobaric floats have been in common use since the pioneering development by John Swallow in the 1950s (Swallow, 1955). Isopycnal studies were initiated in later years (Rossby et al., 1985a).

The problem of locating the neutrally buoyant subsurface drifting float was solved by an ingenious technique of outfitting the float with an acoustic transmitter, known as a *pinger*. In this case, besides having a sufficiently low compressibility, the float must provide enough spare buoyancy to carry this transmitter and must not collapse at the greatest working depth. In the design adopted by Swallow, the acoustic transmitters were capable of sending out a short pulse every few seconds for two or three days. Aluminum alloy scaffold tubing (alloy specification HE-10-WP) has been found to possess the required mechanical properties and can be made into convenient containers for the electrical circuits and batteries. Six meters of such tubing were needed to provide sufficient buoyancy to each float (Figure 6.2), and for ease of handling this tubing was cut into two 3-m lengths laid side by side—one containing the transmitter circuit and batteries and the other providing buoyancy. A simple electronic circuit provided the 10-kHz signal that drove a magnetostrictive nickel scroll sound source, which was wound toroidally and energized by discharging a capacitor through a flash tube. The sound source required for the experiment was available from the Royal Navy. End caps were secured using the then-new O-rings. Properly sealed scaffold tubing has been successfully tested to a 4,500-m depth.

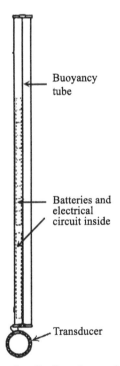

FIGURE 6.2 Sketch of a Swallow float and acoustic transmitter. *(Source: In part from Swallow, 1955.)*

The floats weighed around 10 kg in air but had to be weighed in water so that they could be ballasted to stabilize at their target depth (known as *parking depth*). In the float designed by Swallow (1955), the mean density of each complete float and transmitter was adjusted to an accurately known value by immersing it in a salt solution of known density and temperature and adding weights until it was neutrally buoyant. This adjustment could be made to 1 gm without difficulty, with a float weighing about 10 kg in air. The density could be altered to any desired value by adding or subtracting weights in proportion to the total weight of the float. All the extra weights were put inside the buoyancy tube so that no change in volume had to be allowed for. Before launching any of the floats, temperature and salinity observations were made and the water densities *in situ* calculated from tables (Zubov and Czihirin, 1940). The extra weight required to take the float down to any desired depth could then be determined from the known density at that depth and the calculated compressibility of the float. The floats needed only 38 g of negative buoyancy to stabilize at 1,000 m, so great care was needed with the weighing and density calculations and to eliminate trapped air bubbles.

The method used for tracking the drifting float was to lower two acoustic receivers (known as *hydrophones*) over the sides of a ship, which followed the subsurface drifting float. The two hydrophones were maintained as far apart as possible. The ship track was maneuvered so that the shipborne hydrophone continued to receive the acoustic signals (pings) transmitted by the float-borne pinger and thus tracked the path of the subsurface currents at the depth layer of the drifting float. The acoustic pulses from the neutral buoyancy float will be received at the two spatially separated hydrophones at two different times unless the float's location is just below the ship or at an orientation perpendicular to a plane joining the positions of the two hydrophones. From the magnitude and sign of the observed time difference, it is possible to estimate the bearing of the float with reference to the plane joining the positions of the two hydrophones using known values of the spatial separation between the two hydrophones and the velocity of sound in seawater. To avoid errors in estimation of the bearing of the float, the hydrophones were kept fairly shallow (approximately 7 m below the sea surface) and were weighted to prevent their cables from straying too far from the ship's side. Detection of float position using shipborne hydrophones is schematically shown in Figure 6.3. Because the chasing ship continuously determined its own position by conventional navigational techniques, the speed and trajectory and, therefore, the direction of movement of the current-driven drift of the float at that depth could be determined.

In practice, tracking a drifting subsurface float is more complicated, as indicated in this description given by

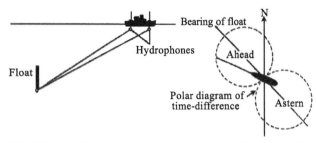

FIGURE 6.3 Method of locating a Swallow float. *(Source: Swallow, 1955.)*

Swallow: With the ship stopped head-to-wind, two hydrophones are lowered over the side, as far apart as can be conveniently arranged. The signals from them are fed via separate tuned amplifiers to a double-beam oscilloscope, the time base of which can be triggered from signals applied to either beam. Pulses from the subsurface drifting transmitter are received at different times at the two hydrophones, and the magnitude and sign of this time difference can be measured on the oscilloscope. As the ship's head falls away from the wind direction, the time difference is observed as a function of the bearing of the line joining the hydrophones. It follows a "figure-eight" polar diagram, with sharp zero values when the bearing is at right angles to the line from the ship's position to the subsurface drifting float. Usually observations over an arc of about 120° are enough to indicate the bearing from the ship to this float. The process is repeated with the ship in other positions, and the intersections of these bearings locate the subsurface float in a horizontal plane. The ship's position is determined by radar range and bearing from an anchored buoy, and the movement of the buoy itself is checked by sounding over small but recognizable nearby features on the sea bed.

Each time a bearing of it is taken, the depth of the subsurface drifting float can be estimated from the size of the "figure-eight" pattern obtained when the time differences are plotted. The ratio of the maximum time difference observed (when the ship is heading directly toward the float) to the direct travel time from one hydrophone to the other is the cosine of the angle between the horizontal and the ray coming from the float to the ship. The depth of the float can then be found when the horizontal distance between it and the ship is known. The direct travel time between hydrophones is measured by floating a transmitter on the surface and observing the maximum time difference at the two hydrophones as the ship is swung round. To avoid errors in bearings and in the time differences, the hydrophones are kept fairly shallow (about 7 m) and are weighted to prevent their cables from straying too far from the ship's side.

The first floats were deployed in June 1955 over the Iberian Abyssal Plain only six months after construction started. The first float was tracked for two and half days by determining its azimuth relative to the ship by using two hydrophones fore and aft and displaying their outputs on a cathode ray oscilloscope as just described. It has to be remembered that at that time, over much of the world's ocean, navigation was by sun and star sights and dead reckoning. (Later on, the availability of LORAN navigation significantly reduced tracking uncertainties). The float could be followed for a period of days or even weeks. Of the six floats deployed, only two worked satisfactorily, but nevertheless the method had been demonstrated, and the results reported by Swallow (1955) were detailed enough to show evidence of tidal variations. As an example, Figure 6.4 shows the temporal sequence of northward and westward movements of a Swallow float deployed near a submarine ridge. A steady drift plus a lunar semidiurnal oscillation has been fitted by least squares to each of these, leaving the residuals shown. Figure 6.5 shows tidal components of

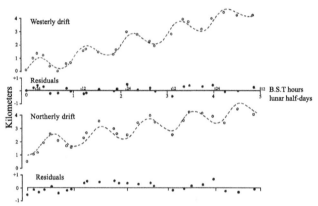

FIGURE 6.4 Temporal sequence of northward and westward movements of a Swallow float deployed near a submarine ridge. *(Source: Swallow, 1955.)*

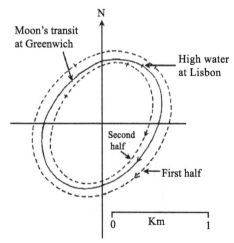

FIGURE 6.5 Tidal components of displacements of a Swallow float combined to form an ellipse. *(Source: Swallow, 1955.)*

displacement combined to form an ellipse. Elaborate deployment of Swallow floats as outlined gave us our first glimpse of the details of deep-water motions and thus contributed significantly to our understanding of several major oceanic subsurface current systems.

While Swallow was engaged in this pioneering development at the National Institute of Oceanography (later renamed the Institute of Oceanographic Sciences) in the United Kingdom, it seems that the neutrally buoyant float concept had also developed quite independently and simultaneously on the other side of the Atlantic (Gould, 2005). Stommel (1955) had called for direct measurement of deep currents and had suggested that it might be done using subsurface neutrally buoyant floats. However, his idea was that they should be tracked through the Sound Fixing and Ranging (SOFAR) channel by the floats creating regular explosions!

According to Gould (2005), at some juncture before 1966, the method of float tracking changed from hydrophones mounted on a ship's hull to two pairs of hydrophones, each pair towed on a cable deployed from the ship's quarter. The "square" of hydrophones separated by about 100 m fore and aft and by the breadth of the ship could be towed at speeds of 3—4 knots (limited by ship noise and hydrodynamic noise) and could determine at what stage a float was abeam of the hydrophone array. This method significantly reduced float-tracking time.

A limitation of Swallow's method is that one ship can follow only a very small number of floats at one time. If several floats are released, e.g., at different depths, it is quite likely that they will perverse and drift off in different directions and the ship will not be able to keep track of them for long. Nevertheless, Swallow floats were really the first oceanographic devices to give us reliable information on the speed and direction of deep currents, and some of the results obtained in this new area have been unexpected. Although several techniques have been proposed for direct measurement of subsurface currents, the most powerful and widely used, perhaps, has been the Swallow float. Further advancements in technology permitted tracking of these neutrally buoyant floats without a ship.

Other than Swallow, only a very small number of researchers used Swallow floats before the late 1960s. Gould (2005) reckons that this was perhaps attributable to a degree of "mystique" about float tracking engendered by Swallow's attention to detail in float preparation and the difficulties of float tracking using low-energy acoustic transmitters. This early limited use of floats contrasts with the larger number of laboratories that were measuring currents using moored instruments at that time. However, because of technological limitations of that era, such as power consumption and data storage capacity, moored current meters were also difficult to use for any period longer than weeks. Consequently, most users were confined to the shelf seas.

6.4. SUBSURFACE FLOATS TRANSMITTING TO MOORED ACOUSTIC RECEIVERS

There are few precise observations of average deep-water motions, primarily because of the limited duration of each experiment. The prevalence of large-velocity fluctuations over short time scales demands long averaging periods before the low-frequency motions with weekly and monthly periods and the net transports can be resolved. Continuous observations over months rather than days would be needed if the mean ocean circulation were to be revealed. Such measurements can, for example, be made with Swallow floats or with current meters, which are designed to operate for months. Both methods are expensive. The latter require two trips to sea, one for their deployment and another for their recovery. A large number of current meters must be distributed in a suitable array if in addition one wants to resolve the spatial structure of the water motions. Such programs are most desirable, but cost considerations, in terms of hardware as well as in ship time, necessarily limit the number and scope of such studies. Similar cost considerations arise in the use of Swallow floats, for which reason these usually cannot be tracked from a surface vessel for longer than perhaps a week. However, the very fact that a Swallow float "tags" and drifts with the water mass makes it ideal for tracing deep-water motions. Thus, if there were a method of locating Swallow floats over long periods of time (months) without the use of a ship, one should hope to obtain, at much less expense, a more accurate estimate of the net drift of the deep waters as well as some information on the character of the water flow, such as, for example, steady, linear, eddy-like motions. It is also interesting to examine what happens as the float approaches or leaves a coastal region.

The prohibitive time factor, expense, and logistical difficulties associated with the method of tracking subsurface floats by ship-borne hydrophones prompted scientists to explore alternate techniques of detection of subsurface drifting floats. M. Ewing's invention of the SOFAR channel during World War II heralded a new era in underwater acoustics, paving the way for development of techniques for making continuous day-by-day measurements of subsurface oceanic currents and their trajectories on a permanent basis. With this development, it was hoped that we should eventually be able to map synoptic charts of currents at various depths in the ocean in the same way as meteorologists keep abreast of the winds. The SOFAR channel permits acoustic signaling over great distances (greater than 1,000 km), provided that the frequency of transmission is low (much smaller than 1,000 Hz). An ingenious method thus evolved for remotely exploring large-scale ocean circulation trajectories is the use of the so-called SOFAR floats. The SOFAR floats are neutrally buoyant subsurface floats that are outfitted with acoustic pingers and released into the oceanic SOFAR channels.

The SOFAR channel, which is an acoustic wave guide, owes its existence to the fact that the speed of sound reaches a minimum value at a certain depth in the ocean. This is because the speed of sound is a function of pressure and temperature (and salinity, to a lesser extent). The speed of sound below the warm surface waters of the ocean decreases with increasing depth due to decreasing temperature. Below the main thermocline, where the thermal gradients become small, the effect of hydrostatic pressure on the speed of sound becomes dominant and thus increases as the depth is further increased. In the western Sargasso Sea, the minimum sound speed is $1,492 \pm 2$ m/s and is located at $1,200 \pm 200$ meters depth. The accuracy with which a float can be located depends directly on the stability of this minimum.

The significance of the minimum velocity becomes clear when it is realized that sound rays that are radiated within certain angles from the horizontal from a source at the depth of the minimum velocity (the sound channel axis) will be refracted back toward this depth. This acoustic energy, trapped in the vertical, radiates horizontally with a geometric rate of attenuation proportional to $1/R$, where R is the distance, instead of $1/R^2$, as in the case of a spherical radiator. This low rate of attenuation of acoustic signals propagating in the SOFAR channel permits acoustic signals to be detected at great distances from the source. In addition to the geometric attenuation, there is also the frequency-dependent attenuation due to scattering and absorption. Thus, the acoustic signal is attenuated much more rapidly at higher frequencies. For this reason, better reception and greater ranges are obtained by transmitting at as low a frequency as possible. The dual virtues of vertical trapping and low rate of attenuation of acoustic energy in the SOFAR channels permit hydrophones to track the SOFAR floats drifting at far-off distances from the hydrophones.

This application of the SOFAR channel was suggested by H. Stommel (1949, 1955). He suggested that the basic instrument required for this purpose is the oceanographic equivalent of the meteorological constant-altitude balloon: an unmanned subsurface buoy or float, devised so as to float at a nearly constant depth or along an isopleth of temperature or density and equipped with a clockwork designed to drop and/or fire SOFAR charges at certain predetermined times, say once a week, for periods up to half a year. The trajectory of each buoy could be determined from the times of arrival of the explosive sound wave at three SOFAR stations. He predicted that if this method proves feasible, the way will then be clear for a major advance in physical oceanography.

Encouraged by this suggestion, Rossby and Webb (1970) initiated an observational program to study abyssal water motions by tracking subsurface floats that were released in the SOFAR channel, and the first Swallow float was launched in the SOFAR channel in January 1968. These floats are now known as *SOFAR floats*.

SOFAR floats are similar in concept to the original Swallow floats, which have been used with great success. However, the technology for SOFAR floats is much more demanding in terms of reliability, economy of power, batteries with high energy density per unit weight, a low-frequency (preferably less than 500 Hz) acoustic projector with low weight and high efficiency, and corrosion resistance. These special demands of SOFAR floats arise because they act as acoustic beacons for tremendous distances (more than 1,000 km), which must operate over long periods of time (over a year).

Usable acoustic ranges that can be achieved depend largely on ambient acoustic noise. Unfortunately, the ocean is not noiseless. Ships, waves, rainfall, earthquakes—all generate noise in various parts of the frequency spectrum. This imposes a limit on what signals can be detected. Most noise is due to surface agitation, although ships can cause serious interference at frequencies below 400 Hz. According to Rossby and Webb (1970), the sea noise to be expected near Bermuda, Eleuthera, and Puerto Rico in the Atlantic Ocean is largely predictable from knowledge of wind conditions.

It has been found that useful acoustic ranges in excess of 1,000 km are possible for tracking SOFAR floats. In this method, at known time intervals, each float transmits a low-frequency acoustic signal. From the time of arrival of this signal at several widely spaced hydrophones (at ranges of approximately 500 to 1,000 km), the positions of the floats can be determined to within a few kilometers.

The SOFAR floats have been tracked by the use of two receivers separated by a large distance. If the travel times of an acoustic pulse from the SOFAR float to these two receivers are known, then the distance between the float and each receiver can be computed from an *a priori* knowledge of the mean sound velocity in that ocean basin. The position of the float will be one of the two intersection points of the two circles whose centers are the positions of the two receivers and whose radii are the two pulse-travel distances between the float and the two receivers (see Figure 6.6).

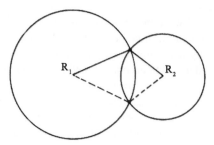

FIGURE 6.6 Method of tracking a SOFAR float with the aid of two acoustic receivers R_1 and R_2.

The ambiguity (i.e., which of the two intersecting points on the two circles represents the true position of the float) can be resolved by knowing its previous location or the location where it was launched. If the pulse-travel time is to be determined precisely, the time of transmission of the acoustic pulse from the float must be known precisely. For this purpose, pulses are transmitted from the float at pre-determined times. This means that the clock in the float must be extremely precise. The crystal clocks are known to drift with time. To circumvent this difficulty, three receivers are usually used rather than two. In this case the float position can be determined even if the clock in the float gains or loses time slightly. This is because of the fact that when three receivers are used, the *difference* in time of arrival between two pairs of receivers gives two hyperbolae (on a sphere), the intersection of which is the location of the float (Rossby and Webb, 1970). Because the "difference" in time of arrival is used, any error due to gain or loss of time (which is common to both of the received signals) gets cancelled out. To get sufficient data for this purpose, many pulses (usually a set of 100) are transmitted 1 minute apart every few hours. Because the deep currents are usually small, this methodology of tracking is not expected to give rise to serious errors in the position determination of the float.

In the receiver electronics, some precautions are needed to keep the influence of noise level to a minimum. This is because the acoustic signal traversing a long distance (1,000 km or more in the horizontal) picks up noise in various parts of the frequency spectrum (noise generated by ships, waves, rainfall, earthquakes, and so on). Any slight nonlinearity of the signal amplifiers in the receiver side can generate harmonics, and this will raise the effective noise level of the system. This problem can be minimized by filtering out the noise components in the received signal and amplifying only the frequency band of interest. Usually the received signals are detected in an approximately 1 Hz wide band-pass filter centered at the transmission frequency.

Rossby and Webb (1970) have described an observational program designed to study abyssal water motions in the western Sargasso Sea by tracking Swallow floats in the SOFAR channel. The development of floats that can be tracked by sound transmissions through the SOFAR channel can be traced back to Stommel's original SOFAR concept; but Rossby and Webb (1970) used a low-frequency (500–600 Hz) sound source rather than Stommel's primitive idea of repeated explosions. The prototype SOFAR float constructed by Rossby and Webb (1970) is a tightly engineered Swallow float with an acoustic transducer that transmits CW pulses according to a precisely timed schedule. Their float consisted of a spherical housing (in two hemispheres) of adequate stiffness and resistance to hydrostatic pressure (Figure 6.7). Each hemisphere of the float was spun from aluminum alloy 6061-T6 and had an internal

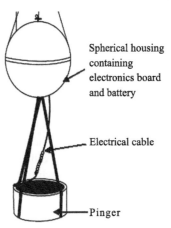

FIGURE 6.7 Prototype SOFAR float. *(Source: Modified from figure 3 of Gould, 2005.)*

Labels: Spherical housing containing electronics board and battery; Electrical cable; Pinger

diameter of 0.93 m, a minimum wall thickness of 25 mm, and a hydrostatic collapse safety factor of two or more. All exposed external surfaces were either machined hard anodized aluminum alloy or nonconductive synthetic material. The float enclosed the required power supply and electronic equipment and supported the electroacoustic transducers suspended about 1.5 m beneath it. The electroacoustic transducer was a hollow cylinder of 750 mm diameter, 325 mm length, and 50 mm wall thickness, which vibrated in the fundamental radial mode. The construction was a composite of active barium titanate and passive aluminum alloy in a barrel-stave configuration pre-stressed into compression by circumferential glass fibers in tension and insulated with polyurethane. The electrical connectors to the transducers were protected against fish bite using nylon tubes.

The computational method of determining the position of the float is to assume an initial position, the last one, say, and then to calculate the travel times. These are compared to the measured values, and the resulting errors are used to estimate the new position. The process is repeated until the error is negligible. The travel time is estimated from the mean velocity of sound and the range, which is calculated using the Andoyer-Lambert correction for the ellipticity of the Earth.

It may be noted that three major international campaigns—POLYGON, MODE, and POLYMODE—were planned and executed with the hope of better understanding the large-scale circulation features in the Atlantic. POLYGON was led by Leonid Brekhovskikh from the Acoustics Institute, Russia, involving six research vessels and an extensive network of current meters. The current meters were deployed in a pattern resembling the shape of a cross, dubbed the "POLYGON", spanning a region of 113 × 113 nautical miles. This experiment provided some interesting results. It looked as though some large-scale eddy or wave disturbances were travelling across the POLYGON site from east to west. Their scales were close to those of the planetary baroclinic Rossby waves.

Mid-Ocean Dynamics Experiment (MODE) was led by Henry Stommel from the United States. According to Walter Munk, the POLYGON experiment "ignited the mesoscale revolution and that MODE defined the new order" and that "oceanography has never been the same" since. POLYMODE was a combination of the Russian experiment POLYGON and the U.S. led MODE. Russian ships came into Woods Hole during this collaboration. Scripps as well as Woods Hole oceanographic institutions from the United States were involved in this campaign. POLYMODE experiment, led by Andrei Monin, took place in 1978.

The U.S. Air force maintained an array of Missile Impact Location (MIL) hydrophones that located the positions of test missiles that dropped SOFAR charges at the end of their flights. These hydrophones could monitor floats in much of the northwest Atlantic. The first two SOFAR floats were deployed in the Sargasso Sea in 1968 and showed that reception of signals was possible at ranges of up to 1,000 km and that float positions could be determined with accuracy of the order of 3−5 km. The premature failure of both floats after one week and after two days, respectively (they were designed for a life of 9−12 months), was worrisome and attributed to biological attacks on the floats.

The availability of the MIL hydrophones was encouraging to carry out a systematic exploration of the ocean mesoscale over a substantial part of an ocean basin and thus paved the way for planning the 1973 Mid-Ocean Dynamics Experiment (MODE), (MODE Group, 1978). In the MODE experiment, a combination of floats, moored current/temperature recorders, hydrographic surveys, and bottom pressure gauges was used in a nine-month collaborative U.S./U.K. experiment. Because the abrupt failure of both spherical-shaped SOFAR floats after only a few days of operation was a cause for concern, Rossby et al. (1975) adopted an alternate design for use in the MODE experiments at Bermuda triangle. This design was similar to the neutral buoyancy float originally designed by Swallow. The SOFAR floats constructed for MODE (see Figure 6.8) consisted of three cylinders of aluminum alloy.

The central cylinder was 5.2 m long and 30.5 cm in diameter, with hemispherical aluminum end closures, and contained the battery and electronic equipment. The two short cylinders shown in the figure are the low-frequency acoustic transducers, which operate freely flooded. The upper end was open and the acoustic driver, which was a bender plate, was fitted to the lower end. These transducer tubes were mounted on the main housing with four heavy stainless steel studs. All 20 SOFAR floats constructed for MODE were equipped with two acoustic signaling systems: a low-frequency system for long-range shore locations and a high-frequency subsystem for shipboard location and command recovery. A 7.3 kg cylinder of lead was connected to the lower end through an electronically

FIGURE 6.8 Photograph of the modified SOFAR float used in the MODE program. *(Source: Rossby et al., 1975. Reproduced with kind permission of Dr. Thomas Rossby.)*

controlled release mechanism and was jettisoned for float recovery on command from the surface. The small, high-frequency (10 kHz) ceramic transducer was connected to the bottom end. All signals were derived from a temperature-compensated quartz crystal oscillator. Electrical energy was furnished by a single battery of 73 kg, 30 volt nominal, and 5 kW hrs. Over 90 percent of the energy was used in the low-frequency signal. The completed float weighed 430 kg. Ballasting was carried out in an enclosure attached to the Woods Hole dock. The floats also had a 10-kHz short-range navigation system to allow their location identification and subsequent recovery by a ship.

For identification of all 20 floats from the same receiver stations, the method used was to have a combination of one of three carrier frequencies (approximately 267, 270, and 273 Hz) and seven different pulse-repetition rates ranging from 1,437 to 1,443 transmissions per day. These three differing carrier frequencies and seven pulse repetition rates provided 21 channels, which was more than sufficient to track 20 floats. Finally, 20 MODE SOFAR floats, each with a one-year design life, were deployed at a target depth of 1,500 m. The floats were tracked from four widely spaced hydrophone sites (Bermuda, Bahamas, Grand Turk, and Puerto Rico) at ranges of approximately 500 to 1,000 km, as indicated in Figure 6.9. After amplification, filtering,

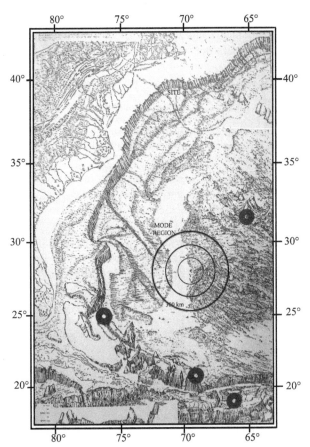

FIGURE 6.9 Physiographic diagram of the western North Atlantic (straddling the rough bottom to the east, the Hatteras Abyssal Plain to the west, and further to the west the Blake-Bahama Outer Ridge), where MODE-I experiment was conducted. The MODE-I region is indicated by three concentric rings of 100, 200, and 300 km radius. The float-monitoring sites are indicated by thick rings. *(Source: Rossby et al., 1975. Reproduced with kind permission of Dr. Thomas Rossby.)*

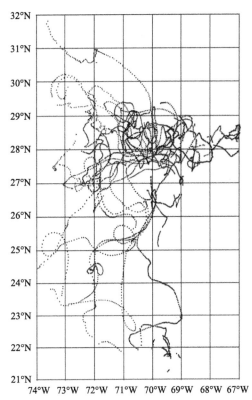

FIGURE 6.10 Trajectories of individual floats, operated at 1,500 m depth in the Bermuda triangle during the MODE experiment (September 28, 1972, to December 31, 1974). *(Source: Rossby et al., 1975. Reproduced with kind permission of Dr. Thomas Rossby.)*

and addition of timing, the signals were recorded and used for analysis.

The navigational resolution for the trajectory data was typically ±500 m and the accuracy was 2 to 3 km. A good fraction of the error was attributed to the assumption of a "universal" speed of sound of 1,492 m/s throughout the Bermuda triangle, where the MODE experiments were conducted. Trajectories of individual floats, operated at 1,500 m depth in the Bermuda triangle during the MODE experiment, are shown in Figure 6.10. The experiment revealed some new information on the structure and variability of the deep-ocean currents. Some floats remained stationary for a year, whereas others covered hundreds of kilometers. The experiments indicated that the Blake-Bahama Outer Ridge had considerable influence on the organization of the eddy field in the MODE area. Regions of sudden swirls and large horizontal shear were also observed.

The MODE marked a quantum leap in our ability to observe the state of the ocean. Although the experiment

was still primarily ship based, the SOFAR floats enabled day-to-day objective mapping of the ocean mesoscale over an area 400 km square and revealed the long-term propagation of these features (Freeland et al., 1975; Freeland and Gould, 1976). The floats were also entrained into the Gulf Stream rings and were thus able to reveal both their rotation rates and propagation (Cheney et al., 1976).

Since the MODE experiment in spring 1973, acoustically tracked SOFAR floats have been regularly used as Lagrangian drifters to study subsurface currents in the western North Atlantic. By emitting acoustic signals on a regular schedule, their positions have been determined as a function of time at ranges in excess of 2,000 km, depending on acoustic propagation considerations, source power level, and ambient noise conditions. Initially, the use of SOFAR floats was restricted to areas of the western North Atlantic within reach of shore-based SOFAR hydrophones at Bermuda, Eleuthera (Bahamas), Grand Turk Island, and Puerto Rico.

In an alternate design, Webb (1977) used a cylindrical float as part of the POLYMODE programs in a major study of mesoscale circulation in mid-ocean. These floats, although used with no protective coatings, were reported to have performed well even after four years of continuous service. The signals were received at deep hydrophones from moorings and transmitted to land-based stations. A multiple address system using a 1,420 Hz phase-encoded signal transmitted from a recovery ship to the float could establish communication to the float. Upon reply from the float, a recovery command sent from the ship caused the jettison of a 7 kg external ballast weight and initiated a special fast-pressure telemetry cycle to verify release and aid recovery. The jettison was accomplished via a simple electrochemical release. For pickup at sea, a buoyant grapnel tethered to the float by a buoyant line was snared with a heaving line from the ship, and the float was hauled into its launch and recovery frame using the buoyant line.

In the POLYMODE experiments, temperature and pressure were also measured as additional parameters. The measurements were suitably averaged to avoid contamination or aliasing of the data by internal waves. Because early observations revealed a tendency of the floats to sink slowly, approximately 0.5 m/day, apparently due to inelastic creep in the main buoyant housing, adequate precautions were taken in subsequent experiments to prevent such sinking.

To ensure that all the POLYMODE floats drifted over the same isobaric surface, these floats were equipped with a controller that maintained a constant operating pressure. The controller consisted of a block of anode-quality zinc mounted externally in the sea water and could be electrically connected to the main aluminum housing via a switch controlled by the pressure measurement and averaging circuit. The controller works as follows: When the circuit is open, the electrochemical couple is quite inactive. When the circuit is closed, under control of the pressure measurement circuitry, a small saltwater battery is formed. When this happens, zinc goes into solution in the seawater and the whole instrument begins to rise. When the float has risen to the required level, the switch is again opened so that dissolution of zinc is arrested. The regularly telemetered pressure signals indicated that all the POLYMODE floats operated within ± 5 decibar of their normal equilibrium depth. An excellent description of the careful preparations needed before ballasting may be found in Webb (1977). This information will be very useful to those involved in deep-sea circulation measurement using SOFAR floats.

From the mid-1970s, acoustically tracked floats were used extensively to further explore the ocean's mesoscale structure and variability. However, the use of floats was almost entirely applied to the North Atlantic. This bias was the result of the regional interests of the laboratories involved (Woods Hole, University of Rhode Island, and the U.K. Institute of Oceanographic Science) and, of course, the existence of the acoustic tracking network for SOFAR floats. The shorter-range MiniMODE floats were applied to study a number of physical phenomena in detail. In the 1980s SOFAR floats were in extensive use in the eastern North Atlantic in U.S., British, and French oceanographic programs. The SOFAR floats, considering their relatively short development and trial phase, were remarkably successful and, therefore, can be used worldwide.

6.5. SUBSURFACE FLOATS LISTENING TO MOORED ACOUSTIC SOURCES

Subsurface floats transmitting to moored acoustic receivers suffered from several limitations. For example, a close look at its size and its deployment reveals that, to be cost-effective, these floats have to be used in large numbers because the listening stations are rather expensive and require a ship for replacement and final recovery. Second, because of large size and weight, these floats require specialized launching equipment. Although this is not a serious problem for large dedicated studies, it becomes a serious handicap for studies where economy is a concern. Third, although neutrally buoyant Swallow floats and their larger counterparts, SOFAR floats, have been used as subsurface drifters in numerous studies of mesoscale circulation (Swallow, 1955; Rossby et al., 1975; Price and Rossby, 1982) and small-scale motion (Voorhis, 1968; Pochapsky, 1963), these floats approximately track isobaric surfaces and thus are not Lagrangian followers of water parcels in a strict sense. Deviations of Swallow floats and SOFAR floats from true isobaric tracking are due in part to small but finite differences in compressibility and thermal expansion from that of seawater. A better representation of Lagrangian motion, particularly in regions of strong baroclinicity and vertical motion, can be obtained from a Swallow float modified to follow isopycnal rather than isobaric surfaces.

Thus, the requirement for a low-cost, lightweight, and ideal Lagrangian drifter for studies in the Gulf Stream (Rossby et al., 1985a) resulted in the development of a new subsurface float, known as a Ranging and Fixing of Sound (RAFOS) float (Rossby et al., 1986). The RAFOS (SOFAR spelled backward) float is a small, neutrally buoyant subsurface drifter that, like the SOFAR float, uses the deep sound (or SOFAR) channel to determine its position as a function of time. The technical difference between the two systems is that whereas the SOFAR float *transmits* to moored receivers, the RAFOS float *listens* to accurately timed signals from moored sound sources to determine its position. (This difference justifies the use of the backward spelling in naming the float!) Since the RAFOS float is lightweight, making deployment is a simple hand operation, and, in fact, there is no reason why these floats could not be adapted for launch from aircraft.

Isopycnal operation of a subsurface float can be accomplished by the addition of a compressible element, or *compressee* as it is called, which adjusts the effective compressibility of the float package to approximate that of seawater. If the coefficient of thermal expansion of such a float is much smaller than that of seawater, it will remain close to a given isopycnal surface because neither salinity nor temperature changes will affect the density of the float. Hence, it becomes an isopycnal float. In isopycnal operation, no Archemedian restoring force is introduced due to pressure changes, as in the case of the standard Swallow float. For the standard (isobaric) float, the compressibility is typically 30–50 percent less than that of seawater (Rossby et al., 1985b). Consequently, if a float is depressed from its equilibrium depth, it compresses less than the surrounding seawater, gains buoyancy, and thus has a relatively larger restoring force than an equivalent parcel of water. This restoring force maintains the equilibrium depth of the float, which in turn may be affected by the difference in coefficient of thermal expansion of the float and seawater. The compressee consists of a spring-backed piston in a cylinder so that an appropriate volume loss is achieved as a function of pressure (Rossby et al., 1985, 1986).

It must be borne in mind that by using a passive mechanical compressee, isopycnal tracking can only be approximated. This is due to the fact that a system with compressibility that exactly equals that of seawater would still not be a perfect isopycnal follower due to a small but finite coefficient of thermal expansion, although this can be reduced by 30 percent or more by fabricating the compressee from nonmetallic materials. In summary, an isopycnal float can be optimized by matching its compressibility to that of seawater and minimizing its coefficient of thermal expansion as much as possible. In addition, the restoring forces will be maximized and the ballasting errors will be reduced when the float is deployed in a region of strong stratification.

The float used by Rossby and co-researchers in the first test, August 1981, was packaged in a standard 25-cm glass sphere. Two seawater connections drilled through the sphere connected the external 12-kHz acoustic transducer with its power source. Temperature was measured by a thermistor, which was in thermal contact with the sphere wall. It was encoded into a 12-digit binary word, which was transmitted by the transducer every five minutes. The signal was displayed in analog form on the ship's graphic recorder. This binary printout was then read and converted to decimal form. The aluminum compressee consisted of a spring-backed piston set in a cylinder made from solid aluminum stock, with an O-ring set in the piston wall. Thus, unlike the isobaric operation scheme employed in the Swallow float, isopycnal operation is employed in the RAFOS float.

Two preliminary field experiments were conducted in the upper main thermocline near the North Wall of the Gulf Stream. During August 1981, the spherical glass float was deployed near the northern edge of the Gulf Stream (200 km downstream from Cape Hatteras) and was tracked for approximately 100 km. The float, having remained neutrally buoyant in the oxygen minimum zone near 380-m depth,

tracked an isopycnal surface as confirmed from CTD data during its passage through a mini-meander. The velocity vectors, based on successive pairs of float positions, showed reasonable values, which compared favorably with the acoustically tracked velocity profiler Pegasus station values for similar depths (see Chapter 10 for a description of Pegasus).

The updated float design (Figure 6.11) used in September 1982 was packaged in a 1.52-m glass pipe with a 7.6-cm nominal ID manufactured by the Corning Glass Co. The RAFOS float is nearly identical to the deep drifter previously developed by Rossby and Dorson (1983). The pipe is made of standard Number 7740 borosilicate glass, which provides the flotation and housing of all electronics. The upper end of the pipe, which floats vertically, is rounded off, and the lower end is sealed with a flat aluminum end-plate where all electrical and mechanical penetrators are located. The glass-wall thickness is ∼5 mm, giving these pipes a theoretical maximum depth (length/diameter » 1) of ∼2,700 m. The internal mechanical assembly consists of a single PVC spar running the length of the pipe. Mounted on it from top to bottom are the radio antenna, the ARGOS transmitter, the microprocessor and memory circuit board, the analog circuit board (pressure, temperature, and acoustic filtering circuits), and at the bottom, for vertical stability, the battery pack. The electronics, flasher, and power packs are

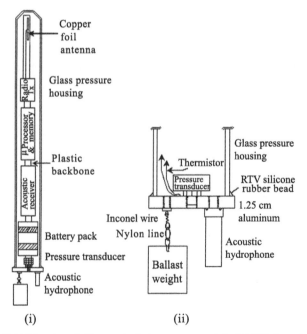

FIGURE 6.11 (i) The mechanical arrangement of the RAFOS float. A glass pipe, rounded at the upper end, is closed at the lower end with an aluminum endplate. All of the components—radio antenna, radio transmitter, microprocessor acoustic receiver, and battery pack—are mounted on a PVC spar prior to insertion in the pipe. (ii) Detailed drawing of the endplate arrangement. *(Source: Rossby et al., 1986, ©American Meteorological Society. Used with permission.)*

seated in foam rubber in order to avoid points of concentrated stress on the inside walls of the glass tube. The batteries are placed at the lower end of the float to facilitate replacement and provide vertical stability, especially at the sea surface. Bolts with washers connecting a polypropylene collar provide sufficient closure to prevent low-pressure leakage. The retrieval bridle is attached to the collar rather than the glass to prevent nonuniform stresses in the glass pipe.

The pressure gauge is rigidly threaded to the endplate and the thermistor is attached to the inside surface. The attachment of the aluminum endplate to the glass pipe is accomplished by applying a bead of silicone rubber around the perimeter where the plate is pressed against the pipe. This is a change from the glass flange-and-clamp arrangement used earlier by Rossby and Dorsan (1983), which was both more expensive and probably less reliable at high pressures. As a precaution, a thin (~ 0.025 cm) sheet of Teflon is inserted between the flat-ground end surface of the pipe and the endplate, to permit the glass pipe to compress radially under pressure without spalling against the metal surface.

The hard-coat anodized aluminum endplate supports the penetrators for the hydrophone, the ballast release wire, and hole for the pressure gauge. The hydrophone is a single ceramic element and is rigidly potted onto a standard three-pin connector using a two-component polyurethane resin compound. The compressee (see Figure 6.12) used in the

1982 experiment contained a spring-backed piston, with diameter 2.25 cm, set in an aluminum cylinder made of tubing and machined endplates. The cylinder had an outer diameter of 7.57 cm and an overall length of 25.35 cm. The piston was made to protrude approximately 2.54 cm from the endplate to provide better alignment at high pressures and had an O-ring set in the piston wall. The diameter of the piston and the spring constant are adjusted so that the compressibility of the float assembly matches that of seawater. All parts are anodized to minimize saltwater corrosion. Hysteresis due to striction of the piston is reduced by choosing O-rings of high durometer. In subsequent experiments, the closure design was improved by grinding flat the open end of the glass tube and using a very thin Teflon gasket. This greatly reduces the volume change associated with cold flow of the Teflon.

The RAFOS electronics consist of four main parts: a set of sensors for collecting temperature, pressure, and tracking information; an ARGOS-compatible transmitter to relay the collected data after surfacing; a clock for time reference; and a microprocessor (CPU) to control the sensors, store the data, and format the data for the transmitter. The acoustic signal detection and storage of data are all handled by the CMOS microprocessor in the float.

For launching, the 10-kg instrument is lowered to near the water line, the ship is headed up into the current, and floats are released manually. In the absence of the now-ubiquitous GPS position-fixing devices, instrument position was determined from LORAN fixes made while the ship was directly overhead of the float, shown schematically in Figure 6.13. The depth of the float was determined

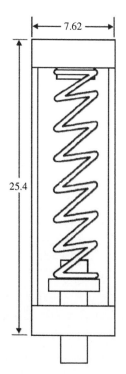

FIGURE 6.12 Schematic diagram of the compressee used in the RAFOS float. *(Source: Rossby et al., 1985.)*

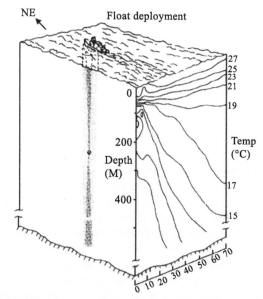

FIGURE 6.13 Schematic diagram of the tracking scenario showing the float, and the direct and bottom-reflected acoustic paths. The upstream temperature section from the August 1981 pilot experiment is also shown. *(Source: Rossby et al., 1985.)*

to an accuracy of 5 m based on the arrival time of the signal reflected from the transducer.

The external ballast (for return to the sea surface) is suspended by a short piece of Inconel wire chosen for its resistance to saltwater corrosion. By dissolving it electrolytically (2 min at 0.3 ampere), the approximately 1-kg ballast is released. A flasher is actuated externally prior to launch and a semi-rigid PVC plastic bridle is used to facilitate recovery. Subsequent to the ballast weight release, the float can be located at the sea surface by the flasher during night, by the brightly colored upper portion during daylight, and by the acoustic signal, which continues to transmit at the surface. Recovery is accomplished by snagging the PVC bridle with boat hooks and lifting it directly on board.

The acquired recorded data are recovered at the end of the mission when the float surfaces and telemeters its memory contents to Systeme Argos, a satellite-borne platform location and data collection system (see Chapter 3). Accordingly, at the preprogrammed end of the mission underwater, the CPU activates a release circuit, which drops a ballast weight, and the float returns to the sea surface. Exactly 30 min after release initiation, under control of the master clock, the ARGOS transmitter in the float starts transmitting the contents of its memory repeatedly to Systeme Argos until the battery in the float is exhausted (a duration of about two weeks). The transmission schedule consists of uploading, approximately every 43 seconds, an ARGOS format message with 32 data bytes. The first byte of each message is a check byte to detect any data errors introduced during transmission and data transfer. The second byte is a message number to keep track of each data block. The remaining 30 bytes are the collected data sequentially transmitted in the order in which they were stored. A satellite pass is within range for less than 15 minutes (there are two satellites), and therefore a number of orbits are required to successfully transfer the complete dataset. Typically, at least three days are required to transfer the 2,835 bytes of data.

All data collected from each 100-min polar orbit are processed at ARGOS headquarters in Toulouse, France, where a few hours later they can be examined via telephone. The data are also forwarded by mail on a regular basis. On calm days at sea, about 90 percent error-free data transfers can be achieved; in severe weather (heavy sea states), this figure can drop to ~50 percent. If necessary, the rejected messages could be examined more closely for salvageable data.

Tracking information is obtained by detecting the time of arrival of SOFAR signals from moored sound sources at the RAFOS float. Just a few sound sources provide navigation for an arbitrary number of floats.

During the September 1982 field experiment, the cylindrical float was tracked for approximately 100 km east

of Cape Hatteras. Deployed in the Gulf Stream near the 250-m level was striking evidence that the float was adjusting depth to maintain its prescribed isopycnal surface during a gradual descent and cross-stream motion over the 24-hr sampling period (Rossby et al., 1985). Thus, the float exhibited its isopycnal-following character in both field tests.

Although the use of the SOFAR channel for tracking subsurface floats is quite straightforward, there are several acoustic considerations that should be kept in mind in order to maximize the area of coverage or insonification. These include (Rossby et al., 1986):

- Acoustic source level
- Ambient noise conditions
- Depth of the float in relation to the sound channel axis

Rossby and co-researchers essentially used standard SOFAR floats as acoustic beacons (Webb, 1977). They are buoyed up from the bottom and provide a CW acoustic source level of about 175 dB relative to 1 μPa. The ambient noise conditions are a strong function of weather and shipping. For example, the ambient noise in the Gulf Stream is quite high due to a combination of heavy shipping and strong winds. Despite this fact, useful detection ranges as great as approximately 2,000 km were possible in the Gulf Stream.

Unlike the SOFAR float system, wherein continuous signal monitoring is possible at land-based stations, the RAFOS floats are limited to storing only the two largest correlations per expected signal. Thus, it is necessary to require a large S/N ratio for positive signal detection. It was found that the shallower floats show somewhat poorer performance than the deep floats, but both groups work over the entire listening range of 1,230 km. Cyclonic meanders of the Gulf Stream and/or cold-core rings may cause additional losses.

The procedures for editing and processing the acoustic travel-time data to reconstruct the float trajectories are very similar to those for SOFAR float tracking. To compute the position of the float, the following information is needed (Rossby et al., 1986):

- The time of signal transmission
- The time of arrival of the signal at the float
- The average speed of sound in seawater
- A Doppler correction

The time of transmission is known from the preset transmission schedule for each sound source. The time of signal arrival is determined relative to the float's clock. The float clock is set "on time" and verified (to within a second) at launch, and the accumulated error at the end of its submergence is indicated by the error in radio transmission schedule, which can be determined to a fraction of a second. The biggest uncertainty in the path-averaged speed of sound arises from the fact that the received signal is really a complex signal of many path arrivals spread out over several seconds. Each one of these has a different average speed of sound. However, sophisticated signal identification analysis procedures allow this uncertainty to be minimized to an acceptable level. Doppler correction can be achieved because the transmitted acoustic signal is a linearly ascending CW tone (chirp signal).

The tracking procedure is iterative. In this procedure, a position is assumed (initially, the launch position) and distances to the sound sources are computed. The difference between these and the measured values are used to improve the initial guess. This process is repeated until the absolute sum of the differences is less than 1 km. This position is then used as the initial guess for the next position in time, and so on.

Several RAFOS floats have been launched in the Gulf Stream since the beginning of 1983. An example of the trajectory of a float is shown in Figure 6.14. The three sound sources, indicated by the numerals 1, 2, and 3, are located south of Cape Hatteras, on the northern slope of the Bermuda Seamount and on top of one of the Southern New England seamounts. The trajectory of the float was determined with sound sources 1 and 2 until year-day 74 and with sound sources 2 and 3 thereafter. The trajectory exhibits the characteristic wavy nature of the meandering Gulf Stream. The mean speed of the float (i.e., the traveled distance divided by the elapsed time) is 55 cm/s. A striking aspect of this and all other float tracks was found to be the tendency for floats to shoal from meander trough to crest and to deepen from a meander crest to the next trough. It was observed that during the float's 2,100 km journey to the east, its lateral displacements relative to the current are less than 100 km. From the pressure record it was noticed

that these motions are not random but clearly a result of the dynamics of curvilinear motion.

It was found that the RAFOS system provides a straight-forward means for studying oceanic variability over a wide range of spatial and temporal scales. Sequentially launched in the Gulf Stream, Rossby and co-researchers used these floats to study the space-time evolution of the meandering current. According to them, deployments of dense clusters of floats would enable study of dispersive processes, and when used in concert with hydrographic surveys, these have the potential for unique studies of mixing, entrainment, and topographic effects. Deployed in large numbers over wide areas, they can also be used effectively to estimate mean flows.

The limitations of the single-mission RAFOS floats (acoustic data downloaded only when the float surfaced at the end of its life) were relaxed by development of a multi-cycle float, the MARVOR (named after the Breton word for a seahorse). MARVOR floats were acoustically tracked but surfaced at regular intervals (typically three months) to transmit the signal arrival times. The process of ascent and descent was achieved by pumping fluid from an internal reservoir to an external bladder (Ollitrault et al., 1994). MARVOR floats were first deployed in 1994 and were used in the South Atlantic SAMBA project and in the Euro float and Arcane projects in the Northeast Atlantic (Gould, 2005).

6.6. ALACE: HORIZONTALLY DISPLACED AND VERTICALLY CYCLING SUBSURFACE FLOAT

Acoustically tracked floats are more logistically efficient when used in dense localized arrays than when providing long records from the widespread low-concentration arrays most appropriate for mapping low-frequency flows. As the number of floats within acoustic range of a single tracking-array element decreases, the fraction of costs devoted to maintaining the tracking array increases.

Motivated by these considerations in exploring large-scale, low-frequency subsurface oceanic currents velocities, Davis et al. (1992) developed a new kind of subsurface float that is autonomous from acoustic tracking networks and is named the Autonomous Lagrangian Circulation Explorer (ALACE, pronounced like Lewis Carroll's Wonderland explorer, Alice). The float periodically increases its buoyancy and rises to the surface, where it is located by the ARGOS satellite system. Subsequently, the float returns to depth to continue water following. ALACE is a subsurface float that cycles vertically from a depth (where it is neutrally buoyant) to the surface where it is located by, and relays data to, System ARGOS satellites. ALACE floats are intended to permit exploration of large-scale, low-frequency currents and to provide repeated vertical profiles of ocean variables. ALACE floats periodically change their buoyancy by pumping

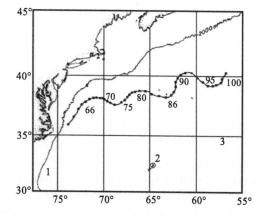

FIGURE 6.14 The 45-day unsmoothed trajectory of a RAFOS float. Elapsed time in year-days (1985) is indicated with one dot per day. The numbers 1, 2, and 3 indicate the locations of the three sound sources. (*Source: Rossby et al., 1986, ©American Meteorological Society. Used with permission.*)

hydraulic fluid from an internal reservoir to an external bladder, thereby increasing float volume and buoyancy. Because positioning and data relay are accomplished by satellite, ALACE floats are autonomous of the cumbersome acoustic tracking networks and are suitable for global deployment in arrays of any size. While providing only a sequence of displacements between surfacing intervals, ALACE floats are efficient in gathering the widely spaced long-term observations needed to map large-scale average flow.

The major challenges involved in the design of ALACE are obtaining high-energy efficiency so that long lifetimes are possible and maintaining adequate surface following so that ARGOS transmissions are reliable. The major components of an ALACE float are shown in Figure 6.15. The three major subsystems of an ALACE float are (1) a hydraulic system to adjust buoyancy, (2) a microprocessor to schedule and control various functions, and (3) the ARGOS transmitter and antenna. An expanded schematic description of the hydraulic system is given in Figure 6.16.

Buoyancy changes are accomplished by moving hydraulic oil from the internal reservoir to inflate an external bladder, thereby increasing float volume and buoyancy, or allowing fluid to flow from the bladder back into the internal reservoir. Ascending to the sea surface is achieved by allowing hydraulic fluid to flow from the internal reservoir through a 25-μm filter to a small motor-driven hydraulic pump, which pumps high-pressure fluid through a one-way

check valve and into the external bladder. The one-way valve prevents reversed flow of high-pressure fluid back through the pump. Descending to the ocean depth is achieved by opening a latching valve, which allows oil to flow from the external bladder (which is at atmospheric pressure or higher) back into the internal reservoir that is maintained at the float's internal pressure of about 0.7 times the atmospheric pressure. The solenoid-actuated latching valve is closed before pumping is restarted. It uses no electrical power to hold in either position. The hydraulic system is completely sealed by the flexible internal reservoir and the external bladder. The external bladder that receives hydraulic fluid is a Buna-N hemisphere. In the float's minimum volume state, the bladder retracts into a hemispheric cavity of approximately the same radius that is machined into the instrument's lower end cap. High reproducibility of the float's minimum buoyancy (and hence equilibrium depth in the submerged state) is obtained because the bladder retracts smoothly into this hemispheric cavity without forming folds or bubbles between the bladder and the cavity surface. The most difficult problem of avoiding loss of prime in the high-pressure pump, caused by air bubbles in the hydraulic fluid, was solved by the use of pumps with a self-priming option. Cleanliness of the hydraulic system remains important for minimizing pump wear and avoiding malfunctions in the latching valve, which must provide a very low leakage rate when closed against high pressure. With the introduction of the self-priming option, the hydraulic system has proven reliable and robust.

The pressure case is a 1-m-long 6061 T-6 aluminum cylinder with 170-mm diameter and 9.5-mm-thick walls. A complete float has a mass of about 23 kg. The upper end cap

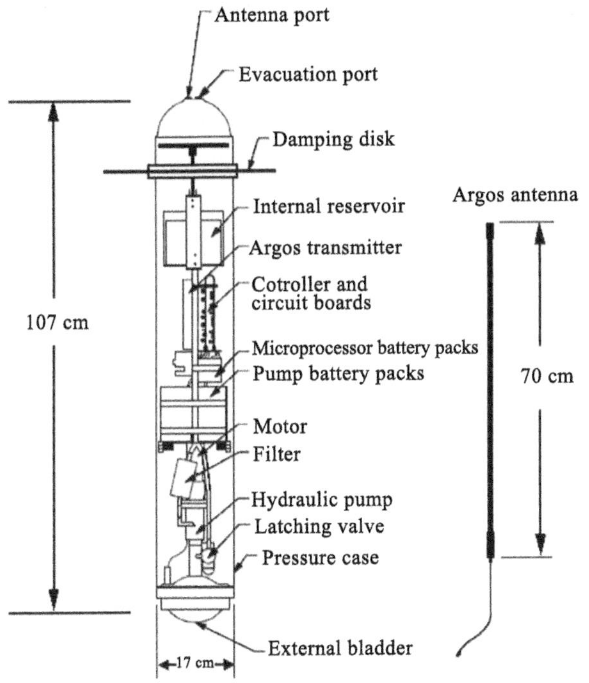

FIGURE 6.15 Schematic of an ALACE float. The antenna shown to the right is mounted on the top hemispherical end cap. (*Source: Davis et al., 1992, ©American Meteorological Society. Used with permission.*)

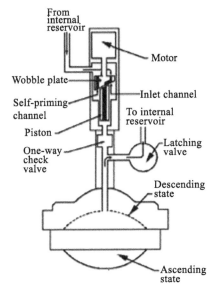

FIGURE 6.16 Expanded schematic of the ALACE hydraulic system. (*Source: Davis et al., 1992, ©American Meteorological Society. Used with permission.*)

is a machined hemisphere (3.8-mm thickness), which in most floats is held in position with epoxy glue, without additional seals. Access to the instrument is through the lower end cap, which forms the hemispheric cavity for the external bladder and is an integral part of the hydraulic subsystem. A plastic disk, of which the diameter is larger than that of the cylindrical float, surrounds the pressure case, perpendicular to the float axis and about 25 cm below the top hemispherical end cap. This disk provides hydrodynamic resistance to relative flow past the float parallel to its axis and damps heaving oscillations when the float is at the surface. This scheme allows the barely buoyant float to follow the wave-disturbed surface and ensures reliable ARGOS communication.

Davis and co-designers of ALACE reckon that a critical component of the instrument is the antenna that must efficiently radiate the 401-MHz ARGOS signal after years of submersion in high-pressure saltwater, dozens of cycles from one to several hundred atmospheres of pressure, and the physical insults of deployment and high sea states. The antenna strength member is a 70-cm-long tapered fiberglass tube that protects the internal radiating element occupying the tube's upper 40 percent. This tube is glued into an aluminum ferrule that screws into the instrument's upper-hemispherical end cap and is sealed with an O-ring. The outside of the fiberglass tube is sealed from the sea by a thermoplastic sheath (shrink tube) bonded to the antenna surface by meltable glue. The electronic and hydraulic components inside the pressure case are mounted to form a single structure that is rigidly attached to the end cap.

The major electronic components include a single-board computer that serves as the ALACE controller, the ARGOS transmitter, and a board for auxiliary functions, such as generating control signals and conditioning analog signals. The microprocessor-based controller's primary function is to schedule all system activities, such as starting the pump motor, opening or closing the latching valve, or starting the ARGOS transmitter. The microprocessor also generates all ARGOS transmitter timing signals and supplies the data message to be transmitted. Reliable ARGOS transmission is critical to achieving the scientific goal of measuring subsurface currents. During operation, when a control function is completed, the microprocessor sets an external programmable alarm clock and puts itself into a low-power dormant state. At the end of the programmed delay, the clock generates a hardware interrupt that wakes up the controller to carry out its next scheduled activity. An internally mounted strain-gauge pressure transducer for which the pressure inlet is exposed to the seawater via a small oil-filled flexible tube mounted on the outside of the end cap measures the submerged depth. An internal thermistor cemented to the lower end cap is used to report at-depth temperature. These data are sent through ARGOS during surfacing of the float.

ALACE is programmed to begin a surfacing cycle by pumping for a specified time, waiting about an hour while the float ascends, and then filling the external bladder to its surface configuration. Typically, about 100 cm^3 more volume than is needed to become neutrally buoyant at the surface is pumped initially, and then the remainder of the internal reservoir's 750-cm^3 capacity is pumped at the surface to provide reserve buoyancy for surface following. The design life of ALACE depends on how far the batteries can be discharged before failing to supply the needed operating voltage.

Due to poor ARGOS satellite coverage, the float positions are not obtained exactly at the surfacing and descent times. This lacuna made it necessary to extrapolate the observed positions to these times. In addition, surface drift must be accurately measured so that measurements of the much slower subsurface motions are not contaminated. Because this requires obtaining as many ARGOS fixes as possible, an important design objective for ALACE was to obtain good surface following so that the ARGOS antenna is rarely submerged. Because ALACE is a float with quasi-linear dynamics, one essential requirement for good surface following is adequate reserve buoyancy. Fortunately, the hydrodynamic resistance to vertical relative flow provided by horizontal damping disks was found to be effective in preventing submergence of the float below the sea surface in the presence of energetic surface waves. Because the reason for good surface following is to make ARGOS communication reliable, the ultimate test of on-surface performance is success in being located by, and sending messages to, Systeme Argos satellites. Quite encouragingly, no connection between wind and either the number of fixes or correct messages could be detected during extensive field trials. It is fortunate that ARGOS communication performance is not influenced by wave state.

Accuracy of subsurface current represented by the subsurface drift of the float, inferred from ARGOS surface fixes, depends on estimating the float positions at the ends of the subsurface sampling period. In addition to extrapolating surface drift to the times of ascent and descent, this depends on rapid transit through shallow regions where current velocities can differ significantly from those at the level of interest. The damping disk used to produce good surface following increases flow resistance and slows vertical motion. Ascent is much more rapid than descent because the buoyant force is intentionally made larger by pumping. On descent, the initial negative buoyancy depends primarily on the equilibrium depth for which the float is ballasted and, to a lesser extent, on the density stratification above that depth. Initial descent is relatively fast, and it takes less than 2 h to reach half the target depth so long as that depth is 2,000 m or less.

A critical element in accurate determination of subsurface drift is extrapolating the observed surface drift to the positions where surfacing first occurs and descent begins. The accuracy

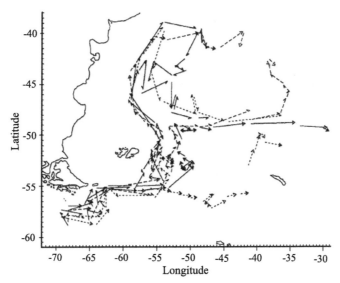

FIGURE 6.17 Displacement vectors from the first year of seven ALACE floats deployed in the Drake Passage. *(Source: Davis et al., 1992, ©American Meteorological Society. Used with permission.)*

with which this can be done depends primarily on the frequency of ARGOS fixes and on the variability of surface currents. Bretherton et al. (1975) used an objective estimation procedure for extrapolation. This procedure is based on treating Lagrangian time series of velocity as samples of an isotropic, stationary random process with known statistics.

After initial field trials in 1988, the first operational use of ALACE floats was carried out in January 1990 for the first World Ocean Circulation Experiment (WOCE) in the Drake Passage (Davis et al., 1996). The floats were ballasted to a nominal depth of 750 m (near the depth of Antarctic intermediate water as it is injected into the South Atlantic) and set to complete a full cycle every 15 days and to remain on the surface for 24 h. The measurements yielded a clear understanding of the circulation route in the region (see Figure 6.17). The WOCE strategy was to make velocity estimates at a common level to provide velocity constraints on the global inverse calculations using hydrography, tracers, and altimetry.

Field experiences have shown that the autonomous float concept is sound and have demonstrated how the data can be used. The potential difficulties in interpretation of data from these floats are (Davis et al., 1992):

1. Trajectory interruptions when the floats rise for satellite locating
2. Uncertainty of the float trajectory between the times it is positioned
3. Contamination of the deep velocity measurements by motion on the surface or during ascent or descent

Fortunately, errors of deep-velocity averages are largely random and much smaller than velocity variability. According to Davis et al. (1992), the procedure of extrapolating surface drift leads to errors of the order 3 km in measuring the displacement between the point at which a float leaves the surface and the point where it resurfaces. The primary contributor to this error is incompletely predictable on-surface motion before the first ARGOS position and after the last fix. These displacement errors correspond to velocity errors of the order 2 mm/s when distributed over the submerged period of 14 days used in Drake Passage and 1 mm/s over the 25-day periods subsequently used in the interior deep ocean. Such errors are largely random because consistent drifts are accurately accounted for by the extrapolation procedure. Developed by Scripps Oceanographic Institute in the mid-1980s and manufactured by Teledyne Webb Research Corporation (WRC), ALACE floats are one of the earlier versions of the floats used in most research today. Although they are still used, there have been improvements in their design.

6.7. DRIFTING PROFILING FLOATS (ARGO FLOATS)

An innovative step initiated in 1999 to take full advantage of the capabilities of drifting profiling floats for prolonged global-scale ocean monitoring gave birth to an international project known as the Array of Real-time Geostrophic Oceanography (Argo). The Argo program was conceived by scientists to greatly improve the collection of observations inside the ocean through increased sampling and increased coverage in terms of time and area via the use of an array of drifting profiling floats known as *Argo floats*. As the name suggests, Argo floats are primarily intended for the study of thermohaline circulation; therefore, apart from drift measurements, temperature, salinity, and pressure are three other essential parameters that are measured by these drifting profiling floats. The drifting and profiling character of the Argo floats allow estimation of geostrophic currents at various depths from repeat profiles of water temperature, salinity, and depth data (density data). Argo's initial aim was to build an array of a real-time, high-resolution monitoring system for upper and middle layers of the world ocean by deploying approximately 3,000 automatic drifting profiling floats, called Argo floats, throughout the ice-free regions of the world ocean, with average spacing of about 300 km.

Three types of drifting instruments that measure ocean temperature and salinity (ALACE, P-ALACE, and SOLO floats) form the backbone of the international Argo program. (P-ALACE is an acronym for Profiling ALACE; SOLO floats are very similar to P-ALACE floats but have better satellite communication and acoustic tracking capabilities). The oceanographic data collected by the Argo floats are primarily intended to understand various oceanographic processes that influence global climate. The original neutrally buoyant floats were designed solely to explore ocean circulation, but Argo floats serve a dual purpose by additionally collecting the CTD profile data. The velocity data from Argo floats have demonstrated enormous potential despite the uncertainties due to their not being acoustically tracked and their departure from being truly Lagrangian due to the time spent at the surface.

As noted in the previous section, Argo floats have the ability to change their own buoyancy, and, when they are deployed, they sink to a prespecified depth, known as *parking depth*. To control the buoyancy of the float, a small amount of oil is contained within it. When the float is submerged, all the oil is kept entirely within the hull. The Argo floats usually remain at the prespecified depth for 7–10 days, drifting with the ambient ocean currents. The float will then rise to the ocean surface, where it communicates its data and position to an orbiting satellite. The float then sinks again, continuing the process. When it is time to rise to the surface, the oil is pumped into an external rubber bladder that expands. Because the weight of the float does not change but its volume increases when the bladder expands, the float becomes more buoyant and ultimately floats to the surface. Similarly, when the float is on the surface and it is time to submerge, the oil is withdrawn from the bladder into the hull of the float and the buoyancy decreases.

Argo floats are built in specialist factories in the United States, France, and Germany. They are built very carefully to work reliably for four years. A float is about 1.1 m tall and weighs around 25 kg. Its body (the pressure case) is made of aluminum tubing sealed at the ends and is strong enough to withstand pressures of more than 200 atmospheres (i.e., the pressure at 2,000 m depth). At the top are the sensors that measure temperature, salinity, and pressure (depth) and an antenna to transmit the data via satellite. At the bottom there is a rubber bladder, which can be deflated to make the float sink or inflated to make it rise under the control of an internal hydraulic system. The pressure case contains electronics,

pumps, and many batteries. The electronics include a microprocessor that stores the data from the sensors until it can be transmitted, a program that controls when the float sinks and rises, and a position-fixing and data transmission system that controls the interaction with the satellite. Each float has a unique number that allows it to be recognized and distinguished from all the other floats. Each float is checked carefully before it is launched. The temperature, salinity, and pressure sensors are calibrated in the laboratory to make sure that the measurements made by the float are accurate. All parts of the system are tested to make sure that the float is working properly. Floats are launched from ships doing scientific research, from large container ships, and sometimes even from aircraft. The floats may be lowered into the water from stationary ships, or they may be packaged into deployment boxes, which protect the floats from water impact when they are launched from moving ships or aircraft. Argo has one float on average in every $3° \times 3°$ area of the ocean that is deeper than 2,000 m and not covered by ice. New floats are needed each year to replace old floats that have stopped working.

An Argo float basically drifts at a depth of approximately 1,000 m at which the float's density is the same as the density of the surrounding water and, therefore, it stays at that level, drifting slowly with the ambient currents. It goes down another 1,000 m every 10 days, and then ascends to the surface, measuring temperature and salinity profiles. At the surface, it transmits the observed data to land-based facilities via the ARGOS satellite system and then submerges again to 1,000 m. The Argo float's measurement cycle is repeated every 10 days. After 150–200 repeats (3–4 years), the batteries are exhausted. With no energy to bring it to the surface, the float drifts until the pressure case corrodes and leaks and the float sinks to the sea floor.

The Argo program allowed, for the first time, continuous monitoring of water temperature, salinity, and velocity of the upper ocean, with all data being relayed and made publicly available within hours after collection. This observation system makes possible real-time monitoring of the ocean conditions. This will greatly contribute to the study of interannual, decadal, and interdecadal variations of the climate system, and it will bring substantial improvement in the performance of long-term forecasts.

The first Argo floats were deployed in 2000. As part of the World Ocean Circulation Experiment (WOCE), 306 autonomous floats were deployed in the tropical and South Pacific Ocean and 228 were deployed in the Indian Ocean to observe the basin-wide circulation near 900 m depth. Davis (2005) conducted a comprehensive analysis of the voluminous dataset obtained from Argo floats using various methods such as area averages, local function fits, and a novel application of objective mapping to estimate the mean circulation. By late 2004, over 1,500 neutrally buoyant floats were drifting at depth throughout the global ocean.

They were approximately 50 percent of the final global Argo array that was scheduled for completion by 2007.

By January 2005, more than 1,600 floats were delivering data. By November 2007 the array was 100 percent complete. The project far exceeded the initial target (3,000 floats), and about 5,000 of them were deployed in different parts of the World Oceans by 2012. Besides float deployment, Argo has worked hard to develop two separate datastreams: real time and delayed mode. A real-time data delivery and quality-control system has been established that delivers 90 percent of profiles to users via two global datacenters within 24 hours. A delayed-mode quality-control system (DMQC) has been established, and 60 percent of all eligible profiles have had DMQC applied.

The accuracy of surface currents estimated from Argo floats is constrained by several factors. First, these floats spend only relatively short periods freely drifting at the sea surface (mostly no more than 15 h in the Pacific). Furthermore, their fixes have relatively large errors (150 m, 350 m, or even 1,000 m), and the interval time between the adjacent fixes varies from several minutes to several hours.

Xie and Zhu (2008) reported the development of a new sequential method of Argo float surface trajectory tracking and extrapolating based on the Kalman filter (KF) method, under the presumption that a surface trajectory of Argo float is dominated by a constant current plus inertial oscillation. This trajectory tracking and extrapolating method is claimed to be able to reduce the positioning uncertainties of Argo surface trajectories and provide error estimations. When this method was applied to extrapolate the Argo float positions at the times of float resurfacing and descending, the estimation error of the mid-depth currents could be reduced. Utilizing this method in the Pacific, surface and mid-depth currents were estimated from surface trajectories of Argo floats from 2001 to 2004, along with their detailed error estimations. The average error for surface currents was found to be about 4.4 cm/s, which is equivalent to the accuracy order (5 cm/s) of the Surface Velocity Program drifters. The estimation error of the mid-depth currents at 1,000 db could be reduced to about 0.21 cm/s without considering the effect of vertical shear. The study of Xie and Zhu (2008) showed that the surface trajectory from Argo float provides a new means to measure surface circulations in the global ocean in real time and that the estimated mid-depth current could be one of the important sources in improving the understanding of ocean dynamics.

6.7.1. Profiling Observations from Polar Regions

Because of a lesser amount of observational data in the Arctic Ocean interior, changes of the Arctic Ocean circulation, oceanographic parameters, and sea-ice conditions remained

unclear. In addition, there is debate over how changes in the Arctic Ocean are affecting global climate, e.g., as related to the global ocean circulation. Even though the Arctic Ocean plays a critical role in global climate (e.g., Morison et al., 2000; Hassol, 2004), sea ice had previously prohibited Argo float observations in the Arctic Ocean. Instead of such instruments as Argo floats, ice-drifting buoys have been the main method of year-round observation in the Arctic Ocean. Compared to hydrographic surveys of the Arctic Ocean by icebreakers or aircrafts, one of the most important advantages of an ice-drifting buoy is that it can obtain data even in darkness and during severe winter conditions. To provide meteorological and oceanographic observation data throughout the year, the International Arctic Buoy Programme (IABP) maintains a network of ice-drifting buoys (e.g., Rigor et al., 2000).

To monitor and better understand the thermohaline conditions in the ocean interior of the polar regions and thus to elucidate Polar Ocean change, Kikuchi et al. (2007) developed a new Argo-type ocean profiling system. This Polar Ocean Profiling System (POPS) is an ice-drifting buoy system that tethers an Argo-type CTD profiler and is deployed in multiyear ice. JAMSTEC and METOCEAN began collaboration to develop POPS in 2004 with the aim of obtaining oceanographic profiling data from beneath the Arctic ice. POPS consists of an ice platform and a subsurface CTD profiler (see Figure 6.18). It also provides

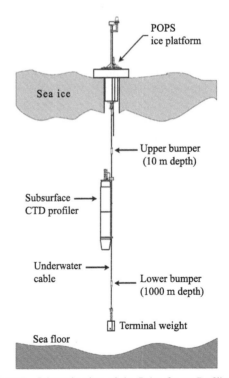

FIGURE 6.18 Schematic view of the Polar Ocean Profiling System (POPS) consisting of an ice platform and a subsurface CTD profiler. *(Source: Kikuchi et al., 2007.)*

meteorological data. Iridium and GPS antennae are located at the top of the meteorological mast, which is placed into a drilled hole in the ice.

The ice platform includes a system controller that manages all data acquisition, processing, formatting, and messaging. The profiler is mounted on an oceanographic cable interfaced to the platform. The profiler moves along the cable between depths of 10 and 1,000 m. The inductive modem system provides data transfer between the ice platform and the profiler. The inductive modem (IM) telemetry system, which includes a Surface Inductive Modem (SIM), an Underwater Inductive Modem (UIM), and two Inductive Cable Couplers (ICCs), is used to establish communication between the ice platform and the subsurface CTD profiler. The SIM is located inside the ice platform, whereas the UIM is inside the profiler. The two ICCs, the plastic-jacketed wire, and the water together make the connection between the SIM and the UIM.

The surface unit power supply consists of two 152-Ah lithium battery packs. The expected lifetime of the ice platform is about 2.68 years for meteorological data acquisition with a GPS position every 3 h and oceanographic profiling data acquisition transferred and processed every three days.

Iridium satellite communication technology sends the observation data and allows remote commands to be sent from the laboratory to the buoy. Data can also be sent to the global telecommunication system (GTS) in real time. Data can easily be accessed from the Argo data server. The system was successfully tested in the Arctic Ocean near the North Pole.

The major difference between POPS and the standard Argo floats is that the POPS profiler is mounted on a cable and slides between a pair of upper and lower bumpers located at 10 and 1,000 m on the cable. The cable is a plastic-impregnated 7×7 strand galvanized wire rope. Figure 6.19 shows how the profiler is mounted on the cable.

The upper and lower riders attached to the profiler are carefully designed to minimize friction and drag when the profiler is moving along the cable. In contrast to the original Argo float, the POPS controller software is modified to cope with the nonsurfacing properties and to interface with the IM system (UIM and ICC) to communicate with the ice platform. Its functions include maintenance of the calendar and internal clock, supervision of the depth cycling process, and activation and control of the hydraulic system.

In addition to the 37-kg weight of the 1,000-m cable itself, a 20-kg terminal weight is used to keep the cable as vertical as possible. Although all these calculations allow full data point collection, much faster ice motion will result in missing data due to tilting of the cable. Placing a heavier terminal weight at the end of the cable is possible in order to minimize the risk of missing data points.

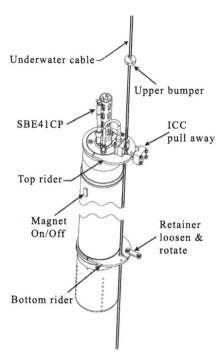

FIGURE 6.19 Sketch of the subsurface CTD profiler and the attachments of the Polar Ocean Profiling System. (*Source: Kikuchi et al., 2007.*)

The total weight of the POPS (ice platform, profiler, cable, and 20-kg terminal weight) is less than 150 kg in air. Therefore, not only a big icebreaker but also a small airplane or helicopter allows accessing the target sites for the POPS deployment on the Arctic multiyear ice. Because of the presence of sea ice, it is necessary to drill a hole through the ice for the POPS deployment. A 10-in. (25.4 cm) diameter hole is enough to put the terminal weight, profiler, and cable into the seawater.

When the profiler has reached and is stabilized at its parking depth, it will establish communication with the platform. The profiler travels at approximately 8–10 cm/s in ascent while traveling at approximately 5 cm/s in descent. The parking and maximum profiling depths are set to 300 and 1,000 m, respectively. This bidirectional communication system provides the users with flexibility and control of the observation after the deployment. For example, it is possible to fine-tune these parameters for optimum performance of the system when ice condition warrants closer attention.

REFERENCES

Bretherton, F.P., Davis, R.E., Fandry, C.B., 1975. A technique for objective analysis and design of oceanographic experiments. Deep-Sea Res. 23, 559–582.

Briscoe, M.G., Frye, D.E., 1987. Motivations and methods for ocean data telemetry. Marine Technol. Society J. 21 (2), 42–57.

Cheney, R.E., Gemmill, W.H., Shank, M.K., Richardson, P.L., Webb, D.C., 1976. Tracking a gulf stream ring with SOFAR floats. J. Phys. Oceanogr. 6, 741–749.

Davis, R.E., 1991. Observing the general circulation with floats. Deep-Sea Res. 38, S531–S571.

Davis, R.E., 2005. Intermediate-depth circulation of the Indian and South Pacific Oceans measured by autonomous floats. J. Physical Oceanogr. 35, 683–707.

Davis, R.E., Webb, D.C., Regier, L.A., Dufour, J., 1992. The Autonomous Lagrangian Circulation Explorer (ALACE). J. Atmos. Oceanic Technol. 9, 264–285.

Davis, R.E., Killworth, P.D., Blundell, J.R., 1996. Comparison of ALACE and FRAM results in the South Atlantic. J. Geophys. Res. 101 (C1), 855–884.

Freeland, H.J., Gould, W.J., 1976. Objective analysis of mesoscale ocean circulation features. Deep-Sea Res. 23, 915–924.

Freeland, H.J., Rhines, P.B., Rossby, T., 1975. Statistical observations of the trajectories of neutrally buoyant floats in the North Atlantic. J. Mar. Res. 33, 383–404.

Gould, W.J., 2005. From Swallow floats to Argo: The development of neutrally buoyant floats. Deep-Sea Res. II 52, 529–543.

Hassol, S.J., 2004. Impacts of a warming Arctic: Arctic Climate Impact Assessment (ACIA). Cambridge University Press, New York, NY, USA.

Kikuchi, T., Inoue, J., Langevin, D., 2007. Argo-type profiling float observations under the Arctic multiyear ice. Deep-Sea Res. I 54, 1675–1686.

Merritt, S.R., Pattullo, J.G., Wyatt, B., 1969. Subsurface currents off the Oregon coast as measured by parachute drogues. Deep-Sea Res. 16, 449–461.

MODE Group, 1978. The mid-ocean dynamics experiment. Deep-Sea Res. 25, 859–910.

Morison, J.H., Aagaard, K., Steel, M., 2000. Recent environmental changes in the Arctic: a review. ARCTIC 53 (4), 359–371.

Niiler, P., Davis, R., White, H., 1987. Water-following characteristics of a mixed-layer drifter. Deep-Sea Res. 34, 1867–1881.

Ollitrault, M., Loaec, G., Dumortier, C., 1994. MARVOR: A multicycle RAFOS float. Sea Technol. 35, 39–44.

Pickard, G.L., Emery, W.J., 1982. Instruments and Methods. In: Descriptive Physical Oceanography: An Introduction. Pergamon Press, New York, NY, USA. pp. 77–124.

Pochapsky, T.E., 1963. Measurement of small-scale oceanic motions with neutrally buoyant floats. Tellus 15, 353–362.

Price, J., Rossby, H.T., 1982. Observations of a barotropic planetary wave in the Western North Atlantic. J. Marine Res. 40 (Suppl.), 543–558.

Rigor, I.G., Colony, R.R., Martin, S., 2000. Variations in surface air temperature observations in the Arctic, 1979–1997. J. Climate 13 (5), 896–914.

Rossby, H.T., Bower, A.B., Shaw, P.-T., 1985a. Particle pathways in the Gulf Stream. Bull. Amer. Meteor. Soc. 66, 1106–1110.

Rossby, H.T., Voorhis, A.D., Webb, D., 1975. A Quasi-Lagrangian study of mid-ocean variability using long range SOFAR floats. J. Marine Res. 33, 355–382.

Rossby, H.T., Dorson, D., 1983. The Deep Drifter: A simple tool to determine average ocean currents. Deep-Sea Res. 30, 1279–1288.

Rossby, H.T., Levine, E.R., Connors, D.N., 1985a. The isopycnal Swallow float: A simple device for tracking water parcels in the ocean. Prog. Oceanog. 14, 511–525.

Rossby, H.T., Levine, E.R., Connors, D.N., 1985b. The lsopycnal Swallow Float: A simple device for tracking water parcels in the ocean. Prog. Oceanogr. 14, 511–525. Pergamon.

Rossby, T., Webb, D., 1970. Observing abyssal motions by tracking swallow floats in the SOFAR channel. Deep-Sea Res. 17, 359–365.

Rossby, T., Webb, D., 1971. The four-month drift of a Swallow float. Deep-Sea Res. 18, 1035–1039.

Rossby, T., Dorson, D., Fontaine, J., 1986. The RAFOS system. J. Atmos. Oceanic Technol. 3, 672–679.

Rossby, T., Levine, E., Conners, D., 1982. The isopycnal Swallow float: A simple device for tracking water parcels in the ocean (abstract). EOS 63. American Geophysical Union.

Stewart, R.H., Joy, J.W., 1974. HF radio measurements of surface currents. Deep-Sea Res. 21, 1039–1049.

Stommel, H., 1949. Horizontal diffusion due to oceanic turbulence. J. Marine Res. 8, 199–225.

Stommel, H., 1955. Direct measurements of subsurface currents. Deep-Sea Res. 2, 284–285.

Sturges, W., 1980. Measurements of currents. McGraw-Hill Encyclopedia of Ocean and Atmospheric Sciences, 329–330.

Swallow, J.C., 1955. A neutral-buoyancy float for measuring deep currents. Deep-Sea Res. 3, 74–81.

Swallow, J.C., Hamon, B.V., 1960. Some measurements of deep currents in the eastern North Atlantic. Deep-Sea Res. 6, 155–168.

Vachon, W.A., 1977. Current measurement by Lagrangian drifting buoys: Problems and potential. Proc. Oceans '77, 46B-1–46B-7.

Voorhis, A.D., 1968. Measurements of vertical motion and the partition of energy in the New England Slope Water. Deep-Sea Res. 15, 599–608.

Webb, D.C., 1977. SOFAR floats for POLYMODE. Proc. Oceans '71 Vol. 2, 44B-1–44B-5.

Weidemann, H., 1966. Oceanic currents and wave recording: Instrumentation. The Encyclopedia of Oceanography 1, 597–599.

Xie, J., Zhu, J., 2008. Estimation of the surface and mid-depth currents from Argo floats in the Pacific and error analysis. J. Marine Systems 73, 61–75.

Zubov, N.N., Czihirin, N.J., 1940. Oceanological Tables. Moscow.

BIBLIOGRAPHY

Anon, 1989. Guide to the Argos system, CLS Service Argos, Toulouse Cedex, France.

Bowditch, N., 1962. American Practical Navigator, Vol. 1. U.S. Naval Oceanographic Office, Washington.

Bryan, G.M., Truchan, M., Ewing, J.I., 1963. Long-range SOFAR studies in the South Atlantic Ocean. J. Acoust. Soc. Am. 35, 273–278.

Frosch, R.A., 1964. Underwater sound: deep-ocean propagation. Science 46, 889–894.

Fuglister, F.C., 1963. Gulf Stream '60, Progress in Oceanography, Vol. 1. Pergamon Press, New York, NY, USA, 265–383.

Hale, F.E., 1961. Long-range sound propagation in the deep ocean. J. Acoust. Soc. Am. 33, 456–464.

Urick, R.J., 1975. Principles of underwater sound. McGraw-Hill, New York.

Webb, D.C., Tucker, M.J., 1970. Transmission characteristics of the SOFAR channel. J. Acoust. Soc. Amer. 48, 767–769.

Chapter 7

Horizontally Integrated Remote Measurements of Ocean Currents Using Acoustic Tomography Techniques

Chapter Outline

7.1. One-Way Tomography   203
7.2. Two-Way Tomography (Reciprocal Tomography)   210
7.3. Acoustic Tomographic Measurements from Straits   212
7.4. Coastal Acoustic Tomography   215
7.5. River Acoustic Tomography   226

7.6. Acoustic Tomographic Measurements of Vorticity   230
7.7. Horizontally Integrated Current Measurements Using Space-Time Acoustic Scintillation Analysis Technique   232
References   235
Bibliography   237

Satellite remote-sensing techniques, employing active and passive optical, thermal, and microwave signals, and coastally operated remote-sensing techniques employing active electromagnetic signals in the HF, VHF, UHF and microwave bands are used on an operational scale for remote detection and quantitative mapping of ocean surface current vectors and circulation patterns. However, the inability of electromagnetic signals to penetrate below the surface layer of the ocean has rendered these techniques unusable for remote measurements of subsurface currents and their circulation features. Physical oceanographers have an interest in subsurface current measurements to gain insight into the water circulation in the ocean layers at various depths and its dependence and possible effects on climatological conditions. Apart from this, subsurface currents are of considerable importance in marine geology because of their influence on the transportation and deposition of sediment. Knowledge of deep currents is also of interest to biologists because of the currents' influence on the dispersal of organisms and the maintenance of supplies of nutrients. Regions of convergence or divergence in the horizontal movements of water mass are of particular interest because of their association with vertical movements in the form of sinking or upwelling.

The traditional means of making observations of subsurface currents was an indirect one, the so-called *dynamical method*, based on highly precise measurements of water temperature, salinity, and depth; the hydrographic tables for computing density; the geostrophic equation; and an assumption regarding the "depth of no motion." The valuable review of Bowden (1954) focuses attention on the assumptions made and uncertainties involved in the dynamical computations regarding the depth of no motion and the mean subsurface current charts. In fact, Stommel's (1955) letter to the editor provides an indication of an almost total lack of knowledge on subsurface currents in the early 1950s.

For lack of proper tools, direct measurements of subsurface currents were limited to those made from current meters tethered from anchored ships, moored current meters, freely sinking/rising vertical profilers, and so on. An example whereby much effort has been expended with a large variety of techniques is the Straits of Florida. Current meter moorings (Lee et al., 1985), Pegasus sections (Leaman et al., 1995), sea-level differences (Maul et al., 1985), and undersea cables (Larsen and Sanford, 1985) have all been employed for measuring the water-current flow there. For a variety of reasons, such limited measurements were not adequate to resolve the spatial structure of deep-water motions. Tracking of subsurface drifters (see Chapter 6) yielded Lagrangian descriptions of subsurface currents. However, these tools were inadequate to provide basin-scale horizontally integrated current measurements.

Copyright © 2014 Elsevier Inc. All rights reserved.

Since the infancy of oceanographic research, oceanographers have been strapped for observations over large regions of the ocean in anything like rapid enough time to get synoptic views or snapshots. If one goes out and surveys a large region of the ocean with a ship, the ocean might have already changed by the time the survey is completed.

The situation faced by oceanographic researchers is quite different from the situation of meteorologists, who can get virtually instantaneous pictures of the atmosphere from either a global measuring network or satellite images of clouds and so forth. They really do get images that are, effectively, snapshots. In contrast, oceanographers are not so fortunate, with the exception of satellite maps of surface temperature, color, and roughness and possibly, satellite altimetry measurements used as sea surface tomography. But none of these measurements extends beneath the surface. It was hoped that acoustic transmissions and receptions could well turn out to be one of the few ways to acquire measurements of what is happening in the ocean over a large spatial scale fast enough that one is not hopelessly mixed up between time variations and space variations.

Ocean acoustic tomography (OAT) is a method employed in measuring the ocean by utilizing the favorable properties of sound propagation through water. The outlines of such a system (i.e., the feasibility of monitoring and ultimately studying the oceans by measuring acoustic transmissions between moorings over large distances) were originally proposed by Walter Munk (Scripps Institution of Oceanography) and Carl Wunsch (Massachusetts Institute of Technology), as provided in Munk and Wunsch (1979). In subsequent years there has been a substantial effort to demonstrate both the practicality of the idea (Spiesberger et al., 1980) and to further analyze the theoretical aspects of sound propagation in this context (Munk and Wunsch, 1982a,b, 1983; Spofford and Stokes, 1984). The theory was further developed and tested to a large part with active contribution from Robert Spindel (Woods Hole Oceanographic Institution).

Ocean acoustic tomography techniques have been developed for remote measurements of large-scale subsurface currents and the associated large-scale circulation features. Similar to the use of X-rays to produce medical computer-assisted tomography (CAT) scans in hospitals to examine the interior of the human body and the use of seismic waves by geophysicists to determine the Earth's internal structure, OAT employs low-frequency sound waves to probe large sections of the oceans. Ocean tomography describes, layer by layer, the interior features of the ocean by transmitting sound waves along many transmitter-to-receiver paths between distantly spaced instrument moorings. The details of the ocean's interior are revealed by interpreting the arrival times of the sounds, since their speed is either accelerated or decelerated by the

temperature and current of the interior ocean. By transmitting sound waves through hundreds of miles of ocean, it becomes possible to make measurements over an area that would otherwise require a fleet of ships working for many weeks. With OAT it becomes possible to take a look at large-scale ocean circulation in a synchronous manner, which is a requirement to better understand a fundamental problem in oceanography. No other technique has the potential to make these ocean-interior measurements. The appeal of acoustical techniques is several-fold (Cornuelle et al., 1985):

1. Tomography techniques are by nature *integrating*, automatically filtering out undesirable small-scale features that "contaminate" normal point measurements. The unwanted features can range from microstructure and internal waves, if one's interest is in the mesoscale (order 100 kilometers), to the mesoscale itself, if one's interest is in the gyre-scale circulation.

2. In principle, the information content of tomographic arrays grows quadratically with the number N of moorings deployed rather than approximately linearly, as with conventional point moorings. It may be noted that over long distances, a source may not be heard by all receivers, and the addition of new instruments thus may not provide information quite as fast as N^2.

3. As a consequence of the waveguide nature of sound propagation over much of the ocean, a single source-receiver pair can provide information about the horizontal average of the vertical structure of oceanic disturbances that would otherwise require large numbers of vertically distributed instruments.

The theory on paper advanced to sea in two tests conducted in the Atlantic Ocean during 1981 and 1983 and subsequent tests funded by the U.S. National Science Foundation (NSF) and the Office of Naval Research (ONR). Each at-sea test became more complicated as the tomography itself grew more sophisticated. There are essentially two kinds of tomography: (1) one-way tomography (called the *zero-order* kind of tomography) and (2) two-way tomography (called *reciprocal tomography*). The former is essentially measuring the temperature of the ocean by measuring the time it takes for sound to traverse from a source to a receiver; the latter measures the difference in time it takes to send a signal from a source to a receiver and back again.

Although the principle of OAT was well founded, its practical realization on an operational scale took a long time because the issues had to do more with engineering developments. The principles had been demonstrated, but it was necessary to develop better, cheaper, very reliable, long-lived sources and receivers. The science that could be developed with the application of acoustic tomography did not advance rapidly enough, because oceanographers had to wait for the more complicated engineering to catch up to the need. The elaborate electronic equipment for the acoustic tomography experiment was not yet commercially available and had to be designed and built by the technologists at various oceanographic research institutions.

7.1. ONE-WAY TOMOGRAPHY

In one-way tomography, the travel time between moorings of the multipaths created by an acoustic pulse is very accurately measured, and the way the multipath travel times change over daily, weekly, and seasonal time periods is monitored. The fundamental simplicity of the idea of acoustic tomography is impressive. Sound propagation in the ocean is described by the equations of classical physics. In acoustic tomography, acoustic ray theory is used to compute sound propagation in a complex environment. It is unique in identifying which part of the ocean (in the intervening space between a source-receiver pair) is sampled by the sound signal corresponding to the arrival of each pulse. With the aid of computers it is economical to simulate measurements prior to the conduct of any field experiment. Ray theory is adopted for computing the trajectory of an acoustic signal based on the assumption that (1) the acoustic wavelength is much smaller than the local water depth, and (2) the change of sound speed is negligible over several wavelengths. Following Arthur et al. (1952), the ray geometry (Figure 7.1) is governed by the following equations:

$$\frac{dx}{ds} = \cos\theta \qquad (7.1a)$$

$$\frac{dy}{ds} = \sin\theta \qquad (7.1b)$$

$$\frac{d\theta}{ds} = \frac{1}{c}\left[\frac{\partial c}{\partial x}\sin\theta - \frac{\partial c}{\partial y}\cos\theta\right] \qquad (7.1c)$$

In these expressions, θ is the angle of emergence of the sound ray, s is the arc length along the ray, and c is the sound speed. ds represents an infinitesimal arc length along the ray, and dx and dy represent the corresponding infinitesimal lengths along the x-axis (i.e., the acoustic ray axis) and the y-axis, respectively. Numerical integration of these equations using available means (e.g., the Runge-Kutta technique) yields the eigen rays. Acoustic rays for various emergence angles are traced using appropriate software packages.

Tomography is based on the notion that if we can understand how sound propagates under specific physical conditions, we ought to be able to understand various aspects of the ocean through which the sound has passed. The concept of time-of-flight acoustic tomography to measure the ocean mesoscale was formulated in the late 1970s. The first demonstration of multipath resolution, stability, and identification through a 900-km propagation test was carried out in 1978. It was found that the individual arriving acoustic rays are identifiable, and that the rays are stable through time. Since then there has been an evolving (and expanding) research program involving at least a half a dozen institutions and many more scientists and engineers to test and evaluate the utility of the idea and to develop practical implementations. The mathematics of "inverse theory" (in many forms) provides us with convenient machinery for understanding the relationship between the measurements and the questions we have about how the ocean behaves (including the uncertainties with which quantities of interest are inferred). Based on these ideas, the measured multipath travel-time data are "inverted" by computer codes to obtain the temperature and current fields of the ocean, between the moorings of acoustic source-receiver pairs. Although the preliminary experiments supported the concept of tomography as a scientific tool, some researchers (e.g., Cornuelle et al., 1985) believed that a three-dimensional test of the method was necessary. It was felt that experience was needed with the second part of tomography—i.e., the "inversion." As outlined by Munk and Wunsch (1979), one needs to take the "data" (i.e., the travel times of individual rays passing through a volume of ocean) and invert it to obtain information about the state of the ocean.

One-way tomography is employed in oceanography mainly to remotely detect mesoscale circulation features such as cold/warm-core eddy structures (known as the *ocean weather*) that are superimposed on a generally sluggish large-scale circulation (known as the *ocean climate*). The ocean mesoscale eddy field is closely analogous in character to weather systems in the atmosphere, but the two differ in terms of sizes and lifetimes. Whereas the oceanic eddy structures are several hundred kilometers in diameter and have lifetimes of a few months, the weather systems in the atmosphere are several thousand kilometers

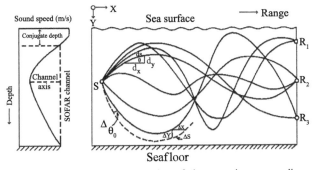

FIGURE 7.1 Schematic presentation of the acoustic ray coordinate system. (*Source: Kumar et al., 1994.*)

in diameter and have lifetimes of a few days. Detection of mesoscale eddy structures in an ocean basin using acoustic one-way tomography relies on the measurement of travel time fluctuations induced by changes in the acoustic field within the ocean by acoustic transmission along many diverse paths. The first test, held in 1981 in areas north and west of Bermuda, involved the making of a three-dimensional, time-evolving map of the area.

In acoustic one-way tomography, sound sources are placed over a specified distance and give off signals. The signals, in turn, are picked up by receivers—essentially omnidirectional hydrophones—which are hung on moorings to record when a signal is received. The received signals can be used to determine water temperature, density, and current. With higher frequencies, bandwidth increases, which, in turn, makes timing of signals more accurate. However, high frequencies tend to be absorbed by the water medium because absorption increases with frequency. This, in turn, diminishes monitoring range. Therefore lower frequencies produce greater range but with less accuracy.

The sound speed in the ocean is predominantly a function of temperature and, to a lesser extent, salinity and water depth (Chen and Millero, 1977). Thus, a cold eddy within the observation region will delay the arrival of any transmission through the eddy, and a warm eddy will cause faster arrival of the acoustic transmission at the receiver. For example, in the Bay of Bengal in the Indian Ocean, an observed cold-core eddy of $5°$ temperature drop that brings about a reduction in the ambient sound speed by 10 m/s delays the travel times by 100 to 200 ms for a mesoscale range (Kumar et al., 1977).

As indicated earlier, the OAT technique involves two aspects, namely (1) the "forward" problem of finding the behavior of sound transmission in an ocean basin over distances of order 100 km, and (2) the "inverse" problem of determining the interior structure of this ocean basin from travel-time measurements of sound waves transmitted through the basin. In the OAT technique, the variable acoustic travel times between all source-receiver pairs of an array of underwater acoustic sources and receivers, moored at spatial intervals of usually more than 100 km, are used to construct the three-dimensional (time-variable) eddy field using inverse theory.

Munk and Wunsch (1979) simulated the inversion of real data by a method chosen primarily for its simplicity. Munk and Wunsch (1983) presented a finite amplitude (i.e., nonlinear) procedure for inverting tomographic data based on Abel transforms. Many superior inversion methods are available, all of which are intimately related (Herman, 1979, 1980). Inversion theories applied in various disciplines of science such as medical tomography, geophysics, and the like may be found in the literature (e.g., Liebelt, 1967; Bretherton et al., 1976; Parker, 1977; Aki and Richards, 1980; Wunsch and Minster, 1982; Cornuelle, 1983; Zlotnicki, 1983), and the possible tradeoffs could only be studied with real data. An experiment conducted by Cornuelle et al. (1985) was the first attempt to use tomography as a full system at sea; it was in large measure an engineering demonstration. They wanted to try this novel system in a region of the ocean that was well understood and where there would be few, if any, surprises that could hamper an evaluation of the procedures. The at-sea experiment was thus deliberately conducted in an area that was congenial to focusing on the technology. Because of the rapid spatial coverage possible with tomography (the entire area was mapped once every three days), the experimenters did gain some valuable insight into the rapid time-evolution possible in mesoscale eddies.

To convert travel-time measurements to sound speed anomaly maps, they used the method often called the *stochastic inverse* in geophysics (Aki and Richards, 1980). Objective mapping in oceanography is a special case of this technique. The method assumes that an *a priori* estimate of the mean and covariance of the unknown field exists, and constructs an estimator which minimizes the square of the difference between the estimate of the unknown field and the true field at each point being mapped. A detailed discussion of this technique is given in Cornuelle et al. (1985).

The main element in any acoustic tomographic experiment is the acoustic transceiver, which functions both as an acoustic source and an acoustic receiver for hearing sounds from other transceivers. Early tomography transmitters were extensions of the technology employed by neutrally buoyant SOFAR floats, namely high-Q, open-end, resonant tubes, approximately one-quarter-wavelength long and driven at one end by a piezoelectric transducer. These devices had sound pressure levels approaching 180 dB re 1 µPa and bandwidths from 16 Hz at a center frequency of about 200 Hz (in 1980) to 100 Hz at a center frequency of about 400 Hz (in 1983). Signal-processing gains of some 35 dB yielded sound-pressure levels equivalent to 215 dB re 1 µPa. However, time resolution with 16-Hz bandwidth is barely adequate, and 100-Hz bandwidths are achievable only at higher frequencies where propagation loss is greater, thus restricting achievable ranges (Spindel and Worcester, 1991). For long-range experiments, hydraulic-acoustic sources, manufactured by Hydroacoustics Inc. (Rochester, New York), with 100-Hz bandwidth centered at 250 Hz, were used. Those sources had sound-pressure levels of 193 dB re 1 µPa, which, together with signal-processing gains, produced an equivalent 228-dB signal. Receivers (which were combined with these transmitters, thus producing transceivers) were equipped with four to six hydrophone vertical arrays to allow multipath discrimination by arrival angle as well as arrival time.

In another design, the sound was produced mechanically by a hydraulic piston that slightly bends an aluminum sheet back and forth, making a 250-Hertz sound equivalent to a low hum. A powerful central microcomputer served as the supervisor of 18 single-chip microprocessors that performed individual tasks. Lithium batteries powered the system. To withstand the tremendous deep-sea pressures, the instruments were housed in inch-thick aluminum cylinders. More technologically advanced instruments were developed in subsequent years.

The great advantage of the OAT technique over conventional point measurements or ship surveys is that the number of data points grows geometrically as the product of the number of sources (S), the number of receivers (R), and the number of resolved acoustic paths (P), compared with the sum ($R + S$) for conventional spot measurements. This concept is demonstrated in Figure 7.2. Additionally, path integration reduces the noise from local fine structure and internal waves that contaminate spot measurements. The superb ability of the OAT technique is achieved essentially because of the transparency of ocean-to-acoustic signals and the presence of an acoustic waveguide known as the *SOFAR channel*. In the ocean there is a minimum sound-speed axis around 1-km depth. The sound speed above this axis gradually increases upward because of increase in temperature, and it increases downward due to the effect of increasing pressure. This gradually increasing sound speed with reference to the acoustic axis corresponds to a gradually decreasing refractive index of the sound channel, because of which the acoustic ray within this channel travels long distances with minimal loss, in a manner similar to the propagation of light rays along an optical waveguide (Figure 7.3). Because the acoustic ray paths oscillate about the axis of the sound channel (i.e., the axis of minimum sound speed) these rays, depending on their inclination, can scan long distances in the vertical plane. It is interesting to note that the steep rays that traverse a large distance of the ocean in the vertical plane arrive at the receiver faster than those flat rays that traverse a much lesser distance in the vertical. Nevertheless, this is an essential peculiarity of acoustic rays traversing in an acoustic waveguide.

The presence of an acoustic waveguide in the ocean helps essentially in the following two ways to simultaneously scan a large ocean volume within a very short time (Behringer et al., 1982): (1) The acoustic rays that are refracted back toward the SOFAR axis before reaching the surface and bottom of the ocean basin lose little energy through the boundaries and, therefore, can be detected over several thousand kilometers; (2) steep rays that sample the entire water column and generally arrive early can be distinguished from flat, late rays that remain nearer the axis, and in this way information can be gathered about the depth dependence of the mesoscale eddy field, which researchers want to detect. Different ray paths give different weights of the water column, and this permits study of vertical eddy structure. Once the travel times of various acoustic rays are obtained (i.e., a direct solution is obtained), the inverse theory is applied to construct a three-dimensional map of the sound-speed field in the scanned ocean basin from *a priori* knowledge of the unperturbed sound-speed field (the reference sound-speed field) in the scanned region.

As part of the Greenland Sea experiment coordinated by the Arctic Ocean Science Board (participating countries included Canada, Denmark, the Federal Republic of Germany, Finland, France, Iceland, Norway, the United Kingdom, and the United States), the Scripps Institute of Oceanography deployed five sets of tomography

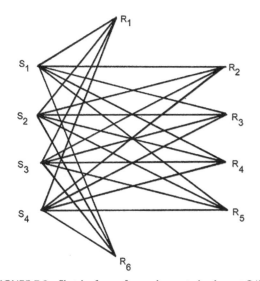

FIGURE 7.2 Sketch of one of several ways to implement OAT.

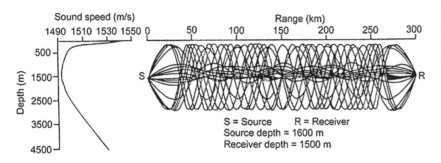

FIGURE 7.3 Propagation of acoustic rays about the axis of minimum sound speed in a SOFAR channel in the Bay of Bengal. *(Source: Kumar et al., 1994.)*

instruments on cables moored to the deep-ocean floor at separations of up to 480 kilometers. Each mooring had a transceiver with four receivers at different depths to increase the number of sound-ray paths. The sounds were emitted in 80-second bursts, which took about five minutes to travel the distance between moorings. Over the course of the year-long experiment, the instruments were designed to transmit for a combined total of about 30 hours. This sampling schedule was expected to be adequate because the large-scale ocean phenomena change slowly.

The acoustic transmission loss is both range- and frequency-dependent. For this reason the transmitted signal frequency employed in OAT measurements involving ocean basin-wide distances generally lies in the range 200–400 Hz and usually has a bandwidth of 2–100 Hz. The acoustic wave source is a resonant tube with a length of approximately one-quarter wavelength (in water), driven at one end and open at the other. The acoustic source transducer usually consists of a cylindrical tube of approximately 0.3 m diameter tuned for resonance at the desired transmission frequency. It is driven at the closed end by a flat circular plate of piezoelectric material (lead zirconate titanate) operated at the fundamental tube resonance. The actual tube length is about 10–20 percent greater than one-quarter wavelength due to finite edge effects (Spindel et al., 1977). The efficiency of the source is roughly proportional to the tube cross-sectional area. The acoustic source transducers are essentially of organ pipe design derived from SOFAR float programs and have lengths of approximately a meter. This type of transducer has severely limited bandwidth and is marginally suitable for tomography because a limited bandwidth of the source transducer will limit the sharpness of the processed received signal, thereby limiting the multipath resolution of the receiver. In the early years of tomographic experiments, there were no alternative broadband transducers available.

Ocean acoustic tomography, the process of deploying spatially distributed arrays of underwater sound source-receiver pairs to produce a picture of the inner sea, has advanced from its relative infancy only in the 1980s to a stage on which it is being proven as a reliable measuring device over long stretches of the ocean. The advancement of this process, thought not too long ago to be theoretically implausible, has proven to be a major advancement in oceanographic research.

Because travel time of acoustic signals traversing over a distance of a few 100 km are to be measured with utmost precision, an essential requirement of the transmitted pulse is that it must contain sufficient power (to traverse long distances) and its width must ideally be very small (so that multipath travel times can be adequately resolved). In practice, there is not really much point in going to narrower pulses because the internal waves in the ocean spread and scatter any pulse that is transmitted. If a perfect delta function is transmitted, the received pulses will still be several milliseconds wide due to internal wave scattering.

Because these two conflicting requirements are difficult to meet with a narrow pulse, an ingenious technique used in OAT is to transmit a signal (sequence) pattern of large width (so that sufficiently large power can be transmitted) for which the autocorrelation function has a very sharp peak with very low side lobes. The width of the correlation peak establishes the achievable multipath resolution of the system. For many purposes the covariance peak can be regarded as though it had in fact been the transmitted pulse. This pulse-compression technique therefore permits pumping sufficiently high power by the transmitter without sacrificing the multipath resolution of the system. In the acoustic tomography experiment conducted by the Ocean Tomography Group in 1982 (Behringer et al., 1982), a single transmission consisting of 24 consecutive sequences lasted for nearly 192 seconds. Each sequence was a 127-digit maximal-length shift register. Thus transmission of each sequence required 8 seconds, and each pulse was of width of approximately 63 ms. This scheme meant that each transmission was equivalent to the transmission of 127 pulses of 63-ms width at intervals repeated 24 times. The carrier was 224 Hz with a bandwidth of 20 Hz, and the transmitted power level was approximately 14 Watts. At the receiver, the arrival time structure of the multipath field between each source and receiver pairs was obtained by cross-correlating the coherently received incoming signal with a stored replica of the transmitted signal. Some receivers perform this cross-correlation *in situ;* others store signal samples for later onshore processing.

The validity of the interpretation regarding mesoscale eddy fields using OAT techniques depends to a large extent on the precision with which the acoustic pulse travel time is measured. For this reason, an essential requirement of OAT instrumentation is the incorporation of a precision time base. Because the time base derived from quartz crystal oscillators cannot provide the required long-term precision and repeatability, rubidium atomic frequency standards must be used. However, the power requirement of rubidium atomic frequency standards is too high for continuous use in a long-term moored instrumentation such as that of OAT measurements. To meet the stringent requirement of high precision in time-base and low-power consumption, a technique that is usually employed is to switch on the highly stable rubidium atomic frequency standard only periodically (so that power consumption is reduced) and compare the frequency of the considerably less stable (but low power-consuming) crystal oscillator clock frequency with the rubidium standard to get the frequency offset of the crystal oscillator clock. The periodically measured frequency offsets of the crystal oscillator clock are then integrated to yield time corrections.

The receivers are usually equipped with a vertical array of hydrophones separated at a suitable spacing so that the ray inclinations of various multipath arrivals can be computed. The pulse arrival pattern for each ray is predicted using ray theory from *a priori* knowledge of the sound speed in the ocean basin being surveyed. The measured deviations are attributed to perturbations in sound speed along the unperturbed ray paths due to the presence of eddies, meanders, fronts, and so on. The three-dimensional shape and nature of the travel-time perturbations can be obtained from analysis of multipath arrivals from many combinations of acoustic transmitter/receiver pairs in the horizontal array, moored near the axis of the sound channel (approximately 1-km depth). The measured travel-time perturbations are then used to generate the perturbations in the sound-speed pattern in the ocean basin and construct a tomographic picture of the perturbations within the array.

In the first major application of OAT, Cornuelle et al. (1985) chose a substantial (300 km × 300 km square) volume of ocean over four months east of Florida in the western North Atlantic and just south of the region in which the MODE-1 program had been conducted. In this experiment, each source was turned on at one-hour intervals for 24 hours and then shut down for two days so that observations were obtained only every third day. Each time the source was turned on, it transmitted a phase-coded, linear maximal shift register sequence on a 224-Hz carrier (62.5 ms long pulses). Several images of this transmitted sequence were received and averaged by each receiver. Each transmission was cross-correlated with a stored replica of the transmitted sequence, and a best estimate was made (see Spindel, 1979) of the time of arrival of each ray. Two corrections were applied to the raw arrival times. Corrections were made to account for clock drift in the sources and receivers (the procedure is described by Spindel et al., 1982). The second correction comes from changes in mooring position. With acoustic instrumentation mounted 3.5 km above the seafloor, the movement of the instruments in the ambient current field introduces changes in distance between sources and receivers, leading to changes in travel time, which can swamp those due to the mesoscale sound-speed changes. Transponders on the bottom (in a system described by Nowak and Mealy, 1981) determined the temporal variations in the three-dimensional positions of the acoustic instruments.

The goal of the experiment was to make the best possible estimate of the three-dimensional sound-speed field in the ocean volume monitored by the acoustic array. Application of appropriate inversion procedures yielded contours of the sound-speed anomaly (Figure 7.4), which primarily corresponds to water temperature anomaly, revealing a clear *pattern* of eddy structure (an important water circulation and vortex feature) in agreement with the

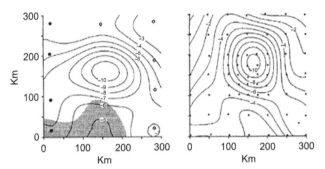

FIGURE 7.4 Contours of sound speed anomaly in a 300 × 300 km square region east of Florida in the western North Atlantic, revealing a clear *pattern* of eddy structure (an important water circulation and vortex feature) in agreement with the direct observations within computed mapping errors. (*Source: Cornuelle et al., 1985, ©American Meteorological Society. Reprinted with permission.*)

direct observations within computed mapping errors. This shows that from daily snapshots of the acoustic travel-time perturbations, it is possible to identify cold and warm eddies and determine their growth, change in shape, speed, and direction of their translatory movement and their ultimate decay. Tomography is fundamentally an integrating measurement, and the computation of the spatial averages is immune to noise.

Several tomographic experiments in the 1980s have indicated that mapping of mesoscale gyres in the ocean can be performed tomographically over large areas. The integrating properties of long-range sound transmissions are now permitting true ocean basin-scale measurements without being perturbed by local small-scale influences. The proponents of the OAT scheme to monitor large ocean basins for mesoscale fluctuations have envisaged real- or near-real-time use of the data for operational ocean-monitoring programs. The spatial resolution could be greatly improved by well-positioned autonomous offshore listening stations and data transmitted periodically, via telecommunication links. With the advent of miniature transmitters and receivers, such data transmissions would be possible.

As with many oceanography techniques, tomography has relied heavily on the internal storage of data in moored instruments, which means that one doesn't know how well the instruments did or what was seen until recovery. To enable a researcher to check on an acoustic tomography experiment, use the data in real time, and talk to the instrument (for instance, giving it global positioning system time), James F. Lynch and his team at Woods Hole (see Lynch, 1995) designed a surface telemetry buoy, known as an *S-tether system*, which is usable with deep-ocean acoustics instruments. The S-tether system can be deployed at any remote deep-ocean site and be amply powered. The concept, in which one uses a flexible S-shaped cable link from a subsurface float to a surface buoy to decouple the surface buoy's motion from the instruments, has been utilized previously with current meters, which are also sensitive to surface-buoy induced motions. S-tether TOMO is a transfer of this design to acoustic devices, which are even more sensitive to motion effects.

The first trial of the system was in the context of the Thetis experiment in the western Mediterranean, which was a large multinational tomography experiment program aimed at both exploring the oceanography of the western Mediterranean and developing large-scale, long-term monitoring capabilities usable in a wide variety of ocean basins. The mooring was deployed successfully, along with six other moorings, in January 1994, for a 10-month period. However, the telemetry system met with only partial success. Local fishing activity off Mallorca severed the telemetry-link electrical connection between surface buoy and subsurface instruments halfway through the experiment. However, both the surface buoy and subsurface buoy still worked autonomously, which allowed achieving the full scientific objectives.

Functionally, the S-tether mooring has four parts: the surface buoy, the S-tether cable connection from the surface to subsurface float, the subsurface float, and the "standard mooring" down from the subsurface float, which contains the instruments. The surface buoy was a 56-inch diameter, 1,228-pound, syntactic-foam float with a buoyancy of 1,895 pounds. It was designed to be submerged for considerable periods (e.g., covering storm periods) and to reemerge in working order. Its survival depth, based on launch and storm conditions, was rated at 300 meters. The float had a 10-inch × 5-foot-long electronics well that accommodated batteries, a computer, and Argos and GPS units. Externally, the float carried Argos and GPS antennae, solar panels, and recovery aids, all of which were protected from seawater leakage and pressure to survival depth.

The S-tether connection between subsurface buoy and surface buoy was formed by a heavy-jacketed electromechanical cable and a section of compliant rubber stretch hose with an internal electrical conductor path. Through careful distribution of flotation along the electromechanical cable, the cable assumes the shape of a tilted letter S, the stretching of which decouples the surface buoy wave motions from the subsurface buoy. In this configuration, the motions of the tomography source and the hydrophone array are minimized. Under the influence of increasing wind and current drag, the S shape of the cable is gradually straightened out. Thus, even an inclined, linear path provides considerable cushioning from the surface buoy's motion.

The Webb Research Corporation (Falmouth, Massachusetts) tomography transceiver, which was positioned 50 meters below the subsurface float, consisted of a 400-Hz organ pipe source, a four-element hydrophone array, internal battery and electronics, and an external mooring motion navigator. This instrument was placed relatively near the ocean surface to take advantage of the upward refracting acoustic propagation conditions (surface sound channel) peculiar to the Mediterranean. The equipment below the transceiver included temperature sensors, an Aanderaa Instruments (Bergen, Norway) current meter, a Benthos Inc. (North Falmouth, Massachusetts) acoustic release, and a 6,000-pound anchor.

To transmit at low power and with high efficiency, the acoustic pulses are actually sent out as extended signals (pseudo-random noise, FM sweeps, etc.) that are replica correlated at the receiver to recreate the sharp-pulse multipath structure.

Additionally, we also need acoustic navigation data, tomography-receiver "housekeeping" data, ambient noise data, GPS position data, and (very important) internal instrument clock versus GPS clock data (very accurate clocks are necessary in tomography). Unaccounted-for clock drift is akin to an ocean travel-time signal, causing error in the estimates of oceanic water temperature and current. One of the big advantages of having a surface buoy is the access to a GPS clock.

Digital communications between the transceiver and surface buoy processors were implemented with a low-power, 1,200-baud FSK modem using two conductors. A simple communication protocol based on the oceanographic standard serial ASCII instrumentation loop (SAIL) was implemented to permit the processors in the transceiver and surface buoy to interact and communicate.

Real-time data telemetry was accomplished by transmitting selected, processed tomography and engineering values via the Argos data collection system. To maximize data throughput, a Seimac Ltd. (Dartmouth, Nova Scotia, Canada) PTT was configured with four IDs under software control, each transmitting four 32-byte multiplex packets. In this way it was possible to transmit 512 bytes of data in about 800 seconds, a time comparable to a satellite pass. By updating the buffer twice per day, a total of 1,024 new bytes of data were telemetered daily. Each data buffer was transmitted 54 times, virtually ensuring that each of the 16 data packets would be received at least once. A checksum was used in each transmission to confirm error-free status. The transmitted data consisted of acoustic correlates and times (tomography results), acoustic navigation data, tomography-receiver housekeeping data along with ambient noise data, time offset (internal clock versus GPS clock), and GPS data. With the advent of high-bandwidth satellite systems (such as the Iridium system), high data rates have become possible. Incorporating such schemes, the thorniest problems in using telemetry systems for high-density data reporting have become past history.

With more and more importance having been attached to acoustic tomography, serious attempts have been made to develop better transducers and the related interfacing instrumentation in place of those borrowed from the SOFAR float era. Because the hitherto existing instruments made by Woods Hole, Webb Research, and the Institute Français de Recherche pour l'Exploitation de la Mer (IFREMER) were not adapted to basin-scale studies (about 1,000 kilometers), in 1994 IFREMER launched the development of a new modular instrumentation for mesoscale OAT measurements. The new autonomous instrument consists of a wideband unlimited-depth acoustic source (Gac et al., 1999) developed and commercialized by ERAMER (Toulon, France), a high-efficiency class-D power amplifier, a programmable multifunction receiver, a long-baseline positioning system, and a low-power, high-stability clock.

In view of the requirement for high-efficiency, low-frequency transducers for operation at great depths, studies conducted in collaboration with the Centre Militaire d'Oceanographie (CMO; Brest, France) indicated that the Janus-Helmholtz technology was the best candidate for basin-scale experiments. For precise measurement of the ocean impulse response, OAT requires a low-frequency acoustic source with a large bandwidth (high time-resolution) and an energetic output sound level. Janus-Helmholtz transducer technology was well suited to long-range OAT applications. With a working central frequency of 400 Hz, 600-km ranges were expected for the initial development. In a simultaneous attempt, a 1-gigabyte storage capacity autonomous receiver capable of operation under various frequency bands was also developed in collaboration with ORCA Instrumentation (Brest, France). Fully compatible with the acoustic source, this receiver system was found to be usable for standard OAT experiments. The new instrument consisted of a 250-Hz Janus-Helmholtz acoustic source, a high-power transmitter associated with many acoustic-processing signal functions, a programmable multifunction acoustical receiver, and a processor with a 1-gigabyte storage capacity.

With very good stability over a large operating temperature range, this new instrument detected and transmitted low-frequency acoustic waves in great water depth. To separate different rays' travel time, the time resolution must be at least 10 milliseconds. Ambient noise in the ocean is a natural limit to the precision that can be attained, but other kinds of effects, such as internal waves scattering and ray interference, would decrease the measurement accuracy. Optimization of the transmit power and the frequency resolution, which define the working characteristic of the transmitter, was found to be the most important criterion in designing and building a Janus-Helmholtz transducer for OAT requirements. The transducer was designed to match, with maximum efficiency, the output level in order to yield a flat bandwidth corresponding to high time resolution. The updated instrument was developed to fulfill the constraints of reliability, low power consumption, and low cost.

The low-frequency Janus-Helmholtz acoustic source (JHAS) is made up of a piezoelectric ceramic stack inserted between two similar head masses. This structure, called a *Janus driver*, is mounted inside a vented cylindrical housing, and the decoupling between head masses and housing is provided by a very thin slit. A fluid with a low-compressibility modulus is inserted inside the cavity in order to satisfy the Helmholtz resonance condition to work at low frequencies and to have a free-flooded device. Because of the coupling of the two resonances, a wide frequency band is available. According to Gac et al. (1999), two frequency bands are usable for OAT applications:

- One between both resonances in order to have a large bandwidth with almost constant impedance values and to be independent of hydrostatic pressure (this was verified from deep-sea measurements in the Mediterranean Sea)
- Another around the second resonance in order to have large transmitting voltage response (TVR) values with high electroacoustic efficiency, allowing small voltage values even if frequency is higher than in the first case

It was found that a working frequency band located between both resonances (frequency bands located around 250 and 400 Hz) was a better way to fulfill bandwidth requirements and to have a constant sound level at any depth, but an improvement of the electrical behavior was then necessary. Due to the length of the ceramic stacks, it was not possible to strongly modify the parallel capacitance (C_p) of the JHAS. The only way to optimize the electrical Q-factor and to minimize the power consumption was to have a better coupling between both resonances, which implies an increase of the TVR values in this frequency band (i.e., a decrease of the voltage values) and a decrease of the parallel resistance values (R_p). Headmass shape and the opening between cylindrical housings were then modified (ATILA finite-element modeling). Because these modifications led to a frequency shift, the length of the driver had been extended. The JHAS TVR was improved: A 7-dB gain was obtained between the resonances, the electrical Q-factor was halved in the working frequency band, and the electroacoustic efficiency was improved. Thanks to the use of more fluid inside the cavity, the in-water transducer weight remained unchanged. A study of the impedance-matching circuit revealed that sound level and bandwidth requirements to reach 1,000 kilometers ranges in the ocean basin with enough time resolution (maximum sound level equaling 190 dB with a 70-Hz bandwidth) could be satisfied with a 1,500-volt-ampere class-D power amplifier. The new transducer and power amplifier were reportedly nearing completion.

The whole system (i.e., transducer and the electronics interface) was proved during the Cambios oceanographic expedition in the Atlantic at a 600-meter depth. The various time delays can be distinguished according to the different travel paths.

7.2. TWO-WAY TOMOGRAPHY (RECIPROCAL TOMOGRAPHY)

Having succeeded in making of three-dimensional, time-evolving map of large-scale circulation features (e.g., gyres), the next phase was to test *reciprocal tomography*. Reciprocal tomography was expected to be a significant development in oceanographic research. It would improve the ability to make acoustic predictions for several operational purposes.

In a broader sense, reciprocal tomography is indicative of oceanographers' attempts to overcome the limitations they traditionally have faced. An approach to the problem of remote measurement of large scale oceanic motion using reciprocal acoustic transmission method was advanced by Stallworth (1973) followed by Rossby (1975). The central element of this scheme is that the line integral of fluid velocity along an acoustic ray joining two points in a fluid flow field is proportional to the difference in travel times of two acoustic signals simultaneously transmitted from these two points in opposite directions. Via this method, the effects of ocean currents on acoustic propagation can be separated from the effects of sound speed structure (i.e., influence of seawater temperature, salinity, and depth). Reciprocal acoustic transmissions can, therefore, be used to measure ocean currents. One of the unique advantages in using acoustic techniques to measure large-scale oceanic phenomena is that they enable an integral or spatially averaged measurement. For many purposes it is the spatial averages that are of interest, and these are extremely difficult to obtain over large ocean areas in any other way.

The basic premise of using reciprocal acoustic transmissions to measure ocean-current flow is that a sound pulse traveling with a current is faster than the one traveling against a current. This method has been routinely used by meteorologists as early as the 1960s in acoustic anemometers (Kaimal, 1980) and later by oceanographers for Eulerian current measurements (Gytre, 1976). Thus, it is well known that the path-averaged sound speed and water current velocity can be measured separately using the reciprocal transmission method. When the direction of water current u is taken from station S_1 to N_1, the travel time of sound propagating from S_1 to N_1 is expressed as:

$$t_1 = \int_\Gamma \frac{dr}{c(r) + u(r)} \qquad (7.2)$$

Similarly, the travel time in the opposite direction is:

$$t_2 = \int_\Gamma \frac{dr}{c(r) - u(r)} \qquad (7.3)$$

In Equations 7.2 and 7.3, $c(r)$ and $u(r)$ are the sound speed in seawater at rest and water current velocity along the ray path Γ, respectively. In Equations 7.2 and 7.3, dr is the increment of arc length measured along the acoustic ray, and the path integrals are taken along acoustic rays. It may be noted that the acoustic ray path is assumed to be overlapped for the reciprocal course. When c_m, u_m, and R are considered as the path-averaged sound speed in seawater at rest, path-averaged water current velocity, and path length, respectively, Equations 7.2 and 7.3 get modified, respectively, as:

$$t_1 = \frac{R}{c_m + u_m} \qquad (7.4)$$

and

$$t_2 = \frac{R}{c_m - u_m} \qquad (7.5)$$

Equations 7.4 + 7.5 and rearrangement yields:

$$c_m + u_m = \frac{R}{t_1} \qquad (7.6)$$

$$c_m - u_m = \frac{R}{t_2} \qquad (7.7)$$

Equations 7.6 + 7.7 and rearrangement yields,

$$c_m = \frac{R}{2}\left(\frac{t_1 + t_2}{t_1 t_2}\right) \qquad (7.8)$$

Taking into account that $t_1 \approx t_2 \approx \bar{t}$,

$$c_m \approx \frac{R}{\bar{t}} \qquad (7.9)$$

Equations 7.6 − 7.7 and rearrangement yields:

$$u_m = \frac{R}{2}\left(\frac{t_2 - t_1}{t_1 t_2}\right) \qquad (7.10)$$

Putting $\Delta t = (t_2 - t_1)$ and taking into account that $t_1 \approx t_2 \approx \bar{t}$,

$$u_m = \frac{R}{2}\left(\frac{\Delta t}{\bar{t} \times \bar{t}}\right) \qquad (7.11)$$

Taking into account (see Equation 7.9) that $\bar{t} = \dfrac{R}{c_m}$, Equation 7.11 gets modified as:

$$u_m = \frac{c^2{}_m}{2R}\Delta t \qquad (7.12)$$

An average sound-speed profile of the ocean basin at which the reciprocal acoustic transmissions are carried out, appropriate for the time during which the experiments are conducted, is usually constructed by combining data from a series of expendable bathythermograph (XBT) casts and salinity measurements taken along the lines joining different moorings of the transceiver array. Measurement of mean flow components along two or more different axes permits estimation of the mean water-current flow vector. In an ocean basin-scale measurement scenario using acoustic techniques, where many pairs of transceivers are usually deployed in a large array, it is possible to estimate the mean water-current flow in an ocean basin using this technique.

As indicated earlier, when the reciprocal acoustic transmission technique is applied for measurement of large-scale oceanic water current flow, the differences in travel times of the oppositely traveling pulses are interpreted in terms of acoustic ray-averaged currents. In the simplest approach, it is assumed that sound speed and water current depend only on the horizontal coordinate along a straight-line ray between a given source-receiver pair. Preliminary investigations by Worcester (1977) suggested that differences in travel times of oppositely traveling pulses can be interpreted in a preliminary manner as ray-averaged currents. Because a number of distinct ray paths with a variety of turning depths exist for each source-receiver pair, and each ray represents a different depth-weighted average of the ocean, the ocean basin currents estimated using this technique represent baroclinic (depth-dependent) spatially averaged current fields. However, methods now exist to estimate basin-scale current fields in several horizontal layers of the ocean depth, which yield barotropic (depth-independent) spatially averaged current fields.

The method employed for generation of acoustic transmission signals and processing of the received signals in the reciprocal acoustic transceivers is similar to those used in acoustic tomography experiments. Before deployment, the source-receiver pairs are usually connected in the same loop for clock synchronization purposes. The transmission signal is usually a pulse stream consisting of period repetitions of a phase-coded linear maximal shift-register pseudo-random sequence. The advantage of such a transmission signal code is that it can be processed at the receiver to yield an output waveform that has minimum side lobes (see Figure 7.5). It has the additional advantage that the processing can be easily implemented on a microprocessor. This scheme enables the signal processing to be performed *in situ* in the instrument itself, thus conserving memory space.

Reciprocal tomography received its first at-sea test in August and September 1983 in a 300-km area below the southern portion of the Gulf Stream. To measure the currents in that area, scientists from Scripps and Woods Hole placed moorings 300 km apart and estimated the intermooring travel time differences. The frequency for the source was set at 400 Hz during the test.

In the reciprocal acoustic transmission experiment conducted by Worcester et al. (1985), the transmission signal code consisted of a carrier frequency of 400 Hz. The transmission

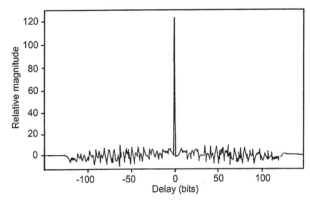

FIGURE 7.5 Autocorrelation of pseudorandom *n*-bit shift register sequences used in reciprocal transmission measurements. The autocorrelation exhibits a triangular peak at zero lag. *(Source: Menemenlis and Farmer, 1992, ©American Meteorological Society. Reprinted with permission.)*

length was 122.64 seconds, consisting of 24 sequences of 5.11 seconds. The phase-coded 400-Hz digital signal is amplified, usually by a constant-power amplifier. After necessary filtering and impedance matching, the signal is impressed on the drivers of the acoustic transducers. The transducers are resonant tubes, driven by a pair of flat piezoelectric elements inserted at the midpoint of the tube, with an effective length of one-fourth the acoustic wavelength.

In any typical ocean-basin experiment, the expected differential travel time is only a few milliseconds. For this reason it becomes necessary to have a clock with nanosecond precision for the several-month duration of the experiment, requiring an oscillator accurate to better than one part in 10^{10}. In the basin-wide experiment of Worcester et al. (1985), a two-oscillator system was employed to achieve this precision at a reasonable level of power consumption. In this system, a low-power (10-mW) temperature-compensated crystal oscillator (TCXO) ran continuously to drive the clock. A high-power (13-W) rubidium (Rb) atomic frequency standard, which returned to its previous frequency to within two parts in 10^{10} within 10 minutes after power is applied, was turned on at 6-h intervals. The frequency offset between the Rb oscillator, after permitting its warm-up, and the TCXO was used in a feedback circuitry to readjust the TCXO frequency. This feedback technique increased the effective stability of the TCXO by approximately one order of magnitude. Any leftover frequency offset that still existed was measured over a 2-minute interval with a precision of one part in 10^{10} using a phase-comparison technique.

At the receiver section, the signal reception is initiated by the processor at preset times, computed by adding to the programmed source transmit times the expected propagation delay for the nominal range. The received signal is amplified and filtered using a band-pass filter centered at the transmission frequency. Two quadrature components of the filtered signals, representing the real and imaginary parts of the complex demodulated signal, are generated, low-pass filtered, and then digitized for further "sharp" processing by the microprocessor, to enable detection of successive peaks and their arrival times. Several peaks in the processed received signal arise from multipath signals corresponding to differing acoustic ray paths through the ocean basin.

To resolve all the multipath arrival peaks, including those arriving simultaneously, a vertical array consisting of several hydrophones is required in the receiving transducer assembly. The vertical distance between adjacent hydrophones is maintained at 1.5 times the acoustic wavelength. In this case, processed data from each hydrophone channel are recorded separately so that beams can be formed at any desired angle during post-recovery processing. This will, in addition to improving S/N ratio, permit estimation of the vertical arrival angles of the several acoustic ray paths that impinged on the receiving hydrophone array, thus assisting multipath ray identification by separating simultaneous arrivals from different angles. To perform an inversion of travel-time data, each arrival must be associated with a particular ray path. Furthermore, ray identification is useful in performing inversions to convert travel-time differences to ocean current structure as well. In this sense, use of a vertical array of several hydrophones rather than a single hydrophone assumes special significance. Usually, all signals arrive within ±15° of the horizontal if mooring motion is negligible. Fortunately, mooring motion does not seriously affect the two-way travel times in reciprocal acoustic transmissions (velocity tomography), although mooring-motion correction is most important in the one-way travel times used in acoustic tomographic measurements (density tomography). This is because, in the case of reciprocal transmissions, the differential travel time is directly proportional to the ray-averaged current with respect to the mean motion of the transceivers (Worcestor, 1977). One probable source of error in the inversion of differential travel times to obtain currents arises from the influence of current shear in causing the acoustic ray paths to differ with and against a current. Sound pulses traveling in opposite directions, therefore, do not sample precisely the same part of the ocean (Worcester et al., 1985). Fortunately, by virtue of Fermat's principle, travel time is unchanged to first order in small perturbations in the ray paths. Effects associated with the nonreciprocity of ray paths are expected to be small if the sound-speed gradient and current gradient are comparable.

7.3. ACOUSTIC TOMOGRAPHIC MEASUREMENTS FROM STRAITS

Water currents in straits (narrow passages of water connecting two large water bodies such as basins of oceans or marginal seas) are typically highly variable in both the horizontal and the vertical. Straits potentially provide important observation sites for various applications. For example, the net transports of mass, heat, and salt through such passages give integrals of the fluxes over the interior basin. Alternatively, the flow through a strait may represent a control for interior processes or forcing and thus is of interest for the functioning of the basin or as a boundary condition for modeling studies (Send et al., 2002). The highly variable nature of currents in straits makes it difficult and expensive to obtain reliable, long-term transport observations using point measurements. An adequate current-meter array, for example, typically requires a sequence of current-meter moorings across a strait, spaced closely enough in the horizontal to resolve the cross-strait current scales and with current meters spaced sufficiently densely on each mooring to resolve the vertical scales.

For long-term observations, one would prefer a shore-based observing system. Among those systems, sea-level differences integrate only the surface currents across the section. Electromagnetic (cable) methods (see Chapter 2) seem promising for observing barotropic currents but are subject to some side effects and are not suitable for observing the exchange in two-layer currents. The acoustic transmission method (e.g., transmission from one side of a strait to the other) offers some promise because this method inherently integrates horizontally over the flow and provides information on the flow along the path bounded between the acoustic instruments, without the need to deploy instruments in the interior of a strait. A variety of methods are theoretically possible for the use of acoustics in this context.

It has been found that in the deep ocean, the acoustic arrivals are stable and the received acoustic pulses can be resolved and identified with ray paths. However, acoustic transmission in shallow-water regions such as straits, coastal water bodies, and estuaries is quite different from that in deep oceans. An axis of minimum sound speed exists in deep oceans, but such an axis does not exist in shallow waters, where sound speed typically decreases with depth so that ray paths are refracted downward, and propagation to moderate ranges (of the order of a few kilometers) necessarily involves bottom bounces (De Ferrari and Nguyen, 1986). Further, because of the loss caused by bottom interactions, long-range acoustic transmission in shallow water is not possible. Although long-range acoustic transmission paths in the deep ocean are mostly refracted/reflected (RR) paths, the transmission paths in shallow-water acoustic transmission are usually refracted/bottom-reflected (RBR) paths and surface-reflected bottom-reflected (SRBR) paths. Those arrivals traveling via the SRBR paths suffer high losses due to high-angle interactions with the ocean surface and bottom so that pulse responses are dominated by the RBR arrivals. The earliest arrivals are those that follow the steepest RBR paths and reach the higher sound-speed region of the upper ocean. The latest arrivals are those that travel along the flattest RBR paths, propagating in the slower sound-speed near-bottom water. These observations indicate that deep-ocean tomography methods may not be successful in shallow-water applications. Deep-ocean tomography inversion techniques require that the individual eigen rays be separable in time, identifiable, and have known paths through the ocean. However, in shallow areas at long ranges, unresolved multipaths will be the rule (Muller et al., 1986).

A major difficulty with acoustic tomographic and reciprocal measurements in shallow water is that multipath interference rapidly changes the pulse shape and phase. Furthermore, acoustic multipath pulse arrivals overlap and form groups. Measurements by Ko et al. (1989) in the Florida Straits showed that these groups are generally not consistent and often cannot be resolved. This means that in shallow water, it is difficult to identify acoustic arrivals with particular ray paths. However, shallow areas are often of great oceanographic interest.

To surmount some of the difficulties associated with long-range acoustic propagation in shallow-water basins and to achieve precision in measurement, one approach is to measure the pulse response of the acoustic channel with high resolution and then attempt to use multipath groupings, correlation methods, or phase information to resolve travel times.

Experiments by De Ferrari and Nguyen (1986) in the Florida Straits indicated that a convenient depth average is associated with those rays forming the late peak, and the arrival time of the late peak is a measurable and consistent feature of the data. The resulting depth average is determined by the source and receiver depths. Thus, information about the depth dependence of current is possible with a single parameter inversion by employing a vertical array of hydrophones. The purpose of having such a hydrophone array in deep-ocean tomographic measurements is to select the acoustic arrivals, but the purpose of such an array in shallow-water acoustic measurements is to select average height of the ray, thus controlling the depth and the extent of averaging the current.

In shallow-water acoustic pulse propagation, the signals arriving earlier than the late pulse generally are less stable and of lower amplitude. Even with longtime averaging, it is difficult to identify characteristic features that can be tracked and used for precise measurement of travel time. However, a consistent feature of the pulse response of a channel is a sharp peak associated with the late-arriving RBR paths. The purpose of averaging several pulse responses is to smooth out interference effects. The stable and sharp cutoff of the late peak is used to align the pulses prior to their averaging. The arrival time of the pulses is estimated using a *threshold cross-time* method, whereby the arrival time of the lagging edge is estimated as the time when the edge crossed an intensity threshold. This threshold is usually an average of several thresholds, which are 10 dB or more below the maximum intensity of the late peak. The aligned pulse responses are averaged, with the moving window over several records (the *sliding average method*).

Ko et al. (1989) found that the pulse responses lined up very well after being aligned and averaged. Because the time corresponding to the maximum intensity of the peak is an imprecise estimate, the centroid estimation is usually used to estimate the travel time of the arrival peak. The time corresponding to the centroid of the late peak is computed using several points around the maximum value. The time variation for the late peak has been found to be a good indicator of the travel-time variation for the latest, and thus the flattest, near-bottom RBR rays. Details of data analysis methods may be found in De Ferrari and Nguyen (1986).

Suesser (1990) examined the properties of the acoustic environment in the Strait of Gibraltar, with the goal of determining the feasibility of using acoustic remote-sensing methods to monitor temperature and/or current structure there. Subsequently, Elisseeff et al. (1999) reported acoustic tomographic measurement of a coastal front in Haro Strait, British Columbia (Canada). In their experiments, each mooring consisted of 16 receivers, a 1.5-kHz tomographic source, and a 15-kHz communication source (see Figure 7.6). A chain of thermistors was added to two of the four moorings. The moving source used in this study consisted of ship-deployed lightbulbs. The lightbulbs were lowered to a specified depth in a casing apparatus, as shown in Figure 7.7. The shot was then triggered by breaking the bulb at depth using an operator-released lead mass that dropped along the cable from the ship to the casing. Lightbulbs generate a short, reproducible bubble-pulse waveform. The spectral peak of the

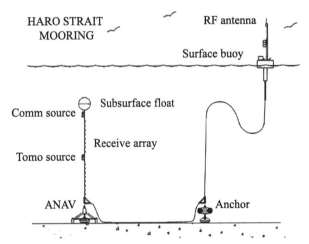

FIGURE 7.6 Haro Strait acoustic tomographic array mooring design. (*Source: Elisseeff et al., 1999.*)

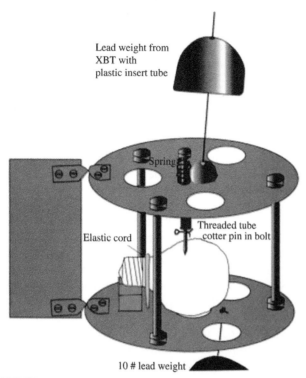

FIGURE 7.7 Design of moving acoustic source consisting of ship deployed lightbulb lowered to a specified depth in a casing apparatus. The lightbulb generates a short reproducible bubble-pulse waveform (shot) when the bulb is pierced at the specified depth using an operator-released lead mass that is dropped along the cable from the ship to the casing. *(Source: Elisseeff et al., 1999.)*

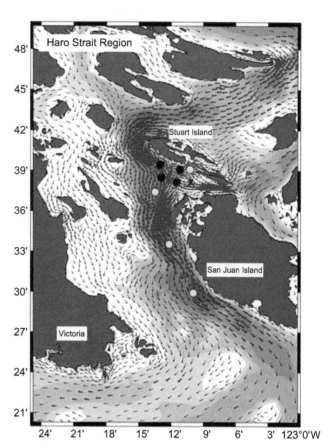

FIGURE 7.8 Topographic map of Haro Strait region with predicted ebb tide currents during tomographic experiment. Tomographic arrays (black circles) were deployed to investigate the front south of Stuart Island. Current meter moorings (yellow circles) and a meteorological surface buoy (black cross) were used to help clarify the large-scale circulation. *(Source: Elisseeff et al., 1999.)*

lightbulb-generated wideband acoustic signals is typically about 500 Hz at a depth of 50 m, with a 3-dB bandwidth of approximately 300 Hz. Although there is significant spectral content to several kHz, the band was limited in this experiment by the relatively low sampling frequency of 1,750 Hz. Source locations were obtained using differential GPS data.

The four 16-element vertical acoustic line arrays were moored south of Stuart Island (see Figure 7.8) around the location of a coastal front driven by estuarine and tidal forcing. Four non-acoustic moorings were also deployed along the Haro Strait channel. Local temperature, salinity, current magnitude, and current direction were recorded by these moorings at discrete depths. These shots were recorded on all the moorings and were all deployed within a time period of approximately 90 minutes. Source depths ranged from 30 to 70 m. Source levels were approximately 160 to 170 dB//1 μPa. The acoustic data acquired on each receiver array were sent back to shore via a surface radio link at a rate of 35 kbauds. Various acoustic signals were transmitted from array to array and from the moving source to the arrays over a period of five weeks. Tomographic signals were transmitted over a wide-frequency band (150 Hz to 15 kHz). The novelty of the Haro Strait

dataset resides in its unusual tomographic features: Ranges are short (less than 3 km), sound-speed perturbations are small (2 to 3 m/s), and currents are relatively strong (3.5 kts).

Individual acoustic time series were match-filtered using a single arrival as a substitute for the actual source signal. Absolute timing was provided by an acoustic acquisition system described by Grund et al. (1997). The magnitude of the matched-filter output was then low-pass-filtered using a zero-phase third-order Butterworth filter with a cutoff frequency of 500 Hz. Direct and surface ray arrival times were subsequently measured by identification of the corresponding local maxima of the filtered time series. Acoustic sensor localization was carried out by minimizing the mean-square difference between the measured arrival times and the arrival times predicted using a sensor model.

Raw temperature, salinity, and water-current time series were measured at the mooring sites at a frequency of 12 samples per hour. Sound speed, C (*m/s*), was estimated from water temperature T (°C), salinity S (psu), and depth D (m) using Mackenzie's formula (Mackenzie, 1981):

$$C = 1448.96 + 4.591T - 5.304 \times 10^{-2}T^2 + 2.374$$
$$\times 10^{-4}T^3 + 1.304(S - 35) + 1.630 \times 10^{-2}D$$
$$+ 1.675 \times 10^{-7}D^2 - 1.025 \times 10^{-2}T(S - 35)$$
$$- 7.139 \times 10^{-13}TD^3$$

$$(7.13)$$

Individual samples were grouped by 50-min bins and averaged. Standard deviations were found to be between 6 and 13 cm/s for the current field and between 1 and 25 cm/s for the sound-speed field.

To combine heterogeneous datasets, a variety of observation models is required. Acoustic sensor displacements are accounted for by a sensor model. This model relates acoustic sensor displacements to array design parameters and local current magnitude at the array. The water-current model provides predictions of the tidal current field at any point in the observed region. The acoustic model relates sound speed to acoustic travel times. Statistical field estimates provide *a priori* information on the actual field estimates. *A posteriori* information on water current is provided by local water-current measurements and acoustic sensor displacements. *A posteriori* sound-speed information is provided by local sound-speed measurements and acoustic tomographic data. In a first stage, an estimate of the oceanic water current field is computed by objective analysis of the available current data (Carter and Robinson, 1987). The water-current estimate is then externally melded with the tidal-current model prediction following a standard data assimilation procedure. The sound-speed field is then objectively analyzed using range-averaged acoustic tomographic estimates and local nonacoustic data.

Elisseeff et al. (1999) combined two types of information in the estimation of the sound-speed field:

1. Point measurements of sound speed at the nonacoustic moorings
2. Acoustic travel times measured at the acoustic moorings

First, the acoustic travel times are used to estimate range-averaged sound-speed profiles and their error covariance along the available acoustic transmission tracks. These sound-speed profiles are then merged with the point measurements and interpolated to yield an estimate of the sound-speed field.

Lightbulb-generated low-frequency wideband acoustic signal transmissions and the received acoustic dataset gathered in Haro Strait in June 1996 were used in conjunction with local nonacoustic measurements in order to image the three-dimensional sound-speed and water-current fields within the water mass of approximately 3 km × 3 km × 200 m enclosed by a group of moored arrays. A fast and robust inversion algorithm combining linear techniques previously used in deep-ocean tomography and oceanographic data assimilation was developed by Elisseeff et al. (1999), yielding the real-time oceanic field estimates required in the context of acoustically driven rapid environmental assessment. The use of data melding and data assimilation techniques allowed them to resolve, to some extent, the current/sound-speed ambiguity inherent to any nonreciprocal acoustic transmission. In addition, the combined use of integral and local data led to a significant decrease of the field estimate uncertainty while maintaining coverage of the area that was not achievable by nonacoustic means.

7.4. COASTAL ACOUSTIC TOMOGRAPHY

As mentioned earlier, most acoustic oceanographers have focused their major interest on the open ocean. Thus, in contrast to the frequent use of reciprocal transmission techniques in the open ocean, no reciprocal transmission experiment has been performed in the coastal ocean for quite some time since the development of OAT techniques. The major reason for this inaction has been a plethora of technical problems associated with the complicated nature of sound transmission in shallow water bodies. In the coastal ocean, long-term measurements of currents has often been prohibited by heavy ship traffic and fishing activities. Despite such difficulties, Zheng et al. (1997) designed and developed a reciprocal sound transmission system to measure currents over a long-term basis in a Japan coastal sea with heavy ship traffic and fishing activities. They successfully demonstrated that reciprocal sound transmission is applicable to water-current velocity measurement in the inland sea with heavy ship traffic and fishing activities. It is well accepted now that, especially around Japan, sound transmission techniques may be a possible method to realize a long-term monitoring of through-flows in coastal channels.

The reciprocal transmission system designed by Zheng et al. (1997) was composed of two stations located on the opposite sides of the coastal channel, where measurements were desired. Both stations were equipped with a transmitter, hydrophone, and GPS receiver. The GPS receiver was used not only to locate the stations but also to synchronize their clocks. In this system, 1-Hz and 1-kHz signals from the GPS receiver were used for timing of the system. Starting signals of transmission were sent by detection of local time and 1-Hz pulse signals. Reception of signals was initiated by counting a delay time with 1-Hz and 1-kHz pulse signals. The transmission clock (CLK_t) and the coherent carrier signals for demodulations were synchronously locked by a phase-locked loop (PLL) circuit using 1-kHz signals from the GPS. As a result, the timing of transmission and reception for each ping was controlled to an accuracy of 0.1 μs. Transmission signals, a carrier of frequency 10.6 kHz modulated by M-sequence (briefly described in the following paragraph), were generated by driving the data written in programmable read-only memory (PROM). First the received signals were processed through a preamplifier, band-pass filter, and amplifier. Next they were demodulated by the multiplier and low-pass filter using the coherent carriers and divided into in-phase (I) and quadrature (Q) components. Finally, the analog signals were digitized by an A/D converter and recorded on the hard disk of a microcomputer.

Coastal oceans frequently suffer from a noisy acoustic environment due mainly to ship traffic. Consequently, a special method of sound transmission and signal processing was needed. A method for extracting the received signals from noisy data is the use of M-sequence and the cross-correlation of received signals with it. The *M-sequence* is a kind of pseudo-random signal by which a phase shift of π radians in the carrier is generated with irregular time intervals (Okujima and Ohtsuki, 1981). Zheng et al. (1997) have reported a typical example of M-sequence, with time series of the original signal and its autocorrelation coefficient. There are two kinds of time scales in the M-sequence, the period (T_p) and the width of one digit (T_r). After the autocorrelation procedure, the S/N ratio can be improved by ($2n - 1$) times for the M-sequence of nth order. Zheng et al. (1997) nomenclatured the improved rate of the S/N ratio as *processing gain* (G_p). The travel time of an acoustic ray propagating between two stations is determined as a peak position of the autocorrelation coefficient of triangular shape in an ideal case without ambient noise. When two acoustic rays with the same intensity arrive successively within a short interval of time, the autocorrelation pattern for the rays is overlapped. For the travel-time difference (Δt) less than T_r, we lose the individual peaks that each autocorrelation pattern possesses. According to Zheng et al. (1997), T_r can be termed as the *time resolution* of the system for multiple arrivals. In the case of two acoustic rays with different intensities, individual correlation peaks may be detectable even for the overlapped arrival because of the difference of the peak height. Smaller T_r is required for a better resolution of time but needs a transmitter with a wide range of frequency response. In the system designed by Zheng et al. (1997), the order of M-sequence is set to 10 for increasing the S/N ratio by $2^{10} - 1$ (30.1 dB), and T_r is taken as three times the period of the carrier. This condition puts the frequency range of transmitted signals into 10.6 ± 3.53 kHz and matches with the frequency response of the transmitter. As a result, T_r and T_p are given as 0.283 ms and 0.290 s, respectively.

A sound level (SL) at the transmitter is set to about 190 dB (relative to 1 μPa at 1 m). The propagation loss (PL) of sound waves against the range (R) is expressed by (Urick, 1983):

$$PL = 20 \log R + \alpha R + L_0, \tag{7.14}$$

where the first, second, and third terms on the right-hand side of Equation 7.14 correspond to the spreading, absorption, and other losses (directivity, reflection, interference, etc.), respectively. The absorption coefficient α is set at 0.00132 dB/m at 10.6 kHz. In Equation 7.14, R is m in dimension. The received sound level (RL) at the hydrophone is given from the sonar equation as:

$$RL = SL - PL = SL - 20 \log R - \alpha R - L_0. \tag{7.15}$$

The signal-to-noise ratio $(S/N)_R$ of received signals obeys the following equation:

$$(S/N)_R = RL + G_p - N_a$$
$$= SL - 20 \log R - \alpha R - L_0 + G_p - N_a, \tag{7.16}$$

where G_p is the processing gain due to the use of M-sequence and taken to be 30.1 dB, and N_a is the total level of the noise composed of the ambient noise (NL) and the system noise. In the Seto Inland Sea, NL is mainly caused by ship traffic, and within the operating frequency range of the hydrophone and the receiving circuit was estimated as 85 dB, greater by 30 dB than that in the open ocean (Wenz, 1962). The total level of the noise (N_a) reduces to about 100 dB in consideration of system noise. Setting $L_0 = 10$ dB for direct rays (not reflected at both the surface and bottom), Zheng et al. (1997) obtained $RL = 97.36$ dB and $(S/N)_R = 27.46$ dB. For the sound transmission system used by Zheng et al. (1997), the accuracy of travel-time measurement depends on T_r and $(S/N)_R$ and is expressed by (Munk and Wunsch, 1979) as:

$$T_a = \frac{T_r}{\sqrt{10^{(S/N)_R/10}}} = 12.0 \mu s, \tag{7.17}$$

where $T_r = 0.283$ ms.

Zheng et al. (1997) carried out a reciprocal sound transmission experiment on July 14, 1995, in the Seto Inland Sea, Japan. The peculiarity of this region is that a pair of vortices is induced by an eastward tidal jet coming out from the Neko Seto channel (Takasugi et al., 1994). Accordingly, the possibility of a clockwise vortex appearing on the proposed sound transmission line was expected when the direction of tidal current turns from the east to west and moves slowly to the southwest in the growth of westward tidal flow. Based on this realization, the sound transmission experiment of Zheng et al. (1997) was planned to start immediately after the tidal current changed its direction from east to west. They expected that water and flow properties in the tidal vortex are homogenized (i.e., well mixed) in the entire depth because the diameter of the vortex of about 4 km is much larger than water depths at the observation site. The main purpose of this study was to develop a reciprocal sound transmission system of a 10-km scale applicable to the coastal ocean with severe ambient noise due mainly to ship traffic.

The system was composed of two stations spaced with a distance of 5.7 km on both sides of a channel in the Seto Inland Sea. Each station was equipped with a transmitter, hydrophone, and GPS receiver. The transmission line of length 5.7 km between the stations N_2 and S_2 was taken to cross a small coastal channel surrounded by Honshu (the biggest main island of Japan) and Kami-Kamagari Island. The bottom slopes steeply down near these two stations and maintains a constant depth of about 70 m in the remaining region except for a bank at the northern part of the line. The subsurface system equipped with a transmitter and hydrophone was suspended down in water through an aluminum shaft from a fishing boat anchored at the stations N_2 and S_2, the depths of which were about 20 m. The transmitter and hydrophone were installed at 5 m and 5.5 m, respectively. A directional transmitter and an omnidirectional hydrophone were used.

As an example, for $R = 5.7$ km and $c = 1,500$ m/s, measurement of the seawater current velocity of 1 cm/s requires that the acoustic travel-time difference (Δt) be measured with an accuracy of 50.67 µs. Taking this into account, a pulsed signal of time width 0.29 s corresponding to a period of the M-sequence was transmitted every minute from the stations N_2 and S_2 with a time difference of 30 s between the two stations. The recording of received signals started in a delay time of 3.75 s from the transmission and continued for 0.983 s. The sampling time of the A/D converter was 20 µs. In this experiment, one-period M-sequence was used as a transmitted signal. The one-period M-sequence method is known to make spurious correlation peaks on both sides of the true correlation peak, which corresponds to an arrival ray (Okujima and Ohtsuki, 1981). Zheng et al. (1997) estimated the level of the spurious peaks to be 14.4 dB lower than that of the true peak. This means that no problem takes place in this study, which deals with only a first arrival ray.

The third boat was operated for ADCP and CTD measurements, which proceeded in parallel with the sound transmission experiment. The RD Instruments 300-kHz broadband ADCP was installed at 1 m below the surface by an aluminum frame mounted on the side of the boat. The bin length and sampling time were set to 2 m and 12 s, respectively. The bottom tracking bins also profiled bottom topography. CTD casts (Alec Electric Corporation AST-200) were done over the whole depth at seven stations, including the stations N_2 and S_2. The correlated signals possessed small peaks behind the first correlation peak, implying the successive arrival of signals. The original

signals also showed a remarkable change before and after the first correlation peak.

Combination of measurements of water temperature, salinity, and flow velocity between stations C_5 and C_7 enabled detection of the anticipated clockwise vortex. Southwestward movement of the vortex was well traced with a sequence of positions where minima of u or zero velocities of v occurred. The correct feature of such moving vortices may be a difficult target to be measured by the slowly moving ship equipped with an ADCP.

Yamoaka et al. (2002) made continuous effort since 1994 to construct a cost-effective multiple set of coastal acoustic tomography (CAT) systems. At the beginning of March 1999, five sets of the moored-type CATs were constructed, and they have been applied to measure vortex structures in the Neko-Seto Channel of the Seto Inland Sea with strong tidal currents. Inverse analysis has been applied for reconstructing horizontal current fields from the travel-time difference data. The internal host computer controls all the clock, transmission, and receiving circuits. Timing of sound transmission and receiving is coherently synchronized with an accuracy of 0.5-µs by GPS pulse signals. The time base module (a SeaSCAN precise crystal clock) becomes a system clock when the GPS signals are interrupted because of unexpected problems. Temperature and salinity are well homogenized due to strong tidal mixing over the entire observation region except the upper 5-m layer. In the upper 5-m layer, the sound speed follows the temperature profile, and below the surface layer it increases with the pressure, forming an adiabatic profile.

Acoustic signals are transmitted every 5 min by an omnidirectional transmitter (ITC2011/ITC2040) with horizontally directed beams of $\pm 30°$. A Gold sequence of the tenth order, which is a kind of pseudo-random sequence, is used for phase modulation of signal transmission (Simon et al., 1994). The Gold sequence is constructed by a product of the preferred pair of M-sequence codes of tenth order with the optimal cross-correlation property, maintaining the original cross-correlation property of the M-sequence. The M-sequence of tenth order has only three sets of the optimal cross-correlation property, whereas the Gold sequence of the same order can generate 1,025 optimal sets. This advantage enables CATs to carry out multistation tomography with a simultaneous transmission. Transmission signals are received by a hydrophone (Benthos AQ-1). The received signals are preamplified and divided into sine and cosine channels for a complex demodulation. After the signals are low-pass-filtered and A/D converted, the digital signal processor (DSP) TMS320c548 promptly calculates the cross-correlation between the processed signal and the transmission codes that are used in the transmission and stores the results to the hard disk. In the observation region, a depth-averaged water-current velocity field can be reconstructed through the inverse analysis of travel-time difference data for all station pairs in the tomography array.

The first coastal acoustic tomography experiment was carried out on March 2–3, 1999, in the Neko-Seto Channel of the Seto Inland Sea, Japan. Several crowded shipping routes are distributed throughout most parts of the Neko-Seto Channel. Five sets of CATs were placed at stations from $S1$ to $S5$ in the periphery of the Neko-Seto Channel. The distance between two neighboring stations ranges from 1.9 to 5.5 km. In this region, the tidal current is directed eastward at the flood tide and westward at the ebb tide. The bottom topography, which is characterized by sand banks and troughs, is formed by the action of a pair of tidal vortices induced by a strong eastward tidal jet flushed out from the narrow western inlet (Takasugi et al., 1994). Previous research has determined that the tidal vortices develop as the eastward current is strengthened and they reach a maximum size of about 2.5 km two hours after the strongest eastward current. The vortices move toward the narrow inlet at the incipient phase of the westward current and diminish rapidly with increasing westward currents.

The rays of sound propagation are simulated by the ray-tracing method in which the CTD data are used to specify the reference sound speed. The direct rays, which have no interaction with the sea surface and bottom, pass the upper 20-m layer, where the near-surface duct exists. The computed travel times for all the direct rays have a scatter of 0.4 ms, which is less than the one-digit width (0.54 ms) of a Gold sequence. Thus the arrivals cannot be resolved, and the first arrival peak is composed of multiple direct rays.

Received signals for each station are cross-correlated with the same Gold sequence code as that used in the transmission signal. Travel-time difference data for the first arrival peak determined in the correlation peaks is used as data in the inverse analysis. In fact, the travel-time measurement is affected by the accuracy of clock timing and the horizontal movement of the transmitter and hydrophone. When the accurate clock timing is available, the relative error of current velocity measurement ($\Delta u/u_m$) may be formulated by $\Delta u/u_m \approx \Delta L/L$, using the relative-positioning error ($\Delta L/L$), where u_m and L are the range-averaged current velocity and the station-to-station range, respectively (Zheng et al., 1997). It would be noteworthy to keep in mind that the relative error of current velocity is only 0.2 percent of the observed current velocity for observation range 5 km and positioning error 10 m.

The travel-time difference (Δt) may be converted into the range-averaged current velocity (u_m) using Equation 7.12, where R is the station-to-station range and c_m the range-averaged sound speed. It has been noted that the oscillation due to the semidiurnal tide is featured for several transmission lines.

In the studies reported by Yamoaka et al. (2002), the inverse analysis was performed only for the first arrival peaks estimated by the ray theory, because the identification of bottom reflected/scattered rays was much more difficult due to the interference of sound waves in shallow water. The high-frequency variability of current velocity was found to have been considerably enhanced with decreasing station-to-station ranges. This high-frequency variability expressed by the standard deviation from the 30-min mean was proposed as a good index of the error bar in the velocity measurement. The relative magnitude of direct to scattered rays in received signals was found to increase with increasing station-to-station ranges because the intensity of scattered rays is more rapidly damped due to bottom bouncing. That was why the high-frequency variability of travel-time difference was found to have been so large for the station pairs with much shorter station-to-station ranges.

It is encouraging to note that the CAT system composed of five moored acoustic stations could reveal the horizontal structure and the temporal variations of the tidal vortices in straits. Although an advanced technique of data analysis is expected, the overall process of growth, translation, and decay of the tidal vortex pair was fairly reconstructed by the conventional inversion method. Based on the studies of Yamoaka et al. (2002), it was found that for the same number of moorings, the spatial resolution of the tomography is considerably improved compared with the conventional point measurement technique. Whereas the spatial resolution for the conventional technique can be improved by increasing the number of moorings, the instrumentation cost is correspondingly increased with the number of moorings. On the other hand, whereas the mapping of coastal surface currents can be performed by the HF radar systems, the acoustic tomography measures depth-averaged currents with the future extension to the vertical profile measurement of current. In the Seto-Inland Sea region, because of the development of residential and industrial areas, it may be difficult to find sufficient shore space for locating the array of antennae needed by HF radar. In contrast, CAT, operated by multiple sets of compact mooring stations and placed near the shore, can be a practically feasible system with more flexibility and potential ability than HF radar.

In the Neko-Seto Channel, the mooring observation has been strictly prohibited by the Japan Maritime Safety Agency, except for the near-shore region, because of the risk of shipping accidents. Limited information on the vortex generation in the Neko-Seto Channel may be provided by shipboard ADCP operating along several lines covering the observation region. However, the rapid processes of growth, transition, and decay of the tidal vortices were first measured by Yamoaka et al. (2002) using an acoustic tomography technique. It is hoped that in the near future, the tomography system may make further progress, with new instruments to measure the phenomena of strong tidal mixing and dissipation in the Neko-Seto Channel and around a huge number of islands located in the Seto Inland Sea without disturbing marine traffic. When this happens, the sophisticated tidal model of the Seto Inland Sea, including the Neko-Seto Channel, will be fully operational. At this stage, real-time data telemetry via satellite or mobile phone is desired for predicting current variability in combination with the ocean model. It was also proposed that a long-term operation at shorter intervals could be conducted by putting solar panels on a surface buoy.

From the preceding examples of acoustic tomographic experiments, it can be expected that an increase in the number of tomographic source-receiver pairs makes it possible to achieve more detailed mapping of current fields in the coastal seas. Yamaguchi et al. (2005) applied multiple CAT systems to map tidal current structures generated at the Hayatomono-Seto of the Kanmon Strait. In Japan, most of the CAT measurements have been carried out in this strait, which is a narrow passage with a width of 1−2 km and a length of about 28 km. This strait is located in the Sea of Japan and it is famous, not only as an important shipping traffic route to China and Korea but also as a dangerous passage with quite strong tidal current exceeding 5 m/s at the narrowest point.

Since the time of tidal current measurements in this strait with the use of drifting floats tracked by many small boats in the 1940s, more comprehensive measurements have been prohibited due to crowded shipping traffic. Although the maximum current and the vertical section structure of current across several transects were well observed with the use of repeat shipboard ADCP measurements, information on the horizontal structure of current and its temporal variation in this strait was rather scarce. The temporal change of volume transport across the strait was also poorly known. Except for continuous measurement of currents routinely performed at one station near the Kanmon Bridge (the narrowest point) by the Japan Coast Guard using an upward-looking, bottom-mounted ADCP and the collected data made available to the public in the form of tidal current charts published by the Japan Coast Guard (1994), no more information on the current structure and its spatial variability was available. This was the reason that primarily stimulated the oceanographic researchers to undertake CAT experiments in this strait.

We have noted that Yamaoka et al. (2002) challengingly measured the two-dimensional tidal vortex fields in the Neko-Seto Channel of the Seto Inland Sea using CAT experiments. It has been seen that the overall features of the tidal vortex fields were reasonably well reconstructed through inverse analysis in which travel-time differences obtained for the pairs of five acoustic stations were used as input data. However, the influences of the open boundary conditions and the complex coastline on current fields were difficult to include in the inversion. A coupling of the tomography data with the ocean model was expected to satisfy the dynamic constraints of current fields and the boundary conditions.

Although CAT technology was found to be a practical means for integrated measurement of coastal currents and their circulation structures, the application of this technology in the Japanese Inland Sea was beset with a couple of practical issues that needed to be resolved. In an attempt to better reconstruct the vortex pair observed in the Neko-Seto Channel of the Seto Inland Sea, Japan, Park and Kaneko (2000) applied the ensemble Kalman filter (EnKF) technique of the data assimilation accompanied by the estimate of error covariance by means of the Monte Carlo method (Evensen, 1994) to analyze the first CAT data. This technique has distinct advantages in the application to strongly nonlinear current fields, and there is also no restriction in its application to the coastal seas (Madsen and Canizares, 1999). Application of the EnKF technique, as stated, was found to be the best choice to the present problem because currents in the tomography region are dominated by the strongly nonlinear tidal vortices. They found that the results obtained with the use of this method were in better agreement with the shipboard ADCP measurements (which were carried out in parallel with the CAT measurements) than those obtained by the inverse analysis of the travel-time difference data for the pairs of acoustic stations.

The simulated current fields were surprisingly improved through the implementation of the new data assimilation technique. The model result showed a tidal vortex pair composed of a western counterclockwise vortex and an eastern clockwise vortex—a result that agrees very well with observations made by aerial photographs (Takasugi et al., 1994). The tomography data are generally less accurate near the periphery of the observation region because of the decreasing number of ray paths there. Taking this limitation into account, it is justifiable to conclude that the data assimilation technique proposed by Park and Kaneko (2000) can be used as a powerful technique that is superior to the inverse analysis in analyzing the CAT data. As a result of the data assimilation, the ocean model thus reaches a level suitable for explaining the strongly nonlinear phenomena in the coastal seas.

As part of a continuing effort to improve CAT measurements in Japan, Yamaguchi et al. (2005) carried out a CAT experiment using eight source-receiver stations during March 17−20, 2003, at the Hayatomono-Seto around the Kanmon Bridge, located at the eastern part of the Kanmon Strait (Figure 7.9). The observational period included a spring tide. The area of the tomography domain was about 2 km × 2 km. The northern part of the region is characterized by a narrow trench with depths of 16−20 m, the existence of which is attributed to a strong tidal jet flushing out westward from the narrow passage where the

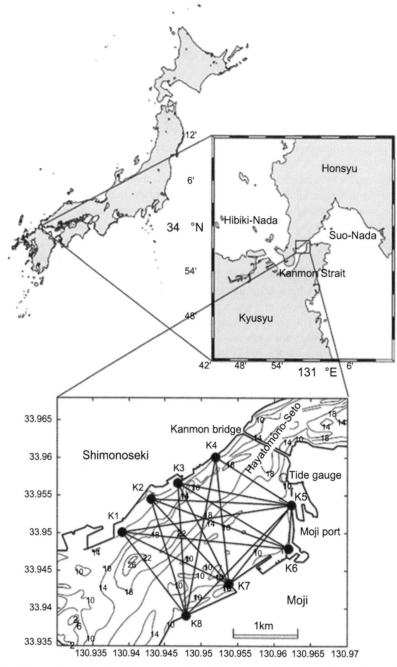

FIGURE 7.9 Location maps of the experimental region (Hayatomono-Seto of the Kanmon Strait, Japan). The positions of the CAT stations K1–K8 and the bathymetric contours are shown in the most magnified figure with solid circles and thin solid lines, respectively. The thick solid lines connecting the CAT stations are the sound transmission lines. An interval of contour lines is 4m. *(Source: Yamaguchi et al., 2005.)*

Kanmon Bridge is constructed. Except for the trench, the average depth of the tomography domain is about 12 m. The seabed is mainly composed of sand.

Eight CAT systems (*K*1 to *K*8) were located at the northern and southern coasts using wharfs and piers. The strong tidal current in the strait and the safety traffic strategy of the Japan Coast Guard prohibited deployment of the CAT system in water off wharfs and piers, even near the coast. Thus the main portions of CAT, such as the electronic housing, battery box, and GPS antenna, were placed on the ground near the edge of wharfs and piers, and only transmitters and four hydrophones were placed in subsurface water, using a steel frame in touch with their vertical walls (see Figure 7.10).

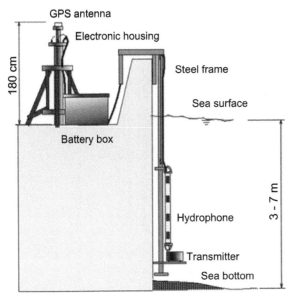

FIGURE 7.10 Schematic diagram of the CAT system deployed at the front of wharves or piers. *(Source: Yamaguchi et al., 2005.)*

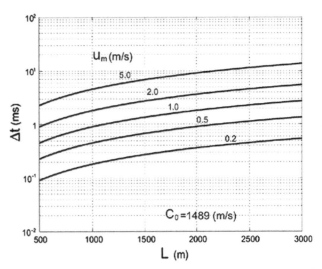

FIGURE 7.11 Relationship between the travel-time difference (Δt) and the station-to-station range (L) with the range-averaged current velocity (u_m) as a parameter. Here, the reference sound speed C_0 is given to be 1,489 m/s from the reference values for temperature, salinity, and water depth: $T_0 = 10.45°C$, $S_0 = 33.03$, and $D = 5$ m. *(Source: Yamaguchi et al., 2005.)*

A pseudo-random signal, called the *10th order Gold sequence*, was transmitted every 10 minutes from the broadband transmitter (ITC2011/ITC2040) of central frequency 5.5 kHz. The 5.5-kHz carrier is modulated by the Gold sequence with a different code for each acoustic station, and one period (0.56 s) of the modulated signal is transmitted. One digit (0.54 ms) of the Gold sequence, which is the minimum unit of the Gold sequence, was set to include three waves of the carrier. The transmission signals were received by the four-hydrophone array located mainly on the opposite side of the strait. The two-way sound transmission and reception on one side of the strait were possible only between the station pairs $K5-K7$ and $K5-K8$. The received signals were preamplified, complex demodulated, cross-correlated with the Gold code, digitized by the A/D converter, and recorded into the memory. In the original schedule, the sampling frequency of the A/D converter was set to 11 kHz, i.e., two samples per wave. The relationship between the travel-time difference (Δt) and the station-to-station range (L) with the range-averaged current velocity (u_m) as a parameter is shown in Figure 7.11. Note that the time resolution (t_r) for multiple arrival rays in the shallow-sea sound transmission is defined by one-digit length of the Gold code (Zheng et al., 1997).

All timings of transmission, reception, and A/D conversion were synchronized by the GPS high-precision clock, which has an accuracy of about ± 500 ns. The operation of the CAT system was controlled by the internal host computer, and various parameters to specify experimental conditions were configurable at any time by connecting the external host computer with the internal one via the infrared data association.

The tomography data are obtained as integral values of current velocity along the ray paths. If sound speed has inhomogeneous distribution in seawater, rays draw a curve obeying Snell's law of refraction. Each eigen ray may be determined by a set of the following ordinary differential equations (Pierce, 1989; Dushaw and Colosi, 1998):

$$\frac{d\cos\theta}{dr} = -\sin\theta\left(\frac{1}{C}\frac{\partial C}{\partial r}\tan\theta - \frac{1}{C}\frac{\partial C}{\partial z}\right)$$
$$+ \left(3 + \tan^2\theta\right)\frac{u}{C^2}\frac{\partial C}{\partial r} - \frac{1}{C}\frac{\partial u}{\partial r} - \tan\theta\frac{u}{C^2}\frac{\partial C}{\partial z}$$

$$(7.18a)$$

$$\frac{d\sin\theta}{dr} = \cos\theta\left(\frac{1}{C}\frac{\partial C}{\partial r}\tan\theta - \frac{1}{C}\frac{\partial C}{\partial z}\right) + 3\frac{u}{C^2}\frac{\partial C}{\partial z} - \frac{1}{C}\frac{\partial u}{\partial z}$$

$$(7.18b)$$

$$\frac{dz}{dr} = \tan\theta\left(1 - \frac{u}{C\cos\theta}\right)$$

$$(7.18c)$$

$$\frac{dt}{dr} = \frac{1}{C\cos\theta} - \frac{u}{C^2}\left(2 + \tan^2\theta\right)$$

$$(7.18d)$$

The acoustic ray paths are determined from the integral of Equation 7.18.

In the original ray-tracing method, no transmission losses of sound are taken into consideration. Yamaguchi et al. (2005) modified the original method to evaluate the acoustic intensity of transmission signals by considering the dominant transmission losses along a ray. The acoustic intensity along a ray is dissipated by transmission losses

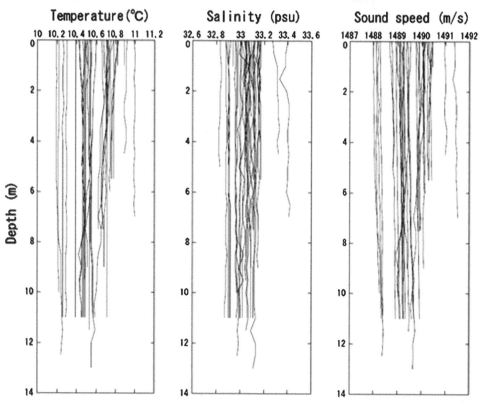

FIGURE 7.12 Vertical profiles of temperature, salinity, and sound speed obtained at all CAT stations. *(Source: Yamaguchi et al., 2005.)*

due to spreading, absorption, reflection, and scattering. The sound transmission losses in shallow water may be written as:

$$TL = 10 \log r + 10^{-3}\alpha r + L_B + L_s \qquad (7.19)$$

In this expression, r is the range (in meters) between the source and receiver, α is the absorption coefficient (dB/km), L_B the bottom loss, and L_S the surface loss. The absorption coefficient, α, may be expressed in terms of the frequency of sound (f) as:

$$\alpha = 3.3 \times 10^{-3} + \frac{0.11f^2}{1+f} + \frac{44f^2}{4100 + f^2} + 3.0 \times 10^{-4}f^2 \qquad (7.20)$$

In shallow water such as the Kanmon Strait, sound waves can propagate in a duct between the sea surface and the seabed. That is why cylindrical spreading of sound is applied in the first term of Equation 7.19. The second term indicates the absorption loss, depending on the square of frequency, and according to Yamaguchi et al. (2005), it is mainly caused by the relaxation of magnesium ions in sound waves. At the third term, the bottom loss depends on the incident angle of sound wave and the seabed roughness and materials. Yamaguchi et al. (2005) used the following formula for interfaces with a random roughness (Jensen et al., 1994):

$$BL = 10 \log\left(\mu(\theta)\exp\left(-0.5\kappa^2\right)\right)^{-1} \qquad (7.21)$$

In this expression, μ is the reflection coefficient and κ is the Rayleigh roughness parameter with the following relationship:

$$\mu = \frac{(\rho_2 C_2/\sin\theta_2) - (\rho_1 C_1/\sin\theta_1)}{(\rho_2 C_2/\sin\theta_2) + (\rho_1 C_1/\sin\theta_1)} \qquad (7.22)$$

$$\kappa \equiv 2k\sigma\sin\theta_1 \qquad (7.23)$$

In these equations, (ρ_1, C_1) and (ρ_2, C_2) are the (density, sound speed) pertaining to (seawater, flat seabed), respectively. Also, θ_1 and θ_2 are the grazing angle to the seabed and the refracted angle measured from the seabed, respectively. Further, k is the acoustic wave number and σ is the RMS roughness of the seabed. In the experiment of Yamaguchi et al. (2005), they chose $k = 23.2$ and $\sigma = 0.3 m$ in a rough estimation.

Figure 7.12 shows the vertical profiles of temperature, salinity, and sound speed obtained by CTD measurement at the eight CAT stations, in which the sound speed was calculated by Mackenzie's formula. Water was well homogenized (i.e., no stratification) at every CAT station due to strong tidal currents. The averaged profile calculated from all the profile data was used in the ray simulation. Figure 7.13 shows the result of ray simulation obtained between the CAT

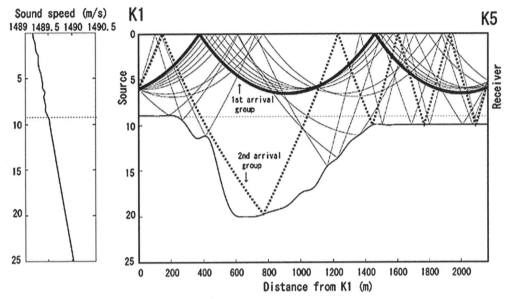

FIGURE 7.13 Ray simulation result between CAT stations K1 and K5 determined by the ray-tracing method. Only the rays with S/N ratios greater than 10 at the receiver position are drawn. The average sound-speed profile used in the simulation is presented left of the figure. *(Source: Yamaguchi et al., 2005.)*

stations K1 and K5 as a typical example. The acoustic intensity (signal level) of received signals at the hydrophone position was evaluated by the modified ray simulation, which can consider the transmission losses along the ray.

The correlation waveforms of signals released from K1 and received at K5 are shown typically in Figure 7.14, together with the arrival signal pattern calculated by the modified ray simulation. The tiny effect of current was not considered in this ray simulation. The simulated ray arrival peaks are located inside the flattened broad peaks in the real data (Figure 7.14a), implying that the broad peaks are composed of multiarrival rays (Figure 7.14c) separated into two groups (Figure 7.14b). The travel time for the typical ray path at the first arrival group (thick solid line) passing the upper 7 m layer was determined at a time (dot) when the S/N ratio was over 10 dB at the upward-sloping front of the broad peak. A typical ray path for the second arrival group is drawn with a thick dotted line in Figure 7.14b. The 10-minute interval data for travel time were further smoothed through a one-hour running mean to reduce high-frequency variations existing in the strong current. Hourly maps of the horizontal current distributions from 6:00 to 17:00 of March 18, 2003, obtained by the inversion analysis, are shown in Figure 7.15. CAT stations with successful sound transmission are connected with solid lines. Hourly plots of the maximum current velocity and the average current velocity across the transect K2-K7 are shown in Figure 7.16.

The correlation diagram between the CAT and ADP data is presented in Figure 7.17. The correlation rates for the east-west current (u) and north-south current (v) are 0.84 and 0.82, respectively. The RMS difference is

0.47 ms^{-1} for u and 0.48 ms^{-1} for v. At an overall view, the CAT velocities are considerably smaller than the ADP velocities. Yamaguchi et al. (2005) reckons that the observed difference in the measurements obtained from the two systems may have been caused by the different averaging procedure: The average along a ship track was adopted for the ADP data, whereas the CAT data were averaged through a resolution window on a horizontal plane.

The agreement between the CAT and ADP data is considered to be satisfactory regarding the maximum tidal current magnitude reaching 5 m/s and the time resolution of 0.54 ms (one-digit width) corresponding to the velocity bias of about 0.3 m/s. As for the operational system of the Kanmon Strait, the use of higher-frequency sound of about 20 kHz is considered optimal to get velocity accuracy better than 0.1 m/s. The acoustic tomographic measurements at Kanmon Strait indicate that the daily mean transport for the upper 7-m layer across the transect K2−K7 is directed eastward; Yamaguchi et al. (2005) estimated this transport to be about 1,470 m^{3}s^{-1} during a spring tide. The daily mean eastward transport at the spring tide is strongly supported by the three-dimensional Kanmon Strait model, which is validated with the just-mentioned CAT and ADP data. The CAT experiment conducted by Yamaguchi et al. (2005) demonstrates the ability to monitor strong tidal current structures in the Kanmon Strait.

The tidal current in the Kanmon Strait is as strong as 5 m/s at the phase of maximum tidal current. It was found that significant numerical errors are produced in the tomographic inversion. The Kanmon Strait acoustic tomography data acquired at 5-min intervals by Yamaguchi

FIGURE 7.14 (a) Stack diagram of the correlation waveforms of signals released from K1 and received at K5 accompanied by (b) the acoustic intensity and (c) ray travel time calculated at the receiver by the modified ray simulation. *(Source: Yamaguchi et al., 2005.)*

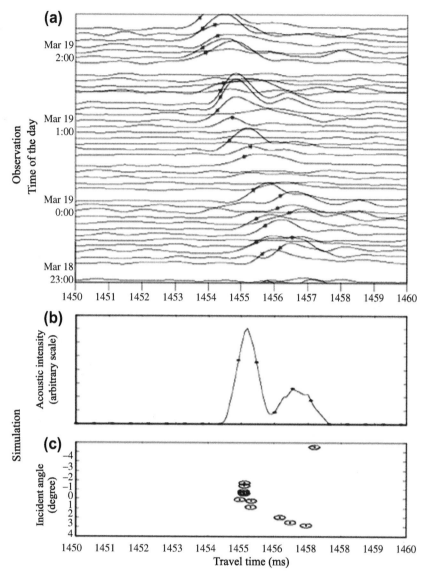

et al. (2005) were assimilated sequentially by Lin et al. (2005) into a 2D ocean model on the basis of the EnKF scheme to image strong tidal current structures occurring in the strait. When the accurate range-averaged currents obtained for the sound transmission lines connecting eight tomography stations were used as assimilation data, the complicated vortex-imbedded currents were imaged with horizontal resolution and accuracy much better than the result of tomographic inversion. The assimilated currents were well compared to the shipboard ADCP data with a RMS difference of about 24 cm/s for both the horizontal velocity components. The assimilated volume transport across the strait also showed good agreement with the transport estimated from the range-averaged current on a pair of transmission lines crossing the strait, making a RMS difference of 3,700 m³/s. It was concluded that the prediction of tidal current structures is useless in the Kanmon Strait with a strong forcing at the open boundaries, and instead their continuous imaging at 5-min intervals is the best policy to be taken.

The success of CAT experiments in the Japan Sea began to show its influence being permeated into the neighboring China Sea. For example, Zhu et al. (2010) successfully carried out a tomography experiment with seven acoustic stations (CAT systems) during July 12–13, 2009, in the Zhitouyang Bay near the Zhoushan Island, facing the mouth of Hangzhou Bay (China). The water depths are deeper than 40 m in the southwestern part of the bay and shallower than 20 m in the northeastern part. The maximum and minimum water depths are 91 m and 3 m, respectively, inside the tomography domain (about 11.4 km × 11.5 km). The inverse results provided continuous maps of horizontal

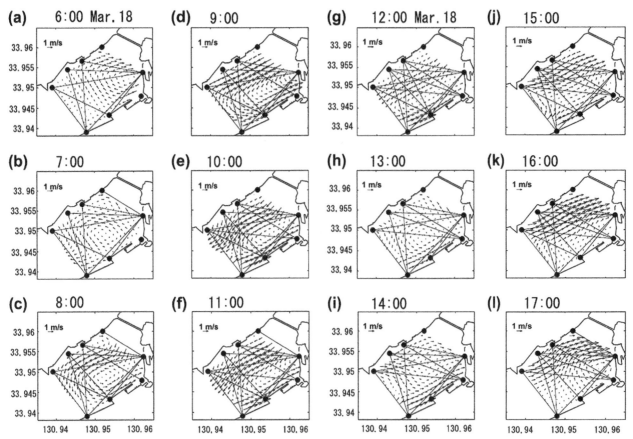

FIGURE 7.15 Hourly maps of the horizontal current distributions during 6:00 to 17:00 of March 18, 2003, obtained by the inversion analysis. CAT stations with successful sound transmission are connected with solid lines. *(Source: Yamaguchi et al., 2005.)*

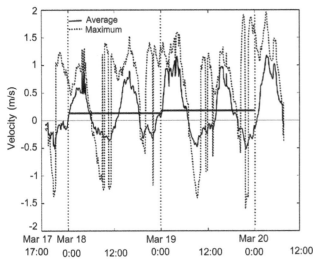

FIGURE 7.16 Hourly plots of the maximum current velocity (thin dotted line) and the average current velocity (thin solid line) across the transect K2–K7. The daily mean velocity for March 17 and 18 is indicated with thick horizontal solid lines. *(Source: Yamaguchi et al., 2005.)*

tidal current distributions at 3-min intervals. It was found that the strong eastward tidal current, with a maximum velocity of 2.34 m/s, entered into the tomography site from the western part of the bay and separated into two branches during the ebb tide. The horizontal tidal current fields (strong eastward tidal currents with two branches during the ebb tide and the reverse currents during the flood tide) were well reconstructed by the inverse analysis of reciprocal travel-time data. The clockwise tidal vortex of size 5 km at the eastern part of the bay in the transition phase from the ebb to flood was also reconstructed. Another piece of information gained from the tomographic experiments was the absence of tidal vortices in the weak tidal currents in the transition phase from the flood to ebb while changing the current direction from westward to eastward.

The observed travel-time difference (Δt) in this relatively shallow region could be well simulated by harmonic analysis, which considered eight major tidal constituents, K1, M2, M3, M4, 2MK5, M6, 3MK7, and M8. This result was found to be so useful not only for interpolating and smoothing the raw Δt dataset but also for predicating Δt values for future applications. The tomographic scheme field-tested by Zhu et al.

FIGURE 7.17 Correlation diagrams between CAT and ADP data. *(Source: Yamaguchi et al., 2005.)*

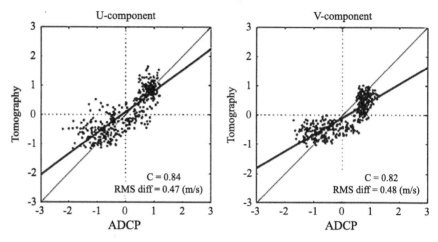

(2010) makes it possible to predict tidal currents inside the tomography domain by the inverse analysis from the known Δt data without relying on any tidal circulation model or data assimilation method. These results suggest that the CAT is a quite powerful instrument for continuously mapping the horizontal tidal current structures in coastal regions with heavy shipping traffic and active fishing.

7.5. RIVER ACOUSTIC TOMOGRAPHY

River discharge is an important hydrological quantity in terms of flood control and the management of water resources. This parameter is also important for river and coastal planning/management, control of water resources, and the like. Therefore, establishment of appropriate methods and technologies for measurement of water discharge assume great practical significance. Evidently, continuous progress of discharge measurement methods is an important issue. Conventional current measurement methods, such as ADCP (Lin, 2003), can only observe the current along a linear array. To acquire long-term data of cross-sectional average velocity and resulting river discharge, arrays of current meters should be deployed over the stream section of the river. Unfortunately, long-term measurement of river streams is often prohibited by heavy shipping traffic, especially in large rivers. As a traditional method, a water-level meter (He, 2008) is often operated to measure the discharge of nontidal rivers. However, in large tidal rivers, water level may not be a good parameter to quantify river discharge because of the nonapplicability to time-varying flows. Unfortunately, it is very difficult to measure cross-sectional mean water velocity in complex flow regimes such as tidal estuaries or during extreme hydrologic events.

Equipment that is available for continuous measurement of water discharge includes acoustic velocity meters (AVMs), horizontal acoustic Doppler current profilers (H-ADCPs), and the like (Catherine and DeRose, 2004;

Wang and Huang, 2005). The main drawback of the conventionally used methods is that often the number of velocity-sampling points in the cross-section of stream is insufficient to realistically estimate cross-sectional averaged flow velocity. Although several methods have been introduced to estimate the flow-velocity distribution (e.g., Chiu and Hsu, 2006), the results are disputable in complex flow fields such as stratified tidal flows. This calls for an innovative method and instrumentation for continuous measurement of water discharge in tidal estuaries. Taking this aspect of the problem into account, Kawanisi et al. (2009) carried out continuous water discharge measurements in a shallow tidal channel (the Ota River diversion channel), which is 120 m wide and 0.3 ~ 3 m deep, with large changes of water depth and salinity, using a new *river acoustic tomography* (RAT) system. The RAT system possesses technical advantage compared to competing techniques in terms of accurate measurement of acoustic travel time using GPS clocks and achievement of high S/N ratio due to the carrier modulation by 10th order M-sequence.

The cross-sectional averaged velocity v_m is estimated from (Kawanisi et al., 2009):

$$v_m = \frac{u_m}{\cos\theta} = \frac{1}{\cos\theta}\frac{1}{M}\sum_{i}^{M} u_{m_i} \qquad (7.24)$$

In Equation 7.24, u_{m_i} is the averaged water-flow velocity along the ray path, and θ is the angle between the ray path and the water-flow streamline. Substituting for u_m from Equation 7.12:

$$v_m = \frac{1}{\cos\theta}\frac{1}{2M}\sum_{i}^{M} \frac{C_{m_i}^2}{R_i}\Delta t_i \qquad (7.25)$$

To estimate the cross-sectional average velocity v_m, the ray paths have to penetrate through all layers between bottom and water surface. If the sound speed has inhomogeneous distribution in water, the acoustic rays follow a curve

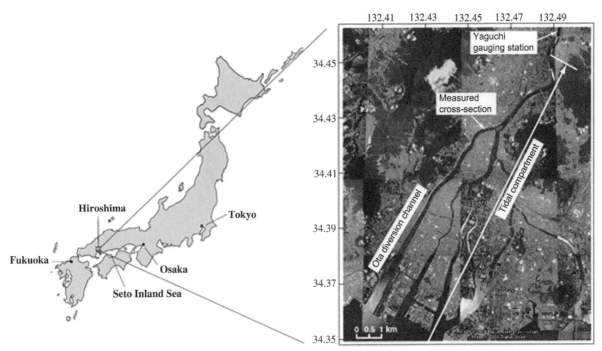

FIGURE 7.18 RAT measurement region and experimental site in Japan. *(Source: Kawanisi et al., 2009.)*

obeying Snell's law of refraction. Kawanisi et al. (2009) implemented acoustic ray simulations by solving the following differential equations (Dushaw and Colosi, 1998):

$$\frac{d\varphi}{dr} = \frac{\partial c}{\partial r}\frac{1}{c}\tan\varphi - \frac{\partial c}{\partial z}\frac{1}{c} \tag{7.26a}$$

$$\frac{dz}{dr} = \tan\varphi \tag{7.26b}$$

$$\frac{dt}{dr} = \frac{\sec\varphi}{c} \tag{7.26c}$$

In these expressions, φ is the angle of the acoustic ray relative to the horizontal axis r; z is the vertical coordinate; and t is the time. Kawanisi et al. (2009) estimated the in-water sound speed c using Medwin's formula (Medwin, 1975) as a function of temperature T (°C), salinity S, and depth D (m). Water discharge can be calculated by the RAT system from the expression (Kawanisi et al., 2009):

$$Q = A(H)v_m \sin\theta = A(H)u_m \tan\theta \tag{7.27}$$

In this expression, A is the cross-sectional area in which sound paths travel (i.e., the cross-sectional area bounded between the acoustic transceivers), and H is the water level. $A(H)$ implies that the cross-sectional area is a function of H, which undergoes temporal variability imposed by tidal height variability.

Kawanisi et al. (2009) reported a RAT experiment that was carried out during June and July 2008 at the Ota River

diversion channel. The upstream border of the tidal compartment in the Ota River estuary is about 13 km upstream, far from the mouth. The tidal range of an extreme spring tide at the mouth is about 4 m. The measurement site was located at 246 m downstream from the Gion sluice gates, as shown in Figures 7.18 and 7.19. The Ota River diversion channel at the site is 120 m wide and the water depth ranges from 0.3 m to 3 m by tide. The saltwater in this river can intrude to about 11 km upstream from the mouth.

A couple of broadband transducers were installed diagonally across the channel as shown in Figure 7.19. The

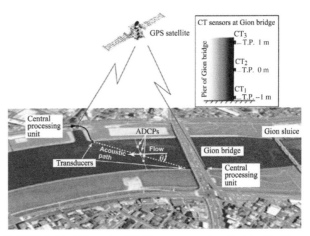

FIGURE 7.19 Schematic of RAT measurement method in Ota River, Japan. *(Source: Kawanisi et al., 2009.)*

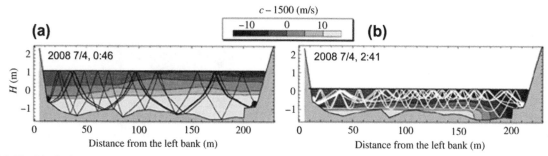

FIGURE 7.20 Distributions of sound speed at the RAT measurement site in Ota River, Japan, and results of ray simulation. *(Source: Kawanisi et al., 2009.)*

central frequency of transducers was 30 kHz. The angle θ between the acoustic ray path and the stream direction was 30°. The transducers were mounted at a height of 0.2 m above the river bottom. The altitudes of left and right transducers were −0.46 m and −0.70 m, respectively. The sound pulses of the RAT system were simultaneously transmitted from the omnidirectional transducers every minute, triggered by a GPS clock.

Three moored ADCPs were used to validate the velocity data obtained from the RAT system. These three ADCPs were arranged along the Gion Sluices in a way that each of two ADCPs were 30 m away from each other while the central ADCP location aligned with the river centerline. The distance between each ADCP and the Gion Sluice was 59.1 m.

Vertical distribution of water temperature and salinity was measured every 10 minutes by CTD sensors attached to the pier of the Gion Bridge at 40 m from the left bank. In addition, cross-sectional distributions of temperature and salinity were measured by CTD casts from the Gion Bridge. The transverse interval of the CTD casts was 20 m and crossing time was about 10 minutes.

Figure 7.20 shows distributions of the sound speed and results of the ray simulation just after HWS and just before LWS as typical examples. The tiny effect of current is not considered in the ray simulation. The sound speed was calculated from projected data of the CTD cast from the Gion Bridge. The salinity increases with depth, and as a result of this distribution, sound speed ranged from 1,515 m/s in deeper layers to 1,485 m/s near the surface.

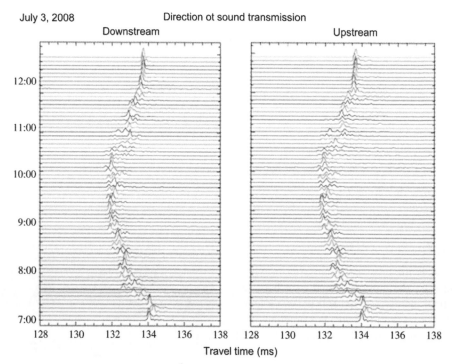

FIGURE 7.21 Stack diagrams of correlation waveforms of 10th order M-sequence modulated acoustic transmissions from upstream and downstream transducers at the RAT measurement site in Ota River, Japan. *(Source: Kawanisi et al., 2009.)*

Most of the time, the sound paths cover the cross-section, as shown in Figure 7.20a. At times a near-bed-established salt wedge caused the sound paths to be reflected; consequently, a part of sound paths were not able to penetrate into the lower layers; a typical condition under strong stratification is depicted in Figure 7.20b. In this case, the cross-sectional averaged velocity is somewhat over-estimated by the RAT system. For this reason, the cross-sectional averaged water-flow velocity had to be modified using velocity distribution of two-layer flow when there was a salt wedge under the transducer.

The cross-correlation waveforms of signals transmitted from the upstream and downstream transducers (plotted every 5 min) are shown typically in Figure 7.21. The fact that the cross-correlations obtained from both sides are of similar form suggests that the two-way path geometry is reciprocal. The broad peaks are composed of multiarrival rays. It seems that the single peak is also composed of multiarrival rays because the experimental site is shallow. The mean arrival time changes because of salinity change. Sometimes there were no clear peaks. In the study reported by Kawanisi et al. (2009), the two-way travel-time difference was calculated when the S/N ratio was over 14.

Figure 7.22 shows temporal variations of the cross-sectional averaged velocity, v_m, and the water level, H. The thick red line denotes the cross-sectional averaged velocity deduced from three moored ADCPs. The broken line denotes the water level. Because of the strong nonuniformity of flow at the observation site, data from just three ADCPs were unlikely to have well represented the cross-sectional averaged velocity. Despite this lacuna, the difference between water-flow velocities acquired by the RAT system and the ADCP is small. The observed comparison implies that the cross-sectional averaged velocity obtained from the RAT system fulfills an

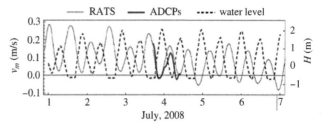

FIGURE 7.22 Comparison of cross-sectional averaged water-flow velocity between RATS and ADCPs at the RAT measurement site in Ota River, Japan. *(Source: Kawanisi et al., 2009.)*

acceptable compliance with the results derived from an array of ADCPs.

Temporal variations of water levels at Yaguchi and Gion and the flow discharge for 44 days are illustrated in Figure 7.23. It was found that the water discharge values modified using velocity distribution of two-layer flow were, on an average, ~10 percent smaller than those without correction for the observation period. It is interesting to note that the pattern of the trend of discharge, denoted by a thick line, closely resembles that of the change of water level at the Yaguchi gauging station. These observations indicate that employing RAT system is a promising method for continuous measurement of water discharge in rivers and estuaries.

Although the reciprocal sound transmission method has been employed as an innovative technology to continuously measure the cross-sectional average velocities in a tidal channel in Japan (Kawanisi et al., 2009), no such experiments were attempted to measure the variation of discharge in tidal rivers in China. However, Zhang et al. (2010) reported such an experiment, which was performed in the Qiantang River as the first application in China.

Zhang et al. (2010) carried out reciprocal sound transmission experiments for current measurement in the

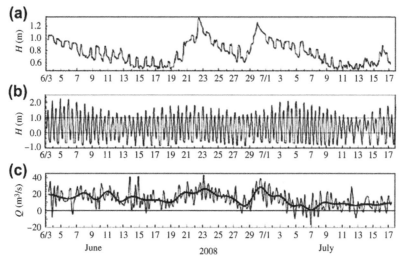

FIGURE 7.23 Temporal variations of water levels at (a) Yaguchi and (b) Gion at the RAT measurement site in Ota River, Japan, and (c) water discharge for 44 days. *(Source: Kawanisi et al., 2009.)*

upstream region of the Qiantang River, about 90 km from the mouth of Hangzhou Bay from April to December 2009. This estuary is notorious for the occasional occurrence of the dreaded tidal bores. Two systems were set up at stations east and west with a distance of 3,050 m across the Qiantang River. The transducer (a Neptune T170) was suspended down to 3 m depth through a rope from the wharf; the system components such as electronic housing, battery, and GPS antenna were placed on the wharf. The 5-kHz sound signal, modulated by one period (0.64 s) of the pseudo-random signal, called the 10th order M-sequence, was transmitted every 3 min from the broadband transducer.

Reciprocal sound transmissions were successfully performed along the sound transmission line. The travel time changed in the range 2.033–2.103 s during the whole experimental period, conforming to the expected travel time of about 2.061 s between the two stations (3,050 m / 1,480 m/s = 2.061 s). The reciprocal travel time decreased suddenly at the arrival of tidal bores. Also, travel time from west to east is larger than that from east to west. The changes in travel time during the arrivals of tidal bores are caused by the sound-speed changes due to the downstream increase of water temperature along the Qiantang River (because salinity is nearly zero).

The time plots of the path-averaged alongstream water-current flow velocity were estimated using the travel-time difference data. The tomographically measured and ship-board ADP velocities were in good agreement, producing a root-mean-square difference (RMSD) of 0.03 m/s. The tomographically measured water-current flow velocity made a dramatic change during the arrival of tidal bores. Before their arrival, freshwater flowed eastward from upstream to downstream with velocities ranging from 0.25 m/s to 0.82 m/s. The eastward freshwater flow became weak and then changed its directions from eastward to westward within a short period of about 20 minutes when the tidal surges arrived at the experiment site. This west-ward flow reached maximum velocities ranging from −0.11 m/s to −1.18 m/s, then kept its direction and magnitude for about 100 minutes constant. This westward flow was gradually replaced by the eastward flow, and this "drama" finished in about 5 hours and 20 minutes after the arrival of the tidal bores. This "drama" occurred twice one lunar day in a synchronized timing with the semidiurnal tide in Hangzhou Bay.

The acoustic travel-time differences $\Delta\tau$ data have been found to be well correlated with the ADP data (Q_{ADP}), with a linear relation. The correlation coefficient is 0.99, and the RMSD from the regression line is 241 m^3/s. It was found that the empirical formula that connects tomographically obtained river discharge (Q_{TOMO}) and $\Delta\tau$ may be expressed by (Zhang et al., 2010):

$$Q_{TOMO} = -2400 \times 10^3 \Delta\tau + 179 \qquad (7.28)$$

Zhang et al. (2010) estimated the river discharge during the whole experimental period in a time series through this empirical formula. It is seen that Q_{TOMO} has a quite good agreement with Q_{ADP}. The variations of the river discharge were very similar to those of the velocity. During the arrival of tidal bores, Q_{TOMO} changed not only in its values but also in its directions, meaning that the river discharge changed from eastward to westward. The eastward freshwater discharge reached a maximum value of 5,096 m^3/s during 0:00–6:30 of August 13, 2009, when a rainstorm attacked the upstream region of the Qiantang River just before the experiment. On the other hand, the maximum westward discharge of −7,626 m^3/s (from downstream to upstream) occurred in the spring tide during 2:0–3:40 h of September 19, 2009, when the tidal bores became the strongest in all the experimental periods. The mean of Q_{TOMO} for the whole experimental period was 1,246 m^3/s and equivalent to the averaged freshwater discharge of the Qiantang River.

7.6. ACOUSTIC TOMOGRAPHIC MEASUREMENTS OF VORTICITY

The concepts of vorticity and especially its conservation have played a central role in furthering our understanding of the meso- and large-scale oceanic and atmospheric circulations (Rossby, 1975; Conzemius and Montgomery, 2009). Through Gauss's theorem, the tendency of circulation for any enclosed area fixed in space can be written in terms of the line integral of the flux component normal to the boundary of the area. Circulation in an ocean basin is often associated with vorticity. Because solid Earth spins, a water mass that resides on it will suffer an additional vorticity associated with the spinning Earth. Thus, the vorticity measured by an observer on Earth is not absolute vorticity but is *relative vorticity*. Estimates of relative vorticity are necessary to determine the long-term balance of energy and momentum in an ocean basin. Muller et al. (1986) pointed out the importance of the vorticity mode of motion that must coexist with internal gravity waves at small scales. Conventional methods of measuring currents do not allow for an accurate calculation of the vertical component of vorticity on geostrophic time scales. Because relative vorticity and horizontal divergence are difficult to measure using conventional instruments, the vortical mode of motion has traditionally been ignored. Muller et al. (1988) made an attempt to measure potential vorticity at small scales in the ocean from measurements at three discrete locations using a triangular array of current meters. However, such discrete measurements can give rise to significant sampling errors because the measurements do not represent a continuous line integral (i.e., path-averaged measurement) along the sides of the triangle.

As we have already noted, reciprocal acoustic transmission methods have proved to be successful in several experiments, and this provides optimism for the feasibility of ocean basin-wide measurements. This technique has indeed been found to be useful for estimation of relative vorticity. In fact, the most exciting application of reciprocal transmissions is considered to be the measurement of relative vorticity. This concept of determining the relative vorticity of oceanic motion in a realistic manner (i.e., from continuous path-averaged measurements) was originally suggested by Rossby (1975), although its practical implementation came much later. According to this concept, acoustic transmissions in opposite directions around a closed loop give circulation directly. This quantity, on basin scale, is quite difficult to measure using traditional approaches. In addition, its unprecedented sensitivity allows detection of phenomena not observable with traditional sensors.

Even in the absence of appropriate technologies, Thomas Rossby of Yale University's Department of Geology and Geophysics was optimistic that as more sophisticated studies of oceanic motion were mounted, increasing stress would be laid on our ability to reveal the higher-order vorticity dynamics. Accordingly, he proposed a new approach to this measurement problem, which is based on an acoustical method for determining the circulation around a closed path (Rossby, 1975). Once the circulation around a closed path is known, Stokes' theorem can be applied to obtain the average vorticity of the "enclosed" fluid. The central element of Rossby's scheme is the notion that the line integral of fluid velocity along a ray between two points is proportional to the difference in travel time of two acoustic signals, transmitted in opposite directions. (This concept has already been elaborated in this book under reciprocal acoustic method of ocean-current measurement.) If v is the velocity vector in a flow field in the direction of the flow stream, and s is a unit vector along the acoustic propagation path (which can be oriented in any direction relative to v), the line integral of fluid motion around an N-cornered polygon (i.e., fluid circulation around the polygon) becomes $\oint v.ds$. Stokes' theorem tells us that this fluid circulation is equivalent to the surface integral of the normal component of vorticity, but this is the average vorticity, ξ, times the area, A, encompassed by the closed loop. That is, fluid circulation around the polygon, by virtue of Stokes' theorem, is equivalent to the areal-average relative vorticity of the enclosed fluid. In mathematical terms,

$$\oint v.ds = \xi \times A \qquad (7.29)$$

Thus, the average vorticity is given by:

$$\xi = \frac{1}{A} \oint v.ds \qquad (7.30)$$

Rossby (1975) has called attention to the fact that for each side of the polygon we also have a line-averaged

current meter (equivalent to reciprocal tomographic current measurement). He suggested that if the N-cornered polygons are placed horizontally in the ocean so that only the vertical vorticity is sensed, it would be possible to determine a numerical relationship between ξ and the travel-time difference around the loop. While conceptualizing the applicability of reciprocal acoustic transmission method for estimation of vorticity, Rossby was optimistic that although the acoustic travel-time difference is very small, it would be possible with very narrow band-pass filters or phase-lock techniques to achieve the desired timing resolution (of the order of μs) using a CW frequency of, say, 10 kHz.

As pointed out under the preceding discussion, estimation of relative vorticity of fluid motion in an ocean basin requires measurement of path-averaged fluid velocity around a closed loop (Figure 7.24). This, in the reciprocal acoustic transmission method, is equivalent to the measurement of travel-time differences of acoustic propagation along this closed loop. From the theory of ATT current meters (see Section 9.2), the travel-time difference Δt_{ij} along an acoustic ray path of range l_{ij} joining points P_i and P_j is given by:

$$\Delta t_{ij} = \frac{(2v_{ij} \times l_{ij})}{(c_{ij})^2} \qquad (7.31)$$

where c_{ij} and v_{ij} are the path-averaged sound speed in seawater at rest and fluid velocity, respectively, along the ray path joining P_i with P_j. Thus,

$$v_{ij} = \frac{\Delta t_{ij} \times (c_{ij})^2}{2l_{ij}} \qquad (7.32)$$

The line integral of the fluid motion between points P_i and P_j can, under the assumption that the flow is momentarily steady, be written as:

$$\int_{P_i}^{P_j} v.ds = v_{ij} \times l_{ij} = \frac{\Delta t_{ij} \times (c_{ij})^2}{2} \qquad (7.33)$$

where v is the velocity vector and s is a unit vector along the acoustic propagation path. Estimation of vorticity, based on

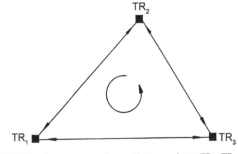

FIGURE 7.24 Arrangement of acoustic transceivers TR_1, TR_2, and TR_3 (minimum configuration) for measurement of vorticity.

the method proposed by Rossby (1975), requires estimation of the line integral of fluid motion around a closed loop. Because a minimum of three lines are required to form a closed loop, a minimum configuration of instrumentation required for estimation of relative vorticity is a set of acoustic transceivers placed on the vertices of a triangle (see Figure 7.24). Thus, the line integral of fluid motion around a triangle $P_1P_2P_3$ becomes:

$$\oint v.ds = (v_{12} \times l_{12}) + (v_{23} \times l_{23}) + (v_{31} \times l_{31}) = \frac{\left[\Delta t_{12} \times (c_{12})^2\right] + \left[\Delta t_{23} \times (c_{23})^2\right] + \left[\Delta t_{31} \times (c_{31})^2\right]}{2} \quad (7.34)$$

Let us make the assumption that the triangle lies on a horizontal plane. Equation 7.30 gets modified as:

$$\xi = \frac{\left[\Delta t_{12} \times (c_{12})^2\right] + \left[\Delta t_{23} \times (c_{23})^2\right] + \left[\Delta t_{31} \times (c_{31})^2\right]}{2A} \quad (7.35)$$

Assuming that c_{12}, c_{23}, and c_{31} are nearly the same and equal to c, we get:

$$\xi = \left(\frac{c^2}{2A}\right)(\Delta t_{12} + \Delta t_{23} + \Delta t_{31}) \quad (7.36)$$

Taking into account a situation in which adequate numbers of acoustic transceivers are configured on an isosceles triangle or square of 3-km/side, placed horizontally in the deep ocean, Rossby discussed possible error budgets in the measurement of acoustic travel-time difference. According to him, although the assumption that c and v are temporarily steady is not strictly valid, if the time difference is based on simultaneously transmitted signals, as it certainly should be, then it is only a variability in the timeframe between transmit and receive that can affect the measurement (~2 seconds for a 3-km path). Variations in the water-current flow speed on this time scale will be negligible. The local speed of sound is especially stable in deep waters and for a given pressure will depend mostly on local water-temperature variations, which are of order 3–10 centi-degrees. That is equivalent to changes in the sound speed of 15–50 cm/s. Such variations result from the advection of water masses, which are large compared to the polygon mentioned earlier.

The temporal variance in temperature on a "seconds time scale" is one of milli-degrees or less. The effect of this on the speed of sound is equivalent to less than 1 cm/s, and when this is also line averaged, it can be confidently stated that the path-averaged sound speed in deep water should have quite negligible variability within the 2-s timeframe (for a 3-km path). Spatial variations in the speed of sound due to temperature and salinity are

by themselves unimportant because the quantity Δt is a differential measurement along the same path. On the strength of these considerations and rapid advances achieved on the technology front, Menemenlis and Farmer (1992) employed this concept to measure path-averaged horizontal current and relative vorticity of water mass in the sub-ice boundary layer of the eastern Arctic.

7.7. HORIZONTALLY INTEGRATED CURRENT MEASUREMENTS USING SPACE-TIME ACOUSTIC SCINTILLATION ANALYSIS TECHNIQUE

An alternative acoustic method developed to measure coastal water-current flows is the *acoustic scintillation method*. The acoustic scintillation technique of flow measurement is similar in principle to the well-known tracer technique wherein a dye, a chemical, or a radioactive isotope is injected into the flowing fluid, and its transportation over a known distance is timed to estimate the flow velocity, using the definition of velocity (i.e., distance traveled in unit time). The essential difference between the tracer method and the acoustic scintillation method is that whereas a foreign substance is injected into the flow field in the tracer method, the "tracer" in the acoustic scintillation method is merely some acoustically detectable random fluctuations already existing naturally in the flow field.

The random fluctuations can take many forms, such as velocity turbulence, presence of a second phase, and temperature fluctuations. Internal waves, layering, and turbulence create inhomogeneities of sound speed in the ocean, and these inhomogeneities induce fluctuations in the ocean acoustic transmission. Through both refractive and diffractive effects, sound waves from a point source are perturbed from the simple spherical or plane-wave geometry into more complicated phase fronts (Duda and Flatte, 1988). A wave propagating through a random medium, such as a turbulent flow field, undergoes distortions in its amplitude and phase. Initially, only the phase of the wave is distorted as it propagates through the turbulent medium. These phase distortions or wrinkles in the wave front redirect the wave's energy, eventually producing both amplitude and phase distortions and, finally, intensity fluctuations (i.e., scintillation) due to interference across some distant receiving plane (Clifford and Farmer, 1983). This results in a complicated diffraction pattern, similar in nature to the stellar scintillations (twinkling of stars) as observed near the focal plane of an astronomical telescope. In analogy, a similar pattern at the receiver of an acoustic transducer is known as *acoustic scintillation*. The pattern of irregular intensity (the scintillation pattern) in a receiving plane, perpendicular to the acoustic propagation axis, evolves with time. The evolution of the scintillation pattern occurs because of advection and/or decay of the density fluctuations or eddies that produce the wave perturbation. If the transit time of the scintillation pattern across the detectors is short compared to the eddy lifetimes, it is possible to derive quite accurate estimates of the intervening transverse flow from a statistical analysis of the scintillation pattern.

Scintillation techniques have long been employed to elucidate the fine scale structure and motion of the turbulent media. The acoustic scintillation analysis technique is, in fact, a spin-off from the applied optics wherein the properties of the intervening medium such as ionospheric, solar, and atmospheric winds were determined from an analysis of turbulence-induced stellar scintillations. Because of the unique spatial distribution of medium irregularities for each of these applications (i.e., concentrated at one position along the path or uniformly distributed), different techniques have been devised to estimate flow information. However, in acoustic scintillation analysis only three methods have so far been applied to estimate oceanic water current flows.

Acoustic scintillation technique has successfully been used for measurement of water flows in straits and channels. The minimum configuration of a flow-measuring device that employs this technique consists of a single acoustic transmitter and two acoustic receivers located in such a way that the acoustic transmission path is transverse to the flow direction (see Figure 7.25). The separation between the transmitter and the receiver ranges from a few meters to a few km. The two acoustic receivers are placed at a known distance apart in the flow field so that the receiving planes of these two hydrophones are normal to the acoustic path. The scintillation pattern at the receiver plane drifts with the flow. Upon reception, the acoustic signal from one receiver is compared with that from the other receiver to find the time lag, T, at which the scintillation patterns at the two receivers have maximum similarity or a best fit in waveform (i.e., *autocorrelation*). It is well known that autocorrelation is a mathematical representation of the degree of similarity between a given time series and a lagged version of itself over successive time intervals. It is the same as calculating the correlation between two different time series, except that the same time series is used twice (i.e., once in its original form and once lagged one or more time periods). The term can also be referred to as *lagged correlation* or *serial correlation*. From the estimated time lag, T, and the known receiver separation, the flow velocity is computed.

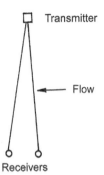

FIGURE 7.25 Sketch of minimum configuration of an acoustic scintillation device, consisting of a single acoustic transmitter and two acoustic receivers located in such a way that the acoustic transmission path is transverse to the flow.

Flow measurement using scintillation analysis technique is based on Taylor's frozen-flow hypothesis, according to which the fluctuations in the refractive index at any point result solely from the advection of a frozen refractive index field past that point by a mean flow. This hypothesis is reasonably well satisfied by atmospheric and oceanic turbulence over short distances, but it is violated by internal waves. This is the reason that the method has not been used for measurement of flows in the open ocean where internal waves may be present. However, in straits and tidal channels, the flow is mainly driven by tide, and acoustic scintillation arises only from turbulence. In these environments Taylor's hypothesis is well satisfied and the acoustic scintillation technique can be applied for measurement of water-current flows. As the refractive index irregularities are advected *across* the acoustic propagation path, the structure in the scintillation pattern drifts across the receiving plane with a speed proportional to the flow speed if weak scattering restriction is satisfied. The value of the proportionality constant depends on the type of incident wave. For example, the pattern drift-velocity V_p for an incident spherical wave is given by $2V_t$, where V_t is the transverse flow velocity. The factor of 2 arises because of the diverging nature of the spherical wave field and the sensitivity of the measurement to refractive eddies at mid-path (Clifford et al., 1990). For an incident plane wave, the pattern drift velocity V_p is equal to V_t.

It has been noted that the acoustic scintillation analysis technique is applicable for flow measurement if weak scattering restriction is satisfied. Over short ranges, characterized well by a single transmission path from a source to any receiver, the wave front may be subjected to effects that are weak enough that the variation of signal amplitude in a plane transverse to the transmission path is a small fraction of the average amplitude. This is the weak scattering or unsaturated regime. Weak scattering theory has been shown to accurately describe wave transmission when the normalized variance of intensity (sometimes called the

scintillation index) is less than 1. For centimeter acoustic
wavelength, the acoustic path length in a flow field is
limited to a few km.

Two techniques that are generally used in the scintillation analysis method of flow measurements are based on
peak delay and slope of covariance at zero lag. Clifford and
Farmer (1983), who originally used acoustic scintillation
analysis for ocean-flow measurement, employed the latter
technique. In this method, a time-lagged covariance function (*C*) at displacement (*d*) is constructed to study the
spatio-temporal variation of the scintillation patterns at the
two receivers. The slope at zero time lag of the normalized
covariance is proportional to the path-weighted average of
the transverse flow (*u*) and is given by (Clifford and Farmer,
1983):

$$m_N = \frac{1}{L} \int_0^L W_y U_y dy \qquad (7.37)$$

where *y* is the constant plane, the refractive index fluctuations along which have given rise to the scintillation drift; *L*
is the acoustic path length; and W_y is a weighting function.
In general, *U* can be expressed as $U = R \times m_N$, where *R* is
a calibration factor in meters. *R* takes different values for
different turbulence spectral power laws and spacing
$B = \left(\frac{d}{R_f}\right)$, where R_f is the Fresnel radius given by the
square root of the product of the acoustic wavelength with
the path length. Values of the calibration factor for different
values of *B* and spectral power laws are given in Clifford
and Farmer (1983).

In the first experiment conducted by Clifford and
Farmer (1983) to test the validity of the scintillation technique for ocean-flow measurement, the acoustic transducers
were mounted on two masts, which were rigidly mounted
on the leading edge of a research barge that could be towed
at various speed steps (see Figure 7.26). The transducers
were maintained at a depth of 2.1 m below the water
surface. The transmitter-receiver separation was 12.4 m.
The separation of the two receiving hydrophones, oriented
along a plane perpendicular to the acoustic transmission
path, was set at 15.7 cm, which was approximately one-half
of a Fresnel radius for the given experimental setup.

The experiment was conducted in Saanich Inlet in 1982.
Both the projector (*P*) and the receiving hydrophones (*R*)
consisted of single hexagonal elements of 6 cm diameter.
The projector was driven at approximately 214 kHz with
5 cm pulses once every 100 ms. For a transmitter-receiver
separation of 12.4 m, the receiving hydrophones detected
only the direct pulse, without contamination by multipath
propagation. The signals received at the hydrophones were
amplified, detected with an rms-to-DC converter, filtered,
digitized, and recorded. When the flow was zero, the
signals received at both the hydrophones displayed a low-

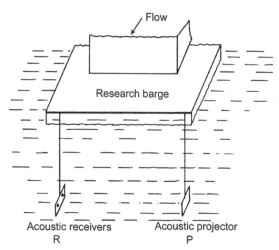

FIGURE 7.26 A sketch of the initial experimental arrangement used to
test the practicability of acoustic scintillation technique of flow measurement. (*Source: Modified from Clifford and Farmer, 1983.*)

frequency fluctuation associated with slowly moving and
changing patterns in the refractive index along the acoustic
path. With increase in flow, the signal frequency greatly
increased, as anticipated on theoretical grounds.

Data analysis begins with breaking the edited time
series of data received at the two hydrophones into 1-min
segments, followed by log normalization and high-pass
filtering. For each segment, the cross-covariance is calculated, together with the first central difference estimate of
the cross-covariance slope at zero lag. Three examples of
cross-covariance functions, corresponding to zero flow and
flows in two opposite directions, are shown in Figure 7.27.
It can be seen that the peak of the normalized covariance
corresponding to zero flow lies on the zero-lag axis,
whereas the peaks corresponding to nonzero flows get
shifted with reference to this axis. The position of the peak
also gives an indication of the flow direction.

The amount of the shift is a measure of the magnitude of
the flow normal to the acoustic path. The magnitude and

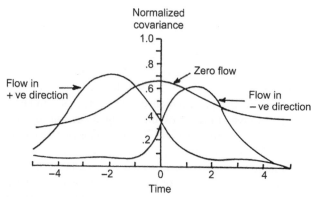

FIGURE 7.27 Cross-covariance as a function of lag for still water and
flows in opposite directions. One unit on the abscissa is 0.1-s time lag.
(*Source: Modified from Clifford and Farmer, 1983.*)

sign of the slope of the normalized covariance at zero lag can also be used for estimation of the magnitude and direction of the flow perpendicular to the acoustic path. Initial experiments have indicated that the time series of path-averaged flow transverse to the acoustic propagation path, derived from a statistical analysis of the scintillation pattern at the two receiving hydrophones, compared favorably with independently determined flow-speed measurements using an Aanderaa current meter, thus demonstrating the viability of the scintillation technique for remote probing of ocean flows. Subsequently, this technique was used to measure flows across a 0.66-km path in a channel, and the results compared well with current measurements obtained from a moored current meter (Farmer and Clifford, 1986). Investigations by Farmer et al. (1987) showed that acoustic scintillation technique can also be used to study the fine structure of turbulent flows. Menemenlis and Farmer (1992) used the scintillation technique to study the ice-water boundary layer. Thus, the path-integral scheme for measurement of turbulence and mixing appears to prove a useful adjunct to the conventional flow measurement techniques. Further investigations have shown that, by spatial filtering of acoustic scintillations from an array of hydrophones, it is possible to obtain horizontal profiles of fine-scale variability and transverse current along short (up to ~2 km) acoustic propagation paths (Farmer and Crawford, 1991; Clay and Medwin, 1977). So far, the application of the acoustic scintillation method is restricted to ranges of less than 1 km because the range is suppressed by ambient noise (Send et al., 2002).

REFERENCES

Aki, K., Richards, P.G., 1980. Quantitative Seismology. vol. 2. Freeman, pp. 932.

Arthur, R.S., Munk, W.H., Issacs, J.D., 1952. The direct construction of wave rays. Trans. Am. Geophys. Un. 33, 855–865.

Behringer, D., Birdsall, T., Brown, M., Cornuelle, B., Heinmiller, R., Knox, R., Metzger, K., Munk, W., Spiesberger, J., Spindel, R., Webb, D., Worcester, P., Wunsch, C., 1982. A demonstration of ocean acoustic tomography. Nature 299, 121–125.

Bowden, K.F., 1954. The direct measurement of subsurface currents in the oceans. Deep-Sea Res. 2, 33–47.

Bretherton, F.P., Davis, R.E., Fandry, C., 1976. A technique for objective analysis and design of oceanographic experiments applied to MODE-73. Deep-Sea Res. 23, 559–582.

Carter, E.F., Robinson, A.R., 1987. Analysis models for the estimation of oceanic fields. J. Atmos. Ocean. Technol. 4, 49–74.

Catherine, A.R., DeRose, J.B., 2004. Investigation of hydroacoustic flow-monitoring alternatives at the Sacramento river at Freeport, California: Results of the 2002–2004 pilot study. Scientific Investigation Report, USGS(2004-5172). Virginia, VA, USA.

Chen, C.T., Millero, F.J., 1977. Speed of sound in sea water at higher pressure. J. Acoust. Soc. Amer. 60, 129–135.

Chiu, C.L., Hsu, S.M., 2006. Probabilistic approach to modeling of velocity distributions in fluid flows. J. Hydrology 316, 28–42.

Clay, C.S., Medwin, H., 1977. Acoustical oceanography: Principles and applications. Wiley, New York, NY, USA.

Clifford, S.F., Farmer, D.M., 1983. Ocean flow measurements using acoustic scintillation. J. Acoustic Soc. Am. 74, 1826–1832.

Clifford, S.F., Farmer, D.M., Lataitis, R.J., Crawford, G.B., 1990. Ocean remote sensing by acoustic scintillation techniques. In: Singal, S.P. (Ed.), Acoustic remote sensing. Tata McGraw-Hill, Uttar Pradesh, India, pp. 189–197.

Conzemius, R.J., Montgomery, M.T., 2009. Clarification on the generation of absolute and potential vorticity in mesoscale convective vortices. Atmos. Chem. Phys. Discuss 9, 7531–7554.

Cornuelle, B., 1983. Inverse methods and results from the 1981 ocean acoustic tomography experiment. Ph.D. Thesis. Massachusetts Institute of Technology/Woods Hole Oceanographic Institution, Cambridge, MA, USA.

Cornuelle, B.C., Wunsch, C., Behringer, D., Birdsall, T., Brown, M., Heinmiller, R., Knox, R., Metzger, K., Munk, W.H., Spiesberger, J., Spindel, R., Webb, D., Worcester, P., 1985. Tomographic maps of the ocean mesoscale: Pure acoustic. J. Phys. Oceanogr. 22, 133–152.

De Ferrari, H.A., Nguyen, H.B., 1986. Acoustic reciprocal transmission experiments, Florida straits. J. Acoust. Soc. Amer. 79, 299–315.

Duda, T.F., Flatte, S.M., 1988. Modeling meter-scale acoustic intensity fluctuations from oceanic fine structure and microstructure. J. Geophys. Res. 93, 5130–5142.

Dushaw, B.D., Colosi, J.A., 1998. Ray tracing for ocean acoustic tomography, Technical Memorandum. Applied Physics Laboratory, University of Washington, Seattle, WA, USA. TM 3–98.

Dushaw, B.D., Colosi, J.A., 1998. Ray tracing for ocean acoustic tomography, Applied Physics Laboratory Report. University of Washington, Seattle, WA, USA.

Elisseeff, P., Schmidt, H., Johnson, M., Herold, D., Chapman, N.R., McDonald, M.M., 1999. Acoustic tomography of a coastal front in Haro Strait, British Columbia. J. Acoust. Soc. Amer. 106, 169–184.

Evensen, G., 1994. Sequential data assimilation with a nonlinear quasi-geostrophic model using Monte Carlo methods to forecast error statistics. J. Geophys. Res. 99, 10,143–10,162.

Farmer, D.M., Crawford, G.B., 1991. Remote sensing of ocean flows by spatial filtering of acoustic scintillations: Observations. J. Acoustic Soc. Am. 90, 1582–1591.

Farmer, D.M., Clifford, S.F., 1986. Space-time acoustic scintillation analysis: a new technique for probing ocean flows. IEEE. J. Oceanic Eng. OE-11, 42–50.

Farmer, D.M., Clifford, S.F., Verrall, J.A., 1987. Scintillation structure of a turbulent tidal flow. J. Geophys. Res. 92, 5369–5382.

Gac, C., Gall, Y.L., Terre, T., 1999. For ocean acoustic tomography new modular instrumentation. Sea Technol. 40 (3), 55–60.

Grund, M., Johnson, M., Herold, D., 1997. Haro Strait tidal front mapping experiment. Technical report, Woods Hole Oceanographic Institution. February 1997.

Gytre, T., 1976. The use of a high-sensitivity ultrasonic current meter in an oceanographic data acquisition system. Radio and Electronic Engineer 46 (12), 617–623.

He, J.M., 2008. The methods to determine stage-discharge relation and examine error. J. Water Conservancy Science, Technology and Economy 14 (9), 700–701.

Herman, G.T., 1980. Image reconstruction from projections: The fundamentals of computerized tomography. Academic Press, Waltham, Massachusetts, USA, pp. 316.

Herman, G.T. (Ed.), 1979. Image reconstruction from projections: Implementation and applications. Springer-Verlag, London, UK, pp. 284.

Jensen, F.B., Kuperman, W.A., Porter, M.B., Schmidt, H., 1994. Computational ocean acoustics. American Institute of Physics, New York, NY, USA.

Kaimal, J.C., 1980. Sonic anemometers. In: Dobson, F., Hasse, L., Davis, R. (Eds.), Air-sea interaction: Instruments and methods. Plenum Press, New York, NY, USA, pp. 81−96.

Kawanisi, K., Watanabe, S., Kaneko, A., Abe, T., 2009. River acoustic tomography for continuous measurement of water discharge. 3rd International Conference and Exhibition on Underwater Acoustic Measurements: Technologies & Results, 613−620.

Ko, D.S., De Ferrari, H.A., Malanotte-Rizzoli, P., 1989. Acoustic tomography in the Florida Strait: Temperature, current, and vorticity measurements. J. Geophys. Res. 94, 6197−6211.

Kumar, S.P., Navelkar, G.S., Murthy, T.V.R., Murthy, C.S., 1977. Acoustical characteristics and simulated tomographic inversion of a cold core eddy in the Bay of Bengal. Acustica-Acta Acustica 83, 847−854.

Kumar, S.P., Murty, T.V.R., Somayajulu, Y.K., Chodankar, P.V., Murty, C.S., 1994. Reference sound speed profile and related ray acoustics of Bay of Bengal for tomographic studies. Acustica 80, 127−137.

Larsen, J.C., Sanford, T.B., 1985. Florida current volume transports from voltage measurements. Science 227, 302−304.

Leaman, K.D., Vertes, P.S., Atkinson, L.P., Lee, T.N., Hamilton, P., Waddel, E., 1995. Transports, potential vorticity and current/temperature structure across Northwest Providence and Santaren Channels and the Florida Current off Cay Sal Bank, J. Geophysical Res. Lett. 27 (18), 2949−2953.

Lee, T.N., Schott, F.A., Zantopp, R., 1985. Florida Current: low-frequency variability as observed with moored current meters during April 1982 to June 1983. Science 227, 298−302.

Liebelt, P.B., 1967. An Introduction to Optimal Estimation. Addison-Wesley, Boston, USA.

Lin, H.Y., 2003. Application of ADCP for river hydrometry. J. China Water Resources 9 (B), 37−39.

Lin, J., Kaneko, A., Gohda, N., Yamaguchi, K., 2005. Accurate imaging and prediction of Kanmon Strait tidal current structures by the coastal acoustic tomography data. Geophys. Res. Lett. 32. L14607.

Lynch, J.F., 1995. Tomography via motion-decoupled surface buoy. Sea Technol 36 (9), 25−31.

Mackenzie, K.V., 1981. Nine-term equation for sound speed in the oceans. J. Acoust. Soc. Am. 70, 807−812.

Madsen, H., Canizares, R., 1999. Comparison of extended and ensemble Kalman filters for data assimilation in coastal area modeling. Int. J. Numer. Math. Fluids 31, 961−981.

Maul, G.A., Chew, F., Bushnell, M., Mayer, D.A., 1985. Sea-level variation as an indicator of Florida Current volume transport: comparisons with direct measurements. Science 227, 304−307.

Medwin, H., 1975. Speed of sound in water: A simple equation for realistic parameters,. J. Acoust. Soc. Am. 58, 1,318.

Menemenlis, D., Farmer, D.M., 1992. Acoustic measurement of current and vorticity beneath ice. J. Atmos. Oceanic Technol. 9, 827−849.

Muller, P., Holloway, G., Henyey, F., Pomphrey, N., 1986. Non-linear interactions among internal gravity waves. Review of Geophysics 24, 493−536.

Muller, P., Ren-Chich, L., Williams, R., 1988. Estimates of potential vorticity at small scales in the ocean. J. Phys. Oceanogr. 18, 401−416.

Munk, J.W., Wunsch, C., 1979. Ocean acoustic tomography: A scheme for large scale monitoring,. Deep-Sea Res. 26A, 123−161.

Munk, W., Wunsch, C., 1982a. Observing the ocean in the 1990s. Phil. Trans. Roy. Soc. London A307, 439−464.

Munk, W., Wunsch, C., 1982b. Up-down resolution in ocean acoustic tomography. Deep-Sea Res. 29, 1415−1436.

Munk, W., Wunsch, C., 1983. Ocean acoustic tomography: Rays and modes. Rev. Geophys. and Space Phys. 21 (4), 777−793.

Munk, W., Worcester, P.F., Wunsch, C., 1995. Ocean acoustic tomography. Cambridge University Press, New York, NY, USA.

Nowak, R.T., Mealy, S.F., 1981. Mooring motion monitoring acoustic navigation system. Proc. IEEE Oceans '81 Conf, 793−796.

Okujima, M., Ohtsuki, S., 1981. Observation of underwater ultrasound propagation in the sea by two-period M-sequence signal method. Japanese J. Appl. Phys. 20, 237−239.

Park, J.H., Kaneko, A., 2000. Assimilation of coastal acoustic tomography data into a barotropic ocean model. Geophys. Res. Lett. 27 (20), 3373−3376.

Parker, R., 1977. Understanding inverse theory. Ann. Rev. Earth Planet. Sci. 5, 35−64.

Pierce, A., 1989. Acoustics: An introduction to its physical principles and application. Acoustical Society of America, New York, NY, USA.

Rossby, T., 1975. An oceanic vorticity meter. J. Marine Res. 33 (2), 213−222.

Send, U., Worcester, P.F., Cornuelle, B.D., Tiemann, C.O., Baschek, B., 2002. Integral measurements of mass transport and heat content in the Strait of Gibraltar from acoustic transmissions. Deep-Sea Research II 49, 4069−4095.

Simon, M.K., Omura, J.K., Scholtz, R.A., Levin, B.K., 1994. Spread-spectrum communications handbook. McGraw-Hill, New York, NY, USA.

Spiesberger, J.L., Spindel, R.C., Metzger, K., 1980. Stability and identification of long range ocean acoustic multi-paths. J. Acoust. Soc. Amer. 67, 2011−2017.

Spindel, R.C., 1979. An underwater acoustic pulse compression system. IEEE Trans. Acoust. Speech and Sig. Proc. ASSP-27, 723−728.

Spindel, R.C., Worcester, P.F., 1991. Ocean acoustic tomography: A decade of development. Sea Technol. 32 (7), 47−52.

Spindel, R.C., Worcester, P.F., Webb, D.C., Boutin, P.R., Peal, K.R., Bradley, A.M., 1982. Proc. IEEE Oceans '82 Conference Record, 92−99.

Spindel, R.C., Porter, R.P., Webb, D.C., 1977. A mobile coherent low-frequency acoustic range. IEEE J. Oceanic Eng. OE-2, 331−337.

Spofford, C.W., Stokes, A.P., 1984. An iterative perturbation approach for ocean acoustic tomography. J. Acoust. Soc. Amer. 75, 1443−1450.

Stallworth, L.A., 1973. A new method for measuring ocean and tidal currents. IEEE Oceans '73, 55−58.

Stommel, H., 1955. Direct measurements of subsurface currents. Deep-Sea Res. 2, 284−285.

Suesser, U., 1990. Acoustic propagation in the Strait of Gibraltar, M.Sc. thesis. Royal Roads Military College. British Columbia, Canada, Victoria.

Takasugi, T., Fujiwara, T., Sugimoto, T., 1994. Formation of sand banks due to tidal vortices around straits. J. Oceanogr. 50, 81−98.

Urick, R.J., 1983. Principles of Underwater Sound, third ed. McGraw-Hill, New York, NY, USA.

Wang, F., Huang, H., 2005. Horizontal acoustic Doppler current profiler (H-ADCP) for real-time open channel flow measurement: Flow calculation model and field validation. In: XXXI IAHR CONGRESS. Seoul, Korea, pp. 319−328.

Wenz, G.M., 1962. Acoustic ambient noise in the ocean: spectra and sources. J. Acoust. Soc. Am. 34, 1936−1956.

Worcester, P.F., 1977. Reciprocal acoustic transmission in a mid-ocean environment. J. Acoustic Society of America 62, 895−905.

Worcester, P.F., Spindel, R.C., Howe, B.M., 1985. Reciprocal acoustic transmissions: Instrumentation for mesoscale monitoring of ocean currents. IEEE J. Oceanic Eng. OE-10, 123−137.

Wunsch, C., Minster, J.-F., 1982. Methods for box models and ocean circulation tracers: Mathematical programming and nonlinear inverse theory. J. Geophys. Res. 87, 5647−5662.

Yamaguchi, K., Lin, J., Kaneko, A., Yayamoto, T., Gohda, N., H-Nguyen, Q., Zheng, H., 2005. A continuous mapping of tidal current structures in the Kanmon Strait. J. Oceanography 61, 283−294.

Yamoaka, H., Kaneko, A., Park, Jae-Hun, Zheng, H., Gohda, N., Takano, T., Zhu, X.-H., Takasugi, Y., 2002. Coastal acoustic tomography system and its field application. IEEE J. Oceanic Eng. 27 (2), 283−295.

Zhang, C., Zhu, X.-H., Kaneko, A., Wu, Q., Fan, X., Li, B., Liao, G., Zhang, T., 2010. Reciprocal sound transmission experiments for current measurement in a tidal river, 2010 IEEE, Solaris, Singapore.

Zheng, H., Gohda, N., Noguchi, H., Ito, T., Yamaoka, H., Tamura, T., Takasugi, Y., Kaneko, A., 1997. Reciprocal sound transmission experiment for current measurement in the Seto Inland Sea, Japan. J. Oceanogr. 53, 117−127.

Zhu, X.H., Kaneko, A., Wu, Q., Gohda, N., Zhang, C., Taniguchi, N., 2010. The first Chinese coastal acoustic tomography experiment. Proc. IEEE, Solaris, Singapore.

Zlotnicki, V., 1983. The oceanographic and geoidal components of sea surface topography. Ph.D. Thesis, MIT/WHOI, Cambridge, MA, USA.

BIBLIOGRAPHY

Bellingham, J.G., Schmidt, H., Deffenbaugh, M., 1996. Acoustically focused oceanographic sampling in the Haro Strait experiment. J. Acoust. Soc. Am. 100, 2612.

Canizares, R., Madsen, H., Jensen, H.R., Vested, H.J., 2001. Developments in operational shelf sea modeling in Danish waters. Estuarine Coastal Shelf Sci. 53, 595−605.

Chester, D.B., Malanotte-Rizzoli, P., DeFerrari, H.A., 1991. Acoustic tomography in the Straits of Florida. J. Geophys. Res. 96, 7023−7048.

Chester, D.B., Malanotte-Rizzoli, P., Lynch, J., Wunsch, C., 1994. The eddy radiation of the Gulf Stream as measured by ocean acoustic tomography. Geophys Res. Lett. 21, 181−184.

Chiu, C.S., Lynch, J.F., Johanessen, O.M., 1987. Tomographic resolution of mesoscale eddies in the Marginal Ice Zone: A preliminary study,. J. Geophys. Res. 92 (C7), 6886−6902.

Chiu, C.S., Miller, J.H., Lynch, J.F., 1994. Inverse technique for coastal acoustic tomography. In: Lee, D., Schultz, M.H. (Eds.), Theoretical and computational acoustics, 2. World Scientific, Singapore, pp. 917−931.

Chui, C.K., Chen, G., 1991. Kalman filtering with real-time applications. Springer, New York, NY, USA.

Clifford, S.F., Farmer, D.M., 1983. Ocean flow measurements using acoustic scintillation. J. Acoustic Society of America 74, 1826−1832.

Cornuelle, B.D., Worcester, P.F., 1996. Ocean acoustic tomography: Integral data and ocean models. In: Malanotte-Rizzoli, P. (Ed.), Modern approaches to data assimilation in ocean modeling. Elsevier, Amsterdam, the Netherlands, pp. 97−115.

Cornuelle, B., Munk, W., Worcester, P., 1989. Ocean acoustic tomography from ships. J. Geophys. Res. 94, 6232−6250.

Crawford, G.B., Lataitis, R.J., Clifford, S.F., 1990. Remote sensing of ocean flow by spatial filtering of acoustic scintillations: theory. J. Acoustic Society of America 88, 442−454.

Davis, R.E., 1985. Objective mapping by least squares fitting. J. Geophysical Res. 90, 4773−4777.

DeFerrari, H.A., Nguyen, H.B., 1986. Acoustic reciprocal transmission experiments, Florida Straits. J. Acoustic Society of America 79, 299−315.

Deutsch, R., 1965. Estimation theory. Prentice-Hall, 269 pp.

Di Iorio, D., Farmer, D.M., 1996. Two-dimensional angle of arrival fluctuations. J. Acoustic Society of America 100, Melville, NY, USA, 814−824.

Dushaw, B.D., Worcester, P.E., Cornuelle, B.D., Howe, B.M., 1994. Barotropic currents and vorticity in the central North Pacific Ocean during summer 1987 determined from long-range reciprocal acoustic transmissions. J. Geophys. Res. 99, 3263−3272.

Elisseeff, P., Schmidt, H., Johnson, M., Herod, D., Chapman, N.R., McDonald, M.M., 1999. Acoustic tomography of a coastal front in Haro Strait, British Columbia. J. Acoust. Soc. Am. 106, 169−184.

Evensen, G., 2003. The ensemble Kalman filter: Theoretical formulation and practical implementation. Ocean Dyn. 53, 343−367.

Ewart, T.E., Reynolds, S.A., 1978. The mid-ocean acoustic transmission experiment-MATE. J. Acoust. Soc. Amer. 63, 1861−1865.

Farmer, D.M., Di Iorio, D., 1991. Two-dimensional acoustical propagation in a stratified shear flow. In: Potter, J., Warn-Varnas, A. (Eds.), Ocean variability and acoustic propagation, Proceedings of the Workshop on Ocean Variability and Acoustic Propagation. Kluwer Academic Publishers, Dordrecht, the Netherlands, pp. 41−55. La Spezia, Italy, June 4−8, 1990.

Farmer, D.M., Crawford, G.B., 1991. Remote sensing of ocean flows by spatial filtering of acoustic scintillations: observations. J. Acoustic Society of America 90, 1582−1591.

Farmer, D.M., Clifford, S.F., 1986. Space-time acoustic scintillation analysis: a new technique for probing ocean flows. IEEE Journal of Oceanic Engineering OE-11, 42−50.

Flatté, S., Dashen, R., Munk, W.H., Watson, K.M., Zachariasen, F. (Eds.), 1979. Sound transmission through a fluctuating ocean. Cambridge University Press, Cambridge, UK.

Flatté, S.M., Stoughton, R.B., 1986. Theory of acoustic measurement of internal wave strength as a function of depth, horizontal position, and time. J. Geophysical Res. 91, 7709−7720.

Flatté, S.M., Stoughton, R.B., Howe, B.M., 1986. Acoustic measurements of internal wave rms displacement and rms horizontal current off Bermuda in late 1983. J. Geophysical Res. 91, 7721−7732.

Foreman, M.G.G., Walters, R.A., Henry, R.F., Keller, C.P., Dolling, A.G., 1995. A tidal model for eastern Juan de Fuca Strait and the southern Strait of Georgia,. J. Geophys. Res. 100 (C1), 721−740.

Gelb, A. (Ed.), 1974. Applied optimal estimation. MIT Press, Cambridge, MA, USA.

Hansen, P.C., O'Leary, D.P., 1993. The use of the L-curve in the regularization of discrete ill-posed problems. SIAM J. Sci. Comput 14, 1487–1503.

Headrick, R.H., Spiesberger, J.L., Bushong, P.J., 1993. Tidal signals in basin-scale acoustic transmissions. J. Acoust. Soc. Am. 93, 790–802.

Howe, B.M., 1987. Multiple receivers in single vertical slice ocean acoustic tomography experiments. J. Geophys. Res. 92 (C9), 9479–9486.

Howe, B.M., Worcestor, P.F., Spindel, R.C., 1987. Ocean acoustic tomography: Mesoscale velocity. J. Geophys. Res. 92, 3785–3865.

Hukuda, H., Yoon, J.-H., Yamagata, T., 1994. A tidal simulation of Ariake Bay; A tideland model,. J. Oceanogr 50, 141–163.

Johnson, L.E., Gilbert, F., 1972. Inversion and inference for tele-seismic ray data. *Methods Comp. Phys.* 12, 231–266.

Kaneko, A., Yamaguichi, K., Yamamoto, T., Godha, N., Zheng, H., Fadli, S., Lin, J., Nguyen, H.Q., 2005. Coastal acoustic tomography experiment in the Tokyo Bay. Acta Oceanologica Sinica 24, 86–94.

Kapoor, T.K., 1995. Three-dimensional acoustic scattering from arctic ice protuberances. Ph.D. thesis. Massachusetts Institute of Technology, Cambridge, MA, USA. June 1995.

Ko, D.S., DeFerrari, H.A., Malanotte-Rizzoli, P., 1989. Acoustic tomography in the Florida Strait: temperature, current, and vorticity measurements. J. Geophysical Res. 94, 6197–6211.

Lafuente, J.G., Vargas, J.M., Plaza, F., Candela, J., Baschek, B., 2000. Tide at the eastern section of the Strait of Gibraltar. J. Geophysical Res. 105 (C6), 14,197–14,213.

Lawrence, R.S., Ochs, G.R., Clifford, S.F., 1972. Use of scintillations to measure average wind across a light beam. Applied Optics 11, 239–243.

Le Groupe Tourbillon, 1983. The Tourbillon experiment: A study of a mesoscale eddy in the eastern North Atlantic,. Deep-Sea Res. 30, 475–511.

Lee, D., 1994. Three-dimensional effects: interface between the Harvard Open Ocean Model and a three-dimensional acoustic model. In: Robinson, A.R., Lee, D. (Eds.), Oceanography and acoustics: Prediction and propagation models. AIP, New York, NY, USA.

Lozano, C.J., Robinson, A.R., Arango, H.G., Gangopadhyay, A., Sloan, Q., Haley, P.J., Anderson, L., Leslie, W., 1996. An interdisciplinary ocean prediction system: assimilation strategies and structured data models. In: Malanotte-Rizzoli, P. (Ed.), Modern approaches to data assimilation in ocean modeling. Elsevier, Amsterdam, the Netherlands, pp. 413–452.

Malanotte-Rizzoli, P. (Ed.), 1996. Modern approaches to data assimilation in ocean modeling. Elsevier, Amsterdam, the Netherlands.

MODE Group, 1978. Mid-ocean dynamics experiments. Deep-Sea Res. 25, 859–910.

Munk, W., Wunsch, C., 1979. Ocean acoustic tomography: A scheme for large-scale monitoring,. Deep-Sea Res. 26A, 123–161.

Munk, W., Worcester, P., Wunsch, C., 1995. Ocean acoustic tomography. Cambridge University Press, New York, NY, USA.

Ocean Acoustic Tomography Group, 1982. A demonstration of ocean acoustic tomography. Nature 299, 121–125.

Paduan, J.D., Graber, H.C., 1997. Introduction to high-frequency radar: Reality and myth. Oceanography 10, 36–39.

Pan, C.H., Lin, B.Y., Mao, X.Z., 2007. Case study: Numerical modeling of the tidal bore on the Qiantang River, China. J. J. Hydraulic Eng. 130–138.

Park, J.H., Kaneko, A., 2001. Computer simulation of the coastal acoustic tomography by a two-dimensional vortex model. J. Oceanogr. 57, 593–602.

Park, J.-H., Kaneko, A., 2000. Assimilation of coastal acoustic tomography data into a barotropic ocean model. Geophys. Res. Lett. 27, 3373–3376.

Richman, J.G., Wunsch, C., Hogg, N.G., 1977. Space and time scales of mesoscale motion in the western North Atlantic. Revs. Geophys. Space Phys. 15, 385–420.

Robinson, A.R., Carman, J.C., Glenn, S.M., 1994. A dynamical system for acoustic applications. In: Robinson, A.R., Lee, D. (Eds.), Oceanography and acoustics: Prediction and propagation models. AIP, New York, NY, USA, pp. 80–117.

Siegmann, W.L., Lee, D., Botseas, G., Robinson, A.R., Glenn, S.M., 1994. Sensitivity issues for interfacing mesoscale ocean prediction and parabolic acoustic propagation models. In: Robinson, A.R., Lee, D. (Eds.), Oceanography and acoustics: Prediction and propagation models. AIP, New York, NY, USA, pp. 133–160.

Sloat, J.V., Gain, W.S., 1995. Application of acoustic velocity meters for gaging discharge of three low-velocity tidal streams in the St. John River Basin, Northeast Florida. U.S. Geological Survey, Water-Resources Investigations Report, 95–4230.

Sørensen, J.V.T., Madsen, H., 2004. Efficient Kalman filter techniques for the assimilation of tide gauge data in three-dimensional modeling of the North Sea and Baltic Sea system. J. Geophys. Res. 109. C03017.

Spindel, R.C., Spiesberger, J.L., 1981. Multipath variability due to the Gulf Stream. J. Acoust. Soc. Amer. 69, 982–988.

Takasugi, Y., Fujiwara, T., Sugimoto, T., 1994. Formation of sand banks due to tidal vortices around straits, *J. Oceanogr.* 50, 81–98.

Takeuchi, T., Taira, K., 1993. Development of multipaths inverted echosounder. In: Teramoto, T. (Ed.), Deep ocean circulation: Physical and chemical aspects. Elsevier, Amsterdam, the Netherlands, pp. 375–382.

Tarantola, A., 1987. Inverse problem theory: Methods for data fitting and model parameter estimation. Elsevier, Amsterdam, the Netherlands.

The ocean tomography group, 1982. A demonstration of ocean acoustic tomography. Nature 299, 121375–382125.

Tiemann, C.O., Worcester, P.F., Cornuelle, B.D., 2001a. Acoustic scattering by internal solitary waves in the Strait of Gibraltar. J. Acoustic Society of America 109, 143–154.

Tiemann, C.O., Worcester, P.F., Cornuelle, B.D., 2001b. Acoustic remote sensing of internal solitary waves and internal tides in the Strait of Gibraltar. J. Acoustic Society of America 110 (2), 798–811.

Wolanski, E., Williams, D., Spagnol, S., Chanson, H., 2004. Undular tidal bore dynamics in the Daly Estuary, Northern Australia. J. Estuarine, Coastal and Shelf Science 60, 629–636.

Worcester, P., 1981. An example of ocean acoustic multi-path identification at long range using both travel time and vertical arrival angle. J. Acoust. Soc. Amer. 70, 1743–1747.

Worcester, P.F., 1977. Reciprocal acoustic transmission in a mid-ocean environment. J. Acoust. Soc. Amer. 62, 895–905.

Worcester, P.F., 1979. Reciprocal acoustic transmission in a mid-ocean environment: fluctuations. J. Acoust. Soc. Amer. 66, 1173–1181.

Worcester, P.F., Dushaw, B.D., Howe, B.M., 1991. Gyre-scale reciprocal acoustic transmissions, pp. 119–134. In: Potter, J., Warn-Varnas, A. (Eds.), Ocean variability and acoustic propagation. Kluwer, Dordrecht, the Netherlands.

Worcester, P.F., Williams, G.O., Flatté, S.M., 1981. Fluctuations of resolved acoustic multipaths at short range in the ocean. J. Acoust. Soc. Amer. 70, 825–840.

Worcester, P.F., Send, U., Cornuelle, B.D., Tiemann, C.O., 1997. Acoustic monitoring of flow through the Strait of Gibraltar. In: Zhang, R., Zhou, J. (Eds.), Shallow-Water Acoustics, Proceedings of the International Conference on Shallow Water Acoustics. China Ocean Press, Beijing, China, pp. 471–477. April 21–25, 1997.

Yamaoka, H., Kaneko, A., Park, J.-H., Zheng, H., Gohda, N., Takano, T., X-Zhu, H., Takasugi, Y., 2002. Coastal acoustic tomography system and its field application. IEEE J. Oceanic Eng. 27 (2), 283–295.

Yanagi, T., 1989. Coastal oceanography. Koseisha-Koseikaku, Tokyo, Japan (in Japanese).

Zheng, H., 1997. Study on development and application of the coastal acoustic tomography system, Ph.D. dissertation. Hiroshima University, Japan.

Zheng, H., Yamaoka, H., Gohda, N., Noguchi, H., Kaneko, A., 1998. Design of the acoustic tomography system for velocity measurement with an application to the coastal sea. J. Acoust. Soc. Jpn. (E) 19, 199–210.

Zheng, H., Gohda, N., Noguchi, H., Ito, T., Yamaoka, H., Tamura, T., Takasugi, Y., Kaneko, A., 1997. Reciprocal sound transmission experiment for current measurement in the Seto Inland Sea, Japan. J. Oceanogr. 53, 117–127.

Zhu, X.H., Gohda, N., Wu, Q.S., Naowa, T., Zhang, C.Z., 2009. Acoustic current meter for measurement of velocity in a shallow channel. J. Electronic Measurement Technology 32 (2), 166–169.

Chapter 8

Eulerian-Style Measurements Incorporating Mechanical Sensors

Chapter Outline

8.1. Eulerian-Style Measurements   242
8.2. Savonius Rotor Current Meters   242
8.3. Savonius Rotor and Miniature Vane Vector-Averaging
      Current Meters   249
8.4. Propeller Rotor Current Meter (Plessey Current Meter)   250
8.5. Biaxial Dual Orthogonal Propeller Vector-Measuring
      Current Meters   253

8.6. Calibration of Current Meters   255
8.7. Graphical Methods of Displaying Ocean Current
      Measurements   258
8.8. Advantages and Limitations of Mechanical Sensors   263
References   263
Bibliography   264

With the growing dependence of man on the oceans, the need to acquire knowledge of ocean currents has also correspondingly increased. Ocean-current measurements are made in many ways. One such method is the so-called *Eulerian method*, in which flow of water in the ocean (i.e., ocean current) past a geographically fixed position is determined as a function of time using a single instrument or a chain of instruments deployed at a specific station. Current measurements in a Eulerian fashion become important when a detailed study of currents in a particular region of interest is required on a long-term basis, either for operational applications or in academic interest. Such time-series measurements are particularly important for the design and installation of offshore structures and underwater pipelines. Current meters of many designs have also been used by biologists and oceanographers to determine the current velocities in marine waters. Present knowledge of currents in the world oceans has accumulated in part from time-series measurements made on different occasions through a variety of current meters deployed at different depths in different regions of the oceans. Long-term time-series measurements using an array of current meters have revealed several interesting phenomena.

In the past, the Eulerian method was mostly employed in situations where local time-series current measurements in the upper few meters of the ocean surface became necessary. At present, this method is used also for current measurements from much larger depths. Near-shore current measurements using current meters mounted on solid bases and near-bottom measurements made with the aid of tripods or similar structures are representative of this method. Short-term measurements using a direct reading current meter (DRCM) hung from the side of country crafts, boats, or ships also come under this category.

Eulerian current measurements in the open ocean are usually accomplished using self-recording current meters attached to a taut-moored, surface, or subsurface buoy system and are capable of operating unattended for long periods of time. Deep-sea moorings are usually deployed with several current meters located at different depths. Although this method provides excellent time coverage, the current-versus-depth information is obviously limited to the number of current meters one can afford. Furthermore, in the design of oceanographic experiments using arrays of moored current meters, the question of their optimum spacing always arises. Because of the high cost, it is desirable to space the current meters as far apart as possible. On the other hand, the current meters must be close enough so that adjacent observations can be correlated. However, the question as to how close the current meters must be deployed in order that the observations can be correlated has not yet been fully answered.

For a given vertical distance, the coherence between the currents measured at a pair of current meters is considered the *correlation coefficient* between the two time series as a function of frequency. Webster (1972) observed that the coherence estimates have the general property of being high at low frequency and dropping as frequency increases. He obtained an empirical relationship between the vertical separation of a pair of instruments and the frequency above which the coherence is less than a chosen critical value. However, the physical mechanism underlying this relationship has not been properly understood.

8.1. EULERIAN-STYLE MEASUREMENTS

This chapter addresses the Eulerian method of ocean current measurements. Eulerian current meters (sometimes referred to as *flow meters*) employ a wide range of sensors such as the Savonius rotor, unidirectional impellers, bidirectional propeller pairs, electromagnetic sensors, acoustic travel-time (ATT) difference sensors, and acoustic Doppler sensors. In some specialized situations, such as measurements of turbulence and very small flows, thermal and laser Doppler sensors are also used. Although all these sensors have varying degrees of merits and limitations, the choice of a particular type of sensor often depends on the type of study desired. A current meter intended for measurements in estuaries and inlets must be able to measure current flows as large as 300 cm/s or more. Because tidal variations in estuaries introduce variations in salinity and water temperature, and some locations have large sediment concentrations, the current meter must also be insensitive to these effects.

Eulerian current-measuring devices may be classified under the following categories:

1. Purely mechanical devices, whereby detection and recording of currents are controlled solely by mechanical devices
2. Current meters incorporating mechanical sensors, whereby detection of currents is achieved by mechanical devices, whereas acquisition and display/recording are controlled by electronic circuitries

3. Current meters whereby detection of currents is achieved by nonmechanical sensors and acquisition and display/recording are controlled by electronic circuitries

The present chapter primarily addresses mechanical sensors, which include Savonius rotor and vane, unidirectional impellers, and bidirectional propeller sensors. Some aspects of tow-tank measurements and calibration procedures are also discussed. Some of these current meters are direct reading types and others are recording types. In the former, current speed and direction as well as deployment depth are displayed on a deck unit in analog or digital display format, and the observations are noted by an observer on board (e.g., country crafts or research vessels) from where the current meters are deployed (Figure 8.1). Direct reading current meters are usually deployed for short-term *in situ* measurements of currents in coastal waters, estuaries, or ports where ship traffic and/or vigorous fishing activities render deployment of current meter moorings difficult. On the other hand, recording current meters on surface or subsurface moorings are usually deployed in offshore areas that are free from disturbances caused by fishing/navigational activities. Long-term time-series current measurements are desirable from shallow waters too but are not easy to obtain for various logistical reasons.

8.2. SAVONIUS ROTOR CURRENT METERS

Although the history of making current measurements in the ocean is long, the advent of user-friendly current meters is fairly recent. The Christian Michelsen Institute in Bergen, Norway, contributed immensely to the initiation and development of various types of current meters. At this institute, Ivar Aanderaa was the main contributor in this field. Aanderaa designed the classic current meter in the late 1950s (Aanderaa, 1964) using a Savonius-like rotor (Savonius, 1931) and a magnetic compass as the speed- and

FIGURE 8.1 (left) Deployment of a direct-reading-type Savonius rotor current meter from a country craft; (right) current speed and direction and deployment depth displayed on a deck unit being noted by an observer on board the country craft. *(Source: Courtesy of CSIR-National Institute of Oceanography, India.)*

direction-sensing elements, respectively. This current meter, universally known as the Aanderaa Current Meter, has thereafter been the "workhorse" of oceanographic current measurements and is traditionally used for deep-ocean, shelf, and estuary researches. The Savonius rotor's insensitivity to variations in the surrounding environmental conditions is its noteworthy feature, making it suitable for current flow measurements in estuaries and inlets. The first-generation Aanderaa current meter senses scalar speed above a threshold of 2 cm/s and relies on a large vertical fin to orient the entire instrument in the direction of current flow. The current meter remains vertical in steady conditions at mooring wire angles of up to 27°; from the vertical. Despite the advent of more sophisticated current meters in the 1980s, rotor current meters continued to be improved by M/s. Aanderaa Instruments and are being used by several oceanographers from across the world.

Measurement of currents using a mechanical sensor such as a rotor is based on the mechanism of physical rotation of the rotor in response to the drag force it experiences from the moving water in a flow field. For steady flows, the average rotational frequency of the rotor is closely related to the speed of the fluid impinging on it. The first-generation Aanderaa Savonius rotor (see Figure 8.2) permitted it to rotate in the same sense, irrespective of the direction of current flow. A small magnet embedded at the bottom end of the rotor stud permits detection of its revolution by magnetically coupling it to a sensing device such as a reed relay, magneto diode, or inductor that is rigidly mounted in the watertight housing immediately beneath the rotor. Speed is sampled by summing the revolutions of the rotor over a defined sampling interval (the sum of pulses

FIGURE 8.3 Aanderaa north-seeking magnetic compass. (*Source: Aanderaa Instruments Model 4 Recording Current Meter Manual.*)

corresponding to rotor revolutions over ~32 seconds in the case of the Aanderaa rotor shown in Figure 8.2 provides an approximate value of current speed in cm/s). The speed of the water current is precisely calculated from the steady-state calibration equation for the sensor. The measured speed is, therefore, an integrated or averaged speed over the measurement interval.

A method popularly used for determination of the direction of water current flow relative to Earth's magnetic north is to rigidly couple a magnetic compass (see Figure 8.3), which is mounted within the current meter (CM) housing, to a large tailfin that is rigidly fixed to the CM housing (see Figure 8.4). With this technique, the

FIGURE 8.2 Initial design of the Aanderaa Savonius rotor, capable of rotating in the same sense irrespective of the direction of current flow. A small magnet embedded inside the bottom end of the rotor stud permits detection of its revolution by magnetically coupling it to a sensing device rigidly mounted in the watertight housing immediately beneath the rotor. (*Source: Aanderaa Instruments Model 4 Recording Current Meter Manual.*)

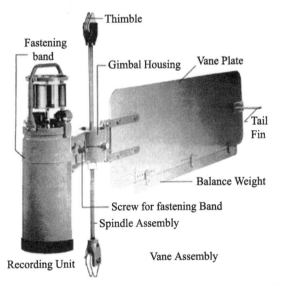

FIGURE 8.4 Aanderaa Savonius rotor current meter housing rigidly coupled to a large tailfin for automatically orienting the current meter in the direction of current flow. (*Source: Aanderaa Instruments Model 4 Recording Current Meter Manual.*)

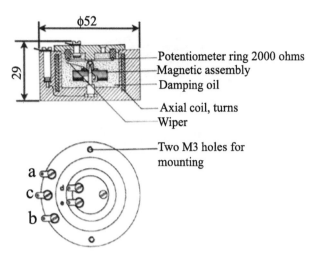

FIGURE 8.5 Constructional details of Aanderaa magnetic compass. *(Source: Aanderaa Instruments Model 4 Recording Current Meter Manual.)*

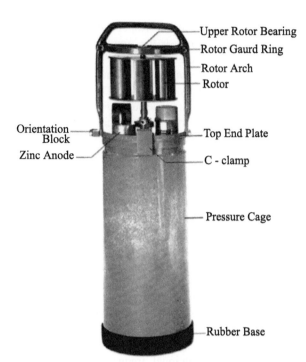

FIGURE 8.6 Constructional details of the Aanderaa Savonius rotor current meter. *(Source: Aanderaa Instruments Model 4 Recording Current Meter Manual.)*

whole CM is constrained to rotate in tune with the changing current direction. The magnetic compass gives the orientation of the fin and, therefore, the direction of current flow relative to Earth's magnetic north, obeying the universal convention of water current direction, namely "current toward." (Note that the universal convention of wind direction is "wind from.") The construction details of the magnetic compass are shown in Figure 8.5.

M/s. Aanderaa Instruments has pioneered in rotor current-meter technology, and several of their devices have been deployed in the world oceans. The current meter (see Figure 8.6) is deployed on a mooring line. Aanderaa rotor current meters are self-contained instruments that can be moored in the sea and can record ocean current, electrical conductivity, and temperature of the water, together with deployment depth (CTD). Construction details of the vane assembly and the mechanism for supporting the Aanderaa CM housing are shown in Figure 8.7. The combination of recording unit and vane assembly is equipped with a rod that can be shackled into the mooring line (see Figure 8.8). This arrangement permits the instrument to swing freely and align with the current. The recording unit contains all sensors, the measuring system, battery, and a detachable, reusable data storage tape unit (see Figure 8.9). In later modifications, the tape was replaced by a solid-state data storage device.

In operation, a built-in clock triggers the instrument at preset intervals and a total of six channels are sampled in sequence. The first channel is a fixed reference reading for control purposes and data identification. Channels 2, 3, and 4 represent measurement of temperature, conductivity, and depth, respectively. Channels 5 and 6 represent the vector-averaged water current speed and direction since the previous triggering of the instrument. The data are sequentially fed to a data storage unit (DSU). Simultaneously, as the reading takes place, the output pulse keys on and off an acoustic carrier emitted by an acoustic transducer. This

allows monitoring of the performance of the moored instrument from the surface by a hydrophone and can be used for real-time telemetry of data. The recording interval of the instrument is set by an interval selector switch. When a 10-minute interval is used, the operating period of the instrument will be two months. Aanderaa current meters are available in different depth ranges.

Until about 1987 a very simple sampling scheme was used in the Aanderaa current meter: The number of rotor revolutions during a predetermined sampling interval was recorded on magnetic tape, together with a single spot measurement of the meter orientation, derived from the mechanically clamped compass. Analog-to-digital conversion of the six available data channels was accomplished by an electromechanical encoder controlled by a crystal clock. The same encoder was used to drive the magnetic tape transport system and to actuate the parameter-selecting switch. In subsequent modifications M/s. Aanderaa Instruments incorporated a vector-averaging sampling scheme and solid-state memory. Samples are taken every 12 seconds and are decomposed into east and north components. Successive east components and successive north components are added separately, and the vector-averaged speed and direction are computed at the end of the measurement interval and then recorded in the data storage media. The capacity of the system is 10,900 sampling cycles, which, for example, provides 75 days of operation at 10-minute sampling intervals.

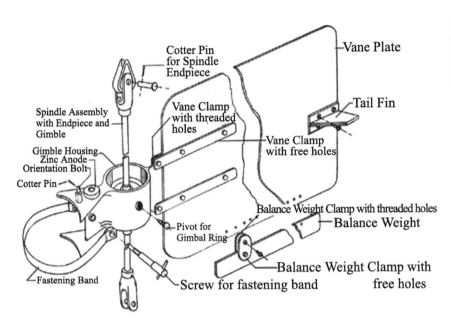

FIGURE 8.7 Constructional details of the vane assembly and the mechanism for supporting the Aanderaa current meter housing. *(Source: Aanderaa Instruments Model 4 Recording Current Meter Manual.)*

Over the years, the first-generation Aanderaa Savonius rotor-type current meter established an enviable reputation for durability and reliability. Reasonable price was another attraction. There is widespread experience of its use in the major oceanographic institutions and in industry. Because the first-generation Aanderaa CMs were used by several oceanographers and environmentalists over a long time, these were also the ones that were evaluated the most. Based on these studies it has been observed that the usefulness and accuracy of a CM incorporating a Savonius rotor for measurement of small currents is related to the sea state, the degree of unsteadiness of the current flow, and the mooring style. Paquette (1962) pointed out that a rotor-type CM exhibits over-registration when the mean flow is weak. Pollard (1973) reported that error currents as large as 20 cm/s could be recorded by surface-moored rotor CMs in severe wave conditions. Karweit (1974) observed systematic errors in Savonius rotor current measurements in unsteady currents as inferred from steady-state calibrations and small overestimates of low-frequency current measurements by a vector-averaging rotor CM in the presence of intense high-frequency oscillatory currents.

It has also been observed that the rotor tends to respond differently at different frequencies of the oscillatory current. Field experiments conducted by Halpern and Pillsbury (1976) revealed that as a result of mooring motion, the rotor is "pumped round" and often senses an excess speed, an effect popularly known as *rotor pumping* or *over-speeding*. Saunders (1980) noted that a Savonius rotor is also sensitive to vortices shed by the CM's body. Comparison of rotor CMs on surface moorings showed that rotor pumping by mooring motion is a major limitation. If

rotor pumping adds a constant speed to the current measurement record, the observed energy will be elevated by an approximately constant proportion across the entire spectrum (Pearson et al., 1981). In high sea-state conditions, there is a high probability of contamination of current measurement records by rotor pumping. The influence of

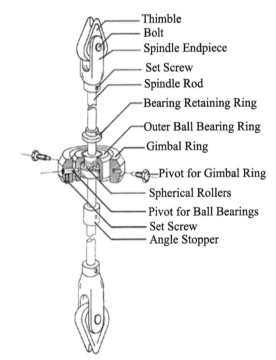

FIGURE 8.8 Constructional details of the rod that is shackled into the mooring line of the Aanderaa current meter. *(Source: Aanderaa Instruments Model 4 Recording Current Meter Manual.)*

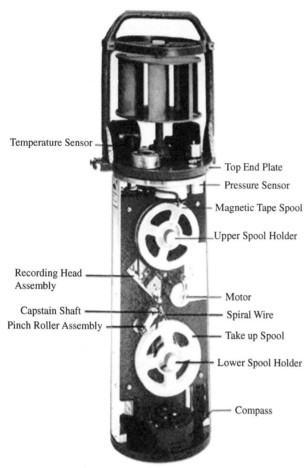

Temperature Sensor

Top End Plate

Pressure Sensor

Magnetic Tape Spool

Upper Spool Holder

Recording Head
Assembly

Motor

Capstain Shaft

Spiral Wire

Pinch Roller Assembly

Take up Spool

Lower Spool Holder

Compass

FIGURE 8.9 Aanderaa Savonius rotor-style recording current meter containing all sensors, the measuring system, battery, and a detachable, reusable data storage tape unit. *(Source: Aanderaa Instruments Model 4 Recording Current Meter Manual.)*

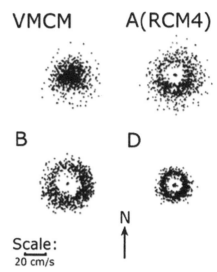

FIGURE 8.10 Presence of "holes" in the Savonius rotor current meter's current velocity distribution at low speeds (less than 50 cm/s). *(Source: Sherwin, 1988.)*

wave motion decreases exponentially with depth; therefore, the depth of the subsurface float is a critical factor in using Savonius rotor-type CMs.

Field experiments conducted by Halpern and Pillsbury (1976), Halpern et al. (1981), and Johnson and Royer (1986) revealed that the surface-moored rotor-type CM exhibits over-registration when the mean current is small and the oscillatory currents are large. Schott et al. (1985) found that the error can be reduced but cannot be fully eliminated by adopting improved mooring techniques.

It was found that even the fin can introduce some errors in the measurements. The large-fin system suffers from long response time and is therefore unsuitable for accurate current measurements under large turbulence. Kenney (1977) analyzed the dynamics of such a direction-orienting fin and noted that the use of a long fin to orient a CM into the "mean" current requires caution. For example, when a Savonius rotor CM is deployed in a turbulent flow regime, the measured "mean" current could be grossly over-estimated, depending on the intensity and spectrum of the

lateral component of turbulence. The orbital velocity of the oscillations at the sea surface due to waves, especially due to swells, penetrates to great depths following an exponential law related to the angular velocity and amplitude of the wave at the sea surface. This indicates that the mean current at depths below a persistent wave- or swell-ridden sea is superimposed by oscillatory motions. If the CM deployed in such an environment employs a large direction-orienting fin rigidly coupled to the CM's body, it is likely that the CM may undergo some degree of angular motion. Such motions result in the rotor's nonlinear response because the axis of the CM swings through an arc (Sherwin, 1988). In a nutshell, it was observed that although well suited to measurements in steady flow regimes, the first-generation Aanderaa CMs are considered unsuitable for applications in or near the surface wave zone or indeed wherever appreciable instrument motion is likely to occur (Saunders, 1976; SCOR WG 21, 1974).

Field intercomparison studies conducted by Johnson and Royer (1986) indicated that speed measurements obtained from rotor CMs are influenced by the effect of rotor pumping. This effect is manifested by the presence of "holes" (see Figure 8.10) in the current-velocity distribution at low speeds (less than 50 cm/s). The observed hole is indicative of the absence of low speeds at rough-sea conditions. With reduced wave activity, the size of these "holes" is also reduced and the holes finally became absent in calm-sea conditions. Sherwin (1988) observed a mean-speed overestimation of 66 percent by the Savonius rotor-type CM when current-flow speeds were less than 10 cm/s and waves and swells were large. Analytical treatment of moored CM motion (Griffin, 1988) indicated that it is necessary to minimize the phase difference between the current-flow velocity and the CM's velocity to minimize the relative oscillatory current detected by the rotor.

Analysis of Savonius rotor performance in varying current-flow conditions, field intercomparison studies, and analytical treatment of mooring motion in oscillatory flows as just mentioned indicate that amplification of rotor CM speed measurements will be greatest when the mean currents are small and high-frequency oscillatory flows are of larger amplitudes. In summary, a surface-moored Savonius rotor CM is unable to record small horizontal currents at high sea states when other types of sensors record small currents under similar sea states.

In an attempt to unravel the mysteries that shroud the motion of a Savonius rotor in an unsteady current, various investigators have performed numerous laboratory experiments. A series of independent experiments conducted by Fofonoff and Ercan (1967) and Saunders (1980) in wave tanks provided contradictory results as to the response of the rotor in accelerating and decelerating flows.

One line of thought attributed current-speed over-registration by Savonius rotors during small current flows to the two end plates, which support the curved blades of the rotor. The argument often put forward is that, in this configuration, the rotor is likely to "trap" a certain amount of water during its rotation and, therefore, increases its inertia. In an attempt to improve the rotor CM's performance, a new version of the Savonius rotor was introduced by M/s. Aanderaa Instruments. This new version is devoid of the end plates, and the curved blades are replaced by a set of flat-bladed paddlewheels that are distributed in diametrically opposite directions (see Figure 8.11). This arrangement (incorporated in Aanderaa RCM 7 current

meter) necessitates covering half the portion of the rotor (in the direction of current flow) to allow its unidirectional rotation in the presence of current flow (see Figure 8.12). Various stages of deployment of an Aanderaa RCM 7 current meter are shown in Figure 8.13.

Replacement of the Savonius rotor by a "flat-bladed" paddlewheel design and modifications to the vane was expected to improve the performance of the CM in the near-surface region, though prior to the introduction of vector averaging the fundamental limitations in directional measurement remained (Woodward, 1985). As yet relatively little information is available on the effectiveness of the more recent changes, though Loder and Hamilton (1990) have reported on some effects of high-frequency mooring vibration.

Obviously, the half-cylindrical cover now used in the Aanderaa RCM, having a possible drawback of shedding of eddies from its edges, could have been avoided if the flat-bladed paddlewheels were replaced by curved-bladed paddlewheels having no end plates. When the author of this book put forward this suggestion to Ivar Aanderaa in March 1995, he went from the visitors' room and came back with a rotor that had curve-bladed paddlewheels devoid of end plates (see Figure 8.14), which resembled the one that I suggested. He told me with a smile, "My wish was to use this rotor; but somehow it did not materialize." Because the two ends are "open" in either of these configurations, the

FIGURE 8.11 M/s. Aanderaa Instruments' modified version of the Savonius rotor. A set of "flat-bladed" paddlewheels without end plates is uniformly distributed in diametrically opposite directions. *(Source: Aanderaa Instruments Current Meter Manual.)*

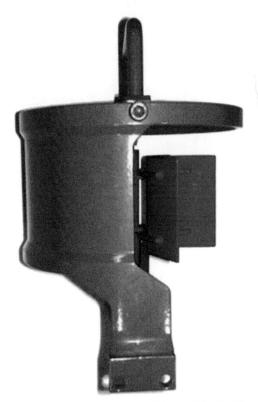

FIGURE 8.12 Half-cylindrical cover of the "paddlewheels" rotor to allow unidirectional rotation of the rotor in presence of current flow. *(Source: Aanderaa Website: www.aanderaa.com.)*

FIGURE 8.13 Various stages of deployment of an Aanderaa RCM 7 current meter. *(Source: Courtesy of Vijayan Fernando, CSIR-National Institute of Oceanography, Goa, India.)*

problem of increase in inertia due to water trapping is expected to be eliminated. To this author's knowledge, no performance evaluation of the modified rotor vis-à-vis the old version of the Savonius rotor in terms of rotor pumping effect had been reported. The fact that a cup anemometer with no end plates also exhibits overspeed registration in a turbulent wind field (Busch et al., 1976) is probably an indication that removal of end plates of the Savonius rotor may not be a remedy against overspeed registration by the Savonius rotor. In any case, there appears to be no experimental evidence as to the real cause of overspeed registration by Savonius rotors, and therefore it may be premature to attribute "overspeed registration" to "rotor overspeeding."

Looking at the problem of overspeed registration (during weak currents) by surface-moored Savonius rotor CMs from a purely instrumentation point of view, Joseph (1991) proposed that one of the reasons for this problem could be slow bidirectional rotation of the rotor, instead of the expected unidirectional rotation, in a weak or no-flow situation. When the mean horizontal current is weak and the Savonius rotor CM moves up and down under the influence of mooring-line motion, it is quite possible that the rotor might tend to make slow bidirectional motions about its axis. This is because the horizontal drag force, which is responsible for the unidirectional rotation of the rotor, is negligible at very weak currents.

FIGURE 8.14 Rotor with curved-bladed paddlewheels devoid of end plates, which would have avoided the use of the half-cylindrical cover. *(Source: Courtesy of the late Ivar Aanderaa, Aanderaa Instruments, Norway.)*

If the rotor executes bidirectional motions, the reed relay (which is magnetically coupled to the rotor and used for detecting the current speed) can close and open whenever it happens to be within the magnetic field of the embedded magnet at the end portion of the rotor's stud. The technique devised by Joseph (1991) to detect and tag the anticipated bidirectional rotation of the rotor consists of three reed relays, each of which is rigidly mounted within the waterproof housing of the CM at an angular separation of 120° and is located within the magnetic field of the rotor. A rotor-following magnet, which is free to rotate about the principal axis of a ball bearing and to freely sweep over the three relays, is so positioned that only one pole of the magnet can effect a closure of the relay. Thus, one full revolution of the rotor in the same sense will result in serial closures of all the three relays, thereby generating three electrical pulses during one full revolution of the rotor.

The electronic circuit incorporating the three relays generates a frequency-shift-keyed (FSK) signal (i.e., closure of each relay is represented by a unique pulse-stream frequency). If the rotor rotates unidirectionally in the correct sense (i.e., as expected in a steady current flow), the circuit registers the correct counts during the current-speed measurement interval. If there is no flow, no count is registered. If, however, the rotor undergoes bidirectional motions about its axis (i.e., the rotor's motion is not due to a steady current flow), the flow-unrelated rotations of the rotor during the current-speed measurement interval are also registered. Thus, the device enables counting the rotor-generated electrical pulses in a form that permits detection of error counts produced by the bidirectional motions of the rotor. In this way, the device prevents attribution of an erroneous higher value to current speed when the mean current is absent or it is actually very weak.

8.3. SAVONIUS ROTOR AND MINIATURE VANE VECTOR-AVERAGING CURRENT METERS

Some of the problems associated with the first generation of the Aanderaa Savonius rotor CM incorporating a large direction-orienting vane arose primarily from the scalar averaging of speed and the inadequate sampling of direction. The CMs that came to be known as vector-averaging current meters, or VACMs (McCullough, 1975), the first of the VACMs, offered substantially improved performance over the first generation of the Aanderaa CMs in nonsteady flows. When this was realized, the vector-averaging procedure was incorporated in the second-generation Aanderaa rotor CMs.

In the VACM the measurement is made of polar components using a Savonius rotor, the rotation of which is sensed eight times per revolution, and a miniature vane (9 × 17 cm) is used to indicate flow direction (see Figure 8.15). The direction orientation of the vane relative to the "zero" of the magnetic compass (which is rigidly mounted inside the in-water housing of the current meter) is measured to determine the direction of water-current flow in the conventional style. At each rotor count, the sine and cosine of the flow direction relative to magnetic north are computed from measurements of compass and vane-follower outputs and are accumulated in east-west and north-south registers over the sampling interval so as to provide a vector-averaged mean. Though representing a considerable improvement over CMs employing simple sampling schemes, the VACM falls short of ideal

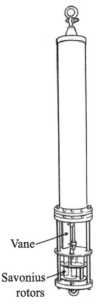

Vane

Savonius
rotors

FIGURE 8.15 Vector-averaging current meter (VACM) with a miniature fin capable of free motion about the axis of the instrument housing. *(Source: Beardsley, R.C. 1986.)*

performance in oscillatory flow, due partly to inadequacies inherent in the rotor and vane dynamic response characteristics (Fofonoff and Ercan, 1967), and partly to rotor pumping. This last effect is the name given to the rectification of eddies shed by supporting structures and the instrument case, particularly when the instrument is subjected to strong vertical motions. Horizontal oscillatory flows can cause either underspeeding or overspeeding, depending on the characteristics of the flow and the amplitude of the fluctuations relative to the mean. The response of the Savonius rotor to a change in flow conditions may be characterized in terms of the filament length of the fluid that must pass through the rotor before adjustment to the new fluid speed is complete. The time constant is thus dependent on fluid speed, the rotor responding more closely to an increase than to a reduction in fluid speed. An estimate of the likely effects of rotor time constants on accuracy has been provided by Saunders (1980), who measured response lengths in both accelerating and decelerating flows. Application of these values in a simple rotor model of the form $\frac{dR}{dt} = \frac{(S-R)}{t}$ subjected to simple harmonic periodic forcing (where R is the rotor speed converted to apparent flow speed, S is the true speed, and t is the rotor time constant) generated maximum fractional errors of ~10 percent when oscillatory and mean current values were comparable in magnitude.

A similar form of model was also used to examine the vane and the vane-follower responses (Saunders, 1976). In this instance, $\frac{d\theta}{dt} = \frac{(\theta_i - \theta)}{t}$, where θ_i, θ are, respectively, the instantaneous current and vane directions, and $t \geq 1.5$ sec is the combined vane and vane-follower time constant. When applied to some representative near-surface current data, an overreading of currents of order 10 percent was predicted by the model. For a discussion of the importance of vane characteristics, refer to Kenney (1977). More recent work by Patch et al. (1990, 1992) considers the dynamic response characteristics of the VACM compass and vane follower.

In the case of a direct reading current meter (DRCM) incorporating a fixed and large direction-orienting vane, the motions of its cable and the current meter's in-water housing may cause additional errors in the measurement of the mean current velocity. These problems can be circumvented by incorporation of a miniature vane that is capable of free movement with respect to the axis of the instrument housing. However, unlike the fixed-vane configuration, a constant heading of the CM's in-water housing relative to the direction of current flow cannot be maintained in a free-vane configuration. Consequently, variation of the horizontal angles between the current-flow direction and the supporting rods of the rotor causes errors in the flow-speed measurements (Pite, 1986; Joseph and Desa, 1994).

8.4. PROPELLER ROTOR CURRENT METER (PLESSEY CURRENT METER)

While the Savonius rotor CMs were already in wide use, searches were in progress for the design of a different style of CM using a different sensing device. Hodges (1967) reported the design and development of a propeller-type rotor current meter. He carried out the work at the Christian Michelsen Institute in Bergen, Norway, under the auspices of the NATO Subcommittee for Oceanographic Research, with cooperation and help from Odd Dalh and Ivar Aanderaa from the Institute in Bergen. (Aanderaa later left Christian Michelsen Institute and started a private company, Aanderaa Instruments.) Subsequent to completion of the basic design and development in Bergen, particularly the novel electromechanical analog-to-digital converter that also drives the magnetic tape transport system and actuates the parameter selecting switch, the basic instrument was then further engineered at the Marine Systems Division of the Plessey Company Ltd. to make use of a propeller-type rotor, in place of the Savonius type rotor fitted to all previous prototypes, to measure current flows.

The Plessey Recording Current Meter (see Figure 8.16) was designed primarily to measure and record data on the speed and direction of water flow. The method by which the instrument measures the current flow is by obtaining the total rotor revolutions over a known period of time. The total revolutions are divided by the time to give an average speed in revolutions per second, which can be directly related to the flow in feet per second from a calibration graph. There is no means of knowing whether the total revolutions measured occurred at the average rate or at some other varying rate. To avoid error due to the assumption of an average rate, the rotor calibration needs to be linear.

The main factors considered in the propeller design were to achieve linearity, low-speed response, good mechanical strength, reliability, and ease of servicing. Linearity is obtained by designing the propeller to give

FIGURE 8.16 Plessey propeller-type current meter. *(Source: www .aagm.co.uk/thecollections/objects/object/Plessey-Current-Meter-And-Fin-Mo21—Serial-Number-267?.)*

a minimum of interference between blades and by ensuring that the blades never operate in a semi-stalled condition. Consideration of a suitably sized propeller makes this a three-bladed design of a length to permit the trailing edge of a blade to be in line with the leading edge of the following blade (see Figure 8.17). By the selection of a suitable length, this results in a convenient nominal pitch of 1 foot. For low-speed response with consistent performance over a long immersed period, the bearing design is based on a watch-type escapement pivot. The assembly is also spring-loaded to give adequate mechanical protection. The pivot material is stainless steel and the bearing bush is loaded nylon. To reduce weight on the bearing, the material chosen for the propeller itself is plastic. Mechanical strength and reliability are achieved by the sprung design of the bearing bush, which permits the whole assembly to move sideways until the hub is supported by the main frame (see Figure 8.17).

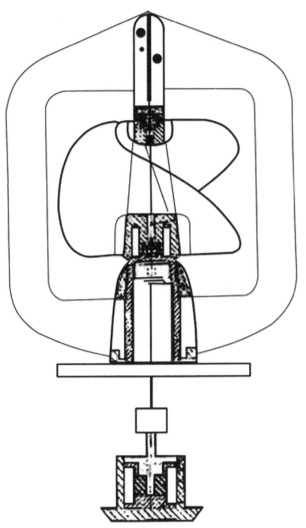

FIGURE 8.17 Details of mounting of Plessey propeller. *(Source: Hodges, 1967.)*

Very heavy side shock loads are supported in this manner and do not damage the pivot. This feature also provides a self-aligning action, which means that performance is maintained even if the outer frame is partially distorted. The hub is also shrouded at both the front and the rear by suitable skirt extensions, which are designed to prevent fouling from weed and growth. The whole propeller is painted with antifoulant. For ease of servicing, the front bearing assembly can slide forward. This releases the propeller, which can be extracted through the side supports and gives immediate access to the bearing assembly. Based on performance data obtained from the Wallingford Hydraulics Research Station, the minimum starting speed is 2.03 cm/s. The stopping speed is 1.98 cm/s. It was found that good linearity is maintained from 4 to 250 cm/s. Coupling of the propeller to the instrument proper is made by magnets inserted within the propeller hub immediately on each side of the bearing housing. This magnet drive operates a concentric gearbox housed within the support pillar, which also carries the propeller bearings and bearing frame. The final driven potentiometer is built as an integral part of the gearbox, and the overall loading on the propeller has been kept to a minimum by this method.

A major change introduced in the Plessey meter is the method of suspension from the mooring cable. It can be appreciated that to make good use of the propeller-type rotor for measuring current flows, it is essential that the propeller face correctly into the direction of the water flow. It is little gain to have a propeller sensitive to low velocities if it is not also able to swing into the flow direction under the same low-flow velocity conditions. One method adopted to give low-velocity alignment response is to have less frontal area, which means that less fin area is needed. An additional scheme is the use of an offline suspension system. In the use of this system, the instrument only is suspended below the pivot. This reduces the overall hung weight, and the load on the pivot is reduced.

To suit practical use of the meter, all the instrumentation and power supplies are contained in a cylindrical pressure-sealed tube that is 5 inches in diameter and 15 inches long. This tube is suspended in the water and has attached to it small directional fins and a propeller-type rotor that drives through a magnetic coupling to the internal instrumentation so that the basic cylinder is completely sealed and self-contained. Standard domestic magnetic tape is used for recording and storing the information from the speed and direction sensors. The information is related to variations in resistance ratios and is converted to 10-bit binary numbers and recorded onto magnetic tape in the form of long and short DC pulses. The binary number code is obtained from an electromechanical analog-to-digital converter, which is generally referred to as the *encoder*.

The current speed is obtained from a change in a resistance ratio over a period of time of a potentiometer, driven by the rotor via a magnetic coupling and a reduction gear as mentioned earlier, with the rotor being turned by the water current. Current direction is obtained from a magnetic compass that is made to include a potentiometer resistance element; at the start of a direction measurement, a floating contact clamps to the potentiometer to provide the required resistance ratio, which is proportional to the angle with reference to the Earth's magnetic north.

A further feature of the design is the inclusion of a fixed resistance ratio within the instrument, the repeated measurement of which is used to identify the individual instrument. This is referred to as the *reference*. The whole system is powered by batteries and controlled by a separate battery-driven clock mechanism that initiates the measuring cycle at preset time intervals.

For telemetry and direct reading of the data, the pulses (which represent the data) fed to the recording head of the tape recorder are propagated through the water as an acoustic signal via a spherical transducer mounted on the top plate of the instrument. These pulses can be picked up and monitored on a suitable hydrophone receiver at ranges up to 500 meters.

The receiving hydrophone is used on board a surface vessel and can accept the pulses and convert these into an audible note that can be listened to and recorded or fed into a pen recorder or special printout unit, which will show the pulses as short and long dashes. The long pulses represent binary 0 and the short pulses binary 1; knowing this, the binary number can be converted to its equivalent decimal number. Thus, a direct-reading facility is made available by this means.

Ramster and Howarth (1975) reported a detailed comparison of the data recorded by Plessey MO21 and Aanderaa Model 4 recording current meters (the ones that incorporated the first-generation Savonius rotor shown in

Figure 8.2) moored in two shelf-sea locations in the North Sea and the Irish Sea, each with strong tidal currents. Schematic diagrams of the moorings of these two current meters are shown in Figure 8.18. Obtained comparison measurements are shown in Figure 8.19. The instruments measured integrated counts from the speed sensor and discrete directions from the direction sensor every 10 minutes. In general, there were no very strong winds during the measurements, so that a comparison of performance in rough seas is not possible. It is apparent that in the calm-sea conditions of the present measurements, there is very good agreement between the data recorded by the two types of current meters every sample interval. Calculation of "stream prediction" residuals (residual current is the measured current minus the predicted tidal current) is expected to highlight the differences in the sets of data not readily apparent in the straight computation of tidal constituent characteristics and shows up the full implications of apparently small and unimportant differences. Based on the use of this scheme, it was found that there is very good agreement between the residuals calculated from the data recorded by the two types of current meters. Fortunately, there was a small window during which relatively strong wind blew during the measurement, and this provided a fortuitous opportunity to examine the residuals from the near-surface meters (see Figure 8.20). In general, the hourly estimates of residual speed for the near-surface meters agree to within 4 cm/s and for the near-bottom meters to within 2 cm/s (see Figure 8.21). The fact that the residuals agree so well in the presence of strong tidal currents is encouraging.

As indicated earlier, a known drawback of the Savonius rotor is that its omnidirectional properties do not allow it to record currents faithfully in the presence of waves

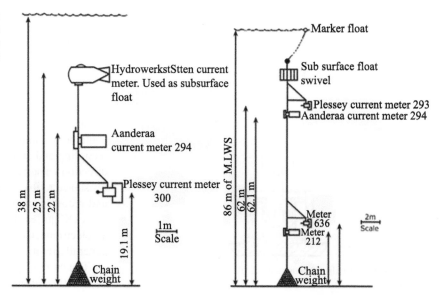

FIGURE 8.18 Schematic diagram of the moorings of the Plessey MO21 and Aanderaa Model 4 recording current meters moored in two shelf-sea locations. *(Source: Ramster and Howarth, 1975.)*

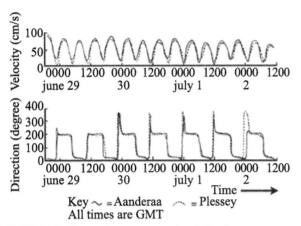

FIGURE 8.19 Time-series current speed and direction measurements obtained from Plessey MO21 and Aanderaa Model 4 recording current meters moored in two shelf-sea locations. *(Source: Ramster and Howarth, 1975.)*

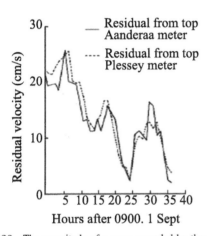

FIGURE 8.20 The magnitude of a surge recorded by the near-surface moored Plessey MO21 and Aanderaa Model 4 recording current meters. *(Source: Ramster and Howarth, 1975.)*

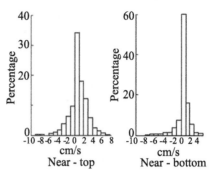

FIGURE 8.21 Percentage-frequency diagrams of the differences (cm/s) between the hourly estimates of residual drift as calculated from the two sets of intercalibration data from the near-surface moored Plessey MO21 and Aanderaa Model 4 recording current meters. *(Source: Ramster and Howarth, 1975.)*

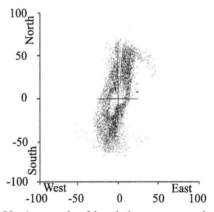

FIGURE 8.22 A scatter plot of the velocity components calculated from the near-bottom Aanderaa data. *(Source: Ramster and Howarth, 1975.)*

(Hansen, 1964). This is because the orbital velocity of water particles due to wave action will always add to the speed seen by the rotor, whereas an impeller's speed of rotation will only correspond to the current flowing axially to it and will reverse if the flow reverses. If the direction of the waves and the tidal stream is the same and if the speed of the tidal stream is in excess of the orbital velocity due to waves, then both the rotor and the impeller will record the same current speed. The rotor, however, will record higher speeds than the impeller for tidal stream speeds that are lower than the orbital velocity. In a moored system where the subsurface float is influenced by the surface waves, the surface-wave action on the subsurface buoy is being taken up by the mooring wire, leading to the movement of even the bottom instruments to and fro in the water. The omni-directional sensitivity of the rotor of the Aanderaa meter would then lead to velocities being recorded that are higher than those measured by the Plessey meter. A scatter plot of

the velocity components calculated from the near-bottom Aanderaa data is shown in Figure 8.22. It may be recalled that appearance of "holes" in the scatter plot of Savonious rotor current meter data was indicated in Section 8.2.

8.5. BIAXIAL DUAL ORTHOGONAL PROPELLER VECTOR-MEASURING CURRENT METERS

The biaxial dual orthogonal propeller current meter measures averaged orthogonal components of the flow with two pairs of orthogonally mounted dual propellers (i.e., two propellers fixed on one axle) with accurate cosine-response characteristics. The term *cosine response* refers to the dependency of the measured flow on the angle of attack of the incident current. The ideal response of a propeller is the variation of its response in proportion to the cosine of the incident flow angle (i.e., full response for flow parallel to the axle and no response to flow normal to the axle). The revolution rate of an ideal propeller sensor is proportional to the magnitude of the flow times the cosine of the angle between the propeller axle and the flow vector. *Horizontal cosine response* refers to the sensor's response to flow in the horizontal plane, whereas *vertical cosine response* refers to the sensor's response to flow in any of the vertical planes through the sensor.

The biaxial dual-propeller current meter is designed such that, while moored in a flow field, the shaft that bears the dual propeller pairs (a total of four propellers) will remain almost vertical (see Figure 8.23). The instrument relies on the propeller sensors' cosine response—a characteristic most desirable for any CM sensor but unfortunately lacking in many of them. It has been observed (Dean, 1985) that poor cosine response can lead to rectification of wave components and therefore large errors in the measurement of mean flows. The cosine-response characteristic is particularly relevant to ocean-current sensors deployed in the wave zones because in this zone the sensor encounters a wide range of flow incident angles. Two cosine-response propeller sensors mounted at right angles to each other directly measure the components of horizontal velocity parallel to the axles of the two propellers.

The rotation of each propeller is detected by magneto-diodes—asymmetrically set so as to indicate the direction of rotation—which sense the passage of four permanent magnets embedded in an epoxy disc rotating with the propeller axle (Collar, 1993). The heading of the CM is provided by a flux-gate compass, the output of which is sampled at a 1-Hz rate. On the production of each pulse pair by the rotation of a propeller, sine and cosine of the heading angle are added to registers storing the east-west and north-south components of the current, thus forming at the end of the sampling period a vector average of the current flow. It is quite evident that if the mean flow of magnitude U makes a horizontal angle θ with a given propeller, this propeller will sense $U\cos\theta$ and another propeller orthogonal to it will measure $U\sin\theta$. A similar pair of orthogonal propellers mounted back to back with the first set measures orthogonal components of currents from the other quadrants. This arrangement ensures that at least one sensor turns when the current comes from any angle in the horizontal plane.

Using heading signals from a suitably aligned magnetic compass within the instrument housing, the measured components are "rotated" into the conventional north-south and east-west components, averaged, and recorded. Summing the components and finding the resultant yield the vector-averaged current measurement. Because two orthogonal components of flow are directly measured, use of a direction-orienting vane is avoided for determination of flow direction. Avoidance of vane, which is an extra moving component, ensures that errors normally associated with ocean-current measurements due to imperfect response of the vane to high-frequency fluctuations are absent. Considering all these features, this current meter is popularly known as a *vector-measuring current meter* (VMCM). Preset sampling intervals between 1 and 15 minutes can be selected, the data then being written to cassette tape. The instrument is designed for inline mooring, the exterior titanium frame—from which the instrument housing is isolated—carrying the mooring tension.

The development of the biaxial dual-orthogonal propeller current meter offers an example of the way in which modeling of complex dynamic sensor response and the feedback of results from tank testing have played a central role in achieving a near-optimum design. A simple model of propeller dynamics such as that developed by Davis and Weller (1980), though neglecting interactions between propeller blades, serves to identify parameters important in determining the propeller response; it also provides insight into the nature of the compromise that must be reached in selecting the values for design parameters such as propeller pitch angle.

In steady axial flow, an ideal, frictionless propeller is accelerated until the lift-and-drag forces on its blades are perfectly balanced at a given radius along the blade, and the sensor is essentially linear. Any off-axis flow component or unsteadiness in the flow modifies the angle of attack of the blade on the fluid, thereby disturbing the dynamic balance, and hence includes contributions from the quadratic forces; a likelihood of nonlinear response then exists. Low pitch

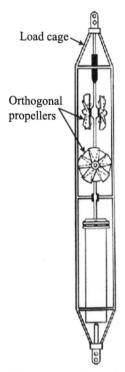

FIGURE 8.23 A biaxial dual orthogonal propeller current meter, which directly measures current vectors and is therefore known as a vector-measuring current meter (VMCM). This current meter does not have a fin. *(Source: Weller and Davis, 1980.)*

Load cage

Orthogonal propellers

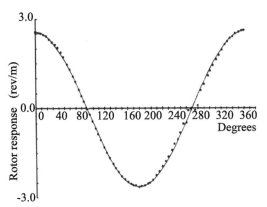

FIGURE 8.24 A typical plot of angular response of the propeller sensor attached to a VMCM, including the load cage, with flow in the horizontal plane at 25 cm/s. *(Source: Weller and Davis, 1980.)*

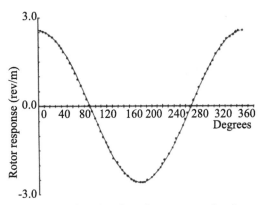

FIGURE 8.25 A typical plot of angular response to flow in a vertical plane parallel to the axle (i.e., the angle of attack of the horizontal component is kept at zero degrees whereas the vertical angle is varied). Flow was 25 cm/s, and the rms deviation of the measured response from cosine was 2.8 percent of the full-scale response. *(Source: Weller and Davis, 1980.)*

angles minimize the effects of off-axis components at the expense of increased bearing loads. However, response to unsteadiness in the axial component of the flow is also important and can be described—as for the VACM—in terms of a distance constant L, which in this case is proportional to the blade inertia and inversely proportional to the sine of the angle of attack. Excellent high-frequency response therefore demands low blade inertia and high pitch angle, and a compromise has to be sought.

Models of the instrument were made with progressively smaller spacings between the two propeller sensors and the pressure case. Tow tests of these models verified that the angular response of the sensors was degraded by placing them too close to the pressure case or to each other. The spacings chosen kept the response to flow in the horizontal plane close to cosine, whereas the response to flow with a vertical component deviated slightly from cosine due to flow disturbance by the pressure case and upstream sensor. The model included a simulation of the load cage so that any effect of flow disturbance by the rods of the load cage is included. Figure 8.24 shows a typical plot of angular response in the horizontal plane; the rms deviation of the propeller response from cosine was usually between 1.0 and 1.5 percent of the zero-degree angle of attack response. Figure 8.25 shows a typical plot of angular response to flow in a vertical plane parallel to the axle (i.e., the angle of attack of the horizontal component is kept at zero degrees, whereas the vertical angle is varied). Flow was 25 cm/s and the rms deviation of the measured response from cosine was 2.8 percent of the full-scale response.

The VMCM has been found to perform well in near-surface conditions. In combination of steady and unsteady currents, a mathematical model for the sensor predicted the possibility of "underspeeding" or underregistration of mean currents by the propeller sensor (Weller and Davis, 1980). This prediction was found to be true in actual tow-tank tests in which the instrument was subjected to combinations of steady and oscillatory flows. Results from some of the

subsequent evaluations (Beardsley, 1987) suggest that the VMCM has the tendency to underread slightly in the presence of intense high-frequency oscillatory flows (−5 percent). It has, however, been reported that the slight underregistration error of the biaxial propeller-style current meter in such an environment is considerably small relative to the overregistration error of the Savonius rotor-style current meter under similar conditions (10−20 percent). The sensor threshold has been reported to be ~1 cm/s, i.e., the CM cannot measure currents below 1 cm/s. It has also been reported that the propeller sensors do not rectify oscillatory motion when no mean motion is present. This characteristic is a remarkable feature of the propeller sensor over Savonius rotor, which does rectify oscillatory motion in the absence of a mean current.

8.6. CALIBRATION OF CURRENT METERS

Calibration is the determination of the sensor's response to steady-state relative motion of water past the sensor. The calibration relation is a best fit of one or more linear segments over the range of interest, expressed as an intercept and slope (Dean, 1985). Whereas laboratory calibration can be carried out at a towing-tank facility under controlled conditions, intercomparison studies are carried out in the real field under the existing natural conditions of a given water body. Laboratory calibration is usually carried out with a single current meter rigidly fastened to a trolley at one time. In a conventional procedure of CM calibration, the current meter is maintained normal to the trolley track. The towing facility includes a towed carriage that moves along two tracks built on the edges of a rectangular tank (see Figure 8.26). The carriage supports the current meter. Some tanks are provided with a shallow-water zone, which allows trouble-free mounting of the current meter on the trolley, and a deep-water zone for

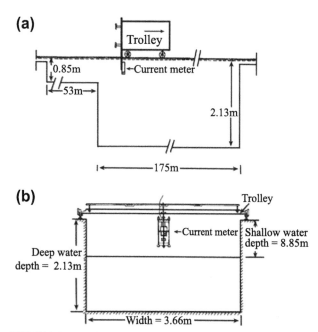

FIGURE 8.26 (a) Tow-tank and trolley facility at Central Water & Power Research Station (CWPRS) at Pune (India), used for conducting current meter calibration; (b) flow calibration setup. (*Source: Joseph and Desa, 1994, ©American Meteorological Society, reprinted with permission.*)

carrying out the actual calibration runs. The tow tank at the Central Water & Power Research Station (CWPRS) at Pune in India, used for conducting CM calibration runs, is part of a long channel extending from a nearby dam; therefore, the residual flow disturbances after a trolley run are rapidly damped.

The calibration setup shown in Figure 8.26b permits the CM to be moved in still water along the center of the tank with minimal flow disturbances. During calibration runs, the trolley is run at specified constant speeds, and the successive CM readings are noted, from which an average CM output value is derived. The CM is calibrated at discrete steps of towing speeds. At the end of the calibration runs at a number of discrete trolley speeds, from minimum to maximum, a calibration graph connecting a sufficiently large number of trolley speed values and the corresponding mean value of the CM outputs are plotted and the least-squares-fitted calibration equation is estimated. If the trolley speed did not maintain constant speed during a given run, the average trolley speed during that run is calculated and considered against the corresponding average value of the CM output for preparation of the calibration graph and estimation of the least-squares-fitted calibration equation.

An instrument's quality parameters such as accuracy and linearity are deduced from the least-squares-fitted calibration graph. Accuracy refers to the closeness of the instrument's output to the true value. In CM calibration, the trolley speed determined from first principles (i.e., speed defined as distance traveled in unit time interval) is taken as the true

value. The error is the difference between the instrument output and the corresponding true value, taken positively if the instrument output is greater than the true value (Doebelin, 1983). More often, accuracy is quoted as a percentage figure based on the full-scale reading of the instrument. Linearity is a specification relating to the degree of conformity of an instrument's calibration graph to the least-squares-fitted straight-line behavior. The linearity is a measure of the deviation of the calibration points from this straight line. Deviation from linearity is the difference between the instrument output and the corresponding least-squares-fitted value, taken positively if the instrument output is greater than the corresponding least-squares-fitted value (Doebelin, 1983). Linearity is usually expressed as a percentage of the full-scale reading of the instrument.

Some current meters are designed to orient into the flow irrespective of the change in current-flow direction. Gimbaled propeller current meters (e.g., Plessey current meter) and gimbaled fixed-vane type Savonius rotor current meters (e.g., Aanderaa recording current meter) are examples of such an ideal design. In such cases, the response of the flow sensor remains more or less unaffected by the changes in the direction of the current flow. However, in some designs (e.g., a free-vane system such as VACM), the CM does not orient itself in a specified direction in relation to the flow direction. If this is applicable to a given current meter, the flow sensor of which may need to be held within the supporting hardware of the current meter, the presence of the supporting hardware may give rise to differences in the CM's response with different azimuthal directions relative to the flow. In effect, the flow obstructions from the supporting hardware tend to modify the current-flow pattern in the vicinity of the flow sensor and, therefore, may affect its response to the flow field. If such a CM rotates in the azimuth for the same incident flow, the flow experienced by the flow sensor will change as the heading of the supporting hardware situated outside the sensor's periphery changes. This will introduce some errors in the current meter's output signal when the meter rotates in the azimuth.

In contrast to a fixed-vane system, the body of a free-vane system (e.g., VACM) does not have a fixed orientation with respect to the flow direction. Consequently, for the same incident flow, the flow pattern in the volume cell bounded by or surrounding the flow sensor of a free-vane system can vary during the measurement interval, resulting in corresponding variations in the flow sensor's directional sensitivity. Flow-pattern modifications imposed by a cylindrical support rod of a free-vane system can be calculated using the formula (Eskinazi, 1965):

$$V = \left(V_r^2 + V_\theta^2\right)^{1/2},$$

where V is current flow at any given point in the volume cell bounded by or surrounding the flow sensor, under the influence of the cylindrical support rod; V_r is the flow component along the line joining a given point and the axis of the given rod; and V_θ is the flow component at this point perpendicular to V_r. Here, V_r and V_θ are given by the expressions:

$$V_r = U_o \left[\left(\frac{a_o}{r} \right)^2 - 1 \right] \cos\theta \qquad (8.1)$$

$$V_\theta = U_o \left[\left(\frac{a_o}{r} \right)^2 + 1 \right] \sin\theta, \qquad (8.2)$$

where U_o is the undisturbed flow approaching the cylindrical rod, a_o is the radius of the rod, r is the distance between the axis of the rod and a given point, and θ is the angle between r and the vector, U_o. The formulas for V_r and V_θ are strictly valid only for rods of infinite length and when the flow is steady. However, these calculations enable first-order estimates to be made of the flow patterns in the volume cell bounded by and in the surrounding of the flow sensor. Joseph and Desa (1994) reported calibration of a "two-support rod" current meter, which employed an Aanderaa curved-bladed Savonius rotor, at two orthogonal orientations. The calibration results supported the inferences drawn from the flow patterns estimated for the volume cell swept by the rotor blades and provided a more quantitative figure of the effective flow deviation. The results are summarized in Figure 8.27. Because flow-pattern variations at two limiting orientations have been considered, it is expected that the flow output of the meter at any other orientation will lie within the two orthogonal calibration limits. Equations 8.1 and 8.2 suggest that the asymmetries in the flow patterns in the volume cell swept by the rotor in two orthogonal orientations can be reduced either by reducing the diameter of the support rods or by increasing the pitch circle of these rods.

One method of reducing the directional sensitivity of the rotor may be the use of a sufficiently large number of support rods instead of two. Detailed tow-tank experiments by Pite (1986) have shown that the flow asymmetries can be reduced to near zero if the number of support rods surrounding the flow sensor is such that near cancellation of positive and negative asymmetries occurs.

As indicated before, in the conventional procedure of current meter calibration, the CM is maintained normal to the trolley track. However, in most cases, while the current meter is deployed in a natural flow regime, it usually undergoes tilt from the vertical. At increasing current-flow speed, the speed-dependent wakes shed from the CM's housing are likely to generate unsteady flows in the vicinity of the flow sensor. This may cause nonlinearities in the tilt responses with increasing flow speeds. It is therefore useful to investigate the true behavior of a current meter while it is tilted from the vertical (i.e., its tilt response) at various towing speeds of interest.

Joseph and Desa (1994) conducted tilt-response experiments in a tow tank. The experimental setup used for investigation of tilt response is shown in Figure 8.28. The

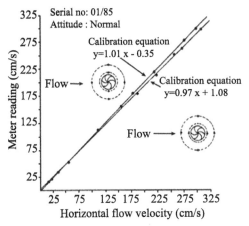

FIGURE 8.27 Calibration limits of a "two-support rod" current meter, which employed an Aanderaa curved-bladed Savonius rotor, at two orthogonal orientations. *(Source: Joseph and Desa, 1994, ©American Meteorological Society, reprinted with permission.)*

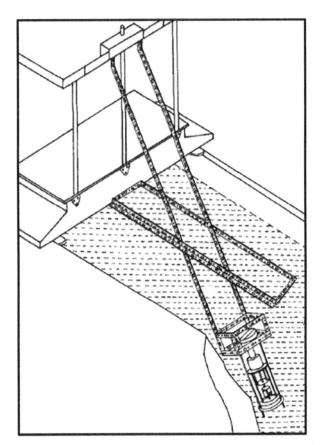

FIGURE 8.28 Experimental set-up for investigation of tilt response of a current meter. The current meter is seen attached to the tilting mechanism. *(Source: Joseph and Desa, 1994, ©American Meteorological Society, reprinted with permission.)*

tilting mechanism is fabricated from simple slotted angles, and the tilt angle is varied at will by simply shifting the mounting bolt from a pair of holes on the slotted angle to another pair. Because the tilting mechanism can be held above the surface of water in the tow tank, the tilting mechanism itself does not affect the flow in the vicinity of the flow sensor. In addition, because all tests can be performed on the current meter as it would be deployed in the real field, with all the hardware surrounding the flow sensor, the test results can be expected to represent the total tilt performance characteristics of the whole meter rather than that of the sensor alone. For analyzing the tilt response, for each towing speed of the current meter in its tilted position, the expected flow output of the current meter in its normal attitude can be derived from the least-squares fit of the flow calibration equation of the current meter in its normal attitude. Different flow sensors may respond differently to unsteady flows. For example, Karweit (1974), who conducted extensive tow-tank tests of a Savonius rotor in oscillating flows of differing frequencies, concluded that in unsteady flows the Savonius rotor may respond differently at different frequencies of the flow constituents. Joseph and Desa (1994) found that compared to a curved-bladed rotor, the flat-bladed rotor assembly is particularly sensitive to tilt. Its large and inconsistent deviations from cosine response have been attributed to the flow distortions caused by the semicircular cover of the rotor assembly.

The performance of a current meter in combinations of steady and oscillatory flows can be tested by adding to the tow cart a platform capable of swinging the current meter back and forth through an arc of variable period. This test set-up is shown in Figure 8.29. The tow cart travels on rails at a preset speed and a motor mounted on the cart with variable-speed drive, and an arm of variable throw provides oscillatory motion by moving the current meter back and forth at the desired period and amplitude. The current meter can be rotated so that testing can be carried out for study of the current meter's azimuth response (e.g., with the axle of each sensor parallel to, at a 45° angle to, or perpendicular to the direction of tow). Shown in Figure 8.29 is a prototype of the VMCM with a mock-up of the load cage so that any flow disturbance caused by the rods of the load cage will be included in the testing.

Calibration of the current meter involves not only speed performance but direction performance as well. According to Appell et al. (1983), a simple and inexpensive calibration facility at a field far from local magnetic influences is adequate for calibration of direction. The current meter can be placed over a graduated circle on a horizontal plane so that the vertical plane passing through the axis of the north-seeking magnetic compass points toward 0°. The current meter is then operated as is usually done in field measurements. After allowing a settling time of 30 s, the average of at least three successive readings may be noted. The current

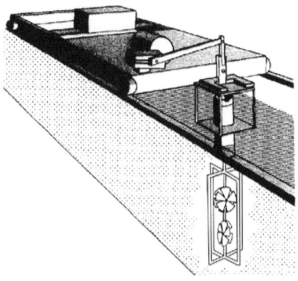

FIGURE 8.29 Test set-up used to create combinations of steady and unsteady flow. The tow cart travels on rails at a preset speed, and a motor mounted on the cart with variable speed drive and an arm of variable throw provides oscillatory motion by moving the current meter back and forth at the desired period and amplitude. Shown here is a prototype of the VMCM with a mock-up of the load cage so that any flow disturbance caused by the rods of the load cage will be included in the testing. *(Source: Weller and Davis, 1980.)*

meter can then be slowly rotated clockwise so that the axis of the direction-orienting vane is positioned in line with the next graduation marking, separated by a discrete angular spacing (say, 10°), and an average of at least three successive readings in this new position can be recorded after allowing for adequate settling time. In this manner, readings over the entire range 0–360° in discrete steps can be recorded, with the current meter rotated in the clockwise direction. Experiments can then be repeated in the same manner with the current meter rotated in the counterclockwise direction. Based on such a direction calibration procedure, Joseph and Desa (1994) found that a fixed-vane CM exhibits superior direction performance compared to a free-vane CM. The comparatively poor direction performance of the free-vane system is due to the poor coupling to the "vane-follower" magnet from the external vane.

8.7. GRAPHICAL METHODS OF DISPLAYING OCEAN CURRENT MEASUREMENTS

Understanding the measured dataset is as important as the measurement itself. The first task to be carried out to retrieve meaningful information from the data (measured or modeled) is to present them in meaningful graphical formats. There are many ways of graphically presenting the water-current measurements obtained from current meters. These include time-series plots indicating the temporal variability (see Figure 8.30), stick diagrams of current

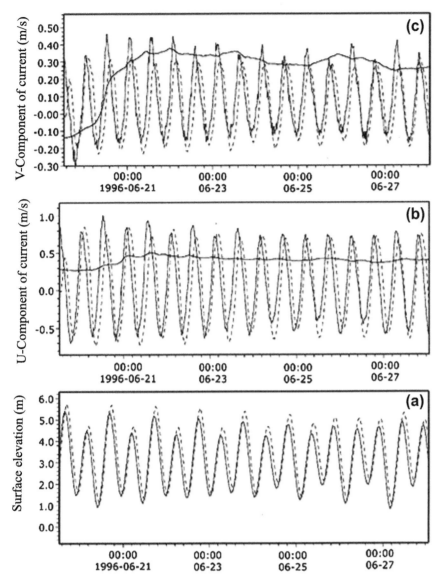

FIGURE 8.30   Time-series plots of measured and modeled current speeds and residual current during southwest monsoon conditions off Sikka in the Gulf of Kachchh in the Eastern Arabian Sea. *(Source: Babu et al., 2005.)*

velocity vectors that provide an indication of the temporal relationship of the current velocity vector (see Figure 8.31), rose diagrams that provide a pictorial description of the directional features of current propagation (see Figure 8.32), scatter plots (also called *dot plots*) indicating the relative spread of *u* and *v* components of currents (see Figure 8.33; see also Figure 8.22), progressive vector diagrams of currents showing the path that a fluid particle would follow (see Figure 8.34), and vertical profiles of horizontal current if current measurements are available from closely spaced depth intervals. If time-series current measurements from a geographically distant array of current meters are available, time-averaged currents for each separate location in the array can be plotted on a geographical map of the current measurements region (see Figure 8.35).

The progressive vector method takes advantage of the fact that the vertical variation in speed and direction of the current flow is usually small. If a region of fluid is imagined to be moving as a solid body, then a progressive vector plot can be viewed as a slice through the advected fluid. A standard progressive vector diagram is drawn showing the direction in which the current is flowing. For instance, if a current is flowing to the west, the point for day 2 will be drawn to the west of the point for day 1. A standard progressive vector plot can be thought of as showing the path that a fluid particle would follow. Thus, a progressive vector diagram provides an indication of the trajectory of the current.

In the case of regions of tidal dominance, plotting tidal current ellipses of various tidal constituents is an efficient

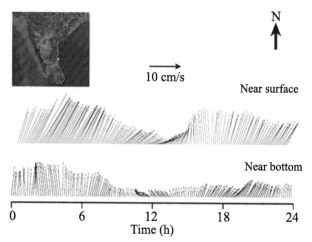

FIGURE 8.31 Stick diagrams of current vectors off Cuddalore on the southeast coast of India on May 22, 1998, at near surface and at 10 m depth. The measurement location is shown by a filled circle. *(Source: Courtesy of V. Kesavadas, P. Vethamony, and K. Sudheesh, CSIR-National Institute of Oceanography, India.)*

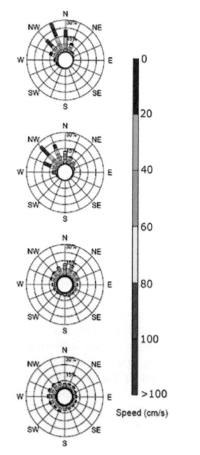

FIGURE 8.32 Rose diagrams of current velocity from five locations off Taiwan. *(Source: Liang et al., 2003.)*

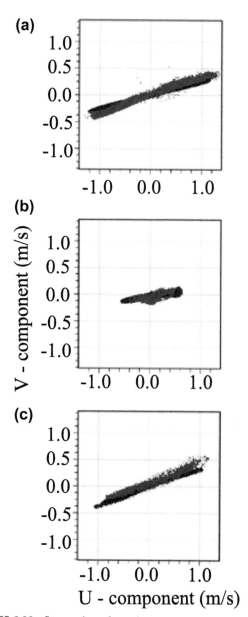

FIGURE 8.33 Scatter plots of u and v components of measured (red dots) and modeled (black dots) currents at (a) Mundra, (b) Vadinar, and (c) Sikka in the Gulf of Kachchh in the Eastern Arabian Sea. *(Source: Babu et al., 2005.)*

method of pictorially illustrating the tidal current features in a region. A harmonic analysis of the CM data yields the amplitude and phase of the east-west and north-south velocity components for each tidal constituent (Foreman, 1977). These four parameters define the tidal current ellipse that is traced out by the tip of the current vector in terms of semi-major (M) and semi-minor (m) axis, angle of inclination or ellipse orientation (Ψ), and Greenwich phase angle (ϕ). Alternatively, the tidal ellipse velocity vector can be represented by the sum of two co-rotating vectors with amplitudes of the rotary components R_+

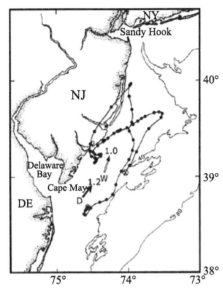

FIGURE 8.34 Progressive vector diagram of low-pass filtered surface current records during summer 1983 at a mooring on the inner continental shelf off Delaware Bay, USA. *(Source: Epifanio et al., 1989.)*

and R_- and phases of ϕ_+ and ϕ_- (Makinson et al., 2005). This idea is shown in Figure 8.36. There are locations in the southern Weddell Sea (e.g., along the Ronne Ice Front, Antarctica) where the M_2 rotary components, R_+ and ϕ_+, show a strong seasonal signal that coincides with the changes in stratification at FR6 (see Figure 8.35), resulting in the amplitude of R_+ at 442 m exhibiting a two-fold increase during summer months with swings of up to

$50°$ in ϕ_+. The changes in ϕ_+ for the other semidiurnal tides can also change by over $40°$, equivalent to $20°$ in ellipse orientation, which, at FR6, is typically around $55°$ during winter. Surprisingly, at this location, the clockwise tidal components of the semidiurnal tides (R_- and ϕ_-) and all components of the diurnal tides remain unaffected by the seasonal changes in stratification induced by intense wintertime heat loss arising from intense winds and sea-ice production. The observed specialties of the tidal ellipses in this region match observations from beneath fixed ice cover in the Arctic (Prinsenberg and Bennett, 1989) and are also predicted by the boundary layer theory (Makinson, 2002). The observed sensitivity of the anti-clockwise rotary components to changes in stratification is interpreted to be the best indicator of changes in stratification after direct observations of density variations. To identify any seasonal changes in the tidal currents, it is necessary to analyze short sections of the current meter time-series data record.

Spatially distributed tidal ellipses over a given oceanic region provide a clear indication of the nature of bathymetric and topographic influence on the tidal currents in that region (see Figure 8.37). Time-series residual currents are extracted from time-series records of current speed and direction with the use of a low-pass numerical filter (Godin, 1967) or any other convenient software packages. The resulting time series (see the residual graph in Figure 8.30) illustrates the nontidal part of the currents in a region. Vertical profiles of residual horizontal currents can also be presented graphically (see Figure 8.38).

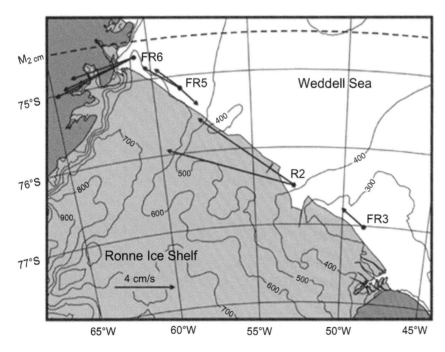

FIGURE 8.35 Time-averaged currents (indicated by arrows) at selected locations (FR3, R2, FR5, FR6) in the ice front region of the Ronne Ice Shelf in the Weddell Sea, Antarctica, plotted on a geographical map. The map shows the contours indicating the bedrock depth below sea level, with a 100-m contour interval (Vaughan et al., 1994). *(Source: Makinson et al., 2005.)*

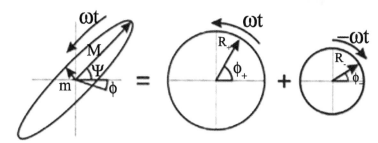

FIGURE 8.36 The basic parameters of a tidal ellipse and its two counter-rotating vectors. (*Source: Makinson et al., 2005.*)

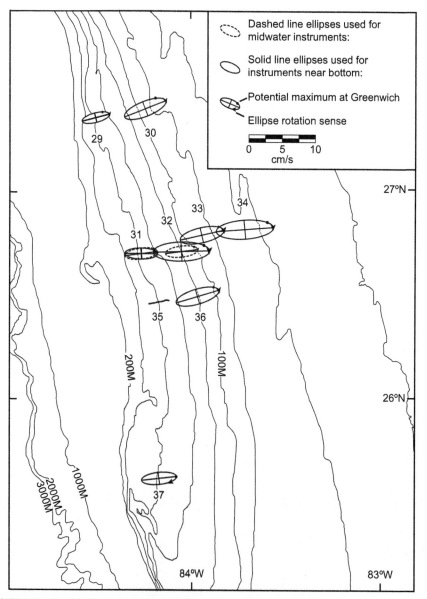

FIGURE 8.37 Spatial distribution of M$_2$ tidal ellipses on the West Florida Shelf. (*Source: Koblinsky, 1981.*)

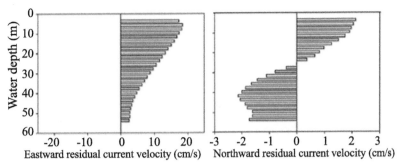

FIGURE 8.38 Vertical profile of residual current components for 24.8 h at a site in the Tachibana Bay, Japan, on July 31, 2006. *(Source: Tamaki et al., 2010.)*

8.8. ADVANTAGES AND LIMITATIONS OF MECHANICAL SENSORS

A unique advantage of mechanical sensors is their insensitivity to changes in the surrounding environmental conditions, such as temperature, conductivity, and sediment concentration. This unique quality of mechanical sensors probably makes them the most suitable for applications in estuaries, where freshwater influx and tidal variations introduce significant salinity variations, and in locations where sediment concentration is large. (The Hugli estuary that joins to the Bay of Bengal in the Indian Ocean is an example of such a situation.) In fact, current measurements in the Hugli estuary have traditionally been carried out using CMs that incorporated mechanical sensors. Because high-frequency wave components are comparatively minimal in an estuary, the drawbacks of mechanical sensors with reference to their sensitivity to wave motions become less severe in most situations. In addition, mechanical sensors tend to be low power consumers, are durable, and are relatively cheap. Furthermore, the inherently digital nature of these sensors permits their easy interfacing to digital systems. Bidirectional propeller sensors possessing good cosine-response characteristics directly measure current vectors.

A major drawback of CMs that incorporate mechanical sensors is their lower threshold (i.e., insensitivity to very weak currents). Furthermore, their bandwidths are poor and their usefulness is limited by their inability to measure the smaller time and length scales of fluctuating turbulence components. This makes them unsuitable for fast turbulence studies. Furthermore, except for the bidirectional propeller sensors, mechanical sensors permit current measurements along only one axis. This means that they need to be oriented into the current, often using large tailfins fixed on their supporting frame. This could be a disadvantage for their application for long-term deployments in fast-reversing flows. In addition, large vanes suffer from large response times and introduce their own errors (Kenney, 1977). Current meters that incorporate mechanical sensors therefore have difficulty measuring currents in which the mean velocities are small relative to the fluctuating currents such as those found in surface wave zones.

REFERENCES

Aanderaa, I., 1964. A recording and telemetering instrument. NATO Subcommittee on Oceanographic Research, Technical Report No. 16, Bergen, Sweden.

Appell, G.F., Mooney, K.A., Woodward, W.E., 1983. A framework for the laboratory testing of Eulerian current measuring devices. IEEE J. Oceanic Eng. 8, 2–8.

Babu, M.T., Vethamony, P., Desa, E., 2005. Modeling tide-driven currents and residual eddies in the Gulf of Kachchh and their seasonal variability: A marine environmental planning perspective. Ecological Modelling 184, 299–312.

Beardsley, R.C., 1987. A comparison of the vector-averaging current meter and New Edgerton, Germeshausen and Grier Inc. Vector-Measuring Current Meter on a surface mooring in Coastal Ocean Dynamics, Experiment I. J. Geophys. Res. 92, 1845–1859.

Busch, N.E., Kristenson, L., 1976. Cup anemometer over-speeding. J. Applied Meteorol. 15, 1328–1332.

Collar, P.G., 1993. A review of observational techniques and instruments for current measurements from the open sea. Report No. 304, Institute of Oceanographic Sciences, Deacon Laboratory, Wormley, Godlming, Surrey, UK.

Davis, R.E., Weller, R.A., 1980. Propeller current sensors. In: Dobson, F., Hasse, L., Davis, R. (Eds.), Air-sea interaction: Instruments and methods. Plenum Press, New York, NY, USA, pp. 141–153.

Dean, J.P., 1985. Problems and procedures associated with the calibration of ocean current sensors. In: Advances in underwater technology and offshore engineering, 4; Evaluation, comparison, and calibration of oceanographic instruments. Graham & Trotman, London, pp. 83–101.

Doebelin, E.O., 1983. Generalized performance characteristics of instruments. In: Measurement systems: Application and design. McGraw-Hill, pp. 37–206.

Epifanio, C.E., Masse, A.K., Garvine, R.W., 1989. Transport of blue crab larvae by surface currents off Delaware Bay, USA. Mar. Ecol. Prog. Ser. 54, 35–41.

Eskinazi, S., 1965. Uniform flow around a cylinder. In: Principles of fluid mechanics. Allyn & Bacon Inc, Boston, USA, pp. 286–290.

Fofonoff, N.P., Ercan, Y., 1967. Response characteristics of a Savonius rotor current meter. Woods Hole Oceanographic Institution Technical Report WHOI-67-33, Woods Hole, USA.

Foreman, M.G.G., 1977. Manual for tidal currents analysis and prediction. Institute of Ocean Sciences, Patricia Bay, Sidney, BC, Canada.

Godin, G., 1967. The analysis of current observations. Int. Hydrogr. Rev. 44, 149.

Griffin, D.A., 1988. Mooring design to minimize Savonius rotor overspeeding due to wave action. Continental Shelf Research 8 (2), 153–158.

Halpern, D., Pillsbury, R.D., 1976. Influence of surface waves on subsurface current measurements in shallow water. Limnology and Oceanography 21, 611–616.

Halpern, D., Weller, R.A., Briscoe, M.G., Davis, R.E., McCullough, J.R., 1981. Intercomparison tests of moored current measurements in the upper ocean. J. Geophys. Res. 86 (C1), 419–428.

Hansen, P., 1964. Note on the omni-directionality of the Savonius rotor current meter. J. Geophys. Res. 69, 4419.

Hodges, G.F., 1967. The engineering for production of a recording current meter. International Hydrographic Review 44 (2), 151–168.

Johnson, W.R., Royer, T.C., 1986. A comparison of two current meters on a surface mooring. Deep-Sea Res. 33 (8), 1127–1138.

Joseph, A., 1991. A technique for detection of Savonius rotor oscillation. Proc. Fourth Indian Conference on Ocean Engineering, Dona Paula, Goa, India, pp. 165–168.

Joseph, A., Desa, E., 1994. An evaluation of free-and fixed-vane flow meters with curved- and flat-bladed Savonius rotors. J. Atmos. Oceanic Technol. 11 (2), 525–533.

Karweit, M., 1974. Response of a Savonius rotor to unsteady flow. J. Marine Res. 32, 359–364.

Kenney, B.C., 1977. Response characteristics affecting the design and use of current direction vanes. Deep-Sea Res. 24, 289–300.

Koblinsky, C.J., 1981. The M2 tide on the West Florida Shelf. Deep-Sea Res. 28A (12), 1517–1532.

Liang, W.D., Tang, T.Y., Yang, Y.J., Ko, M.T., Chuang, W.S., 2003. Upper-ocean currents around Taiwan. Deep-Sea Res. II 50, 1085–1105.

Loder, J.W., Hamilton, J.M., 1990. Degradation of paddlewheel Aanderaa current measurements by mooring vibration in a strong tidal flow. In: Maryland, G.F., Appel Curtin, T.B. (Eds.), Proc. IEEE Fourth Working Conference on Current Measurement, April 3–5, 1990. IEEE, New York, NY, USA, pp. 107–119.

Makinson, K., 2002. Modeling tidal current profiles and vertical mixing beneath Filchner-Ronne Ice Shelf, Antarctica. J. Phys. Oceanogr. 32 (1), 202–215.

Makinson, K., Schroder, M., Osterhus, S., 2005. Seasonal stratification and tidal current profiles along Ronne Ice Front. FRISP Report No. 16.

McCullough, J.R., 1975. Vector averaging current meter speed calibration and recording technique. WHOI Technical Report, WHOI-75-44.

Paquette, R.G., 1962. Practical problems in the direct measurement of ocean currents, Marine Sciences Instrumentation. Instrument Society of America Vol. 2, 135–146.

Patch, S.K., Dever, E.P., Beardsley, R.C., Lentz, S.J., 1992. Response characteristics of the VACM compass and vane follower. J. Atmos. Oceanic Technol. 9, 459–469.

Patch, S.K., Beardsley, R.C., Lentz, S.J., 1990. A note on the response characteristics of the VACM compass. In: Maryland, G.F., Appel Curtin, T.B. (Eds.), Proc. IEEE Fourth Working Conference on Current Measurement, April 3–5, 1990. IEEE, New York, NY, USA, pp. 129–133.

Pearson, C.A., Schumacher, J.D., Muench, R.D., 1981. Effects of wave-induced mooring noise on tidal and low-frequency current observations. Deep-Sea Res. 28A (10), 1223–1229.

Pite, H.D., 1986. The influence of support rods on the rotation speed of Savonius rotors. J. Atmos. Oceanic Technol. 3, 487–493.

Pollard, R.T., 1973. Interpretation of near-surface current meter observations. Deep-Sea Res. 20, 261–268.

Prinsenberg, S.J., Bennett, E.B., 1989. Vertical variations of tidal currents in shallow land fast ice-covered regions. J. Phys. Oceanogr. 19 (9), 1268–1278.

Ramster, J.W., Howarth, M.J., 1975. A detailed comparison of the data recorded by Aanderaa Model 4 and Plessey MO21 recording current meters moored in two shelf-sea locations, each with strong tidal currents. Deutsche Hydrographische Zeitschrift 28, 1–25.

Saunders, P.M., 1976. Near-surface current measurements. Deep-Sea Res. 23, 249–257.

Saunders, P.M., 1980. Overspeeding of a Savonius rotor. Deep-Sea Res. 27 A, 755–759.

Savonius, S.J., 1931. The S-rotor and its applications. Mechanical Engineering 53, 333–338.

Schott, F., Bedard, P., Haldenbilen, K., Lee, T., 1985. The usefulness of fairings for moored subsurface current measurements in high currents. J. Atmos. Oceanic Technol. 2, 260–263.

Sherwin, T.J., 1988. Measurement of current speed using an Aanderaa RCM-4 current meter in the presence of surface waves. C. Shelf Res. 8 (2), 131–144.

Tamaki, A., Mandal, S., Agata, Y., Aoki, I., Suzuki, T., Kanehara, H., Aoshima, T., Fukuda, Y., Tsukamoto, H., Yanagi, T., 2010. Complex vertical migration of larvae of the ghost shrimp, Nihonotrypaea harmandi, in inner shelf waters of western Kyushu, Japan, Estuarine. Coastal and Shelf Science 86, 125–136.

Vaughan, D.G., Sievers, J., Doake, C.S.M., Grikurov, G., Hinze, H., Pozdeev, V.S., Sandhager, H., Schenke, H.W., Solheim, A., Thyssen, F., 1994. Map of subglacial and seabed topography 1:2000000 Filchner-Ronne-Schelfeis. Antarktis, Institut fur Angewandte Geodasie, Frankfurt am Main, Germany.

Webster, F., 1972. Estimates of the coherence of ocean currents over vertical distances. Deep-Sea Res. 19, 35–44.

Weller, R.A., Davis, R.E., 1980. A vector measuring current meter. Deep-Sea Res. 27A, 565–582.

Woodward, M.J., 1985. An evaluation of the Aanderaa RCM-4 current meter in the wave zone. In: Proc. IEEE Conference: Ocean Engineering and the Environment. IEEE, San Diego, CA, USA. November 1985, pp. 755–762.

BIBLIOGRAPHY

Beardsley, R.C., 1986. An inter-comparison between the VACM and new EG&G VMCM on a surface mooring. IEEE 1986, 52–62.

Beardsley, R., Briscoe, M., Signell, R., Longworth, S., 1986. A VMCM S4 current meter intercomparison on a surface mooring in shallow water. IEEE 1986, 7–12.

Gaul, R.D., Snodgrass, J.M., Cretzler, D.J., 1963. Some dynamical properties of the Savonius rotor current meter. In: Marine sciences instrumentation. Plenum Press, New York, NY, USA.

Gould, W.J., 1973. Effects of non-linearities of current meter compasses. Deep-Sea Res. 20, 423.

Halpern, D., 1986. Preliminary results of four VACM-VMCM dyads in the upper ocean of the equatorial Pacific. IEEE 1986, 13–19.

Henning, B., Gandelsman, V., Cope, K., 1981. Vector-measuring current meter (VMCM). IEEE 1981, 522–525.

Liang, W.D., Tang, T.Y., Yang, Y.J., Ko, M.T., Chuang, W.S., 2003. Upper-ocean currents around Taiwan. Deep-Sea Res. II 50, 1085–1105.

Pingree, R.D., Maddock, L., 1985. Stokes Euler and Lagrange aspects of residual tidal transports in the English Channel and the Southern Bight of the North Sea. J. Mar. Biol. Assoc. UK 65, 969–982.

Prandle, D., 1982. The vertical structure of tidal currents and other oscillatory flows. Cont. Shelf Res 1, 191–207.

Prandle, D., 1985a. On salinity regimes and the vertical structure of residual flows in narrow tidal estuaries, Estuarine. Coastal Shelf Sci. 20, 615–635.

Robinson, I.S., 1979. The tidal dynamics of the Irish and Celtic Seas. Geophys. J. Roy. Astron. Soc. 56, 159–197.

SCOR Working Group 21, 1974. An intercomparison of some current meters. UNESCO Technical Papers in Marine Sciences. No. 17, Paris, France.

Srinivas, K., Revichandran, C., Thottam, T.J., Maheswaran, P.A., Asharaf, T.T.M., Murukesh, N., 2003. Currents in the Cochin estuarine system (southwest coast of India) during March 2000. Indian J. Marine Sci. 32 (2), 123–132.

Veth, C., Zimmerman, J.T.F., 1981. Observations of quasi-two-dimensional turbulence in tidal currents. J. Physical Oceanogr. 11, 1425–1430.

Chapter 9

Eulerian-Style Measurements Incorporating Nonmechanical Sensors

Chapter Outline

9.1. Electromagnetic Sensors 267
    9.1.1. Electromagnetic Current Meters 269
9.2. Acoustic Travel-Time Difference and Phase Difference Sensors 281

9.3. Acoustic Doppler Current Meter 291
References 294
Bibliography 296

Understanding the dynamics of a number of important ocean processes, such as frontogenesis, mixed-layer evolution, inertial-interval wave generation, and coastal/shelf wave development, requires long-term (weeks to years) high-resolution (meters to tens of meters vertically, hundreds of meters to kilometers horizontally) measurement of the three-dimensional velocity field in the frequency band from 0 to 1 Hz. There are also fine and microstructures in the ocean current regime; therefore, sensors used for studies of such currents must respond fast enough to follow most of the high-frequency components in the flow signal and must be sensitive enough to detect minute variations in the flow field. Measurements of low-frequency, small-amplitude mean currents in the presence of turbulent currents as well as surface wave- or internal wave-induced orbital currents also require sensitive sensors. Nonmechanical speed and direction sensors best satisfy the required frequency response and ruggedness criteria. Purely mechanical sensors' inherent limitations, such as high thresholds, poor high-speed endurance, poor sensitivities, and limited frequency response, render rather difficult the measurement of very weak and very strong currents as well as high-frequency constituents that are present in the surface wave-induced or internal wave-induced orbital currents.

These limitations of mechanical sensors led to the development of rapid-responding nonmechanical sensors. Collective efforts expended by several technologists led to the successful implementation of such sensors into the subsequent generation of current meters. Thus, over a period of time, Eulerian-style measurement methods gradually progressed from those relying on purely mechanical sensors to those incorporating advanced electromechanical, acoustic, laser, and thermal sensors. These efforts resulted in the availability of highly sensitive sensors for measurements in turbulent current regimes and, in turn, stimulated the oceanographers to undertake fine-scale measurements that were not possible before for lack of suitable instrumentation. In situations in which high-frequency processes need to be examined, these specialized sensors' output can be sampled rapidly enough in time to match the processes of interest. Where low-frequency processes are to be studied, the averaging time can be chosen appropriate to the type of motions involved.

This chapter addresses three types of nonmechanical sensor: electromagnetic (EM) sensors, acoustic travel-time (ATT) difference sensors, and acoustic Doppler (AD) sensor.

9.1. ELECTROMAGNETIC SENSORS

EM sensors have been used for several decades for measurement of water flows in open water bodies such as oceans, lakes, and rivers as well as in constrained environments such as pipes. The basic principle dates from the days of Faraday. The EM current meter was developed and has been used for a number of years by the Institute of Oceanographic Sciences, Wormley, England. In the 1950s it was employed to measure turbulent fluctuations in tidal currents in the Irish Sea (Bowden and Fairbairn, 1956). Subsequently, it was developed as a ship's log, and Tucker et al. (1970) described its use for this purpose and its

Copyright © 2014 Elsevier Inc. All rights reserved.

principle of operation. As a current meter it has several advantages over more conventional devices. Its output is linear down to zero flow and independent of temperature, salinity, and pressure. It has no moving parts and is not particularly prone to damage during launching operations nor to fouling or corrosion. Use of two orthogonal pairs of electrodes permits simultaneous measurement of two orthogonal components of flow signals and therefore avoids the need for a current direction-orienting vane.

EM current meters also have the advantage of having high sensitivity, excellent linearity, and good directional response but no moving parts (Olson, 1972). They can measure two components of water current; this property may be useful, for example, if Reynolds stresses are to be measured (Thorpe et al., 1973). An EM sensor also possesses large linear dynamic range and excellent dynamic response and keeps a constant calibration. Kanwisher et al. (1975) reported an interesting example to illustrate the rapid dynamic response of EM sensors. In an EM sensor, the electrical output changes sign with the change in direction of the water current. This means that a moored EM current meter, unlike a Savonius rotor current meter, will not tend to produce a net output as a result of "rectification" of apparent currents in response to pumping motions.

The theory of EM current meters belongs to the subject of magneto-hydrodynamics, formed by the combination of the classical disciplines of fluid mechanics and electromagnetism (Shercliff, 1962; Bevir, 1970). Fluid moving in a magnetic field experiences electromotive force acting in a direction perpendicular to both the motion and the magnetic field. According to Faraday's principle of electromagnetic induction (EMI), an electrical potential gradient is developed across any conductor that moves perpendicularly through a magnetic field (see Figure 9.1). The magnitude of this potential gradient (E) depends on the magnetic flux density (B), the conductor velocity (V), and the conductor length (L), according to the relationship:

$$U = (V \times B)L \qquad (9.1)$$

Thus, for any constant B and L, the potential gradient (U) across the conductor is directly proportional to the conductor velocity, V. When Faraday's principle is

employed for water current measurements, the "conductor" is the water. Fortunately, the river and ocean waters possess sufficient conductivity to permit the use of the EMI principle for water current measurements. Apart from the primary application of EM current meters for measurement of motional velocities of electrical conducting fluids, there are several potential applications for EM flow sensors in industrially important multiphase flows in which the continuous phase has a relatively high electrical conductivity (e.g., water) whereas the dispersed phase or phases have a much lower conductivity. These include (1) "rock particle in water" flows that occur during the hydraulic transportation of rock in mineral-processing applications; (2) "rock cuttings in water-based drilling mud" flows that occur during oil-well drilling operations; (3) oil-in-water flows that occur down-hole in offshore oil wells at pressures that are too great to allow dissolved gases to come out of solution; (4) oil-and-gas-in-water flows that can occur at the wellhead in oil production operations, and (5) solids-in-water flows that occur in wastewater treatment applications (Wang et al., 2007). It has also been shown that in vertical, bubbly gas-water flows, EM flow meters can be used to measure the mean velocity of the water (Bernier and Brennen, 1983). These characteristics of EM flow meters render them suitable for current measurements in the surf zone of the oceans, where the seawater is often impregnated with air bubbles and minute sand particles thrown by the breaking waves.

After some assumptions were made, Shercliff (1962) proposed a weight function that describes the contribution of different parts of the EM flow meter measurement to the total signal. This function shows that the effect of velocity is strong near the electrodes and decreases when moving further away from the electrodes. The weight function therefore describes spatial variations in the EM flow sensor's sensitivity, which is a function of the magnetic flux density and the size, shape, and positions of the electrodes. According to Shercliff's weight function, the induced voltage across the electrode pair can be expressed as follows (Shercliff, 1962):

$$U = \frac{2a}{\pi a^2} \iint B(x,y)v(x,y)W(x,y)dxdy \qquad (9.2)$$

where U is the induced voltage when the electrode pair is located at various positions (x, y), B is the magnetic flux density, v is the flow velocity at each point of the flow region in the cross-section of the flow channel bounded by the magnetic field, W is the "weight" value at each point of the flow region in the cross-section of the flow channel bounded by the magnetic field, and a is the radius of the circular cross-section bounded by the magnetic field. Wang et al. (2007) reported a numerical approach to the determination of EM flow meter weight functions. Although the induced voltage is "conductivity-weighted," laboratory

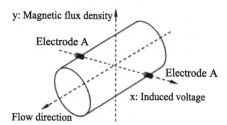

FIGURE 9.1 Geometry illustrating the principle of electromagnetic flow meters. *(Source: Wang et al., 2007.)*

experiments have shown (Olson, 1972; Sanford et al., 1975) that changes in "zero" (i.e., output in the absence of currents) and sensitivity due to water composition changes can be neglected in circumstances in which the electrode separation (L) is not appreciably large.

EM sensors employed for detection of ocean currents can broadly be categorized into two classes according to the source of the magnetic field used to detect the water-motion signal: (1) those that utilize the Earth's magnetic field, and (2) those that locally generate their own magnetic field. Eulerian-style EM current meters fall under the latter category.

9.1.1. Electromagnetic Current Meters

In the design of EM current meters meant for Eulerian-style current measurements, the geomagnetic electro kineto-graph (GEK) method mentioned in Chapter 2 was modified to generate a sufficiently strong local magnetic field using a coil that functions as an electromagnet. Uniformly distributed local magnetic field (i.e., homogeneous magnetic field) can be created using a pair of Helmholtz coils. Helmholtz coils are usually used as the excitation coils in the design of EM flow meters intended for measurement of fluid flows through large pipes, wherein the flow velocity away from the axis of the pipe decreases with increasing distance from the pipe's axis because of boundary effects imposed by the wall of the pipe. A Helmholtz coil consists of a parallel pair of identical circular coils spaced one radius apart and wound so that the electrical current used for exciting the coils flows through both coils in the same direction. This winding results in a uniform magnetic flux density between the coils, with the primary component parallel to the common axes of the two coils.

The material of Helmholtz coils is copper. Based on measurements using a magnetic camera, consisting of a rectangular array of closely spaced Hall sensors, Wang et al. (2008) experimentally demonstrated that Helmholtz coils generate uniform magnetic flux density. From Equation 9.2, a uniform magnetic flux density simplifies the computation of water current flow velocity based on induced voltage measured by the data acquisition system. Figure 9.2 shows a typical example of the placement of the Helmholtz coils used with a flow meter for water pipes, together with the orientation of the magnetic field generated by the Helmholtz coil, positioning of electrode pairs, and the water flow-induced voltages.

For logistic reasons, current meters need to be small in size, and therefore the sensor of an EM current meter must also be small in size. This requirement calls for the use of a multiturn single-coil bunch rather than a pair of Helmholtz coils for generation of a strong magnetic field. A pair or pairs of electrodes situated within this relatively strong

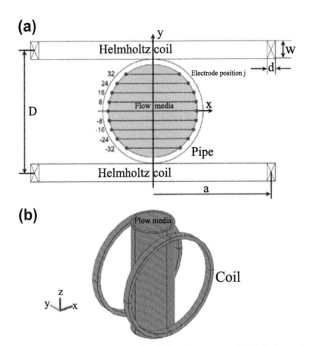

FIGURE 9.2 A typical example of the placement of Helmholtz coils used for flow measurements in pipes. *(Source: Wang et al., 2008.)*

magnetic field give rise to measurable potentials with small (less than 10 cm) electrode separation. Thus, in this type of EM current meter, the magnetic field is impressed on the water using a multiturn coil buried in the sensing head.

EM current meters are usually configured for measurement of two mutually orthogonal horizontal current components. In this case, measurements of the potential gradients are made from two orthogonally mounted pairs of electrodes. A steady direct current (DC) could be used to drive the Helmholtz coils, resulting in a steady, nonvarying magnetic field. However, this technique would superimpose the resulting microvolt-level DC signal voltage (i.e., flow-induced potential gradient across the electrodes) on the wideband electrical noise generated from a variety of sources such as electrochemical actions at the electrode surface, variable polarization potentials at the electrodes, and stray DC potentials in the water. In the case of ocean currents, the electrochemical voltages generated at the measuring electrodes are typically 10–1,000 times larger than the water-motion-induced voltages (McCullough, 1980). Consequently, it would be very difficult to accurately detect a DC signal; therefore, this calls for some sort of frequency separation of the low-frequency electrochemically induced voltages. As a result, most designs of local magnetic field type EM current meters drive the electromagnet with an alternating current (AC) or switched DC, thereby creating an alternating signal (voltage) of known frequency. Furthermore, application of an alternating magnetic field permits drift-free amplification of the microvolt-level alternating potential generated across the electrodes and largely reduces long-period noise and electrode effects. Because the magnetic field is modulated at a higher frequency relative to the undesired electrode potential variations, synchronous detection of the water-current-induced voltage signal becomes possible. This technique greatly simplifies the instrument design. Practical designs typically evolve from sinusoidal to chopped-DC excitation.

The water current flow field in the immediate vicinity of a sensor head, established to a large extent by the head geometry, is of critical importance in that it determines the degree of linearity that can be achieved as well as the nature of the directional response. Characteristics of the flow that are important include the thickness and velocity dependence of the surface boundary layer, especially in the proximity of the embedded electrodes, as well as the existence of flow separation effects that create complex flows capable of generating substantial nonlinearity in the sensor output. Modeling techniques, when applied to flow conditions and sensor response, can provide a basis for understanding sensor behavior and for prediction when seeking an optimum head design (Collar, 1993). Cushing (1976) constructed analytical models in order to compare the responses of different head geometries.

Forms of head shape that have been considered or used include "open" electrode configuration and various solids of revolution, such as spheres, cylinders, and ellipsoids (Figure 9.3). Although the hydrodynamic performance weighs heavily in choice of shape, this may be balanced in individual cases by considerations such as ease of fabrication and durability.

EM current meters of initial designs employed an "open" electrode configuration (Olson, 1972). This scheme minimizes the disturbance to the flow as far as is practicable. This approach has been adopted at the National

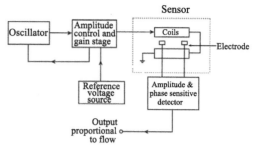

FIGURE 9.4 Block diagram of a conventional electromagnetic flow meter system. *(Source: Desa and Desa, 1980.)*

Institute of Oceanography (NIO) in India (Desa and Desa, 1980; Joseph, 1981) and Institute of Oceanographic Sciences (IOS) in the United Kingdom in work on the measurement of mean current in the wave zone (Collar, 1993). A block diagram of the conventional electromagnetic flow meter system is shown in Figure 9.4. In this flow meter, a pair of Helmholtz coils is driven by a low-frequency sinusoidal current. The induced flow signal is sensed by two electrodes located in the plane midway between the two coils. In operation, the output of a Wien bridge oscillator is passed through a fixed gain stage, and thence to a driver, which powers the coils. The electrode output is sensed by a field effect transistor (FET) input instrumentation amplifier (a near-ideal differential amplifier). The output contains both the flow signal and the induced quadrature voltages. The quadrature voltages are cancelled by adjusting the phase-sensitive detector waveform to sample equal amounts of positive and negative excursions of the quadrature signal. The remaining flow signal is also fed to the detector, which averages out noise and other unwanted signals. The detector consists of an FET switch by a suitably phased signal derived from the driver. The FET output is coupled to a buffer amplifier for transmission and display. The output of the detector is linearly related to the water flow velocity. This EM current meter was tested in a freshwater tow tank of dimensions 213 m × 3.6 m × 2.7 m. The results are shown in Figure 9.5. A root-mean-square (rms) deviation of 0.009 m/s

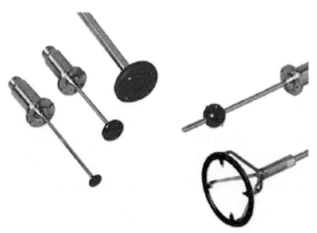

FIGURE 9.3 Samples of some electromagnetic sensors incorporated in electromagnetic current meters. *(Source: Valeport brochure; reproduced with kind permission of Kevin Edwards, Sales and Marketing Manager, Valeport Ltd., UK.)*

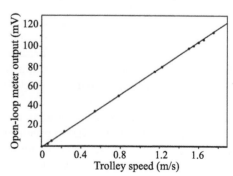

FIGURE 9.5 Calibration curve of a conventional EM current meter output with trolley velocity. *(Source: Desa and Desa, 1980.)*

over a range from 0.06 to 1.96 m/s (a factor of approximately 33) was obtained.

A flexible configuration of a variant of the open electrode configuration for an array of sensors has been developed by Sorrell et al. (1990) for measurements in energetic regions of the ocean where the current flow has high spatial and temporal variability. Central to the system is a two-axis EM current meter that measures the water current flow through a ducted volume containing a uniform magnetic field and nonprotruding electrodes flush with the duct surfaces. This geometry minimizes the electrode boundary-layer effect on the output, thus improving calibration and reducing fouling and damage potential. A patented inexpensive amplifier for low S/N ratios is employed to produce very low "zero drift" during operation. Electromagnetic sensors in various compact configurations have been fabricated previously, and many have worked acceptably. A central problem with this type of sensor has been the inherently low induced-output voltage, when power is limited for magnetic field generation. The low S/N ratio necessitates high-gain amplification, leading to problems with zero drift. In addition, the voltage signal-sensing electrodes have inherent, transient, electrochemical potential differences as well as other types of potential differences related to local boundary layer development/destruction. The relatively slow drifting electrochemical differences can be effectively eliminated by AC driving the magnetic field generating coil. The boundary layer-related differences can be controlled by physical isolation of the electrodes from the flow and/or with the use of geometry with reproducible boundary layer effects related to the flow-field measurement volume.

In the EM sensor reported by Sorrell et al. (1990), two improvements address the criteria of low zero drift and good high frequency as well as angular response on and off axis. First, a new low-cost amplifier circuit was designed for low S/N ratio and a high common mode rejection ratio (CMRR), thus providing very low zero drift. Second, a better geometric configuration (i.e., silver-silver chloride electrodes mounted flush in the face of the coil housing) was developed to control boundary layer and fouling problems as well as improve ruggedness and deployment flexibility. The magnetic field is generated by a Helmholtz coil arrangement, as reported by Olson (1972); however, the electrodes and coils are configured differently. For example, the coil units are solid rather than open, thus providing a ducted flow measurement volume within a highly uniform field. Identical sets of electrodes are flush-mounted on opposing faces of the sensor (coplanar with the coils) and connected in parallel. The complete configuration is mechanically robust, hydrodynamically clean, and symmetric, providing reliable precision measurements.

The electrodes have diameter of 2 mm, and two pairs of electrodes are located at 90° to each other. Each component of the water flow velocity is measured by the two electrodes; the axis connecting them is perpendicular to the flow. By locating two sets of electrodes 90° to each other (i.e., orthogonally), the velocity components 90° to each other can be measured. To average the flow velocity in the flow-sensing volume, electrode pairs are located in both the top and bottom coil housing. These electrode pairs provide two sources of induced voltage output, and the output from the pairs is averaged. An electrode located at the center of each coil housing is used to establish an "electrical signal ground" for the electronics. This arrangement of the electrodes not only averages the induced voltage over the flow-sensing volume, but it also provides redundancy if one of the two pairs of electrodes is damaged or fouled. The interior location of the electrodes minimizes the likelihood of biological fouling or mechanical damage from external impact.

The coil housings are held in place by a support structure, which consists of a hub over each coil housing connected by a support frame of four tubular members. One of the tubular frame members can be removed for access to the coil housings. Each coil housing can be removed individually from the structure. For shallow-water applications, the sensor is attached to the mooring by plastic piping that fits over the support hub. The pipe is attached to the hub by a locking ring and sealed with O-rings. The electronics for each sensor can be housed in a sealed portion of the mooring pipe, next to the sensor. Another alternative is to run electrical cables through the mooring pipe and the tubular support members. The usual practice is to locate the amplifier electronics next to each sensor and cable the output from several sensors in an individual mooring to a central power supply/data logger/compass unit located at the bottom of the mooring. The tubular support members are specifically designed to be large enough in internal diameter to carry all the cables necessary for most applications. With this flexibility in electronic component location, and because the sensor is an integral load carrying part of the mooring, the sensors can be configured as a vertical array. The sensors can also be electronically connected to common components or used as a single-point mooring. In situations where a number of closely spaced sensors are needed, such as in boundary layer phenomena, several of them can be located in a vertical array.

The sensor electronics supplies a stabilized square-wave excitation to the Helmholtz coil system. A synchronous demodulation system is then used to measure the flow-induced voltage signal. The square-wave frequency is maintained accurately at 15 Hz to cancel the large amounts of 60-Hz interference often encountered in the tow tanks used for calibration. The high-gain front-end amplifier uses low-noise operational amplifiers configured into a unique differential scheme. The circuit is specifically designed for the decoupling of both DC and low-frequency electrode offset potentials due to galvanic electrochemistry, noise, or other sources. A capacitor is used in the feedback loop to decouple DC and low-frequency voltage offset on the electrodes. Sorrell et al. (1990) used this patented scheme (Sorrell et al., 1985) in place of the usual method of AC coupling of the electrodes to the amplifiers. It has the advantage of giving high common-mode rejection when the electrical characteristics of the electrodes are not identical. This advantage is missing with direct AC coupling. Two such differential amplifiers are employed in a cascade to permit high gain with good bandwidth and minimum phase errors.

The synchronous demodulator uses a refractory period after each transition of the coil excitation for magnetic field generation. This scheme permits the coil current to stabilize and allows the large electrical transients, which result from the high rate of change of the magnetic field, to decay. Sorrell et al. (1990) observed that these transients are more pronounced in the presence of boundaries that disturb the magnetic field symmetry. Therefore, for example, these transients tend to be larger near the wall in a tow tank. The modulator samples the amplified signal only during the second half of each magnetic field phase. Signal conditioning for the resulting demodulated flow signal varies with the specific application. For the near-shore applications, which have a measurable surface gravity-wave (wind-generated wave) velocity component, the signal is filtered by a second-order Butterworth filter, usually having a 3-dB roll-off at 1 Hz.

As noted in Chapter 8, the performance of current meters needs to be established based on the results of tow-tank tests and field intercomparison studies. Sorrell et al. (1990) conducted all tow-tank tests in the Environmental Protection Agency's (EPA) Fluid Modeling Facility. Calibration of their EM sensor was carried out by mounting the sensor on the carriage of the tow tank and then towing the carriage down the tank at a number of selected constant towing speeds. The results of the test show that the sensor's output is essentially linear over the entire range of carriage speeds (the maximum carriage speed possible was 49 cm/s), with an rms error (in speed) of 0.3 cm/s. The maximum percent error in speed was 4%, with an average error of 2%. The range of testing not only indicated the proper operation of the sensor electronics but also lent confidence that changes in the boundary layer over the electrodes, for the speed range investigated, did not alter the linear response of the sensor.

A series of detailed tests on the angular response of the EM sensor was carried out by Sorrell et al. (1990) by repeating the full-calibration tests by rotating the sensor in normal attitude through various selected measured positive and negative horizontal azimuthal angles. The test results show that the sensor output is slightly higher at negative horizontal azimuthal angles than at positive angles. The reason for the observed difference in the azimuthal angular

response has been attributed to the sensor's misalignment at the start of the angular response tests. However, the azimuthal response is reasonably consistent and, therefore, reduction of data to determine the flow angle should be straightforward.

Sorrell et al. (1990) conducted dynamic response tests by mounting a mechanical oscillating device on the carriage. The oscillator was driven by a Scotch yoke driver, which was intended to produce an oscillating horizontal motion with a single harmonic component. The sensor was mounted on a string attached to the mechanism, and the entire system is referred to as an *oscillating arm*. The arm position was measured with a linear displacement transducer. The arm position was then differentiated to yield the arm velocity. This was then added to the carriage velocity to provide the instantaneous sensor velocity. The sensor was mounted with one axis parallel (parallel channel) to the flow motion and the other axis perpendicular (cross channel) to the flow. Output from both sensor channels and the oscillating arm position was digitally sampled at 0.01-s intervals. In each individual test, a total of 2,048 data points were recorded for each channel. The conditions for these tests were a mean flow (carriage) velocity of 30.7 cm/s and a maximum oscillatory velocity of 44 cm/s with a frequency of approximately 1 Hz. This is a relatively severe test in that the sensor is subjected to a high-frequency (1 Hz) unsteady flow and to both positive and negative velocities. Output from the parallel channel of the sensor indicates good dynamic response at frequencies of 1 Hz. The cross-channel output exhibits a higher frequency oscillation component. This component becomes quite large at the time when the oscillating arm has a relatively large deviation from single-frequency harmonic motion (referred to as a *jerk* in the arm motion). Similar higher-frequency oscillations were observed in both the parallel-channel and the cross-channel outputs when the sting was visually observed to vibrate. The frequency of these higher oscillations was found to be at approximately the natural frequency of the sting, which adds to the belief that they are due to the sting vibration. The higher-frequency oscillations in the cross-channel output did not average to zero, which is believed to be due to slight misalignment of the cross-channel axis with the flow motion (i.e., the cross-channel is not exactly perpendicular to the arm plus carriage flow motion).

Often, in the unsteady flows of interest, the flow velocity is not confined to two dimensions. The EM current sensor developed by Sorrell et al. (1990) is designed to measure two planar-flow components that are typically, but not necessarily, in the horizontal plane. To be able to do this, it is important that any flow component that is in the third dimension, or "out-of-plane" for a particular sensor orientation, not alter or contaminate the "in-plane" measurement. A flow component at an angle α should not alter the flow in the plane of the sensor housings. To investigate this idea, the entire sensor and the semi-rigid mooring pipe on which it was attached were mounted at an angle of 45° to the vertical. The sensor, so inclined, was then oscillated and translated horizontally down the tank, as in the prior tests. This produced a flow component that was inclined 45° to the sensor, or $\alpha = 45°$. Thus, in this dynamic response test, the flow was 45° out of plane of the sensor, i.e., with the sensor inclined at an angle of 45° to the vertical. The sensor's dynamic response (i.e., its tilt response) at high angle of flow inclination (45°) appears to be encouraging.

In addition to the extensive steady-state flow-calibration tests and dynamic tests of the EM sensor in the tow tank, which demonstrated its linear response over a moderate range of velocities and its reasonably good cosine angular response, Sorrell et al. (1990) reported separate *in situ* tests. In one of these tests, the sensor was installed from a bridge in an estuary. The flow had few dynamic characteristics that challenged the sensor's response characteristics, but the situation tested the sensor's long-term reliability. Data were collected for a period of over a year, indicating the sensor's reliability under field conditions for at least this long. The sensor required no maintenance during the entire year, but by the end of the year the sensor had become so biologically fouled that there was partial blockage of the flow sensor volume.

It was shown that the EM sensor has a sufficiently good dynamic response so that the high-frequency parameters of even the highly unsteady oceanic flows can be accurately measured, even in relatively severe environments. In addition, the sensor was found to maintain linear response and low zero drift required to accurately measure the mean flow properties.

As indicated earlier, the EM sensors have undergone several development phases. The drawback of an increase of noise level due to vortex shedding from the Helmhotz coils and their supports employed in the "open" electrode configuration necessitated alternate sensor designs. An ellipsoidal "head" is a result of such design initiatives. For example, the EM current meter designed by Thorpe et al.

(1973) consists of an energizing coil with a vertical axis (Figure 9.6), which is encapsulated in an ellipsoidal head. The electrical current passing through the coil produces a magnetic field. Conducting seawater moving through the magnetic field produces a potential voltage gradient at right angles to both the magnetic field and the direction of seawater flow, and this is sensed by two pairs of electrodes mounted on the face of the head. The voltage ratios bear a direct relationship to the horizontal direction of the water flow past the head. Electrochemical effects associated with DC and direct electromagnetic and capacitive coupling induced by AC are avoided by driving the coil by a switched DC. A steady current (governed by the switched DC coil-excitation voltage waveform) is passed through the coil and, after a fixed interval of time to cover the decay of transients, the amplified voltages from each pair of electrodes are briefly switched to storage capacitors. The current passing through the coil is then reversed (again governed by the switched DC coil-excitation voltage waveform) and the process is repeated using second capacitors. The potential difference between the two pairs of stored voltage levels is then a direct measure of water current speeds normal to the two pairs of electrodes.

The sensing head (Figure 9.7) used by Thorpe et al. (1973) is based on the design described by Tucker et al. (1970) for the ship's log and consists of a bifilar wound energizing coil and four electrodes contained in a solid epoxy resin molding with a maximum external diameter of 11.2 cm. Considerable research has been done to obtain the

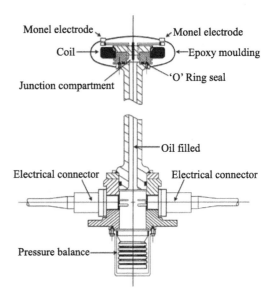

FIGURE 9.7 The cross-section of the sensing head of an EM current meter sensor, which consists of a bifilar wound resin-impregnated energizing coil and four electrodes contained in a solid epoxy resin molding. The supporting strut and the junction compartment at the back of the head are filled with oil and pressure-balanced by means of a diaphragm. *(Source: Thorpe et al., 1973.)*

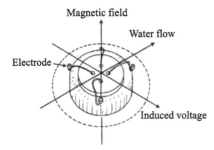

FIGURE 9.6 An EM sensor's bifilar wound resin-impregnated energizing coil and two pairs of electrodes across which water-flow-induced potential voltage gradients are generated. *(Source: Thorpe et al., 1973.)*

best configuration for the coil and electrodes and for the shape of the head. For example, the existing sensors were found to contain coils with an air gap between the windings, which inevitably resulted in a collapse of the outer molding when the sensor was subjected to high pressures, and new heads were therefore made with resin-impregnated coils to overcome this problem. Stresses on the head were further reduced by filling the supporting strut and the junction compartment at the back of the head with oil and pressure-balancing this by means of a diaphragm. This oil filling also helped with the insulation of the leads where they emerge from the head casting in the junction compartment. The head has been tested hydrostatically to a depth of 3,000 m.

Figure 9.8 shows the assembled instrument. The sensing head (A) with the plane of the coil horizontal is mounted on the supporting strut (B) 55 cm above the aluminium sphere (C), which houses the electronics and data recorder. The sphere has a wall thickness of 3.81 cm and external diameter of 71.1 cm. It is made up of two hemispherical forgings bolted on either side of a central ring through which electrical leads are taken. The forgings are manufactured from high-strength aluminium and heat-treated, both in the forged state and after rough machining, to obtain maximum-strength properties with good

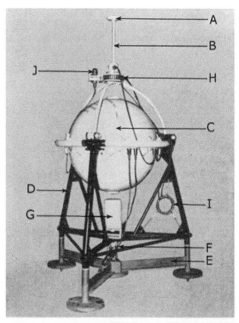

FIGURE 9.8 The assembled EM current meter. (A) Sensing head. (B) Supporting strut. (C) Aluminum sphere that houses the electronics and data recorder. (D) Framework to support the sphere. (E) Triangular base plate used for anchoring the current meter to the seabed. (F) Modified Van-Dorn release to which the base plate is held. (H) Scroll that receives signal from surface for firing the release mechanism. (G) Tilt- and direction-indicating device. (I) Transmission scroll used as a location device. (J) Flashing light used to aid location finding during recovery at night. (*Source: Thorpe et al., 1973.*)

dimensional stability. The sphere has a working depth of 5,000 m and a theoretical collapse depth of 7,500 m.

The sphere is supported on a framework (D) and is anchored to the seabed on a triangular base plate (E), to which it is held by a modified Van-Dorn release (F). This release is triggered by a "Pyro" release fired from a 6-V mercury cell. A 10-kHz frequency-modulated signal is transmitted from the surface, received by scroll (H), and fed into a receiver, where the signal is amplified and fed through a discriminator. The signal then passes through a high-Q filter switching a relay and firing a release. Ranges of 4−5 miles are obtainable, and the system demonstrated an operating life of eight months at temperatures down to −4°C. The Pyro release has no explosive force and, on igniting, burns at a temperature of 1,500°C, melting an aluminum tensile rod holding the jaws of the Van-Dorn release. Two Pyro releases are used, the second being triggered via a pre-set clock and acting as a fail-safe device for recovery. On release, the base plate is left on the seafloor and the remainder of the instrument rises to the surface. It has a buoyancy of 14 kg, which gives it an ascent speed of approximately 0.8 m/s, taking 1.75 hours to rise to the surface from a depth of 5,000 m. It floats with about 10 cm of the sphere clear of the water.

The sensing head is 2.1 m above the bottom of the base plate and will be at this height above the seafloor if no sinking occurs. The orientation of the head and the tilt of the instrument on the seafloor have been successfully measured by a simple but effective tilt- and direction-indicating device (G) shown in Figure 9.9. The solid polythene cylinder has a hemispherical upper face marked with 5° rings and radial lines that have a known orientation relative to the electrodes on the sensing head. Above it hangs a compass on a torsionless thread passing through a hollow brass pendulum rod. The solid cylinder can slide within the hollow cylinder, which is shown cut away in Figure 9.9, and is held away from the pendulum against the force of the spring by the magnesium-nickel alloy rod. When the device is in the sea, this rod corrodes and breaks in about two to three hours (some time after the instrument has settled on the seafloor), thereby allowing the spring to force the solid cylinder upward, to trap the compass and pendulum, thus providing a record of the tilt and orientation for examination when the instrument is recovered. The magnet is affected by the base plate and tripod supports, and therefore the direction can only be read to within 4°, but at worst the orientation is known to within 10°, the tilt to 1°, and the direction of tilt to 20°.

A 10-kHz 1-second acoustic pinger with the transmission scroll mounted on the vehicle framework (I in Figure 9.8) is used to accurately time the record-switching sequence and to act as a location device. The instrument, attached to the base plate, is released at the sea surface and allowed to free-fall to the seafloor. A flashing light

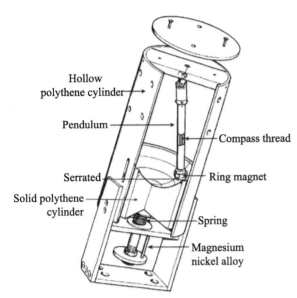

Hollow
polythene cylinder

Pendulum

Compass thread

Serrated

Ring magnet

Solid polythene
cylinder

Spring

Magnesium
nickel alloy

FIGURE 9.9 Tilt and direction indicator. *(Source: Thorpe et al., 1973.)*

(J in Figure 9.8) aids location finding during recovery at night.

A leak detector was housed in the sphere. This consisted of three electrode pairs set in a triangle inside the bottom of the sphere. A short circuit between any of these pairs of electrodes due to the presence of seawater caused a change of 5% in the pinger rate, which would immediately be detectable on a facsimile recorder on board a surface ship.

Current measurements made with this EM current meter in the Gulf of Cadiz revealed rapid and irregular fluctuations characteristic of a highly turbulent flow. Comparisons between records of currents made by the EM current meter and by other meters showed that the EM current meter is a useful instrument for use in the benthic boundary layer.

Head shapes such as the discuss form, developed earlier for use in a ship's log (Tucker et al., 1970) and used successfully also for current measurement in near-rectilinear flows (Thorpe et al., 1973), do not perform adequately in three dimensional flows (Griffiths et al., 1978). In the open-head construction, the magnetic field coil is contained in an epoxy molding of circular or ellipsoidal cross-section so as to minimize disturbance when flow is mainly in the plane of the coil. Four electrodes are mounted within the annulus, and screened electrode leads and coil-drive conductors are led away through the tubular support arms. It has been the practice in sensor heads designed at the Institute of Oceanographic Sciences (IOS) in the United Kingdom to use Monel as the electrode material, although other materials used elsewhere include carbon and silver. Variants of this form of head construction, which was developed at IOS for a vector-averaging current meter (Clayson, 1983), include a 45-cm diameter sensor designed for mounting beneath a surface buoy (Collar et al., 1988);

a miniature 4-cm sensor for small-scale resolution; and an experimental sensor incorporating two field coils and pillar-mounted electrodes intended for simultaneous measurement at four levels within 0.5 m of the sea surface. For each of these open forms, extensive tow-tank tests (Griffiths et al., 1978) have shown excellent linearity and off-axis response. For example, the maximum departure from linearity for the 15 cm diameter head is less than ± 2 mm/s in a 1.5 m/s speed range. Perhaps the main relative disadvantages of this form of construction compared with solid heads are the complexity of construction and their fragility, which often requires some form of protective ring or cage to be mounted around the sensor.

Olson's "open" electrode configuration gives rise to vortex shedding from the Helmhotz coils and their supports, thereby adversely affecting the sensor's angular responses. One way of enhancing the sensor's angular response is the choice of a spherical shape to the sensor head. The shape and dimensions of the sensor head are clearly important in the performance of the sensor in a flow field. The flow regime around a solid body depends on the dimensionless Reynolds number, $R_e = \dfrac{VL}{\nu}$, where V is fluid velocity, L is a characteristic dimension, and ν is the kinematic viscosity. Therefore, the design must avoid the critical range of Reynolds numbers if changes in sensitivity and response are not to occur (Griffiths et al., 1978).

Sensor head design affects other aspects of performance. For example, head sensitivity (which is of the order 20 μV/m/s for 1 Watt dissipation) increases as the square root of both the power dissipation in the coil and a characteristic linear dimension of the coil. Head shape rather than size, on the other hand, is generally more important in determining noise level. With the possible laminar flow around streamlined head forms such as the ellipsoid, flow-induced noise is likely to provide the dominant contribution to sensor noise level. Wherever in the spectrum this is true, S/N ratio cannot be improved merely by increasing coil input power. This situation is true in the case of the open sensor. A series of tow-tank measurements (Griffiths et al., 1978) showed that this has associated with it an rms noise level of ~ 0.6 cm/s at 50 cm/s with a nominal bandwidth of 5 Hz. The thermal noise within the same bandwidth is of order $(1.7 \times 10^{-20} \times RB)^{1/2}$, where B is the bandwidth and R is the electrode source impedance (Collar, 1993). In fresh water, $R \approx 3$ kΩ and in seawater it is $\approx 70\ \Omega$; the corresponding thermal noise that creeps into the current signal is ≈ 0.07 cm/s and ≈ 0.01 cm/s, respectively. Noise levels are reduced by taking vector-averaged values over a minute or two, yielding a current resolution typically <1 mm/s.

EM current meters are produced and marketed by several agencies such as Valeport; Cushing Engineering, Inc. (CEI); Marsh-McBirney, Inc. (MMB); InterOcean Systems, Inc.; Applied Microsystems Ltd., and so on.

FIGURE 9.10   Valeport-make Midas ECM. *(Source: Valeport brochure, reproduced with kind permission of Kevin Edwards, Sales and Marketing Manager, Valeport Ltd., UK.)*

A Valeport-made Midas Electromagnetic Current Meter (ECM) is shown in Figure 9.10. This small solid-state sensor has been designed specifically for use in open channels where fouling by weed or sewage can be a problem. Valeport's experience in electromagnetic technology has ensured that the Midas is a high-precision instrument that can be relied on to give accurate readings ($\pm0.5\%$ of reading plus zero stability) over a wide flow range (±5 m/s). The control display unit provides a choice of averaging modes, standard deviation of the data, and an optional logging facility. Valeport's latest electronics architecture allows multiple additional sensors and a variety of communications options, making it one of the few multiparameter current meters that allows real-time operation over several thousand meters of cable as well as autonomous deployments. A choice of titanium or acetal housing gives depth rating up to 5,000 m.

Cushing Engineering, Inc. (CEI), employs orthogonal electrodes. This is a concept that has been in use for several years. However, the techniques of application are different. The CEI uses a 7-inch cylindrical flow transducer. Its pressure housing is 64 inches long, with an outside diameter of 7 inches; the total length is 89.5 inches. It weighs 95 pounds in air. This design has a cylindrical flow sensor large enough to house all subsystems. The sensor incorporates an alternating electromagnet and two orthogonal pairs of flow-detection electrodes located near the center of the housing and attached flush with the outside container. Each pair of electrodes senses a voltage proportional to the component of velocity perpendicular to the plane, determined by the axis of the electrode pair and the sensor's longitudinal axis of symmetry. These voltages are conditioned and resolved into east and north components of current by a vector rotator using the output of a flux-gate compass. The current velocity components are then processed into mean values and recorded incrementally on magnetic tape.

Some degree of immunity from the problems caused by changes in flow regime around a solid of revolution can be obtained by mounting the electrodes out from the surface. Furthermore, locating the sensor head away from the sensor's housing will avoid vortex-induced deterioration of the sensor's performance. These two techniques have been adopted in the EM current meters manufactured by Marsh-McBirney, Inc. (MMB), an assessment of which has been made by Aubrey and Trowbridge (1985). Like the CEI

current meter, the MMB current meter also uses pairs of orthogonal electrodes mounted along the equatorial plane of its 4-inch spherical transducer. Figure 9.11 shows the MMB current meter (designated Model 585 Adaptive Recording Current Meter). Its pressure housing is 42 inches long, with an outside diameter of 7 inches; the total length is 60 inches. It weighs 95 pounds in air. This sensor has

FIGURE 9.11   Marsh-McBirney, Inc. (MMB) current meter, designated Model 585 Adaptive Recording Current Meter. *(Source: Marsh-McBirney, Inc. brochure.)*

a spherical electromagnetic sensor with two orthogonal pairs of flow detection electrodes. A digital geomagnetic compass measures the orientation of the current meter with respect to the magnetic meridian. A microprocessor-based data acquisition system resolves the current vectors into east and north components of currents, computes the mean values of these components, and transfers them to an incremental tape unit for recording.

In current flows near the air-sea interface, where wave orbits are circular in the vertical plane, the vertical cosine response is also important. With poor vertical cosine response, the vertical components of flow can reduce or augment the horizontal component in one horizontal direction either more or less than that in the orthogonal horizontal direction. This can lead to an error in direction of the horizontal component of flow as well as the magnitude. Though a spherical sensor by itself has excellent horizontal and vertical cosine response characteristics, the vertical cosine response of a current meter that incorporated a spherical EM sensor has been reported to have been degraded (Appell, 1979), probably due to current flow distortion as the flow passes over the pressure housing for the electronic circuitries. Performance degradation as a result of current flow masking associated with the electrode housing and the instrument supports indicated the need for refining the instrument's vertical and horizontal angular responses (i.e., tilt and azimuthal responses).

One method of reducing the distortion of the current flow field, generated by the instrument's supports, is the use of advanced designs where the instrument package is mounted within a spherical housing and the electrodes mounted on the equator of this sphere (Lawson et al., 1983). The sphere has proved particularly attractive in that it possesses three-dimensional symmetry and might therefore be expected to exhibit good responses to off-axis flows. However, any sensor head inevitably disturbs the free stream flow and introduces the risk of flow separation. Spherical design has been incorporated into EM current meters commercially manufactured by InterOcean Systems, Inc. (Figure 9.12). Two orthogonal pairs of electrodes and an internal flux gate compass provide the current vector, without the need for a vane. Improved azimuthal response takes better advantage of the inherent dual-axis capability of the sensor. On the other hand, improvement in the tilt response gives rise to a considerable enhancement in the current measurement accuracy when the sensor is used for particle velocity studies in the surface-wave field (Mulcahy, 1978). Ideal cosine-response characteristics are a highly desirable feature when one considers the very high accuracies required to extract mean velocities and, in particular, mean velocity shear in the presence of surface orbital motions. These are the major reasons for the preferential application of the spherical EM current meter for current measurements in the wave zone. As indicated

FIGURE 9.12 InterOcean S4 current meter. *(Source: InterOcean Systems, Inc., brochure.)*

earlier, the InterOcean S4 current meter incorporates the entire instrument within a spherical housing, which can be inserted directly into a mooring line. The resulting instrument dimensions would normally yield critical Reynolds numbers at some point within the working range, but this is forestalled by the use of surface ribs so as to induce a fully turbulent boundary layer throughout the working speed range. Good linearity is thereby achieved.

Blanton et al. (2002) reported the application of an InterOcean S4 current meter to study tidal asymmetry in three tidal channels with different morphology (e.g., [1] in a 100-km long coastal plain estuary, [2] a 15-km long tidal creek closed at the end, and [3] a small side-channel of a coastal plain estuary closed at low water) by comparing axial velocity as a function of water level (stage-velocity diagrams). Water current and water depth data were harmonically analyzed for the semidiurnal tide M2 and the over-tides M4 and M6, and the results were displayed in a plot of tidal current versus water level. The effects of the M2, M2 + M4, and M2 + M4 + M6 were compared in order to show how over-tides affect tidal asymmetry in the different systems. Stage-velocity diagrams illustrating the spatial dependence of the relative strength of influence of the over-tides along the axis of the just-mentioned three tidal creeks are shown in Figure 9.13. Progressive addition of M4 and M6 to M2 illustrates how over-tides affect the current velocity asymmetry observed in the different tidal channels. The stage-velocity diagrams for stations along

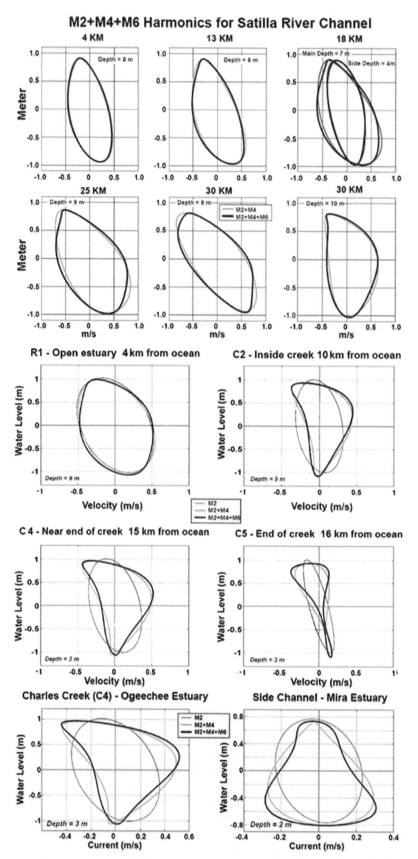

FIGURE 9.13 Temporal progression of water level and tidal current in various estuaries and creeks diagrammatically illustrated in clockwise direction. *(Source: Blanton et al., 2002.)*

the Satilla River show a gradual increase in the strength of the flood and ebb currents. It is seen that the presence of over-tides in the estuary changes the duration of ebb and flood currents. The results show that processes that generate the M4 and M6 over-tides in shallow estuaries and tidal creeks distort the normal sinusoidal character of the tide, as has been recognized by several investigators. The changing relative phase is manifested by the changing orientation of the stage-velocity diagrams, which is related to a steady dissipation of tidal energy along the channel.

For many purposes, power dissipation levels need not exceed a few tens of milliwatts, although at these levels difficulties can sometimes be experienced in calibration in freshwater tanks, where the sensing head source resistance is relatively high and stray potentials are often present in the water as a result of electrical activity in surrounding buildings. In such cases, careful attention to electrical-earthing arrangements is essential but is usually enough to overcome any problems. The mains supply frequency can be intrusive in this respect and is usually evident in the form of an unwanted low-frequency beat at the output of the instrument. For this reason, the choice of head drive frequency should avoid the mains frequency, its harmonics, and its subharmonics; otherwise a spurious DC offset may result in the calibration (Collar, 1993).

Unlike current meters incorporating mechanical sensors, EM current meters have no nonzero velocity threshold. However, response down to zero velocity is of little value unless the output is stable. Potential causes of instability include electrode contamination, biological fouling of the sensor producing a major disturbance of the boundary layer, and changes in the insulation resistance between the magnetic coil and the water. Experience suggests that electrode contamination is generally not a problem in measurement at sea, although care is required when the sensor heads are first put into the flow-tank water for calibration. The water surface in a tow tank frequently carries a thin, almost imperceptible greasy film. The prior addition of a drop or two of detergent to the tow-tank water has been found to eliminate this nuisance. Biological fouling is site and time dependent, but in the open sea it has not generally been found to present any problem.

Instability of the current meter output due to the changes in the insulation resistance between the magnetic coil and the water arises because the coil-water resistance path forms a potential divider with the electrode source resistance, thus permitting a proportion of the coil drive voltage to appear at the output. The need for adequately high ($\geq 10^9$ Ω) coil insulation resistance is particularly acute in the form of construction adopted for the annular head, but with care stabilities to within a few mm/s over periods of several months can be realized.

Overall, electromagnetic sensors potentially possess advantages such as fast response and good linearity,

although sensor head design probably represents a more critical area. Interaction of the flow with the sensor head is clearly of fundamental importance. Although much evaluation work has been done in laminar flow conditions, knowledge of response to turbulent conditions is less well developed. This is a difficult area in which to work, partly because turbulence characteristic of the sea is difficult to create in the laboratory tow tank and partly because the EM sensor yields a weighted, volumetric average. Evidence from the few investigations reported in turbulent flow is conflicting, though this may result from the nature of the flow around differing head shapes. Bivins (1975) found up to 20% changes in sensitivity when a cylindrical sensor was subjected to grid-induced turbulence. Results obtained by Aubrey and Trowbridge (1985) using a Marsh McBirney spherical sensor were not conclusive but showed some evidence of a decreased sensitivity. Griffiths (1979), on the other hand, found no evidence for change in experiments using an annular sensor in turbulence created either by a towed cylinder or by a submerged jet.

A disadvantage, perhaps, that can be stated about an EM sensor is that a large magnetic field is required to achieve an adequate S/N ratio for detecting oceanic turbulence (Jones, 1980). This means that it needs a comparatively high electrical power supply to drive the coil. As a compromise, an ingenious technique of conserving coil-drive power in relation to the flow speed has been reported by Desa and Desa (1980). In this technique, a closed-loop circuit enables the coil power of the current meter to be continuously adjusted for current speeds exceeding a preset threshold value. The closed-loop circuit permits less power to be fed to the Helmhotz coils when the electrode potential is large.

In the closed-loop system shown in Figure 9.14, the buffer-amplified output from the electrode-pair is compared with a standard precision voltage reference V_{ref}. The comparator output is used to modify the drive to the coils via the amplitude control stage. The amplitude controller consists of an integrating comparator wherein a DC reference level is compared with the full-wave-rectified output of the oscillator. The output of the integrating

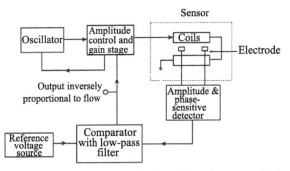

FIGURE 9.14 Block diagram of a closed-loop electromagnetic flow meter system. (*Source: Desa and Desa, 1980.*)

comparator drives the gate of an FET so as to maintain the AC amplitude of oscillation equal to the DC reference level. The integrating comparator output thus alters the coil power so that the difference between the buffer amplifier output and V_{ref} is reduced to a very low level. The closed-loop action thus maintains the electrodes at a constant voltage by altering the coil power when there are changes in water flow velocity.

Because the closed-loop system operates to maintain a constant voltage at the electrodes, Equation 9.1 can be rewritten as (Desa and Desa, 1980):

$$V_{ref} = (V \times B)L \tag{9.3}$$

Because B is directly proportional to the voltage across the coils, V_c, then from Equation 9.3,

$$V \propto \frac{V_{ref}}{V_c} \tag{9.4}$$

The coil voltage will be a maximum at the minimum flow velocity. This minimum is set by the capabilities of the amplification and detection circuitry. In the circuit tested by Desa and Desa (1980), the output of the comparator in Figure 9.14 is limited to a +2 V swing, at which level the output to the coil just begins to clip the rail voltages. At higher water currents than the minimum, the comparator output decreases, thus reducing the coil voltage as outlined. The reference source was equal to 1.600 V with a temperature coefficient of 35 ppm/°C.

Tow-tank results of the flow meter in its open-loop configuration (see Figure 9.5) yielded the sensitivity figures for the current meter. Thus, the coil power and also the sensor output for a known velocity having been determined, it was possible to set the range of flow velocities for the closed-loop system. The sensor output after amplification was approximately 20 mV/(m/s) for a coil power of 2 W. This sensitivity figure was subsequently increased by a factor of 3 for all further measurements.

To set up the closed-loop system for tank calibration in the velocity range 0.12−1.96 m/s, the reference source used for comparison was stepped down to 8 mV, corresponding to 0.12 m/s. The output of the comparator to the amplitude control stage was then equal to about 1.65 V, and this after the gain stage produced an oscillator output of 10 V rms just short of clipping the rail voltages. The upper end of the flow range now automatically adjusts itself to the requisite lower value. Results of calibration of the closed-loop EM current meter in the form of a graph of the reciprocal of the closed-loop output are shown in Figure 9.15. The correlation coefficient was found to be 0.998. The errors observed in this measurement could be largely attributed to the averaging of the closed-loop output, because a filter with a large time constant had to be incorporated into it. The low excitation frequency of 20.8 Hz required a large time constant in the phase-sensitive detector stage to obtain

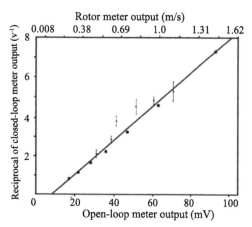

FIGURE 9.15 Reciprocal of the closed-loop EM current meter output with the open-loop meter output in a tank (shown with the best fit through the filled square points) and the corresponding field points (shown with error bars) using a Savonius rotor current meter. *(Source: Desa and Desa, 1980.)*

a low-ripple DC level for error comparison with V_{ref}. This time constant enforced a larger time constant in the comparator to maintain reasonably steady oscillator outputs. The overall time constant was kept equal to several seconds because the type of oceanographic measurements for which the current meter was designed did not demand faster response characteristics.

The location chosen for the field trials of the closed-loop system was an estuary where a large volume of outflow or freshwater to the open sea created layers of differing flow velocities. The reciprocal of the closed-loop output was recorded against the flow velocity as measured by a mechanical rotor current meter with a time averaging of 25 s. The rotor current meter had previously been calibrated in a tow-tank facility and subsequently against the open-loop EM current meter in the field. The field data points and their associated error bars are shown in Figure 9.15 for comparison with the tank calibration points. Data over only a limited range were obtained because these were the prevalent conditions during the field trials. The agreement is surprisingly fair despite the large uncertainty in the rotor current meter measurements brought about by its own inherent limitations and the natural variations of flow normally encountered in the field.

The closed-loop EM current meter feeds more power to the sensor coils when the water-flow velocity is low and less power when the flow signals are large. This type of current meter is therefore useful where long-term monitoring in a high-flow regime is required. In situations where a large range of current velocities is encountered, this system is especially advantageous because power saving of approximately a factor of 1,000 can be achieved for a velocity range of 33 cm/s. Because the S/N ratio is higher at large current speeds, reduced magnetic field as a result of reduced coil power does not degrade the sensor's ability to detect oceanic turbulence.

The inverse relation between water current velocities and coil voltages is easily converted to read as directly proportional quantities by the use of appropriate inverting circuits. One satisfactorily tested method was to convert the coil voltage to DC and feed it to a function generator integrated circuit chip (a voltage-to-frequency converter). The output frequency of this setup is then directly proportional to the water-flow velocity. This signal is readily transmitted over reasonable cable lengths, as would be required in the case of a submerged flow sensor and an onboard read-out unit.

9.2. ACOUSTIC TRAVEL-TIME DIFFERENCE AND PHASE DIFFERENCE SENSORS

When precise measurement of ocean currents is desired, one seeks linearity, accuracy, stability, freedom from flow disturbance, sensitivity, range, and cosine response (Thwaites and Williams III, 1996). Depending on the application, some features would be more important than others. Unfortunately, hardly any current meter has all the features that are desired. An acoustic travel-time (ATT) difference sensor is not an exception, but it has some admirable qualities, such as linear response through zero velocity, three-dimensional measurements, the ability to measure currents along many axes using one set of circuits, the ability to measure water current flow velocity vectors at a rapid rate over a small volume with little flow obstruction, omnidirectional response, the absence of lower speed threshold, and sensitivity to resolve turbulent fluctuations in the sea. These specialties render ATT current meters valuable for precision current measurements in diverse oceanic environments.

An Eulerian-style current meter incorporating an ATT difference sensor operates based on the principle that the resultant velocity of an acoustic wave propagating in a moving fluid is the vector sum of the fluid velocity and the sound velocity in the fluid at rest. The time required for an acoustic pulse to travel from one transducer to another, or vice versa, is the ratio of the path length to the speed of sound in the fluid. With no fluid flow, the travel-time difference between the transit times of the pulses in opposite directions is zero. When there is a flow, the sound velocity is aided in one direction and retarded in the opposite direction.

In designs where differential travel time is directly measured, two piezoelectric transducers of a given pair are simultaneously excited by voltage steps from 100 to 400 volts at a chosen repetition rate. For every excitation, each of the transducers generates a burst of exponentially damped high-frequency acoustic oscillations (in megahertz range) at the same repetition rate. A mathematical expression relating the current flow speed component (v) along the acoustic path, the acoustic path length (l), the speed of sound in the water at rest (c), and the acoustic pulse travel-time difference (Δt) can be derived as follows: Let the time required for an acoustic pulse to travel from transducer #1 to transducer #2 be T_1. Let us assume that during this travel along the acoustic path length (l), the sound velocity (c) is aided by the current flow speed component (v) along the acoustic path. In this case,

$$T_1 = \frac{l}{(c + v)} \qquad (9.5)$$

Let the time required for an acoustic pulse to travel in the opposite direction (i.e., from transducer #2 to transducer #1) be T_2. In this case, the sound velocity (c) is retarded by the current flow speed component (v) along the acoustic path, and therefore T_2 can be expressed as:

$$T_2 = \frac{l}{(c - v)} \qquad (9.6)$$

It is obvious that $T_2 > T_1$. Let $T_2 - T_1 = \Delta t$. From Equations 9.5 and 9.6,

$$\Delta t = \frac{2lv}{c^2 - v^2} \qquad (9.7)$$

In practice, v is very much smaller than c so that $(c^2 - v^2) \approx c^2$. Based on this approximation, when the transducers stay in the moving fluid, the difference between the up- and downstream sonic travel times (i.e., the travel-time difference, Δt) given by Equation 9.7 is simplified as:

$$\Delta t = \frac{2lv}{c^2} \qquad (9.8)$$

From Equation 9.8,

$$v = c^2 \left(\frac{\Delta t}{2l} \right) \qquad (9.9)$$

As indicated earlier, l is the straight line distance between the transducers of a given pair, v is the mean current flow velocity component along the acoustic path, and c is the sound velocity through the fluid at rest. Equation 9.8 shows that the differential travel time depends on the square of the sound velocity in water. The value of c is known to vary with the water temperature, pressure, and salinity (Urick, 1975). To compensate for variations in c, one method that has been adopted is to determine the mean (T) of the upstream and downstream acoustic travel times and to compute sound velocity using the relation $c = \frac{l}{T}$, thereby adjusting the current flow speed to the correct value (Lowell, 1979). Another method relies on measurements of temperature, salinity, and pressure and application of the required correction to the assumed sound velocity based on accepted formulae connecting the described three measured parameters. Measurement of current flow components along two orthogonal axes permits computation of the mean flow vector. Inclusion of a magnetic compass provides a magnetic north reference, and the two orthogonal flow components can be rotated into the magnetic components. Use of a third axis orthogonal to both the first and the second axes permits three-dimensional current measurements.

In a self-recording current meter for which the acoustic path length (l) is typically in the range 10−15 cm, it follows from the expression for Δt that the resolution of currents to better than 1 cm/s requires time discrimination to about 10^{-9} s. This emphasizes a need for highly stable, wideband detection of pulse arrivals, which is a critically important factor in achieving a successful design. Aspects of the design and performance of ATT sensors are discussed by Gytre (1976, 1980).

Apart from aspects pertaining to ATT difference discrimination, hydrodynamic considerations are also critically important in achieving the best possible accuracy of current measurements. The principal difficulty that arises in the case of ATT current meters is that of providing rigid mounting arrangements for the transducers, which are typically of order 1 cm in diameter, at each end of an acoustic path of perhaps 10−15 cm without significantly disturbing the free stream flow. Resolution of high-frequency processes requires that any noise contamination in the current meter be significantly reduced. To minimize vortex shedding (i.e., wakes) behind the leading transducers (which modulates the flow velocity field, thereby contributing to errors in the current flow measurements), an acoustic reflector is often used to deviate the ultrasonic beam between the transducers into a V shape (McCullough and Graper, 1979). By thus folding the sound path (see Figure 9.16), most of the travel of the acoustic signal is out of the shadow of the transducer and reduces the effect of turbulence around the sensor probes. In practice, ultrasonic signals are emitted at an angle of 45° to the horizontal plane (when the current meter is deployed with its major axis vertical). The transducers' tips are shaped to make the local currents around them cross the ultrasonic signals at an angle of 90° so that flow signal contamination does not take place. Though the acoustic signal folding technique reduces the primary interference, it introduces wakes from the mirror and its supports. Latterly the development of microprocessor technology has enabled use of redundant paths, so that for a given instrument orientation, the least disturbed paths can be selected in subsequent processing.

The basic ATT technique has been implemented in various forms for a range of applications. Examples of commercially available *in situ* recording ATT current

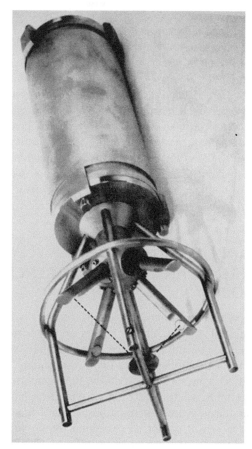

FIGURE 9.16 ATT current meter (original Christian Michelsen Institute design). Hatched line shows one of the two orthogonal acoustic paths via a reflector. *(Source: Collar, 1993, reproduced with kind permission from National Oceanography Centre, Southampton SO14 3ZH, United Kingdom.)*

meters in common use are the Simrad Ultrasonic Current Meter (UCM), manufactured by EG&G Inc.; the Acoustic Current Meter (ACM), manufactured by Neil Brown Instrument System (NBIS); and the one manufactured by Falmouth Scientific, Inc. These current meters are excellent instruments, but, like many other current meters, they are likely to be limited by their horizontal response characteristics. Saunders (1980) investigated the errors due to noncosine response in the horizontal plane (i.e., azimuthal response) in the case of the NBIS ACM. According to the manufacturer's specifications, the overall accuracy of the vector magnitude of the current is ±0.5 cm/s or 3%, whichever is greater. The direction accuracy is specified as ±5° for currents greater than 10 cm/s. At a current speed of 50 cm/s, the vector magnitude should be accurate to about ±1.5 cm/s. The manufacturer specifies horizontal cosine response error of ±2% of the vector magnitude over a 360° rotation.

Deviation from the ideal azimuthal response usually arises due to the presence of the support rods surrounding the sensor. For example, deviation as great as 5% in magnitude due to the presence of support rods has been reported in the case of VACM (Woodward and Appell, 1973). The NBIS ACM also has four support rods; therefore, investigation of a possible departure from a "cosine response" due to the support rods in the case of NBIS ACM is justified. This source was confirmed by tests of the horizontal response of an earlier model of the NBIS ACM by Appell (1978) and McCullough and Graper (1979). Tow-tank tests were carried out at David Taylor Naval Ship Research and Development Center (DTNSRDC) at an angular increment of 15° in the horizontal plane at tow speeds of 12.7 cm/s (nominal) and 31.1 cm/s (nominal) and at the National Space Technology Laboratories at an angular increment of 2° in the horizontal plane at tow speed of 10 cm/s. The direct voltage outputs from both axes of the current meter were fed to digital voltmeters operating on an IEEE-488 data bus. The tow carriage speed values and the ACM voltages were averaged over 1-minute intervals.

Because the carriage speed could not be held exactly constant, the output of each axis was normalized by the carriage speed. This quantity is designated as the *gain* of each axis. The unit of the gain is mV/(cm/s). The nominal gain for NBIS ACM is about 7 mV/(cm/s). In general, this value may vary slightly with the axis under consideration and the orientation of the current meter housing. Modification of the wake structure behind the support rods and the transducer supports may also render the gain a weak function of speed.

The theoretical gain curves assuming a maximum gain of 7 mV/ (cm/s) are plotted for comparison. Certain deviations from the theoretical cosine and sine curves are apparent, but in general, the observed points seem to fit the theoretical curves rather well.

The relative gain amplitude error $\varepsilon(\theta)$ is defined by (Saunders, 1980):

$$\varepsilon(\theta) = \frac{(G_x^2(\theta) + G_y^2(\theta))^{1/2} - \bar{G}}{\bar{G}} \qquad (9.10)$$

where $G_x(\theta)$ and $G_y(\theta)$ are the gains at the orientation, θ, of the current meter case (i.e., current meter housing) and

$$\bar{G} = \frac{1}{N}\sum_{i=1}^{N}\left\{G_x^2\left(\theta_i\right) + G_y^2\left(\theta_i\right)\right\}^{1/2} \qquad (9.11)$$

Here θ_i, $i = 1, \ldots, N$ are all the case orientations for which the gains are measured. The direction error (ϕ) is defined as:

$$\phi(\theta) = \theta - \tan^{-1}\left(\frac{G_y(\theta)}{G_x(\theta)}\right) \qquad (9.12)$$

The relative gain amplitude error is a measure of fractional error in speed when the current meter case is oriented at an angle θ, whereas the direction error is the error in direction

that will be made when the case is oriented at angle θ. These parameters will approximate the error when it is assumed that the case is oriented at the angle defined by $\tan^{-1}\left(\frac{G_y}{G_x}\right)$, providing $|(\phi)| \ll 1$. It should be noted that ε achieves local maxima at odd multiples of 45° and minima near the multiples of 90°. The maximum deviations of ε are of the order of $\pm6-8\%$, whereas the maximum direction errors are of the order of $\pm5°$.

Depending on the direction of current flow, the wakes from the tension rods can interfere differently with the acoustic sampling volume, thus deteriorating the cosine response of the current meter. When the current meter is rotated 45°, the wake from the support rod directly crosses the acoustic sensing volume, thus indicating a speed deficit. The role played by the wakes shed by the tie rod is distortion of the theoretical cosine response. As expected, the largest contribution is the "cosine" response of the first harmonic. It is seen that at $\theta = 0°$, the third harmonic contribution is positive (enhanced observed speed). By the time $\theta = 30°$, there is no contribution, because the effects of the wake deficit and streamline compression cancel one another. From 30° to 90°, the wake effect dominates, thereby producing the speed deficit. The same description applies to the other quadrants. Another expected effect is due to the wakes of the transponder support struts.

It appears that if the orientation of the current meter housing (case) relative to the current is known throughout the sampling period, it may be possible to correct for the imperfect horizontal response of the current meter. Because the case orientation is obtained only once every eight samples, the described requirement necessitates that the case orientation and the current direction both be slowly varying functions of time with respect to an 8-minute time scale. In general, however, this situation will not probably exist except for mooring in rather weak flows far from surface. The orientation is not a slowly varying function of time with respect to the sampling interval. However, it is conceivable that sufficiently smooth conditions may be more common.

Based on evaluation tests on four NBIS ACM current meters, Saunders (1980) found that if a calibration is to be used to correct for case orientation effects, a separate calibration must be used for each current meter. His studies gave an indication that the horizontal response may be speed-dependent as well as instrument-dependent. It was also inferred that the major source of error in horizontal current measurements utilizing the NBIS ATT current meter is due to its imperfect horizontal response. If the torsional motion of the current meter is absent or varies slowly enough, the amplitude error due to the horizontal response could be reduced by about a factor of 5 and the directional error reduced by a factor of 2.

Trivett et al. (1991) and Thwaites and Williams III (1996) conducted extensive studies of ATT current meters, and according to them the error sources for an ATT current meter include:

- Time-average flow obstruction from wakes
- Time-varying disturbances from wakes, vortex shedding
- Potential flow disturbance
- Zero offset bias
- Electronic nonlinearity
- Electronic noise

Electronic nonlinearity is described by Williams III (1995). Vortex shedding from bluff structural members is a significant contributor to measured velocity noise. The dominant frequency of vortex shedding for cylinders is given by the expression (Blevins, 1977):

$$F_s = \frac{0.2v}{d} \qquad (9.13)$$

Here F_s is the frequency in Hertz, v is the fluid velocity relative to the cylinder, and d is the cylinder diameter. ATT measurement of flow velocity can be thought of as measuring a line integral of velocity along the acoustic path. If the acoustic path is downstream and at an angle to the cylinder, the measured flow noise will scale with:

$$\text{Velocity noise } \alpha \; \frac{vd}{L} \qquad (9.14)$$

In this equation, L is the acoustic path length. If the structure upstream of an acoustic path is streamlined but stalled, its wake will also form a vortex street (Abernathy and Kronauer, 1962). To reduce the measured flow noise associated with vortex streets caused by the sensor, the ratio of cylinder diameter to acoustic path length needs to be minimized. The path length cannot be made large and still measure small-scale turbulence. To avoid Strouhal resonance, it is desirable to minimize the cross-sectional thickness of the sensor structure that supports the acoustic transducers consistent with structural considerations of strength and stiffness.

Falmouth Scientific, Inc. (FSI), reported a different configuration of the acoustic current meter—acoustic phase shift current meter (Figure 9.17). The current measurement technique FSI used was invented by Brown (1992) at Woods Hole Oceanographic Institution. The theory of operation of an acoustic phase shift current meter is as follows (Brown, 1992): If we consider an acoustic path with two transducers at points A and B where each transducer is alternately transmitting and receiving, the total phase shift between the received and transmitted acoustic signals is as follows.

$$\Theta_{ab} = \Theta_{ta} + \Theta_{rb} + \Theta_{ttab} + \Theta_{rec} \qquad (9.15)$$

FIGURE 9.17 Falmouth Scientific, Inc. 3-D Acoustic Current Meter, incorporating acoustic phase-shift measurement principle, in protective frame. *(Courtesy of Falmouth Scientific Inc.; www.falmouth.com/sensors/currentmeters.html.)*

$$\Theta_{ba} = \Theta_{tb} + \Theta_{ra} + \Theta_{ttba} + \Theta_{rec} \qquad (9.16)$$

In these expressions, Θ_{ta} and Θ_{tb} are the phase angles between the applied voltage and the resulting acoustic pressure wave for transducers A and B acting as transmitters. Similarly, Θ_{ra} and Θ_{rb} are the phase angles between the output voltage and the arriving acoustic pressure wave for transducers A and B acting as receivers.

Θ_{ttba} and Θ_{ttab} are the phase shifts due to the acoustic travel times from $A \rightarrow B$ and from $B \rightarrow A$, respectively. Θ_{rec} is the phase shift through the receiver. For any piezo-electric transducer driven by an essentially zero impedance generator or loaded by an essentially zero impedance receiver, the transmitting and receiving phase angles between the electrical current and the acoustic pressure wave are identical. Therefore, $\Theta_{ta} = \Theta_{ra}$ and $\Theta_{tb} = \Theta_{rb}$. Substituting these in Equations 9.15 and 9.16, and subtracting Equation 9.16 from Equation 9.15 yields:

$$\Theta_{ab} - \Theta_{ba} = \Theta_{ttab} - \Theta_{ttba} \qquad (9.17)$$

As the acoustic waves travel from one transducer to the other along the acoustic path length l, the sound speed c in the flow medium is aided or retarded by the current flow speed component v along the acoustic path, depending on the direction of v relative to that of the acoustic wave propagation. Thus, as indicated in Equations 9.5 and 9.6, the time T required for an acoustic wave to travel from one transducer to the other and vice versa along the acoustic path length l differs as a function of the water flow velocity component along the acoustic path. This difference in travel times of the received acoustic signals introduces corresponding phase shifts, given by $\Theta_{ttab} = \dfrac{\omega l}{c - v}$ and $\Theta_{ttba} = \dfrac{\omega l}{c + v}$, where ω is the angular frequency (radian/s) of the acoustic wave and v is the water flow velocity (cm/s) component along path $A \rightarrow B$. Thus, Equation 9.17 can be modified as:

$$\Theta_{ab} - \Theta_{ba} = \frac{\omega l}{c - v} - \frac{\omega l}{c + v} = \frac{2\omega l v}{c^2 - v^2} \qquad (9.18)$$

In Equation 9.18, $(c^2 - v^2) \approx c^2$, so that,

$$\Theta_{ab} - \Theta_{ba} = \frac{2\omega l v}{c^2} \qquad (9.19)$$

From Equation 9.19,

$$v = c^2 \left(\frac{\Theta_{ab} - \Theta_{ba}}{2\omega l} \right) \qquad (9.20)$$

Use of the acoustic phase shift principle eliminates the need for very high speed circuits by heterodyning the two received high frequency carrier signals to obtain a significantly smaller difference frequency (i.e., *beat frequency*) and by measuring the phase difference at the beat frequency with low power circuits.

The FSI acoustic phase shift current meter has four "fingers". Each finger houses two acoustic transceivers. The transceivers are used to create four acoustic paths. The flow velocity is measured by comparing phase shifts of sound pulses traveling along three of the four acoustic paths. One path, which is contaminated by the wake from the center support strut, is always disregarded.

To be able to deploy the instrument for several months using battery power only, the power consumption of the instrument needs to be very low (on the order of 10 mA). For this, first the measurement technique uses direct acoustic paths and not reflected sound (for example, from a mirror). This reduces the power dissipated into the water and thus the power required to operate the instrument. Second, the instrument measures phase shifts of the acoustic signals along the paths, not the time of travel along the individual paths. The advantage of measuring phase shift is that it can be accomplished with slower circuits than measuring the time of travel. Slower circuits in turn require less power to operate. The accuracy claimed for current measurement using this device is $\pm 2\%$ of reading or ± 1 cm/s. The dynamic range is ± 600 cm/s. The accuracy of the directional measurement is specified as $\pm 2°$ (Kun and Fougere, 1999). The instrument is protected from mechanical damage by its protective frame. The frame can also be used to tie the instrument to a mooring cable.

In near-bed benthic boundary layer studies, Reynolds stress is an important parameter. Correlation of fluctuations in velocity components in a boundary layer flow represents turbulent exchange of momentum across streamlines. Averaged over many burst and sweep events, the correlation of velocity fluctuations normal to the boundary with stream-wise velocity fluctuations is Reynolds stress (Williams III and Beckford, 1999). Exchange of fluid is the dominant stress-carrying process from the viscous sublayer to the outer boundary layer. In this region, millimeters to meters, the stress is constant, and a single measurement of Reynolds stress represents the shear stress at the boundary. Sediment transport and boundary layer mixing require estimates of the bottom shear stress. Reynolds stress measured within the inertial sublayer represents the boundary layer shear stress because within this region, stress is carried by turbulent exchange of momentum. In stationary, homogeneous flow, the stress in this region is constant, so a measurement at one height represents the stress at the boundary, even if the boundary is rough. If a current meter can measure this turbulent exchange of momentum, it will be a useful tool for boundary layer research.

With a view to achieving improved performance to meet the needs of near-bed benthic boundary layer studies, Williams III (1985) designed a novel ATT current meter known as a *Benthic Acoustic Stress Sensor* (BASS). BASS uses a braced metal frame of small-diameter rods to minimize flow noise, but this requires external electrical cables that are thicker than the structure cage rods. The BASS sensor has no undisturbed flow directions, but all horizontal flow directions are pretty good (Thwaites and Williams III, 1996).

BASS is a pulse travel-time acoustic current meter with four axes arranged about a 15-cm diameter measurement volume. It measures current flow velocity vectors with an accuracy of 0.3 cm/s and precision of 0.03 cm/s and can sample the flow speed at a 5 Hz rate. An assembly of sensors in a vertical array permits profiles of mean velocity, turbulent kinetic energy, and Reynolds stress to be obtained. BASS arrays have been deployed to depths of 4,800 m on instrumented tripods to estimate bottom stress and to monitor the inner boundary layer. The omnidirectional response, absence of velocity threshold, nonobstructive design, and high resolution of BASS renders it ideal for measurements in locations where current flow speed ranges from 0.5 to 70 cm/s and direction varies through all horizontal angles.

The important design criteria for BASS were that it measure water current flow velocity vectors at a rapid rate over a small volume with little flow obstruction, be omnidirectional in response, have no lower speed threshold, and be sensitive enough to resolve turbulent fluctuations. In addition, the constraints of field deployments required it to have low power consumption, be durable, and have redundancy to recover a vector if an axis were disturbed. The BASS system achieves these goals by using small transducers to measure flow speeds via unobtrusive acoustic pulse travel-time measurements. Four pairs of transducers surround a 15-cm-diameter water volume to measure speed along four axes, thus giving a complete velocity vector with an additional redundant axis. The sampling rate can be set as high as 5 Hz and the power consumption is low enough.

As shown in Figure 9.18, the acoustic transducers in the sensor are carried on two vertically separated rings. Each axis is inclined 45° to the planes of the two rings and spaced 90° in azimuth. BASS uses a lead titanate–zirconate piezoceramic transducer that is 9.5 mm in diameter, 1.27 mm thick, and resonant at 1.75 MHz. The central lobe of the acoustic beam is about 6° wide, giving a spot size of 15 mm

at the location of the opposite transducer. This requires alignment within 2°. This may not be the optimum size. According to Williams III et al. (1987), a 5-mm-diameter transducer would be more tolerant of alignment errors and have lower flow disturbance than the existing one and might still have enough gain for a good S/N ratio over a 150-mm path. However, the 9.5-mm transducer is a reasonable compromise for gain, flow disturbance, and alignment sensitivity as well as manufacturability and cost.

As Figure 9.19 shows, the transducer is cast in a rigid urethane (Conap DP-10767) and faced with a soft urethane matching the acoustic index of water (Conap EN-4). It is exposed to hydrostatic pressure and transmits a pressure wave equally from both faces when electrically driven. The mold accurately aligns the transducer with its faces elevated 45° from the mounting flange and aligned with its center 2.3 mm from the inner corner of the rectangular cross-section support ring. Because a burst of many cycles is transmitted and reflections from the early part of the burst transmitted from the back face could add to the later portion of the burst transmitted from the front face, it is important to minimize the amplitude of the reflections along the acoustic axis. The avoidance of any surface parallel to the transducer in the back does this, and the scattering from edges is too diffuse to contribute a measurable effect at the opposite transducer. Little refraction or reflection occurs at the EN-4 facing, and because this is slightly convex, what

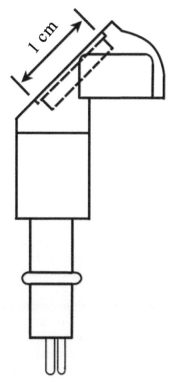

FIGURE 9.19 BASS sensor. *(Source: Williams III et al., 1987; ©American Meteorological Society. Reprinted with permission.)*

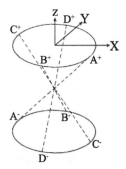

FIGURE 9.18 BASS sensor geometry. Four acoustic transducers (each 1 cm in diameter) are spaced 90° apart in azimuth at the locations indicated by the letters on each of the upper and lower 10-cm diameter rings. Under typical deployment geometry, the measurement path is inclined 45° to the horizontal. The wakes of the support rings and of the transducers lie outside the measurement volume for near-horizontal flows. *(Source: Williams III, 1985.)*

reflection does occur is defocused and thus attenuated at the opposite transducer. The transducer is connected through an integrally molded connector to urethane-jacketed RG-316 coaxial cable and then to the electronics pressure housing end cap.

Because of the particular geometry used in BASS, flow disturbance to the measurement volume is minimal for near-horizontal flows. The sensor is linear in flow velocity along any axis up to 120 cm/s and free of wake effects. Deviation from cosine response is less than 5% of the flow speed for all horizontal flows, and for flows out of the horizontal plane it is within 20° of an acoustic axis. This is possible because the transducers are small (~ 1 cm^3) and their supporting structures are thin (0.3 cm diameter). Thus, the wake of the structure is not great, and what wake there is crosses the measurement path only for a small percentage of its length. The major flow obstruction is the supporting ring for the sensor cage. This does not shed a wake into the measurement volume until the flow is tipped to about 50° from horizontal. If an individual axis lies within 20° of the flow (mean stream more than 25° out of the horizontal), the wake of one transducer influences the velocity along that axis significantly, and that axis cannot be used for the vector velocity determination. Fortunately, only one such disruption can occur at a time. System redundancy leaves three axes, which can provide all the information needed to measure the velocity vector.

Differential travel-time measurements can be made with pulses of acoustic energy, in which case time differences are measured, or with continuous waves (cw), in which case phase differences are measured. The former method is selected in BASS (a pulse consisting of 16 cycles at 1.75 MHz is transmitted from both transducers of a selected path driven in parallel) because, although it requires more power, it permits simple multiplexing so that many axes can be measured with one set of circuits and each axis can be measured with the transducers connected normally and reversed. The measurement of a single-flow component requires a little over 200 μs, which is the travel time for pulses crossing the measurement volume with the transducers connected normally and again with the transducers connected in reverse. By electrically reversing the transducers, offsets in the electronics can be subtracted. Doing this reduces the noise and drift of the measurements.

Williams III (1985) has provided a detailed description, with schematic illustration, of how precision measurements are accomplished in BASS. Deviations of 0.3 cm/s between sensors can result due to capacitance differences of different sets of transducers and cables. This offset limits the absolute accuracy of the velocity measurement to 10 times the pulse-to-pulse variation, but because the offset is fixed for each deployment of the system, it does not affect the turbulent quantities measured. To avoid aliasing (i.e.,

leakage of high-frequency energy present in the signal into a lower-frequency portion of the spectrum), the signal must be sampled at twice the maximum frequency present. The 15-cm diameter of the BASS sensor requires a sample rate of 1.6 Hz for a maximum water current speed of 25 cm/s, and a 2-Hz sampling rate is generally used. A sample volume of 15 cm diameter and a sampling rate of 2 Hz capture the turbulent scales of interest.

Good performance is achieved by minimizing the sensitivity of the system to the large and the variable capacitance of the long transducer cables and by isolating each timing circuit from the disturbances created by the other when a pulse is detected. These are described electronically as lowering the impedance of the first electronic stage, the multiplexer, and minimizing crosstalk. Electronic drifts are removed by reversing the connections to the transducers. Finally, the time differences between the first pulse arrival and the second pulse arrival are measured and digitized.

According to Williams III (1985), the horizontal and vertical cosine response, as determined from tow-tank test results, is close to ideal, with a maximum error of 5% of the speed along either axis until the vertical angle causes the flow to come within 20° of an acoustic axis. With regard to the sensitivity of the sensor to oscillatory flows, it was found that there is hardly any change in sensitivity or "zero offset" due to oscillations from less than a sensor diameter to several times the diameter.

Because the sensors are cable-connected to the electronics housing with 6-m-long cables, a tower of sensors and spacer cages can be constructed to obtain vertical distributions of mean and turbulent velocities. The tower must be rigid and fixed; to achieve this structure it is supported within a bottom tripod, as illustrated in Figure 9.20. The flow disturbance of the tower is not significantly different from that of a sensor cage (see Figure 9.21), and this prediction has been found to be true based on tow-tank test results. The tripod would add significant flow disturbance only near the legs, which are sufficiently away from the tower except for the topmost sensor. No large structures such as electronics housings or buoyancy modules are at the same height as a sensor. Also, the sensors are at least 2.5 buoyancy spheres and pressure-cylinder diameters away in the vertical.

With the bottom tripod illustrated in Figure 9.20, water-flow velocity measurements can be made to within 30 cm of the seabed. The ability to measure Reynolds stress means that no dependence on logarithmic velocity profiles is required to estimate the total stress above 30 cm. For a wide range of current speeds, BASS can measure turbulent scales extending into the inertial subrange. The instrument mounted on a deep-sea tripod provides the inner boundary layer measurements required to describe the benthic boundary layer dynamics.

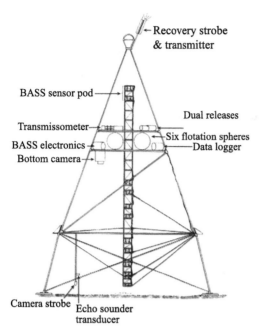

FIGURE 9.20 BASS tripod and sensor tower. The tower at the center of the tripod contains velocity sensors at 0.25, 0.5, 1, 2, 2.5, and 5 m. The instrument housings, buoyancy, and transmissometer are concentrated at 4 m. The lower frame is jettisoned for instrument recovery. *(Source: Williams III, 1985.)*

For deep-water measurements, the heavy lower portion of the tripod is jettisoned by acoustic command when recovery is desired. The upper portion carrying the tower and electronics floats to the surface, and the rising tower assembly is tracked acoustically for pickup. In shallow water, a somewhat different tripod system is used, which is recovered using an acoustically released float and recovery

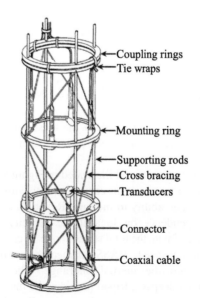

FIGURE 9.21 Sensor cage of BASS. *(Source: Williams III et al., 1987; ©American Meteorological Society. Reprinted with permission.)*

cable. BASS has been very successful at measuring currents, shear, and Reynolds stress (the turbulent transport of momentum).

In ATT current sensing, the "zero point" is not inherent to the sensor and therefore must be determined by calibration in still fluid. This is an important but troublesome chore in the cable-connected sensors of BASS in which flexure of the cables, when dressed on the BASS tripod, changes their capacitance (Williams III et al., 1987). In BASS, after the tower of two to seven sensors is assembled, the urethane-jacketed coaxial cables are tie-wrapped to wire guys and the legs of the tripod. Only then can the zeros be measured. These values are determined by casting the sensors in carrageenan gel or by bagging them in plastic film and deploying the tripod while measuring the flow inside the bags (Morrison et al., 1993).

Decomposition of current velocity fluctuations into vertical and horizontal components has been effective since the first BASS deployments. Sensitivity, linearity, and freedom from flow disturbance have allowed BASS to measure Reynolds stress. Figure 9.22 shows a short segment of an event record from a 1984 deployment at the HEBBLE site (Grant et al., 1985). Following the success of BASS, Thwaites and Williams III (1996) started developing its derivative, known as Modular Acoustic Velocity Sensor (MAVS), to achieve better performance by removing some of the obstacles found in the BASS. MAVS is also a three-axis ATT current meter that measures differential-acoustic travel time in a small measurement volume (see the photograph of a MAVS sensor in

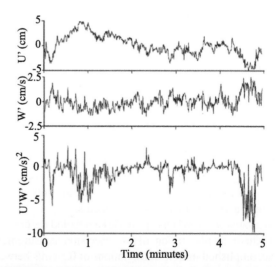

FIGURE 9.22 BASS data. The horizontal velocity fluctuations in the top plot, when multiplied by the vertical velocity fluctuations in the middle plot, give the bottom plot. The average value of the correlation weighted by the density of the fluid is the Reynolds stress. Reynolds stress is an average of this bottom plot, but structure in the flow is revealed by the predominant negative correlation in sweeps and ejections. *(Source: Williams III et al., 1987; ©American Meteorological Society. Reprinted with permission.)*

Chapter 2). Each acoustic axis on MAVS is oblique, with the axes 45° to the plane of the rings and passing from one ring to the other. The paths are spaced 90° in azimuth around the rings, one path going obliquely up, the next obliquely down, the third obliquely up, and the fourth obliquely down. Opposite paths—1 and 3, for example—when subtracted give one horizontal component of flow and when added give the vertical component of flow. The other opposite paths, 2 and 4, when subtracted give the other horizontal component of flow and when added give a second realization of the vertical component of flow. Because the vertical components are the same, if one axis fails for some reason, the full vector flow can be reconstructed using the redundant vertical components. In the normal rotation into Earth coordinates (or into instrument coordinates), all axes are used for each Cartesian component. U, V, and W components are estimated as follows (as per Dr. Albert J. Williams III, Woods Hole Oceanographic Institution, personal communication):

$$U = (-V_a + V_b + V_c - V_d) \times \beta \qquad (9.21)$$

$$V = (-V_a - V_b + V_c + V_d) \times \beta \qquad (9.22)$$

$$W = \frac{(V_a + V_b + V_c + V_d) \times \beta}{\sqrt{2}} \qquad (9.23)$$

In these equations, β is the scaling factor. Thus, MAVS measures the instantaneous flow velocity vector along four acoustic paths resolved into Cartesian Earth coordinates and thus the 3-D vector velocity in a cylindrical volume 10 cm in diameter and 9.5 cm high. This small region can be assumed to be uniform in velocity for purposes of resolving Reynolds stress down to a scale of 20 cm, although there are certainly turbulent fluctuations down to the dissipation scale of millimeters that are not resolved by the 10-cm acoustic paths in MAVS. Such a measurement is valuable for studies of sediment transport in which the shear stress on the bottom is the forcing function for erosion and a barrier to deposition. It may be noted that the energy content of the fluctuations is less at the 1-cm scale than at the 10-cm scale of the sensor if the boundary that generates the Reynolds stress is several sensor diameters away. The goal of developing MAVS was to give oceanographers access to the ATT technology in an affordable and easy-to-use form. The sensor was designed to reduce the amount of labor required in its manufacture compared to a BASS sensor, together with achieving reduction of flow disturbance error when flows are steeper than 30° from the horizontal. Other attributes of MAVS are low cost, small size, higher accuracy, good cosine response to current direction (immunity to off axis flow and gain independent of attitude), lack of bias in a wave environment, ability to measure turbulence and the Reynolds stress, resistance to

fouling (helped by no moving parts), ability to measure near a boundary, and enhanced accuracy at low current speeds (linear response through zero flow). In MAVS, the horizontal cosine response is excellent due to symmetry, and the vertical cosine response has been improved by fairing the rings that support the transducers. If the cost and size of the sensor can be kept small enough, arrays of current sensors could be deployed to measure flow fields, not just the time series of current at one or two points. Measurement and understanding of organized structures, such as Langmuir cells, could benefit greatly from such an array.

Whereas a BASS sensor requires a great deal of labor to build, thereby making it expensive, the MAVS sensor requires fewer labor-intensive potting steps of molding the transducers and cables in polyurethane. MAVS is only a single-velocity sensor that outputs a serial datastream to a user-supplied logger. It has been noted that one of the limitations of BASS is that cable flexing alters the "zero offset" of the velocity measurements by slightly changing the cable capacitance (Morrison et al., 1993). When the cables are flexed or retied, the zero offset typically changes by one-third to one-half cm/s. Based on this realization, in past BASS deployments, careful calibrations of the zero offsets were typically done after all the cables had been secured. In contrast, the advantage with the MAVS sensor is that it has internal, rigidly fixed wires that do not flex, and therefore its zero offset does not change as much, thereby avoiding the need to recalibrate the zero offsets (Thwaites and Williams III, 1996). In fact, a single "zero" calibration should suffice, barring dimensional changes with temperature and age. In any case, MAVS is relatively easy to "zero" because the sensor is at the end of a tube and can be immersed in a bucket of still water or a pot of hot carrageenan that is allowed to cool and gel. If MAVS is powered up while so immersed, it automatically obtains an average of 16 measurements of velocity and stores them in an array for subtraction from subsequent measurements. This is essential for doing Earth coordinate rotations with an ATT current meter (Williams III and Thwaites, 1998).

MAVS has other advantages over BASS. For example, the structural rings holding the acoustic transducers have streamlined cross-sections to reduce drag when the flow is 45° from perpendicular to the central tube, to reduce flow obstruction at this attitude, and to improve the cosine response. The geometry is simpler to make, because larger pieces of the sensor are moldable, to reduce labor and cost of production. The MAVS uses four acoustic paths of 10 centimeters to reduce sensor size and is operated by modified BASS electronics. The production version uses surface-mount technology electronics to reduce the size of the sensor electronics. The central tube is large, to allow use as the tension member in a mooring.

Reynolds stress and turbulent kinetic energy derive solely from fluctuations in velocity components (Williams III and Beckford, 1999). In the case of Reynolds stress, the mean is subtracted from the time series of velocity before the product of the downstream component with the vertical component is averaged. The average self-product of the fluctuations gives the turbulent kinetic energy, again rejecting the mean velocity. Consequently, "zero-offset" error is not a concern for these measurements. Only a very high rate of zero-offset drift would be a concern. However, for other purposes, the loss of vector truth and the loss of accuracy in speed are concerns. Thus, evaluation of performance of the instrument under various conditions that are expected in real field situations needs to be carried out.

Thwaites and Williams III (1996) reported the comparison of the measured sensor performance of BASS and MAVS for cosine response and flow noise. Williams III and Beckford (1999) reported results of additional tests of BASS and MAVS such as bucket zero, tank tests, and tow-tank tests. ATT sensors have electronic offsets in general that can only be corrected up to a point without a physical zero-point measurement. For such measurement, the sensor is placed in still fluid, such as water, seawater, gelatin, or some other stationary fluid, and a measurement is made. This becomes the zero-point calibration, and it is subtracted from all subsequent determinations of velocity. As noted earlier, BASS and MAVS both measure the vector velocity components in a cage 10 cm in diameter. The essential constructional difference between BASS and MAVS is that whereas the acoustic paths of BASS cross the center of the measurement volume, the acoustic paths of MAVS are on chords surrounding a central support. An ideal ATT current meter measures the component of flow speed along an acoustic axis. Combining several axes, three or more for a 3-D current meter, produces a flow vector. If one or more axes have a zero-point error, the resulting vector will be skewed. A 10 cm/s current with a zero-point error of 1 cm/s can have a vector direction error of 0.1 radian or 5.7°. According to Williams III (2001), the ratio of the zero-point error to the velocity is the critical determinant of this error. Because the offset is a constant, the direction error increases as the speed decreases. Cosine response measures sensitivity to off-axis velocity and any change in sensor gain with attitude relative to the flow. An ideal sensor measures the undisturbed flow, which is independent of sensor attitude. The rings in the BASS sensor appear to partially obstruct flow that is steep, thereby causing the sensor to measure less than the undisturbed flow, when the flow is more than 30° from the horizontal. Thus the BASS sensor provides good cosine response for horizontal flows but imperfect cosine response for steep flows. The BASS cosine response to steep flows is thought to be impacted by the rings. Streamlining the rings for these steep angles

in the MAVS sensor reduced the flow blockage and improved the sensor's cosine response.

The measured flow noise for both BASS and MAVS has been found to be largely linear with relative velocity, as expected. Unfortunately, the MAVS sensor has a larger cylinder-diameter-to-acoustic-path length ratio, and therefore its flow noise is much larger than that of BASS. The ratio of noise was found to be consistent with the diameter over path-length ratio of Equation 9.13. In a nutshell, performance evaluation studies conducted by Thwaites and Williams III (1996) provided an indication that an ATT current meter such as MAVS, with a central strut large enough to take a mooring load, despite good cosine response to steep flow angles, will not be good at measuring small-scale turbulence due to its large ratio of strut diameter to acoustic path length. However, fairing of the transducer support rings achieved reduction of flow distortion by the sensor to an acceptable level (Williams III and Thwaites, 1998). According to them, the main error in flow-velocity measurement using acoustic current meters is loss of a received acoustic pulse. An acoustic reflector such as a large air bubble or a bone can block an acoustic path and prevent one or more components of the velocity from being measured. This is detected by requiring that both transducers defining an acoustic path have received signals by the end of the measurement interval. If either or both have not received signals, a flag word is stored in place of the velocity. This word is detected when the normal and reversed measurements are subtracted. If either the normal or the reversed measurements contains the flag word, the flag word replaces the result of subtraction. When the velocities in the sensor frame are rotated into Earth coordinates, the presence of a flag word is checked for each component of the velocity and, if it is present, the Earth coordinate velocities are all replaced by flag words.

Because the zero offset can be measured and applied as a correction, zero offsets themselves are not a problem. It is the changes in zero offset that create error. Williams III (2001) reported study of zero-offset drift of MAVS in laboratory as well as field environments. In MAVS and its predecessor, BASS, the connection of each pair of transducers to the measurement circuits is reversible with a transistor switch. The electronic circuits after the switch drift with temperature, age, and voltage variations, but these effects cancel out when two measurements are made sequentially, first with the transducers connected normally and then with the transducers connected in reverse. The results are subtracted. The drift in electronic characteristics of the cascade amplifier, voltage comparator, charge integrator, voltage follower, differential amplifier, and A/D converter are all cancelled when the second member of the pair is subtracted from the first. In extended zero tests conducted by Williams III (2001) in a bucket in the lab, the "zero" was not seen to drift. Cooling the electronics with a circuit cooler did not have an effect either, until condensation occurred on the circuit board, degrading the insulation of the off circuits.

Three MAVS were deployed at 2,300 meters for up to three months at the Endeavour Field of the Juan de Fuca hydrothermal vents (Williams III and Tivey, 2001). Organisms covering the instruments were an interesting feature of long deployments of current meters. It was thought that these organisms might impede the flow, but fortunately they did not in this deployment.

Morrison et al. (2002) reported the development of a modified MAVS, known as MAVS3, which is the third version of MAVS. To make a measurement of the component of flow along a single acoustic axis, MAVS transmits a short pulse (15 cycles at 1.7 MHz) simultaneously from both ends of the acoustic path. The received signals are amplified in a cascade circuit and passed through a Schmitt trigger to a counter. This process provides significant immunity from noise (Williams III et al., 1987). In each receive channel, the counter detects the 15th negative-going zero crossing of the received signal and switches an active current source to a capacitor that has previously been fully discharged. The two acoustic signals arrive differentially as a function of flow speed along the axis, so the integrating capacitors in the receiver channels begin charging at different times. After the arrival of the second signal, MAVS simultaneously switches both current sources away from the capacitors. The capacitor voltages are differentially amplified, and the analog difference voltage is digitized with a 12-bit analog-to-digital converter (ADC).

To further improve accuracy, each measurement is repeated with the receiver channels electrically exchanged. The results of the normal and reversed measurements are differenced in software to remove receiver bias. The result is a differential travel-time measurement, in digital form, with accuracy in time of 40 picoseconds. Over the 10 cm MAVS acoustic path length, this means a velocity accuracy of 0.05 cm/s. The full linear range of the measurement is ± 180 cm/s. With this level of measurement accuracy, the limiting factor becomes the disturbance of the flow created by the sensor. The sensor head was designed to minimize the distortion.

Real-time MAVS3 measurements are available to an operator as a serial stream of ASCII data using standard RS-232 or RS-485 protocols. Any generic PC with a serial port and running a simple terminal emulator can display and log MAVS3 data in real-time. The problem of facilitation of simultaneous real-time data recording from an array of multiple MAVS3 sensors without the use of multiple PCs has been well addressed by adding an analog output capability to MAVS3. The advantage of doing this was that multichannel analog-to-digital acquisition and logging systems are standard equipment in many hydrodynamics laboratories, and such systems are well suited to sampling a large array of sensors; they provide the operator with real-time datastreams that have been referenced to a common time base. The digital-to-analog conversion of MAVS3 velocity measurements provides stable analog outputs that can be sampled asynchronously by the external system. Alternatively, a "data-ready" pulse permits sampling synchronized to the MAVS3 measurement interval (Morrison et al., 2002).

9.3. ACOUSTIC DOPPLER CURRENT METER

Acoustic Doppler water-flow measurement technologies have a close similarity to the laser Doppler technologies discussed in Chapter 2, and both can measure oceanic currents remotely. The major differences between these two technologies lie in the differences in spatial resolution and the maximum achievable range for remote measurement. The considerably short wavelength of the laser beam permits flow measurements with fine spatial resolution, but large attenuation of the laser beam in water does not permit remote measurements of water currents from distances more than a meter or so away from the sensor. Whereas seawater is rather opaque to electromagnetic radiation such as laser and radio waves, it is almost transparent to the mechanical form of energy known as *acoustic radiation*. This unique property of the acoustic radiation permits the acoustic Doppler current meter to nonintrusively measure ocean currents from large distances from the sensor. This superb capability of acoustic Doppler current meters circumvents one of the major limitations of the conventional Eulerian current meters, namely obstruction of the flow field by the body of current meter and the resulting dependence of the accuracy of current measurements on the current meters' azimuthal and tilt responses, which are generally far from the ideal cosine responses.

Acoustic Doppler current meter's operation relies on the mechanism of acoustic volume scattering from a cloud of moving scatterers in the flow field and the well-known phenomenon of the Doppler effect. The term *scatterer* encompasses any kind of inhomogeneity in water, including suspended particulate matter, biological organisms, minute bubbles, and so on. These scatterers serve as passive tracers of the flow field. In operation, acoustic sonar (which is the sensor) transmits a narrow-beam acoustic pulse into the flow field in a given direction, and the Doppler shift of the volumetric echo returned from the scatterers in the flow field is used to determine the relative radial velocity of the current flow between the volume bounded by the scatterers and the sonar. If the stationary transmitter, stationary receiver, and flow velocity vector *v* are in the same plane, the

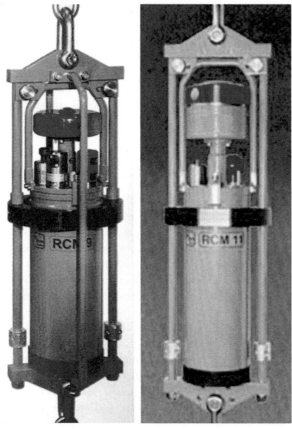

FIGURE 9.23 Eulerian-style Aanderaa RCM 9 MkII and RCM 11 Doppler acoustic current meters. *(Source: Aanderaa RCM 11 brochure.)*

frequency f_r of the received acoustic wave differs from the transmitted frequency f_t according to the relation (Clay and Medwin, 1977):

$$f_r = f_t \left(\frac{c + v\cos\theta_t}{c - v\cos\theta_r} \right) \qquad (9.24)$$

where c is the velocity of sound in the water at rest, and θ_t and θ_r are the angles made by the flow velocity vector v with the transmitter beam and the receiver beam, respectively. For a given geometry, the Doppler shift $(f_r - f_t) = f_D$ is +ve or −ve, depending on the direction of the flow. From Equation 9.24, we get,

$$\frac{(f_r - f_t)}{(f_r + f_t)} = \frac{(v\cos\theta_t + v\cos\theta_r)}{(2c + v\cos\theta_t - v\cos\theta_r)} \qquad (9.25)$$

For ocean currents, the received frequency differs only slightly from the transmitted frequency, so that $(f_r + f_t) = 2f_t$. For a symmetrical geometry, $\theta_t = \theta_r = \theta$. Based on these considerations, Equation 9.25 can be written as:

$$f_D = \frac{(2f_t v\cos\theta)}{c} \qquad (9.26)$$

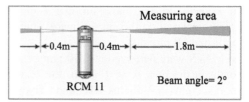

FIGURE 9.24 Illustration of the acoustic beam geometry and the active measuring window (starting at ∼0.5 m and ending at ∼2 m from the transducer) of the Aanderaa Doppler acoustic current meter. *(Source: Aanderaa manual for RCM 11 Doppler acoustic current meter.)*

If the receiving transducer is the same as the one used for transmitting the ping, then $\theta = 0$ so that Equation 9.26 gets simplified as:

$$f_D = \frac{(2f_t v_r)}{c} \qquad (9.27)$$

In this expression, v_r is the radial current velocity, i.e., the component of the current speed along the acoustic transmission path.

An important factor that influences the development of acoustic Doppler current meter is the transmission frequency f_t. From Equation 9.27 it is evident that for a given flow velocity, higher transmission frequencies are desired to produce larger Doppler shifts. The transmission frequency also influences the transducer design, its weight, and its size. However, attenuation of the acoustic wave increases as the transmission frequency increases, and the choice of frequency will, therefore, be a compromise.

The backscattered signal, normally of microvolts in amplitude, is composed of the random sum of the individual scattering amplitudes, each having a Doppler frequency shift associated with the radial velocity (i.e., water-flow velocity along the acoustic beam) of the scatterer. Because the micro-inhomogeneities that cause the scattering are of random sizes, the received signal is subject to amplitude modulation. Phase incoherencies also arise as a result of different ranges of the individual scatterers from the acoustic receiver as well as their random motion due to turbulence within the defined scattering volume. The signal is also corrupted by an additive white Gaussian noise. Thus, the received signal is a highly distorted version of the transmitted signal wherein its deterministic properties are no longer valid; rather, the signal is statistical in nature with Gaussian distributions. For this reason, a realistic approach for estimation of Doppler shift information is some form of statistical methods. Most of the Doppler shift information is contained in the first and second moments of the Doppler spectrum. One method of obtaining these moments is to Fourier transform the received echoes. These can also be obtained from estimates that do not involve the Fourier transform operation.

Based on the experience gained by past researchers in Eulerian-style Doppler acoustic current meters (see Chapter 2) and taking due account of various geometries vis-à-vis their advantages and limitations, Aanderaa Instruments of Norway developed a current meter, known as the RCM 9, and subsequently its modified model, known as the RCM 11 (Figure 9.23). In these current meters, a Doppler measurement starts by sending out a pulse of acoustic energy, called a *ping*. As mentioned earlier, due to particles and bubbles in the water, a fraction of the transmitted sound is reflected backward. This echo is then picked up by a receiver, which, in the present case, is the same transducer used for transmitting the ping. The uncertainty of a single Doppler measurement is very dependent on the transmitted frequency (f_t), the length (L) of the transmit pulse (which is assumed to be the same as the length of the active measuring window) and the beam geometry, and a proportionality constant (k_b), which is dependent on beam angle and several other factors. The standard deviation of one measurement (σ_s) is expressed as:

$$\sigma_s = \left(\frac{k_b}{f_t L}\right) \qquad (9.28)$$

By averaging several of these Doppler measurements, the standard deviation can be minimized. It was found that the accuracy would be acceptable by using a transmission frequency of 2 MHz, a beam angle of $\pm 2°$, a measuring window of ~ 1.5 m, and averaging 150 pings. To receive as much reflected acoustic energy as possible, the transmit pulse is set to be the same length as the measuring window. The pulse length of the ping is therefore set to 1 ms, which corresponds to a distance of 1.5 m. By measuring the incoming echo frequency from 0.67 ms to 1.67 ms after the ping is sent out, the active measuring window will start at ~ 0.5 m and end at ~ 2 m from the transducer (Figure 9.24). The water current velocity vector in the horizontal plane can be determined by the use of two mutually orthogonal transducers on a horizontal plane. However, to make the sensor as symmetrical as possible, Aanderaa chose to use four transducers, two opposite at each axis (see Figure 9.24). By pinging one at a time, the same electronics are used for transmitting and receiving. The circuit incorporates two switches: one for selecting one of the four transducers and the other to switch between transmitting and receiving.

When a Doppler sampling starts, the correct transducer is selected and the ping is transmitted. After the first part of the ping has traveled 0.5 m and back again, the active transducer is switched to receiver mode. The echo signal received in the receiver electronics is amplified and filtered. The signal is then mixed down to 455 KHz. This signal is then filtered and amplified until digital level is reached. The amplified signal is then fed to a programmable logic device

(PLD) where the frequency is measured very accurately. Both the transmit pulse and the measurement are based on the same 20 MHz crystal oscillator. This methodology minimizes the risk of temperature influence of the measurements.

A microprocessor is used for controlling the different timing and for calculating the current velocity vectors. The main component of current in the sea is usually horizontal. Because a moored instrument is likely to be inclined, the measuring axis will not be in the horizontal plane. For this reason, the current sensor is equipped with an electrolytic tilt sensor for both axes. The acoustic transducers, a Hall effect compass, and the tilt sensors are all incorporated in one molded unit, called a Doppler current sensor. After a cycle of four pings, one in each direction, the tilt is measured and two horizontal current components are calculated. For current direction to be meaningful, these components must be referred to Earth coordinates. On the basis of the readings from the internal Hall effect compass, the two current components are converted to north and east components. These components are accumulated for each cycle of four pings. At the end of the measurement interval, the vector average of these components is calculated and sent out as current speed and current direction at the sensor output.

Most of the electronics are turned off between the Doppler samplings. This makes the average power consumption very low, almost proportional to the ping rate. The sensors are mounted on the top endplate of the instrument case, with the Doppler current sensor on a central stud. All the sensors except for the temperature are molded in polyurethane with excellent antifouling properties. The measuring interval of the instrument can be selected by setting a rotary switch on the electronic board inside the pressure tube. This board controls the ping rate so that 600 pings are taken during each measurement interval. This feature makes the accuracy independent of the measuring interval and the power consumption per measuring interval almost constant. Figure 9.25 shows typical examples of preparations for deployment of an Aanderaa Doppler acoustic current meter in the Indian Ocean.

One of the main applications of this current meter is monitoring the large-scale circulation of the oceans. Because this current meter is free of the threshold found in several mechanical sensors, weak currents such as those found in the deep oceans can be monitored more accurately and, due to the low power consumption, data can be collected for up to two years (Hovdenes, 1998).

In tropical water bodies, the upper few meters of the water column (the photic zone) are subject to severe biological fouling, which includes algal growth and heavy incrustation with barnacles and mussels. In higher latitudes, such fouling is very common during the summer season. Figure 9.26 shows an example of such fouling

FIGURE 9.25 Photographs showing preparations for deployment of the Aanderaa Doppler Acoustic Current Meter in the Indian Ocean. *(Source: Courtesy of Dr. Sanil Kumar, National Institute of Oceanography, Goa, India.)*

that occurred on a deployment at the Gulf of Maine. The Doppler acoustic current meter has no moving parts, and it is immune to even the severe fouling and mooring motion that plague mechanical current meters under these conditions. Aanderaa's proprietary forward-pinging algorithm especially assures that the data are collected upstream and free from contamination from the instrumental wake.

FIGURE 9.26 Photograph showing severe biological fouling on the Aanderaa RCM 9 MkII Acoustic Doppler Current meter deployed at the Gulf of Maine and still giving reliable data. *(Source: Aanderaa magazine Sensors & Systems, No. 21, p. 1–4, June 2004.)*

REFERENCES

Abernathy, F.H., Kronauer, R.E., 1962. The formation of vortex streets. J. Fluid Mech. 13, 1–20.

Appell, G.F., 1978. A review of the performance of an acoustic current meter. In: Woodward, W., Mooers, C.H.K., Jensen, K. (Eds.), Proc. Working Cong. on Current Measurement. Tech. Rep. DEL-SG-3-78 University of Delaware, Newark, DL, USA, pp. 35–58.

Appell, G.F., 1979. Performance assessment of advanced ocean current sensors. IEEE J. Oceanic Eng. OE-4 (1), 1–4.

Aubrey, D.G., Trowbridge, J.H., 1985. Kinematic and dynamic estimates from electromagnetic current meter data. J. Geophys. Res. 90, 9137–9146.

Bernier, R.N., Brennen, C.E., 1983. Use of the electromagnetic flow meter in a two-phase flow. International Journal of Multiphase Flow 9 (3), 251–257.

Bevir, M.K., 1970. The theory of induced voltage electromagnetic flow meters. J. Fluid Mech 43 (3), 577–590.

Bivins, L.E., 1975. Turbulence effects on current measurement. M.S. thesis. University of Miami, Coral Gables, FL, USA.

Blanton, J.O., Lin, G., Elston, S.A., 2002. Tidal current asymmetry in shallow estuaries and tidal creeks. Continental Shelf Research 22, 1731–1743.

Blevins, R.D., 1977. Flow-induced vibration. Van Nostrand.

Bowden, K.F., Fairbairn, L.A., 1956. Measurements of turbulent fluctuations and Reynolds' stresses in a tidal current. Proc. R. Soc. (A) 237, New York, NY, USA. 422–438.

Brown, N.L., 1992. A simple low-cost acoustic current meter, Vol. 2. Proc. Oceanology International '92, Brighton, UK. March 10–13.

Clay, C.S., Medwin, H., 1977. Doppler effects for moving objects, sea surface and ships. In: Acoustical oceanography: Principles and applications. Wiley Interscience, John Wiley & Sons Singapore Pte. Ltd., Singapore, pp. 334–338.

Clayson, C.H., 1983. A vector averaging electromagnetic current meter for near-surface measurement. In: Proc. IEEE Third Working Symposium on Oceanographic Data Systems. IEEE Computer Society, New York, NY, USA, pp. 81–87.

Collar, P.G., 1993. A review of observational techniques and instruments for current measurements from the open sea. Report No. 304, Institute of Oceanographic Sciences, Deacon Laboratory, Godalming, Surrey, United Kingdom.

Collar, P.G., Hunter, C.A., Perrett, J.R., Braithwaite, A.C., 1988. Measurement of near-surface currents using a satellite-telemetering buoy. Journal of the Institution of Electronic & Radio Engineers 58, 258–265.

Cushing, V., 1976. Electromagnetic water current meter. Proc. Oceans '76, MTS-IEEE Conference. IEEE, Washington, DC, USA, September 1976.

Desa, E.S., Desa, E., 1980. A closed-loop electromagnetic flow meter. J. Phys. E: Sci. Instrum. 13, 233–235.

Grant, W.D., Williams III, A.J., Gross, T.F., 1985. A description of the bottom boundary layer at the HEBBLE site: low-frequency forcing, bottom stress and temperature structure. Mar. Geol 66, 219–241.

Griffiths, G., 1979. The effect of turbulence on the calibration of electromagnetic current sensors and an approximation of their spatial response. Institute of Oceanographic Sciences Report No. 68, Godalming, Surrey, United Kingdom.

Griffiths, G., Collar, P.G., Braithwaite, A.C., 1978. Some characteristics of electromagnetic current sensors in laminar flow conditions. Institute of Oceanographic Sciences Report No. 56, Godalming, Surrey, United Kingdom.

Gytre, T., 1976. The use of a high sensitivity ultrasonic current meter in an oceanographic data acquisition system. Radio and Electronic Engineer 46, 617–623.

Gytre, T., 1980. Acoustic travel time current meters. In: Dobson, F., Hasse, L., Davis, R. (Eds.), Air-sea interaction: Instruments and methods. Plenum Press, New York, NY, USA, pp. 155–170.

Hovdenes, J., 1996. The RCM 9: A unique new instrument for measuring ocean currents and other oceanographic parameters. Proc. Oceans '96 Conference Vol. 1. Issue 1 (revised January 1998).

Jones, I.S.F., 1980. Electromagnetic current meters. In: Dobson, F., Hasse, L., Davis, R. (Eds.), Air-sea interaction: Instruments and methods. Plenum Press, New York, NY, USA, pp. 219–229.

Joseph, A., 1981. Electromagnetic flow meter. NIO Technical Report, Dona Paula, Goa, India, pp. 1–22.

Kanwisher, J., Lawson, K., 1975. Electromagnetic flow sensors. Limnology & Oceanography, Alberta, Canada, 174–182.

Kun, A.L., Fougere, A.J., 1997. New wave direction and spectrum measurement technique. Proc. Third International Symposium on Ocean Wave Measurement and Analysis. Virginia Beach, VA, USA, Nov. 3–7, 1997, 1278–1281.

Kun, A.L., Fougere, A.J., 1998. Applications of an acoustic current measurement technique. Ocean News and Technology. September/October 1998, 36–37.

Kun, A.L., Fougere, A.J., 1999. A new low-cost acoustic current meter design. Proc. IEEE 1999, 150–154.

Lawson, K.D., Lessieux, B.J., Luck, J.M., Woody, D.C., 1983. The development of a spherical electromagnetic current meter. Proc. IEEE Oceans '83 Conference. New York, NY, USA, 187–193.

Lowell Jr., F.C., 1979. Acoustic flow meters for pipeline flow rate. Water Power & Dam Construction, 39–46.

McCullough, J.R., 1980. Survey of techniques for measuring currents near the ocean surface. In: Dobson, F., Hasse, L., Davis, R. (Eds.), Air-sea interaction: Instruments and methods. Plenum Press, New York, NY, USA, pp. 105–126.

McCullough, J.R., Graper, W., 1979. Moored acoustic travel time (ATT) current meters: Evolution, performance and future design. Woods Hole Oceanographic Institution Report No. WHOI-79-92, Woods Hole, MA, United States.

Morrison III, A.T., Williams III, A.J., Waterbury, A.C., Tierney, C.M., 2002. Analog output from a differential travel-time current meter. IEEE Oceans '02, 708–712.

Morrison III, A.T., Williams III, A.J., Martini, M., 1993. Calibration of the BASS acoustic current meter with carrageenan agar. Proc. Oceans '93, IEEE/OES Vol. III, 143–148.

Mulcahy, M., 1978. A solid state water current meter for wave direction sensing. Sea Technol. July 1978.

Olson, J.R., 1972. Two-component electromagnetic flow meter. Marine Technology Society Journal 6 (1), 19–24.

Sanford, T.B., Flick, R.E., 1975. On the relationship between transport and motional electric potentials in broad, shallow currents. Journal of Marine Research 33, 123–139.

Saunders, K.D., 1980. , Horizontal response of the NBIS acoustic current meter. Proc. Oceans '80. Seattle, WA, USA, September 8–10, 1980, 220–225.

Shercliff, J.A., 1962. The theory of electromagnetic flow measurement. Cambridge University Press, Cambridge, United Kingdom.

Sorrell, F.Y., Curtin, T.B., Feezor, M.D., 1990. An electromagnetic current meter-based system for application in unsteady flows. IEEE J. Oceanic Eng 15 (4), 373–379.

Sorrell, F.Y., Curtin, T.B., Feylor, M.D., 1985. Electromagnetic shear flow meter for remote oceanic applications. U.S. Patent # 4543822.

Thorpe, S.A., Collins, E.P., Gaunt, D.I., 1973. An electromagnetic current meter to measure turbulent fluctuations near the ocean floor. Deep-Sea Res 20, 933–938.

Thwaites, F.T., Williams 3rd, A.J., 1996. Development of a modular acoustic velocity sensor. Proc. Oceans '96, 607–612.

Trivett, D.A., Terray, E.A., Williams III, A.J., 1991. Error analysis of an acoustic current meter. IEEE J. Ocean. Eng 16 (4), 329–337.

Tucker, M.J., Smith, N.D., Pierce, F.E., Collins, E.P., 1970. A two-component electromagnetic ship's log. J. Inst. Navig. 23, 302–316.

Urick, R.J., 1975. Propagation of sound in the sea: transmission loss, I. In: Principles of underwater sound. McGraw-Hill, Columbus, OH, United States, pp. 93–135.

Wang, J.Z., Tian, G.Y., Simm, A., Lucas, G.P., 2008. Simulation of magnetic field distribution of excitation coil for EM flow meter and its validation using magnetic camera. In: 17th World Conference on Nondestructive Testing. Shanghai, China, Oct. 25–28, 2008.

Wang, J.Z., Tian, G.Y., Lucas, G.P., 2007. Relationship between velocity profile and distribution of induced potential for an electromagnetic flow meter. Flow Measurement and Instrumentation 18, 99–105.

Williams III, A.J., 1985. BASS, an acoustic current meter array for benthic flow-field measurements. Mar. Geol 66, 345–355.

Williams III, A.J., 1995. Linearity and noise in differential travel time acoustic velocity measurement. Proc. IEEE Fifth Working Conference on Current Measurement, IEEE/OES. Feb. 7–9, 1995, 216–221.

Williams III, A.J., 2001. Acoustic current meter zero offset drift. Proc. IEEE Oceans '01, 916–921.

Williams III, A.J., Beckford, C., 1999. Reynolds stress resolution from a Modular Acoustic Velocity Sensor. Oceans '99, 386–390.

Williams III, A.J., Thwaites, F.T., 1998. Earth coordinate 3-D currents from a modular acoustic velocity sensor. Proc. IEEE Oceans '98, 244–247.

Williams III, A.J., Tivey, M.K., 2001. Tidal currents at hydrothermal vents, Juan de Fuca Ridge. Sea Technology. June 2001, 62–64.

Williams III, A.J., Tochko, J.S., Koehler, R.L., Grant, W.D., Gross, T.E., Dunn, C.V.R., 1987. Measurement of turbulence in the oceanic bottom boundary layer with an acoustic current meter array. J. Atmos. Oceanic Tech 4 (2), 312–327.

Woodward, W.E., Appell, G.F., 1973. Report on the evaluation of a vector averaging current meter. NOAA Technical Memorandum NOAA-TM-NOS-NOIC-1. pp. 13−16.

BIBLIOGRAPHY

Chriss, T.M., Caldwell, D.R., 1982. Evidence for the influence of form drag on bottom boundary layer flow. J. Geophys. Res. 87, 4148−4154.

Cushing, V., 1976. Electromagnetic water current meter, Paper 25C. Proc. MTS-IEEE Oceans '76 Conference. IEEE, 1976; 25C-1-25C-17, Washington, DC, USA, Sept. 13−15, 1976.

Hardies, C.E., 1975. An advanced two-axis acoustic current meter. Proc. Offshore Technology Conference II, 465−476.

Heldebrandt, K.E., Trest, L.S., Michelena, E.D., 1978. The development and testing of current meters for long-term deployment on the continental shelf. Proc. Oceans '78. Washington, DC, USA, 308−314.

Lawson, K.D., Brown, N.L., Johnson, D.H., Mattey, R.A., 1976. A three-axis acoustic current meter for small scale turbulence. ISA, 501−508.

Morrison, A.T., Williams III, A.J., Martini, M., 1993. Calibration of the BASS acoustic current meter with carrageenan agar. Proc. Oceans '93, 143−148.

Robbins, R.J., Morrison, G.K., 1981. Acoustic direct reading current meter. Proc. IEEE Oceans '81 Conference, 506−511.

Thwaites, F.T., Williams, A.J., 2001. BASS measurements of currents, waves, stress and turbulence in the North Sea bottom-boundary layer. IEEE J. Oceanic Eng 26, 161−170.

Trivett, D.A., Williams, A.J., 1994. Effluent from diffuse hydrothermal venting. 2. Measurements of plumes from diffuse hydrothermal vents at the southern Juan de Fuca Ridge. J. Geophys. Res 99, 18417−18432.

Wang, J.Z., Lucas, G.P., Tian, G.Y., 2007. A numerical approach to the determination of electromagnetic flow meter weight functions. Measurement Science and Technology 18 (3), 548−554.

Chapter 10

Vertical Profiling of Horizontal Currents Using Freely Sinking and Rising Probes

Chapter Outline

10.1. **Importance of Vertical Profile Measurements of Ocean Currents**   297

10.2. **Technologies Used for Vertical Profile Measurement of Ocean Currents**   300

    10.2.1. Freely Sinking and Wire-Guided Relative Velocity Probes   300

    10.2.2. Bottom-Mounted, Winch-Controlled Vertical Automatic Profiling Systems   302

    10.2.3. Acoustically Tracked Freely Sinking Pingers   305

    10.2.4. A Freely Sinking and Rising Relative Velocity Probe (Cyclesonde)   308

        10.2.4.1. Operation   309

        10.2.4.2. Advantages   310

    10.2.5. A Freely Falling Electromagnetic Velocity Profiler   311

    10.2.6. Free-Falling, Acoustically Tracked Absolute Velocity Profiler (Pegasus)   318

    10.2.7. Freely Falling, Acoustically Self-Positioning Dropsonde (White Horse)   320

    10.2.8. Freely Rising Acoustically Tracked Expendable Probes (Popup)   322

10.3. **Technologies Used for Vertical Profile Measurements of Oceanic Current Shear and Fine Structure**   325

    10.3.1. Free-Fall Shear Profiler (Yvette)   326

    10.3.2. Acoustically Tracked Free-Fall Current Velocity and CTD Profiler (TOPS)   326

10.4. **Technologies Used for Vertical Profile Measurements of Oceanic Current Shear and Microstructure**   329

    10.4.1. Towed Acoustic Transducer   329

    10.4.2. Free-Fall Probe (PROTAS)   330

    10.4.3. Free-Falling Lift-Force Sensitive Probes   332

10.5. **Merits and Limitations of Freely Sinking/Rising Unguided Probes**   335

References   335

Bibliography   337

10.1. IMPORTANCE OF VERTICAL PROFILE MEASUREMENTS OF OCEAN CURRENTS

Qualitatively, the ocean is regarded as a deck of cards (water layers) wherein each card moves horizontally at a given velocity. The difference in horizontal velocity of neighboring water layers in the ocean gives rise to what is known as *vertical shear of horizontal current velocity* (defined as the derivative of current with respect to depth). Although current shears are most frequently found in the upper layers of the water column, primarily because different time responses by water layers at different depths to a fluctuating wind regime induce high lateral shear, they are also found within the main thermocline and within 500 m of the seafloor.

By the early 1960s it had become increasingly evident that not only the surface waters of the oceans but also the deep waters are active over a wide range of frequencies (e.g., Crease, 1962; Pochapsky, 1966). With the advent of continuously profiling seawater temperature and salinity systems, oceanographers became aware of a new world of fine and microstructure in the density field of the oceans. A wide spectrum of small-scale, predominantly stratified structure was found at all depths in tropical as well as polar oceans (Stommel and Federov, 1967; Neshyba et al., 1971). It could be inferred from basic physical considerations that this complex density field cannot exist without a corresponding richness in the velocity field. What was missing was the ability to resolve the fine-scale velocity simultaneously with the density field in order to explore and quantify the causal relationship between the two.

There has been evidence to suggest that kinetic energy propagates downward, perhaps as isolated events or packets of energy. Based on measurements from a few sites, Rossby and Sanford (1976) showed that the inertial frequency-band energy propagates downward at a group velocity with

Copyright © 2014 Elsevier Inc. All rights reserved.

a vertical component of about 0.5 mm/s. In general, current variability is observed throughout the water column. Depth dependence of water currents at different regions of the oceans is influenced by various factors, such as the presence of ridges, gorges, valleys, and banks. It was suspected that a current velocity field in the deep ocean along the Continental rise contains significant spatial gradients in the vertical. For instance, measurements of the vertical profile of currents carried out by Rossby (1969) on the southern slope of the Plantagenet Bank southwest of Bermuda showed that the Bank has a pronounced influence on the local flow. In this region, at depths less than 500 m, the currents are found to be swift at more than 30 cm/s, whereas at this depth a sudden transition to a slow drift of the deep water along the topography was observed. Changes in horizontal velocities with depth are not gradual but occur in regions of limited vertical extent where current velocity differences of 10–20 cm/s in 10 m were observed. The effective blocking of the deep water by the Bank causes a local depression of the main thermocline deeper than is typical for the Bermuda area. Influence of ridge valleys on the structure and evolution of vertical profiles of horizontal motions is particularly interesting. Garcia-Berdeal et al. (2006) reported fascinating vertical structures of time-dependent water currents in the axial valley at the Endeavor Segment of the Juan de Fuca Ridge.

Oceanic water current velocity shear has great importance in turbulent mixing and dispersal effects. An interesting application of vertical profile measurements in oceanography is in the study of fine structures and microstructures and velocity shear in the water column. Ocean-current velocity profile observations have revealed considerable complexity to the vertical structure or distribution of currents in the deep sea. An individual vertical profile is composed of a superposition of a wide spectrum of different spatial and temporal scales of motions operating simultaneously. In a study of low-frequency variability of the velocity profile at one site, Rossby (1974) found that monthly or even weekly sampling was inadequate to follow the time evolution of the velocity profile. Low-frequency (quasi-geostrophic) structures can be advected past the observation site in a matter of days. Moreover, high-frequency contributions, such as those due to internal tidal or inertial period motions, are important constituents to an individual profile. Thus, repeated current-velocity profiles at one site are needed in order to estimate the energy contributed by a variety of time-dependent motions.

General knowledge of oceanic motions has increased considerably during the past decades, primarily due to the successful use of moored current meters. The technology of mooring and maintaining chains of current meters in the deep ocean has steadily improved; moorings became maintainable for even a year or more, yielding time series long enough to cover the greater part of the spectrum of oceanic motions. Observation of poor vertical coherence at inertial frequency for current meters separated 100 m in the vertical, based on examination of an extensive data bank at the Woods Hole Oceanographic Institution (Perkins, 1970), gave rise to the conjecture that some of the most energetic temporal frequencies are badly sampled (aliased) in the vertical by typical fixed instrument spacing.

Considering further what is already known about temperature and salinity distributions in the oceans—to cite an example, data from the San Diego Trough showed conductivity structures with vertical wavelengths down to less than a centimeter (Gregg and Cox, 1971)—it is amply clear that vertical profiles of scalar or vectoral quantities of interest cannot be obtained with an economical number of fixed instruments. The often observed extremely low correlation between motions at different depth levels (Webster, 1968, 1969), together with the great expense of current meter chains, indicated the necessity for additional tools (say, vertically moving sensor packages providing high-resolution profiles) to measure ocean currents effectively over a variety of space and time scales. Obviously, it would be interesting to explore how these horizontal motions on different time scales couple in the vertical. This calls for fine-resolution measurements of horizontal motions over both small and large vertical separations. Besides being suitable for directly verifying coupled thermal and circulation models in the ocean and large lakes, the dataset obtained from vertical profilers possessing fine spatial and temporal resolution should provide an understanding of a number of geophysical phenomena and processes, such as the evolution of the surface mixed layer; the generation and decay of internal waves and seiches; and the turbulent transport and mixing of heat, momentum, and water masses.

With rapid improvements in oceanographic instruments in post World War period, large fluctuations in the vertical and horizontal distributions of seawater temperature and salinity were found. These fluctuations have come to be called *fine structure* or *microstructure*, depending on their wavelengths. Such fluctuations were expected to occur in the ocean currents as well, in which wavelengths λ, shorter than 1 m, came to be known as microstructure, and wavelengths in the range 1–100 m are known as fine structure.

The study of oceanic microstructure is really a study of oceanic turbulence and mixing. The source of the turbulent energy and the mechanisms of extracting the energy must be determined. It may then be possible to understand the effect of the turbulence on the temperature and salinity profiles. Observations and photographs by Woods and Fosberry (1967) in the thermocline off Malta indicated that shear instability is a source of turbulent energy. Gregg and Cox (1971) reported measurements of vertical profiles of temperature microstructure to the millimeter range in the ocean. Osborn (1969) noticed oceanic regions of intense gradient activity several meters thick, which are suggestive of the turbulent patches observed by Woods (1968) in the thermocline. The local Richardson number (Ri) is considered the most important parameter that controls the stability of a stratified fluid and thus occurs as the essential ingredient in a wide range of schemes for parameterizing the turbulent fluxes. A necessary criterion for the growth of infinitesimal shear instability is:

$$Ri = \frac{g}{\rho}\frac{\partial \rho}{\partial z}\bigg/\left(\frac{\partial U}{\partial z}\right)^2 < \frac{1}{4} \qquad (10.1)$$

To evaluate the Richardson number (Ri), it is necessary to know the vertical component of the density gradient $\left(\frac{\partial \rho}{\partial z}\right)$ and the vertical shear of the horizontal current velocity $\left(\frac{\partial U}{\partial z}\right)$. The vertical profile of the density can be calculated from vertical profiles of the temperature and electrical conductivity of the ambient ocean water (i.e., CTD measurements). Techniques for measuring these parameters are given in the literature (e.g., Osborn and Cox, 1972; Gregg and Cox, 1972). A quantitative study of the roles of shear instability in generating oceanic microstructure requires detailed vertical profiles of the shear of the horizontal velocity.

In isotropic turbulence, the energy dissipated by viscosity can be estimated from (Hinze, 1959):

$$\varepsilon = 7.5\nu\left(\frac{\partial u}{\partial z}\right)^2 \qquad (10.2)$$

In this expression, ε is the viscous energy dissipation, ν is the coefficient of kinematic viscosity, and u is one horizontal component of the velocity. There is evidence that turbulence in the ocean is not generally isotropic (Grant et al., 1968; Nasmyth, 1970) and much work needs to be done to determine the three-dimensional nature of oceanic turbulence. It may be argued that the vertical shears are a significant part of the energy dissipation, and if the turbulence is not isotropic, it will change the numerical coefficient from 7.5 (in Equation 10.2) to a lower value, which will still be greater than 2. The turbulence is more likely to be isotropic where ε is large, on the order of 10^{-3} erg/cm^3/s or more, in which case the rms shear is $\sim 10^{-1}$/s. When the energy dissipation is low, the isotropic formula probably overestimates the dissipation rate. It might, therefore, be appropriate to change the factor 7.5 (in Equation 10.2) to 5 ± 2.5 (Osborn, 1974). There is an increasing belief that it is the catastrophic events and not the mean situations that are important in turbulent transfer in the ocean. Therefore, more interest lies in the regions with high values of ε rather than low values.

In the past, measurements of small-scale velocity structures in the ocean were attempted, in limited depths only, by the use of moored current meter chains. Maintaining such moorings at large depths in the open ocean is prohibitively expensive. Furthermore, in situations where

intense current are to be measured, it is difficult to maintain these moorings. Additionally, these current profiling installations using a series of Eulerian current meters fastened to a tether are not conveniently mobile, nor are the installations readily adaptable to changing situations. Ideally, a single package scanning the entire water column repeatedly to sense the vertical distribution of horizontal motions with high vertical resolution is the preferred choice.

Single-probe profiling techniques have emerged as an attractive alternative tool to effectively measure vertical profiles of horizontal currents over considerable vertical distances while retaining high resolution of the small-scale features without being contaminated by mooring-line motions. Ideally, one might want a single package scanning the entire water column in order to sense the vertical distribution of horizontal motions with high vertical resolution. This basic necessity resulted in the development of single-probe profiling techniques. In fact, awareness of many complex features of oceanic motions, such as sharp and discrete shear zones, internal waves, and eddies in the ocean, has been enhanced by current velocity profile observations.

Measurements of the vertical structure of ocean currents are important for a wide variety of studies and operational applications. These include estimating depth-dependent variability of inertial and near-inertial motions in a stratified rotating fluid under the influence of topographical constrictions (e.g., over ocean ridges and valleys); study of layered current structure (attributed to the mixing process); quantifying/examining the velocity structure in hydrothermal plumes within a few hundred meters of the bottom (Thomson et al., 1989); estimation of volume transport; understanding the stratification of current jets; and so forth.

Deep-water current-velocity profiles are essential for studying interesting phenomena such as inertial oscillations and equatorial currents. The 1950s and subsequent years witnessed a rapid evolution of new methods of measuring ocean currents. The generally most useful instrument remained the self-recording current meter, which is capable of operating unattended for long periods of time. These meters are commonly used on deep-sea moorings, usually with several Eulerian current meters located at different depths. Whereas this method provides excellent time coverage, the current-versus-depth information is obviously limited to the number of instruments that can be afforded.

Fundamental discoveries have been made with the first application of new profiling technologies. These include, for example, the ratio of upward to downward energy flux in inertial waves (Leaman and Sanford, 1975) and the stack of alternate eastward and westward currents on the equator below the thermocline, known as the *deep jets* (Luyten and Swallow, 1976). Current profiles in mid-ocean ridges and their valleys are particularly interesting. For example, within the axial valley of the Endeavor Ridge (well known as a hydrothermal vent region), located off the coasts of Washington and Oregon, a significant part of hydrothermal fluid and heat is released in the form of low-temperature diffuse plumes that rise 10–50 m above the bottom (Trivett and Williams, 1994; Rona and Trivett, 1992) and therefore remain within the confines of its valley. The vertical structure of water flows within the valley has important implications for the transport of hydrothermal vent fluid and the dispersal of larvae of vent organisms (Mullineaux and France, 1995).

Clearly, free-fall profilers are useful in strong currents where current meter moorings cannot be applied. Other applications include the study of internal waves and local dynamic investigations of ocean fronts and processes that are intermittent with respect to depth, such as lateral intrusions of different water masses. Obviously, the wide variety of applications of vertical profile measurements of horizontal currents in the ocean attracted the design and development of a number of vertical profiling instruments over a period of time, based on different principles of operation. These tools are addressed in the following sections.

10.2. TECHNOLOGIES USED FOR VERTICAL PROFILE MEASUREMENT OF OCEAN CURRENTS

The traditional method of measuring vertical profiles of horizontal currents is to use conventional fixed-point current meter (CM) moorings. However, moorings with fine vertical resolution (of the order of a few cm or tens of cm required for understanding a number of geophysical phenomena) are impractical and uneconomical. Furthermore, measurements using CM moorings are usually limited to depths far below the sea surface because of the disturbing influence of waves on the subsurface float. To circumvent such difficulties associated with moored CM chain measurements, various ingenious technologies have been developed from time to time for vertical profile measurement of ocean currents using a single package of freely moving sensors or a fixed remote sensing probe.

Basically, two methods are employed in the single-probe current profiling scheme. One method employs free-falling, wire-guided, or free-falling/rising autonomous probes as a means for measuring the vertical profiles of horizontal currents in the ocean. Launched from a ship, the probe falls freely, with no connections to the sea surface. The probe, in some designs, is capable of penetrating deep into the water column. The other method is acoustic Doppler profiling to remotely measure velocity profiles.

The first method, the free-falling probe, utilizes two basic techniques to measure current profiles. In one

technique, a pinger is acoustically tracked while it sinks or rises, much as a weather balloon is tracked using electromagnetic signals and the pinger's trajectory is thus determined. In the other technique, the trajectory of the probe is determined without the use of acoustic tracking. In either case, vertical profiles of horizontal current velocities are estimated from the trajectory described by the freely moving probe. Before the advent of ADCPs, freely moving sensor packages were the only available tools for high-resolution measurement of vertical profiles of horizontal currents in the ocean. This chapter addresses technologies using such moving packages. (ADCPs are discussed in chapter 11.)

10.2.1. Freely Sinking and Wire-Guided Relative Velocity Probes

Several attempts have been made in the past to measure vertical profiles of horizontal currents. One of the most successful techniques is the *free-drop* technique (Richardson and Schmitz, 1965; Richardson et al., 1969). This method is ideally suited for fast surveys over considerable horizontal distances, yielding the mass transport through a cross-section. Its vertical resolution, however, is limited. A very high resolution has been achieved by Plaisted and Richardson (1970), who used a small winch mounted on a drifting sparbuoy to lower two current meters, thus obtaining relative profiles. A disadvantage of this method was the need for a highly accurate navigation system in order to convert the measured current to absolute current.

Duing and Johnson (1972) reported a simple, relatively inexpensive method for repeated observations of high-resolution current profiles, in which a profiling current meter (PCM) is constrained to roll down an anchored taut wire, with the roller attachment providing the necessary decoupling of the instrument from vertical wire motions. In this method, current profile measurements are carried out from anchored vessels. Figure 10.1 shows the principle of the profiling method. The profiler consists of a self-contained Aanderaa CM attached to a cylindrical hull. The density of the instrument package is slightly greater than that of the surrounding water. The Savonius rotor extends out from the bottom side of the cylindrical hull when it is in its horizontal working position. The entire package is attached by a roller to a taut wire suspended beneath the anchored ship and allowed to descend slowly through the entire water column. During operations in the Florida Current, Duing and Johnson (1972) found that the descent rate of the instrument was most favorable when it had approximately 1,000 g mean negative buoyancy. The principle of the ballasting procedure Duing and Johnson adopted is the following: The effective density of the instrument package can be measured and adjusted to the desired range by submerging it in a saltwater tank at

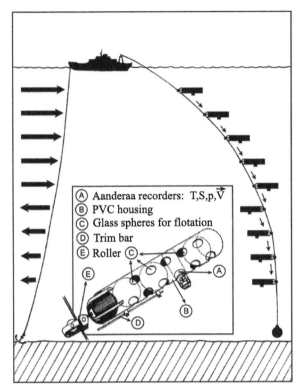

FIGURE 10.1 Principle of the current-profiling method in which a profiling current meter (PCM) is constrained to roll down an anchored taut wire, with the roller attachment providing the necessary decoupling of the instrument from vertical wire motions. *(Source: Duing and Johnson, 1972.)*

a known salinity and temperature and by measuring its overweight with a scale (say, a spring balance) mounted above the tank. The adjustment is made with the use of the following formula (Duing and Johnson, 1972):

$$\rho = \rho_w + \frac{\Delta m_{ow}}{V} + \frac{\Delta m_a}{V} \qquad (10.3)$$

In this expression, ρ is the desired effective density of the instrument; V is the volume of the instrument; ρ_w is the density of the tank water; Δm_{ow} is the measured overweight; and Δm_a is the mass added or subtracted. The mean buoyancy is $(\bar{\rho} - \rho)g$. In this, $\bar{\rho}$ is the estimated mean density of the oceanic water column to be sampled. Ballast is added or subtracted by changing the amount of water in the forward and aft glass spheres inside the profiler hull. A certain amount of ballast is added near the instrument's center of buoyancy in the form of variously sized shackles to allow for ballast corrections on board ship.

Because the flow of water has to be perpendicular to the axis of the Savonius rotor, the profiler must be horizontally trimmed. This is done during the ballasting procedure by injecting water into the aft or forward glass spheres and by moving the additional shackles fore or aft. Once the instrument is horizontally trimmed, the center of buoyancy

of the profiler is determined by using a fulcrum to balance the instrument in water to a horizontal position. During operations at sea, it is thus possible to vary the sink rate of the instrument without changing the horizontal trim of the profiler by adding or removing weight from the center of buoyancy.

In operation, the PCM is attached to the taut wire above the sea surface by means of the roller and then lowered into the water and released by tripping a hook attached to its forward end. The hollow glass spheres inside the profiler hull provide an excellent acoustic target, and it is thus possible to follow the descent of the PCM on the ship's precision echo sounder. By this means, it is possible to monitor the sinking speed of the instrument on board ship and to change it, if necessary, by adding or subtracting weight from the center of buoyancy of the profiler hull.

Due to the catenary shape of the suspended wire and due to the vertically varying current profile, the sinking speed of the profiler varies with depth. Typical values are 15−20 cm/s in the upper 100 m and 5−8 cm/s in the deeper layers. A profile to 500 m is completed in approximately one hour. Because of high wire angles and larger currents in the upper layer, a horizontal motion of the profiler occurs that produces a velocity error, which needs to be corrected, as described later. The advantages of this method are the following (Duing and Johnson, 1972):

- A single instrument yields high-resolution vertical profiles.
- The axis of the Savonius rotor (i.e., a current-measuring sensor) remains always perpendicular to the flow of water, as expected, because of the horizontal trimming of the hull.
- The 2-meter-long profiler hull acts as a stable current vane.
- The roller attachment decouples the instrument from vertical wire motions.

If the profiler is deployed in a high-speed flow regime, there is an advantage of stabilization of the heading of the anchored ship. However, there are also some disadvantages. The greatest of these is the occurrence of large angles in the suspended wire along which the PCM descends. As a result of the wire angle, the PCM has a horizontal component to its motion as it descends; this component subtracts from the actual current speed. To correct for this, the product of the descent rate and the wire angle as a function of depth is calculated and added to the measured current. The wire shape as a function of depth is measured by taking the slant range to the PCM, measured by the sonar, together with its depth as measured by the PCM's pressure sensor. These two values in the *x-z* plane provide a locus of points that is an image of the wire shape. To check this correction and to obtain rough information on the accuracy of an instantaneous downward profile, the profiler may be stopped at discrete depth intervals while being raised. It must, however, be borne in mind that this check is only approximate, because the instrument hangs by its forward section at these stops and, consequently, is tilted. The tilt should be small in the high-intensity currents.

It is difficult to quantify the errors involved in the current profiles that have been measured by this technique. *In situ* comparison is one possible option. The individual sources of errors may be broken into two categories: those that are introduced by the profiling technique and those that are common with moored current meters. Accordingly, the following errors contribute to the total error crept into the measurements made by this method (Duing and Johnson, 1972):

1. Errors introduced by the profiling technique:
   - Deviation of the profiler from its horizontal trim position
   - Induction of rotations or slowdown of the Savonius rotor due to vertical descent
   - Errors in determining the wire shape
   - Depth errors due to error of pressure sensor
2. Errors common with moored current meters:
   - Inertia of the Savonius rotor in the presence of rapidly changing currents
   - Wire vibrations in intense currents
   - Horizontal motions of the supporting platform
   - Basic calibration errors

The effects of rotor inertia are greater in the profiling technique than in a moored current meter because the profiling meter is likely to pass through regions of large current shear. In addition, very little is known about wire vibrations in intense currents. Anchored ships in the deep ocean are known to exert swaying motions of the order of 15 cm/s. Figure 10.2 provides two selected profiles obtained using the PCM method from the same location in the Florida Straits in two different months in 1970. These two profiles of considerably different patterns from the same location emphasize the importance of the time-dependent aspect of current profiles in the Florida Straits.

10.2.2. Bottom-Mounted, Winch-Controlled Vertical Automatic Profiling Systems

The Vertical Automatic Profiling System (VAPS) is a profiling device specifically designed for vertical profile measurements in low-current regimes (e.g., typical of deep lakes). However, it can be deployed in coastal waters as well if the horizontal currents are not too large. Ward-Whate (1977) described a device that is designed to

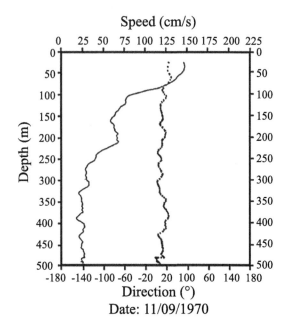

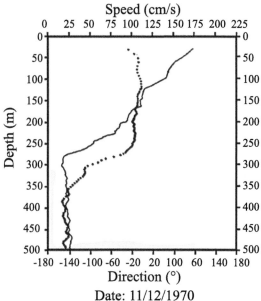

FIGURE 10.2 Two selected profiles obtained using the PCM method, from the same location in the Florida Straits on two different months in the year 1970. *(Source: Duing and Johnson, 1972.)*

measure continuous profiles of horizontal current and temperature in small current regimes over long periods of time while unattended. It is reasonable to assume that vertical spacing of 10 cm should sample the mean properties of the velocity and density field sufficiently well to define a meaningful local Richardson number and yet not be unduly contaminated by the motions of the turbulent eddies.

One possible method of measuring vertical profiles of horizontal current and temperature with fine temporal and spatial resolution is the use of a slowly descending/

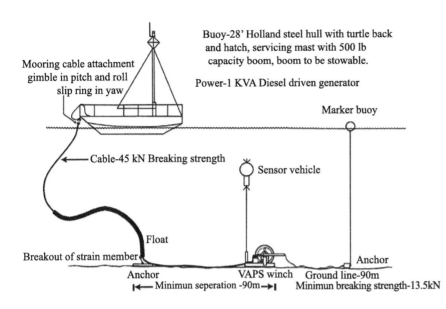

Mooring cable attachment gimble in pitch and roll slip ring in yaw

Buoy-28' Holland steel hull with turtle back and hatch, servicing mast with 500 lb capacity boom, boom to be stowable.

Power-1 KVA Diesel driven generator

Marker buoy

Cable-45 kN Breaking strength

Sensor vehicle

Float

Breakout of strain member

Anchor

Anchor

VAPS winch

Ground line-90m
Minimun breaking strength-13.5kN

|◄— Minimun seperation -90m —►|

FIGURE 10.3 Schematic of the VAPS system. *(Source: Royer et al., 1987.)*

ascending probe. The automatic profiling system (VAPS) reported by Ward-Whate (1977) consists of a buoyant sensor vehicle that carries a suite of sensors, capable of traversing 150 m of water column in the vertical via a lightweight tether cable and a bottom-mounted winch. A schematic of the VAPS system is shown in Figure 10.3. A 2-km-long cable from shore carries electrical power and control signals to the profiler system. X and Y components of current speeds, compass heading, time, depth, temperature, roll, and pitch signals are telemetered back to a recorder on shore.

The 360-kg bottom-mounted aluminum winch is powered through a regenerative (four-quadrant), ¼-hp, SCR-based speed controller located on shore. The drum speed is sensed by a tachometer and fed back to the shore-based controller, which corrects the ascent or descent rate to within several percent of the nominal values. The tether is wound onto a 64-cm-diameter drum, grooved in a single layer, which is guided by a level-wind assembly. Limit switches at the two ends of the level-wind assembly are preset manually to determine the reversing points of the winch. The reversing mechanism is controlled automatically by electronic logic circuits. The automated mode can be overridden manually to stop the float or to measure profiles at different rates or between various depth limits.

Connecting the winch and vehicle to shore is 2 km of 10-conductor, 2.5-cm-diameter cable, which is split on to two reels for easier deployment from a 28-m boat. The system is powered by either a conventional line source or an on-site generator. If power fails, the internal clock continues to function for about six days, a normal end-of-profile marker is recorded at shore, and the controller logic is set to begin a new profile starting at the bottom when the power resumes.

A standard 22-cm-diameter electronics canister mounts and interfaces the following sensors (Ward-Whate, 1977):

- *Current sensor.* Two-axis acoustic travel-time difference (ATT) type, mounted on the lid of the electronics can ($\pm$100 cm/s full scale with 0.1 cm/s sensitivity).
- *Direction sensor.* Optically scanned/encoded magnetic compass gimbaled; oil damped; 1 s recovery time; $0-360\pm3°$ with digital-to-analog conversion.
- *Depth sensor.* Strain gauge type; $0-200\pm0.6$ m.
- *Temperature sensor.* High-purity, platinum-resistance thermometer; fast response; 0 to $25\pm0.04°$C.
- *Tilt sensor.* Pendulum type; gimbaled; dual axis; oil damped; 1 s recovery time; $45\pm1°$.

A 69-cm-diameter syntactic foam sphere surrounds the electronics package, giving a net positive buoyancy of 600 N (150 lb) force. This shape has been tow-tested and found to be stable in pitch, roll, and yaw at velocities up to 100 cm/s and has moderately low drag, thereby minimizing lateral drift.

The winch is lowered from a boat or pontoon tender suspended from a pivoting frame. Based on experience, the entire operation requires four people and three hours of calm weather.

Of paramount concern to a moving current sensor is the influence of self-motion on measurement accuracy. In all but shallow water moorings, the two-axis tilt sensor showed very little change ($<1°$). In shallow moorings, the following three effects have been found to dominate with increasing wave activity: (1) near the surface, wave particle motion is impressed on the current sensor; (2) the unsteady wave flow can initiate a yaw and roll instability in the vehicle as well as a surging action; this activity is at a high enough frequency that the compass readings become very erratic; and (3) near-bottom wave motion can excite a pendulum-type oscillation of the body and tether above the winch.

An inherent limitation results from hydrodynamic effects due to the position of the current sensor atop the vehicle. Measurements are made on the up cycle only, and this constant upward velocity of the vehicle is added vectorially to the natural horizontal current. Lateral drift of the vehicle may be neglected due to the high buoyancy relative to the drag. The current sensor detects a simple, skewed resultant, which, in the case of low natural currents, impinges on the vehicle beneath, the presence of which displaces the flow streams through the measurement plane. The size and location of the current meter's acoustic mirror and its support posts add nonlinearly to output noise. Calibration of the current sensor, which is mounted on the vehicle, and the results obtained have been reported by Ward-Whate (1977). It has been found that the effective threshold for the vehicle-borne current sensor is 10 times its actual sensitivity of 1–2 mm/s, primarily because of its location atop the vehicle and its own physical shape.

Hamblin and Kuehnel (1980) successfully used this profiler for long-term unattended measurements of profiles of horizontal current and temperature at fine vertical resolution in a deep lake of moderate size (Kootenay Lake in British Columbia, Canada). Based on the excellent profiles they collected, it has been legitimately concluded that the disadvantages of the solid-state current sensor—namely, the care required to establish the environmental influences on the zero stability and gain and to establish its nonlinear calibration—are outweighed by its ability to detect the weak flows found in deep lakes that cannot be sensed by conventional mechanical devices on account of their high speed threshold.

Royer et al. (1986) conducted an intercomparison study of the VAPS in Lake Erie with drogues and current meters. Figure 10.4 shows a comparison between currents sampled by VAPS and currents measured at 21-m depth with a current meter at a nearby mooring. The agreement between the current speeds is extremely good, and, for most of the time, so is the agreement between the current directions. The few cases in which the directions disagree correspond to times of such low velocity that the directional thresholds of the fixed CM are being approached. According to Royer et al. (1986), the correspondence between instruments at 10-m depth is also as convincing as the one at 21-m depth for speeds above 2.2 cm/s, which is the threshold speed of the Plessey current meter.

Figure 10.5 shows a comparison between currents sampled by VAPS and currents measured with drogues at 3-m depth. The velocities can be extracted from successive positions and corresponding times of each drogue. A comparison of Lagrangian drift to an Eulerian current is strictly valid only for a uniform flow, but it is reasonable to expect that horizontal uniformity in the comparison

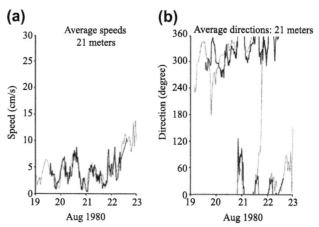

FIGURE 10.4 Comparison between currents sampled by VAPS (solid line) and currents measured at 21-m depth (dotted line) with a Geodyne current meter at a nearby mooring. (a) Compares speed; (b) compares direction. *(Source: Royer et al., 1986.)*

experiment reported by Royer et al. (1986) would be sufficient to allow comparison of the resulting drogue speeds and directions to the VAPS data as being characteristic of the whole area. The drogues chosen for the comparison were usually closer to the VAPS site; therefore, the good correspondence between the speeds and directions of the two current estimators is not surprising. The near-surface speeds are usually higher than those at lower depths, thus giving rise to a better definition of the direction of the current.

The successful comparison with conventionally measured currents and temperatures establishes that these parameters can be accurately measured from a moving profiler in an exposed large water body. The agreement between the uppermost VAPS current and the surface wind field suggests that the profiler would be capable of resolving the surface and bottom boundary layers, provided that some way could be found to maneuver the sensors more closely to the surface and bottom.

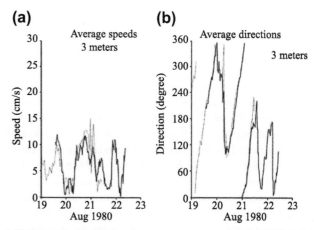

FIGURE 10.5 Comparison between currents sampled by VAPS (solid line) and currents measured with drogues (dotted line) at 3-m depth. (a) Compares speed; (b) compares direction. *(Source: Royer et al., 1987.)*

The VAPS record provided a fascinating glimpse of the intimate coupling between the temperature and velocity distributions in Lake Erie. In particular, the velocity profiles indicated a surface current swinging in direct response to the observed winds. It was also observed from the VAPS vertical profile measurements that "transient" thermal structure exerts an important influence on the distribution of momentum during periods of relatively light winds, which is an observation in agreement with that reported by Price et al. (1986). The VAPS measurements provided a clear indication that only during periods of very strong winds does Lake Erie assume a truly two-layered structure. The tantalizing glimpse of the richness and variety of the subsurface velocity and temperature structure suggests that only the data from a profiling instrument can adequately resolve the structural details of a current regime.

10.2.3. Acoustically Tracked Freely Sinking Pingers

The pinger-tracking method used for measurement of vertical profiles of horizontal currents in the deep ocean is the oceanic analog of the meteorological pilot balloon. The acoustic method obtains a current-velocity profile from the acoustical determination of the trajectory of a free-falling probe. In this method, an instrument package containing an acoustic pinger is ballasted to descend to the ocean floor, where it can be released by an acoustic release mechanism, a corrodible link, or a preset timed-release mechanism controlled by the electronics in the instrument package. The pinger's trajectory is determined by acoustically tracking it relative to an array of at least three units of spatially separated surface- or bottom-moored acoustic transponders.

While the pinger slowly descends to the seafloor and then ascends to the surface, it transmits a continuous-wave (CW) pulse (i.e., an acoustic ping of a few milliseconds' temporal width) at selected time intervals. On reception of the ping, the transponders reply at separate frequencies. The round-trip travel time of pulses between the pinger and each transponder (i.e., the time elapsed between the interrogation and reply pulses) is recorded in the electronics package of the pinger. From the different round-trip travel times, the trajectory of the pinger in three-dimensional space is determined. As the pinger falls and rises, it encounters a variety of forces that result from its relative velocity with respect to the surrounding water mass. The vertical profile of horizontal flow velocities is estimated from the horizontal displacements of the freely falling or rising pinger relative to the moored transponders. Thus, from the tracking data, it is possible to compute the horizontal velocity as a function of depth.

Rossby (1969) described an experiment to measure the overall vertical profile of horizontal ocean currents on the southern slope of Plantagenet Bank southwest of Bermuda

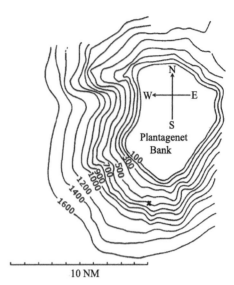

FIGURE 10.6 Portion of topography around Plantagenet Bank with contour lines shown at 100-fathom intervals. The Bank is situated 25 nautical miles (NM) southwest of Bermuda. The sinking float was released at the point marked by a cross. *(Source: Rossby, 1969.)*

(see Figure 10.6) while retaining high resolution of the small-scale structure by tracking acoustically a slowly sinking pinger (a standard Swallow float). The float, equipped with slanted vanes causing it to rotate as it sinks, was released from a research vessel. The design for rotation was meant to eliminate or considerably reduce any inherent tendency for the float to veer or coast in any preferred direction. The particular location for the experiment was dictated by the existence of a set of hydrophones on the southern slope of the Bank.

The successive positions (i.e., three-dimensional geographical coordinates) of the free-falling pinger were determined based on a principle that is somewhat analogous to that used for position fixing (i.e., determining the precise position) of a ship surveying in the offshore regions. Before the advent of the now ubiquitous GPS, the geographical coordinates of a survey ship were determined based on the known positions of a navigation transit satellite (i.e., a satellite that passes overhead) in near-polar orbit at an altitude of ∼1,100 km and with an orbital period of ∼107 min, moving at a speed of ∼20,000 km/h. In this satellite navigation (SatNav) scheme, a given navigation satellite transmitted a set of signals at successive discrete time intervals and a Doppler Navigation System receiver located in the ship received the signals when the satellite remained on the horizon. The position fixing of the ship was accomplished based on an estimate of the intersection of the two surfaces of the space hyperboloid (constructed from the successive range differences between the satellite and the ship) with the sea surface, and judiciously choosing one of the two mathematically admissible locations of the ship, the alternate location being on the other side of the track swept by the satellite's footprint on the surface of the Earth (Joseph, 2000).

Position fixing of the free-falling pinger, based on the principle of construction of a hyperboloid as indicated, requires determination of successive range differences between the pinger and a two- or three-dimensional array of spatially distant hydrophones located on the seafloor. This is because hyperbolae are loci of constant differences in distance to two points. The float (i.e., pinger) used by Rossby periodically transmitted a ping, which was received by four hydrophones located on the seafloor. The theory for tracking is as follows: Assume that the four receivers have the locations (x_i, y_i, z_i), where $i = 1, 2, 3, 4$. The distances r_i from the pinger at (x, y, z) to the receivers are $((x - x_i)^2 + (y - y_i)^2 + (z - z_i)^2)^{1/2}$ and are equal to the effective sound speed v_i times the travel time t_i. Assuming that the effective sound speeds v_i are the same and known, it is possible to construct three difference equations to be solved for the three unknowns x, y, and z from the four range equations (note that $i = 1, 2, 3, 4$):

$$v_i t_i = \left((x - x_i)^2 + (y - y_i)^2 + (z - z_i)^2\right)^{1/2} \quad (10.4)$$

In practice the solution is obtained by an iterative procedure. Allowance is made for the variation in effective sound speed, which corrects for ray bending and is defined as the equivalent sound speed for an acoustic signal along a straight path. It can be computed from knowledge of the vertical sound-velocity profile, together with the angle from the vertical of the equivalent straight ray. Expansion of Equation 10.4 yields the following:

$$v_i^2 t_i^2 = x^2 - 2xx_i + x_i^2 + y^2 - 2yy_i + y_i^2$$
$$+ z^2 - 2zz_i + z_i^2 \quad (10.5)$$

Putting $i = 1$ in Equation 10.5 yields the following:

$$v_1^2 t_1^2 = x^2 - 2xx_1 + x_1^2 + y^2 - 2yy_1 + y_1^2$$
$$+ z^2 - 2zz_1 + z_1^2 \quad (10.6)$$

Determination of range differences for construction of hyperbolae (which is a requirement for determining the successive positions of the free-falling pinger in three-dimensional space) can be accomplished by forming three difference equations. This can be achieved by subtracting Equation 10.5 from Equation 10.6. Accordingly, Eq. (10.6) − (10.5) and appropriate rearrangement yields the following expression:

$$2x(x_i - x_1) + 2y(y_i - y_1) + 2z(z_i - z_1)$$
$$= (x_i^2 + y_i^2 + z_i^2) - (x_1^2 + y_1^2 + z_1^2)$$
$$- v_i^2 t_i^2 + v_1^2 t_1^2 \quad (10.7)$$

where $i = 2, 3, 4$, and $x, y, z, t_1, t_2, t_3, t_4$ are unknown. However, t_i can be expressed as $t_1 + (t_i - t_1) = t_1 + \Delta t_i$, where Δt_i are the measured time differences. If these are substituted into the three equations, we have three unknowns (x, y, z) as a function of t_i. The solution is found by assuming a value for t_1 and perturbing it until a set of coordinates (x, y, z) is found, which, when inserted into the four range equations, will yield the same time differences as those measured. In a three-dimensional situation, with the slowly sinking float (i.e., the transmitter) hanging in the ocean and the four spatially separated hydrophones (i.e., receivers) fixed on the seafloor, the slant-range differences (equivalently, the measured time differences) define a hyperbolic surface. The transmitter can be anywhere on the surface of the hyperboloid. Thus, there is more than one solution to the problem, but only one solution will satisfy the expectation that the transmitter be at a reasonable point in the water and not in the Earth's crust or atmosphere!

In the method adopted by Rossby (1969), two passes are made through the data. In the first pass, (x, y, z) as a function of time are obtained. In the second pass, the depth is treated as a known smooth function of time and only (x, y) are computed. This permits one to select the best three of four receivers for the final calculations. The only assumption made is that the float sinks at a steady rate. Systematic and random errors affect the accuracy of the velocity estimates. The errors are those associated with the acoustic navigation of the pinger and those involved with the interpretation of the horizontal velocity of the pinger as oceanic motion. Some sources of possible errors in the position estimates are (Rossby, 1969):

- Uncertainty in timing resolution
- Amplification of errors by the geometry of hyperbolae
- Variation in effective sound velocity due to internal waves
- Uncertainty of the sound-velocity profile and receiver locations

Of these, the first two are the most significant. In Rossby's experiment, the time difference to 0.1 ms was measured. Advancement in electronics technology easily reduces such uncertainties.

The difficulty associated with the hyperbolic geometry is that hyperbolae are loci of constant differences in distance to two points. Given a distance difference with a certain possible error, the error band for the locus will grow rapidly as the distance from the axis between the two points increases or if the solution lies close to the baseline extension between the two points. The result of these two errors is that for each of the time differences, two time measurements are required, both of which introduce timing errors that together become amplified if the solution is far from the baseline of the receivers. The error arising from variation in effective sound velocity is believed to be minor

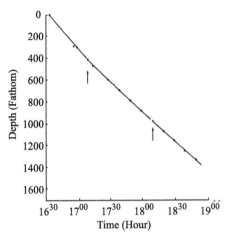

FIGURE 10.7 Depth of the float as a function of time. It is approximated by three linear sections for the purpose of ease in analysis. *(Source: Rossby, 1969.)*

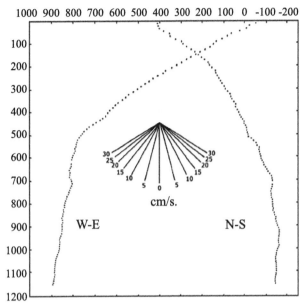

FIGURE 10.8 The trajectory of the float separated into its N-S and W-E components as a function of depth. The horizontal velocity at any depth is obtained from the slope of the trajectory and the nomogram in the center of the figure. The vertical scale is in meters from the sea surface and the horizontal scale is in meters N-S or W-E with respect to an arbitrary origin. The dots indicate the position at one-minute intervals. *(Source: Rossby, 1969.)*

as long as the acoustic ray has not been totally refracted. In this case there is little scatter and a single path arrival is achieved. Internal waves will perturb the path. However, if the mode number is high, the combined effect should remain small. Also, if the thermocline as a whole moves up or down, the resulting change in effective sound speed is estimated to be less than 1 m/s. The consequence of change in travel time arising from this change in effective sound speed is considered negligible. Uncertainty of the sound-velocity profile can be minimized by knowing the water-temperature profile in the region of interest, because the sound-velocity profile can be estimated from the measured or known temperature profile.

A depth-versus-time curve for the float, approximated by three linear sections and obtained from the first analysis of the tracking data, is shown in Figure 10.7. The sinking rate was about 17.5 cm/s. The second analysis gave the results shown in Figure 10.8, in which the trajectory is composed into its north-south (N-S) and west-east (W-E) components. In interpreting this figure, it must be borne in mind that the horizontal velocity at any depth is given by the deviation of the trajectory from the vertical. The nomogram in the center of the figure gives the velocity as a function of angle. Similarly, the curvature of the trajectory is a measure of the vertical shear. In principle it is obtained from the second derivative $\frac{d^2}{dz^2}$ of the trajectory. In practice it is difficult to measure this quantity, because it is limited to thin regions where the velocity changes almost discontinuously.

The dots along the trajectory are the computed positions of the float at one-minute intervals. Although the ping repetition rate was one every 2.4 s, the times of arrivals were averaged in 15-s groups from which the float position was calculated. It is evident from the trajectory in Figure 10.8 that high velocities dominate from the surface

down to the upper limit of the main thermocline, where there is a sudden transition to slowly moving deep water. Rossby (1969) proposes that it is not possible to state with certainty whether this transition is typical of the flow far from the Bank or whether it is peculiar to the flow around the Bank. According to Crease (1962), such low velocities are not typical of deep water in the Sargasso Sea. According to the profile measurements by Rossby (1969), the deep water appears to be moving to the west, suggestive of a weak tangential flow along the bottom topography. Based on several hydrographic measurements and current profile measurements, it is inferred that the Bank has a pronounced influence on the water-current flow in its vicinity.

The abrupt change in N-S flow at 510 m corresponds to a velocity difference of 20 cm/s. It is inferred that this particular high shear zone is an interface between turbulent patches of water agitated by the flow around the Bank. Noteworthy information that could be gleaned from the profile measurements of Rossby (1969) is poor indication of the regularly spaced steps reported by Cooper and Stommel (1968) in this region.

In an analysis by Rossby and Sanford (1976), the trajectories in the vertical and east plane and the vertical and north plane during an interval of 200 s are separately fitted in a least-squares sense to parabolic curves. The fitted curve is differentiated with respect to time to determine the east, north, and vertical-velocity components of the probe's motion. It is assumed that the horizontal velocity components of the float are equal to those of the surrounding water. The rms difference between the measured and fitted trajectories is typically 1 m. Thus, if the instrument is sinking at 0.25 m/s in the presence of a 0.10 m/s horizontal current, it will be advected 20 m horizontally during a 50 m vertical interval. According to Rossby and Sanford (1976), a single end-point determination of the advection rate without any curve fitting would lead to a maximum 10 percent, or 0.01 m/s, uncertainty in horizontal speed. The least-squares curve fit reduces this level of expected error.

10.2.4. A Freely Sinking and Rising Relative Velocity Probe (Cyclesonde)

Van Leer et al. (1974) reported the design and use of a vertically scanning instrument package called the *Cyclesonde* (see Figure 10.9). The Cyclesonde derives its name from the fact that it profiles the upper ocean periodically, in much the same way as a radiosonde station profiles the lower atmosphere. Furthermore, both these tools are buoyancy-driven systems.

The Cyclesonde consists of a gas-operated, buoyancy-driven probe with a recording package containing sensors for measurements of water-current velocity, temperature, and depth. It makes repeated automatic round trips up and down a taut-wire subsurface mooring at selected vertical speeds of between 2 and 20 cm/s while scanning the horizontal currents at successive depth layers. The combination of a ball-bearing roller block and smooth plastic-coated wire gives a low coefficient of friction (μ) of about 0.02 on the wire, with no noticeable increase in friction for periods of immersion of a week or more. A block designed with three overlapping wheels prevents the plastic wire from chafing on the block's cheeks and reduces friction.

FIGURE 10.9 A complete Cyclesonde being deployed for ocean-current profile measurements. *(Source: Van Leer et al., 1974.)*

This block opens on one side to allow the Cyclesonde to be removed from the wire without breaking the wire. A safety wire holds the wire, sphere, block, and hull together. The Cyclesonde's motion is controlled by ballast, buoyancy, and drag adjustments.

The sensors (water-current velocity, temperature, and pressure) are located in two protecting cages mounted symmetrically on both sides of an aluminum ball on the head of the Cyclesonde. The side location gives the sensors unobstructed exposure for both the up and down profile of each round-trip cycle. A variable buoyancy device (VBD) package and a scuba tank of the Cyclesonde trail downstream of the taut wire, enclosed in a low drag (drag coefficient $C_D = 0.7$) plastic hull that acts as a 1.5-m-long current vane. The long vane greatly reduces vane flutter due to wave and mooring motion. The hull orients the current speed sensor (Savonius rotor) so that the vector sum of the vertical speed of the Cyclesonde and the horizontal component of current lies in a plane perpendicular to the rotor's axis of rotation. The whole Cyclesonde, including the magnetic compass for current direction measurement, turns with the vane, and thus no vane follower is required for speed-direction estimation. The device measures and records these parameters as a function of time. The use of the unattended Cyclesonde on a taut-wire mooring is limited to current regimes with speeds of less than 90 cm/s.

The Cyclesonde may be operated from a ship or deployed on a mooring. When a ship is anchored in a region where currents are less than 2 knots (1.02 m/s) or it is drifting slowly, a Cyclesonde will run up and down a ship-lowered wire held taut by a heavy weight. Because the wire is run in or out only when the Cyclesonde requires service, a crude winch or capstan is sufficient, freeing the winches for other sampling programs. Several hundred profiles of current speed and direction, temperature, and depth have been recorded in this mode of operation.

When the Cyclesonde is deployed on a mooring, the device is constrained by a closed pulley to ride up and down a smooth plastic-coated wire (about 4-mm diameter). The wire is held taut between a seafloor-mounted anchor weight (e.g., a railroad wheel) and a subsurface float of ~350-lb buoyancy to minimize the wire angle θ. At the bottom of the wire there is an auxiliary float and a weak link. Should the anchor become fouled, the weak link will break when the mooring is recovered, thereby saving the Cyclesonde wire from breaking. A ground-line auxiliary anchor and slack-wire surface float may be used in typical continental shelf depths. If the weather permits, divers may exchange Cyclesondes in a short period of time, giving minimum data interruption. Using this technique, Van Leer and colleagues (1974) regularly serviced Cyclesonde moorings in an hour in 80-m water depth, and several hundred profiles were made on moorings in the Gulf of Mexico and Oregon shelf regions.

10.2.4.1. Operation

The Cyclesonde's cyclic vertical motion is controlled by changing the mean density of the instrument package a few percent. These density changes are accomplished by an inflatable bladder, which is used to change the displacement of the Cyclesonde so that its buoyancy is made negative at a pre-selected low pressure and positive at a pre-selected high pressure. This causes the instrument package to cycle between these two specific water pressures. The original design covers a maximum depth of 200 m. Figure 10.10 shows the variable buoyancy device. Its component parts are an aluminum scuba tank, pressure regulator, deep valve, differential pressure valve, shallow valve, and bladder assembly (see Figure 10.11). The swim bladder assembly, shown in Figure 10.10, consists of a rubber tube closed at one end and fastened to an inlet port at the other. The tube is placed over a PVC rod to maintain its volume when deflated. The swim bladder is enveloped by a Plexiglas tube so that its inflated dimensions are the

same for each inflation. The neoprene bladder can be inflated to 400 cm^3.

The variable buoyancy device is powered by compressed helium contained in an aluminum scuba tank. Aluminum was chosen because of its lightweight and nonmagnetic properties. A Décor Model 300 first-stage regulator valve was modified to maintain a pressure of 25 psi ± 10 over the ambient water pressure to inflate the bladder. A detailed drawing of the valve assembly is shown in Figure 10.12. The stainless-steel heads of the deep and shallow control valves protrude from the valve body and are used to set the upper and lower turnaround pressures. Each of the control valves has a calibration curve of turns versus pressure that is used for accurately selecting the turnaround depths.

Six ports, P_1 to P_6, are drilled into the valve body. P_1 allows gas to enter the control valve assembly from the regulator, P_2 supplies gas at a controlled pressure to the swim bladder, and P_3 to P_6 are vented to the seawater. A deep turnaround depth between 0 and 200 m and a shallow turnaround between 0 and 26 m can be selected without changing the spring. To prevent overpressure inside the swim bladder during ascent, a pressure-relief valve opens to P_4 when the pressure differential across the swim bladder and seawater exceeds 37 psi.

A loss of gas occurs each time the gas in the bladder is expelled to the sea. Because of its low molecular weight, helium was used to minimize the changes in trim arising from this loss. Up to 300 cycles can be obtained using helium before weight compensation becomes necessary.

A complete Cyclesonde system is composed of the following components: (1) a variable buoyancy device (VBD), (2) mooring or ship-lowered wire, (3) an instrument package, (4) a pulley block, and (5) propulsion housing. In a successful Cyclesonde system, the change in buoyancy, $\Delta B/2$, must always exceed that portion of the

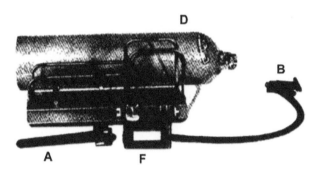

FIGURE 10.10 Variable buoyancy device. A: Swim bladder. B: Tank regulator valve. D: Scuba tank. F: Differential pressure valve. *(Source: Van Leer et al., 1974.)*

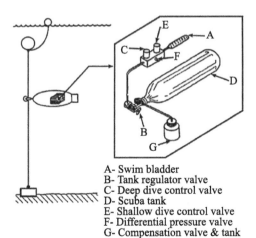

A- Swim bladder
B- Tank regulator valve
C- Deep dive control valve
D- Scuba tank
E- Shallow dive control valve
F- Differential pressure valve
G- Compensation valve & tank

FIGURE 10.11 Principal parts of the variable buoyancy device. A: Swim bladder. B: Tank regulator valve. C: Deep-dive control valve. D: Scuba tank. E: Shallow-dive control valve. F: Differential pressure valve. G: Compensation valve and tank. *(Source: Van Leer et al., 1974.)*

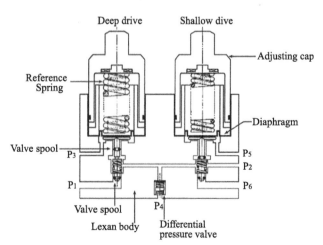

FIGURE 10.12 Valve body arrangement. *(Source: Van Leer et al., 1974.)*

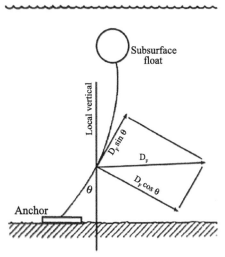

FIGURE 10.13 Schematic diagram of acting forces. *(Source: Van Leer et al., 1974.)*

drag force acting along the wire plus the component perpendicular to the wire, multiplied by the coefficient of friction of the roller running on the wire (see Figure 10.13). Thus, the condition to be satisfied is the following: $(\Delta B/2) > D_F (\mu \cos \theta \pm \sin \theta)$, where ΔB is the buoyancy change due to the swim bladder; $D_F = (C_D A \rho/2) \, V_H^2$; θ is the wire angle, and μ is the coefficient of friction. $A =$ Cyclesonde frontal area; $V_H =$ horizontal velocity; C_D is the horizontal drag coefficient; and ρ is the local seawater density. The sign of the sine term depends on whether the direction of travel is the same as the direction of the drag component acting along the wire.

For a moored Cyclesonde, the surface or subsurface float usually trails downstream (see Figure 10.13) so that the component of drag in the layer of strongest current acting along the wire usually acts to oppose the friction force when the Cyclesonde moves upward. The friction force always acts to oppose the motion. Because the horizontal velocity component V_H and the local seawater density ρ are determined by the ocean, the variables available to the designer are C_D, A, μ, and θ. To make $C_D A$ as small as possible, the displacement of the Cyclesonde must be minimized. Each kilogram of excess wet weight in the instrumentation or VBD system must be supported by about 2 kg of syntactic foam, giving a total increase in the displacement of about 3 kg. The additional area added by such an increase in Cyclesonde displacement is multiplied by the square of the velocity. This aspect of the problem results in a greater drag force, which must be overcome by increasing the change in buoyancy ($\Delta B/2$) used to move the Cyclesonde vertically. For a given number of profiles, the gas supply would need to be increased, which again increases the volume displaced. To avoid this "snowballing" effect of increasing total displacement, every possible effort must be made to keep the component weight to a minimum.

Based on field experience, some modifications in the design became necessary. For example, the VBD had to be used with some modifications as follows: The heavy stainless steel mounting frame had to be omitted to save weight. Surgical rubber tubing was substituted for the neoprene swim bladder to extend bladder life to approximately 300 inflations using a displacement of $\Delta B = 600$ cm³. A new longer-life swim bladder using toroidal geometry, developed at the University of Miami, survived without failure for over 5,000 inflations (at 600 cm³) in a prototype model (Van Leer et al., 1974).

Because the axis of rotation of the Savonius speed sensor is horizontal, it responds with equal sensitivity to both the horizontal current speed and the vertical motion of the Cyclesonde. This means that the vertical velocity of the Cyclesonde must be subtracted from the observed speed. This is done by using the rate of change of water pressure with time as a vertical velocity estimate. One advantage of the sensitivity of the horizontally mounted Savonius rotor to the sum of the horizontal current speed and the vertical motion of the Cyclesonde is that the rotor operates outside its nonlinear threshold region. Figure 10.14 presents unedited, unsmoothed raw data in the format of time series as recorded by the Aanderaa current meter, which formed the sensing and recording device of the Cyclesonde. Figure 10.15 presents unsmoothed, original vertical profile observations obtained during February 1973 in the Gulf of Mexico.

10.2.4.2. Advantages

A noteworthy advantage of the Cyclesonde is that the roller decouples the Cyclesonde from the vertical heaving motions of a ship or mooring. This makes relatively noise-free measurements possible because the measuring instrument moves slowly, smoothly, and monotonically through the water. Noise problems such as "salinity spiking" and "rotor pump up" are much reduced relative to conventional techniques. Furthermore, the Cyclesonde has nearly a cosine response to currents not in the plane of the fin (hull), as a result of which horizontal wave-induced mooring motions perpendicular to the fin tend to average. This leaves the fore and aft motions (in the plane of the fin) of the mooring as a major source of wave-induced rotor error.

In regions of weak currents, Cyclesonde measurements from taut-wire, subsurface moorings are preferable to ship-lowered measurements because of uncertainties introduced by the ship's motion on its anchor, the ship's changing magnetic influence on the compass used in the current meter, and the ship's effect on flow in its immediate vicinity. The magnetic properties of a Cyclesonde mooring must be given special consideration because the instrument can move close to the iron anchor and subsurface float. The large hull makes a very sensitive vane, allowing accurate

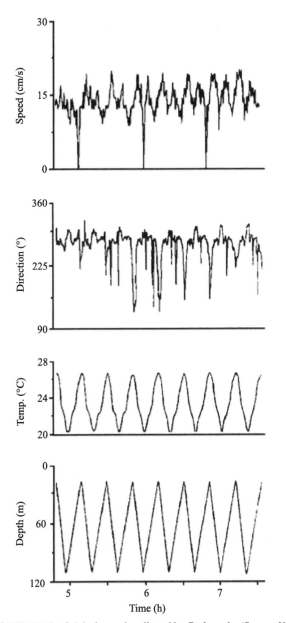

FIGURE 10.14   Original records collected by Cyclesonde. *(Source: Van Leer et al., 1974.)*

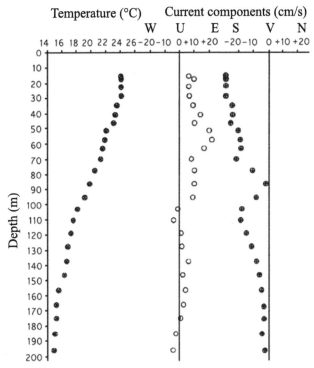

FIGURE 10.15   Unsmoothed, original vertical profile observations obtained by Cyclesonde during February 1973 in the Gulf of Mexico. *(Source: Van Leer et al., 1974.)*

current-direction data at very small horizontal water speeds. Evidence for this was found by Van Leer et al. (1974) in their shallow-water tests in a protected region. Limitations of Cyclesonde include its limited vertical range and limited life for a given design.

10.2.5. A Freely Falling Electromagnetic Velocity Profiler

Various methods have been employed to determine the vertical profile of horizontal current $V(z)$ at depth z. One approach, based on the measurement of motionally induced electric effects, was developed by Drever and Sanford

(1970). They showed that the weak electric fields and currents generated by the motion of the seawater through the geomagnetic field could be measured and related to the velocity profile. The instrument measures the voltages induced by the motion of the sea and the instrument through the geomagnetic field. The electric field is interpreted in terms of ocean-current velocity. Comparisons with an acoustically tracked probe demonstrated the electromagnetically inferred velocity profile to be accurate to about ±1 cm/s (Rossby and Sanford, 1976). Sanford et al. (1978) developed an improved technique to determine the variations of horizontal flow velocity between the sea surface and the seafloor based on the measurement of electric currents generated by the motion of the seawater through the Earth's magnetic field. In this device, a freely falling *electromagnetic velocity profiler* (EMVP) senses the currents as a function of depth. The device is released from the surface, falls to a preset depth or to the seafloor, and then returns to the surface. Both descent and ascent take about 90 min in water 6,000 m deep. The design requirements of the EMVP were the following (Sanford et al., 1978):

- Velocity resolution of 1 mm/s, to observe fine structure
- Fall speed of about 1 m/s, for rapid profiling
- Rotation rate of once each 5 to 10 seconds, to separate motional signal from electrode offset
- Real-time acoustic telemetry and tracking

- Depth capability to 6000 m
- CTD functions
- Internal digital storage
- Rugged construction
- Orthogonal electric measurements every ½ s, for redundancy and fine vertical resolution
- Autonomous release from the sea surface
- Recovery aids including radio, flashing light, and acoustics

The EMVP differs from conventional electromagnetic (EM) current meters in that the geomagnetic field, rather than a locally generated magnetic field, is utilized. The response of the EMVP results from the motion of seawater relative to the geomagnetic field, whereas the EM current meter responds to the motion of water relative to its own magnetic field. In a sense, the geomagnetic field establishes a reference frame and the electrical response arises from motion relative to this reference. The physics of motional induction was discussed by Longuett-Higgins et al. (1954).

The EMVP is allowed to free-fall and is advected horizontally by the local flow velocity $V(z)$ at depth z. It thus records a potential difference $\Delta\phi = B \times (V(z) - \overline{V}^*)$, where B is the local magnetic flux density of the Earth, and $\overline{V}^*$ is the barotropic (largely depth-independent) contribution to the current. The local magnetic field of the Earth is available from the world magnetic field charts. The EMVP yields a profile in the form of $[V(z) - \overline{V}^*]$ (Sanford, 1971). In the expression $\overline{V}^*$, the overbar denotes the vertically averaged character of the quantity. The asterisk denotes that there is an electrical conductivity weighting involved and that other contributions exist. In the EMVP, the electric current is measured as the voltage difference between two points on opposite sides of a vertical, insulating cylinder. The use of a large diameter outer skin not only increases the electrical sensitivity but also provides room inside for the electrodes, electronics pressure vessel, and floatation. In addition, by keeping the length down and the body streamlined, the profiler vehicle (see Figure 10.16) is easier to handle at sea. The vehicle is built up around a 7,075 aluminum alloy tube, 5 ft (1.52 m) long, with an outside diameter of 7.5 in (19.1 cm) and wall thickness of 0.75 in (1.91 cm). The tube and end caps are designed to withstand a pressure equivalent to 6,000 m.

The forward half of the tube is surrounded by five discs of syntactic foam, each of which has buoyancy in seawater of about 10 lb (4.5 N). Covering the foam and the remaining tube is a 14-in. (35.6-cm) cylindrical medium-density polyethylene shell, 0.25 in. (0.64 cm) thick. Beneath the lower portion of this shell are electrode housings connected by hollow plastic tubes to four holes equally spaced around the circumference (see Figure 10.17). A collar holding eight pitched fins, which

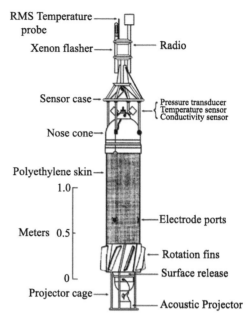

FIGURE 10.16 Free-fall EMVP vehicle. A balloon keeps the profiler on the surface until the preset time when it is released by the surface release mechanism and the profiler begins its down profile. (Source: Sanford et al., 1978.)

cause the instrument to rotate as it falls, is mounted over the lowest part of the skin. Cages are attached to each end of the tube to protect the sensors and to provide places to lift and handle the probe. On the upper end of the sensor cage are mounted the recovery aids: a radio transmitter and a xenon flasher. These units are self-contained, having no electrical connection with the main instrument.

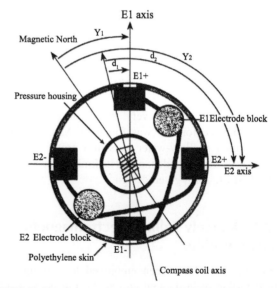

FIGURE 10.17 Cross-section of the free-fall EMVP vehicle at the level of the electrodes showing the arrangement and orientation of the electrode arms (E1 and E2) and compass coil (CC). (Source: Sanford et al., 1978.)

Between the radio and flasher is a thermistor bead that is used to sense the high-frequency variations (0–16 Hz) of seawater temperature. The assembly is mounted on a frame that is able to flex relative to the sensor cage. An extra 13 N (30 lb) of expendable weights overcomes the positive buoyancy of the instrument (6.5 N) and forces it to fall toward the bottom. The weights are released either on contact with the bottom or at a preset depth. The release mechanisms are similar to those used by Richardson and Schmitz (1965). As a safety measure, a small corrosive magnesium link is placed between the release mechanism and the weights. During a 5,000-m drop, the fall and rotation rates gradually change by about 8 percent because of changes in buoyancy and viscosity. The ratio of fall and rotation rates changes by only 3 percent.

The sensors used in the EMVP are the following (Sanford et al., 1978):

- Two electrode assemblies (*E*1 and *E*2)
- A solenoid coil wound on a high-permeability core (*CC*)
- A bonded strain-gauge pressure transducer (*P*)
- A pressure-protected platinum resistance temperature sensor (*T*)
- An inductively coupled electrical conductivity sensor (*C*)
- A three-axis flux-gate magnetometer (*F*1, *F*2, and *F*3)
- A thermistor in the magnetometer (*TM*)
- An external thermistor (*TF*)

The major component of the profiler is the electrode and salt bridge system, which senses the microvolt potential difference across a diameter of the instrument's surface. The potential difference is the integral of potential gradients distributed on the path from one electrode along its salt bridge, around the skin through the surrounding seawater, and down the salt bridge to the other electrode. In any EM current sensor, offset voltage is an issue, and this needs to be tackled in the most optimal manner. The simplest way to make good use of the EM sensors is to protect the electrodes from environmental changes and to modulate the desired signal at a large frequency compared with the rate of electrode drift. This is achieved by the use of salt bridges, as proposed by Mangelsdorf (1962), and by the rotation of the entire instrument. Even with the use of Ag-AgCl (silver-silver chloride) electrodes, a voltage of ~350 µV will be generated for each degree in temperature difference between the electrodes of a pair.

To guard against this problem, the salt-bridge arms and electrode blocks are designed to retard heat and salt fluxes into electrodes, and the data are processed to remove the remaining slowly changing electrode offset. Taking all these into account, the electrode system (see Figure 10.18) uses large Ag-AgCl electrodes in a PVC block. The orientation of the electrode arms is determined from the signal of the compass coil (CC). Temperature gradients in the thermocline produce negligible errors, but bubbles in

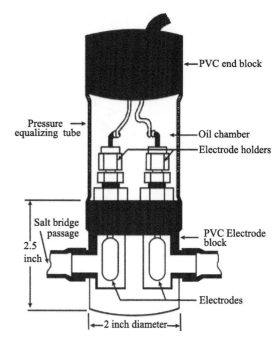

FIGURE 10.18 Construction of electrode block of EMVP. *(Source: Sanford et al., 1978.)*

the tubes disturb operation within 30 m of the surface. The acoustic telemetry system provides real-time information about the current velocity and seawater temperature structure, slant range, and depth. A typical plot of the current velocity components (E-W and N-S) indicating the shear features as produced by standard processing is shown in Figure 10.19. Rapid changes of current velocity with depth in the thermocline are quite evident.

A disadvantage of the method is that it yields a profile relative to an unknown, depth-independent velocity contribution. It reveals how the ocean-flow velocity varies with depth, but the depth-averaged or barotropic component is not observed. Advantages are that the measurements are made from a self-contained, free-falling instrument. It can be used from small vessels, does not interfere with other shipboard operations, and is mobile. There are several ways by which the unknown barotropic contribution can be determined so that the velocity profiles can be made absolute. As indicated earlier, the profiler has a cylindrical form and is equipped with angled fins so as to induce steady rotation at 0.15 Hz about the vertical axis throughout its descent, thereby modulating the potential difference detected at the electrodes. This rotation enables the direction-of-flow vector to be established. It also permits the wanted signal (typically in the range 1–100 µV) to be extracted from much larger offset DC potentials (up to millivolts in magnitude) arising from electrochemical effects and temperature differences. A performance level of ±1 cm/s at a vertical resolution of 10 m was indicated. This level is expected in the absence of

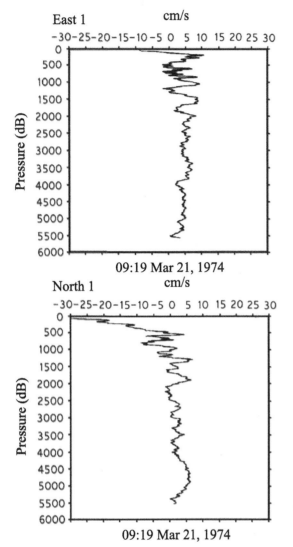

East 1   cm/s

North 1   cm/s

09:19 Mar 21, 1974

FIGURE 10.19 A typical plot of the current-velocity components (E-W and N-S) indicating the shear features as produced by standard processing. (*Source: Sanford et al., 1978.*)

velocity) as a function of depth. This is because the velocity of the profiler itself was unknown. Difficulties exist in interpreting velocity profiles from freely sinking instruments because the instrument is affected by the flow it is attempting to measure, and depending on the size, shape, and buoyancy of the probe, its trajectory may also vary from one probe to another for the same current profile. With the aid of mathematical models for the probe and from the sensor performance, the measured profile has to be interpreted as an estimate of the actual current profile (Hendricks and Rodenbusch, 1981). One method of correcting the drift-contaminated profile data would be to use commercially available acoustic positioning. This technique would, however, be tedious and restrictive in its area of operation.

The EMVP has undergone substantial refinements since its introduction, particularly in the resiting of its electrode ports for improved sensitivity. The difficulty in measuring "absolute" velocity profiles using freely falling/rising probes has been circumvented by an ingenious method reported by Sanford et al. (1985). In this modified device, known as the *absolute velocity profiler* (AVP), in addition to the relative velocity profile measured by the moving probe, an acoustic Doppler probe determines the absolute velocity of the profiler with respect to the seafloor. In construction, two acoustic transceivers on the probe, in Janus configuration, "look" toward the ocean bottom. As the probe sinks or rises, the transceivers transmit ultrasonic pings and receive the Doppler-shifted backscattered signals from the seabed. As the probe sinks toward the seafloor the returned signals are shifted "up-Doppler" in frequency, and while it rises up, the returned signals are "down-Doppler" in frequency. From measurements of the Doppler frequency shifts in the returned signals, the probe velocity relative to the fixed seafloor is independently determined. From knowledge of the acoustically measured probe velocity and the relative current-flow velocity measured by the velocity probe, the vertical profile of the absolute horizontal flow velocity is determined without depending on the hydrodynamic characteristics of the probe or additional equipment. This development thus included a measurement of $\overline{V}^*$. This has been achieved by mounting downward-looking Doppler sonar in the nose of the instrument (see Figure 10.20).

The transducers are arranged in a Janus configuration (see Figure 10.21). In this configuration, two transceivers of equal elevation angle point in opposite directions in the UW plane. This configuration, named after the Roman god of gates and doors who is represented as having two faces, allows determination of the vertical and horizontal components of the probe's velocity relative to the seafloor over the last 60–300 meters of the downward excursion. The advantages of the Janus arrangement are that it reduces the influences of W and small tilt errors. For a periodic

strong magneto-telluric currents, which can produce errors as large as 10 cm/s during infrequent (10 to 20 times per year) periods of strong temporal fluctuations of the geomagnetic field. The data (flow velocity, water temperature, electrical conductivity, pressure, and other variables) are recorded at fine temporal resolution (twice each second). The instruments have been profusely used in the western North Atlantic to describe the deep-ocean velocity profiles (Sanford, 1975); to investigate the structure and propagation of inertial period motions (Leaman and Sanford, 1975; Leaman, 1976); and to study low-frequency, baroclinic Rossby waves (Hogg, 1976).

The ocean-current profiler just mentioned and those addressed in the previous sections could not measure absolute velocity profiles; they only measured relative profiles of horizontal velocity (i.e., relative to the profiler's

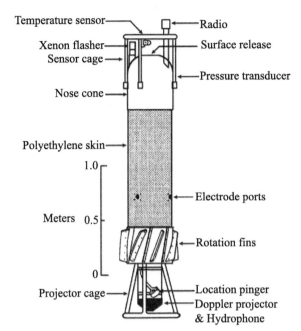

FIGURE 10.20 Absolute velocity profiler (AVP). *(Source: Sanford et al., 1985, ©American Meteorological Society, reprinted with permission.)*

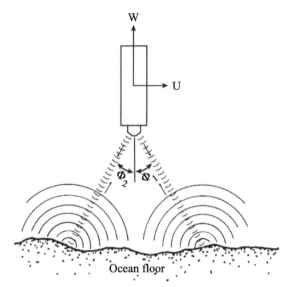

FIGURE 10.21 AVP in which the acoustic Doppler transducers are arranged in a Janus configuration. The freely falling profiler moves with velocity components of U and W in the horizontal and vertical directions, respectively, while projecting ultrasonic beams in the directions ϕ_1 and ϕ_2 relative to the vertical. *(Source: Sanford et al., 1985 ©American Meteorological Society, reprinted with permission.)*

signal transmitted and returned along a ray making the angle ϕ relative to the vertical in the UW plane, the measured Doppler shift is (Berger, 1957):

$$f_{D1} = \frac{2f_1}{c}(U\sin\phi_1 - W\cos\phi_1) \qquad (10.8)$$

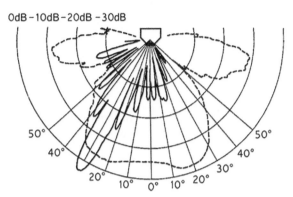

FIGURE 10.22 Directivity patterns for the projector (solid line) and the hydrophone (dashed line) used in the AVP. *(Source: Sanford et al., 1985, ©American Meteorological Society, reprinted with permission.)*

where

f_{D1} = Doppler frequency shift on beam 1
f_1 = transmitted frequency on beam 1
c = speed of sound in seawater at rest

Given $f_1 = 300$ kHz, $\phi_1 = 30°$, $U = 10$ cm/s, $W = -100$ cm/s, and $c = 1.5 \times 10^5$ cm/s, estimates of the expected Doppler-frequency shifts are obtained from the previous expression as $f_{D1} = 367$ Hz. Vehicle rotation necessitated the use of two separate frequencies for the two acoustic beams in the AVP because the narrow-beam transducers could not be used for both transmission and reception. (See the directivity patterns for the projector and the hydrophone in Figure 10.22.)

The AVP determines *V* at several levels, averaged over some vertical distance near the seafloor. Consequently, the combination of the acoustic Doppler (AD) and EM datasets near the seafloor yields:

$$AD - EM = V(z) - \left[V(z) - \overline{V}^*\right] = \overline{V}^* \qquad (10.9)$$

Thus, subtraction of the electromagnetically derived velocities from those obtained using the Doppler sonar yields $\overline{V}^*$, which can then be added to the EM profile to provide the absolute current profile *V(z)*. Once $\overline{V}^*$ is known, the whole EMVP profile can be adjusted so that it is an absolute velocity profile. Use of the absolute velocity profiler has shown that it can provide measurements to within 1 cm/s. A limitation of this probe, however, is that an absolute flow velocity profile can be measured only within the specified maximum slant range of the acoustic Doppler transceivers relative to the seafloor.

Electromagnetic profiling measurements clearly show promise. Duda et al. (1988) extended the technique to an autonomous instrument that can conduct multiple profiles to 1-km depth in the ocean (see Figure 10.23). Its name, *Cartesian diver*, is drawn from the toy made from an inverted bottle in which the buoyancy can be controlled by compression of an enclosed volume of gas while it is

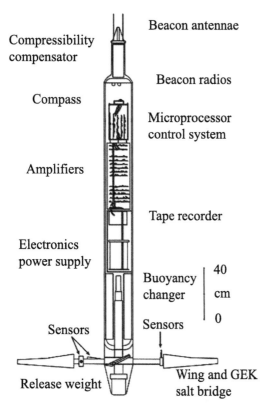

Compressibility compensator

Compass

Amplifiers

Electronics power supply

Sensors

Release weight

Beacon antennae

Beacon radios

Microprocessor control system

Tape recorder

Buoyancy changer

Sensors

Wing and GEK salt bridge

40

cm

0

FIGURE 10.23 Side view of the Cartesian diver profiler. *(Source: Duda et al., 1988, ©American Meteorological Society, reprinted with permission.)*

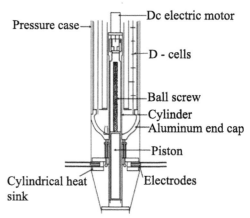

Pressure case

Dc electric motor

D - cells

Ball screw
Cylinder
Aluminum end cap

Piston

Cylindrical heat sink

Electrodes

FIGURE 10.24 Cross-section of the Cartesian diver profiler showing the electrodes within two salt bridges within the disk-like heat sink. The ball-screw-driven volume-changer piston is shown in extended position. *(Source: Duda et al., 1988, ©American Meteorological Society, reprinted with permission.)*

submerged in a liquid. The device was used prior to Descartes to illustrate the incompressibility of water relative to air, but Descartes' name has since been associated with it. The oceanic Cartesian diver has been engineered primarily to record continuous profiles of horizontal velocity using a variation of the method of geomagnetic induction, such as those reported by Sanford et al. (1978) and Sanford et al. (1982).

The assumption made in the estimation of the horizontal velocity of the diver is that the ocean electric field E_s, measured in a stationary coordinate system, does not vary with depth. Consequently, the electric field measured in the reference frame of a drifter, $E' = E_s + V \times B$, yields the drifter velocity V. Here B is the local magnetic induction. The diver records the horizontal components of the apparent electric field, which includes the effect of all three components of the diver's velocity. If the instrument moves horizontally at the same speed as the surrounding fluid and if its vertical velocity is steady or known, then vertical shear of horizontal velocity is directly determined from the vertical variations of the measured horizontal electric field.

The vertical velocity of the water is estimated by observing the vertical velocity of the instrument relative to the surface with a pressure gauge. Because the package profiles at terminal velocity, changes in its falling or rising

rate can be attributed to the vertical motion of the surrounding fluid.

The *Cartesian diver* is an untethered autonomous device that alternately falls and rises freely through multiple cycles, recording data internally. The diver propels itself vertically by adjusting its volume while its mass is unchanged, thereby changing its buoyancy. The instrument rotates once in 3 m vertically.

One feature of untethered profilers is that they have constant and reproducible dynamics. Because they fall and rise at a characteristic terminal velocity, the force required to propel them is equal to the drag at that velocity. Buoyancy of the diver is changed by moving a piston, which seals a cylinder extending through the bottom end cap. The recovery release assembly is attached to the lower end of the cylinder, which is open to seawater. Because the density of seawater changes rapidly through the thermocline and the instrument is virtually incompressible, a passive compensation piston at the top end of the diver is used to give the entire instrument roughly the same compressibility as seawater as both seawater temperature and pressure vary in that region. Without the compensation, the Cartesian diver would fall at an ever-decreasing rate throughout its descent and would rise at an increasing rate. The diver is equipped with beacon radios and a programmable underwater acoustic transponder to facilitate recovery.

The diver's buoyancy change is implemented by a ball-screw-driven volume-changer piston at the lower end of the diver (see Figure 10.24). The piston is sealed with a sealing ring. A DC electric motor with a reduction gear drives the piston through a ball-screw jack. A servo-controlled electromechanical brake must be energized to physically release the motor for motion. Both the motor and brake voltage regulators are of the switching type in order to conserve battery energy. The buoyancy changer is controlled by an onboard computer, and the decision making can be based on any measured parameter such as seawater pressure, temperature, or time. The buoyancy is always changed to its maximum or minimum value so that the instrument can only rise or fall at its maximum speed. The logic behind this choice is that to simplify statistical analysis of repetitive profiles, it is desirable for the Cartesian diver to move at a constant vertical velocity.

The buoyant forces that propel the instrument are determined by the difference in density between the instrument and the water surrounding it. Density adjustments are made with the buoyancy changer only as the instrument changes direction, but during the course of a profile the density difference between the instrument and the seawater must remain constant. The density of seawater increases more rapidly with pressure than that of the aluminum pressure case. To compensate for this, a highly compressible volume of gas is added to the instrument to make its overall compressibility more similar to that of seawater. The gas volume is in the form of a cylinder sealed with a piston. One side of the piston is exposed to ambient pressure. The piston is captured when the gas has expanded to a predetermined level. The O-ring seal of the two-piece, hollow piston is lubricated with oil forced out from within the piston. The gas volume is pressurized with a regulated gas bottle before the instrument is launched.

Silver-silver chloride electrodes with 5–25 cm² of exposed silver are used to sense the induced voltages. They provide a stable, low-noise electrical connection to the surrounding seawater. Unfortunately, the noise spectrum of the measurement electrodes is quite similar in shape and magnitude to the signal spectrum that would be sensed if the profiler simply profiled and recorded the potentials due to its horizontal drift, without spinning. The electrodes also have a thermal coefficient, which will create error voltages if the electrodes change temperature relative to one another or if that coefficient varies between members of an electrode pair. To record the induced voltage signal despite the red noise and to achieve 5-m resolution of water-current velocity, the electrodes must be physically interchanged at 0.03 Hz or faster in as short a vertical distance as possible (i.e., at a high wave number). This is a modulation of the signal spectrum into the frequency (wave number) of rotation. Higher modulation frequency and higher modulation wave number each give smaller effective rms noise velocity and better resolution of the water-current velocity. The interchange of the electrodes is accomplished by spinning the instrument with four wings, which have a span of 1.60 m. Adjustment of the wings allows profiling velocity variation over a range ±20 percent of the nominal velocity. Having two pairs of electrodes provides redundancy and a method for evaluating electrode noise levels.

The wings are made of epoxy-Fiberglas skin over polyethylene pipe. The foil sections are symmetrical in order to have the same lift and drag characteristics for both rising and falling motions. To minimize the voltages induced by temperature differences and to provide insulation from the changing temperature of the passing seawater, the electrodes are mounted within a 0.05-m-tall, 0.11-m-diameter cylindrical heat sink made of aluminum and PVC. The electrical conducting paths from the ocean to the electrodes extend through the polyethylene wing pipes, filled with agar and seawater, which serve as salt bridges. The resistance of each salt bridge is about 500 ohms, contributing insignificant thermal agitation noise of 3 nV rms in each of the two voltage channels over the 0.25 Hz bandwidth of the instrument. The lattices of agar suppress noise by preventing water of varying conductivity from flowing into and out of the salt bridges. The salt bridges extend the effective electrode positions away from the pressure case. Effective 1.6-m electrode separation, rather than the 0.22-m case diameter, provides voltage gain and allows a measurement of the motionally induced voltage uninfluenced by the presence of the instrument body (Sanford et al., 1978). Thus the Cartesian diver differs from the geomagnetic inductive profilers of Sanford et al. (1978) and those of Sanford et al. (1982) by measuring the electric field away from the body of the instrument instead of one perturbed by the presence of the instrument.

With the end assemblies in place and with an expendable ballast "drop weight" attached at the lower end, the instrument has a total length of about 3 m and a total mass of 90 kg. The drop weight is released by venting a vacuum chamber that holds on the weight. The venting is performed at a prearranged time by two redundant explosive devices activated by countdown timers. At this time, profiling is suspended and the instrument floats to the surface, extending two beacon radio antennae well out of the water.

The Cartesian diver is a quasi-Lagrangian drifter in the horizontal in the sense that it moves with the mean velocity of the water through which it profiles. Because it profiles through a sheared horizontal velocity field, it cannot follow the water at any depth in a horizontally Lagrangian manner, even if instantaneously perfectly Lagrangian in the horizontal. The Lagrangian character distinguishes the diver from moored instruments, past which there is continual advection, and from instruments that profile from a ship moving through the water. One appropriate use for the diver is study of the evolution of properties of water mass. Another is the determination of statistical differences between separate water masses, such as variations of the internal wave field on either side of a front (Duda and Cox, 1987).

The Cartesian diver, which is an untethered and unguided, axially spinning and horizontally drifting, self-profiling device that has the capability to alternately fall

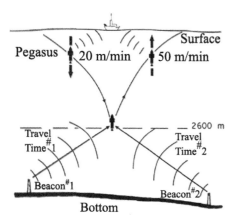

FIGURE 10.25 Schematic diagram showing acoustic tracking of free-falling current velocity profiler (Pegasus) relative to two expendable beacons. *(Source: Spain et al., 1981.)*

and rise freely through multiple cycles at almost equal down- and up-characteristic terminal velocity, is perhaps a marvel in the realm of electromechanical engineering design in the field of oceanography. The ability to make Lagrangian measurements is an important feature of this diver-probe. However, this very feature also renders it unsuitable to make repeated vertical profiling in a given location of interest.

10.2.6. Free-Falling, Acoustically Tracked Absolute Velocity Profiler (Pegasus)

As noted in Section 10.2.3, measurement of absolute ocean-current velocity profiles by the method of acoustic tracking of a free-falling pinger was first introduced by Rossby (1969). In this method, the differences in the time of arrival of an individual acoustic pulse at four or more hydrophones were measured. Hyperbolic tracking was then used to fix the position of the pinger, and the trajectory was differentiated to compute the current velocity profile. The hardware positioning was later reversed (Pochapsky and Malone, 1972) so that acoustic transponders were bottom-mounted and the hydrophone was contained within a free-falling float. The acoustic travel times from the sources to the receiver, as well as water pressure and temperature at the hydrophone, were telemetered to a support vessel. When travel times were measured, the spherical tracking method (as opposed to the hyperbolic tracking method used by Rossby, 1969) could be used to compute the trajectory of the moving hydrophone. In later instruments (Luyten and Swallow, 1976), the free-falling hydrophone was used as self-recording probe for monitoring travel times, water pressure, temperature, and salinity in the close vicinity of the probe.

Spain et al. (1981) reported a free-falling current velocity profiler, which is acoustically tracked relative to two expendable beacons (see Figure 10.25). In this scheme,

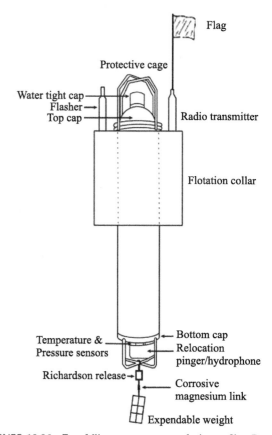

FIGURE 10.26 Free-falling ocean current velocity profiler (Pegasus). *(Source: Spain et al., 1981.)*

regular acoustic transmissions from the profiler enable its movement to be monitored from a ship. Every 8 seconds, water pressure (depth) and temperature, together with the two travel times from the beacons, are logged in the profiler Pegasus (see Figure 10.26). The profiler sinks at a speed that enables the fine structure in the vertical to be resolved accurately.

Pegasus is a compact and lightweight probe. The pressure case is an 80-cm length of 6061-T6 aluminum tubing with an outer diameter of 17.5 cm and wall thickness of 1.25 cm. A hemispherical top cap is permanently fixed to the tube, whereas the flat bottom cap is removable. The tube and end caps can nominally withstand a pressure equivalent to 3,000-m seawater depth. A syntactic foam floatation collar is fastened to the outside of the pressure case to provide 10 kg of positive buoyancy. Two pitched fins are fixed on opposite sides of the collar. As the profiler descends and ascends, the fins cause it to rotate, thereby removing any effects of an asymmetric weight distribution on the profiler's motion.

Standard recovery aids such as flag, radio, and flasher are inserted into the top of the collar. Cages are attached at either end of the probe to protect the sensors and watertight seals. A recovery line is tied to the top cage, and a Richardson hydrostatic weight release (Richardson and Schmitz, 1965) is secured to the lower cage. Pegasus weighs 45 kg in air. Four kilograms of expendable weight sink the profiler at a constant vertical velocity of 20 m/min and the expendable weight is dropped at a preselected depth by the Richardson release. A corrosive link is the backup weight release. The recovery aids and 5 cm of the floatation collar are visible when the instrument is floating at the surface.

In Pegasus, all the sensors are mounted on the bottom end cap. Whereas water temperature, pressure, and oxygen probes penetrate the end cap, the acoustic transducer is simply mounted to it. The acoustic transducer includes a diode transmit-receive switch that allows the transducer to serve as both the relocation pinger and the hydrophone. An 8-ms acoustic pulse is transmitted every 8 seconds at 10 kHz so that the movement of the profiler can be monitored from the ship. These transmissions can also be used for the interrogation pulses for transponders. During the remaining time, the transducer is used as a hydrophone across the 11- to 13-kHz band. The low-level input from the hydrophone is fed to a preamplifier, which also bandpass-filters the signal between 11 and 13 kHz. The signal is then sent to three signal detector circuits, each of which is centered on a specific beacon frequency. The detectors band-pass-filter the signal in a wide and narrow band. When a signal is received from a given beacon, the rms level in the relevant narrow band rises much more than in the corresponding wide band. If the ratio of the rms levels in the two bands exceeds a threshold for a duration of 4 ms, a beacon signal is considered to have been received. An output pulse is sent to the stop gate of the corresponding travel-time counter. Water temperature is measured by a thermistor with an accuracy of 0.03°C. A strain-gauge pressure transducer with a nominal accuracy of 0.5 percent makes pressure measurements. The oxygen sensor (optional) measures the dissolved oxygen in water. The acoustic travel times and the profiler-borne sensor outputs are logged in the memory bank of the profiler. At a typical descent speed of 20 m/min, the profiler will complete a 2,500-m cast in 3 hours.

The expendable beacons used with Pegasus are constructed from a 30-cm length of 7075-T6 aluminum tubing with an outer diameter of 12.5 cm. The beacons weigh 6 kg and are sealed by two 2-cm-thick flat end caps. A transducer with an acoustic output power of 91 dB re 1 μbar at 1 m is mounted to the top cap of the beacon. With this power, the beacon signals can be heard at distances in excess of 7 km. At a profiling site, two of the three operational frequencies (11.5, 12.0, and 12.5 kHz) are used; the third is available for a backup beacon. The beacons, sealed by evacuation, can nominally withstand a pressure equivalent to 5,500-m seawater depth.

The beacons are deployed while the ship is underway at a constant speed and heading so that a reasonable estimate

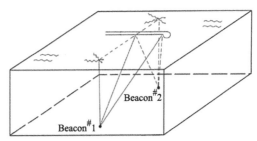

FIGURE 10.27 Beacon survey method. *(Source: Spain et al., 1981.)*

of their juxtaposition can be made. More accurate measurements of the separation distance and orientation are made in a survey of the beacons. The acoustic travel times from the beacons are continuously monitored on a graphic recorder. In the survey, the ship makes several passes at each beacon to try to pass directly overhead. During a pass, the ship is continuously maneuvered to minimize the travel time from the beacon to the ship. When the ship passes directly over each beacon, a minimum acoustic travel time is recorded. After the beacon depths have been ascertained, the ship steams steadily on a heading to intersect the baseline perpendicularly. When the baseline is crossed, the minimum travel times for that heading are measured (see Figure 10.27). Using Pythagorean geometry, the baseline length is determined from the minimum slant ranges and beacon depth measurements. A reciprocal heading is sailed to make a second crossing of the baseline. The ship's velocity on each heading is computed in both the geographical and beacon reference frames. Vector subtraction of these velocities removes the effect of the ocean currents and enables the baseline direction to be ascertained. The just-mentioned technique does not depend on electronic navigational aids.

When the beacons are deployed, they ping synchronously with each other, with Pegasus, and with a timing reference aboard the ship. However, if a profiling site is later revisited, any drift in the transmission time of the beacons must be determined before absolute travel times can be recorded. To determine this drift, the ship must travel directly over each beacon, and a minimum acoustic travel time must be measured relative to the timing reference, as indicated earlier. Because the travel time corresponding to the water depth is already known, the transmission time of each of the beacons relative to the reference can be computed. Pegasus's timing must also be similarly referenced so that the offset between the clocks in the beacons and the profiler can be measured.

Spain et al. (1981) reported the profile data collected by Pegasus. In their measurements, a constant descent rate was assumed for Pegasus. This approach correctly resolved the fine structure. When Pegasus was tracked, the measured water pressure determined its vertical position. Its horizontal position was fixed by the two slant ranges to the

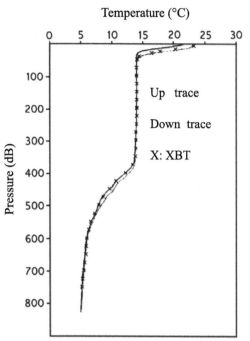

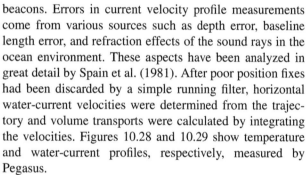

FIGURE 10.28 Water-temperature profiles measured by Pegasus. *(Source: Spain et al., 1981.)*

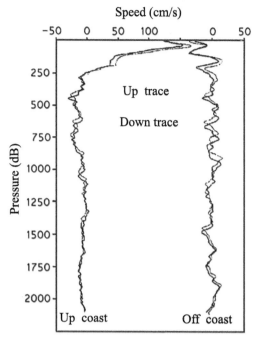

FIGURE 10.29 Water-current profiles measured by Pegasus. *(Source: Spain et al., 1981.)*

beacons. Errors in current velocity profile measurements come from various sources such as depth error, baseline length error, and refraction effects of the sound rays in the ocean environment. These aspects have been analyzed in great detail by Spain et al. (1981). After poor position fixes had been discarded by a simple running filter, horizontal water-current velocities were determined from the trajectory and volume transports were calculated by integrating the velocities. Figures 10.28 and 10.29 show temperature and water-current profiles, respectively, measured by Pegasus.

A glass-housed version of Pegasus was reported by Cole (1981). This modified version incorporates a number of advantages over earlier versions in aluminum housings. These advantages include operation to full ocean depth, integral floatation, no corrosion, optical synchronization where desirable, and large battery capacity. The enhanced battery power supply allows 100 deployments without opening; a very ample safety factor for a normal cruise is usually about 20 deployments. As in the case of the aluminum units, the glass units are also small, easy to handle, and easily shipped. Enhancement was also achieved in the bottom-mounted acoustic array and the shipboard data handling and storage unit. The electronics of Pegasus consist of two subsystems: (1) a data collection unit, which measures water pressure and temperature and up to three acoustic travel times; and (2) a microprocessor-controlled data storage package, which also controls the operation of the Pegasus instrument itself.

The acoustic tracking system used with the modified Pegasus instrument has the feature of either bottom transponders or beacons. Both methods have been used successfully. When a bottom array of free-running beacons is used, precise synchronization of the beacons and the Pegasus instrument clocks is required. Because of the glass housing, this synchronization can be carried out without opening the housing. The Pegasus instrument acoustic transmitter (in the transpond mode) and receiver (both modes) are based on a novel approach (Hayward and Ferris, 1979) that provides extremely low jitter, thereby giving good resolution and accuracy of position. It is of great value to be able to preview the data immediately on return of the instrument without opening it. A simple software package has been developed to present the data in graphic form so that the success or failure of a particular deployment can be assessed immediately before moving to the next station.

10.2.7. Freely Falling, Acoustically Self-Positioning Dropsonde (White Horse)

Luyten et al. (1982) have reported a dropsonde system known as *White Horse*, which was designed and constructed at the Woods Hole Oceanographic Institution by W. J. Schmitz, Jr., and R. Koehler and first deployed in 1972 (Gould et al., 1974) to determine vertical profiles of horizontal velocity. This device is a freely falling instrument package that interrogates bottom-moored acoustic transponders (see Figure 10.30) and records internally the

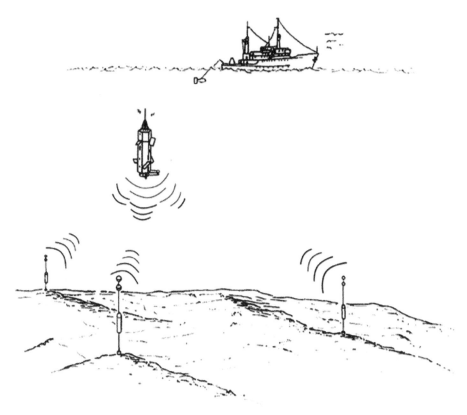

FIGURE 10.30 Diagram showing the freely moving instrument package known as White Horse, which interrogates three bottom-moored acoustic transponders. *(Source: Luyten et al., 1982.)*

time elapsed between the interrogation and reply pulses. The interrogation cycle is initiated every 20 s. The White Horse system also includes a Neil Brown microprofiling CTD operated at 1 Hz. The instrument is ballasted to fall and rise at approximately 1 m/s. The acoustic navigation package was designed around an underwater navigation package marketed by EG&G Sea Link using its transponding releases. It is a four-channel system in which the instrument transmits a 20-ms pulse at 11 kHz and the bottom-moored transponders reply at 9.0, 9.5, and 10.5 kHz. The fourth channel provides a spare in case one of the other channels becomes unusable. (Standard use of four transponders provides redundant navigational data.) When the instrument detects a reply pulse in one of the receivers, the appropriate time-interval counter is stopped. The time intervals are stored in a shift register, which is transferred onto the Sea Data cassette when the next interrogation pulse is transmitted. The CTD samples data (16 bits) once in a second. At a fall or rise rate of 1 m/s, this corresponds to a nominal 1-m vertical resolution for the CTD. Data are logged on the cassette every second.

The dropsonde is ballasted to descend to the ocean floor, where it can be released by any of three independent release mechanisms: an acoustic release, a preset timed release controlled by the internal master clock, or a corrodible link. The former two methods have been found to be very reliable.

The White Horse consists of four instrument packages: (1) the navigator-controller, (2) CTD, (3) an acoustic release, and (4) a data recorder. Each unit is contained in a standard 5-in. ID anodized aluminum deep-sea pressure case. The cases are cable-connected and placed in the vehicle, constructed of rigid plastic with syntactic foam for buoyancy. There is sufficient reserve buoyancy for the instrument to remain slightly buoyant if all the pressure cases become flooded. The assembled vehicle (see Figure 10.31) has a hexagonal cross-section of 0.7 m², an

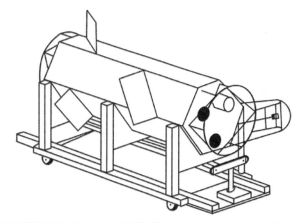

FIGURE 10.31 Drawing of White Horse, depicting its hexagonal cross-section of 0.7 m² and an overall length of 1.75 m. *(Source: Luyten et al., 1982.)*

overall length of 1.75 m, and a mass of 475 kg. To increase the stability and avoid erratic skating motions, fins have been added to make the instrument rotate clockwise as it falls at approximately one revolution per instrument length.

To record a profile at a given site, three acoustic transponders are deployed in a triangular array on the seafloor and surveyed before the dropsonde probe is deployed. This generally requires 4–5 hours if a towed fish is used for the survey. The dimensions of the triangular array must be comparable with the ocean depth. Although a longer baseline array reduces the navigation errors somewhat, the S/N ratio is also reduced due to longer acoustic paths, and consequently more data are lost. The transponders are tethered approximately 20 m above the ocean floor, resulting in some mooring motion, normally 10 to 15 cm, but there is less chance of the refracted acoustic rays intersecting the ocean floor.

The relative coordinates of the transponder array are determined by interrogating the array from the sea surface. The White Horse system is used for such interrogations with a transducer suspended 15 m below the freely drifting ship. To reduce errors in the determination of the coordinates of the transponder array, a pattern of 13 interrogation sites is used. These sites are equally spaced around the perimeter of a triangle, which is twice the size of the transponder net. An additional site is chosen in the center of the net. The orientation of the transponder net relative to North can be determined by using absolute navigation fixes, together with simultaneous acoustic ranges. The dropsonde is deployed from the ship's crane near the center of the transponder net. The dropsonde descends to the seabed and is released after a few minutes on the bottom, either acoustically or with a timed release.

The slant range between the dropsonde and the ship is monitored using a deck clock, which is synchronized with the instrument to drive a precision graphic recorder. Because the ascent rate of the dropsonde is nearly constant and the time of anchor release from the bottom is known, the slant range can be converted to an approximate horizontal range from the ship to the instrument. With this information it is possible to be near the instrument when it surfaces (typically 200 m). A submersible radio and light are used as recovery aids and are particularly useful if acoustic contact with the instrument is lost.

As in the case of any freely falling/rising probe, the profile of local horizontal current velocity in the ocean is estimated from the horizontal displacement of the White Horse relative to the bottom-moored transponders. Systematic and random errors affect the accuracy of the current velocity estimates. The errors are those associated with the acoustic navigation of the instrument and those involved with the interpretation of the horizontal velocity of the instrument as oceanic motion. The latter are generally hydrodynamic in origin. The individual sources of error in position and velocity determination are given in Luyten et al. (1982). Despite various sources of error, intercomparison measurements against a vector-averaging current meter (VACM) have indicated that the average difference between the White Horse and the VACM measurements are 0.5–0.7 cm/s, with standard deviations (2.0, 1.6) cm/s for the (east, north) components. Although no direct simultaneous comparisons have been made between the White Horse and other similar profiling instruments, Rossby and Sanford (1976) reported comparisons between the EMVP and an acoustic dropsonde similar to the White Horse. They concluded that the observed differences between the EMVP and the acoustic dropsonde are not significantly different from the expected errors in each technique. Evans and Leaman (1978) reported further comparisons between the White Horse and the University of Miami PCM. The differences between the two sets of observations were reported to be within the margins of errors appropriate for the two systems. Detailed assessments of the observed errors in the vertical profile of horizontal currents have indicated the importance of using the correct local sound-velocity profile (a climatological average may suffice in some circumstances) for the current profile and survey calculations. The key to the success of the dropsonde technique is considered to be matching the accuracy of the navigational technique to the particular environment. For the pulsed acoustic navigation system, the dominant error arises from uncertainties in determining the leading edge of the acoustic pulse. As with any dropsonde technique used in the measurement of the vertical profile of horizontal currents, the accuracy is related to the vertical resolution by an uncertainty principle $\Delta u \Delta z = w \Delta x$, where Δu and Δz represent the average horizontal velocity error (due to position errors) and sampling depth interval, respectively. In this expression, w and Δx represent the vertical motion speed (i.e., fall rate) of the profiler and the error in the horizontal position of the profiler, respectively. It has been found that using the system in the White Horse, current-velocity profiles can be determined with an accuracy of ± 4 cm/s over 25-m depth intervals.

10.2.8. Freely Rising Acoustically Tracked Expendable Probes (Popup)

A limitation of the acoustic dropsonde methods discussed in the previous sections was that the spatial separations of the transponder array needed to be comparable with the ocean depth. Accurate determination of the coordinates of the transponders was also difficult. Furthermore, even the bottom-moored transponders needed to be tethered approximately 20 m above the ocean floor in order to avoid the refracted acoustic rays intersecting the ocean floor

(Luyten et al., 1982). This resulted in some motion of the transponders, causing errors in the estimation of the vertical profiles of the horizontal current flow velocities.

These limitations have been circumvented in a design reported by Voorhis and Bradley (1984). This device is a self-contained, bottom-mounted instrument for measuring vertical profiles of horizontal current in the deep ocean over long time periods (up to a year). It employs an interferometric technique to track small, expendable, buoyant probes containing CW acoustic beacons, which are released on a preset schedule. A carrying capacity of 60 probes is feasible. Tracking data are stored in the bottom instrument and retrieved after recovery. The overall profiler system consists of an open pyramidal frame with a triangular base (5 m on each side) outfitted with a canister of simple, expendable, buoyant acoustic probes and three tracking hydrophones located at the three corners of the pyramid base (Figure 10.32).

The pyramidal frame is constructed with aluminum tubing that can be unbolted easily to permit air shipment and field assembly. Beneath the main frame is a releasable anchor frame that supports the entire unit on the seafloor and remains on the bottom when the unit is recovered. Recovery buoyancy is provided by nine glass floatation spheres attached to the sides of the main frame. The complete instrument, including anchor frame (250 kg), weighs about 570 kg in air and 114 kg in water. Without the anchor frame, the instrument has a positive buoyancy of ∼120 kg.

The three tracking hydrophones are mounted rigidly on the outer corners of the main frame base, allowing an undisturbed acoustic "view" of the ascending probe. The receiving hydrophones are identical to the probe transmitter

element and are a piezoelectric crystal type contained in oil-filled boots. The three bottom hydrophones receive a continuous incoming signal from the rising monochromatic probe, which transmits a 15-kHz CW tone with an acoustic wavelength of 10 cm. Each hydrophone has a preamplifier with band-pass filtering at 15±2 kHz. The hydrophones are directional and reject acoustic radiation from below.

The primary electronics, which include a complete recovery command system, tracking receivers, data logger, internal magnetic compass, and instrument tilt sensor, are contained in a single pressure case suspended vertically from the apex of the main frame. Mounted on top of the main pressure case are the instrument recovery beacon and a command receiver/transducer. Thus, all components associated with recovery are part of this single pressure housing. Additionally, the release mechanism is a shared double release that can be activated by a command, either to the main command receiver/transducer or, as a backup, to a completely independent release in a second tube that is mounted alongside the main electronic housing.

The distance from the ascending probe to each hydrophone is determined by measuring and keeping track of the phase difference between a fixed, precision frequency source within the bottom unit and each incoming signal from the ascending probe, the frequency of which has been Doppler downshifted by the probe's upward motion. Signal processing required to achieve better precision in distance measurement is accomplished as follows: The electrical signal from each hydrophone goes to a superheterodyne receiver with a noise bandwidth of approximately 10 Hz. In this receiver, the hydrophone signal is multiplied by a constant 15.1515 kHz signal, giving sum and difference frequencies, termed by Voorhis and Bradley (1984) as *interferometric technique.* A narrow-band phase-lock loop tracks the difference frequency of ∼151 Hz and multiplies it by 512. This ∼77 kHz signal goes to a free running counter, the contents of which are recorded at each sampling interval. The least-phase count is therefore equivalent to an acoustic phase resolution of $2\pi/512 = 1.2 \times 10^{-2}$ rad at each hydrophone. Thus, the signal phase is measured with a resolution of 1/512 cycle so that in the absence of significant noise, the slant range to the probe is measured with a resolution of $\lambda/512$. This scheme made it possible to have a much shorter baseline compared to the pulse system, where the baseline needs to be comparable to the ocean depth. The advantage of using phase-lock loops is that it permits the tracker to "flywheel" over occasional dropouts in the acoustic signal if they occur.

The launch of Popup is shown in Figure 10.33. A schematic of the expendable acoustic probe assembly (Hoyt, 1982) is shown in Figure 10.34. The expendable probe is constructed from thick-walled cylinders (7.5 cm outer diameter, 2.5 cm inner diameter) of high-density

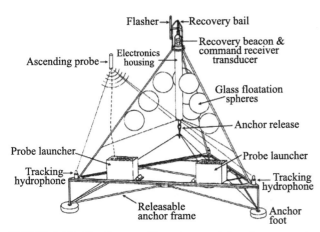

FIGURE 10.32 The Popup profiler system consisting of an open pyramidal frame with a triangular base (5 m on each side) outfitted with a canister of simple expendable buoyant acoustic probes and three tracking hydrophones located at the three corners of the pyramid base. *(Source: Voorhis and Bradley, 1984, ©American Meteorological Society, reprinted with permission.)*

FIGURE 10.33 Launch of Popup. *(Source: Voorhis and Bradley, 1984, ©American Meteorological Society, reprinted with permission.)*

syntactic foam (0.6 gm/cm³) enclosed by syntactic foam end caps. The overall height of a probe is 47 cm, and it ascends to the sea surface in a vertical orientation. Within the central cavity are the battery pack (four alkaline C cells) and electronics. The latter consist of a 15-kHz crystal-controlled oscillator, amplifier, and matching transformer to drive the acoustic transducer, which is mounted in an oil-filled boot on the bottom of the probe. The drive circuit

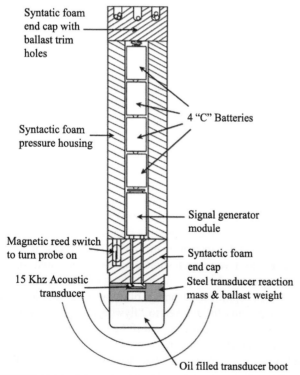

FIGURE 10.34 Schematic of the expendable acoustic probe assembly. *(Source: Voorhis and Bradley, 1984, ©American Meteorological Society, reprinted with permission.)*

automatically shuts down after 5 or 10 hours to prevent interference with subsequent probes. The transducer is a piezoelectric crystal type tuned to 15 kHz. A heavy metal disk sandwiched between the lower end cap and the acoustic transducer serves the following purposes (Voorhis and Bradley, 1984):

- Increasing the probe weight to near-neutral buoyancy in seawater
- Lowering the center of gravity and increasing vertical stability
- Improving the acoustic radiation pattern (i.e., decreasing the ratio of upward to downward radiated power); this minimizes phase interference from surface reflection as the probe approaches the sea surface

Syntactic foam construction suffers from two limitations: (1) relatively low compressive strength, which must be compensated for by the thick wall construction of the probes, and (2) buoyancy loss due to water absorption over long periods at high pressures.

The probe begins its ascent at the preset time after a warm-up period of 30 min, during which the receivers in the main housing lock on and the probe frequency stabilizes. A reasonable ascent rate for the probes is 25 cm/s so that 5–6 hours are required to reach the surface.

The bearing of the rising probe with respect to the bottom unit is calculated from the difference in slant ranges to the three hydrophones. The trajectory of the ascending probe is measured and recorded within the bottom unit. The bottom unit is recovered at the end of the deployment mission and the probe trajectories analyzed. Successive determinations of positions of the probe are used to estimate the oceanic current field. Figure 10.35 shows east and north components of trajectory and horizontal velocity of the probe released on May 28, 1983, at 39°00′N, 69°00′W. Probe ascent speed was 35 cm/s and water depth was 3,110 m. Data were sampled every 2 s, smoothed by an 80-s triangular window and plotted every 40 s (14-m depth intervals).

It may be noted that no field measurements are free from errors. The probe just described also suffers from such limitations. Tracking error is one such error. This arises from random variations in acoustic phase and hence in the measured travel times to the probe. Although the probe is designed for downward radiation of acoustic signal, some degree of upward radiation is also present. Such radiations from the ascending probe get reflected from the sea surface and are added to the direct downward radiated signal, producing a phase uncertainty at the hydrophones. This uncertainty is small when the beacon is deep, because of the large distance to the surface and back (spherical spreading). The uncertainty becomes significant, however, as the probe nears the surface, where it becomes the major source of random error in the Popup tracking system. The phase

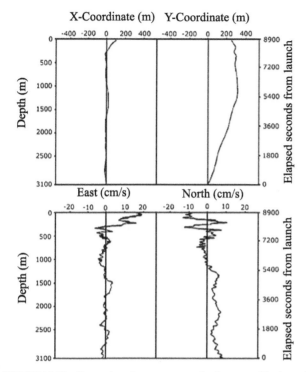

FIGURE 10.35 East and north components of trajectory and horizontal velocity of the probe released on May 28, 1983, at 39°00'N, 69°00'W. *(Source: Voorhis and Bradley, 1984, ©American Meteorological Society, reprinted with permission.)*

uncertainty depends on the acoustic directivity of the probe (upward to downward radiated power). Voorhis and Bradley (1984) found that digital processing of the acoustic phase data with a triangular weighting filter over 80 s reduces the near-surface probe position error to about 70 cm and horizontal current velocity errors to about 2 cm/s for a 14-m vertical resolution. They also noticed occasional large phase fluctuations that cause probe tracking to jump one or more phase "lanes" (i.e., change by multiples of 2π radians), particularly when the probe was near (or on) the sea surface. However, because the phase jumps must be exactly an integer number of cycles, it was possible to add or subtract the missing cycles to generate a continuous trajectory. This strategy can be successful only if the phase lock loops are out of lock a small fraction of the time.

Apart from the previously mentioned problem, another issue that may deteriorate the performance of the tracking system is the changes in the probe signal frequency during its ascent, primarily due to thermal drift of the resonant frequency of the crystal oscillator in the beacon as it ascends into warmer layers of the upper ocean. Yet another aspect that might deteriorate the measurement accuracy is the uncertainty in converting the travel time to slant range. The conversion requires knowledge of the nonfluctuating sound velocity along the acoustic ray paths, including the effects of refraction, from probe to hydrophones. However,

in the absence of sound-velocity measurements by the probe, errors arising from poor knowledge of sound velocity profile can be ameliorated by acquiring it independently, either from hydrographic measurements at the site or from historical data appropriate to the site.

An improved design that does not require an array of bottom-moored transponders is a phased-array hydrophone and an expendable, free-falling, and depth-measuring pinger, reported by Riggs et al. (1989). In this method the pinger's slant range, depth, and azimuth relative to a single hydrophone deployed below the sea surface from a vessel are determined at regular time intervals throughout the descent. A crystal-controlled clock in the shipboard electronics is synchronized with a similar clock in the probe prior to the deployment. The probe's range is then obtained from a synchronized pulse, which is transmitted by the probe. As the slant range of the probe to the hydrophone increases, the time delay of the incoming signals relative to time of synchronization is measured and the range calculated. The probe depth relative to the sea surface is determined by a pressure transducer, which is housed in the base of the probe, and the depth data are transmitted to the surface. The angle of arrival of the incoming signal is measured by the hydrophone's ultra-short-baseline phased array. The measurement of changes in the phase of the incoming signals as they pass over the elements of the array allows for an accurate determination of the azimuth. Once the three-dimensional trajectory of the probe is determined, the speed and direction of horizontal flows at various depths are estimated from the predicted dynamic response of the probe relating its horizontal displacement and the oceanic current velocity. Comparative statistics indicated (Riggs et al., 1989) that such a probe can measure speed and direction of horizontal flows accurate to better than ±4 cm/s and ±10°, respectively, for current speeds of 30 cm/s or greater.

10.3. TECHNOLOGIES USED FOR VERTICAL PROFILE MEASUREMENTS OF OCEANIC CURRENT SHEAR AND FINE STRUCTURE

It was found to be difficult to measure the detailed velocity structure of important western boundary currents, such as the Kuroshio and the Gulf Stream, by means of conventional moored current meters, primarily because of the motion of mooring lines due to strong currents and the deployment cost of a mooring system over a wide area. In view of such difficulties, freely dropped ocean current profilers were developed (Halkin and Rossby, 1985; Leaman et al., 1987) for the study of vertical fine-structure regimes ($1m < \lambda < 100m$). Subsequent research in this area gave rise to the development of more advanced technologies. These are addressed in the following sections.

10.3.1. Free-Fall Shear Profiler (Yvette)

The first requirement for an instrument intended to resolve small-scale ocean-current velocity and shear is that its own motion be quiet so that it can fall freely through the water column rather than be cable-lowered from a ship. Second, both the water-current velocity and water-density sensors must have fine resolution, because it is necessary to detect differences in these properties over small vertical scales. Third, the velocity and density measurements should be made simultaneously, at the same point in space. To fulfill these requirements, Evans et al. (1979) reported the design of a free-fall instrument, known as Yvette, which simultaneously resolves the local current velocity field relative to the instrument as well as the density field. Yvette was designed primarily to investigate the relations between shear and density gradient on small scales.

The Yvette comprises a 4-m, ~17-cm circular long tube (which provides the basic framework and flotation), the lower end of which houses a current meter and sensors for measurements of water conductivity, temperature, and pressure (CTD). The electronics main frame and batteries are all located at the lower end to give the instrument vertical stability against shear-induced tilting (see Figure 10.36). A protective cage surrounds the sensors and supports the pressure-actuated ballast-release systems. At the upper end of the tube is a 10-kHz pinger with 4-s repetition rate, flasher, and radio as well as all aids required for relocation and recovery. The current-sensing device used in this profiler is an ATT difference current meter developed by Trygve Gytre at the Christian Michelsen Institute of Bergen, Norway. In the present application, the CM sensor design sought a compromise between the need to obtain a good S/N ratio and the need to keep the acoustic transducers as small as possible, to avoid disturbance to the current flow, and their separation large enough to obtain a high acoustic pulse travel-time difference value. The compromise was to choose the probe

distance about twice the diameter of the pressure housing. With this arrangement and careful electronic design, the current meter is capable of measuring correctly both static and dynamic currents from ~1 mm/s to ~1 m/s.

A master clock in the profiler controls the entire operation at a basic sampling rate of 2.5 Hz. A problem that seems to characterize long tubes such as this is a tendency for them to behave like pendulums. The Yvette exhibits a small-amplitude oscillation with a period of ~5 s. By working backward from a worst-case current-velocity measurement, where the 5-s oscillation had an amplitude of 0.25 cm/s, the amplitude of the oscillation was estimated to be less than 0.02°. Such oscillation was found to introduce a resonant peak in the shear data. Vortex-shedding surfaces of the profiler were found to be responsible in part for exciting these oscillations and could be attenuated by reducing such surfaces. The observed resonant peak in the shear data could be removed using a digital notch filter. The Yvette rotates slowly while sinking, and its body motion effectively high-pass-filters the velocity profile to scales smaller than the rotation scale.

Yvette was subjected to comparison studies by attaching a pinger to it, and a number of profiles were made in the U.S. Navy underwater tracking range at St Croix, U.S. Virgin Islands. Figure 10.37 shows a comparison of two components of velocity obtained by differentiating the tracking data (tracked) and reconstructing the profile from Yvette (computed). The overall agreement on scales up to about 100 m is good, with the difference less than ± 3 cm/s. Figure 10.38 shows two components of vertical shear from St Croix tracking range data. It has been found that Yvette can resolve the high wave-number end of the internal wave band and the transition region to smaller three-dimensionally turbulent scales. The instrument has been deployed for study of the relation between mesoscale features and small-scale mixing.

10.3.2. Acoustically Tracked Free-Fall Current Velocity and CTD Profiler (TOPS)

As mentioned earlier, the dropsonde proved to be a convenient tool for vertical profiling of horizontal ocean currents. However, in situations in which one's interest is in the study of vertical fine-structure regimes ($1m < \lambda < 100m$), the translation of the dropsonde itself must be adequately accounted for because at long vertical wavelengths, the dropsonde nearly follows the fluid motion and the measured relative velocity is small. In addition, the profiler body often has a natural oscillation frequency, which, at reasonable fall rates, induces spurious current velocity measurements in the fine-structure region.

Thus, to understand the relative velocity measurements, it is necessary to model the response of the dropsonde to the imposed shear flow. The actual ocean current velocity is

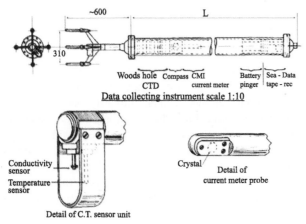

FIGURE 10.36 Diagram of Yvette. *(Source: Evans et al., 1979.)*

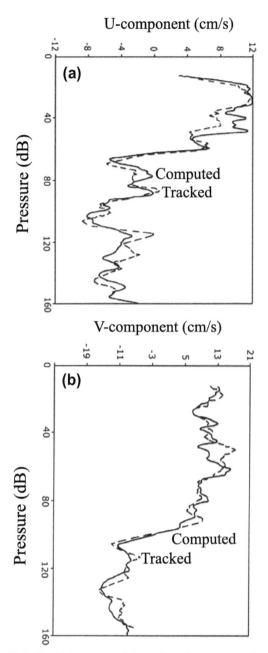

FIGURE 10.37 (a) East-west and (b) north-south components of current measured from St Croix, U.S. Virgin Islands, by the inversion routine (computed) and by acoustic tracking with bottom-mounted hydrophones. An arbitrary zero offset has been added to the computed values because Yvette makes only a relative measurement. *(Source: Evans et al., 1979.)*

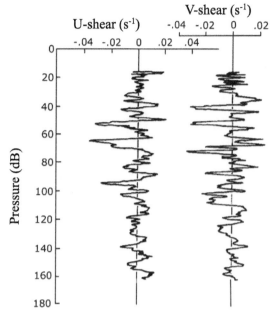

FIGURE 10.38 Two components of vertical shear from St. Croix tracking range data. *(Source: Evans et al., 1979.)*

then extracted by a transfer function acting on the measured velocity. Limitations on this velocity computation occur at long wavelengths because of sensor drift and at short wavelengths because of sensor size and sensitivity. Techniques that measure ocean-current velocity by tracking a freely falling body's trajectory relative to some fixed coordinate system are the natural complement to onboard CM measurements. Deriving ocean-current velocity from

the trajectory of a freely falling body requires that the body be in equilibrium with the current flow in the vicinity. Thus, this technique well resolves the long vertical wavelength structure but smoothes the small-scale fluctuations. However, simultaneously tracking a dropsonde with an onboard current meter yields a profiler capable of resolving vertical scales from the fine-structure range up to the total water depth. Hayes et al. (1984) reported one such instrument, named TOPS (Figure 10.39). In this case, relative velocity measurements are provided by an onboard two-axis acoustic current meter, and the trajectory of the free-fall profiler is measured by acoustically tracking it relative to an array of bottom-moored transponders. Whereas estimation of vertical profiles of horizontal flow velocity using a simple acoustic dropsonde depends entirely on its trajectory, the acoustic CM incorporated into the free-fall instrument, TOPS, measures ocean-current velocity relative to the profiler. In other words, the CM measures the relative velocity between the dropsonde and the surrounding water masses at various depths. Motions of the profiler are monitored with a two-axis accelerometer and flux-gate compass. TOPS' profiling capability extends throughout the full water column (6,000-db pressure limitation). The relative velocity measured by the CM provides additional data to the estimates of the vertical profiles of horizontal flow velocities. From the trajectory described by the dropsonde as well as the relative velocity measured by the CM, the absolute flow velocities at various depths are inferred using an appropriate hydrodynamic model, resolving velocity fluctuations that have vertical wavelengths as small as 0.2 m. Additional

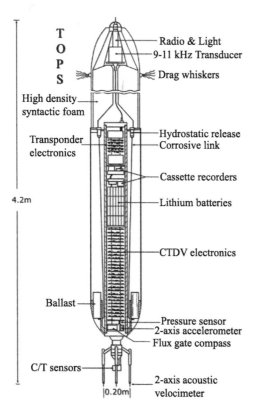

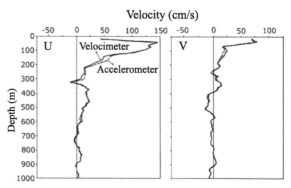

FIGURE 10.40 Ocean-current velocity profiles obtained by inverting the TOPS model using relative velocity measured by the acoustic current meter (solid) and those estimated from the measured acceleration (dashed). In the latter case, the low-frequency form of the transfer function was used. *(Source: Hayes et al., 1984, ©American Meteorological Society, reprinted with permission.)*

FIGURE 10.39 Schematic drawing of TOPS. The lower portion contains the onboard sensors: two-axis acoustic current meter, flux-gate compass, CTD and two-axis accelerometer. The upper portion contains the acoustic tracking system. *(Source: Hayes et al., 1984, ©American Meteorological Society, reprinted with permission.)*

parameters measured by TOPS are water conductivity and temperature.

The acoustic tracking system associated with TOPS is similar to that described by Luyten et al. (1982). In operation, an array of three bottom-moored acoustic transponders is deployed. The relative location of the array elements is determined by a ship survey. During profiling, TOPS is dropped near the center of the net. If for some depths or at some locations, replies from only two bottom transponders are available, then the trajectory and velocity of the TOPS are estimated from these replies, combined with the measured pressure from the CTD. Because TOPS usually rotates slowly (1 revolution per 100 m) about its axis of symmetry, the compass goes through several oscillations. Plotting the x and y components against each other should describe a circle. If necessary, the components are adjusted in offset and amplitude to yield a circle centered on zero.

As TOPS falls through the ocean, it is subjected to a variety of forces that result from the relative velocity between the vehicle and the water. Consequently, interpretation of the TOPS onboard velocity measurements requires a model that describes the response of the profiler to the ocean shear flow. If motions of the vehicle can be

accurately predicted given only the relative water velocity at the nose (which is measured), it is possible to infer the actual water velocity by combining vehicle and relative velocities. In practice, there is always an arbitrary constant velocity that cannot be obtained solely from the relative velocity measurement. However, this barotropic component is measured by the acoustically tracked velocity profile.

Models of the response of freely falling long cylindrical profilers (similar to TOPS) to shear flows have been discussed in several studies, but none of them account adequately for the distribution of forces along the body or the inclination of the body relative to the vertical. To account for these effects, Hayes et al. (1984) developed a two-dimensional model for TOPS. Figure 10.40 shows ocean-current velocity profiles obtained by inverting the TOPS model using relative velocity measured by the acoustic current meter and those estimated from the measured acceleration. Measurements have shown that TOPS and the dynamic model that describes its motion are efficient tools that permit study of the vertical structure of oceanic velocities over vertical wavelengths from 0.20 m to the total water depth. This spectrum encompasses the mesoscale and fine-structure regimes. The principal components are the combination of acoustic tracking and onboard CM measurements (both of which had been used independently in several previous instruments) and a linear model that permits reconstruction of current velocities through the fine structure region where vehicle motions potentially contaminate ocean-current velocity estimates. In addition, the full-depth capabilities of the instrument permit study of fine-structure velocities in the weakly stratified abyssal ocean. These measurements have provided new insights into the relationships between large and small vertical-scale ocean currents and the underlying physical processes.

10.4. TECHNOLOGIES USED FOR VERTICAL PROFILE MEASUREMENTS OF OCEANIC CURRENT SHEAR AND MICROSTRUCTURE

Several types of free-fall instrument packages that provide vertical profiles of horizontal currents have been addressed in the previous sections of this chapter. Apart from such measurements, oceanographers have shown keen interest in examining the vertical microstructure (i.e., wavelengths λ, shorter than 1 m), where the major concern is shear over vertical length scales that are much less than the instrument length. At such scales, only the high-frequency vibrations of the profiler affect the measurements; the lateral translation of the body can be neglected. Technologies used for vertical profile measurements of oceanic current microstructure are addressed in the following sections.

10.4.1. Towed Acoustic Transducer

Probably the first experiment conducted to examine the feasibility of measuring ocean-current shear is the one reported by Thomas (1971) using acoustic backscattering. Measurements of reverberation statistics carried out by Thomas with a towed transducer indicated that volume backscattering can be used to measure ocean-current shears over the interval occupied by deep scattering layers—normally from near the surface down to depths of 400 to 1,000 m. A series of experiments performed by Thomas employed an acoustic transducer towed behind a ship and with a searchlight beam pattern, as indicated by the broken line in Figure 10.41. An impeller speedometer attached to the towed body gave a continuous record of the transducer speed relative to the water with an accuracy of about 0.1 knot. Tone pulses of temporal width $\tau = 0.5-1$ s and 9.5-kHz frequency were transmitted, and the backscattered acoustic energy was received by the same transducer. Velocity of sound profile and corresponding ray plots of acoustic propagation indicated downward refraction of acoustic energy through the deep scattering layers (see Figure 10.42). The rays correspond to energy leaving the transducer horizontally and at $\pm 5°$. Downward refraction eliminates surface backscattering at distances beyond 0.4 km.

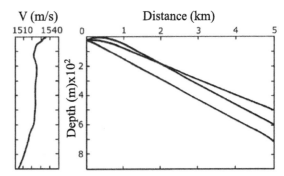

FIGURE 10.42 Examples of velocity of sound profile (left) and ray plots showing acoustic energy refracted downward through the deep scattering layers (right). *(Source: Thomas, 1971.)*

In several of the experiments, conditions were such that the reverberation was entirely due to volume scatterers beyond some relatively short distance where the surface was no longer insonified. Most of the volume backscattering in deep-water areas is from relatively small organisms that are presumably carried along by the water current. If water current is primarily in a horizontal direction, one might expect the current shear structure to remain much the same over relatively large areas in the deep ocean. Downward refraction of the acoustic energy means that the received volume backscattering corresponding to increasing time steps (i.e., from increasing radial distances) arrive from increasing scatterer depths. The relative velocity between the acoustic transducer and the moving scatterers give rise to Doppler shift in the received acoustic energy. Thus, the frequency spectrum of reverberation at any time after transmission, corresponding to some distance interval, gives the effective radial velocity distribution of the scatterers over the water volume insonified at that distance and depth.

Figure 10.43 indicates that the average frequency or Doppler shift of the reverberation changes with distance due to changing radial velocity components of the scatterers. Thus, spectra of successive samples of reverberation of the form shown in Figure 10.43 enable estimation of the horizontal current velocity vectors as a function of depth. All that is required are the mean reverberation Dopplers (ensemble averages of from 30 to 50 samples of spectra in order to obtain statistically significant estimates), the transducer headings (preferably 90° apart), and the calculated angles from the horizontal at which acoustic energy reaches the scatterers at different depths. During periods of vertical migration of scatterers, which usually occur near sunset and sunrise, some errors could exist in measurements, depending on the rate of migration and the percentage of the relevant scatterers migrating.

Although the transducer used in Thomas's (1971) experiments was not optimum for current shear measurements, the results obtained were sufficient to demonstrate

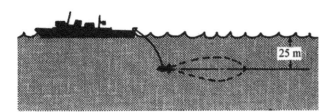

FIGURE 10.41 Schematic diagram depicting the use of an acoustic transducer towed behind a ship for measurement of the vertical profile of ocean-current shear. *(Source: Thomas, 1971.)*

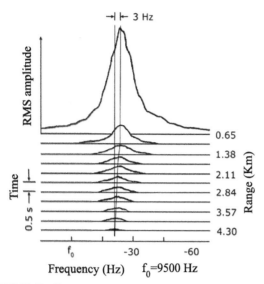

FIGURE 10.43 Root mean power spectra of reverberation at half-second time intervals corresponding to successive scattering volumes separated by 380 m. The average frequency or Doppler shift of the reverberation changes with distance due to changing radial velocity components of the scatterers. *(Source: Thomas, 1971.)*

the effect of horizontal current shears. Better measurements should be possible using transducers with narrower stabilized vertical beams and higher acoustic frequencies whereby pulse lengths could be shortened.

The technique Thomas (1971) used is a valuable tool for both continuous monitoring and sampling of ocean-shear currents. The transducer could be mounted on the seafloor or, with suitable stabilization, on a buoy, in a towed body, or on a ship's hull. Perhaps the greatest attraction is that, combined with velocity of sound measurements using expendable bathythermographic equipment, this acoustic technique could provide relatively quick sampling of near-surface ocean-shear currents without lowering complex equipment through the water column.

10.4.2. Free-Fall Probe (PROTAS)

Knowledge of small-scale flow-velocity structure in the ocean is meager in relation to what is known of the density structure. Techniques for the observation of temperature and salinity (and therefore density) in the ocean are well established and have been widely exploited. For instance, Woods (1969a) described a probe measuring the temperature difference over a vertical spatial interval of 10 cm, which falls down a vertical wire, and observations on vertical scales of less than 10 cm have been reported by several other researchers. This method overcomes the difficulties associated with vertical and horizontal ship motions transmitted to conventional TSD systems by the supporting cable.

Often, the water-flow velocity gradient in the ocean is concentrated into the regions of intense water-density

gradients. For closely examining such velocity gradients, it is helpful to have an instrument that is capable of resolving velocity differences as small as 1 mm/s over vertical separations down to ~30 cm. Because of the generally small velocity changes observed in the deep oceans, even weak coupling to ship motions would be unacceptable; therefore, a completely free-fall system was considered imperative. Such a free-fall system makes only relative velocity measurements unless it can be provided with a reference frame by, for example, an acoustic navigation system—a requirement that would severely restrict observations. A technique for establishing a reference frame has been described by Plaisted and Richardson (1970), who used a conventional current meter lowered at a slow uniform rate from a free-floating spar buoy. To obtain an absolute velocity profile, the CM velocities were combined with the velocity of the spar, which was determined by repeated plotting on Hifix.

An ingenious method of measuring velocity shear developed by Drever and Sanford (1970) made use of the electrical currents that flow in response to water movement in the Earth's magnetic field. The method is not absolute, however, but determines the horizontal velocity, with a vertical resolution of 8 m, relative to its value averaged over the entire depth.

A further difficulty arises from the small magnitude of the velocity differences to be detected. Even in a relatively high-shear region such as the seasonal thermocline, Woods (1969b) found from observations of dye tracers that the largest shear corresponds to a change of velocity of ~1 cm/s in an interval of 10 cm. His results suggested that useful observations should be able to resolve velocity to ~1 mm/s at speeds of 1 cm/s or less, a requirement that cannot be met by any of the conventional instrumentation used in oceanography.

Simpson (1972) reported a probe for measuring small, vertical scale variations in the horizontal velocity field. This probe, named Probe Recording Ocean Temperature and Shear (PROTAS), designed and constructed at the National Institute of Oceanography in the United Kingdom, was intended to overcome the problem of measuring small flow-velocity changes by the use of a neutrally buoyant vane. In this system, the vane is attached by a low-friction universal joint to a framework that protrudes from the main body of the probe (see Figure 10.44). As well as being neutrally buoyant, the vane is ballasted so that there is no static moment about its center of gravity. Consequently, the attitude of the vane is controlled entirely by the dynamic forces associated with the flow past it. The vane responds to changes in the horizontal velocity in a distance approximately equal to its own length of 30 cm.

In contrast, the body of the probe, which consists of a long circular cylinder, is subject to a large static righting moment of ~6 N m per degree of deflection from the

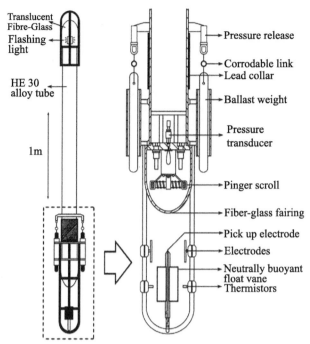

FIGURE 10.44 Construction of PROTAS and details of the sensing head. *(Source: Simpson, 1972.)*

vertical. It falls at between 20 and 40 cm/s under a net weight of ~100 gwt. Under the action of the largest dynamic moment envisaged, it deflects less than 0.1° from its equilibrium position. Deflection of the vane relative to the vertical defined by the probe is therefore an indicator of any horizontal velocity difference between the probe and the water.

Any change in horizontal velocity encountered by the probe in its descent will cause a displacement of the vane. It will also cause the body to experience a horizontal acceleration, with the result that the observed velocity shear will differ from its absolute value. If u and v are the absolute horizontal velocity components and u' and v' the values relative to the probe, this may be expressed as (Simpson, 1972):

$$\frac{du}{dz} = \frac{du'}{dz} + \frac{\dot{p}}{W} \qquad (10.10)$$

$$\frac{dv}{dz} = \frac{dv'}{dz} + \frac{\dot{q}}{W} \qquad (10.11)$$

In these equations, $\dot{p}$ and $\dot{q}$ are the components of horizontal acceleration and W is the free-fall vertical velocity. By making W large, the correction term can apparently be suppressed. Unfortunately, this is not the case in practical terms, because $\dot{p}$ and $\dot{q}$ both are dependent on W, increasing almost in proportion to it.

In addition, because the vane displacement sensitivity is inversely proportional to W, making W large does not make

sense. In practice, the correction term is found to be small and its value may be estimated from the known relative velocity profile. To do this, the flow past the cylinder is considered an axial flow with an independent two-dimensional cross-flow. It was found that if the length of the body can be made to exceed the dominant vertical length scale of the velocity structure, the accelerations $\dot{p}$ and $\dot{q}$ would be reduced. This observation suggests that the cylindrical body of the probe must be as long as is practical. In computing the acceleration of the probe, it is assumed implicitly that the probe is not rotating or at least that it does not rotate significantly in its own length. This assumption about PROTAS was found to be reasonably valid in field trials at sea in which it was fitted with a recording compass. The rate of rotation was then found to be ~1 revolution per 100 meters of depth.

Movement of the shear-detecting vane is sensed by an electrode on the end of the vane, which picks up two alternating voltages provided by pairs of fixed electrodes. The pick-up electrode is, in effect, the wiper of two resistive potentiometers formed by the pairs of fixed electrodes. The two signals are at different frequencies and can, therefore, be separated by tuned amplifiers and converted by phase-sensitive detection to voltages that, for small displacements, represent the rectangular coordinates of the tip of the vane. At a fall speed of 30 cm/s and with a vane of 30-cm length, a cross-flow of 1 mm/s produces a vane movement of 1 mm, which is a displacement readily detectable by the pick-up circuitry.

Justifying the name of the probe (i.e., Probe Recording Ocean Temperature And Shear), the PROTAS incorporated a fast-response temperature gradient measurement as well so that velocity and density gradients may be compared in fresh water. Since it is a profiler, its depth measurements are also made. Temperature and depth measurements are made by conventional thermistor and strain-gauge techniques and simultaneously recorded. The thermistor is a Fenwal GB32, which has a time constant of $\tau \sim 150$ ms and hence a vertical resolution of ~5 cm for $W = 30$ cm/s. A still faster response thermistor GC32 ($\tau = 80$ ms when mounted in a pressure-proof unit) will allow a slight improvement on the vertical resolution.

The temperature gradient is estimated by differentiating the output from a single thermistor. In freefall, where the vertical velocity W is accurately known, the time rate of change of temperature may be interpreted as a spatial gradient. To improve the S/N ratio, the conventional active differentiator circuit is preceded by a low-noise amplifier.

In construction, the body of the probe consists of an HE 30 WP alloy tube of wall thickness 6.35 mm and diameter 18 cm, which can operate to a depth of 500 m with a safety factor of 2. Adequately thicker-walled tubes that can overcome collapse due to elastic instability and yielding of the metal will be capable of reaching the deepest oceanic depths. The ends of the tube are streamlined by Fiberglas fairings, which also house the recovery aids. At the lower end of the tube there is a 10-kHz acoustic beacon; in the upper fairing, a discharge tube flashing light is provided to assist in locating the probe at night. The righting moment is further enhanced by the positioning of the ballast weight and the lead collar, which is used for ballast adjustments. To achieve the necessary degree of static stability, the internal payload (electronics, batteries, and recorder) is carried in the lower half of the tube. When the probe reaches its terminal depth, the ballast is released by one of the two tension-pin pressure releases. The second weight is detached by a corrodible link, which also acts as a back-up time release in the event of a failure in the hydrostatic system.

The first successful trials of the prototype system were made in 1970 from a research vessel in Loch Ness in Scotland, which is a freshwater lake 30 km long by 1.5 km wide with an abyssal depth of more than 200 m. Based on previous temperature structure studies of this lake by Simpson and Woods (1970), at this time of the year (September 1970) there is a surface mixed layer of about 30 m at a temperature of $\sim 12°C$, below which a thermocline rich in microstructure extends down to ~ 100 m. At greater depths the temperature decreases only slightly to a value of $5.3-5.7°C$ at the bottom. An example of the raw data record collected by PROTAS is shown in Figure 10.45. Note that the two components of relative velocity u' and v' vary little in the surface layer down to 30 m. On reaching the thermocline, however, they exhibit marked fluctuations of up to ±1 cm/s that are clearly associated with the temperature gradient. Below the thermocline, the shear diminishes, with relative velocities of <1 mm/s in the deep water. The corrected profiles of the velocity shear components and density gradient, computed based on the profiles of the measured data digitized at intervals of 0.5 s, revealed (Simpson, 1972) a marked relation between the two

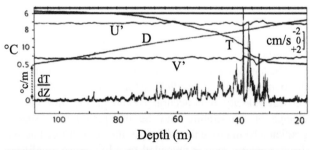

FIGURE 10.45 Example of data record collected by PROTAS profiler probe from the Loch Ness freshwater lake. All parameters are recorded against time, which is indicated by marks at intervals of 10 s at the upper edge of the record. An equivalent depth scale is given at the lower edge. The meanings of symbols are as follows: D = depth; T = temperature; $\frac{dT}{dz}$= temperature gradient. u' and v' are the relative velocity components. Fall speed in this example is 33.3 cm/s. *(Source: Simpson, 1972.)*

gradients, with the strongest shear closely coinciding with the region of largest temperature gradient. The shear exhibited a maximum value of 1.6×10^{-2}/s, which was found to be typical for all the records analyzed and of the same order as that observed by Woods (1969b). Results from sea trials that were undertaken in the stratified area of the western Irish Sea indicated the viability of PROTAS for observation of small-scale velocity shear in the ocean (Simpson, 1972).

10.4.3. Free-Falling Lift-Force Sensitive Probes

It has been seen that the vane used in the free-fall instrument PROTAS (which profiles the horizontal velocity fluctuations by sensing the motion of a neutrally buoyant vane) reported by Simpson (1972) responds to changes in the horizontal velocity in a vertical distance approximately equal to 30 cm, i.e., the vertical resolution achievable by PROTAS is on the order of 30 cm. However, dye studies of Woods and Fosberry (1967) indicate that the shear is concentrated into regions on the order of 10 cm thick. Osborn (1974) reported an instrument that is capable of completely resolving a vertical profile of the vertical shear. The advantage of this high-resolution probe is that from the variance of the vertical shear, one can estimate the turbulent energy dissipation.

The velocity sensor described by Osborn (1974) is an adaptation of the two-component airfoil probe (Siddon, 1965, 1971a, b) to the oceanic environment (see Figure 10.46). The probe tip is an axisymmetric solid of revolution aligned with the major axis of the freely falling body. Variations u' in the horizontal velocity represent a fluctuating angle of attack of the total velocity vector, thereby causing a fluctuating lift force on the probe tip. Embedded in the probe tip are two piezoceramic bimorph beams from a ceramic phonograph cartridge that sense the components of the lift force into two perpendicular directions. To reduce the effect of the lead wires, two preamplifiers, one for each velocity component, are situated directly behind the probe in a small pressure case.

The oceanic measurements require an adaptation of Siddon's probe that has greater sensitivity and can operate at pressures on the order of 20 atm (= 200-m seawater depth). The probe, as far back as the preamplifier, is filled with epoxy. The epoxy tip is molded in place from the same epoxy, thereby waterproofing all the electrical contacts. Osborn and colleagues have routinely operated the probe to a depth of 230 m with no leakage. Tests they conducted in a pressure chamber showed that the probe is insensitive to changes in ambient pressure, even sudden changes on the order of 2 atm. Sufficient sensitivity was achieved by using a urethane-based epoxy with a hardness specification of 45

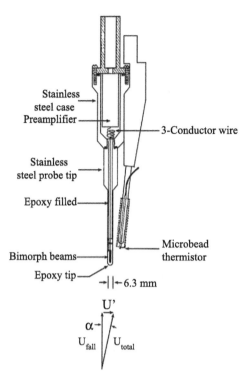

FIGURE 10.46 Diagram of Osborn's velocity and temperature microstructure profiler probes. Vectors show that a change in the horizontal velocity relative to the probe tip appears as a change in the angle of attack of the overall velocity vector. *(Source: Osborn, 1974, ©American Meteorological Society. Reprinted with permission.)*

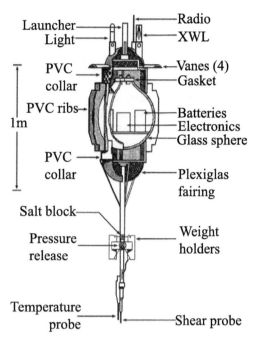

FIGURE 10.47 Schematic diagram of Osborn's velocity and temperature microstructure profiler. *(Source: Osborn, 1974, ©American Meteorological Society, reprinted with permission.)*

on the Shore A scale. An identical tip molded from an epoxy with Shore A hardness of 70 is 60 percent less sensitive. Unfortunately, because most urethane-based epoxies deteriorate in water, the probes have a lifetime in water on the order of 10 hours. However, accelerating technological advancement shows promise of superior-quality epoxies that are insensitive to water. The beams are mounted in a jig while being epoxied to the support and wire leads in order to make the channels orthogonal and to increase the reproducibility of the manufacturing process.

Because the probe is responding to the lift force, the probe's output voltage is proportional to the angle of attack and the mean velocity squared, i.e.,

$$V \approx S\left(\frac{1}{2}\rho\bar{u}^2\right)\sin\alpha, \qquad (10.12)$$

In this expression, V is the output voltage of the preamplifier, S is a calibration constant, $\bar{u}$ is the mean velocity along the central axis of the probe, and $\sin\alpha$ can be approximated by u'/u, where u' is the fluctuating horizontal velocity. Calibration of the probe gave a value of S as 4×10^{-3} V cm sec^2 gm^{-1}, with an estimated error of ± 15 percent. The resolution is equivalent to 0.1 cm/s at a fall speed of 25 cm/s.

The piezoceramic beams are inherently AC devices. The signals from the preamplifiers are sent to the glass

sphere of the profiling instrument (see Figure 10.47) where, due to low-frequency drift in the preamplifiers, there is another low-frequency filter (3-db point at 5×10^{-2} Hz) before the final amplification. The response above 30 Hz is believed to be uniform to beyond 100 Hz, because the resonant frequency of the tip of the probe is between 1 and 2 kHz. The resonance is well damped.

Seawater temperature is sensed with a Veco 43A401C micro-bead thermistor, which is glass-coated and has a nominal diameter of 0.013 cm. It is mounted 2 cm from the velocity probe, and the maximum exposure to the flow is achieved by suspending the bead from its 0.0018-cm-diameter wire leads that extend on opposite sides of the bead.

The probe's fall speed is determined from a record of water pressure versus time sensed with a 0–500 psia Vibrotron pressure transducer mounted inside the glass sphere of the profiling instrument. Typical values of the fall speed are in the range 20–25 cm/s.

The body of the profiler rotates as it descends through the water column. Experience had shown that this rotation must be very slow, at a period of about ~20 second per revolution, for the flow velocity data to be interpretable. With the short wings used, the lift force is small, and therefore the profiler is adjusted to be only slightly heavier (500 gm) than that of the displaced water. Another necessity for rotation of the profiler is determination of the orientation of the velocity sensors. Rotation is monitored as a function of time with a 3,000-turn coil wrapped around a permeable iron core. With this technique, the coil produces a sinusoidal voltage as the body of the profiler rotates through Earth's magnetic field. The orientation of the coil relative to the bimorph beams of the probe is determined in the laboratory.

The two velocity signals, ambient seawater temperature, and its time derivative are converted from voltages to frequency-modulated (FM) signals using voltage-controlled oscillators operating on four standard IRIG frequencies. The Vibrotron pressure transducer is inherently an FM device operating on a standard IRIG channel. The five FM signals are multiplexed together and transmitted from the profiler to its support vessel along a Sippican Expendable Wire Length, which is essentially a Sippican Expendable Bathythermograph without a thermistor or nose weight. It consists of two spools of very fine two-conductor wire (39 gauge), one spool being attached to the instrument and the other remaining on the support vessel. The wire is free spooling, i.e., it comes off the spool like a line off a spinning reel, and causes essentially no drag on the instrument as it falls through the water. In the frequency range used, there is large capacitive coupling between the two conductors, and therefore seawater is used as the signal return path.

The rotation rate of the profiler is telemetered directly up the wire as a slow variation on the offset voltage of the wire. An operational amplifier on the surface ship presents a high impedance load on the wire. The FM signals are recorded directly following high-pass filtering, and the rotation rate is recorded following low-pass filtering. The instrument descends until the weights are released by a Richardson-type stretched-pin release or by the dissolving of a salt block that acts as a back-up release. Upon return to the surface, the instrument is located with the aid of the flashing light and radio transmitter. An STD profile is taken immediately after each successful drop.

Figure 10.48 shows the data for an 8-m-thick region. The fall speed was 20.7 ± 0.5 cm/s and the rotation period 26 ± 0.3 s. There is a turbulent region 5 m thick wherein both the velocity traces and the temperature gradient show intense activity. Some activity is seen in the water above and below. According to Osborn, these data show a very complicated situation, the understanding of which would require information about the horizontal extent of the turbulent region. Figure 10.49 shows the spectrum of data from velocity channel 1 in Figure 10.48 taken at a fall speed of 20.7 cm/s. Temporal frequency times the spectral density of velocity is plotted against the logarithm of the temporal frequency. The plot is variance-preserving in the sense that the areas under the curve are proportional to the variance. Frequency times the spectral density of the vertical current shear, which is frequency-squared times the velocity spectrum, is also plotted. The rise in the shear spectrum above 14 Hz is due to noise.

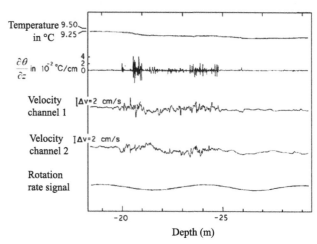

FIGURE 10.48 Data collected by Osborn's velocity and temperature microstructure profiler on August 2, 1972, from Howe Sound. (*Source: Osborn, 1974, ©American Meteorological Society, reprinted with permission.*)

The velocity data are subject to low-frequency contamination due to pendulum-like oscillations of the instrument body. Some of the energy in the frequency bands below 0.1 Hz is from body motions. However, the spectrum of the shear shows that this low-frequency energy does not contribute to the estimate of the variances of the vertical current shear.

High-frequency contamination of the velocity data can come from the acceleration sensitivity of the probe. Vibration of the probe tip places an inertial load on the bimorph beams that is proportional to the amplitude and frequency of the vibration and the mass of the epoxy tip. These high-frequency vibrations can contribute significantly to an estimate of the variance of the vertical shear. The velocity data are differentiated in the instrument, thus giving the velocity shear directly and reducing the noise level in the shear spectra at high frequencies.

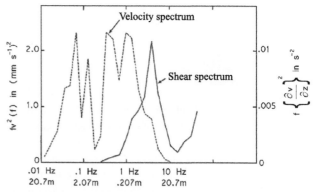

FIGURE 10.49 Spectrum of data from velocity channel 1 shown in Figure 10.48, taken at a fall speed of 20.7 cm/s. The spectrum of the shear is frequency squared times the velocity spectrum. (*Source: Osborn, 1974, ©American Meteorological Society, reprinted with permission.*)

To ameliorate many of the difficulties found in the just-described probe, the instrument housing has been replaced by a long aluminum tube, similar to that of Simpson (1972), in order to (1) increase the stability of the free-fall body, (2) reduce oscillation of the body due to the eddies that are shed off the uppermost part of the housing, and (3) provide sufficient buoyancy. Osborn (1974) concludes that the airfoil probe combined with free-fall instrument housing is ideal for studying the vertical current shear in the ocean. According to him, the resolution and sensitivity are sufficient to provide estimates of the energy dissipation directly. In his measurements, the maximum value of the shear was 0.7/s, which is considerably larger than those reported by Simpson (1972) or Woods and Fosberry (1967).

10.5. MERITS AND LIMITATIONS OF FREELY SINKING/RISING UNGUIDED PROBES

It has been noted that freely sinking probes are an effective means for determining profiles of horizontal current velocity in the ocean. In comparison to a moored instrument, the freely sinking probe has been found to have certain advantages in that an entire profile is obtained from a single record and the measurements are not contaminated by mooring-line motions. However, there are difficulties in interpreting current-velocity profiles derived from freely sinking/rising probes, because the vehicle is affected by the flow it is attempting to measure.

Two basic techniques have been used to measure current profiles. In one method, the probe is acoustically tracked while it sinks or rises, much as a weather balloon is tracked by electromagnetic means. This method has been found to be effective in measuring the slowly varying component of the profile or relatively low wave-number variability. The assumption usually made is that the probe instantaneously follows the horizontal current; but high shears, where the horizontal current velocity changes abruptly, will result in high wave-number variability that the vehicle cannot follow.

In the second type of freely sinking/rising current-measuring probe, there is a current-velocity sensor that measures the current flow relative to the vehicle. Relative velocities are dominated by high wave-number variations because the vehicle will tend to follow low wave-number variations in the current-velocity profile, leaving only the higher wave-number variations in the measured relative-velocity profile. In other words, high-frequency velocity fluctuations are well represented in measurements from a velocity sensor on a freely sinking/rising vehicle, but lower frequencies become increasingly attenuated below the characteristic frequency of order $1/\tau$, where τ is the time scale for the vehicle response.

Hendricks and Rodenbusch (1981) analyzed this problem and concluded that to increase the frequency resolution, it is necessary to minimize τ by decreasing the mass or by increasing the sink/rise rate. According to them, the last option is probably the easiest technically, but the improvement in frequency resolution will not yield proportional gains in vertical wave number. Small vehicles are most effective as tracked current profilers to obtain maximum bandwidth in vertical wave number, whereas larger bodies will give greater wave-number resolution for vehicles measuring relative horizontal velocity with a velocimeter. In most cases, the velocity measurements are ultimately interpreted as a profile or vertical wave-number spectrum rather than a time series or frequency spectrum.

Another effect that may become significant for instruments of the second kind arises when the current-velocity sensors are offset from the main body of the vehicle. In many applications, the current-velocity sensor is not located near the vehicle's center of mass but is placed some distance above or below the main body. Often a placement is chosen to minimize hydrodynamic flow distortion at the sensor. Although it is desirable to minimize such distortion, offsetting the flow sensor introduces an artifact that must be accounted for in the interpretation of the velocity measurements. For example, when the sensors are located below the main body of a sinking vehicle, they measure the relative horizontal velocity before it is felt by the vehicle and before the vehicle can respond to the changes in the drag force. As a result, the sensed velocity difference will, generally, be greater than that which would be measured had the sensors been at the center of mass and the vehicle had accelerated in response to the velocity difference. When the flow sensor is offset as just mentioned, the sensors measure the velocity of a layer that is different from the one occupied by the bulk of the vehicle, thereby introducing a phase shift between the measured velocity and the vehicle velocity.

A third effect arises when the drag force on the body is distributed over a vertical length that is not short compared to the vertical length scales of the measured velocity variation. In this case, the actual drag force distribution must be accounted for to properly interpret the ocean-current velocity measurements.

REFERENCES

Berger, F.B., 1957. The nature of Doppler velocity measurement. IRE Transactions on Aeronautical and Navigational Electronics, ANE-4. September, 3, 103–112.

Cole, D., 1981. A new glass deep-sea acoustically tracked current profiler. Proc. Oceans '81 Conference 1, 257–260.

Cooper, J.W., Stommel, H., 1968. Regularly spaced steps in the main thermocline near Bermuda. J. Geophys. Res. 73, 5849–5854.

Crease, J., 1962. Velocity measurements in the deep water of the Western North Atlantic. J. Geophys. Res. 67, 3173–3176.

Drever, R.G., Sanford, T.B., 1970. A free fall electromagnetic current meter instrumentation. Proc. I.E.R.E. Conf. on Electronic Engineering in Ocean Technology. Swansea, 19, 353–370.

Duda, T.F., Cox, C.S., 1987. Vorticity measurement in a region of coastal ocean eddies by observation of near-inertial oscillations. Geophys. Res. Lett. 14, 793–796.

Duda, T.F., Cox, C.S., Deaton, T.K., 1988. The Cartesian diver: A self-profiling Lagrangian velocity recorder. J. Atmos. Oceanic Technol. 5, 16–33.

Duing, W., Johnson, D., 1972. High resolution current profiling in the Straits of Florida. Deep-Sea Res. 19, 259–274.

Evans, D.L., Rossby, H.T., Mork, M., Gytre, T., 1979. YVETTE: A free-fall shear profiler. Deep-Sea Res. 26/6A, 703–718.

Evans, R., Leaman, K., 1978. A comparison between the Duing profiling current meter and the White Horse acoustic profiler. J. Geophys. Res. 83, 5515–5520.

Garcia-Berdeal, I., Hautala, S.L., Thomas, L.N., Johnson, H.P., 2006. Vertical structure of time-dependent currents in a mid-ocean ridge axial valley. Deep-Sea Res. I 53, 367–386.

Gould, W.J., Schmitz Jr., W.J., Wunsch, C., 1974. Preliminary field results for a Mid-Ocean Dynamics Experiment (MODE-O). Deep-Sea Res. 21, 911–931.

Grant, H.L., Hughes, B.A., Vogeland, W.M., Molliet, A., 1968. The spectrum of temperature fluctuations in turbulent flow. J. Fluid Mech. 34, 423–442.

Gregg, M.C., Cox, C.S., 1971. Measurements of the oceanic microstructure of temperature and electrical conductivity. Deep-Sea Res. 18, 925–934.

Gregg, M.C., Cox, C.S., 1972. The vertical microstructure of temperature and salinity. Deep-Sea Res. 19, 355–376.

Halkin, D., Rossby, T., 1985. The structure and transport of the Gulf Stream at 73°W. J. Phys. Oceanogr. 15, 1439–1452.

Hamblin, P.F., Kuehnel, R., 1980. An evaluation of an unattended current and temperature profiler for deep lakes. Limnol. Oceanogr. 25, 1128–1141.

Hayes, S.P., Milburn, H.B., Ford, E.F., 1984. TOPS: A free-fall velocity and CTD profiler. J. Atmos. Oceanic Technol. 1, 220–236.

Hayward, G.G., Ferris, S.P., 1979. A new acoustic command system for recovery of oceanographic instruments. Marine Technology Society Proceedings, 139–144.

Hendricks, P.J., Rodenbusch, G., 1981. Interpretation of velocity profiles measured by freely sinking probes. Deep-Sea Research 28A (10), 1199–1213.

Hinze, J.O., 1959. Turbulence. McGraw-Hill, New York, NY, USA.

Hogg, N.G., 1976. On spatially growing waves in the ocean. J. Fluid. Mech. 78, 217–235.

Hoyt, J.K., 1982. The design of syntactic foam pressure housings for expendable acoustic beacons, one facet of a prototype current profiling instrument. M.S. thesis. Massachusetts Institute of Technology, Cambridge, MA, USA.

Joseph, A., 2000. Applications of Doppler Effect in navigation and oceanography, Encyclopedia of Microcomputers, 25. Marcel Dekker, Newyork, USA, 17–45.

Leaman, K.D., 1976. Observations on the vertical polarization and energy flux of near-inertial waves. J. Phys. Oceanogr. 6, 894–908.

Leaman, K.D., Sanford, T.B., 1975. Vertical energy propagation of inertial waves: A vector spectral analysis of velocity profiles. J. Geophys. Res. 80, 1975–1978.

Leaman, K.D., Sanford, T.B., 1975. Vertical energy propagation of inertial waves: a vector spectral analysis of velocity profiles. J. Geophys. Res. 80, 1975–1978.

Leaman, K.D., Molinari, R.L., Vertes, P.S., 1987. Structure and variability of the Florida Current at 27°N: April 1982–July 1984. J. Phys. Oceanogr. 17, 565–583.

Longuett-Higgins, M.S., Stern, M.E., Stommel, H., 1954. The electric field induced by ocean currents and waves, with applications to the method of towed electrodes. In: Papers in Physical Oceanography and Meteorology, 13. Woods Hole Oceanographic Institution and the Massachusetts Institute of Technology, Woods Hole, MA, USA, 1–37.

Luyten, J.R., Swallow, J.C., 1976. Equatorial undercurrents. Deep-Sea Res. 23, 999–1000.

Luyten, J.R., Swallow, J.C., 1976. Equatorial undercurrents. Deep-Sea Res. 23, 999–1001.

Luyten, J.R., Needell, G., Thomson, J., 1982. An acoustic dropsonde: Design, performance, and evaluation. Deep-Sea Res. 29 (4A), 499–524.

Mangelsdorf Jr., P.C., 1962. The world's longest salt bridge. In: Marine Sciences Instrumentation, Vol. 1. Plenum Press, Newyork, 173–185.

Mullineaux, L.S., France, S.C., 1995. Dispersal mechanisms of deep-sea hydrothermal vent fauna. In: Humphris, S.E., Zierenberg, R.A., Mullineaux, L.S., Thomson, R.E. (Eds.), Seafloor hydrothermal systems: Physical, chemical, Biological, and geological interactions. American Geophysical Union, Washington, DC, USA, pp. 408–424.

Nasmyth, P.W., 1970. Oceanic turbulence. Ph.D. dissertation. University of British Columbia, BC, Canada.

Neshyba, S., Neil, V., Denner, W., 1971. Temperature and conductivity measurements under Ice Island T-3. J. Geophys. Res. 74, 8107–8119.

Osborn, T.R., 1969. Oceanic fine structure. thesis. University of California, San Diego, CA, USA.

Osborn, T.R., 1974. Vertical profiling of velocity microstructure. J. Phys. Oceanogr. 4, 109–115.

Osborn, T.R., Cox, C.S., 1972. Oceanic fine structure. Geophys. Fluid Dyn. 3, 321–345.

Perkins, H., 1970. Inertial oscillations in the Mediterranean. Ph.D. thesis. Massachusetts Institute of Technology and Woods Hole Oceanographic Institution.

Plaisted, R.O., Richardson, W.S., 1970. Current fine structure in the Florida Current. J. Mar. Res. 28 (3), 359–363.

Pochapsky, T.E., 1966. Measurements of deep water movements with instrumented neutrally buoyant floats. J. Geophys. Res. 71, 2491–2504.

Pochapsky, T.E., Malone, F.D., 1972. A vertical profile of deep horizontal current near Cape Lookout, North Carolina. J. Marine Res. 30, 163–167.

Price, J.F., Weller, R.A., Pinkel, R., 1986. Diurnal cycling: Observations and models of the upper ocean response to diurnal heating, cooling, and mixing. J. Geophys. Res. 91, 8411–8427.

Richardson, W.S., Carr, A.R., White, H.J., 1969. Description of a freely dropped instrument for measuring current velocity. J. Mar. Res. 27, 153–157.

Richardson, W.S., Schmitz Jr., W.J., 1965. A technique for the direct measurement of transport with application to the Straits of Florida. J. Mar. Res. 23, 172–185.

Richardson, W.S., Schmitz Jr., W.J., 1965. A technique for the direct measurement of transport with application to the Straits of Florida. J. Marine Res. 16, 172–185.

Riggs, N.P., 1989. Current profilers: Comparing three different techniques. Sea Technology, March issue, 52–54.

Rona, P., Trivett, A., 1992. Discrete and diffuse heat transfer at ASHES vent field Axial Volcano, Juan de Fuca Ridge. Earth Planetary Science Letters 109, 57–71.

Rossby, H.T., 1974. Studies of the vertical structure of horizontal currents near Bermuda. J. Geophys. Res. 79, 1781–1791.

Rossby, H.T., Sanford, T.B., 1976. A study of velocity profiles through the main thermocline. J. Phys. Oceanogr. 6, 766–774.

Rossby, H.T., Sanford, T.B., 1976. A study of velocity profiles through the main thermocline. J. Phys. Oceanogr. 6, 766–774.

Rossby, T., 1969. A vertical profile of currents near Plantagenet Bank. Deep-Sea Research 16, 377–385.

Royer, L., Hamblin, P.F., Boyce, F.M., 1986. Analysis of continuous vertical profiles in Lake Erie. Atmosphere-Ocean 21, 73–89.

Royer, L., Hamblin, P.F., Boyce, F.M., 1987. A comparison of drogues, current meters, winds, and a vertical profiler in Lake Erie. J. Great Lakes Res. 13 (4), 578–586.

Sanford, T.B., 1971. Motionally induced electric and magnetic fields in the sea. J. Geophys. Res. 76, 3476–3492.

Sanford, T.B., 1975. Observations of the vertical structure of internal waves. J. Geophys. Res. 80, 3861–3871.

Sanford, T.B., Drever, R.G., Dunlap, J.H., 1978. A velocity profiler based on the principles of geomagnetic induction. Deep-Sea Research 25, 183–210.

Sanford, T.B., Drever, R.G., Dunlap, J.H., 1985. An acoustic Doppler and electromagnetic velocity profiler. J. Atmos. Oceanic Technol. 2 (2), 110–124.

Sanford, T.B., Drever, R.G., Dunlap, J.H., D'Asaro, E.A., 1982. Design, operation, and performance of an expendable temperature and velocity profiler (XTVP). APL-UW 8110. Tech. Rep, Appl Phys Lab. University of Washington, Seattle, 88.

Siddon, T.E., 1965. A turbulence probe using aerodynamic lift, Tech. Note 88, Institute for Aerospace Studies. University of Toronto, Toronto, Canada.

Siddon, T.E., 1971a. A miniature turbulence gauge utilizing aerodynamic life. Rev. Sci. Instr. 42, 653–656.

Siddon, T.E., 1971b. A new type of turbulence gauge for use in liquids. preprint of paper presented at Symposium on Flow, Inst. Soc. Amer., Pittsburg, PA, USA.

Simpson, J.H., 1972. A free-fall probe for the measurement of velocity microstructure. Deep-Sea Res. 19, 331–336.

Simpson, J.H., Woods, J.D., 1970. The temperature microstructure in a freshwater thermocline. Nature 226 (5248), 832–834.

Spain, P.F., Dorson, D.L., Rossby, H.T., 1981. Pegasus: A simple, acoustically tracked velocity profiler. Deep-Sea Res. 28A (12), 1553–1567.

Stommel, H., Federov, K., 1967. Small-scale structure in temperature and salinity near Timor and Mindanao. Tellus 19, 306–325.

Thomas, R.S., 1971. Measurement of ocean current shear using acoustic backscattering from the volume. Deep-Sea Res. 18, 141–143.

Thomson, R.E., Gordon, R.L., Dymond, J., 1989. Acoustic Doppler current profiler observations of a mid-ocean ridge hydrothermal plume. J. Geophys. Res. 94, 4709–4720.

Trivett, D.A., Williams, A.J., 1994. Effluent from diffuse hydrothermal venting. 2. Measurements of plumes from diffuse hydrothermal vents at the southern Juan de Fuca Ridge. J. Geophys. Res. 99, 18,417–18,432.

Van Leer, J.C., Duing, W., Erath, R., Kennelly, E., Speidel, A., 1974. The cyclesonde: An unattended vertical profiler for scalar and vector quantities in the upper ocean. Deep-Sea Res. 21, 385–400.

Voorhis, A.D., Bradley, A.M., 1984. Popup: A prototype bottom-moored long-term current profiler. J. Atmos. Oceanic Technol. 1, 166–175.

Ward-Whate, P.M., 1977. A vertical automatic profiling system. Proc. MTS-IEEE Oceans '77 Conf. Record, 25B1–25B5.

Webster, F., 1968. Observations of inertial-period motions in the deep sea. Rev. Geophys. 6, 473–490.

Webster, F., 1969. On the representativeness of direct deep-sea current measurements. Prog. Oceanogr. 5, 3–15.

Woods, J., 1968. CAT under water. Weather 23, 224–235.

Woods, J.D., 1969a. On designing a probe to measure ocean microstructure. Underwater Sci. Technol. J. 1 (1), 6–12.

Woods, J.D., 1969b. On Richardson's number as a criterion for laminar-turbulent-laminar transition in the ocean and atmosphere. Radio Sci. 4 (12), 1289–1298.

Woods, J.D., Fosberry, G.G., 1967. The structure of the thermocline. Report 1966–1967, Underwater Association, London., 5–18.

BIBLIOGRAPHY

Chapman, R.P., Marshall, J.R., 1966. Reverberation from deep-scattering layers in the Western North Atlantic. J. Accoust. Soc. Am. 40, 405–411.

Dahl, O., 1969. The capability of the Aanderaa recording and tele-metering instrument. Progr. Oceanogr. 5, 103–106.

Drever, R.G., Sanford, T., 1970. A free-fall electromagnetic current meter—instrumentation, Electronic Engineering in Ocean Technology. Inst. Electronic Radio Engrs. Swansea, Sept. 21–24, 1970, IERE Conf. Proc. 19, 353–370.

Duing, W., Johnson, D., 1971. Southward flow under the Florida Current. Science 173, 428–430.

Duing, W., Johnson, D., 1972. High resolution current profiling in the Straits of Florida. Deep-Sea Res. 19, 259–274.

Graefe, V., Gallagher, B., 1969. Oceanographic profiling with improved vertical resolution. J. Geophys. Res. 74, 5425–5431.

Grant, H.L., Hughes, B.A., Vogeland, W.M., Molliet, A., 1968. The spectrum of temperature fluctuations in turbulent flow. J. Fluid. Mech. 34, 423–442.

Greg, M.C., Cox, C.S., 1972. The vertical microstructure of temperature and salinity. Deep-Sea Res. 19, 355–376.

Katz, E.J., Witzell Jr., W.E., 1979. A depth controlled tow system for hydrographic and current measurements with applications. Deep-Sea Research 26A, 579–596.

Knight, P.J., 1995. Current profile, pressure and temperature records from the North Channel of the Irish Sea. Proudman Oceanographic Laboratory, Natural Environment Research Council, Bidston Observatory, Birkenhead, Merseyside, United Kingdom. Report No. 39, July 1993–October 1994.

Knight, P.J., Harrison, A.J., Lane, A., Wilkinson, M., Collen, D.G., 1993. Current profile and river bed pressure and temperature records, July 1992. River Mersey, Liverpool-Wallasey transect, Proudman Oceanographic Laboratory, Natural Environment Research Council, Bidston Observatory, Birkenhead, Merseyside, United Kingdom. Report No. 27.

Miles, J.W., Howard, L.N., 1964. Notes on a heterogeneous shear flow. J. Fluid Mech. 20 (2), 331–336.

Osborn, T.R., Cox, C.S., 1972. Oceanic fine structure. Geophys. Fluid Dyn. 3, 321–345.

Pollard, R.T., Millard Jr., R.C., 1970. Comparison between observed and simulated wind-generated inertial oscillations. Deep-Sea Res. 17, 795–812.

Rooth, C., Duing, W., 1971. On the detection of "inertial" waves with pycnocline followers. J. Phys. Oceanogr. 1, 12–16.

Sanford, T.B., 1986. Recent improvements in ocean current measurement from motional electric fields and currents. Proc. IEEE Third Working Conference on Current Measurement, 22–24 January 1986. In: Airlie, Virginia, G., Appell, F., Woodward, W.E. (Eds.). Institute of Electrical & Electronic Engineers, New York, NY, USA, pp. 65–77.

Schott, F., Lee, T.N., Zantop, R., 1988. Variability of structure and transport of the Florida Current in the period range of days to seasonal. J. Phys. Oceanogr. 18, 1209–1230.

Takematsu, M., Kawatate, K., Koterayama, W., Suhara, T., Mitsuyasu, H., 1986. Moored instrument observations in the Kuroshio south of Kyushu. J. Oceanographical Soc. Japan 42, 201–211.

Trump, C.L., Okawa, B.S., Hill, R.H., 1985. The characterization of a midocean front with a Doppler shear profiler and a thermistor chain. Journal of Atmospheric and Oceanic Technology 2, 508–516.

Webster, F., 1968. Observations of inertial-period motions in the deep sea. Rev. Geophys. 6, 473–492.

Webster, F., 1969. On the representativeness of direct deep-sea current measurements. Prog. Oceanogr. 5, 3–15.

Chapter 11

Vertical Profiling of Currents Using Acoustic Doppler Current Profilers

Chapter Outline

11.1. Basic Assumptions and Operational Issues   340
11.2. Principle of Operation   341
11.3. Profiling Geometries   344
   11.3.1. Bottom-Mounted, Upward-Facing ADPs   344
      11.3.1.1. Combination of Surface Currents and Gravity Wave Orbital Velocities   345
      11.3.1.2. A Single Upward-Facing ADP for Measuring Currents and Surface Waves   346
      11.3.1.3. Current Profile and Wave Measurements for Operational Applications   350
11.4. Trawl-Resistant ADP Bottom Mounts   353

11.5. Horizontal-Facing ADPs   354
11.6. Subsurface Moored ADPs   358
11.7. Downward-Facing Shipboard ADPs   360
11.8. Towed ADPs   363
11.9. Lowered ADCP (L-ADCP)   367
11.10. ADPs for Current Profiling and AUV Navigation   368
   11.10.1. AUV-Mounted ADPs for Current Profiling   368
   11.10.2. ADPs for AUV Navigation   369
11.11. Calibration of ADPs   371
11.12. Intercomparison and Evaluation   372
11.13. Merits and Limitations of ADPs   374
References   375
Bibliography   378

In the past, measurements of vertical profiles of horizontal current flows in the ocean were carried out by the use of moored current meter (CM) chains. Maintaining such moorings at large depths in the ocean was found to be difficult and rather expensive. Vertical profile measurements of oceanic currents were indeed a daunting task for several years. Ideally, one might want a single probe that can scan a large water column and remotely measure the vertical distribution of horizontal motions with high vertical resolution. Although freely sinking/rising probes were successfully used for specific studies, time-series measurements with the aid of such probes proved a difficult task. Various difficulties associated with freely sinking/rising probes and the increasing requirement for time-series measurements of vertical profiles of currents led to the development of acoustic Doppler techniques, borrowed from the radar techniques used in meteorology for wind-velocity profile measurements. Although Doppler sonar techniques were investigated several decades ago and were implemented for measurement of ship's speed based on "bottom track" in shallow waters and "water track" in deep waters (Joseph, 2000), power requirements proved a serious impediment to incorporation of the Doppler sonar techniques in battery-powered self-recording CMs. Availability of low-power microprocessors removed this hurdle, and Doppler signal processing in self-recording moored instruments thus became feasible. This paved the way to the success of stand-alone self-recording of acoustic Doppler current profiler (ADP) development, giving great relief to a wide spectrum of oceanographic communities who are interested in the measurement of high-resolution vertical profiles of ocean currents.

An ADP is a type of hydroacoustic device that measures and records water-current velocities over a range of distances both horizontally and vertically over a range of depths. ADPs are now receiving considerable stimulus from requirements for research in upper ocean processes in the context of climate research, whereas in the shallow-shelf seas, current profiles are needed in connection with the proving of three-dimensional circulation models, navigation, sediment transport, and fluid loading of structures. Gordon (1990) provides a useful bibliography relating to a variety of ADPs up to 1990.

Measuring Ocean Currents. http://dx.doi.org/10.1016/B978-0-12-415990-7.00011-9
Copyright © 2014 Elsevier Inc. All rights reserved.

Apart from HF Doppler radar systems (see chapter 4), ADPs are probably the single most significant innovation in oceanography since the mid-1970s. The first commercial ADP, produced in the late 1970s, was an adaptation of a commercial speed log (Rowe and Young, 1979). The speed log was redesigned to measure water velocity more accurately and to allow measurement in range cells over a depth profile. Refined and improved during the 1980s and 1990s, these instruments can now remotely measure both horizontal and vertical profiles of water currents in sequential layers of water columns. The ADPs break up the water-velocity profile into uniform segments or depth cells, often called *bins*. The fact that ADPs can measure flow velocities from such discrete distances (i.e., bins) from the sensor face has led to an almost universal adoption of the instrument within the oceanographic community. Bin lengths and locations are determined by various parameters that set the transmit pulse length, receive window length, and blank after-transmit length. At present, ADPs are used in several offshore renewable energy applications, coastal applications, and offshore oceanographic applications and can be configured for side-looking into rivers and canals for long-term continuous discharge measurements, mounted on boats for instantaneous surveys, and moored on the subsurface and seabed locations for long-term current and wave studies.

11.1. BASIC ASSUMPTIONS AND OPERATIONAL ISSUES

The basic philosophy embodied in the working of ADPs is based on the mechanism of acoustic volume scattering from a cloud of moving scatterers in the flow field and the well-known phenomenon of the Doppler effect. The term *scatterer* encompasses any kind of inhomogeneity in water, including suspended particulate matter, biological organisms (zooplankton), and minute air bubbles (near the sea surface). Any or all of these scatterers are present almost everywhere in the sea. They remain suspended in the water and, on an average, they move at the same velocity as the water particles. Because the suspended particles move along with the water motion, they are assumed to serve as passive tracers of the flow field. Motions of schools of fishes are an exception to this basic assumption, and therefore the presence of fish schools in the acoustic beam gives errors in the water-velocity measurements. However, such errors can be detected from the backscattering strength measurements made by the ADPs, and the erroneous water velocity data can, therefore, be corrected or the erroneous data removed from the dataset.

The physical properties of the transmission medium have a profound effect on system design. Besides simple spherical spreading, a number of frequency-dependent loss mechanisms are involved in acoustic signal propagation—the effect being to produce a rapid increase of attenuation with frequency (Urick, 1982). Thus, although it is desirable to operate at high frequency to improve Doppler frequency resolution, the working frequency is generally set by the maximum range required.

Transducer quality is essential for data quality. The transducers must be directional (including both narrow-beam and suppressed side lobes) and efficient. The ADPs use large transducers (relative to the wavelength of sound in water) to obtain narrow acoustic beams (typically 3° for 400 kHz transducer or 1.7° for 2 MHz transducer). The acoustic beams focus most of the energy in the center of the beam, but a small amount leaks out in other directions. Because sound reflects much better from the sea surface than it does from the subsurface water layers, the small signals that travel straight to the sea surface through the side lobes of the ADP can produce sufficient echo to contaminate the desired signal from the subsurface water layers.

Echo intensity (EI) is a measure of the signal strength of the echo returning from the scatterers. EI depends on transmitted power, backscatter coefficient, sound absorption in the ambient water body, and beam spreading as given by (Urick, 1996):

$$EI = SL + SV - 20 \log (R) - 2\alpha R + K \qquad (11.1)$$

where SL is the transmitted source level (dB//1μPa at 1 m); SV is the mean volume backscatter strength (dB//m^{-1}); R is the range along the beam (m); α is the absorption coefficient (dBm^{-1}). The term K is an instrument- and frequency-specific constant, and $20 \log (R)$ accounts for beam spreading (outgoing and incoming spherical spreading and increased scattering volume). EI is now routinely measured by ADPs as a bonus parameter to survey the concentration of zooplankton or suspended sediment—useful for biologists and sedimentologists.

Our understanding of the way in which the properties and behavior of individual scatterers determine backscattering properties is as yet incomplete in several respects. Acoustic scatterers at depths in the ocean well below the surface are primarily of biological origin, and scattering strength is therefore dependent on biological productivity. This can vary spatially and temporally, being enhanced in regions of upwelling and in the well-defined scattering layers. Entrained air bubbles (known as *microbubbles*) can provide a substantial scattering mechanism within 10–20 m of the sea surface (Thorpe, 1986), whereas in areas of strong tidal mixing in continental shelf waters, suspended sediments may provide the greatest source of backscatter (Collar, 1993).

Small-scale motions of scatterers relative to each other are of fundamental importance, for these produce changes in the relative phases of individual echoes, with the consequence that the phase signature from the ensemble will evolve with time, eventually becoming decorrelated from the initial situation. The characteristic time τ_c during which this process occurs is defined as the *scattering correlation time* and it is a function of the acoustic wavelength, scattering volume, and velocities of the insonified scatterers. In most practical measurement situations, the two-way pulse travel time greatly exceeds τ_c as a result of the relatively slow acoustic propagation velocity, and this initially caused noncoherent processing to be adopted, i.e., the signal frequency is estimated each time from the echo generated from individual transmitted pulses (Collar, 1993). Coherent processing is, however, possible in principle and techniques based on multiple pulses, or coded transmissions have since been developed.

As noted in the preceding introductory remarks, ADPs use acoustic energy directed along narrow beams. The backscattered acoustic energy from suspended material in the water column is analyzed to determine the Doppler shift due to the relative motion between the ADP and the suspended material. The beam width and side-lobe energy levels are important aspects of the performance of the ADP. Beam width is characterized at the −3-db level and is measured in degrees of arc. For example, a typical RD 1,200-kHz transducer has a beam width of 1.4°. The main beam lobe contains most of the energy emitted. However, other transducer characteristics such as size and vibration modes generate beam side lobes. These side lobes are typically −40 dB from the main lobe. The characteristics of these side lobes are significant in determining near-surface and near-bottom measurement accuracy.

The ADP beam angles are typically 30° relative to the principal axis of the transducer, and, therefore, the side-lobe energy at 30° from the main beam points toward the surface or bottom when it is deployed in an upward-looking or downward-looking configuration, respectively. The echo return from the sea surface or the seafloor is considerably larger than that from the suspended material in the water. Therefore, even at relatively low side-lobe energy levels, the return signal is strong enough to contaminate the ADP measurements. This energy is received at the same time as the energy from ∼85 percent of the profiling range (Appell et al., 1991). Accordingly, RD Instruments recommends in their manual not to use measurements acquired beyond 85 percent of the ADP's maximum specified range when surface or bottom reflections are present.

Contamination from side lobes has the effect of biasing the velocity measurements towards zero. The worst-case condition is a mirror surface that would reflect all the energy back toward the receiver. In this case, the surface is not moving and therefore does not Doppler-shift the signal. In actual *in situ* conditions, the results are dependent on sea surface conditions. The angle of reflection with the surface

wave is constantly changing. This means that the amount of backscattered energy received is varying and the moving surface creates a Doppler shift in the signal. The footprint of the beam at the surface and wavelength of the signal may also have an effect. In a downward-looking mode of the ADP, contamination from bottom echoes is a function of bottom characteristics.

11.2. PRINCIPLE OF OPERATION

ADPs operate on the same principle as the meteorological Doppler radar, but the transducers, transmission signals, and the transmission media in these two devices are much different. Accordingly, the complexities in these two systems are also of different nature. The principle on which Doppler profilers operate is as follows: A sonar (which is an active acoustic transducer) transmits a narrow-beam acoustic pulse (of frequency f_t and temporal pulse length τ) through the water in a given azimuthal direction at a known angle to the horizontal: downward from a ship-based installation, upward or downward from a moored instrument, upward from a seabed instrument package, or horizontally from a vertical structure. The transmitted sound pulses from the ADP scatter in all directions from the sound scatterers; most of the sound propagates in the forward direction, unaffected by the scatterers and the small amount that reflects back (i.e., echoes) is Doppler-shifted. After transmission, the same transducer operates as a receiver to receive the echoes from the scatterers, which remain suspended in the water body. The backscattered acoustic signals are received as a function of time after transmission.

The ADP's automatic gain control (AGC) circuit begins to increase the gain (i.e., amplification) of the incoming signal after transmission of a pulse is completed. This time-varying gain (TVG) compensates for the decreasing strength of the return signal with increasing radial (i.e., along-beam) distance from the ADP. Thus, a single transceiver (i.e., transmitter-cum-receiver) detects the components of water-flow velocity resolved along the acoustic beam of the transceiver. The flow-velocity component along the acoustic beam is known as *radial velocity*.

During the reception sequence, the backscattered acoustic beam, which is oriented in a given direction, is divided into different slant range-gated cells, known as *bins* (see Figure 11.1). By time-gating the stream of received backscattered signals, different radial velocity components v_i

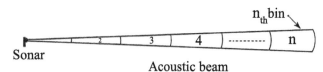

FIGURE 11.1 Range-gated cells, known as bins, of a backscattered acoustic beam, from which radial current measurements are made.

are estimated corresponding to each i_{th} bin. The water motion in a given bin (which is assumed to be the same as the motion of the scatterers in that bin) introduces a Doppler shift $f_D = (f_r - f_t)$ corresponding to that bin, where the transmitted sonar frequency is f_t and the received sonar frequency is f_r. Because the same transducer is used for transmission as well as reception (known as *monostatic configuration*), the Doppler shift f_D can be reduced to a form (see Section 9.3 for details):

$$f_D = \frac{2f_t}{c}v \qquad (11.2)$$

where c is the velocity of sound in water at rest ($\sim 1,500$ m/s). From Equation 11.2, the radial flow velocity v in a given bin is estimated as:

$$v = \frac{cf_D}{2f_t} \qquad (11.3)$$

The Doppler shift from a given bin (say, i_{th} bin) is proportional to the radial velocity v_i whose sign is positive or negative depending on whether v_i is toward or away from the transducer. Thus, the Doppler shift of the volumetric echo from the scatterers in the successive bins in the flow field is used to determine the relative radial velocity of the water-current flow in the successive bins along the acoustic beam. Thus, by examining the backscattered signal within discrete preselectable time intervals (range-gated bins) and measuring the Doppler shift within each bin, a time series of the radial components of the water velocity within the maximum attainable range of the ADP can be deduced.

ADP records current-flow velocity as an average of many velocity estimates (called *pings*). The uncertainty of each ping is dominated by the short-term error. Averaging multiple pings as in Equation 11.4 reduces errors (*Aquadopp Current Profiler User Guide*, Nortek AS, Rud, 2005):

$$\sigma_{Vmean} = \frac{\sigma_{Vping}}{\sqrt{N}} \qquad (11.4)$$

where σ is the standard deviation, and N is the number of pings averaged together.

Each acoustic beam measures the flow-velocity component along the beam and does not sense the velocity perpendicular to the beam. ADP measures the full three-dimensional velocity with a minimum of three beams, all pointed in different angular directions relative to the principal axis of the ADP. Under an implicit assumption that currents are horizontally homogeneous (i.e., uniform currents across layers of constant depth), the current velocity vector can be obtained using appropriate trigonometric relations, taking into account the specific beam geometry of the ADP.

The backscattered signal, normally of microvolts in amplitude, is composed of the random sum of the individual scattering amplitudes, each having a Doppler frequency shift associated with the radial velocity (i.e., velocity along the acoustic beam) of the scatterer. Because the micro-inhomogeneities that cause the scattering are of random sizes, the received signal is subjected to amplitude modulation. Phase incoherencies also arise as a result of different ranges of the individual scatterers from the acoustic receiver as well as their random motion due to turbulence within the defined scattering volume. The signal is also corrupted by an additive white Gaussian noise. Thus, the received signal is a highly distorted version of the transmitted signal wherein its deterministic properties are no longer valid, and therefore application of nonstatistical methods for extraction of Doppler-shift information has invariably failed. For example, Crocker (1983) investigated an axis-crossing counting technique in which the returned signals were first heterodyned with closely spaced, digitally synthesized local oscillators to produce a narrow base-band frequency, but the method was found to be unsatisfactory. Because the backscattered signal is statistical in nature with Gaussian distributions, a realistic approach for estimation of Doppler-shift information is some form of statistical method. Most of the Doppler-shift information is contained in the first and second spectral moments of the Doppler spectrum.

Extraction of the mean Doppler frequency from the backscattered signal can in principle be achieved in several ways. However, the need for real-time processing in low-powered, self-contained instruments has tended in practice to encourage the application of either fast Fourier transform (FFT) or complex autocorrelation methods on the received echoes. The FFT algorithm provides a description of the spectrum of the received signal (the mean Doppler frequency shift is assumed to be the first moment of the discrete power spectrum, and the second moment is assumed to be a measure of turbulence in the flow field) but requires rather more computation than does the complex autocorrelation method, which is implemented relatively easily. The complex autocorrelation method appears to perform better, particularly at low S/N ratios (Sirmans and Bumgarner, 1975). Thus, realization of an optimum ADP is very dependent on the nature of the problem under investigation. Discussion of the fundamental constraints can be found in Pinkel (1980).

There are many ways to process the velocity data, each with its own advantages and drawbacks. Until now there existed three broad classes of ADP in terms of pulse transmission and processing of the received backscattered acoustic signals: (1) incoherent, (2) coherent, and (3) broadband systems.

In the incoherent system, each of the acoustic transducers of the ADP transmits an ultrasonic acoustic pulse of temporal length τ and receives the backscattered return echoes as the pulses propagate through the water column along the respective slant ranges defined by the acoustic beam. The return signals are slightly shifted in frequency relative to that of the transmitted signal (i.e., Doppler-shifted) by an amount proportional to the relative radial velocity of the water current. The Doppler-shifted echoes, received at the ADP as a function of time after transmission, yield the relative radial velocity in sequential layers at each of the "range-sliced," insonified (i.e., acoustically illuminated) volumes of the overlaying water column. By adjusting the transmitted pulse length τ, the portion of the water column insonified at any given instant can be controlled in order to choose the desired range resolution.

For a single-pulse system, as just mentioned, slant-range resolution and radial velocity resolution are closely related. The Fourier transform of a transmitted rectangular pulse of temporal length τ has width $(1/\tau)$. Increasing the pulse length improves the resolution of water-current measurements but simultaneously reduces resolution in range. A limit to this process occurs when the pulse length τ equals the scatterer correlation time; thereafter, range resolution decreases with increasing τ, without any corresponding benefit in velocity resolution. However, it may be noted that the resolution product improves with increasing transmission frequency f_t because, for a given radial velocity v of the flow field, the Doppler shift is proportional to f_t.

A more optimistic lower bound on the attainable measurement accuracy has been derived by Theriault (1986a) based on the Cramer-Rao criterion. Analyses of the performance bounds for practical systems have been made by Hansen (1985), applying several types of velocity estimation algorithms to a series of measured and simulated datasets. The use of coded pulses as a means of reducing velocity variance has been proposed by Pinkel and Smith (1992) and independently by Trevorrow and Farmer (1992). The incoherent method (Doppler processing in terms of changes in frequency) is relatively simpler to implement but suffers from limited space-time resolution.

Unlike the just-mentioned noncoherent processing, coherent processing works in the time domain (not frequency). This method is rather new in ADPs, although it was well established in atmospheric sounding radars (Mahapatra and Zrnic, 1983). Coherent processing ADPs use the phase differences (propagation delays), which are exactly proportional to the particle displacement, and with known speed of sound in the sea and time lag between sound pulses, particle velocity can be computed. In coherent processing, a multiplicity of pulses is transmitted while maintaining phase coherence in the transmitted signal. Doppler shift is evaluated from the pulse-to-pulse change in phase of the received signal, with a frequency resolution that depends on the total dwell time.

Because Doppler resolution is no longer dependent on individual pulse length, in contrast to noncoherent processing, pulse lengths can be chosen to provide a desired range resolution, system, or transducer bandwidth, with S/N considerations alone providing a practical lower limit.

Consequently, coherent systems provide greatly enhanced velocity and range resolution. The ADPs developed from the 1980s onward, incorporating coherent processing method (e.g., Rowe et al., 1986; Lohrmann et al., 1990), are capable of providing precision in oceanic current measurements of a fraction of a cm/s, thereby offering an improvement of two orders of magnitude over noncoherent systems.

Multipulse coherent systems are, however, restricted by ambiguities in both velocity and range. The velocity ambiguity arises because the returned signals are effectively sampled at the pulse repetition frequency, f_{pr}, and the usual sampling theorem considerations apply, i.e., aliasing takes place if the maximum Doppler frequency f_{max} exceeds the Nyquist limit $(f_{pr}/2)$. The maximum velocity that can unambiguously be observed is given by the condition:

$$|V_{max}| < \left| \frac{f_{pr}\lambda}{4} \right| \qquad (11.5)$$

Range ambiguity results from the inability to distinguish between the return from a given pulse and the returns from earlier pulses scattered from greater ranges. Maximum unambiguous range is given by:

$$R_{max} = \frac{c}{2f_{pr}} \qquad (11.6)$$

The combination of unambiguous range and velocity thus has the upper limit given by:

$$V_{max}.R_{max} = \frac{c\lambda}{8}, \qquad (11.7)$$

which is independent of pulse repetition frequency.

The enhanced temporal and spatial resolution available from coherent techniques offers possibilities for determining the parameters of small-scale turbulence, and it is in this context in particular that much pioneering work was done (Lhermitte, 1985; Lhermitte and Serafin, 1984). The major effect of turbulence is to increase the spectral variance of the Doppler signal. Consequently, measurements of the spectral width (given by second spectral moment) as well as the mean (given by first spectral moment) become necessary. Interpretation, however, may be complicated by other contributions to spectral broadening from current shear and by any limitation in target residence time in the acoustic beam (Collar, 1993). The coherent method provides improved space-time resolution compared to narrowband processing by a factor of 2−5, but this method is complex to implement. Coherent processing techniques are discussed by Pinkel (1980) and Rowe et al. (1986).

The use of coded transmissions and application of broadband signal-processing techniques provide a compromise between coherent and the simple incoherent systems. Two such systems have been implemented. Brumley et al. (1991) use correlated pulse pairs; the method adopted by Pinkel and Smith (1992) involves the transmission, within a pulse, of a number of repeats of a broadband subcode. Doppler shift is estimated from the complex autocovariance of the return at a lag equivalent to the subcode length. The transmission of multitone and of frequency-shift-keying (FSK) waveforms has been investigated by Andreucci et al. (1992). The broadband method allows the number, composition, and spacing of pulses to be varied to achieve different values of velocity measurement precision, but each variation imposes restrictions on profiling range, instrument velocities and dynamics, vertical resolution, and power consumption that must be considered. Improvements in the standard operating firmware and data-recording capabilities and a decrease in overall instrument size and weight also favor the broadband ADP over the narrowband if equivalent performance can be achieved.

11.3. PROFILING GEOMETRIES

Both vertical and horizontal profiles of water-current measurements can be obtained using ADPs of various geometries. These typically include two-, three-, four-, and five-beam systems, wherein the individual transducer heads (and therefore the acoustic beams emitted from them) are divergent (typically 20° to 30°) relative to the principal axis of the ADP (see Figure 11.2). The ADP measures the current relative to itself; therefore, it is necessary to correct the data for ADP attitude and motion. In the case of stationary transducers, the Doppler shifts directly yield the profiles of absolute radial velocity components. If the

transducers are nonstationary (e.g., mounted on a buoy or at the bottom of a ship), the absolute water-velocity profile can be obtained from knowledge of the motion of the transducers. Incorporation of a magnetic compass enables the ADP to convert velocity measurements to Earth coordinates (i.e., *east, north, and up*, or ENU, components). To obtain the ENU components, the ADP first converts the data to XYZ coordinates (i.e., orthogonal coordinate system relative to the ADP) and then converts them to the Earth coordinates using tilt and heading data.

The accuracy of water-current speed measurement depends on the incorporation of correct measurement of the speed of sound in still water, the transmission angle, and the accuracy with which the Doppler shift is estimated. The speed of sound c (m/s) in seawater is a function of water temperature t (°C), depth d (m), and salinity s (parts per thousand), as approximately given by:

$$c = 1449.3 + 4.572t - 0.0445t^2 + 0.016d$$
$$+ 1.398(s - 35) \tag{11.8}$$

Thus, an incomplete understanding of the *in situ* sound speed in seawater at rest directly affects the water-current velocity estimation, often dominating the error sources. The ADP usually obtains the speed of sound in seawater by assuming a nominal salinity and computing the sound speed based on the measured temperature and transducer depth. The process works relatively well because sound speed in seawater at rest is more sensitive to water temperature than it is to salinity. A pressure sensor is also needed for the measurement of deployment depth for a moored ADP. Because the velocity of sound in seawater at rest is about 1,500 m/s, a water-current velocity profile over the entire range of the profiler is obtainable within a few minutes, even after averaging over many samples.

11.3.1. Bottom-Mounted, Upward-Facing ADPs

Remote monitoring of seawater currents and sea surface waves in the near-shore region is of great interest, both academically and to the general public, because of the role of these currents and waves in coastline erosion and their impact on recreational activities. Making measurements in coastal regions rich in tidal energy is a challenging and potentially dangerous task. Not only do tides run at such high rates that environmental windows—i.e., the only times when it is possible to deploy and recover measurement instrumentation—are extremely narrow, but regions of large tidal currents commonly have whirlpools and standing waves due to huge velocity shearing, which makes safe navigation particularly difficult. In a vertically sheared current profile in coastal waters, the current may be very strong at the surface and much slower at the bottom.

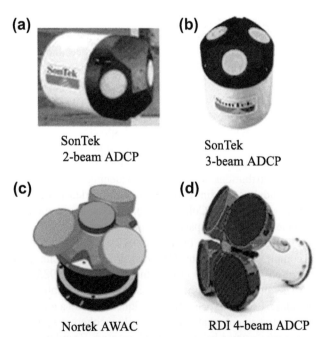

(a)

SonTek
2-beam ADCP

(b)

SonTek
3-beam ADCP

(c)

Nortek AWAC

(d)

RDI 4-beam ADCP

FIGURE 11.2 ADCPs of various geometries, wherein at least two of the acoustic beams emitted by the individual transducers are divergent (typically 20° to 25°) relative to the principal axis of the ADCP.

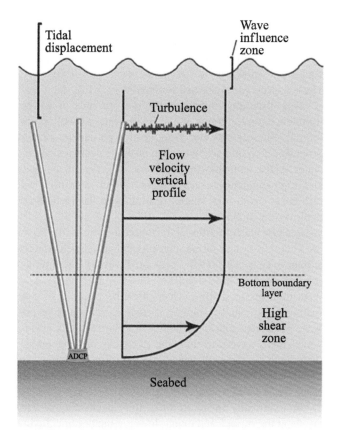

FIGURE 11.3 Schematic diagram showing the vertical current velocity profile in shallow coastal waters. *(Source: Wilson, 2009,* International Ocean Systems, *www.intoceansys.co.uk.)*

Furthermore, highly chaotic current/wave conditions can prevail where there is wind against tidal propagation, where there are dramatic variations in bathymetry, or during storm-surge events. As the current profile in shallow coastal

regions (see Figure 11.3) is influenced by several factors, such as bottom boundary layer, tidal displacement, wave influence, and the like, high-frequency ADPs capable of high-resolution profile measurements (see Figure 11.4) need to be deployed for capturing all these variabilities occurring in a relatively small water column. Placing equipment on the seabed in the hostile conditions of the coastal waters is challenging. Therefore, a single instrument package that can measure multiple parameters is of great practical utility to both technologists and user communities.

11.3.1.1. Combination of Surface Currents and Gravity Wave Orbital Velocities

Water currents at the topmost layers of the ocean are a combination of surface currents and orbital velocities of surface gravity waves (wind-driven waves). In shallow water, the relationship between the vertical and horizontal components of water-flow velocity changes with depth, because wave orbits are approximately circular near the surface (i.e., the magnitudes of the vertical and horizontal components of velocity are approximately equal), but near the bed the orbits are horizontal (the vertical velocity is zero or near zero). In the past, bin-averaged, Earth-referenced vector-current measurements provided by shallow-water ADPs led to some confusion among some inexperienced users who were previously familiar with water-current measurements provided by conventional Eulerian-style CMs only, which do not have the ability to measure sea surface wave orbital velocities, although the water-current measurements provided by such CMs are often contaminated in varying degrees by wave motions. Fortunately, probably in an attempt to provide more clarity on shallow-water ADP measurements, Howarth (1999) reported the raw measurements (individual

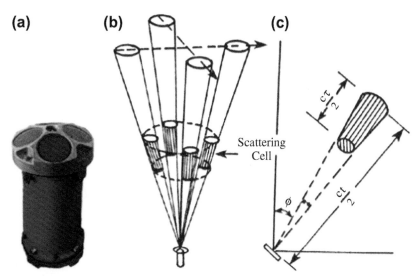

FIGURE 11.4 (a) A four-beam ADCP in Janus configuration; (b) upward-looking slanted beams from the ADCP and the scattering cells in a horizontal plane; and (c) expanded view of a typical scattering cell. High spatial resolution is achieved by the use of small value for τ.

measurements from each bin of each beam) obtained from a standard four-beam, 1.2-MHz ADP deployed in a seabed frame in Red Wharf Bay, off Anglesey in the Irish Sea, in a water depth of 23 m at low water. In this dynamic (tidally stirred) environment, there were sufficient scatterers (sediment) for there to be a good signal in all 20 bins. However, data quality in the uppermost two bins (bins 19 and 20) was affected by side-lobe interference from the sea surface return around low water. The instrument sampled every 2 seconds for 25 hours, recording beam data in 20 1-meter bins from 3.6 to 22.6 m above the bed (the beams were at 20° to the vertical).

To gain a quick impression of the data content and quality, preliminary data analysis was carried out by splitting the records from each beam and for each bin into 10-min blocks (300 samples) and calculating means and standard deviations. Ten-minute blocks have been shown to be a sensible partitioning of the record so that each block includes sufficient high-frequency events (waves and turbulence) for analysis, at the same time as being stationary with regard to the tides (Soulsby, 1980). The "means" clearly shows the tidal signal. It is interesting to note that the standard deviations also exhibit an identifiable signature in which amplitude increases with height above the bed. This contrasts with the deployments in deeper water, where tidal currents were weak, which showed no differences between the bins, or with time, and standard deviations were mainly in the range 10−15 mm/s, as would be expected given the ensemble standard deviation. The patterns illustrated in respect of beam 3 were exhibited by all four beams, both in the vertical and with time. The signal was highly coherent with depth and oscillatory with a period of about 5.5 s.

During a stormy period, the standard deviations decreased exponentially with depth, in which all bins are plotted. The deployment and measurement schemes adopted by Howarth (1999) revealed the finer intricacies of measurements actually performed by the ADP in shallow waters. For example, wave activity, with peak periods of between 5 and 6 seconds, was detected in the top bins throughout the record. The largest wave activity occurred during low water, when wave orbital velocities were twice as large as for the rest of the deployment. During low water, wave activity was detected in all current bins, decreasing exponentially with depth.

To some pessimistic users who lacked acumen and have no nose for knowing the unknown, the ADP measurements as described would have been "instrument noise" or a result of "deployment error." However, to some erudite technologists and devoted researchers with a nose for mining the hidden treasures from the deep, the "noise" was indeed an invaluable signal. These technologists and researchers realized that, clearly, sea surface wave orbital velocities were being measured, and the shallow-water ADP measurements are indeed a combination of currents, wave orbital velocities,

and turbulence. This was realized in the late 1980s, and this realization ultimately culminated in the invention of the now-famous seminal piece of oceanographic remote sensing equipment known as a *directional wave-measuring ADP*. These aspects are addressed in Section 11.3.1.2.

This discussion indicates that the presence of waves, turbulence, and currents in the wave zone, as well as the mutual influence between them, need to be understood and properly taken into account when profilers such as shallow-water ADPs with the capability for detection of currents and high-frequency events (wave orbital velocities and turbulence) with high spatial resolution in the vertical are deployed in shallow waters.

Because water motions at the topmost layers of the ocean are a combination of surface currents and orbital velocities of surface gravity waves (wind-generated waves), fast-sampled (of order 1 Hz) high-frequency (1 MHz or higher) ADP measurements in shallow water offer the prospect of estimating high spatial resolution surface currents and several wave parameters that are difficult to measure by other means: wavelength, wave direction, and the decay of wave energy with depth as well as the standard time-series determination of wave period from individual bin measurements (Terray et al., 1990; Gordon et al., 1998).

In a vertically sheared profile, the current may be very strong at the surface and much slower at the bottom. Taking these into account, it is appropriate to obtain the estimates of an integrated mean surface current by creating a weighted average of the ADP current profile relative to the depth penetration of the wave. The depth-average current U can be estimated from the vertically sheared current $U(z)$ using the expression (Kirby and Chen, 1989):

$$U = \frac{\int_{-H}^{0} U(z)E(\omega, z)dz}{\int_{-H}^{0} E(\omega, z)dz} \qquad (11.9)$$

where ω is the bottom-relative radian frequency of the wave, H is the water depth below the surface, and E is the wave energy. The specialty of the ADP is that it is particularly useful in measuring the current profile over the region in which wave energy is concentrated.

11.3.1.2. A Single Upward-Facing ADP for Measuring Currents and Surface Waves

Historically, the technology for measuring ocean currents and sea surface gravity waves has been distinct, requiring separate instrumentation for each. For instance, it is quite common to measure sea surface waves from a surface-floating buoy (wave rider buoy). However, there are some locations that preclude such measurements on practical grounds, as well as many other locations that present challenges to the survivability of surface buoys. The challenges include shipping traffic, vandalism, and ice cover. One alternative to wave measurements is to measure pressure from below the sea surface. The method has shown good results; however, it is limited to shallow coastal waters (because short gravity waves attenuate with depth) or to measurements of only long waves, such as tides and tsunami waves. Notable developments have improved on the earlier shortcomings.

Fortunately, research initiated in the 1980s and continuing into the 1990s and beyond (e.g., Krogstad et al., 1988; Terray et al., 1990; Herbers et al., 1991; Zedel, 1994; Visbeck and Fischer, 1995; Terray et al., 1997; Terray et al., 1999; Pedersen and Lohrmann, 2004) showed that apart from the well-established capability of the ADPs for remote measurement of horizontal and vertical water-current profiles, they can also be used for measuring sea surface wave height and direction from a single upward-looking multibeam ADP package deployed in shallow depth relative to the surface.

As indicated earlier, ADPs intended for vertical profiling of water currents employ acoustic beams inclined at an angle relative to the vertical (typically by 20−30°). The sonar measures the instantaneous water-velocity component projected along each beam, averaged over a range cell. Because the "mean" current is typically horizontally uniform over the beams (horizontal homogeneity is assumed, at least), its components can be recovered as linear combinations of the measured along-beam velocities.

Application of ADPs for Sea Surface Gravity-Wave Measurements

The ADP measures sea surface waves using three independent techniques for redundancy (Terray et al., 1999). The primary method of wave measurement is profiling the orbital velocity of the wave. The ADP uses the along-beam component of the orbital velocity for each depth cell to construct a virtual array (the collection of ADP range cells constitutes the array). In a four-beam ADP, there are six different combinations of pairs of beams, any two of which allow the calculation of the wave direction and the wavelength for each bin. Array-processing methods are then used to estimate the direction of wave arrival.

Although traditional current profiling expects horizontal homogeneity, wave measurement capitalizes on the phase differences inherent in a spatially separated array of sensors. Cross-spectra are calculated between every sensor, and the directional spectrum is assumed to be linearly related to the cross-spectra. The cross-spectra will give the phase differences between the various beams. From the phase differences, the wave direction and wavelength can be calculated by simple trigonometry based on the beam geometry of the ADP. The procedure used here for wave-direction estimation is similar to that used in HF Doppler radar technology for current direction estimation (see chapter 4). Linear wave theory is used to translate the measurements from velocity spectrum at various depths to surface displacements (Note that surface displacement is the time-integral of orbital velocity). Orbital velocity provides a measure of both directional and nondirectional waves. Because the ADP wave gauge automatically chooses depth cells below the surface, even when the surface moves with tides, wave measurements based on the orbital velocity estimates are of high quality, even during hurricane events. This was noticed in the ADP-based wave measurements during hurricane Lili.

Although fast-sampled, high-frequency ADP measurements of wave orbital velocities in shallow water offer the prospect of measurements of surface gravity wave parameters in a novel perspective, interpretation of such measurements is difficult, both because the ADP measures the component of speed along the beam toward the ADP (neither the vertical nor the horizontal component, but one at an angle to the vertical) and, of greater consequence, the ADP beams diverge. Whereas the separation of the beams is not significant in measuring currents averaged over periods of a minute or longer, beam separation is important for studies of waves, because the separation at the surface can be a significant fraction of the wavelength, hence making it impossible to resolve the pairs of opposing beam velocities sensibly into the vertical and horizontal components of the wave orbital velocity. This also means that to study waves, velocities should be recorded in beam coordinates and not Earth coordinates. A consequence of the angling of the beams is also that vertical profiles are not directly measured. The separation of the beams is, however, turned to advantage, and the array of multiple beams is used to estimate wave direction and wavelength from the phase differences between the beams. Because the lower bins (i.e., those closer to the ADP) are closer together horizontally, the phase differences between the beams should be smaller there.

The second technique is the surface track. This is simply echo location of the range to the surface and is an excellent nondirectional ground truth because it is a direct measure of the surface and does not depend on wave number.

The last measure of nondirectional spectra is the pressure sensor. This method has a long history but is very sensitive to wave number. Having three independent sources of wave spectra in one instrument allows one to identify error sources and ensure data quality. The three spectra should agree very closely.

The conventional slant-beam configuration can be used to determine the water column vertical height above the ADP and, therefore, the wave height. Based on several field measurements, it was realized that the slanted-beam configuration of the ADP for wave-height measurements does not work well during periods of swell. It is generally agreed that for wave-height determination by an ADP, one of its acoustic beams needs to be oriented vertically (Zedel, 1994; Visbeck and Fischer, 1995). The reason for this preference is that a vertical beam for acoustic surface tracking (AST) provides reliable direct measurement of the distance to the surface and thus an estimate of the surface position under almost all wind and wave conditions. In particular, for the so-called "Bug-eye" 3+1 beam configuration (see Figure 11.2c), the vertical wave velocity (which is nondirectional) can be used to estimate the wave-height spectrum (Zedel, 1994) with a slight improvement in performance over conventional Janus type slant-beam ADPs (Terray et al., 1999).

AST is central in the ADP-based wave measurement scheme, to which there are several benefits (Lohrmann and Siegel, 2010):

- Wave processing is much simpler because the estimate of distance to the surface (AST) is free of any complex transfer functions and therefore it provides a direct estimate of the sea surface variations.
- Unlike sea surface wave-related parameters such as pressure and velocity, the AST retains high-frequency wave information because it is not lost by attenuation.

A direct estimate also means that a time series of the sea surface elevation is available to calculate estimates of $H_{1/3}$, $H_{1/10}$, H_{max}, and T_{mean}. Furthermore, it is unaffected by mean currents that impose a Doppler shift on the surface waves (see the principle of HF Doppler radar technique used for surface current measurements)—a result requiring special consideration when calculating the frequency-dependent transfer function for the velocity and pressure. Perhaps the most important benefit is the simple fact that wave resolution (shorter wave information) is only mildly affected by increased deployment depth.

However, it must be noted that although the response is mild, some high-frequency wave information is lost at greater depths. The extent of the high-frequency filter is determined by the size of the sea surface area ensonified (sampled) by the AST pulse. For a given beam width, the diameter or "footprint" of the ensonified area increases as the vertical distance (i.e., depth) from the surface increases. When the diameter of this footprint becomes similar in size to a wavelength of the sea surface wave, then the structure of the waveform (crest and trough) cannot be well resolved. There is an averaging of the distance over the waveform and thus a "smearing" of the true features. A "cut-off" frequency is assumed when the footprint diameter is equivalent to half a wavelength (of the sea surface wave).

Within the limitations just indicated, the AST solved the depth limitation for nondirectional wave measurements, but the limitations for direction waves still remained. Estimation of wave direction is more complicated than wave-height determination. This complication stems from the fact that, unlike tsunami waves, the surface gravity waves are not spatially coherent and the waves at any given instant arrive from multiple directions.

In the state-of-the-art acoustic Doppler method, the directions of arrival of sea surface waves (i.e., wave directions) are estimated by using several measurements of wave orbital velocity, together with the AST. As mentioned earlier, the collection of ADP range cell measurements form an array projected from the ADP to just below the sea surface. A number of methods have been invented to estimate the direction of multiple arrivals using sparse arrays (e.g., Johnson and Dudgeon, 1993). Terray et al. (1999) used an iterative version of Capon's Maximum Likelihood Method (Capon, 1969), known as the Iterative Maximum Likelihood Method, or IMLM (Krogstad et al., 1988; Pawka, 1983). This technique has the advantage that it does not require that the array be uniformly spaced. Iteration improves the consistency of the MLM estimate in the sense that the estimated directional spectrum produces cross-spectra in closer agreement with the observations. The result is to sharpen the directional resolution and decrease the side lobes. Like all high-resolution, direction-of-arrival estimators, the MLM is model-based and requires a relation between the frequency-direction spectrum and the array covariance. For example, Strong et al. (2003) estimated the wave directional spectrum by the IMLM, which requires a model of the ADP response, $H(k)$, to a monochromatic wave propagating at an arbitrary angle. Consequently, at each frequency they computed $k(\omega, \alpha)$ over all angles from 0 to $360°$ (typically at $4°$ increments) and use this to determine the response function H. In this, k is the bottom-relative wave number; ω is the bottom-relative radian frequency; and α denotes the included angle between the directions of wave propagation and the current.

The array-processing method (the MLM) exploits the time lags between spatially separated measurements of the array in order to estimate direction. Alternatively, the same data may be used in the form of "classic" triplet processing (SUV), in which the triplet is formed by the AST (i.e., distance to the surface, S) and two mutually orthogonal horizontal components of orbital velocity measurements (U and V). Both of these directional-processing methods rely on orbital velocity measurements. The frequency-dependent response of orbital velocity with depth determines the frequency resolution for wave directions.

Orbital velocities attenuate exponentially with depth, and this behavior is more severe for higher-frequency waves (short waves). This means that the further down in the water column that the orbital velocities are measured, the less high-frequency information is available. This is the classic problem faced by bottom-mounted instruments, and it may be noted that even the ADP class of instruments suffers from this challenge if it is not managed effectively.

Managing the response means positioning the measurement cells as close to the surface as possible while ensuring that there is no contamination from the surface, either directly from the cells touching the surface or indirectly from side-lobe energy that leaks off the main beam. This can be managed by positioning the cells just below the surface by a fraction of the measured depth; 10 percent of the depth has proven to provide a good signal response without contamination (Lohrmann and Siegel, 2010).

It is important to note that the limitation for the directional estimates is also imposed on the nondirectional estimates if orbital velocities are used to estimate the energy density spectrum. An accurate directional estimate is required if the surface wave is to be estimated accurately for each individual beam. This is why the AST remains the primary estimate and the orbital velocity a secondary estimate for energy.

A second limitation imposed on the resolution for directional estimates is associated with the spatial separation of the measurements. Wave-directional estimates become ambiguous when the horizontal separation is equal to half a wavelength. The result is that waves at the associated frequency cannot be accurately estimated. One perceived solution is to position the measurement cells closer to the ADP such that the horizontal spatial separation between the cells located on a given horizontal plane is reduced; consequently, the surface wave frequency at which this ambiguity occurs is higher. Unfortunately, moving the measurement cell further down in the water column means that the orbital velocity signal disappears due to the well-known sea surface wave attenuation with increasing depth. The result is that there is no performance gain by drawing the cells in closer to the ADP.

In practice, at any instant the wave velocities vary spatially across the array. As a result, except for very long waves that remain coherent during their passage through the array, it is not possible to separate the horizontal and vertical wave-velocity components. However, the wave field is statistically stationary in time and homogeneous in space, and therefore the cross-spectra between velocities measured at various range cells (either beam-to-beam or along each beam) contain information about wave direction. Terray et al. (1999) demonstrated that wave height and direction spectra compare well with a co-located array of pressure gauges.

Although it is true that the collection of ADP range cells constitutes an array, it is a somewhat peculiar one (Terray et al., 1999). First, the S/N ratio in the measured wave velocities varies with depth. At a particular frequency, this variation is due primarily to the vertical decay of the wave energy and (to a lesser extent) on the acoustic S/N ratio, which itself is a function of range. Second, the velocity measured by the ADP is a linear combination of horizontal and vertical wave velocities, with relative weights that depend on both the propagation direction of the waves and the height of the measurement cell above the bottom. For

example, in deep water, as the surface wave direction varies from 0 to 90°, the measured along-beam velocities vary by only 25 percent. Finally, because the range cells closest to the surface have the highest wave velocity S/N ratio, the usable array lag distribution is quite sparse, consisting of long lags from beam to beam and a second group of shorter lags between range cells along each beam. It is quite clear from these considerations that unlike conventional point sensors, the "sensor array" made up of the range cells associated with an upward-looking ADP does not have a simple relationship with the low-order circular moments of the wave directional distribution, and therefore recovering wave direction is considerably more complicated in this case. In any event, the limitations for directional waves require addressing the issue of getting the measurement cells (for orbital velocities) closer to the surface, where the signal is less attenuated by depth.

Strong et al. (2003) observed that the standard ADP wave-processing software (WavesMon) provides accurate spectra and parameters by applying the Doppler-shifted dispersion relationship:

$$(\omega - kU\cos\alpha)^2 = gk\tanh(kH) \qquad (11.10)$$

where ω and k are the bottom-relative radian frequency and wave number (i.e., those observed in a fixed reference frame), H is the water depth, U is the current, and α denotes the included angle between the directions of wave propagation and the current U. Note that Equation 11.10 is just the usual dispersion relation for arbitrary water depth applied to the frequency observed in a reference frame moving with the current. In contrast, other methods of wave measurements showed significant wave-height errors greater than 50 percent for hurricane Lili. This indicates that caution should be exercised in interpreting historical wave height data because it is unlikely that storm surge was taken into account.

Based on persistent research, combined wave- and current-profiling ADPs have now been operationalized to a great level of success. The combined acoustic wave and current (AWAC) profiler system developed by Nortek (see Figure 11.2c) is one such instance. This variant of the traditional ADP has managed to circumvent the classic limitations of measuring short surface gravity waves in deep waters by introducing a vertical beam that directly measures the height of the water-air interface (waves) above the instrument. In the combined wave and current profiling ADPs, acoustic Doppler measurements (standard for an ADP) are used for construction of current profiles; a dedicated vertical narrow beam for AST and near-surface Doppler velocity measurements are together used for wave measurements. The AST method traces the surface wave height profile as it passes through its field of view.

The AWAC, which is intended for greater deployment depths and profiling ranges, is a dual-frequency instrument whereby the off-vertical beams used for current profile measurements transmit at 400 kHz and the vertical beam used for AST transmits at 600 kHz. The vertical beam transmits at a higher frequency in order to maintain the beam's narrow opening angle (1.7°). This same vertical beam has also demonstrated that it is capable of measuring the distance to the water-ice interface, and as a result it can be used as a means to estimate ice draft or ice thickness as well from more extreme latitudes, where the presence of ice is more common (Lohrmann and Siegel, 2010). The wave-burst measurement contains detection methods for both water-air and water-ice interfaces, thereby allowing it to transition seamlessly from wave measurements in the summer to ice measurements in the winter. The AWAC can measure waves and currents over a full depth of 100 meters.

Lohrmann and Siegel (2010) have reported the depth-response evaluation test results of 400-kHz and 600-kHz AWAC wave measurements for both the AST and wave-directional processing, wherein the former and the latter were deployed at 90 and 19 m water depth, respectively, at two spatially separated locations south of Oslo Fjord, which opens to the North Sea. According to them, the AST proved to perform remarkably well. The test period of 10 days provided 240 wave bursts, each of which contained 2,048 samples. Most bursts had no false detects, and the maximum number of false detects was three samples in a single wave burst. The surface wave-frequency resolution associated with the ensonified footprint on the sea surface was consistent with the estimates from the AST beam width and range.

It was observed that there is a low-frequency noise floor for the wave measurements. One explanation for the perceived low-frequency energy is that there are greater fluctuations of the speed of sound (induced primarily by water-temperature fluctuations) as different masses pass over the AWAC during the wave-burst measurement. These fluctuations manifest themselves as variations in the overall range, albeit small relative to the surface wave variations.

Ice Thickness Estimation

Ice thickness estimates from the subsurface require an accurate depth estimate and an accurate distance measurement (AST). To fulfill this requirement, the AWAC now has a temperature-compensated pressure sensor to reduce the uncertainty in the pressure measurements. The AST measurements have been modified by including a second ranging estimate by using a special filter for the water-ice interface, which is different than that used for detecting the water-air interface. Either ice or wave processing can be performed because both estimates are reported within the same wave burst. This makes it ideal for year-long deployments at extreme latitudes where the same measurement scheme is used for waves in the summer, ice thickness in the winter, and both during the transitional periods in the spring and fall.

If an absolute pressure sensor (one that measures subsurface pressure and barometric pressure) is used for subsurface pressure measurements, care has to be exercised while interpreting the ice thickness estimates, because an absolute pressure sensor does not discriminate between subsurface pressure and barometric pressure. Thus, barometric pressure needs to be subtracted from the absolute pressure measurements to obtain the subsurface pressure of interest. Failure to do this will result in contamination of wave and ice thickness measurements.

The 400 kHz AWAC's estimates of significant wave height (H_{m0}) and peak period (T_{peak}) showed good agreement with that of the nearby 600 kHz AWAC. Peak periods are as short as 2 s and the wave heights as little as 30 cm. The directional estimates show a greater variability and not quite as good agreement. There is a notable difference in the first and second half of the deployment. The first half shows a consistent direction from the south-southwest, whereas the second half exhibits a random distribution of the estimates. The reason for this is that the peak period shortens from 8 s to less than 4 s. When it reaches 4 s, the orbital velocities at depth are no longer measurable. Here we witness the frequency limit for directional estimate at approximately 4 s for a deployment depth of 90 meters. The two AWACs have an apparent bias of the wave directions for the first half of the deployment, for which the most plausible explanation is refraction that occurs between the spatially separated AWACs.

11.3.1.3. Current Profile and Wave Measurements for Operational Applications

The unique capability of upward-looking near-surface-deployed ADPs for measurement of vertical profiles of horizontal velocities and surface wave parameters can be effectively utilized for several academic investigations and operational applications. However, for such measurements from shallow coastal regions, ADPs in seabed-mounted frames are required. Use of frame-mounted ADPs requires careful attention because the equipment needs to be located on a stable region of the seabed and in a stable configuration, and it needs to stay in the spot where it is deployed. The ADP needs to remain vertical, which may require frames of specialized design (see Figure 11.5) and possibly divers during deployment to finalize positioning. To ensure recovery, acoustic release units (which can be actuated from a boat or a vessel) should always be integrated into the frame. Bottom-mounted platforms have proven to work well on condensed bottom sediments in relatively shallow water depths.

Requirements are also increasing for real-time current and directional wave measurements at offshore sites in support of oil exploration, wind, wave, and tidal energy production as well as commercial and research ocean

FIGURE 11.6 Platform-mounted AWAC with asymmetric transducer head for deployment on an offshore structure. *(Source: Siegel, 2007,* Ocean News & Technology, *www.ocean-news.com.)*

FIGURE 11.5 A Nortek AWAC profiler instrument being deployed on a bottom frame in the Gulf of Mexico. *(Source: Siegel, 2007,* Ocean News & Technology, *www.ocean-news.com.)*

observing systems. There are many challenges to operational systems that provide real-time data. Siegel (2007) has reported on new hardware products and measurement techniques that have been developed by Nortek (USA) to provide robust solutions for these demanding offshore requirements.

In shallow, coastal environments (less than 50 m depth), a common solution for current and directional wave measurements is to deploy an ADP on the seafloor. Keeping the equipment away from the ocean surface provides several advantages, such as reduced exposure to harsh storms, security from theft or vandalism, and protection from ships, ice, or drifting debris. Historically, wave-rider surface buoys and fixed-mounted equipment such as wave staffs and wave radar systems have been used for making sea surface wave measurements in offshore regions (deeper than 50 m). At this depth, bottom-mounted ADPs do not provide the directional resolution necessary for research and commercial wave-measurement requirements. Mounting an acoustic system on a subsurface buoy or underwater directly to an offshore platform would permit the instrument to be close enough to the surface for high-quality wave measurements, yet be removed from the dangers of exposure at the surface. There was a dearth of commercial off-the-shelf solutions to meet this requirement.

The AWAC is designed for the special purpose of being mounted directly to a subsurface structure on an offshore platform. The resulting design, commonly known as the Platform Mount AWAC, employs four acoustic transducers asymmetrically arranged on one hemisphere of the system (see Figure 11.2c) to point away from the offshore platform. The three slanted beams are used for current profile and wave

directional estimates, and the fourth (vertical) beam is used to measure wave height. This permits the Platform Mount AWAC to be deployed underwater directly to an offshore structure while at the same time measuring the waves and currents away from the structure (see Figure 11.6).

The upward-looking Platform-Mount AWAC is typically deployed 15–30 m below the sea surface. A downward-looking ADP can be placed below the AWAC to extend the current measurement range into deeper waters (see Figure 11.7). A cable running to the Platform Mount

FIGURE 11.7 Platform-mounted AWAC being lowered into place on an oil platform. Downward-looking AWAC is used for current profile measurements into deeper water. *(Photo: B. Magnell, Woods Hole Group. Source: Siegel, 2007,* Ocean News & Technology, *www.ocean-news.com.)*

AWAC provides power to the system and transmits data from the AWAC to a computer or telemetry node on the platform. The first such unit was deployed in 2006 on the Ambrose Light Coastal-Marine Automated Network (C-MAN) Station (offshore New York harbor) operated by NOAA's National Data Buoy Center (NDBC); it transmits processed data to the shore by satellite. Real-time wave data from this AWAC are available on the NDBC Website.

Real-time water current and surface gravity-wave measurements from offshore structures may provide critical decision-making data for safe personnel and vessel operations as well as long-term wave-loading information for facility design, maintenance, and repair planning. Future applications could include real-time dynamic feedback loops for optimizing wave energy-harvesting systems, such as real-time modifications of the harvest system based on parameters such as wave height or peak period.

Offshore wave measurements are often needed for site surveys during the planning stages of petroleum and renewable energy projects as well as for fixing boundary conditions for wave models. Sea surface wave buoys have been a common solution to meet these requirements; however, surface buoys may be damaged by storms, ice, and ships. In addition, there are many places in the world that surface buoys would be vandalized or stolen. The ability to mount an ADP for wave measurements on a subsurface buoy would permit the instrument to be close enough to the surface for high-quality wave measurements yet be removed from the dangers of exposure at the surface (see Figure 11.8). However, a subsurface buoy will both

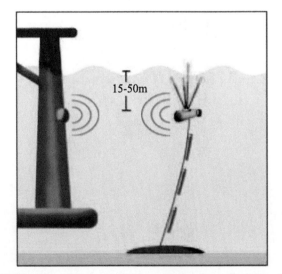

FIGURE 11.8 AWAC mounted on subsurface buoy. Real-time data telemetry is provided to the offshore structure via underwater acoustic modems. *(Source: Siegel, 2007, Ocean News & Technology, www.ocean-news.com.)*

rotate and translate during a wave-sampling period of 10—30 minutes. For this reason, it is not possible to measure waves with a typical ADP deployed on a subsurface buoy without modifying the wave-processing algorithms developed for stationary systems. Nortek managed to resolve this problem by introducing a method called *SUV* (Siegel, 2007), addressed in Section 11.3.1.2.

Real-time current and wave data are critical for safe operations at sea and assimilation with nowcast and forecast models. The two most common methods of underwater data telemetry are cables and wireless underwater acoustic modems. Rugged cables may be used to transfer data from a Platform Mount AWAC up to a data collection station on the platform. This cable may also provide power to the AWAC for long-term measurements without the need to change batteries.

Wireless underwater data telemetry solutions are required for instruments mounted in areas where cables are not suitable. This includes regions where bottom trawling is common, on subsurface buoys, and in other situations where cables are impractical because of time or cost constraints. Nortek uses a modified version of the Benthos modem to transmit data from the AWAC to a receiving station such as an offshore platform or surface buoy. The horizontal distance between the acoustic modems can range up to 1—3 km, depending on several environmental factors such as local bathymetry, hydrography, and environmental noise.

Underwater acoustic modems operate at low data transfer rates (typically 300—9,600 baud). To address the increasing need for wave data passing through low-bandwidth channels, Nortek developed the Nortek Internal Processor (NIP). The NIP is a scaled-down PC running a Windows CE operating system. Small enough to fit within the AWAC, it processes the raw AWAC data into a user-selectable set of current-velocity profiles, wave parameters, and energy spectra. These processed data are considerably reduced in size (0.1—1.0 kilobytes) compared with the raw wave data (25—50 kilobytes) and can be easily transmitted through the Nortek underwater acoustic modems. Data formats can be adapted to the user requirements and can be scaled for satellite transmissions.

It has been indicated that the power of the ADP can now be used to obtain wave data while also measuring current profiles. This has been capitalized on by several competing manufacturers. For example, RD Instruments Inc. reported the availability of a new patented capability as an upgrade for RDI WorkHorse ADCPs (Strong and Devine, 2000). They reported results demonstrating high-quality output from the ADP wave analysis. In particular, it has been found that ADP wave data do not suffer from biases in significant wave height and direction that arise in traditional wave analyses that ignore wave interaction with strong tidal currents. Directional resolution of the ADP wave spectra is improved; wave packets of similar frequency from different directions are distinguished. Because of its remote sampling capability and four acoustic beams, the RDI WorkHorse ADCPs can be deployed safely on the bottom yet act as a surface and near-surface array to measure waves.

Users of the RDI WorkHorse ADCP directional wave gauge can compare statistics from three independent data types measured simultaneously. The first data type is direct tracking of the water surface variation along each of the four beams. The second data type measured by the ADCP is wave orbital current fluctuations. These fluctuations are measured with each beam at 3–5 m depths near the water surface. Once these data are transformed to the surface using linear wave theory, the array has at least 12 elements spaced at longer beam-to-beam lags and shorter bin-to-bin lags. This combination produces wave-directional information across a wide bandwidth, including high-frequency waves in deeper water (e.g., measuring 4-s waves in 40 m of water). The third data type is obtained via a pressure sensor. The pressure sensor is used first to establish the depth of the ADCP below the sea surface, from which the separation of the ADCP array elements is determined. So the extra benefit of these pressure data is a third independent measurement of waves.

Wavelength is shortened when the phase speed of the wave is opposed by a background current. Consequent to this shortening is an enhanced decay with depth of the deep-water wave signature. Without including this influence in the data processing, the surface wave fluctuations inferred from pressure measurements at depth will be underestimated. Wave direction, too, can be distorted by wave-tidal flow interaction. Through the beam-former capability of the RDI WorkHorse ADCP directional wave gauge, the ADCP wave array resolves wave direction into 90 segments around the compass. The ADCP Directional Wave Gauge surpasses the traditional trade-off in wave gauge choices. The advantage comes from the ADCP's capability to measure remotely. In short, direct measurements of surface and near-surface variability are significant advantages compared with pressure-based wave gauges. For example, direct tracking of the surface increases the bandwidth of the wave gauge's frequency response. Whereas the frequency cut-off for the pressure-based wave gauge is about 5 seconds, the surface track ADCP detects periods shorter than 1.5 seconds. The ADCP measurements of wave orbital current fluctuations are near surface and have a high frequency cut-off near 2.5 s. By including the simultaneous current profiles in the wave analysis, the ADCP wave results do not suffer from biases in significant wave height and direction that arise in traditional wave analyses. There is also the improved directional resolution of the ADCP wave spectra,

distinguishing wave packets of similar frequency from different directions.

In summary, for situations where the performances of other wave gauges are limited, the ADCP Directional Wave Gauge continues to measure waves accurately. In particular, the ADCP has been demonstrated to have an impressive bandwidth from short local seas to long swells, eclipsing the performance of other common choices for wave gauges (Strong and Devine, 2000).

In addition to the application of ADPs for port operations and navigational safety, bottom-mounted, upward-looking ADPs have begun to be used for routine real-time reporting of current profiles in hazardous navigation channels (Froysa et al., 2007). For example, Aanderaa Data Instruments (AADI) of Bergen, Norway, installed a real-time environmental data observation station in the notoriously hazardous Bergen Fairway, which is the gateway to Bergen Port, on Norway's rugged West Coast. The majority of Bergen's maritime traffic must pass through these unfriendly straits, where complicated currents are driven by tides, storm surges, and wind. In January 2004 the motor ship *Rocknes* capsized in the straits with the loss of 18 lives, presumably by the force of strong currents and conflicting winds, causing the ship to strike an uncharted reef. The coastal observation station provides hitherto unavailable real-time graphical displays of oceanographic and meteorological data at all states of the tide, aided particularly by AADI's ADP.

Ocean current and surface wave data are measured by the RDCP 600 profiler, which is seabed-mounted in 22-m water depth within a gimbaled frame and cable-connected (75 m) to the host station. The unique feature of AADI's RDCP 600 profiling CM is its ability to measure current "bins" relative to both the sea surface and the seabed. The "relative to sea surface" current data ensures that measurements of current speeds are always available at critical "below surface" depths (e.g., ship's draught, irrespective of the state of the tide).

Acquired data are published to the Web. Results have shown that currents conflict at different depths and are additionally influenced by the prevailing wind conditions. The AADI coastal observation station has considerably improved the safety of vessels navigating this complex and hazardous waterway.

11.4. TRAWL-RESISTANT ADP BOTTOM MOUNTS

Uninterrupted data collection from areas of vigorous fishing activity is a very difficult task. In the areas with heavy trawling activities, a trawl-resistant seabed platform is usually used for protection of the ADP. ADPs' capability for high-resolution remote current profile measurements has created new research opportunities in oceanography. This is especially true in shallow water, where instruments mounted on the bottom in trawl-resistant bottom mounts (TRBMs) can provide time series of current measurements with high vertical resolution from near surface to near bottom in water depths that include most of the world's continental shelves. Application of this technique was accelerated in the late 1990s by a new generation of ADPs that are smaller, lighter, and less expensive than their predecessors. Further encouraging the trend has been an increased concern for the ecological health of the oceans' shallow regions. At the same time, commercial and recreational use of near-shore waters is at a level that puts oceanographic equipment at high risk, especially for long-term deployments. TRBMs for such an application require careful engineering because a single high-value unit is exposed to a large number of risks, such as damage by trawlers, storms, corrosion, or bio-fouling.

Keeping the just mentioned requirements in view, Perkins et al. (2000) reported the development of a TRBM called *Barny* (so called because of its near barnacle-like shape) to meet the needs of long-term deployment of ADPs in water depths up to 300 m. The modified name *Barny Sentinel* is derived from a larger version developed several years ago at SACLANT Center and later extensively redesigned to incorporate an RDI Sentinel ADP and a Sea-Bird Wave-Tide Gauge. In this platform, a concrete ring surrounding the Fiberglas instrument housing serves as a ballast and impact protection. The ballast ring is made of concrete, reinforced with AISI316 stainless steel. This structure, like the rest of the Barny, is entirely nonmagnetic to prevent interference with the ADP compass. Barny's overall smooth profile minimizes the risk of being fouled by fishing gear.

The unit is lowered by electromechanical cable. Through this cable, real-time information regarding instrument pitch, roll, and depth is passed from a sending unit, which is temporarily attached during launch, to a PC-based display on the ship. In operation, two independent recovery modes may be activated by acoustic commands, one of which can operate whether or not the platform is upright. Once satisfactory placement on the bottom is confirmed, the external release is activated and comes away together with the sending unit. Two floats on the launch line provide buoyancy for the release. The acoustic releases also function as transponders so that bottom position of the platform can be determined acoustically. Features have also been included to facilitate recovery by remotely operated underwater vehicle (holes provided on the ballast ring permit recovery by this means). A series of tests and applications have validated all basic functions and trawl resistance. Platform cost has been kept below that of the onboard instruments while still maintaining the required high level of recoverability.

Skelin et al. (2008) reported two deployment configurations of trawl-resistant platforms for bottom-mounted ADPs. The "Light" deployment configuration is suitable in areas with no trawling activities. This configuration comprises an ADP, acoustic transponding release, lead weight with retrieving Kevlar rope mounted on the release, and two underwater buoyancies.

The "Heavy" deployment configuration with trawl-resistant seabed platform Mosor comprises an ADP, a battery canister for longer deployments, double acoustic transponding releases with adjacent buoyancies, and retrieving Kevlar rope. A trawl-resistant seabed platform protects the ADP and other equipment from the trawling nets due to polyester body slanted at typically 32°. This configuration is deployed with retractable inclinometer and remotely operated vehicle (ROV) to make sure that the trawl-resistant platform is set tightly to the seabed.

11.5. HORIZONTAL-FACING ADPs

There are many narrow straits in the world where tidal currents are strong. Most of them are of nautical importance, so real-time information of current velocity in the straits is required for the safety of navigation. In addition, the hydrodynamics of the very near-shore have been the subject of several decades of research for the simple reason that they have a large impact on the morphology and sediment transport on a daily scale. Morphology in this zone is far from uniform, and the formation (and destruction) of near-shore rhythmic patterns such as beach cusps, sandbars, erosion hotspots, and sand waves along the coast remains unclear to the present. Moreover, surf-zone currents play an important role in the transport of nutrients and bacteria in the coastal zone. From a beach tourism point of view, information on currents in the surf zone is of enormous importance to the safety of swimmers.

Precision measurements of currents are also required from the vicinity of offshore oil and natural gas production platforms. However, there are situations in which it is very difficult to measure water-current velocity in the center of a sailing route or in the vicinity of offshore production platforms by previously existing means. One of the reasons, especially pertaining to straits, is that strong currents and traffic congestion prevent placing an ADP on the seabed. Horizontal-looking ADPs capable of being side-mounted on bridges, canal walls, riverbanks, or other vertical structures in the coastal and offshore environments circumvent this difficulty. In horizontal-looking ADPs designed to meet these requirements, two transceivers, which are inclined to each other, are often employed so that the acoustic beams are divergent from each other. This scheme enables estimation of horizontal flow velocity vectors in sequential horizontal layers from the ADP.

FIGURE 11.9 Pictorial view of a typical installation of a SonTek/YSI-make side-looking ADP on a wall, with an upward-looking transducer for wave measurement. The two acoustic beams are divergent from each other to enable estimation of horizontal flow velocity vector in sequential horizontal layers from the ADP. *(Source: SonTek YSI Inc., www.sontek. com/pdf/news/newslrelease_apr07.pdf, reproduced with kind permission of Christina Iarossi, SonTek Marketing Communications, San Diego, CA 91921, USA.)*

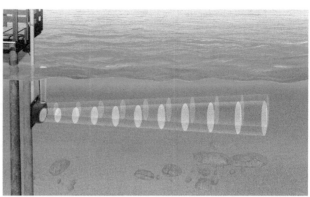

FIGURE 11.10 Side view of a typical installation of a SonTek/YSI-make side-looking ADP on a wall, with an upward-looking transducer for wave measurement. Successive bins in a horizontal layer of water columns are illustrated. Water-current measurement corresponding to each bin (radial current) in sequential layers is computed, and the horizontal flow velocities in sequential layers from various horizontal distances from the ADP are estimated using simultaneous radial current measurements from the respective bins of the two beams. *(Source: SonTek YSI Inc., www. sontek.com/pdf/news/newslrelease_apr07.pdf, reproduced with kind permission of Christina Iarossi, SonTek Marketing Communications, San Diego, CA 91921, USA.)*

Several manufacturers have produced different designs of horizontal-looking ADPs. A typical installation using a SonTek/YSI-make horizontal-looking ADP is illustrated in Figure 11.9. Because radial flow components are to be measured from various layers in the water column, the transceiver system is to be operated in a multiplexed mode, i.e., configuring it for transmission at one instance followed by reconfiguring it as a receiver before the backscattered signals have reached the transducer. In an incoherent system, each transceiver emits a stream of narrow-beam acoustic pulses, insonifying the water-borne scatterers along its path, and receives the Doppler-shifted back-scattered return echoes as the transmitted pulse continues its propagation through the water column. The duration of the pulse equals the gating time τ. The Doppler shift, as a function of time after transmission, yields the relative radial flow velocity as a function of the slant range in sequential layers at each of the "range-sliced" insonified volumes of the successive bins in a horizontal layer of water column (see Figure 11.10). The mean horizontal water-flow velocity vector at a given horizontal water layer, which is located away at a given distance from the ADP, is estimated from measurements of the two divergent radial flow components in this layer. Figure 11.11 provides the photograph of a more recent version of the Argonaut-SL family of SonTek/YSI-make side-looking ADP. Figure 11.12 provides an illustration of the scheme used for estimation of mean horizontal flow vectors in sequential horizontal layers of water using radial currents from these layers, which are

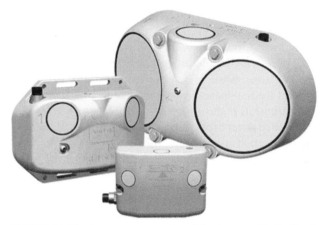

FIGURE 11.11 A more recent version of the Argonaut-SL family of SonTek/YSI-make side-looking ADP. *(Source: Reproduced with kind permission of Christina Iarossi, SonTek Marketing Communications, San Diego, CA 91921, USA.)*

bounded by two acoustic beams. Each measurement is an averaged value over a distance into the medium along the beam, expressed in terms of the two-way travel time. If a pulse of duration τ seconds, representing a time burst, is transmitted, the echo observation (integration) time must be equal to this duration. By adjusting the transmitted pulse length, the portion of the water column insonified at any given instant can be controlled and the desired range resolution can be achieved.

Teledyne RD Instruments Inc. launched its Horizontal ADCP (H-ADCP), which "looks" horizontally to distances

FIGURE 11.12 Illustration of the scheme used for estimation of mean horizontal flow vectors in sequential horizontal layers of water using radial currents from these layers, which are bounded by two acoustic beams.

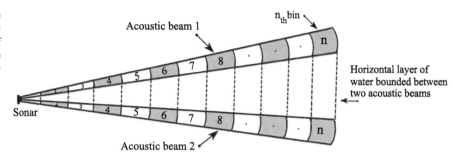

of up to 200 m through the water to measure currents up to 5 m/s at up to 128 individual points (see Figure 11.13). Its applications include in rivers, where it monitors discharge rates and obtains horizontal profiles of water flow; in ports and harbors, where a pier-mounted instrument monitors shipping channels to aid navigation and safety; in estuaries, for defining complex circulation patterns and lateral mixing; on exploration rigs and seismic vessels; and in hydroelectric and tidal power plants.

The Teledyne RDI H-ADCP uses broadband signal processing. According to RDI, this enables it to obtain the best combination of range, resolution, and data quality compared with narrow-band solutions. The H-ADCP is self-contained and can be integrated with a telemetry module for relaying real-time information from remote sites.

Horizontal ADPs have begun to be employed recently in the surf zone, where it is very difficult to carry out water-current measurements by conventional means. Installation of instruments in this zone is not possible with boats due to the limited water depth. A horizontal ADP facing the cross-shore direction installed at low tide can provide a solution in these situations. Because the surf zone is relatively shallow, horizontal-looking ADPs with all beams aligned in a

horizontal plane are required to be used to prevent the acoustic beams from intersecting with the seafloor or the sea surface.

The first test of a horizontal ADP at a macro tidal beach in France showed promising results, especially for the mean flow velocities, revealing strong shear in flow in the vicinity of a rip channel. de Schipper et al. (2010) deployed a horizontal ADP (H-ADP) and a High Resolution upward-looking ADP (HR profiler) on a single measurement frame installed around 1 m below the low water line. The horizontal-looking ADP was a Nortek Aquadopp 600-kHz instrument, recording velocities in 9 bins along the beam axis with a bin size set at 5 m. The two transducer beams have a separation angle of 50° and a blanking distance of 0.5 m. Individual beams fan out with a total beam angle of 4°. The HR profiler was a Nortek Aquadopp HR 2000-kHz instrument. It was set up to measure in 44 bins of 1 cm in the upper part of the water column around high tide. The blanking distance was set at 10 cm. Its internal pressure sensor was used to record the water level above the instrument. Heading and tilt recordings allow examination of the motion of the ADPs.

Because the ADPs are in the very near shore, measurements are likely to be corrupted to some extent by bubbles and the instrument getting dry during low tide. Both instruments record the signal strength. Good velocity measurements are characterized by a high signal strength, which monotonously decreases in the bins further from the ADPs. Data having low signal strength are removed. Surf zone currents should be examined, taking into account the fact that the majority of these currents are induced by wave breaking and variations therein. Dissipation of wave energy generates strong currents of the order of 1 m/s in both cross- and alongshore directions. These wave-induced currents can be characterized by different time scales. Distinction can be made into four different time scales: (1) short-wave, high-frequency motion, (2) infra-gravity, low-frequency motion, (3) very-low frequency (VLF) motion, and (4) mean flow fluctuations. The perceived water-current motion in the surf zone is a superposition of all these elements.

The first category, the short-wave, high-frequency motions, has periods less than 25 s. These rapidly varying

Workhorse H-ADCP-600

FIGURE 11.13 Teledyne RD Instruments horizontal ADCP. *(Source: Reproduced with kind permission from Margo Newcombe, Teledyne RD Instruments, USA.)*

motions are governed by the orbital motion of individual waves. The next category, the infra-gravity, low-frequency motions, has periods in the range of 25 to 250 s. These infra-gravity motions are generated by the wave-height variations over time. High waves are often clustered in wave groups, and these wave groups are accompanied by infra-gravity waves (Longuet-Higgins and Stewart, 1962). Although the velocities and amplitudes of these infra-gravity waves are very little outside the surf zone (i.e., outside the wave-breaking zone), they become of greater relative importance inside the surf zone. The third category, very low-frequency (VLF) fluctuations, has wave periods in the range 250 s to 30 minutes. Water-motion velocity fluctuations with such large wave periods can be generated by instability of the along-shore current-induced or wave-group-induced surf zone eddies (MacMahan et al., 2010). These slow motions have been hypothesized to be important for the initiation or development of near-shore rhythmic patterns (Reniers et al., 2004) as well as the exchange of transported material between the surf zone and the inner shelf (Reniers et al., 2009). The magnitude of VLF velocities and especially the surf zone eddies is influenced by the characteristics of the wave field (i.e., wave period, directional and frequency spreading). The fourth category is the mean flow fluctuations with wave periods larger than 30 minutes. This flow pattern is typically governed by tidal currents.

Based on the just-mentioned considerations of the inner and outer surf zone water-motion characteristics, de Schipper et al. (2010) "dissected" their horizontal and vertical ADP measurements in the very near shore at the Dutch coast in the North Sea. Measurements of both instruments showed the presence of substantial VLF oscillations on one of the field days. The observed root-mean-squared velocity of the oscillations in the VLF frequencies (30 min > T > 250 s) was 0.31 m/s, which is substantial compared with that of the short-wave motion. This finding implies that the VLF oscillations are not only reserved to ocean coasts with medium to high period (T > 10 s) waves. de Schipper et al. (2010) found that a horizontal-looking ADP is most valuable when deployed outside the surf zone, where (1) the flow is more uniform, and (2) the beams intersect less with the sea surface or the seafloor.

In any commercial system, size and weight matter greatly in view of easier transportation and installation in wider, deeper environments. Thus, the design considerations have often resulted from a good deal of customer input on how the products are used. In particular, ergometric features are likely to open up more applications and thus save on installation and maintenance time.

A limitation of the previously mentioned two-beam system is the requirement to make an assumption that the flow field in the horizontal water layer bounded between the two horizontal beams is uniform. This assumption need not necessarily be true if there are abrupt depth discontinuities below this layer or if the water channel is of nonuniform width. To overcome the difficulty of real-time measurements using bottom-mounted ADPs and simultaneously circumvent the limitation of two-beam geometry, the Hydrographic Department and the Japan Hydrographic Association developed a new side-looking H-ADP (Sato et al., 1999) for measuring water-current speed and direction in the center of a narrow strait by transmitting acoustic pulses along a horizontal plane and receiving the echoes at the coastline. The main purpose of the development was to measure strong currents in narrow channels such as the Kanmon Strait in Japan, which is an important sailing route connecting the Sea of Japan and the Seto Inland Sea, which many vessels pass through. The ADP is designed to be placed on a vertical wall beneath the sea surface.

The Horizontal ADP operates by transmitting very narrow acoustic beams horizontally from only one transducer. The beams are transmitted in seven directions, one after another. The angle between successive beams is 10° and the interval of transmission is a second, so that it takes 7 s to transmit beams in seven directions. A single acoustic beam measures a single water-flow velocity component, which is the component parallel to the beam. Radial currents measured by two neighboring divergent beams, which are at an azimuthal separation of 10°, are used to compute the vectorial mean horizontal velocity within the water layer bounded between these two beams. The seven acoustic beams provide the horizontal profile of the water-current velocity in a fan-shaped horizontal plane every 7 s.

When horizontal beams are to be transmitted in shallow water, it is necessary that the beams be of narrow width. If the beam is wide, the echoes from the surface or the bottom are likely to return to the transducer simultaneously with the echoes from the target distances. Sato et al. (1999) generated the required narrow beam using an array of 6,912 elements (144 vertically × 48 horizontally) in the acoustic transducer. The cell size is 10 to 99 m. The transducer's axis must be located more than 5 m below the sea surface so as to measure the surface currents.

The Horizontal ADP consists of four parts: transducer, amplifier, signal processor, and data processor. The amplifier forms the beams and amplifies the signals. The signal processor detects the Doppler frequency shifts and computes the horizontal water-flow velocity components. The data processor displays the distribution of horizontal water-current speed and direction. The transducer was rigidly fixed on an iron frame and mounted on the side wall of the quay by a crane.

Sato et al. (1999) deployed and field-tested the Horizontal ADP at the Kanmon Strait in Japan, which is an important sailing route connecting the Sea of Japan and the Seto Inland Sea. The length of the strait is about 10 km and its width varies from 500 m to 2 km. More than 700 vessels pass through the Kanmon strait every day. The maximum tidal current in this strait reaches about 5 m/s at the narrowest region. The depth at the quay is about 12 m, and the transducer was set at 5 m below the chart datum (CD) level. Water-current speed is expressed in color, and current direction is shown by an arrow. The center of the sailing route is about 500 m away from the quay. Field tests carried out by Sato et al. confirmed that current speed and direction within 500 m of the transducer can be measured by this new ADP. Analysis of the measured data revealed that the phases of the two most dominant harmonic constants of M2 (principal lunar semidiurnal tidal current) and K1 (lunisolar diurnal tidal current) propagate from the coastline to the sailing route, with amplitudes increasing toward the center of the channel. According to Sato et al. (1999), this observation agrees with the theory of the tidal current in a channel.

11.6. SUBSURFACE MOORED ADPs

Dissemination of tide and tidal current predictions is a critical part of several maritime nations' efforts toward promoting safe navigation in their waterways. To assure that the tidal current predictions are reliable, new observational data must be collected periodically, requiring a variety of CM platforms suitable for different environments. For example, the Center for Operational Oceanographic Products and Services (CO-OPS) of NOAA's Ocean Service (NOS) manages a Current Observation Program (Earwaker and Zervas, 1999) with a main objective of improving the quality and accuracy of the annually published Tidal Current Tables (NOS, 2001). For this purpose, CO-OPS strives to use off-the-shelf technologies,

when possible, to help reduce costs and improve efficiency of field operations. Alternative platforms to the commonly used bottom-mounted platforms are subsurface mooring buoys of various shapes (e.g., spherical; streamlined).

Notwithstanding excellent performance of any oceanographic instrument under laboratory conditions, it is necessary to examine its performance under the field environments where it will be deployed. Taking this requirement into account, Bourgerie et al. (2002) carried out several field experiments at the mouth of the Delaware Bay, which is relatively deep (45 m) and the currents there are fairly strong (> 100 cm/s). The bottom-mounted and the subsurface buoy-mounted ADPs used for the experiments (RDI Workhorse Sentinel, WH-300; see Figure 11.14) transmit sound pulses along four narrow beams. The platform was a subsurface, torpedo-shaped, streamlined buoy (SUBS), which was developed and field-tested at the Bedford Institute of Oceanography (Hamilton et al., 1997). This buoy is relatively lightweight (just under 40 kg with ADP) and is considerably easier to handle than the 350-kg bottom-mounted platforms typically used for current observation programs. Use of a streamlined buoy in place of a spherical buoy effectively reduces drag and greatly reduces mooring vibration induced by vortex shedding (Hamilton, 1989). The test consisted of analyses of pressure, tilts, and heading of the subsurface, moored, upward-looking ADP to determine how well its platform performs in high flows. The dynamics of the SUBS were determined from the ADP's pressure sensor, tilt sensors, and compass. It was found that the buoy's tilts fluctuated regularly with the reversing tidal current; the pitch ranged from 0 to −6°, and the roll ranged from −3 to −6°. Although the roll of the SUBS improved with increased current speeds, the pitch of the

Sentinel   Sentinel-03

FIGURE 11.14 Teledyne RD Instruments Workhorse Sentinel. *(Source: Reproduced with kind permission from Margo Newcombe, Teledyne RD Instruments, USA.)*

SUBS became larger at higher speeds. However, these tilts are internally corrected by the ADP and are well within acceptable limits.

There are field-specific issues for which no internal corrections are practical. For example, the water current applies a force along the entire mooring length, causing the mooring to incline and lowering the SUBS down in the water column. This vertical displacement is an important factor in analyzing and interpreting the current velocity profile data. If the SUBS is regularly displaced beyond the cell size, the mapping of the cell depths becomes overly complicated.

The vertical displacement can be isolated from the ADP's pressure sensor measurements by removing the tidal signal observed at a nearby sea-level gauge. Assuming a small time lag between the two locations and referencing both measurements to the same datum (e.g., mean lowest low-water level), the water levels at the sea-level gauge station can be subtracted from the pressure-sensor-derived water-level measurements to arrive at a value referred to as ΔH. The ΔH values represent the distance from the pressure sensor to the water surface minus the tidal elevation; therefore, the ΔH values approximate the vertical movements of the buoy. Accumulation of sediments within the cavities of the SUBS and settling of the anchor into bottom sediments have been offered as two possible explanations for the observed vertical shift of the subsurface buoy (Bourgerie et al., 2002). A reference level from which to quantify vertical displacement due strictly to the current can be derived by averaging all of the ΔH values occurring at slack water (speeds less than 5 cm/s), after the anchor had settled into bottom sediments. Subtraction of this average from all ΔH values yields the vertical displacement of the SUBS.

During slack water, the buoy remains relatively stable in the vertical, but it is suppressed downward with increasing current speed. It was found that the flood currents at the experimental site (Delaware Bay) had a greater effect on the mooring, causing a lowering of the SUBS by as much as 1.9 m. The slightly weaker ebbs produced a maximum vertical displacement of 0.6 m. The flood currents consistently caused a relatively larger vertical displacement of the buoy than the ebb currents.

The amplitudes and phases for six of the largest tidal current constituents (comprising more than 93 percent of the total current) obtained from time-series measurements of SUBS-borne and bottom-mounted ADPs and subjected to similar post-processing and quality-control procedures, showed minor differences. The differences seen in the amplitudes and phases of the two datasets have been attributed in part to a changing water-density structure of the Bay and other seasonal changes in water conditions and flow over the course of the year. (The reference bottom-mounted ADP observations were collected in the summer, whereas the SUBS-borne ADP observations were collected

in the winter.) Tidal current predictions for the 33 days of the SUBS-borne ADP measurement period, generated from each set of 24 constituents (SUBS-borne and bottom-mounted ADPs), yielded good agreement between all three series over a representative period of time. Based on tidal current predictions for 34 days, SUBS-derived tidal current predictions seemed to slightly underestimate the current near maximum flood. However, the overall RMS of the residual current (bottom-mounted ADP observations − SUBS predicted) was 15.3 cm/s, which has been stated to be within the typical range of values seen for flows of this magnitude.

Analyses such as these permitted assessing the quality of subsurface deployed ADP current measurements against those made from stable, bottom-mounted platform. It is seen that there is a certain degree of data degradation due to subsurface mooring dynamics. Bourgerie et al. (2002) exposed the SUBS-borne ADP mooring to maximum current speeds of approximately 120 cm/s. At speeds of this magnitude, the mooring angle was found to be steep and vertical displacement approached 2.0 m. Many areas in the world oceans have significantly stronger currents than those observed at the mouth of the Delaware Bay; hence the subsurface buoy may not be a suitable platform for these locations.

There are four major set-up parameters for moored ADPs: deployment depth, space between measurements in depth and time, data averaging, and deployment duration. These parameters can be modified via professional software, which ADP manufacturers provide. Professional software for deployment planning shows the consequences of set-up choices in terms of power consumption, memory requirements, and velocity precision. For optimizing ADP setup, the importance of the three parameters from the so-called *trade-off triangle* is crucial. These parameters are range, resolution, and random noise (Skelin et al., 2008). Range is associated with transmission frequency. Lower frequency results in longer profiling range. Doubling the cell size will inject two times more acoustic energy into the sea. Usually, profiling range is enhanced by colder and fresher water. Smaller cell size will provide finer resolution but lesser profiling range and more random noise. Therefore, ADP will require a larger measuring period. Doubling the number of cells doubles the power consumption and the random noise. Broader bandwidth and lower frequency will provide finer resolution. More dynamic motions and more turbulence will increase the random noise. More suspended material in the sea will provide lower noise as a result of enhanced echo intensity and, therefore, will improve velocity precision.

Schott (1986) located an ADP at depths of 470 and 610 m on subsurface mooring lines deployed in the Florida Current and obtained detailed vertical profiles over five days.

11.7. DOWNWARD-FACING SHIPBOARD ADPs

Strong ocean currents can disrupt various deep-water activities and cause downtime for several commercial activities such as hydrocarbon exploration and production activities. Thus, accurate and timely observations can help with planning and assuring safe operations. Remote sensing cannot always be counted on to provide timely observations of the sea surface away from the coast. Drifting buoys have proven to be a cost-effective means of collecting ocean current observations; however, drifters report only near-surface currents and the Lagrangian techniques cannot target and monitor a specific site or area. ADPs mounted on moorings and offshore platforms can provide real-time current-flow information but are not always optimally located. ADPs deployed on ships provide a viable way to survey ocean-current profiles in a specific region of interest. Low-frequency ADPs operating at 38 kHz or 75 kHz are needed to provide synoptic coverage of the ocean-flow features from relatively deep-sea regions. Low-frequency operation demands large ADPs, which have to be hull-mounted systems mostly deployed on research vessels.

The development of shipboard ADPs—a logical extension from Doppler navigation systems, which operate on the signals returned from the seabed—began in the 1980s, and a number of successful applications of the technique have since been described in the literature. Surveys using shipboard profilers provide detailed information about the spatial structure of currents in a way that cannot be matched by moored instrument arrays. Such systems were expected to provide a powerful tool in large-scale international campaigns such as the World Ocean Circulation Experiment (WOCE), particularly if they could be made suitable for use on ships of opportunity (Cutchin et al., 1986).

Shipboard hull-mounted ADP systems are found in two geometric configurations: three- or four-beam systems. In a typical three-beam system, the beams are spaced 120° apart in the azimuth, and each beam points down at an angle of $\phi°$ from the vertical. One beam (say, beam 1) is generally oriented forward along the ship's longitudinal axis, and the other two beams (say, beams 2 and 3) lie along the port and starboard quarters (see Figure 11.15). Based on

this geometry, the along-track and the cross-track relative velocity components at various depth layers may be estimated in terms of the in-beam (i.e., radial) Doppler shifts for the three beams from the corresponding depth layers. The mean relative horizontal flow velocity vector may be considered to consist of a horizontal flow component (say, V_y) along the longitudinal axis of the ship (i.e., along the ship's direction of motion) and an orthogonal horizontal flow component, say, V_x. These components are given by the relation (Theriault, 1986a):

$$V_y = \left(\frac{c}{6\omega\sin\phi}\right)(\omega_{D(2)} + \omega_{D(3)} - 2\omega_{D(1)}) \quad (11.11)$$

$$V_x = \left(\frac{c}{2\sqrt{3}\omega\sin\phi}\right)(\omega_{D(3)} - \omega_{D(2)}) \quad (11.12)$$

where c is the velocity of sound in water, ω is the transmission frequency in angular notation, ϕ is the angle (in degrees) made by each beam from the vertical, and $\omega_{D(1)}$, $\omega_{D(2)}$, and $\omega_{D(3)}$ are the Doppler shifts (in angular notation) measured by the beams 1, 2, and 3, respectively. Because only three components of velocity are present, i.e., two horizontal and one vertical component, only three beams are sufficient in principle to fully describe the flow field. However, if the beams are more than three, the performance of the profiler correspondingly increases. For example, in the three-beam system just mentioned, the along-track horizontal flow component V_y is coupled to the Doppler shifts from all the three beams (see Equation 11.11). Furthermore, lack of a coplanar companion beam makes the three-beam system sensitive to pitch and roll of the vessel.

To circumvent some of the limitations inherent in a three-beam Doppler profiler system, a four-beam system has been introduced. This system employs two "Janus configurations" consisting of a total of four beams oriented 90° apart in the azimuth, each beam making an angle of $\phi°$ to the axis of symmetry of the transducer system (see Figure 11.16).

Joyce et al. (1982) described a pulsed, 300-kHz, four-beam, hull-mounted acoustic transducer, which was used to obtain vertical profiles of upper ocean currents relative to a ship (*R. V. Oceanus*), the lateral motion of which was determined by changes in LORAN-C position. The

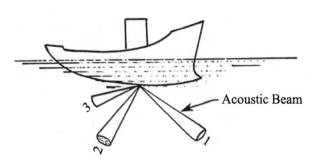

FIGURE 11.15 Geometry for a hull-mounted three-beam ADP.

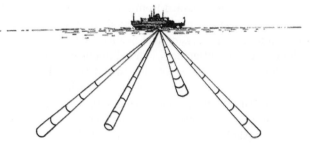

FIGURE 11.16 Geometry of a hull-mounted four-beam ADP.

acoustic transducer and current-profiling electronics were produced by Ametek-Straza (Rowe and Young, 1979) and were successfully used in the western North Atlantic.

The acoustic profiling system consisted of two main parts: the Ametek-Straza Doppler current profiler and a controller-data logger built at the Woods Hole Oceanographic Institution. In this system, a 300-kHz acoustic pulse is transmitted from a transducer mounted in the hull of the ship in four beams down through the water column. As the acoustic signal travels through the seawater, it is scattered by small particles, temperature microstructure, or turbulence. By measuring the frequency of the return signal as a function of time or, equivalently, acoustic range, a profile of water velocity along the acoustic path can be obtained. Combining the data from the four beams with the ship's heading and subtracting the ship's translation over the Earth from navigational data yield a vertical profile of ocean currents.

In this measurement, a 254-cm-diameter transducer assembly had four 10-cm-diameter flat-plane acoustic radiators-receivers mounted at 30° angles to the vertical; two pairs of transducers defined two orthogonal vertical planes. Each radiator transmitted a 3° conical beam at 300 kHz. The beams could be approximately aligned in the fore, aft, port, and starboard directions relative to the ship. The four radiators were driven from one common 40-W power amplifier that produced a sound pressure level of +225 db relative to 1 μPa at a distance of 1 m along each beam. Preamplifiers mounted in the transducer assembly amplified the return signals and sent them via a shielded multiconductor cable to a second set of amplifiers before the signal processing stage. The transducer assembly also included a thermistor temperature sensor to measure seawater temperature to compensate for sound-speed variations due to temperature fluctuations.

Pulses of 300-kHz sound of 10 or 20 ms duration (pings) were emitted at regular intervals from the four radiators. The returned signal was detected by the Ametek-Straza electronics. Following transmission of an acoustic pulse, the returned, amplified signal was detected in a frequency-locked loop, the frequency of which was measured in each of 31 time intervals. These intervals were switch-selectable for 5 or 10 m corresponding to vertical depth bins of 3.2 or 6.4 m, allowing for the 30° beam angles from the vertical. The signal processor measured the amplitude of the return signal and set a quality flag for each Doppler determination in which the signal strength was too low. Profile repetition was once every 1.2 s. The signal processors converted each good Doppler count to a 12-bit binary data word and a good-bad status bit for each depth bin of each beam. Resolution of Doppler-induced currents for a single ping was 1.2 or 2.4 cm/s, depending on the averaging time of 10 or 5 ms in the frequency-locked loop. A more detailed discussion of the detection electronics is given by Rowe and Young (1979).

To record information about the ship's heading and navigation, interface modules were built to convert data from the ship's gyrocompass, a Northstar 6000 LORAN-C receiver, and a Magnavox 706 satellite navigation receiver into a serial datastream that was transmitted over a 20-mA current loop. Each module was assigned a unique address and responded when the data controller sent out the proper address code. The current loop was selected because it required only two simple wires to connect modules located at various places on the ship. It also had a high level of noise immunity. Typically, this serial ASCII information loop (SAIL) transmits data at a rate of 2,400 bits per second (bps).

If flow homogeneity is assumed at a given horizontal plane bounded by the radial bins of the acoustic beams at that horizontal plane, the horizontal flow component u and the vertical flow component w at a given depth bin in a forward-aft-looking Janus arrangement are given by (Joyce et al., 1982):

$$u = \left(\frac{\Delta\sigma_1 - \Delta\sigma_2}{4\sigma}\right)\left(\frac{c}{\sin\phi}\right) \qquad (11.13)$$

$$w = \left(\frac{\Delta\sigma_1 + \Delta\sigma_2}{4\sigma}\right)\left(\frac{c}{\cos\phi}\right) \qquad (11.14)$$

where $\Delta\sigma_1$ and $\Delta\sigma_2$ are the Doppler shifts measured by beams 1 and 2, respectively, in the forward-aft-looking Janus arrangement corresponding to a given horizontal plane bounded by the respective radial bins; σ is the acoustic frequency; c is the sound speed in the seawater medium; and ϕ is the angle made by the forward-aft-looking beams 1 and 2. Since the same argument can be made with the orthogonal pair of beams 3 and 4, the horizontal velocity vector can be estimated along with two independent estimates of the vertical velocity. The difference in the two vertical velocities is therefore related to acoustic noise, imperfect estimation of Doppler shift, and spatially variable currents.

Near the ocean surface, the particle motions due to surface waves are greater, but the physical separation of the two ensonified regions is less. Therefore it is not obvious which of these factors dominates at any depth. Unlike the stable platform *Flip*, ship motion due to surface waves is a significant signal in the Doppler returns and must be considered. The instantaneous decomposition into fore, aft, and athwartship velocities must be put into geographical coordinates because the ship is yawing as it pitches, heaves, and rolls. This is done by measuring the ship's heading for each data cycle. One method of reducing the error induced by ship motion is to vector-average the Doppler currents over many wave cycles and estimate lateral ship velocities from LORAN-C data. However, the resolution of LORAN-C is usually insufficient to detect subtle changes in ship speed. Thus, wave-induced noise will be present, but it will be reduced by averaging over many wave cycles.

ADPs are mounted on moving vessels to map the vertical profiles of upper ocean horizontal currents, but the measured relative velocity profiles (with respect to the velocity of the vessel) can be transformed into an absolute velocity profile using the ship's speed and heading information for each data cycle. In locations where good LORAN-C coverage is available, the vessel's lateral motions can be determined by measurements of changes in LORAN-C position within an uncertainty of 5 to 10 cm/s. This data can, in turn, be used to correct the inherent errors that tend to creep into the velocity field map as a result of the ship's lateral motion during measurements of velocity profiles (Joyce et al., 1982). When both the LORAN-C and the Doppler profiler data are suitably filtered, the resultant absolute velocity profiles are estimated to be accurate to within ± 2 cm/s (Trump et al., 1985). Although this procedure cannot remove the wave-induced noise, these errors can be reduced to a certain extent by vector-averaging the measurements over many wave cycles.

As mentioned earlier, when mounted beneath a vessel, the system is generally operated with one pair of mutually diverging beams (say, beams 1 and 2) oriented along the longitudinal axis of the ship and the other pair of beams (say, beams 3 and 4) oriented along an orthogonal plane so that the profiler resolves the relative velocity field at any given depth layer into along-track and cross-track components. If $\omega_{D(1)}$, $\omega_{D(2)}$, $\omega_{D(3)}$, and $\omega_{D(4)}$ are the Doppler shifts measured along the beams 1, 2, 3, and 4, respectively, the relative horizontal flow component V_y along the longitudinal axis of the ship is given by the relation (Theriault, 1986a):

$$V_y = \left(\frac{c}{4\omega \sin\phi}\right)(\omega_{D(1)} - \omega_{D(2)}) \qquad (11.15)$$

and the relative horizontal flow component V_x perpendicular to the longitudinal axis of the ship is given by:

$$V_x = \left(\frac{c}{4\omega \sin\phi}\right)(\omega_{D(3)} - \omega_{D(4)}) \qquad (11.16)$$

The sign of ω_D depends on the direction of flow relative to the respective axis of the transducer system. In this four-beam geometry, each horizontal flow component is estimated as a mean of the Doppler shifts from the two symmetrically oriented beams in each of the Janus configurations in the two mutually orthogonal vertical planes. It is thus evident that, in the four-beam geometry, the Janus configuration tends to minimize the sensitivity to pitch and roll of the vessel. In addition to these obvious differences in performance, statistical analysis by Theriault (1986b) suggests that, for similar operating conditions, the four-beam geometry yields an enhanced performance

compared to that of the three-beam system. Despite the limit on the maximum allowable ship speed (Theriault, 1986b), the ship-mounted ADPs possess a remarkable feature of large spatial coverage.

Joyce et al. (1982) succeeded in measuring the upper 100-m layer of the Gulf Stream along a traverse line of length about 200 km with an ADP mounted on a ship hull. Vertical profiles of horizontal currents for three selected 10-min segments (see arrows in Figure 11.17) show the change in vertical structure across the section (see Figure 11.18).

Gawarkiewicz et al. (2004) have reported shipboard ADP measurements from the northern half of the South China Sea. Whereas geostrophic calculations from individual sections gave extremely poor results due to aliasing of both tides and high-frequency motions, the ADP measurements were found to be in agreement with the anticyclonic winter circulation in the measurement region from numerical model calculations.

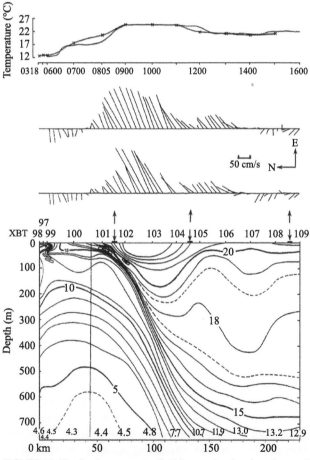

FIGURE 11.17 Section across the Gulf Stream showing thermistor and bucket temperatures (note the 30-min lag of the thermistor). Upper panel: Vector-averaged currents at depths of 28 and 99 m (second and third panels, respectively). XBT section with XBT numbers and three 10-min segments selected for current profiles and denoted by the arrows. *(Source: Joyce et al., 1982.)*

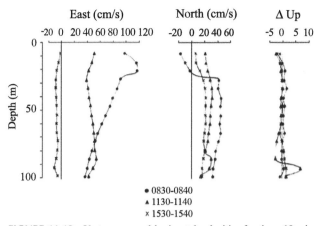

FIGURE 11.18 Vector-averaged horizontal velocities for three 10-min time intervals shown in Figure 12.19 from the Gulf Stream north wall (●), south wall (◐), and Sargasso Sea (×). Difference between two vertical velocity estimates is plotted at the right on an expanded scale. Only data judged "good" by Ametek-Straza interface cards were used. The large shear at 20 m in the north wall profile is across the base of the mixed layer and could be due to inertial waves. *(Source: Joyce et al., 1982.)*

11.8. TOWED ADPs

As indicated earlier, vessel-mounted ADPs have become a standard tool on many research vessels for horizontal and vertical current profile measurements. Their operation has been studied by Kosro (1985), Didden (1987), Chereskin et al. (1987), and Joyce (1989). However, beyond the limitations inherent in the ADP system itself, some limitations have been identified with vessel-mounted systems (Chereskin et al., 1989; Chereskin and Harding, 1993). For example, vessel-mounted ADP systems utilize the ship's inertial or gyro compass. A misalignment generally exists between the reference direction used by the ADP and that of the ship's gyro. Joyce (1989) details a calibration routine to determine this constant misalignment. Pollard and Reed (1989) found that abrupt changes of the ship's course cause yet another compass error as the turn excites persistent Schuler oscillations at periods of 20 and 80 min that bias the compass readings. Although a calibration routine can resolve the misalignment problem, no recourse is available for the Schuler oscillations, the amplitude of which can reach several degrees.

Entrainment of air bubbles is another issue. Air bubbles near the transducer heads and compass biases often degrade shipboard ADP observations (King and Cooper, 1993). New (1992) reported that as the ship heaves, a cloud of air bubbles entrain under the hull of the ship. The bubble cloud then sweeps past the transducer heads in a layer about 1 m thick. This layer degrades ADP current estimation because air bubbles have different acoustic properties than the seawater below. New (1992) found, however, that the problem can be overcome when the transducer heads extend 1.5 m beneath the hull of the ship. Thus, in the use of

vessel-mounted ADP systems, utmost care needs to be taken to ensure that they operate in a low-noise and bubble-free environment.

Probable solutions to overcome the problems related to the hull-mounted ADPs might be to deploy the ADP over the side of the ship or tow the ADP. Low-frequency ADPs are required to profile currents over a large depth extent; therefore, such ADPs would necessarily be large in size. However, deploying a large ADP over the side of a ship would require a large mounting system to be welded to the vessel to withstand structural loading while underway. Thus, it requires modification of the deployment vessel hull. In addition, it would be difficult to design a mount that would position the transducer head deep enough and isolated enough to assure a quiet, bubble-free environment.

A completely submerged, towed ADP system cannot entrain bubbles. In addition, it exhibits neither the Schuler oscillation problem nor the direction misalignment problem because it generally relies on a magnetic compass. The ship's hull and the tow platform, however, may induce magnetic variations that can be large and must be removed through a calibration routine. Hence, both ship-mounted and towed ADP systems require careful *in situ* calibration to account for compass biases. Taking all these into account, it is considered that towed ADPs offer an attractive alternative to vessel-mounted ADPs.

There are other relative benefits in the use of a submerged towed system. A towed system can be used from ships of opportunity, allows easy maintenance, and is expected to provide remedies to several problems of vessel-mounted systems. Kaneko and Koterayama (1988) first introduced a towed ADP. They mounted a 150-kHz ADP (RD Instruments RD-SC0150) on a fish (sled) made of fiber-reinforced plastic (see Figure 11.19), and the sled/fish

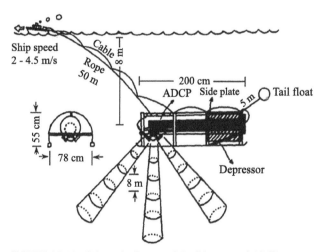

FIGURE 11.19 Schematic diagram of the fish-mounted ADCP system. The ADCP, located inside the cylindrical fish body, is shown in solid black. The hatched region indicates the location of the side plates. On the left is a front view of the fish. *(Source: Kaneko and Koterayama, 1988.)*

was towed with a 50-m-long nylon rope behind a research ship. The nylon rope is expected to dampen the direct response of the fish to ship motion. When the ship was on station, the ADP fish floated up to just below the sea surface because of the buoyancy of a tail float. The trailing buoy also stabilized the fish, much as a tail stabilizes a kite. At ship speeds of 2.0–4.5 m/s, the fish was designed to be submerged up to a constant depth of 6–8 m beneath the sea surface by the downward lift force acting on a depressor.

The front portion of the fish was rounded to reduce hydrodynamic forces. The geometrical positions of the towed rope and the depressor were carefully selected to stabilize fish motions. In particular, the tail float and the side plates served to suppress the yaw and roll motions of the fish. During the traverse survey, the fish could not be turned over because of the heavy transducers and the tail float. A flux-gate compass and dual-tilt sensors located inside the ADP transducer housing were used to convert the measured flow velocities from the instrument coordinates to Earth coordinates (north, east, and vertical velocity components). The ADP fish was covered by thin stainless frames to protect it from collision with the ship's hull during deployment and recovery.

Power to the ADP was supplied from the ship through a cable that also contained wires for an RS-422 serial link. The cable was loosely tied to the nylon rope. Internal ADP sensors measured the heading, pitch, and roll of the sled. Four transducers were arranged in a downward-facing Janus configuration. Water temperature was also measured with a thermistor placed in a space between the ADP transducers. The ADP transmitted 10.6-ms acoustic pulses every 1.8 s. The 10.6-ms pulse width corresponds to a depth resolution of 8 m. The measured data were transferred automatically to an onboard microcomputer through the cable.

Unlike the conventional ship-mounted ADP, the fish-mounted ADP is expected to exhibit the following benefits (Kaneko and Koterayama, 1988):

- Ocean currents are accurately measured, even under severe sea surface conditions.
- Cavitation noises generated by the ship's propeller are much reduced around the fish.
- Interference problems caused by bubbles entrained under the ship's hull do not occur for the fish.
- The fish-mounted ADP can be operated from any unspecified research ships.

Keeping ship speed and heading smooth and constant serves to improve the accuracy of observation. Fish motions measured by the flux-gate compass and the dual-tilt sensors over the whole traverse survey are shown in Figure 11.20. Kaneko et al. (1990) found that when the water depth was shallower than 480 m, the ADP was able to track the sea bottom to obtain fish speed relative to the

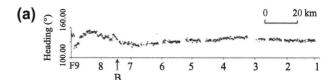

FIGURE 11.20 Fish motions measured by the flux-gate compass and the dual-tilt sensors over the whole traverse survey. (a) Heading. (b) Roll angle. (c) Pitch angle. *(Source: Kaneko et al., 1990.)*

bottom. They also found that when the bottom is within range of the ADP, there is a near-bottom layer in which good data cannot be obtained. This limitation of ADP measurements is caused by transducer side lobes reflecting directly from the bottom. Kaneko et al. (1990) obtained detailed velocity measurements of the Kuroshio Current with their sled-mounted, towed ADP system. The vector time series (stick diagrams) of the Kuroshio current velocities obtained from the ADP measurement are shown in Figure 11.21, with the time plot of water temperature obtained from the thermistor. The Kuroshio Current exhibits a profile like a jet flow, with a width of about 80 km and the maximum current of about 1.3 m/s at the center of the jet (near the middle of the transect) near the surface. The dominant direction of the current was northeastward. The current velocity decreased slowly with depth; it was near 1.0 m/s at 400 m depth. Based on these measurements, Kaneko et al. (1990) concluded that the magnitude of the Kuroshio Current in this region is significantly underestimated through conventional geostrophic calculations that use a level of no motion at 700 or 800 m depth. It was also concluded that there are many similarities between the Florida Current (flowing between Florida and the Bahamas) and the Kuroshio Current (flowing between the East China Sea shelf edge and the Ryukyu Islands). Therefore, many of the mechanisms of these two current systems should be the same.

Towed ADP systems have undergone several design modifications with a view to improving their performance. For example, Munchow et al. (1995) reported a towed ADP platform, which contains both an upward- and a downward-facing ADP (see Figure 11.22). This system differs in design, construction, and operation from that of Kaneko and Koterayama (1988) and Kaneko et al. (1993). The towed fish reported by Munchow et al. (1995) consists of a hydrodynamic body, a seven-conductor torque-balanced

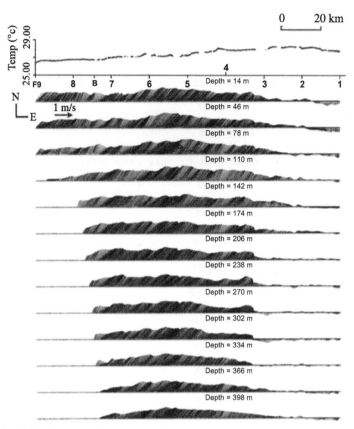

FIGURE 11.21 Vector time series (stick diagrams) of the Kuroshio horizontal current velocities obtained from towed ADCP measurement at 13 different depths. The depth scale for each diagram is inscribed just above the corresponding one. The horizontal variations of the water temperature at depth 8 m is presented at the top of the figure. *(Source: Kaneko et al., 1990.)*

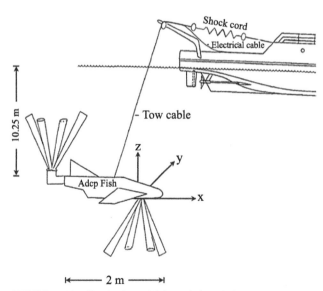

FIGURE 11.22 Sketch of towed fish consisting of a hydrodynamic body, a conducting tow cable, and two independent upward- and downward-facing ADCP systems. *(Source: Monchow et al., 1995, ©American Meteorological Society, reprinted with permission.)*

tow cable with 22 inner steel wires and 42 outer steel wires with an added zipper tubing cable fairing to minimize strumming, and two independent ADP systems manufactured by RD Instruments. The cable is rolled onto one of the winches aboard the towing vessel and paid over the stern or the side through a 0.52-m-diameter block on the aft A-frame or side crane. A shock-cord assembly, consisting of 10 strands of 0.02-m-wide and 3-m-long bungee cord, is fastened to the cable using Yale Minigrip Kevlar braid grips. The shock cord takes the load of the tow between the winch and the block in order to reduce peak loads on the cable due to the differential motion between the towed body and the heaving ship.

The ADPs are mounted inside an Edo Western, Inc., model 1019 hydrodynamic body. The body is about 1.5 m long and has a wing span of 1.5 m. The instrument's cavity is 0.5 m deep and 0.8 m wide and includes a 400-kg lead weight in the front side. The ADP electronics pressure cases are placed horizontally inside the cavity of the tow body. The transducer heads are coupled to the case through 90° adapters so that the 153- and 614-kHz transducers are oriented downward and upward, respectively. The pitch and roll angles are measured using tilt sensors fixed to the ADP, whereas the heading angle is measured with two KVH fluxgate compasses, one un-gimbaled and the other gimbaled. The ungimbaled compass is part of the ADP; the gimbaled one is not. A pressure sensor indicates the vertical displacements of the platform. Sampling frequencies for all these instruments are higher than 0.5 Hz, high enough to resolve motions that are induced by surface gravity waves.

This towed ADP system was deployed from ships of opportunity and towed at depths between 5 and 25 m. The instrument platform (fish) was found to be stable in most operating conditions at ship speeds up to 4.5 m/s. Munchow et al. (1995) identified two separate sources of errors in their tow system: surface gravity waves and compass biases. Surface gravity waves add noise into the compass record dominantly through the rolling motion of the platform. Other variables such as pitch and vertical displacement appear to be unimportant. Under most operating conditions, however, variations in pitch, roll, and vertical position do not exceed 0.5°, 0.4°, and 20 cm, respectively. These variations are random and can thus be removed by averaging. Thus, the towed ADP system shown in Figure 11.22 constitutes a very stable system. Only in the presence of very choppy seas at ship speeds exceeding 3 m/s does the platform become unstable and the data it returns become unusable. Reducing the ship speed to 3 m/s alleviates the problem.

A more serious problem relates to compass biases. It was found that both the ship and the tow platform induce magnetic fields that bias the towed ADP compass. To overcome this problem, an *in situ* compass calibration scheme using GPS data (Joyce, 1989) was found to be necessary. Munchow et al. (1995) found errors in excess of 50 cm/s in the absence of such a calibration. Experience shows that it is important to apply the compass calibration in post-processing, prior to any averaging. Collecting single-ping data in beam coordinates is recommended, even though this increases the volume of data that needs to be stored during an experiment.

It is generally agreed that the ADP system must be towed as deeply as possible in order to minimize the effect of the ship's hull. The use of an upward-looking ADP allows deep tows while still resolving the surface layer. The ease of deployment from ships of opportunity and the capacity of the tow system to provide profile measurements from above and below the platform with different frequencies and thus different vertical resolutions are found to be of much practical value. These features enhance its flexibility and usefulness, especially to study surface and bottom boundary-layer processes.

A towed ADP system that is much different from those discussed so far and that was used to support operational monitoring of deepwater currents was reported by Anderson and Matthews (2005). They used a unique tow-fish system employing a 75-kHz RDI Long Ranger ADP packaged inside of a large Endeco/YSI type 850 V-Fin. The V-Fin is approximately $1.3 \times 1.4 \times 0.7$ m and weighs 185 kg in air. This large towed body has a proven functional design, and similar systems have been used to deploy acoustic equipment for bathymetric and biological surveys. In the present case, the first beam of the ADP is aligned with the nose of the towed body. The ADP has an embedded compass to detect the system orientation, along with pressure and tilt sensors. An electromechanical tow cable provides both data and power. The tow fish is deployed at nominally 20 m depth with tow speeds of 1 to 3 m/s.

The tow cable is deployed through a sheave on the A-frame. The towed-body configuration provides a quiet and stable sensor platform that could be readily relocated and deployed from different vessels. During sea trial by Anderson and Matthews (2005), wind speed was in the range 4–6 m/s and significant wave height was 0.5 m. Under such meteorological and sea-state conditions, the mean (and standard deviation) of the pitch and roll of the tow-fish system were −11.0 (1.1) and −2.5 (0.6) degrees, respectively. On a later deployment, sustained wind speeds and significant wave height reached 11.0 m/s and 1.6 m, respectively. In this case, the mean (and standard deviation) of the ADP pitch and roll were −6.0 (2.9) and −1.1 (0.6) degrees, respectively. The nose was tilted slightly downward, with more motion in the pitch. The roll statistics were almost the same as under light wind conditions. Thus, the platform was found to be very stable, but the ADP system was tilted slightly with the nose down. It was observed from various deployments that the mean pitch varies with ship speed. Improved performance is expected if the V-Fin can be flown completely level.

Anderson and Matthews (2005) successfully deployed the V-Fin towed ADP system in the Gulf of Mexico to track the strong currents associated with the Loop Current and Loop Current eddies (LCE), which are mesoscale features in near-geostrophic balance. LCE currents usually extend to 400 m or deeper. In this operation the ship cruised at around 3 m/s. The tow fish was deployed at 20 m depth, and the first bin was centered near a depth of 44 m. The ADP provided continuous along-track profiles with good data down to the instrument's maximum range of 500 meters.

However, the range performance is linked to the pitch of the V-Fin. Raw data were telemetered to shore for processing and integration with satellite imagery and other *in situ* observations to provide a real-time synoptic analysis. The example includes strong currents from an anticyclonic eddy. It also shows layers of fine-scale shear that are particularly noticeable at the edge of the eddy between 35 and 65 km along the track. These layers are coherent for 5 or more kilometers and might indicate an enhancement of near-inertial internal wave active in the thermocline along the edge of the eddy. It is considered that the ADP mounted on this towed-body system is able to resolve such fine scale variability because it is so stable.

A disadvantage with this towed-body approach is that the ADP would be deployed deeper than a hull- or side-mounted system, and as a result, a single downward-looking ADP would not be able to sample the very near surface. Placing a second ADP configured in an upward-looking direction would allow profiling to the surface. Fortunately, there is room on the V-Fin to add a small (high-frequency) upward-looking ADP, if required.

ADPs operating at 300 kHz to 1.5 MHz have been successfully used in towed bodies for some time now and have performed quite well. In general, high-frequency ADPs are used in towed bodies for shallow-water applications. According to some reports, towed ADPs are not a panacea for all the problems observed in vessel-mounted systems. For example, in a towed ADP system, Monchow et al. (1995) found two separate sources of error: surface gravity waves and compass biases. They concluded that the towed ADP system returned data of the same quality as a vessel-mounted ADP system. From the divergent experiences reported by different users, it appears that the success of towed ADP systems depends primarily on the hydrodynamic stability of the towed system (and therefore on the design of the tow system) and implementation of proper *in situ* calibration procedures.

11.9. LOWERED ADCP (L-ADCP)

Technologies available until the late 1980s for profiling currents over ranges greater than 1,000 m all suffered from several disadvantages that severely limited their use. Hence, there was a major scientific need for a new profiling method with desirable characteristics such as ∼5,500 m depth range, 20-m vertical resolution, accuracy of 2 cm/s or better, requiring little or no additional ship time, and available at a reasonable cost.

As noted in the preceding sections, ADPs have conventionally been used primarily in fixed applications (in side-looking or bottom-mounted configurations, or moored CMs) or hull-mounted on ships or tow bodies, with horizontal ship/tow-body motion removed through analysis of surface navigation data.

It is believed that the first use of a self-contained ADP attached to a CTD and lowered with a standard hydrographic wire was for a study of the hydrothermal plumes in 2,200 m of water at Juan de Fuca Ridge (Thomson et al., 1989). This arrangement is known as *lowered ADCP*, nicknamed *L-ADCP*. Although the main interest of these researchers was in examining the velocity structure in the plume within a few hundred meters of the bottom, it was recognized that the velocity profiles collected while the instrument was being lowered could be patched together into a 2,000-m profile. In principle, the L-ADCP provides

full water-column absolute velocity profiles, coincident with CTD measurements. It also has the advantage of not requiring any extra ship time and being independent of auxiliary bottom-mounted hardware.

Close on the heels of the successful experiment by Thomson et al. (1989), Firing and Gordon (1990) mounted a downward-looking 300-kHz ADP on a rosette sampler with a CTD frame and collected valuable full-water-column vertical profile data of horizontal currents. In this arrangement, the L-ADCP measures the velocity of water current relative to the profiler, obtaining a vertical profile over a range of 100 m or so about twice per second. The motion of the ADP relative to the ground (fixed reference) is unknown, but the first difference of the velocity profile (the shear profile) is invariant (Firing and Gordon, 1990).

Averaging all such measurements of shear in depth bins as the package moves through the water column yields a shear profile for the whole water column. Integration of this vertical shear profile then gives a velocity profile relative to an unknown constant of integration. The constant can be determined if the absolute velocity of the ADP is measured by tracking the seafloor at the deepest part of the profile or by using accurate navigation such as GPS to fix the position of the ADP at launch and recovery times.

Firing and Gordon (1990) reported the results of theoretical evaluation and extensive field testing of the L-ADCP during 1989 via the Hawaii Ocean Time series (HOT) set of approximately monthly cruises north of Oahu and on a Line Islands Array (LIA) cruise from Hawaii southward across the equator. Prior to these cruises, Firing and Gordon carried out error simulation that indicated that the theoretical lower bound on velocity errors could be reduced to 2 cm/s or less, as desired. In these field measurements, the L-ADCP was equipped with a flux-gate compass and tilt sensors. The rosette frame was balanced so that the typical tilt was about 3°. Ensemble-averaged tilt and heading and their standard deviations were recorded with each ensemble and were found to be quite stable. The L-ADCP measured and recorded vector-averaged velocity in Earth coordinates for 10-s ensembles. The deepest cast was about 5,000 m.

Apart from some failures and a "mysterious problem" that caused intermittent loss of data (low percentage of data accepted in all depth bins in some ensembles) in the deepest four profiles, the L-ADCP worked normally at all depths. The profile measurements made by the L-ADCP clearly showed the "equatorial deep jets" that, compared with past profile measurements (Firing, 1987), led to the conclusion that these jets change their vertical positions over several years.

Since the successful experiments by Firing and Gordon (1990), the L-ADCPs have collected hundreds of profiles (Firing and Hacker, 1994). Wilson (1994) used an RD Instruments (RDI) 150-kHz broadband ADCP (BB-ADCP) by mounting it on a CTD frame to measure current-velocity shear (i.e., rate of change of current velocity with depth) below the instrument during casts as deep as 5,500 meters. However, it was found that some practical aspects needed to be considered before using the broadband ADP. For example, initial tests with a low-power (40 Watt acoustic power/beam) 150-kHz BB-ADCP in June 1992 proved unsatisfactory, with range decreasing to nearly zero below 1,500 m depth. RDI subsequently introduced a high-power (240 W/beam) module for the 150-kHz BB-ADCP. Although this modification increased the range to around 150 m at the low scattering levels found below 2,000 m, it drained a standard alkaline battery pack in around 4 hours, which is the duration of a typical deep CTD cast.

Accordingly, the standard RDI pressure housing had to be modified to accept an external rechargeable battery pack. Thus, the enhanced electrical power consumption of the required high-power module necessitated modification of the standard instrument to utilize external rechargeable battery packs. Despite such difficulties, using GPS navigation data and the measured velocities, the profiles could be referenced to produce absolute velocity profiles for both the up and down casts after making correction for sound velocity using ADP temperature and a standard salinity of 35.0 ppt. Wilson (1994) found that data collection and processing methods have improved so much that accuracies comparable to those of the acoustically tracked dropsonde Pegasus (Spain et al., 1981) can be achieved.

A method proposed for referencing L-ADCP profiles is using bottom tracking to measure absolute velocities in the lower part of the profile. The broadband firmware allows "intelligent" bottom tracking so that the instrument does not waste time with bottom pings while out of range. Fischer and Visbeck (1993) have described advances in lowered profiling and associated data-processing techniques using deep profiles collected with a 150-kHz ADP as examples.

11.10. ADPs FOR CURRENT PROFILING AND AUV NAVIGATION

Applications of ADPs have extended beyond the domains of remote measurements of subsurface currents and sea surface waves. AUV navigation is another feather added to the cap of ADPs. These are briefly addressed in the following sections.

11.10.1. AUV-Mounted ADPs for Current Profiling

Apart from a variety of mobile platforms discussed so far for ocean-current profile measurements, *autonomous underwater vehicles* (AUVs) have also begun to be used for such measurements. By merging data acquired by the ADP

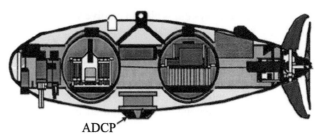

FIGURE 11.23 AUV Xanthos with downward-facing RDI ADCP in its midsection. (*Courtesy of Dr. James Bellingham. Source: Y. Zhang and J. S. Willcox, Current Velocity Mapping Using an AUV-Borne Acoustic Doppler Current Profiler*, Proc. 10th International Symposium on Unmanned Untethered Submersible Technology, *pp. 31–40, Durham, NH, USA, Sept. 1997.*)

with those acquired by other AUV-borne instruments as well as a long baseline (LBL) acoustic navigation system, it is possible to better observe and interpret the underwater processes of interest by utilizing AUVs. This capability was demonstrated by Zhang and Willcox (1997) in a field experiment at Haro Strait, British Columbia, Canada, in the summer of 1996. They developed ADP data acquisition and processing methods for AUV-mounted use.

The method Zhang and Willcox adopted in their measurements is as follows: An ADP is mounted at the midsection of an AUV in a downward-looking configuration, as shown in Figure 11.23. A mission-specific ADP setting is written into the AUV mission file. Just prior to vehicle launch, a software command wakes up the ADP for pinging via an additional circuit board. The ADP is powered by the vehicle's battery pack, and communication with the vehicle's computer system is through an RS-232 port. The ADP is programmed to map a specific column of water (say, 100 m) during the AUV flight. This water column is subdivided into several depth bins. With measurements from its four beams, water current velocities to the specified depth relative to the flying AUV are recorded in the ADP's internal memory and downloaded after missions onto a PC via an external RS-232 cable.

Because the ADP's platform (the AUV) is moving, the vehicle's velocities must be removed vectorially to obtain the current velocities relative to the Earth. If bottom track is not available, LBL navigation data may be used for estimating the vehicle's position and horizontal velocity. For the LBL navigation using four sonar beacons, the position accuracy is better than 10 m, and the precision is better than 2 m when the vehicle's distance from the center of the four-beacon array is less than the array aperture. The ADP clock is automatically synchronized with the vehicle clock at the beginning of each mission, thereby removing the need for data synchronization in post-processing. The vehicle's vertical velocity is estimated from the time derivative of the depth measurements made by its depth sensor. The vehicle's depth sensor (Paroscientific Model 8B-4000) has an

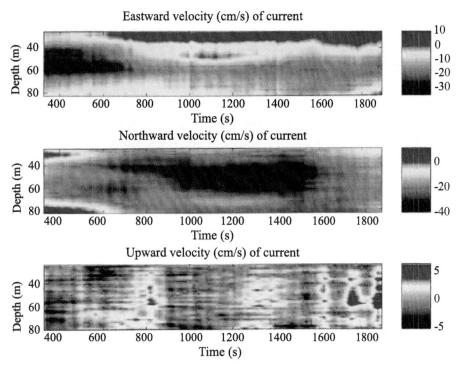

FIGURE 11.24 Earth-referenced current velocities made by an AUV (Xanthos)-mounted ADCP probe. *(Source: Y. Zhang and J. S. Willcox, Current Velocity Mapping Using an AUV-Borne Acoustic Doppler Current Profiler, Proc. 10th International Symposium on Unmanned Untethered Submersible Technology, pp. 31–40, Durham, NH, USA, Sept. 1997.)*

accuracy of 0.4 m, and its precision is better than 0.2 m. The length of the smoothing window is determined based on the measurement errors introduced by the ADP, the LBL navigation system, and the vehicle's depth sensor. The window length is adjustable to achieve a good trade-off between the temporal resolution and the estimation error relative to the maximum current velocity.

At Haro Strait, the ADP was programmed to map a 100-m column of water that was subdivided into 50 2-m depth bins. During a yo-yo mission, the AUV crossed a front, and significant contrasts of temperature and salinity between the two sides of the front were detected. Figure 11.24 shows the Earth-referenced current velocities after removing the vehicle's velocities. Processed ADP data show that the water flowed mostly southward, with the maximum southward velocity of 40 cm/s. The eastward velocity plot demonstrates a layered structure: The upper 40 meters of the water column flowed to the east at about 10 cm/s, whereas the water below flowed to the west with velocity up to 30 cm/s. This kind of layered current structure is attributed to the mixing process.

In Figure 11.25, the vertical current velocity, the vehicle's depth, and the measured temperature and salinity are compared. At time 600 s, the vehicle crossed an oceanic frontal system, entering a lower-temperature and higher-salinity water mass. Subsequently, at time 1,200 s, the vehicle turned around, and at time 1,700 s it crossed the

front again and came back to the higher-temperature and lower-salinity water mass. These signatures were well recorded by the temperature and conductivity sensors. The vertical current velocity shows that within the lower-temperature and higher-salinity region, there was downwelling of up to 5 cm/s, whereas in the higher-temperature and lower-salinity region, upwelling on the order of 5 cm/s existed. The observed alternating upwelling and downwelling events provide insight into the complexity of the mixing process. Current velocity mapping and ancillary measurements utilizing state-of-the-art technology and the AUV's role as a high-performance mobile instrumentation platform provide insight into temporal scales of the mixing process.

11.10.2. ADPs for AUV Navigation

Frequently, AUVs need to operate in ocean environments characterized by complex spatio-temporal variability. This spatio-temporal complexity is induced by the turbulent nature of the ocean, described by the continuous change of a wide range of spatial and time scales. Energetic flows induced by tides and topographic perturbations, as well as instabilities and currents induced by local wind effects, are only a few examples of ocean variability. Such variability can strongly perturb safety conditions and development of AUV operations. In particular, AUVs usually encounter

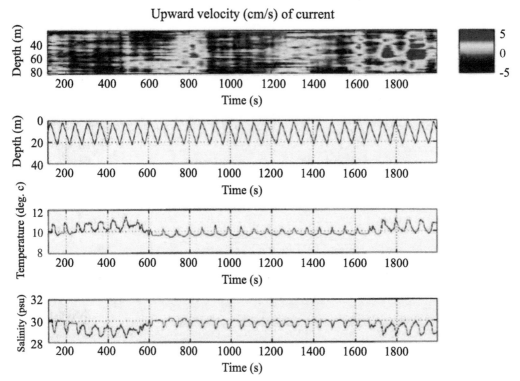

FIGURE 11.25 Earth-referenced vertical current velocity and CTD measurements made by an AUV (Xanthos)-mounted ADCP and CTD probes. *(Source: Y. Zhang and J. S. Willcox, Current Velocity Mapping Using an AUV-Borne Acoustic Doppler Current Profiler, Proc. 10th International Symposium on Unmanned Untethered Submersible Technology, pp. 31–40, Durham, NH, USA, Sept. 1997.)*

strong current fields in the marine environment that can jeopardize their missions. Robustness of AUVs to this strong environmental variability is a key element to carry out safety and optimum operations.

Determining and predicting ocean currents is a fundamental requirement to increase AUV robustness to ocean variability. *A priori* knowledge of this ocean variability would allow us to adequately plan AUV missions, thereby minimizing the possible negative effects of the environment on its operation. Unfortunately, present oceanographic technology is only capable of forecasting the large and slow components of ocean variability, since there are fast and small scales of variability that are still unpredictable. This partial knowledge about future environmental conditions degrades AUV robustness, thereby reducing its autonomy and safety. To circumvent this problem, Garau et al. (2006) proposed a class of real-time algorithm to support AUV navigation in strong, turbulent ocean environments characterized by unpredictable small-scale variability. The algorithm infers the two-dimensional structure of the current field in a limited region in front of the AUV, using the one-dimensional current flow information obtained from a horizontal ADCP known as *H-ADCP*. Subsequently, a planner locally optimizes the performance of the AUV in the inferred field by reducing the traveling time. Based on simulated turbulent environments, it was

found that the solutions provided by the algorithm required significantly less traveling time than straight-line paths.

To obtain the required one-dimensional current-flow information in a limited region in front of the AUV for optimizing its performance, a long-range H-ADCP is incorporated in the AUV's payload. The H-ADCP acoustically measures the current flow velocity profiles in a horizontal line up to ~300 meters in front of the AUV in more than 100 horizontal cells. Apart from the possible use of the water-current velocity profiles gathered from the H-ADCP for a variety of applications, the profiles are primarily employed by the AUV to infer the local spatial pattern of the current field to introduce the required course correction to locally optimize the AUV's navigation and safety in the region by avoiding areas with strong and complex current fields that could jeopardize the AUV mission.

In a first step, the algorithm infers from the H-ADCP profile the current field pattern in an area in front of the AUV. The inferred current field is obtained by considering that the turbulent field results from the superposition of a distribution of vortices known as *viscous Lamb vortices* (Lamb, 1932). The number, centers, and intensities of the vortices closest to the AUV are determined from the H-ADCP profile. Superposition of the current fields generated by the different vortices provides an estimation of the spatial current distribution. Subsequently, based on the reconstructions from the H-ADCP profile and a searching procedure for finding the local minimum-time paths, the AUV plans the trajectory with minimum traveling time in the reconstructed region, trying to take full advantage of the current field that locally favors its mission and to support the AUV navigation to find the way to reach the goal destination, avoiding the hazardous high-speed currents around the goal area. The procedure is repeated when the AUV reaches the border of the inferred current field or when a need to update the inferred current field is detected.

11.11. CALIBRATION OF ADPs

Experiments under controlled conditions are needed to establish the absolute performance of a measuring instrument. In the case of an ADP, calibration in a tow-tank facility enables estimation of its performance. However, unlike conventional Eulerian-style CMs, ADPs have some limitations for being calibrated in a tow tank. These limitations arise primarily from the relatively small size of the tow tank in terms of its width and depth. First, the beams of an ADP are divergent, because of which they intersect with the side walls of the tow tank beyond a certain slant range. Second, the bin length needs to be sufficiently small in size so that the bin is comfortably confined well within the tow tank. Third, the measurements tend to be contaminated by side-lobe interference from the bottom/water surface of the tow tank.

The first limitation is insurmountable, given the large depth range of most of the ADPs. However, the second problem can be resolved by the use of high-frequency ADPs (in effect, shallow-water ADPs) so that pulse length τ is considerably small. In the case of the third limitation, solutions have been found to take care of the problems arising from side-lobe interference. It has been theorized that the acoustic side-lobe energy could be deflected and/or absorbed. The idea embodied in this theory is to develop a device that would allow the main beam to progress undisturbed but restrict the propagation of the side lobes.

NOAA has conducted experiments during field tests with prototype baffles, yielding encouraging results (Appell, 1984; Appell et al., 1985). These tests were conducted in the early 1980s with an AMETEK Straza 300-kHz system. The device consisted of a cone-shaped baffle that was placed between the beams. Side lobes would strike the baffle and be reflected in a horizontal direction. Stray acoustic energy coming from the surface would be reflected and/or absorbed by the top surface of the baffle.

In subsequent tow-tank studies, Appell et al. (1991) subjected RD 1200-kHz and 600-kHz ADPs to tow-tank tests in the David Taylor Research Center (DTRC) towing basin to investigate ADP accuracies. In these experiments, the ADPs could be calibrated in the first four bins (bin 5 indicated signs of contamination) in a downward-looking mode. The ADP parameters were set during DTRC tests (in upward-looking configuration) to provide 1-m bin lengths. During those tests it had been observed that side-lobe returns from the concrete basin bottom biased the measurements at 85 percent (bin 5) of the range. It was also noted that the next bin (bin 6) was not contaminated. In the DTRC tow-tank tests, a 1,200-kHz ADP with a right-angle head adapter was mounted on a bottom sled in an upward-looking mode. The sled was guided by a drainage trough that ran the length of the basin. The sled-borne ADP was towed from the carriage using two tow lines for stability. One end of the tow basin contained a pneumatic wave maker. Maximum amplitude waves of 1-m peak-to-peak were achievable at a 3-s period. An acoustic wave gauge was mounted on the carriage to acquire wave height measurements simultaneously with ADP current measurements.

The bin length (governed by pulse length τ) was set to 1 m and blanking distance to 0.5 m. The tests were conducted with beams 3 and 4 aligned in the towing direction and beams 1 and 2 cross-channel. Data were recorded in beam coordinates so that performance of each beam in the preferential orientation of the ADP could be evaluated on an individual basis. Simultaneous inputs from the tow carriage-speed measurement system and wave gauge were recorded.

Tests were conducted under a variety of tow-speed and surface-wave conditions. A cone-shaped baffle attached to the ADP head prevented side-lobe contamination of the ADP measurements. Tow-tank test results yielded the basic features of the ADP. For instance, zero-flow conditions showed a positive 1 to 2 cm/s bias in all beams and bins. Tow speeds of up to 100 cm/s contained beam errors of within ±2 cm/s in the first three bins. Bin 1 had occasionally higher errors of up to 4 cm/s. Bin 4 (estimated to be at 85 percent of the range and therefore expected to receive the side-lobe energy) consistently had negative errors of 4 to 10 cm/s in the absence of baffle. However, the results in bin 4 were dependent on the size of the baffle. For instance, small baffle did improve the performance of bin 4. However, the larger size baffle showed marked improvement.

These results indicate that the baffle must block sufficient side-lobe energy to be effective. The small baffle did not provide sufficient blockage. With the larger baffle, errors in bin 4 were reduced to normal system errors, showing no signs of the negative bias. Bin 1 showed a slight negative bias compared to the no-baffle data, the reason for which remains to be unknown. With the incorporation of the baffle, bin 5 returned to a normal range of errors.

The tow-tank tests under different artificially generated surface-wave conditions revealed that the frequency content of the wave spectra was produced accurately in the profiler data. Tow-tank wall reflections of wave energy produced velocities in the cross-channel beams. However, wave-particle velocities in the water column had no adverse effect on the ability to determine the "mean" flow.

In a nutshell, based on tow-tank test results of Appell et al. (1991), the 1,200-kHz ADP demonstrated the ability to accurately measure mean currents in a wave-laden flow field. It has been found that side lobes can bias the measurements at 85 percent of the range when bottom or surface boundaries are present. The amount of bias is strongly dependent on surface wave characteristics. Side-lobe bias can be eliminated with a properly designed baffle system. Appell et al. (1991) also demonstrated the ADP's ability to measure water wave particle velocities with a properly configured system (e.g., high transmission frequency and fine sampling interval).

Tow carriages employed for the purpose of calibration of water-flow measuring devices are of various constructions. The drainage trough as available in the DTRC tow basin is not a common feature of every tow basin. Furthermore, the mounting arrangement for the ADP under calibration needs to have the facility to be tilted and rotated so that the acoustic beams can be properly oriented in any desired direction to accomplish different test conditions.

Joseph et al. (2003) reported the design and successful application of a mounting device for a Doppler Velocity Log (see Figure 11.26), which permitted its tilting at roll angles, pitch angles, and combinations of a multitude of chosen roll and pitch angles. Choice of minimum possible projected areas for the components used in the mounting device substantially reduces motion-induced drag force during towing. The edges of all the components of the mounting mechanism are rounded or chamfered to reduce

flow separation and shedding. The mechanism is easy to assemble and mount and is amenable to quick changes of angles. A protractor, mounted on a plane, which is perpendicular to the axis of the support rod, is used for measurement of the horizontal azimuthal directions of the device under calibration. The mounting device was successfully used to examine the performances of a 500-kHz, three-beam downward-looking Sontek Argonaut acoustic Doppler velocity log (DVL).

Figure 11.27 shows a tow tank and towing facility at Central Water & Power Research Station (CW&PRS) at Pune, India, together with the DVL and its mounting attachment. This mounting arrangement can be conveniently used for calibration and performance evaluation tests of an ADP in downward-looking, side-looking, or inclined directions.

11.12. INTERCOMPARISON AND EVALUATION

As in the case of other types of CMs, several aspects of acoustic profiler system performance can, in principle, be evaluated by applying several different techniques. Most of the performance evaluation studies have involved comparison with independent forms of measurement. However, an important step forward has been the development of system-modeling techniques (Chereskin et al., 1989; Chereskin and Harding, 1993). As noted by Chereskin and Harding (1993), this approach, applied in conjunction with signal simulation, enables the performance of individual parts of the system to be analyzed and allows the relationship between system parameter choice and measurement error to be investigated.

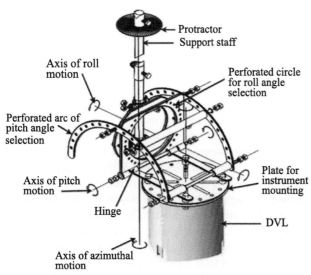

FIGURE 11.26 Support-mechanism for tow-tank experiments on an Acoustic Doppler Velocity Log. *(Source: In part from Joseph et al., 2003.)*

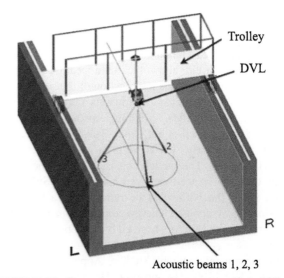

FIGURE 11.27 Tow tank and towing facility, together with the DVL and its mounting attachment. *(Source: Joseph, A., CSIR, National Institute of Oceanography, Goa, India.)*

A number of intercomparison studies have been carried out on ADPs by several researchers, and most of them have been associated with upward-directed ADPs deployed on the seabed in shelf seas, particularly in coastal areas, although Schott (1986), Schott and Johns (1987), and Johns (1988) investigated the use of profilers in deep water for the purpose of obtaining long time series of currents.

For the most part, intercomparisons have been made using data from moored CMs of various types, although Griffiths (1986) has also compared current profiles derived from a Doppler sonar with the predictions of a shallow-water tidal model and found an excellent measure of agreement (Collar, 1993). Figure 11.28 shows current amplitude profiles measured at 3 hours after high water over a number of successive tidal cycles. These have been normalized to water depth D and depth-mean current $\bar{u}$ and clearly show the classical form associated with a logarithmic bottom boundary layer, for which:

$$u(z) = \frac{u^*}{k_o}\ln(z/z_o) \qquad (11.17)$$

where:

k_o = von Karman's constant (= 0.4 in clear water)
z_o = a roughness length constant
z = height above bottom
u^* = shear velocity

Close agreement is evident among the measured profile, the output from an analytical tidal model, and independent bottom observations. The slight offset between the ADP output and the model has been attributed to inaccuracy in extracting a true depth-mean current from the ADP data.

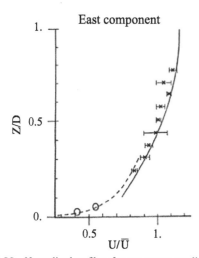

FIGURE 11.28 Normalized profiles of water-current amplitude obtained from a bottom-mounted ADP in shallow water. (*) ADCP with 95-percent confidence limits; (—) analytical model; (-----) theoretical bottom boundary layer; ($\bar{U}$) depth-mean current. *(Source: Collar, 1993, reproduced with kind permission from Jane Stephenson, National Oceanography Centre, Southampton SO14 3ZH, UK.)*

The published results of other instrumental comparisons broadly suggest that below the surface wave zone, incoherent acoustic Doppler provides comparable accuracy of measurement to that of vector-averaging discrete instruments, although Magnell and Signorini (1986) found some discrepancies between data from an ADP and moored instruments at periods outside the main tidal bands. In another study, Pettigrew et al. (1986) found generally excellent agreement between a bottom-mounted ADP and moored VACM and VMCM data. In these studies, the differences observed within 20 m of the surface were consistent with motion-induced errors previously identified in laboratory studies on VACM and VMCM instruments. A subsequent study (Freitag et al., 1992) showed that in some circumstances, ADP speeds can be biased low in comparison with speeds recorded by moored instruments. This is attributed to the presence of fish school in the ADP beams.

Cochin et al. (2006) reported intercomparison studies between an ADP and pulsed VHF Doppler radar systems (VHF COSMER system) in Normand Breton Gulf, France, where the tidal elevation range is about 11 m and the maximum water current velocity is around 2 m/s. During the experiment, the ADP measurements were collected at 10-min intervals. The ADP was configured to make measurements at 28 bins, each with a length of 1 m. Each measurement represented current that was integrated over two bins (i.e., 2 m). On peak spring tide, the sea surface was located above the upper bin (more than 30-m depth).

For intercomparison studies, measurements from a fixed immersion were extracted, taking into account the blind zone of the instrument, which was 4 m below the sea surface. Evaluation of subsurface ADP current measurements was carried out against surface currents obtained from the VHF COSMER pulse-Doppler radar system (the precision of which is around 2 cm/s) and simulation from hydrodynamical numerical model TELEMAC 2-D, developed by the Laboratoire National d'Hydraulique et Environnement (EDF-LNHE), France. Numerical model simulations represent estimation of surface currents, with a mean error of 10 percent. For a fair comparison between the ADP and radar measurements, tidal currents have been extracted from both measurements by subtracting the mean residual current (which is primarily due to meteorological influences) from the observed datasets. Time series of the vector currents (U is the east component and V is the north component) showed that the maximum sea surface currents measured by the instruments yielded different speeds and directions. The observed differences are mainly due to the different techniques of acquisition (Graber et al., 1997).

For example, the electromagnetic method (VHF COSMER) yields a spatially and temporally averaged surface current measurement and is limited to observation near the sea surface, whereas the acoustic method (ADP) produces a subsurface point measurement into bins of the water column. Furthermore, higher time integration and temporal resolution might smooth the currents (which can change rapidly, particularly in spring tides). The observed differences can also be explained by geophysical variability, especially close to the coast (ADP position), where currents are spatially variable (horizontal and vertical). The presence of an island in the area (east of Pointe du Grouin) creates a channel with strong currents during ebb, oriented toward the ADP position. Those currents are measured by the ADP (one measurement every 10 min), whereas the radar measurements smooth them (spatially over approximately 1 km^2 and temporally over 9 min, with one measurement every 30 min). This comparison clearly reveals the complementarities of two different technologies.

Based on examination of numerous intercomparison studies carried out in the initial decades of ADP measurements, Collar (1993) concludes that if adequate leveling of the instrument can be arranged, the use of bottom-mounted ADPs provides substantial advantages over strings of moored CMs in determining vertical profiles of currents. The reasons include the following (Collar, 1993):

- The flow is completely unobstructed by the sensor.
- Measurements are not contaminated by instrument motion.
- Any compass-related errors are common at all levels (bins).
- Relative errors arising from this cause should be small.
- The instrument is probably more secure than a moored string of CMs.
- Costs, including maintenance, are lower than those relating to an equivalent series of discrete instruments.

Instrument tilt can give rise to measurement errors. However, this problem can be taken care of by incorporating tilt sensors in the ADP and by accounting tilt measurements in the velocity calculations. Errors resulting from instrument tilt have been considered by Pulkkinen (1993).

The main problems arise in making measurements close to and within the surface wave zone, in which fluid motions are sheared and are highly variable in magnitude and direction. The instantaneous surface provides a strong acoustic target, as do the air bubbles carried down by the breaking waves (Thorpe, 1986). Backscattered returns may in consequence suffer spectral broadening or distortion, and if the transducer has a near-vertically directed side lobe, which is inadequately suppressed, signals at longer ranges may be masked by strong, unwanted surface returns. Errors may also be generated in determining Doppler shift in high-shear flows (Pullen et al., 1992). In

the initial years of ADP deployments and performance evaluation studies, relatively little effort seems to have been devoted to their near-surface applications, although some preliminary work has been done at the Institute of Oceanographic Sciences (IOS) in the United Kingdom using an upward-looking ADP moored at about 20-m depth (Collar, 1993). Several downward-looking, buoy-mounted systems began to be developed in subsequent years.

Shipboard ADPs have also been subjected to performance evaluation studies. Such studies have included both error analyses and comparative observations made with profiling CMs and moored instruments in both shallow and deep waters (Collar, 1993). These studies (e.g., Chereskin et al., 1987; Magnell and Signorini, 1986; Didden, 1987; Pettigrew et al., 1986; Schott, 1986; Appell et al., 1985; Griffiths, 1986; Delcroix et al., 1992) have generally shown good agreement between the respective techniques, though data on performance under severe sea-state conditions could be limited.

Apart from the physical characteristics of the acoustic medium discussed earlier, factors that have been considered in assessing the accuracy attainable in current shear measurement using shipboard ADP can include various components of ship motion, imperfections in the acoustic geometry (beam pattern, beam orientation), and acoustic noise generated by the ship itself (Joyce et al., 1982; Crocker, 1983; Didden, 1987; Kosro, 1985; Kosro et al., 1986; Joyce, 1989). Velocity quality screening techniques are an important aspect in obtaining maximum accuracy from shipboard ADPs. Real-time methods have been discussed by Zedel and Church (1987). In measurement of absolute current, the rate of change of ship position with time over the Earth's surface must be known to a higher precision than that required of the current measurement. This requirement represents a major constraint (Trump, 1986), although with full implementation of improved satellite-based position-fixing systems such as GPS, the level of attainable accuracy should show marked improvement.

11.13. MERITS AND LIMITATIONS OF ADPs

Probably the most important characteristic of the ADP is its ability to provide a substantial increase in the vertical density of current data obtainable from a single instrument. Furthermore, because the majority of measurements are made remotely, only a small part of the current field—immediately surrounding the instrument—is perturbed. Current measurements are therefore made in a theoretically ideal way. An added advantage, which is not obtainable with older-generation current profilers, is that the acoustic system can also create profiles of acoustic back-scattering strength (Peynaud and Pijanowski., 1979). This information provides an additional window into the environment by which researchers can synoptically study the vertical distribution of biological scatterers such as plankton. The backscattered signal can also be useful to fisheries scientists, assisting them in estimating fish population. The different time responses by water layers at different depths to a fluctuating wind regime induces high lateral shear (Blanton et al., 1974). Oceanic water-current velocity shear has great importance in turbulent mixing and dispersal effects. An interesting application of ADP in oceanography is in the study of fine structures and velocity shear in the water column.

Fast profiling such as that available from an ADP is not possible with free-fall probes. Further, the ADP can map the horizontal (restricted to its field of view) as well as vertical distribution of the current field and its temporal variations. In fact, field experiments using such a profiler could yield three-dimensional current distribution patterns (Okuno et al., 1983). Results of inter-comparison experiments showed striking agreement between results from conventional Eulerian CMs and a bottom-mounted acoustic Doppler current profiler (Pettigrew et al., 1983). Current profiling over vertical ranges of hundreds of meters in the deep ocean has been achieved by an acoustic profiler mounted on top of a subsurface mooring (Schott, 1986). Special precautions must be taken, however, to obtain reliable data from near the sea surface.

Unlike the freely sinking current profiler probe addressed in the previous section, the ADPs can be made to "look" in any direction, and their deployments may be on fixed or moving platforms or on surface, bottom, or mid-depth moorings (Christensen, 1983). Furthermore, the acoustic Doppler technique permits continuous, unattended, fast profiling of currents. Remote sensing with Doppler sonar, therefore, provides an attractive alternative to all previous methods of current profile measurements. Furthermore, in situations where only short-term current measurements are needed, the capability of hull-mountable, downward-looking remote profilers to make measurements from moving vessels is a remarkable feature that permits large spatial coverage and virtually eliminates the tedious and time-consuming logistics in the deployment and retrieval, if conventional moorings were to be used. Mounted beneath a moving vessel, an ADP can make current profile measurements without the motion of the ship significantly affecting the current measurements, because the profiler can measure the speed of the ship with respect to the bottom if the survey area is sufficiently shallow that the ADP's depth range is larger than the local depth of the seafloor beneath the vessel.

REFERENCES

Anderson, S.P., Mathews, P., 2005. A towed 75 kHz ADCP for Operational Deepwater Current Surveys. In: Proc. IEEE/OES Eighth Working Conference on Current Measurement Technology. IEEE, New York, NY, USA, pp. 46–49.

Andreucci, F., Camporeale, C., Fogliuzzi, F., 1992. Acoustic wideband techniques for the remote estimation of the sea current velocity. In: Proc. IEEE Conference Oceans '92. IEEE, New York, NY, USA, pp. 614–619.

Appell, G.F., Gast, J.A., Williams, R.G., Bass, P.D., 1988. Calibration of acoustic Doppler current profilers. Proc. Oceans '88.

Appell, G.F., Bass, P.D., Metcalf, M.A., 1991. Acoustic Doppler Current Profiler performance in near surface and bottom boundaries. IEEE J. Ocean Eng. 16 (4), 390–396.

Appell, G.F., Mero, T.N., Sprenke, J.J., Schmidt, D.R., 1985. An intercomparison of two acoustic Doppler current profilers. In: Proc. IEEE Conference on Ocean Engineering and the Environment, November 1985, San Diego. IEEE, New York, NY, USA, pp. 723–730.

Appell, G.F., Mero, T.N., Sprenke, J.J., Schmidt, D.R., 1985. An intercomparison of two acoustic Doppler current profilers. Proc. Oceans '85.

Blanton, J.O., Murthy, C.R., 1974. Observations of lateral shear in the near-shore zone of a great lake. Journal of Physical Oceanography 4, 660–663.

Bourgerie, R.W., Garner, T.L., Shih, H.H., 2002. Coastal current measurements using an ADCP in a streamlined sub-surface mooring buoy. IEEE 2002, 736–741.

Brumley, B.H., Cabrera, R.G., Deines, K.L., Terray, E.A., 1991. Performance of a broad-band Acoustic Doppler Current Profiler. IEEE J. Oceanic Eng. OE-16, 402–407.

Capon, J., 1969. High-resolution frequency-wavenumber spectrum analysis. Proc. IEEE 57, 1408–1418. also Corrigendum in Proc. IEEE (Lett.), 59, 112 (1971).

Chereskin, T.K., Harding, A.J., 1993. Modelling the performance of an acoustic Doppler current profiler. J. Atmos. Oceanic. Technol. 10, 41–63.

Chereskin, T.K., Harding, J., 1993. Modeling the performance of an acoustic Doppler current profiler. J. Atmos. Oceanic Technol. 10, 42–63.

Chereskin, T.K., Halpern, D., Regier, L.A., 1987. Comparison of shipboard acoustic Doppler current profiler and moored current measurements in the equatorial Pacific. J. Atmos. Oceanic. Technol. 4, 742–747.

Chereskin, T.K., Firing, E., Gast, J.A., 1989. Identifying and screening filter skew and noise bias in ADCP current profiler measurements. J. Atmos. Oceanic Technol. 6, 1040–1054.

Chereskin, T.K., Firing, E., Gast, J.A., 1989. Identifying and screening filter skew and noise bias in acoustic Doppler current profiler measurements. J. Atmos. Oceanic Technol. 6, 1040–1054.

Christensen, J.L., 1983. A new acoustic Doppler current profiler. Sea Technology, Feb. issue, 10–13.

Cochin, V., Mariette, V., Broche, P., Garello, R., 2006. Tidal current measurements using VHF radar and ADCP in the Normand Breton Gulf: Comparison of observations and numerical model. IEEE J. Ocean. Eng. 31 (4), 885–893.

Collar, P.G., 1993. A review of observational techniques and instruments for current measurements from the open sea. Report No. 304, Institute of Oceanographic Sciences, Deacon Laboratory, UK.

Crocker, T.R., 1983. Near-surface Doppler sonar measurements in the Indian Ocean. Deep-Sea Res. 30, 449–467.

Cutchin, D., Christensen, J., Knoop, R., Stillman, J., Woodward, W., 1986. Acoustic Doppler current profiling from a volunteer commercial ship. In: Appell, G.F., Woodward, W.E. (Eds.), Proc. IEEE Third Working Conference on Current Measurement. IEEE, New York, NY, USA, pp. 98–105. January 22–24, 1986, Airlie, VA.

de Schipper, M.A., de Zeeuw, R.C., de Vries, S., Stive, M.J.F., Terwindt, J., 2010. Horizontal ADCP measurements of waves and currents in the very near-shore. Proc. IEEE/OES/CWTM Tenth Working Conference on Current Measurement Technology, 159–166.

Delcroix, T., Masia, F., Eldin, G., 1992. Comparison of Profiling Current Meter and Shipboard ADCP measurements in the Western Equatorial Pacific. J. Atmos. Oceanic. Technol. 9, 867–871.

Didden, N., 1987. Performance evaluation of a shipboard 115 kHz acoustic Doppler current profiler. Cont. Shelf Res. 7, 1231–1243.

Earwaker, K., Zervas, C., 1999. Assessment of the National Ocean Service's Tidal Current Program, NOAA Technical Report NOS CO-OPS 022. Center for Operational Oceanographic Products and Services, NOS, NOAA, Silver Spring, MD, USA, pp. 70.

Firing, E., 1987. Deep zonal currents in the central equatorial Pacific. J. Marine Res. 45, 791–812.

Firing, E., Gordon, L., 1990. Deep ocean acoustic Doppler current profiling. Proc. IEEE Fourth Working Conf. on Current Measurements. Clinton, MD, Current Measurement Technology Committee of the Oceanic Engineering Society, August 1990, pp. 192–201.

Firing, E., Hacker, P., 1994. Equatorial subthermocline currents across the Pacific. EOS Trans., American Geophys. Union, 75(3).

Freitag, H.P., McPhaden, M.J., Pullen, P.E., 1992. Fish-induced bias in acoustic Doppler current profiler data. In: Proc. IEEE Conference, Oceans '92. IEEE, New York, NY, USA, pp. 712–717.

Froysa, K.G., Almroth, E., Andersson, S., Apler, A., Tengberg, A., Hall, P., 2007. Environmental monitoring: Two field cases. Ocean Systems 11 (2), 9–11.

Garau, B., Alvarez, A., Oliver, G., 2006. AUV navigation through turbulent ocean environments supported by onboard H-ADCP. Proc. 2006 IEEE Int. Natl. Conf. Robotics and Automation, Orlando, FL, USA. May 2006, 3556–3561.

Gawarkiewicz, G., Wang, J., Caruso, M., Ramp, S.R., Brink, K.H., Bahr, F., 2004. Shelf-break circulation and thermohaline structure in the northern South China Sea – Contrasting spring conditions in 2000 and 2001. IEEE J. Oceanic Eng. 29 (4), 1131–1143.

Gordon, R.L., 1990. A review of interesting results obtained with acoustic Doppler current profilers. In: Appell, G.F., Curtin, T.B. (Eds.), Proc. IEEE Fourth Working Conference on Current Measurement. IEEE, New York, NY, USA, pp. 181–191. April 3–5, 1990, Maryland.

Gordon, R.L., Brumley, B., Terray, E.A., 1998. Observing wave height and direction with conventional (Janus-style) ADCPs. Oceanology International '98 2, 261–269.

Graber, H.C., Haus, B.K., Chapman, R.D., Shay, L.K., 1997. HF radar comparisons with moored estimates of current speed and direction: Expected differences and implications. J. Geophys. Res. 102 (C8), 18,749–18,766.

Griffiths, G., 1986. Intercomparison of an acoustic Doppler current profiler with conventional instruments and a tidal flow model. In: Appell, G.F., Woodward, W.E. (Eds.), Proc. IEEE Third Working Conference on Current Measurement. IEEE, New York, NY, USA, pp. 169–176. January 22–24, 1986, Airlie, VA.

Hamilton, J.M., 1989. The validation and practical applications of a subsurface mooring model. Canadian Technical Report on Hydrography and Ocean Sciences 119. Ottawa, Ontario, Canada, pp. 49.

Hamilton, J.M., Fowler, G.A., Belliveau, D.J., 1997. Mooring vibration as a source of current meter error and its correction. J. Atmos. Oceanic Technol. 14 (3), 644–655.

Hansen, D.S., 1985. Asymptotic performance of a pulse to pulse incoherent Doppler sonar in an oceanic environment. IEEE J. Oceanic Eng. OE-10, 144–158.

Herbers, T.H.C., Lowe, R.L., Guza, R.T., 1991. Field verification of acoustic Doppler surface gravity wave measurements. J. Geophys. Res. 96 (9), 17,023–17,035.

Howarth, M.J., 1999. Wave measurements with an ADCP. IEEE 1999. New York, NY, USA, pp. 41–44.

Johns, W.E., 1988. Near-surface current measurements in the Gulf Stream using an Upward Looking Acoustic Doppler Current Profiler. J. Atmos. Oceanic. Technol. 5, 602–613.

Johnson, D.H., Dudgeon, D.E., 1993. Array signal processing. Prentice Hall, Upper Saddle River, New Jersey, United States, pp. 533.

Joseph, A., 2000. Applications of Doppler effect in navigation and oceanography, Encyclopedia of Microcomputers. Marcel Dekker Inc 25, 17–45.

Joseph, A., Madhan, R., Mascarenhas, A., Desai, R.G.P., Kumar, V., Dias, M., Tengali, S., Methar, A., 2003. Evaluation of performance of SonTek Argonaut acoustic Doppler velocity log in tow tank and sea, Proc. of the National Symposium on Ocean Electronics: SYMPOL '2003. Department of Electronics, Cochin University of Science and Technology, Cochin, 3–12.

Joyce, T.M., 1989. On in situ "calibration" of shipboard ADCPs. J. Atmos. Oceanic Technol. 6, 169–172.

Joyce, T.M., Bitterman, D.S., Prada, K.E., 1982. Shipboard acoustic profiling of upper ocean currents. Deep-Sea Res. 29 (7A), 903–913.

Joyce, T.M., Bitterman Jr., D.S., Prada, K.E., 1982. Shipboard acoustic profiling of upper ocean currents. Deep-Sea Res. 29, 903–913.

Kaneko, A., Koterayama, W., 1988. ADCP measurements from a towed fish. EOS 69 (23), 643–644.

Kaneko, A., Gohda, N., Koterayama, W., Nakamura, M., Mizuno, S., Furukawa, H., 1993. Towed ADCP fish with depth and roll controllable wings and its application to the Kuroshio observation. J. Oceanogr. 49, 383–395.

Kaneko, A., Koterayama, W., Honji, H., Mizuno, S., Kawatate, K., Gordon, R.L., 1990. Cross-stream survey of the upper 400 m of the Kuroshio by an ADCP on a towed fish. Deep-Sea Res. 37 (5), 875–889.

King, B.A., Cooper, E.B., 1993. Comparison of ship's heading determined from an array of GPS antennas with heading from conventional gyrocompass measurements. Deep-Sea Res. 40, 2207–2216.

Kirby, J.T., Chen, T.-M., 1989. Surface waves on vertically sheared flows: approximate dispersion relations. J. Geophys. Res. 94 (C1), 1013–1027.

Kosro, P.M., 1985. Shipboard acoustic current profiling during the Coastal Ocean Dynamics Experiment. Scripps Institution of Oceanography, Research Theses and Dissertations, California Sea Grant College Program, UC San Diego, CA, USA, Ref, 85–8, pp. 119.

Kosro, P.M., Regier, L., Davis, R.E., 1986. Accuracy of shipboard Doppler current profiling during CODE. In: Appell, G.F.,

Woodward, W.E. (Eds.), Proc. IEEE Third Working Conference on Current Measurement. IEEE, New York, NY, USA, p. 97. January 22−24, 1986, Airlie, VA.

Krogstad, H.E., Gordon, R.L., Miller, M.C., 1988. High resolution directional wave spectra from horizontally-mounted Acoustic Doppler Current Meters. J. Atmos. Oceanic Technol. 5, 340−352.

Lamb, H., 1932. Hydrodynamics. Cambridge University Press, 738.

Lhermitte, R.M., 1985. Water velocity and turbulence measurements by pulse coherent Doppler sonar. In: Proc. IEEE Conference Oceans '85: Ocean Engineering and the Environment. IEEE, New York, NY, USA, pp. 1159−1164. November 1985, San Diego.

Lhermitte, R.M., Serafin, R., 1984. Pulse-to-pulse coherent Doppler sonar signal techniques. J. Atmos. Oceanic Technol. 1, 293−230.

Lohrmann, A., Siegel, E., 2010. Waves in the summer ice in the winter. In: Proc. IEEE/OES/CWTM Tenth Working Conference on Current Measurement Technology, pp. 150−158.

Lohrmann, A., Hackett, B., Roed, L.P., 1990. High resolution measurement of turbulence, velocity and stress using a pulse to pulse coherent sonar. J. Atmos. Oceanic Technol. 7, 19−37.

Longuet-Higgins, M.S., Stewart, R.W., 1962. Radiation stress and mass transport in surface gravity waves with application to surf beats. J. Fluid Mech. 13 (4), 481−504.

MacMahan, J.H., Reniers, A.J.H.M., Thornton, E.B., 2010. Vortical surf zone velocity fluctuations with O(10) min period. J. Geophys. Res. 115, C06007. p. 18.

Magnell, B.A., Signorini, S.R., 1986. Fall 1984 Delaware Bay Acoustic Doppler Profiler Intercomparison Experiment. In: Appell, G.F., Woodward, W.E. (Eds.), Proc. IEEE Third Working Conference on Current Measurement. IEEE, New York, NY, USA, pp. 122−152. January 22−24, 1986, Airlie, VA.

Mahapatra, P.R., Zrnic, D.S., 1983. Practical algorithms for mean velocity estimation in pulse Doppler weather radars using a small number of samples. IEEE Trans. Geosci. Remote Sensing GE-21, 491−501.

Munchow, A., Coughran, C.S., Hendershott, M.C., Winant, C.D., 1995. Performance and calibration of an acoustic Doppler current profiler towed below the surface. J. Atmos. Oceanic Technol. 12, 435−444.

New, A.L., 1992. Factors affecting the quality of shipboard acoustic Doppler current profiler data. Deep-Sea Res. 39, 1985−1992.

Okuno, K., Tsuji, Y., Hisamoto, S., Okino, M., Emura, T., 1983. Three-dimensional current and scattering strength distribution mapping system. Proc. IEEE Oceans '83. New York, NY, USA, 301−305.

Pawka, S.S., 1983. Island shadows in wave directional spectra. J. Geophys. Res. 88, 2579−2591.

Pedersen, T., Lohrmann, A., 2004. Possibilities and limitations of acoustic surface tracking. Proc. Oceans 2004. Kobe, Japan.

Perkins, H., de Strobel, F., Gualdesi, L., 2000. The barny sentinel trawl-resistant ADCP bottom mount: Design, testing, and application. IEEE J. Oceanic Eng. 25 (4), 430−436.

Pettigrew, N.R., Irish, J.D., 1983. An evaluation of a bottom mounted Doppler acoustic profiling current meter. Proc. IEEE Oceans '83, 182−186.

Pettigrew, N.R., Beardsley, R.C., Irish, J.D., 1986. Field evaluations of a bottom mounted acoustic Doppler profiler and conventional current meter moorings. In: Appell, G.F., Woodward, W.E. (Eds.), Proc. IEEE Third Working Conference on Current Measurement. IEEE, New York, NY, USA, pp. 153−162. January 22−24, 1986, Airlie, VA.

Peynaud, F., Pijanowski, J., 1979. An acoustic Doppler current meter, OTC 3457. Proc. Offshore Technology Conference, 863−867.

Pinkel, R., 1980. Acoustic Doppler techniques. In: Dobson, F., Hasse, L., Davis, R. (Eds.), Air-sea interaction: Instruments and methods. Plenum Press, New York, NY, USA, pp. 171−199.

Pinkel, R., Smith, J., 1992. Repeat sequence codes for improved performance of Doppler sounders. J. Atmos. Oceanic Technol. 9, 149−163.

Pollard, R.T., Reed, J.F., 1989. A method of calibrating ship-mounted acoustic Doppler current profilers and the limitations of gyro compasses. J. Atmos. Oceanic Technol. 6, 859−865.

Pulkkinen, K., 1993. Comparison of different bin-mapping methods for a bottom-mounted acoustic profiler. J. Atmos. Oceanic. Technol. 10, 404−409.

Pullen, P.E., McPhaden, M.J., Freitag, H.P., Gast, J., 1992. Surface wave induced skew errors in Acoustic Doppler Current Profiler measurements from high shear regimes. In: Proc. IEEE Conference, Oceans '92. IEEE, New York, NY, USA, pp. 706−711.

Reniers, A.J.H.M., et al., 2009. Surf zone surface retention on a rip-channeled beach. J. Geophys. Res. 114.

Reniers, A.J.H.M., Roelvink, J.A., Thornton, E.B., 2004. Morphodynamic modeling of an embayed beach under wave group forcing. J. Geophys. Res. 109.

Rowe, F.D., Young, J.W., 1979. An ocean current profiler using Doppler sonar, IEEE Proceedings. Oceans '79, 292−297.

Rowe, F.D., Deines, K.L., Gordon, R.L., 1986. High resolution current profiler. In: Appell, G.F., Woodward, W.E. (Eds.), Proc. IEEE Third Working Conference on Current Measurement. IEEE, New York, NY, USA, pp. 184−189. January 22−24, 1986, Airlie, VA.

Sato, S., Nakamura, T., Kawanabe, M., 1999. Horizontal ADCP, new instrument for measuring current speed in a horizontal plane. Proc. IEEE, 121−124.

Schott, 1986. Medium-range vertical acoustic Doppler current profiling from submerged buoys. Deep-Sea Res. 33, 1279−1292.

Schott, F., 1986. Medium-range Vertical acoustic Doppler current profiling from submerged buoys. Rosenstiel school of Marine and Atmospheric Science, University of Miami, Miami, FL, USA. presented orally at the IEEE Third Working conference on current measurement.

Schott, F., 1986. Medium-range vertical acoustic Doppler current profiling from submerged buoys. Deep-Sea Res. 33A, 1279−1292.

Schott, F., Johns, W., 1987. Half-year long measurements with a buoy mounted acoustic Doppler current profiler in the Somali Current. J. Geophys. Res. 92, 5169−5176.

Siegel, E., 2007. New methods for subsurface wave measurements at offshore locations. Ocean News & Technology 13 (6), 38−39.

Sirmans, D., Bumgarner, B., 1975. Numerical comparison of five mean frequency estimators. J. Applied Meteor. 14, 991−1003.

Skelin, D., Bublic, I., Vukadin, P., 2008. Current profile measurement using moored acoustic Doppler current profiler. Proc. 50th International Symposium ELMAR-2008, September 10−12, 2008, Zadar, Croatia, 411−414.

Soulsby, R.L., 1980. Selecting record length and digitization rate for near-bed turbulence measurements. J. Physical Oceanography 10 (2), 208−219.

Spain, P.F., Dorson, D.L., Rossby, H.T., 1981. Pegasus: A simple, acoustically tracked velocity profiler. Deep-Sea Res. 28A, 1553−1567.

Strong, B., Devine, P., 2000. Now use ADCPs for waves as well as currents. International Ocean Systems 4 (5), 4−8.

Strong, B., Brumley, B., Stone, G.W., Zhang, X., 2003. The application of the Doppler shifted dispersion relationship to hurricane wave data from an ADCP directional wave gauge and co-located pressure sensor. Proc. of the IEEE/OES Seventh Working Conference on Current Measurement Technology, 119–124.

Terray, E.A., Brumley, B.H., Strong, B., 1999. Measuring waves and currents with an upward-looking ADCP. In: Proc. IEEE Sixth Working Conference on Current Measurement. IEEE, pp. 66–71.

Terray, E.A., Krogstad, H.E., Cabrera, R., Gordon, R.L., Lohrmann, A., 1990. Measuring wave direction using upward-looking Doppler sonar. In: Appell, G.F., Curtin, T.B. (Eds.), Proc. of the IEEE 4th Working Conf. on Current Measurement. IEEE Press, New York. (IEEE Catalog No. 90CH2861–3), 252–257.

Terray, E.A., Gordon, R.L., Brumley, B.H., 1997. Measuring wave height and direction using upward-looking ADCPs. Proc. of Oceans '97. MTS/IEEE, 287–290.

Theriault, K.B., 1986a. Incoherent multibeam Doppler current profiler performance: Part I—Estimated variance. IEEE J. Oceanic Eng. OE-11, 7–15.

Theriault, K.B., 1986b. Incoherent multibeam Doppler current profiler performance: Part II, Spatial response. IEEE J. Oceanic Eng. OE-11, 16–25.

Thomson, R.E., Gordon, R.L., Dymond, J., 1989. Acoustic Doppler current profiler observations of a mid-ocean ridge hydrothermal plume. J. Geophys. Res. 94, 4709–4720.

Thorpe, S.A., 1986. Measurements with an automatically recording inverted echo sounder; ARIES and the bubble clouds. J. Phys. Oceanogr. 16, 1462–1478.

Trevorrow, M.V., Farmer, D.M., 1992. The use of Barker codes in Doppler sonar measurements. J. Atmos. Oceanic Technol. 9, 699–704.

Trump, C.L., 1986. Estimating absolute current velocities by merging shipboard Doppler current profiler data with Loran C data. In: Appell, G.F., Woodward, W.E. (Eds.), Proc. IEEE Third Working Conference on Current Measurement. IEEE, New York, NY, USA, pp. 177–183. January 22–24, 1986, Airlie, VA.

Trump, C.L., Okawa, B.S., Hill, R.H., 1985. The characterization of a midocean front with a Doppler shear profiler and a thermistor chain. J. Atmos. Oceanic Technol. 2, 508–516.

Urick, R.J., 1982. Principles of Underwater Sound, third ed. McGraw-Hill, New York, NY, USA.

Urick, R.J., 1996. Principles of underwater sound, third ed. Peninsula Publishing, Los Altos, CA, USA.

Visbeck, M., Fischer, J., 1995. Sea surface conditions remotely sensed by upward-looking ADCPs. J. Atmos. Oceanic Technol. 12, 141–149.

Wilson, W.D., 1994. Deep ocean current profiling with a lowered broadband acoustic Doppler current profiler. In: 1994 IEEE. I-660–I-665.

Zedel, L., 1994. Deep ocean wave measurements using a vertically oriented sonar. J. Atmos. Oceanic Technol. 11 (1), 182–191.

Zedel, L.J., Church, J.A., 1987. Real time screening techniques for Doppler current profiler data. J. Atmos. Oceanic. Technol. 4, 572–581.

Zhang, Y., Willcox, J.S., 1997. Current velocity mapping using an AUV-borne acoustic Doppler current profiler. In: Proc. 10th International Symposium on Unmanned Untethered Submersible Technology. Durban, NH, USA, Sept. 1997, pp. 31–40.

BIBLIOGRAPHY

Alderson, S.G., Cunningham, S.A., 1999. Velocity errors in Acoustic Doppler Current Profiler measurements due to platform attitude variations and their effect on volume transport estimates. J. Atmos. Ocean. Technol. 16, 96–106.

Anderson, S.P., Matthews, P., 2005. A towed 75 kHz ADCP for operational deepwater current surveys. Proc. IEEE/OES Eighth Working Conference on Current Measurement Technology, 46–49.

Appell, G.F., 1984. A real-time current measurement system. Sea Technol. 25.

Appell, G.F., Mero, T.N., Sprenke, J.J., Schmidt, D.R., 1985. An intercomparison of two acoustic Doppler current profilers. Oceans '85 Conference Record. IEEE, New York, NY, USA, 723–730.

Beck, S., Pinkel, R., Morison, J., 1986. Doppler acoustic velocity profiling in the Arctic. In: Proceedings of the IEEE Third Working Conference on Current Measurement. January 22–24, 1986, Airlie, VA, IEEE, New York, NY, USA.

Chern, C., Wang, J., 2003. Numerical study of the upper-layer circulation in the South China Sea. J. Oceanogr. 59, 11–24.

Derecki, J.A., Quinn, F.H., 1985. Use of current meters for continuous measurement of flows in large rivers. EOS 66, 907.

Didden, N., 1989. Performance evaluation of a shipboard 115-kHz acoustic Doppler current profiler. Cont. Shelf Res. 7, 1232–1243.

Fischer, J., Visbeck, M., 1993. Deep velocity profiling with self-contained ADCPs. J. Atmos. Oceanic Technol. 10, 764–773.

Gordon, R.L., Skorstad, H.E., 1985. North Sea data will upgrade platform designs. Ocean Industry. July 1985, 35–39.

Irish, J.D., Pettigrew, N.R., Deines, K., Rowe, F., 1983. A new bottom-mounted, internally recording Doppler acoustic profiling current meter. IEEE Proceedings of the Third Working Symposium on Oceanographic Data Systems. IEEE, New York, NY, USA, 109–113.

Johns, W.E., 1988. Near-surface current measurements in the Gulf Stream using an upward looking Acoustic Doppler Current Profiler. J. Atmos. Oceanic. Technol. 5, 602–613.

Kosro, P.M., 1985. Accuracy of shipboard Doppler current profiling during CODE. In: Proceedings of the IEEE Third Working Conference on Current Measurement. January 22–24, 1986, Airlie, VA, IEEE, New York, NY, USA.

Kosro, P.M., 1985. Shipboard acoustic current profiling during the Coastal Ocean Dynamics Experiment. Ph.D. thesis. Scripps Institution of Oceanography, SIO, Reference No. 85–8, La Jolla, CA, United States.

Kosro, P.M., 1985. Shipboard acoustic Doppler current profiling during CODE. SIO Ref. 85–5, La Jolla, CA, United States.

Krogstad, H.E., Gordon, R.L., Miller, M.C., 1988. High resolution directional wave spectra from horizontally mounted acoustic Doppler current meters. J. Atmos. Ocean. Technol. 5, 340–352.

Leaman, K.D., Sanford, T.B., 1975. Vertical energy propagation of inertial waves: a vector spectral analysis of velocity profiles. J. Geophys. Res. 80, 1975–1978.

Luyten, J.R., Swallow, J.C., 1976. Equatorial undercurrents. Deep-Sea Res. 23, 999–1001.

Magnell, B., Signorini, S., 1986. Proceedings of the IEEE Third Working Conference on Current Measurement. Performance of an RDI acoustic Doppler profiler in an intercomparison experiment in Delaware Bay. January 22–24, 1986, Airlie, VA, IEEE, New York, NY, USA.

National Ocean Service, 2001. Tidal Current Tables 2002, Atlantic Coast of North America. U.S. Department of Commerce, National Oceanic and Atmospheric Administration.

Pettigrew, N.R., Irish, J.D., 1983. An evaluation of a bottom-mounted Doppler acoustic profiling current meter, Proceedings of Oceans '83. IEEE, New York, NY, USA.

Pettigrew, N.R., Irish, J.D., Beardsley, R.C., 1983. An intercomparison of a bottom-mounted Doppler acoustic profiling current meter and a conventional current meter mooring. EOS 69, 251.

Pettigrew, N.R., Beardsley, R.C., Irish, J.D., 1986. Field evaluations of a bottom-mounted acoustic Doppler profiler and a conventional current meter moorings. In: Proceedings of the IEEE Third Working Conference on Current Measurement. IEEE, New York, NY, USA. January 22–24, 1986, Airlie, VA.

Pinkel, R., Smith, J.A., 1987. Open ocean surface wave measurement using Doppler sonar. J. Geophys. Res. 92, 12,967–12,973.

Porter, D.L., Williams, R.G., Swassing, C.M., 1986. An intercomparison of surface circulation in the Delaware Bay obtained from CODAR, drifting transponders, and RADS. In: Proceedings of the IEEE Third Working Conference on Current Measurement. IEEE, New York, NY, USA. January 22–24, 1986, Airlie, VA.

Rowe, F.D., Deines, K.L., Gordon, R.L., 1986. High resolution current profiler. In: Proceedings of the IEEE Third Working Conference on Current Measurement. IEEE, New York, NY, USA. January 22–24, 1986, Airlie, VA.

Simpson, M.R., 1986. Evaluation of a vessel-mounted acoustic Doppler current profiler for use in rivers and estuaries. In: Proceedings of the IEEE Third Working Conference on Current Measurement. IEEE, New York, NY, USA. January 22–24, 1986, Airlie, VA.

Smith, J.A., 1989. Doppler sonar and surface waves: Range and resolution. J. Atmos. Ocean. Technol. 6, 680–696.

Stanway, M.J., 2010. Water profile navigation with an Acoustic Doppler Current Profiler. In: 2010 IEEE.

Strong, B., Brumley, B., Terray, E.A., Kraus, N.C., 2000. Validation of the Doppler shifted dispersion relation for waves in the presence of strong tidal currents, using ADCP wave directional spectra and comparison data. Proc. 6th Intl. Workshop on Wave Hindcasting and Forecasting. November 6–10, 2000, Monterey, CA, USA.

Strong, B., Brumley, B., Terray, E.A., Stone, G.W., 2000. The performance of ADCP-derived wave spectra and comparison with other independent measurements. Cesena (FC), Italy. www.rdinstruments.com/library.html.

Terray, E.A., Brumley, B.H., Strong, B., 1999. Measuring waves and currents with an upward-looking ADCP. Proc. IEEE 6th Working Conference on Current Measurement, 66–71. IEEE Press.

Terray, E.A., Gordon, R.L., Brumley, B.H., 1997. Measuring wave height and direction using upward-looking ADCPs, Proc. Oceans '97, MTS / IEEE, 287–290.

Trump, C.L., 1986. Estimating absolute current velocities by merging shipboard Doppler current profiler data with Loran-C data. In: Proceedings of the IEEE Third Working Conference on Current Measurement. January 22–24, 1986, Airlie, VA, IEEE, New York, NY, USA.

Wilson, P., 2009. Measurement of the marine energy resource: Practical solutions for an expectant industry. Part 1: Tidal energy. International Ocean Systems 13 (2), 22–25.

Wolff, P.M., Konop, D., 1984. Predicting water circulation in Delaware Bay and River: NOAA's new approach. Sea Technology. Sept. 1984, 18–22.

Work, P.A., 2008. Nearshore directional wave measurements by surface-following buoy and acoustic Doppler current profiler. Ocean Engineering 35, 727–737.

Chapter 12

Remote Measurements of Ocean Currents Using Satellite-Borne Altimeters

Chapter Outline

12.1. Oceanic Currents and Associated Features
Generated by Sea Surface Slope   382
12.2. Determination of Seawater Motion from Sea
Surface Slope Measurements   385
12.3. Technological Intricacies in Realizing Satellite
Altimetric Measurements   386
12.4. Correction of Errors in Satellite Altimeter Data   389

12.4.1. Correction of Satellite Orbit Errors   389
12.4.2. Correction of Geoid Errors   390
12.4.3. Null Methods for Obtaining Topographic
Height Variability Independent of Geoid   391
12.5. Evolution of Satellite Altimetry   392
References   394
Bibliography   395

Out of academic interest and for a variety of operational applications, it is often necessary to obtain a synoptic view of the spatial dimensions and patterns of water motion over several regions in the ocean under all weather conditions. Seawater motion mapping is useful in establishing large as well as comparatively small circulation routes and in detecting and identifying major ocean gyres. Direct measurement of large-scale ocean currents with current meter arrays is difficult and costly on basin scales. Unfortunately, oceanic water-motion features detected by visible and infrared imagery are sometimes contaminated by the influence of cloud, wind, and sea surface waves. Although passive microwave radiometry is free of cloud effects, higher sea states affect the emissivity. Another difficulty is a weak thermal signature in certain areas in some seasons.

The altimeter, on the other hand, senses oceanic water-motion features based on sea surface height variability and is, therefore, less affected by the mentioned meteorological effects. Furthermore, whereas sea surface images fundamentally provide a qualitative visual indication of the sea surface water-motion features from which water-motion velocity can be estimated using various algorithms, satellite altimetry directly provides quantitative information on various forms of seawater motion.

While we consider the altimetric method of oceanic current measurements, our interest is restricted primarily to boundary currents; variability of spatial length scales of less than 1,000 km, which is typical of most geostrophic

circulation phenomena; and mesoscale phenomena such as gyres and rings (100–300 km range). In this case, the very long length-scale time variability in height caused by tides may interfere with the detection of these circulation features. This error can, however, be corrected by judicious use of tide models. Further, because tides have periodicities of approximately 12 hours, which is very much shorter than the lifetime of most circulation features, the effects of tides may be filtered out. Although the principle of sea-level measurement using altimeters may appear to be simple, much ancillary information as well as highly sophisticated mathematical algorithms are needed to arrive at error-free measurements. The altimetry data have begun to significantly improve our ability to understand oceanic water motion, which strongly influences weather and climate.

In satellite altimetry the sea level relative to a reference ellipsoid that best approximates the shape of the Earth is measured along the satellite ground track (see Wunsch and Gaposchkin, 1980). Because the reference ellipsoid best approximates the geometrical shape of the Earth, with the minor axis of the ellipsoid passing through the poles of the Earth and the major axis of the ellipsoid along the Earth's equatorial plane, the reference ellipsoid surface generally lies above the mean sea level (MSL) and below the mountain levels. For example, the ellipsoid height at the Lakshadweep Island region in the Arabian Sea in the Indian Ocean is ~92 m above the local MSL. In the context of examining satellite altimetric measurements for

Copyright © 2014 Elsevier Inc. All rights reserved.

oceanographic studies, a frequently used term is *marine geoid*. The marine geoid is a surface that would be assumed by the ocean surface if it were a motionless and uniformly dense fluid that is gravitationally bound to a rotating Earth. In other words, the geoid is an equipotential surface of the Earth's gravity field, to which a motionless ocean would conform. Often, the geoid lies far away from the Earth's reference ellipsoid, which is the smooth geometric surface approximating the shape of the entire Earth. Because of Earth's gravity variations, the shape of the geoid is often irregular and in some cases significantly departs from the reference ellipsoid. For example, across the narrow (200 km) Puerto Rico trench, the sea surface sharply dips downward 20 meters relative to the reference ellipsoid (Townsend, 1980; Cheney and Marsh, 1981). Likewise, there is a sharp rise in geoid over the Muir seamount north of Bermuda at about 33°N (Cheney et al., 1984).

The geoid surface is obtained quantitatively from models based on gravity measurements and long-term satellite data. For example, the model, called the PGS-S4, is based on the Goddard Earth Models but also contains Geos-3 and Seasat altimeter data (Cheney et al., 1984). The map of residual sea level (i.e., dynamic topography) is derived by subtracting the local geoid from the altimetric measurement (see Figure 12.1).

A major goal of satellite altimeter missions is the determination of the large-scale dynamic topography of the ocean. Important phenomena include the mean dynamic topography, changes in the mean height of the global ocean, and regional changes of dynamic topography on seasonal or interannual time scales. Knowledge of the mean structure of the dynamic topography is necessary for the determination of the general circulation of the ocean. Satellite altimeter measurements currently provide the only means for determining the level of the sea surface relative to the geoid on a global basis.

Interpretation of satellite altimeter measurements for oceanographic studies is limited by the errors in the marine geoid model. The ocean's surface, while appearing flat, is actually covered with a mosaic of hills, valleys, bumps, and dips caused by variations in Earth's gravity field, currents, winds, and tides. After subtraction of the contribution from Earth's gravity field (i.e., geoid), the residuals are usually smoothed so that only features of interest are retained. Wherever shipboard measurements of temperature and salinity, e.g., those by Levitus (1982), are available, the altimeter-derived dynamic heights can be cross-checked with those calculated from such *in situ* measurements. This will give us a better understanding of ocean circulation field and how it affects climate. The marine geoid was not well known locally before about 2004; therefore altimeters were usually flown in orbits that have an exactly repeating ground track.

Satellite altimeters profile the sea surface with extraordinary precision (a few centimeters) and provide new ways of studying ocean circulation. Altimetric maps of mean sea height have revealed the intricate surface expressions of meanders and eddies. Temporal variability of the sea surface due to meandering currents and eddies has been determined with an uncertainty of only a few centimeters, providing the first comprehensive view of the global eddy field. Even the broad, basin-scale circulation has been observed to a certain degree by subtracting the modeled gravimetric topography from global altimetric mean surfaces.

Superimposed on the static geoid topography is dynamic topography due to ocean circulation. Temporal variability of dynamic height due to oceanic eddies can be determined from time series of repeated altimeter profiles. Maps of sea-height variability and eddy kinetic energy derived from altimetry in some cases represent improvements over those derived from standard oceanographic observations. Measurement of absolute dynamic height imposes stringent requirements on geoid and orbit accuracies, although models and data have been used to derive surprisingly realistic global circulation solutions. Further improvement will be made only when advances are made in geoid modeling and precision orbit determination.

12.1. OCEANIC CURRENTS AND ASSOCIATED FEATURES GENERATED BY SEA SURFACE SLOPE

Geostrophic surface currents are maintained by horizontal pressure gradients and are expressed as sea-height slopes relative to the geoid. Although it is difficult to separate this dynamic component of sea height from the static geoid signal, an altimeter can readily detect the time-dependent variations in height. The most energetic fluctuations have typical scales of 100–300 km and are attributable to the ocean eddy field: meandering of narrow currents and migration of detached vortices. Eddying motions in the ocean are believed to be the dominant mechanism for transferring energy and momentum. Obtaining a basic description of the global distribution of mesoscale eddy

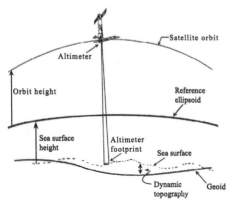

FIGURE 12.1 Concept diagram of satellite altimetry method of dynamic topography measurement.

variability is fundamental to an improved understanding of ocean dynamics (Cheney et al., 1984).

Piling up or piling down of the sea surface relative to the neighboring water body results in characteristic slopes of the sea surface relative to the local marine geoid. Even in a homogeneous ocean, where density is constant, these slopes produce horizontal pressure gradients. The resulting seawater motion is generally called *slope current*. Because it is the pressure field that ultimately drives the slope current, in the literature these motions are also termed *near-surface geostrophic currents*. Because large-scale currents are very nearly in geostrophic balance, their velocity can be calculated from the pressure gradient on an equigeopotential surface. The surface geostrophic current therefore can be calculated from the deviation of sea level from the equigeopotential at the ocean surface (the marine geoid).

Because the primary contribution of altimetry is in the determination of sea surface-height variability, altimetric measurements can be employed to obtain a realistic view of ocean surface current. Measuring sea level from space by satellite altimetry thus offers a unique opportunity for determining the global surface geostrophic ocean circulation and its variability. Coupled with knowledge of the geoid and the ocean density field, satellite altimetry provides the only feasible approach for determining absolute geostrophic currents in the global ocean. Repeated altimetric observations of sea level at the same locations can resolve the variability of surface geostrophic currents, without the requirement for an accurate geoid model (Fu et al., 1988). The most energetic fluctuations of the slope current have typical scales of 100–300 km and are attributable to the ocean eddy field, meandering of narrow currents, and migration of detached vortices. Obtaining a basic description of global mesoscale eddy variability is fundamental to an improved understanding of ocean dynamics. The magnitude of dynamic sea surface elevation ranges from 10 cm (gyre-scale currents) to over 1 m (western boundary currents and eddies), with variance spread over a wide range of wave numbers and frequencies. The required measurement accuracy is thus dependent on the phenomenon of interest. Generally, an overall accuracy of better than 10 cm is required for observations of the dynamic sea surface elevation to be useful.

Temporal variability of the large-scale circulation is strongly linked to weather and climate. An example is the El Niño/Southern oscillation phenomenon in the Pacific. At approximately five-year intervals, patterns of atmospheric pressure and wind in the tropics undergo dramatic change. Easterly trade winds normally create westward currents along the equator and set up a zonal gradient of sea height as water accumulates in the west. During El Niño, trade winds may actually reverse, resulting in a redistribution of water back toward the east. During the 1982–83 event, sea-level changes of up to 40 cm were observed near the equator. Even as far north as 45°N along the North

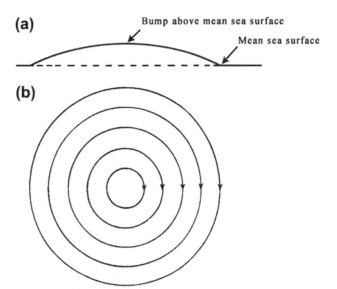

FIGURE 12.2 Concept diagram of a clockwise rotating warm-core circular eddy. (a) Vertical cross-sectional view of sea surface topography of the eddy; (b) plan view.

American coast, 35-cm monthly mean deviations were observed. These anomalous sea-level signals apparently propagate eastward across the central Pacific over a period of a few months. A satellite altimeter with a precision of a few centimeters can, in principle, provide a complete description of sea-level change on all scales.

In the open ocean, slopes of the sea surface can be generated if the wind distribution is not uniform. In such situations the water must pile up in the convergence zone, producing a slope of the sea surface in this zone. Boundaries of two distinct water masses (i.e., frontal region) also exhibit characteristic deviations from the mean sea surface, resulting in water-current motion along the front. Warm-core and cold-core eddies are also associated with characteristic sea surface slopes. For example, a warm-core eddy is associated with a "bump" of the sea surface at the core region (see Figure 12.2), whereas a cold-core eddy is characterized by a "dip" of the sea surface at the core region (see Figure 12.3).

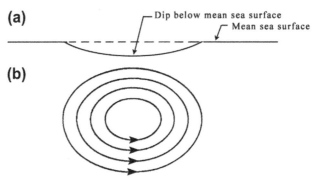

FIGURE 12.3 Concept diagram of a counter-clockwise rotating cold-core elliptical eddy. (a) Vertical cross-sectional view of sea surface topography of the eddy; (b) plan view.

The bumps and dips are caused by spatial variations in seawater density across the eddy region. In reality, several energetic oceanic regions are luxuriant in time-varying sea surface bumps and dips as a consequence of the temporal and seasonal evolution, growth, and decay of a variety of geostrophic currents and eddies. Such a fascinating model-simulated mosaic of mean sea surface heights (SSH) at contour intervals of 8 cm from the Atlantic Ocean, including the Intra-Americas Sea zoomed into the Gulf Stream region between Cape Hatteras and the Grand Banks (see Figure 12.4), has been reported by Hurlburt and Hogan (2008). The ocean eddy fields (typical scales of 100–300 km) have lifetimes ranging a few weeks to a few months. This topic was addressed in more detail in Chapter 1.

In coastal water bodies, interference by the coast to the slow westward drift of water mass results in its piling up on the western boundaries of the oceans. A current is generated to arrest continuous piling up. Because water mass piling up at the western boundaries of the oceans takes place continuously, as a counteracting mechanism the slope currents also persist without break. The slope currents generated along the western boundaries of the oceans are often termed *western boundary currents* (WBCs). The most prominent features in the Southern Hemisphere are the

Agulhas Current off the tip of South Africa, the confluence of the Falkland and Brazil Currents east of Argentina, and the Antarctic Circumpolar Current, represented as a series of sea-level variability maxima surrounding Antarctica. Because of westward intensification of the wind-driven ocean currents, higher sea-level variability is most prevalent in the western parts of the oceans. Seamounts play an important role in controlling the water circulation in large ocean basins. For example, there is a marked contrast between the highly variable western North Pacific and the quieter eastern basin, with the dividing line occurring near the Emperor Seamount chain. This line of seamounts apparently acts as an efficient barrier to the high variability generated by the Kuroshio in the west.

Sea surface slopes have important implications in oceanography. For example, downward sea surface slope along the equator is responsible for the worldwide existence of a remarkably strong current, known as the *equatorial undercurrent*. This current, centered on the equator and flowing from west to east with a maximum speed of 100–150 cm/s, is ~200–300 km wide and ~150–300 m deep. It has been observed that the width, depth, and speed of this current can vary depending on the ocean and, probably, the season (Neumann, 1968). Similarly, the

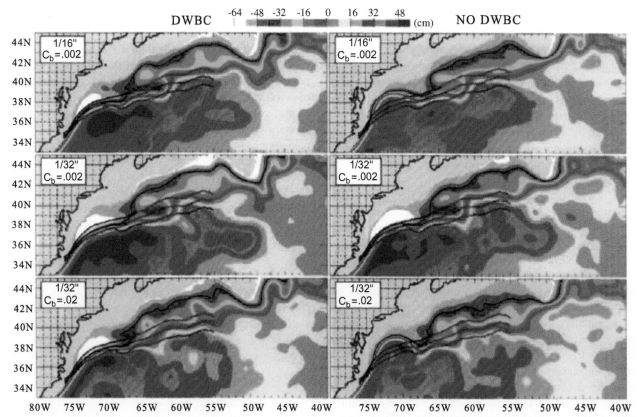

FIGURE 12.4   Model-simulated mosaic of mean sea surface heights (SSH) at contour intervals of 8 cm from the Atlantic Ocean, including the Intra-Americas Sea zoomed into the Gulf Stream region between Cape Hatteras and the Grand Banks, with and without Deep Western Boundary Current (DWBC). *(Source: Hurlburt and Hogan, 2008, Dynamics of Atmospheres and Oceans, Elsevier.)*

development of the Gulf Stream and the Kuroshio, which are two outstanding examples of WBCs, is associated with a westward displacement of the center of rotation of the subtropical gyres and the resulting piling up of water mass on the western boundaries of the respective oceans. In fact, detailed investigations have shown that the Gulf Stream is clearly detectable as a 100–200-cm step in dynamic height (Cheney et al., 1981).

12.2. DETERMINATION OF SEAWATER MOTION FROM SEA SURFACE SLOPE MEASUREMENTS

Slope currents, maintained by horizontal pressure gradients, are manifested as gradients of sea surface relative to an equipotential surface (i.e., the marine geoid). These currents can, therefore, be determined from precise measurements of sea surface slopes relative to the local marine geoid. The sea-level slope, $\tan \beta$, is related to the associated slope current C by the relation (Neumann, 1968):

$$\tan \beta = (2\omega C \sin \varphi)/g \qquad (12.1)$$

In this expression, β is the angle made by the instantaneous mean sea surface with the local marine geoid; ω is the angular velocity of Earth's rotation in radian/s (7.29×10^5 radian/s); φ is geographical latitude; and g is acceleration due to Earth's gravity. In this, ($2\omega \sin \varphi$) is popularly known as the *Coriolis parameter, f*. Thus,

$$C = (g \tan \beta)/f \qquad (12.2)$$

If dz is the difference of the sea surface heights at two locations, separated by a horizontal distance dx, then the slope is given by $\tan \beta = dz/dx$, so that

$$C = (g/f)(dz/dx). \qquad (12.3)$$

If the height of the sea surface relative to the marine geoid is ζ, then the u-component (u_s) and the v-component (v_s) of the surface geostrophic current can be related to surface elevation ζ by the expressions:

$$u_s = -\frac{g}{f}\frac{\partial \zeta}{\partial y} \ ; \qquad (12.4)$$

$$v_s = \frac{g}{f}\frac{\partial \zeta}{\partial x} \qquad (12.5)$$

In practice, the residual height, $\partial\zeta$, may be in cm or a few m, whereas the corresponding horizontal distance will be in hundreds of km. Because we can measure the surface topography from a satellite altimeter, we can calculate the slope of the sea surface and, in turn, the surface geostrophic currents. Typical slopes are $\approx$ 1–10 micro-radians for $v = 0.1$–1.0 m/s at mid-latitudes.

Sea surface slope across straits and channels can be determined using data from sea-level gauge records. This is possible because in coastal regions the absolute geodetic "zero" point can be established by precise leveling, and the mean sea level can be found from long-term sea-level observations. In fact, such observations, in conjunction with simultaneous current measurements in straits and channels, were used to test the validity of Equation 12.3. Using sea-level records from several islands and coastal stations in the equatorial Pacific Ocean, Wyrtki (1977) studied the response of the Pacific equatorial circulation to the 1972 El Niño. In fact, long-term sea-level records are considered an important source of data for the long-term monitoring of coastal currents and their variability.

Although it was known that currents generated by variations in sea surface slope due to actual piling up or removal of mass could be derived from precise sea surface leveling (Sverdrup et al., 1942), in the absence of necessary tools, slope-current measurements from the open oceans continued to remain a mere theoretical concept until satellite altimetry methods were introduced in the 1970s. If the satellite's instantaneous geographical position is independently determined, the sea surface dynamic topography can be inferred from the altimeter height measurements. With the availability of detailed geoid models, satellite altimetry has the potential for generation of near-synoptic maps of seawater motions and thereby to play a significant role in ocean circulation studies through its capability for rapid global observation of sea surface topography. Radar altimetry is identified as an important technology for the acquisition of data on oceanic large-scale water motions. In fact, the Gulf Stream path depicted by altimeter data agreed closely with analysis of simultaneously acquired satellite infrared imagery. Satellite altimetry can provide estimates of global surface geostrophic currents using surface height anomaly measurements (Fu and Chelton, 2001; Bonjean and Lagerloef, 2002). In fact, satellite altimetry has begun to demonstrate its capability to contribute significantly to the reliable mapping of geostrophic circulation of the oceans.

Apart from the well-known slope currents such as the WBCs and the Equatorial undercurrent, the other important oceanographic phenomena that influence the ocean surface topography include fronts, tides, swell waves, storm surges, and the more recently identified spatially fluctuating and circulating currents such as meanders and eddies, respectively. Determination of dynamic topography permits detection and quantification of these oceanic circulation features.

The most direct method of obtaining dynamic topography is to subtract a geoid model from the measured altimetric profiles. Although the detection of dynamic features from the altimeter profiles is straightforward in principle, accurate determination of near-surface geostrophic circulation velocities (through the geostrophic relations 12.4 and 12.5) requires generation of a multitude of sea surface topographic maps using data from different passes of the altimeter in a given region of interest. The axis of water-motion trajectory can be estimated accurately from these multidirectional maps.

The launch of altimeter-borne polar-orbiting satellites was a major milestone in the remote measurement of sea surface parameters, including geostrophic circulation on an all-weather basis. The nadir (the region directly beneath the satellite) viewing radar altimeter measures the altitude of the satellite above the sea surface as follows:

The altimeter transmits a microwave signal of very short pulse width (a few nanoseconds) toward the terrestrial surface. The signal reflected off the sea surface is received back at the altimeter and the round-trip travel time (Δt) of the microwave pulse is measured. Assuming no significant radial motion of the satellite in the interval (Δt), the distance h between the altimeter's measurement point and the sea surface immediately below the altimeter is given by the relation:

$$h = C_m(\Delta t)/2 \qquad (12.6)$$

where C_m is the speed of the electromagnetic wave in the medium that lies between the satellite and the sea surface bounded within the footprint of the altimeter beam. Subtraction of the detailed geoid (along the trajectory of the altimeter footprints) from the altimeter-measured height profiles at close spatial intervals along the satellite ground track yields profiles of residual sea heights (i.e., topographic maps) along that track. The residual sea-height variability, in part, represents the geostrophic height variability that generated the slope currents. Because the Earth spins toward the east, the satellite orbit has a westward drift relative to the Earth's surface. This drift facilitates generation of topographic maps along adjacent tracks and thereby ensures that the satellite altimeter provides a global coverage every few days.

Such maps at close grids contain data on sea surface slopes relative to the local marine geoid, which, in turn, contain information on ocean circulation dynamics. Combined with detailed marine geoid models, the altimeter data can therefore provide descriptions of the near-surface geostrophic current field. This information, together with subsurface measurements of the density structure, may also permit estimation of the velocity field at all depths (Marsh et al., 1982). While addressing the satellite altimetry method of ocean circulation measurements (schematically shown in Figure 12.1), it would be worthwhile to realize that the sea surface is characterized by oscillations and fluctuations associated with long-wavelength ocean tides, short-wavelength gravity waves and ripples, wind set-up/set-down, and inverse barometric effects. Tides in the open ocean are of comparatively small amplitudes (less than a meter). However, waves and inverse barometric effects can have large amplitudes (one to several meters). Fortunately, because the footprint of the altimeter on the sea surface runs to a few square kilometers, errors resulting from the presence of short-wavelength phenomena such as waves and ripples are smoothed out spatially. In effect, the altimeter measures, therefore, the distance to the "instantaneous mean" sea level.

12.3. TECHNOLOGICAL INTRICACIES IN REALIZING SATELLITE ALTIMETRIC MEASUREMENTS

Altitude measurement using a radar altimeter is simple in principle. However, considering the large distance between the satellite and the sea surface (~800–1,300 km), attaining a precision better than 10 cm (required for circulation studies) is a difficult task. Better precision in altitude measurement demands that the transmitted pulse be of extremely narrow width and that the pulse travel time be measured with utmost precision. Because the frequency bandwidth necessary for a pulse is the reciprocal of the pulse width, a sharper transmission pulse necessitates a wider frequency band to carry it. Even if a narrow pulse can be transmitted, with all the attendant constraints, a wave-laden sea surface causes the returned signal to be stretched out. The short pulse of the altimeter is "stretched" by the waves because scattering occurs from crests through troughs as the spherical pulse progresses downward (Barrick et al., 1980). Such degradation of the radar pulse shape leads to errors and adversely affects the altimeter's range resolution. Errors can creep into the calculation of the altitude also because of the variability of the speed (c) of the electromagnetic signal in the ionosphere and in the troposphere. Corrections must be applied to reduce all these errors.

Reduction in radar transmission pulse width to achieve better range resolution prevents injecting sufficient power into the transmission signal. On the other hand, sufficient power has to be injected into the narrow transmission pulse so that the returned signal is sufficiently stronger compared to the ambient noise. The return signal level is proportional to the amount of energy transmitted. The S/N power ratio at the input of the satellite-borne receiving system can be made considerably larger than unity by transmitting very high power. Unfortunately, it is often inconvenient to generate a very high peak power when the source is electronically driven, because the peak power is limited by the amplifier. These conflicting requirements of a high-resolution pulse radar altimetry are met by ingenious techniques such as pulse compression and pulse-limited modes of operation.

In the pulse compression process, a narrow pulse is dispersed by a filter into a very much longer "chirp" waveform (linearly varying frequency; see Figure 12.5a) in the transmission section, and the signal received from the sea

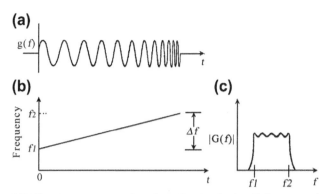

FIGURE 12.5 Representation of (a) chirp signal pulse, (b) frequency of chirp signal as a function of time, and (c) power spectrum of chirp signal pulse. *(Source: Clay and Medwin, 1977.)*

surface is compressed in the receiver section. Mathematical treatment of chirp pulse may be found in Biggs and Jordan (1989). As discussed here, the longer chirp signal permits transmission, with a lower peak power of an equivalent energy in a high-amplitude, narrow sine-wave pulse and achieving comparable fine-range resolution of a narrow sine-wave pulse. By linearly varying the frequency of the signal during transmission, it becomes possible to send a long pulse, keeping the bandwidth large. The transmission starts at frequency f_1 and then sweeps to f_2. The frequency bandwidth of this signal is $(f_2 - f_1) = B$. The superb capability of the long chirp signal to achieve fine-range resolution arises from the wider bandwidth associated with the frequency modulation to this pulse (Ulaby et al., 1982). Because the bandwidth, B, of the chirp signal is considerably larger than the reciprocal of its width τ, better temporal resolution and therefore better range resolution are achievable from a chirp signal relative to that from a sine-wave pulse of the same

width τ. It may be recalled that the bandwidth of a sine-wave pulse of width τ is equal to $(1/\tau)$.

Often, a dispersive delay line (DDL) generates the basic linear FM (chirp) pulse. The heart of the DDL is a surface acoustic wave (SAW) device fabricated on a lithium tantalate substrate. In operation, an impulse of the desired high frequency is applied to the SAW filter, which then generates an expanded chirp pulse with the characteristic of linearly varying frequency. The frequency of the chirp signal as a function of time is shown in Figure 12.5b. The power spectrum of this chirp signal is shown in Figure 12.5c. The turn-on and turn-off transients broaden the spectrum outside f_1 and f_2. The total pulse duration is τ, and its frequency bandwidth after modulation is B. In the transmit mode, the chirp pulse is amplified to the required power level. In the receive mode, the dispersed pulse reflected from the terrestrial surface is compressed in a way that the signal at the output of the receiver system is sufficiently strong. This "de-chirping" process is accomplished by passing the received signal through a delay line-cum-adder system of which the time delay is a function of frequency. The time delay is such that the frequency element of the chirp signal first received from the terrestrial surface is delayed long enough so that it arrives at the output of the receiver system at the same time as the frequency element received last. All the frequencies in between also arrive at this time so that they are superimposed at a single instant of time in the receiver output. This is shown in Figure 12.6.

A particular combination of time delays matches the waveform of a particular signal. This combination is called a *matched filter*. If the filter is truly a matched filter, the time delays suffered by each point of the received chirp signal

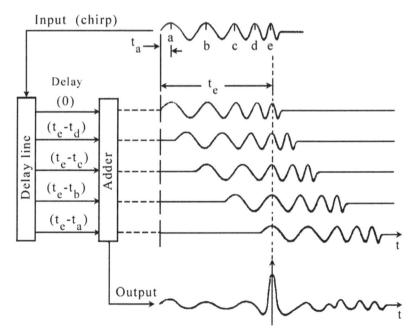

FIGURE 12.6 Schematic diagram illustrating the generation of a narrow peak signal from a long chirp signal when the chirp signal is delayed and added. *(Source: Clay and Medwin, 1977.)*

within the delay line will be such that addition of the various time-delayed signals would generate a very narrow peak with a fixed time delay relative to a fixed temporal position of the received signal. If a different signal, such as a down-swept chirp, is passed through an up-swept chirp matched filter, the output will be low-amplitude wiggles and no peak (Clay and Medwin, 1977). Referring to Figure 12.6, addition of signals with delay times $(t_e - t_d)$, $(t_e - t_c)$, $(t_e - t_b)$, and $(t_e - t_a)$ aligns the peaks of all the delayed signals in the receiver section in such a way that the desired narrow pulse is generated. The matched filter thus compresses the long transmission chirp pulse into a narrow peak. The low-amplitude wiggles, before and after the narrow peak, are called the *side lobes* of the matched filter. If B is the band-width of the transmitted chirp pulse, the effective width of the de-chirped signal at the receiver filter output is $(1/B)$. If the amplitude of the reflected linear FM chirp pulse is constant, the de-chirped signal takes the form of a $((sin\ x)/x)$ pulse. The amplitude of the de-chirped signal is increased from 1 for the signal at the filter input to $\sqrt{(B\tau)}$ for the de-chirped waveform. The de-chirping is, therefore, an effi-cient way of achieving a narrow signal of higher amplitude for a given bandwidth of the transmission chirp signal. An excellent mathematical treatment of the chirp signal and a matched-filter implementation technique for a linear-FM chirp signal may be found in Ulaby et al. (1982).

Returning to the individual measurement of altitude to be representative of a small area, it is necessary that the altimeter footprint on the sea surface be of a minimal size. When one considers the large altitude of the satellite above the terrestrial surface, generation of a radar beam that is sufficiently narrow so that it is sharply focused onto a small area on the sea surface would require far too large an antenna aperture to be practical. Because operation in this ideal situation (i.e., beam-limited geometry) is difficult to achieve, an innovative technique known as *pulse-limited geometry,* which effectively achieves the function of beam-limited geometry, began to be used in precision altimeters. The pulse-limited geometry permits pulse transmission in a wide beam. This technique also minimizes the effects of antenna pointing errors, induced by variations in the atti-tude of the spacecraft, on height measurement precision (MacArthur, 1976). In this technique, a pulse is transmitted from a satellite and is radiated in a spherical shell, of which the intersection with the ocean surface defines an instan-taneous illuminated area.

The leading edge of the transmitted pulse strikes the sea surface first immediately beneath the satellite and then moves out in a circular front. The trailing edge does the same a little later, resulting in the illuminated area being first a circle of growing area, from zero to a maximum when the rear of the pulse just touches the sea surface and then an annulus of constant area. Because the energy reflected from the sea surface is proportional to the illuminated area, the

average return energy exhibits a linear initial rise and is followed by a plateau region resulting from the constant area of the expanding (but thinning) annulus. Each indi-vidual return is similar to the noise that results from the interference between reflections from multiple facets. Ultimately, the overall antenna pattern causes the signal return to attenuate. Geometries of pulse-width-limited mode of operation are given by MacArthur (1976) and Raney (1998). Altitude measurement is implemented based on leading-edge tracking of the pulse-width-limited sea surface return signal. The timing algorithm fits a curve to the leading edge of the return signal up to its maximum. This scheme ensures that the effective area of which the height is sampled is the illuminated circle just before it becomes an annulus. For a calm sea and a pulse length of t_p, this illuminated circle has a radius given by (Robinson, 1985):

$$R_a = \sqrt{(2hct_p)} \qquad (12.7)$$

In this expression, h is the height of the satellite above the mean sea level and c is the speed of electromagnetic signal in the medium of travel. The area, given by $\pi(r_a)^2$, is substantially smaller than the best beam-width-limited footprint achievable by the present technology. For example, for Seasat with h and t_p 800 km and 3 nanosec-onds, respectively, the value of r_a works out to be 1.2 km. With a beam as narrow as $1.59°$ at -3 dB, the beam-width-limited circular footprint associated with the Seasat-A altimeter radar had a radius of 11.1 km (Townsend, 1980). It can be seen that the area of the effective pulse-width-limited footprint (4.52 km^2) for a smooth surface is substantially smaller than that of an appreciably good beam-width-limited footprint (386.88 km^2). For a rough sea with "significant wave height" H, the radius of the illumi-nated circle just before it becomes an annulus is given by:

$$R_a = \sqrt{(2hct'_p)} \qquad (12.8)$$

In this expression, $(t'_p)^2 = (t_p)^2 + (16\ H^2 \ln 2)/c^2$. For very calm seas, the leading edge is sharply defined. For rough seas, where the crest-to-trough distances are substantially greater than the altimeter's basic resolution, the leading edge of the incoming signal echo is stretched out in time, and the point on the leading edge corresponding to the mean sea level is not clearly defined. Thus, the sea state-related alteration of the characteristic shape of the sea surface return signal introduces an error in the determi-nation of a spot value of altitude. These errors are, however, reduced by advanced signal tracking, modeling, and averaging techniques such as those described by MacArthur (1976) and Townsend (1980). The average of many samples acts to reduce the statistical fluctuations inherent in the individual waveform samples and mini-mizes noise.

12.4. CORRECTION OF ERRORS IN SATELLITE ALTIMETER DATA

Although the satellite altimetry method of near-surface oceanic current measurements might seem simple in principle, the actual implementation involves numerous technical hurdles that need to be taken care of before reliable measurements can be achieved. The measured height of the satellite above the sea surface at any instant in time must be corrected for a myriad of instrumental and geophysical effects. These include:

- Satellite height errors resulting from atmospheric-related effects in the electromagnetic pulse travel time
- Sea-state related errors
- Random errors resulting from long wavelength uncertainty in the orbital radius
- Errors arising from the unknown marine geoid

These aspects are briefly addressed in the following sections.

12.4.1. Correction of Satellite Orbit Errors

The orbit of the satellite (800 to 1,300 km above Earth's surface) does not remain precisely stable due to many atmospheric, astronomical, and geophysical forces that act to disturb the dynamics of the satellite. Other small orbit distortions are introduced by changes of the satellite's mass after maneuvers, by gravitational attractions of the moon and the sun, and even by the changes of gravity due to the water mass movements of the ocean tides. Apart from these uncertainties, because the microwave pulse transmitted from the satellite travels through the atmosphere, its round-trip travel time (Δt) must be corrected for propagation delays to accurately obtain the radial height (h) of the altimeter above the sea surface. The atmospherically induced delay in the radar pulse travel time is one of the primary corrections that must be applied to the radar altimeter measurements. The atmospheric effect can be separated into tropospheric and ionospheric effects. The tropospheric effects result in a decrease in the local speed of the electromagnetic signal due to the refractive index changes of the medium of propagation, caused by water vapor and other gases in the troposphere. Similar effects result from the presence of free electrons in the ionosphere. Tapley et al. (1982) discussed the errors due to these effects. It has been noticed that errors in sea-level estimates and water vapor content in the atmosphere are strongly correlated. The amplitude of the water vapor correction to the altimeter path length ranges from approximately 35 cm in the tropics to near zero at the poles and can be modeled with an uncertainty of ~5 cm for monthly means (Cheney et al., 1991).

Most of the atmosphere-related errors are corrected using data collected by other sensors on board the satellite (e.g., wet tropospheric correction to the measured altitude achieved using data collected by the satellite-borne scanning multichannel microwave radiometer [SMMR]). The SMMR measures the integrated mass density of the atmospheric water vapor along a line of sight from the SMMR antenna to the sea surface, thereby obtaining a measure of the increase in the effective radio-frequency path length (Tapley et al., 1982). In a satellite-based system, another source of water vapor data is the Special Sensor Microwave Imager (SSMI). The SSMI data enable derivation of global water vapor fields that provides an altimeter height accuracy of ~2 cm. In fact, for certain oceanographic applications, there is no substitute for a radiometer on board the altimeter satellite.

Errors resulting from ionospheric delay of the microwave signal can be corrected by the use of a dual-frequency altimeter (Lorell et al., 1982). This is possible because, for a given value of the columnar electron content along a ray path, the delay is inversely proportional to the square of the frequency.

Another atmospheric-related error, arising from rain and cloud effects, is caused by the spatial inhomogeneity of the medium of travel of the electromagnetic pulse. This inhomogeneity causes energy in different parts of the altimeter footprint to be attenuated differently, resulting in a distorted return waveform signature. This effect, in turn, causes mean-sea-level errors when standard tracking algorithms, which are dependent on pulse shape, are employed. To avoid misleading interpretation, the use of some kind of a sensor to detect the presence of rain is recommended so that such data can be flagged as questionable.

Separation of the variability in sea surface height, relative to the marine geoid, from the altimeter-measured height requires an independent determination of the radial component of the satellite orbit relative to a common-center-of-mass coordinate system. A dominant contributor to the random errors that creep into a satellite altimeter system is the long-wavelength uncertainty in the orbital radius. Another error is the deviation of the satellite track from the expected ground track. One of the reasons for changes in the satellite orbit is the atmospheric drag. One method of reducing this error component is to increase the altitude of the orbit. At higher altitudes, the atmospheric drag is comparatively lower. Above 1,300 km, the atmospheric drag is small enough that orbital errors may almost be neglected.

Another possibility to reduce satellite orbital error is to carefully design the satellite so as to minimize the effects of surface forces such as drag and solar radiation pressure. These methods can be implemented in a satellite, which is dedicated to altimetry alone. However, in a satellite where various sensors with conflicting altitude requirements are to be installed, these methods of orbit error minimization are difficult to be fully incorporated. Consequently, some sort of a compromise will have to be reached. One method of measuring the orbital radius error is frequent altimeter calibrations. A rather simple method is the use of sea-level gauge data. It has been shown by Wunsch (1986) that calibration by a comparatively modest network of sea-level gauge systems can considerably reduce the overall error in the global estimates of large-scale oceanic water motions.

Because of these several uncertainties, the satellite positions and velocities must be tracked at intervals relative to fixed ground stations. An effective method of satellite altimeter calibration is by comparison with laser altimeter measurements as the satellite passes directly over ground-based laser-tracking stations, which are installed at selected locations (Joseph, 2000). During the calibration process, suitable corrections are applied to the laser-tracking station elevation based on the known local marine geoid and a myriad of optical and microwave propagation path-length effects. The calibration is repeated for many satellite passes over the laser-tracking stations. The microwave propagation effects are corrected using meteorological data acquired around the time of each pass. Because no ground-tracking stations are installed in the open ocean, the orbit of the satellite in these areas must be estimated using the calibration data as one of the inputs.

Laser- or microwave-tracking systems, which measure the range or Doppler shift using signals that pass back and forth between the satellite and ground stations along slant paths, usually track altimeter-bearing satellites. Coincident measurements from at least three such stations are required if the three-dimensional position of the satellite in space, and particularly its altitude, is to be known at the maximum resolution of the tracking system. The distribution of tracking stations is such that a three-dimensional location is seldom possible. The data from an individual tracking system must be used with those data from other stations to construct a global orbit. This provides the required orbit information generalized on a global basis. For this reason, at a particular site and time, the inferred orbit altitude can be in error by a few cm.

Another method employed for precise orbit determination is the use of transponders (Powell, 1986). This allows the altitude of orbit arcs, several thousands of km in length and spanning major oceans, to be located with the same precision so that once every few days the height of the ocean surface under that arc can be measured relative to transponder sites on land with an accuracy of about ± 3 cm. In this method, the predicted orbit of the satellite is used to set the tracking window for the altimeter to view the land-based transponder. When an altimeter passes over a transponder, the latter is illuminated for a few seconds. In this time, a large number of pulses are transmitted and received. The range between the altimeter and the transponder varies in a highly predictable manner and can be expressed as a simple parabolic function of delay time between pulse transmission and reception, and satellite velocity (Powell, 1992). Because the form of the function is known precisely, all the measured pulses can be used to calculate, with very high accuracy, the minimum of that parabola. This minimum corresponds to the delay of pulses transmitted between the altimeter and the transponder at the time of closest approach. The altimeter datastream contains not individual returned pulses but waveforms that are the result of accumulating many returned pulses. The parabola defined by the individual pulses is the same as that defined by the centroid of the waveforms so that the centroid can be used as the measured data. The parabola-fitting process most generally appropriate is an iterative, least-squares-fitting process. The delay measurement is then converted to a range measurement to an estimated accuracy of ± 0.5 cm if the total electron content of the ionosphere, ground-level pressure, and columnar water vapor content of the atmosphere in the region of the transponder are known or if their effects are compensated for.

12.4.2. Correction of Geoid Errors

Several methods have been used for determination of the geoid. One method used accurate tracking of the orbits of various satellites over a long period. Another method relied on measurements of gravity from ships using precision gravimeters. This method provides good spatial details. However, the areas covered by such surveys are limited. In situations in which detailed geoid models are not available, attempts have been made to substitute mean altimetric surfaces. Global mean sea surfaces such as those computed by Marsh and Martin (1982) using this method were not sufficiently accurate for sea-level studies. However, regional surfaces can be generated with a precision appropriate for some applications.

Researchers at NASA's Goddard Space Flight Center have used a portion of the three-month dataset obtained from the Seasat altimeter to derive the topography of the global sea surface with much greater accuracy and detail than has ever been accomplished before. The elevations and depressions have been measured relative to a reference ellipsoid of revolution. The topography resolved small-scale geoidal features due to deep ocean trenches and island arcs (these with signatures of 2–20 meters) as well as the larger geoidal undulations obtained earlier from satellite orbit perturbations and spherical harmonic analyses. This topography includes oceanographic "signals" due to tides and sea-level perturbations arising from a plethora of oceanographically (e.g., currents, gyres) and meteorologically induced causes (e.g., atmospheric pressure loading, wind set-up, and other long-term forces) as well as the geophysical "noise" arising from the gravitational elevations and depressions (i.e., geoidal undulations). The very obvious correlations between the surface topography and the underlying bathymetric features (e.g., deep-sea trenches, island arcs, the continental shelf edge, large seamounts, mid-ocean ridges, and so forth) often seen on the topography map are a manifestation of or contribution from Earth's uncompensated gravitational elevations and depressions. It is clear, therefore, that much information can be gleaned on the dynamics of the sea surface from the altimetric height measurements if the gravity field of the Earth at close spatial intervals is known precisely. If it were possible to arrive at independent estimates of the marine geoid (say, assembled from marine gravity data) or other means, it could be possible to measure variations in the sea surface elevation with greater precision. The intrinsic accuracies of radar altitude measurements and satellite orbit determination were expected to ultimately allow the goal of obtaining sea surface topography to an accuracy of a few cm. The most severe problem that remained for long was the independent determination of a reference geoid to a precision of a few cm.

12.4.3. Null Methods for Obtaining Topographic Height Variability Independent of Geoid

Detailed knowledge of the marine geoid eluded the oceanographers for a long time, and construction of its model was not an easy task due to lack of knowledge about Earth's gravity field. In the absence of accurate geoid models, two "null" methods have been employed for obtaining the topographic height variability independent of geoid. These are (1) collinear passes and (2) crossover points (i.e., intersections of the satellite ground track with itself).

The most effective of these methods is the collinear-track approach, in which altimeter profiles with nearly identical ground tracks are differenced, tending to eliminate their common geoid signal and revealing time-dependent changes in dynamic height (Brown and Cheney, 1983; Douglas et al., 1987). By subtracting sea surface height from one traverse of the ground track from the height measured on a later traverse, changes in topography can be observed without knowing the geoid. The geoid is constant in time, and the subtraction removes the geoid, revealing changes due to changing currents, such as mesoscale eddies, assuming tides have been removed from the data. It must be noted that altimetric determination of surface ocean circulation is inherently more difficult because of the relatively small signal amplitude as well as the need to separate the dynamic topography from undulations of the geoid as well as from apparent undulations due to radial orbit error. However, for the case of mesoscale eddies, these obstacles have been completely overcome using repeated tracks of altimeter profiles from which geoid and orbit signals can readily be removed. This has resulted in highly accurate maps of sea-height variability attributable to the ocean eddy field. Determination of absolute current velocities will require advances both in geoid modeling and in orbit determination. Some success has been attained, however, using existing geoids and global altimetric surfaces to compute the large-scale surface circulation. Cheney et al. (1983) used the techniques of collinear differences to derive the global mesoscale variability from Seasat altimeter data. The altimeter-measured sea height data appeared to be more reliable than those derived from the XBT-inferred sea heights. However, there are some errors that are not easily removed by the collinear difference method. Increased solar activity causes strong atmospheric density-related altitude measurement fluctuations that are difficult to model.

A technique similar to the collinear tracks uses only the points at which the ascending passes intersect the descending passes (crossovers). Analyses of altimeter crossovers are based on the concept that the geo-potential component of sea level is the same at the intersection of the two tracks at a given point and thus cancels out in the crossover difference. Techniques using crossover differences as input data therefore do not require geoid models. This is of critical importance in the application of altimetry to ocean dynamics because uncertainty of marine geoid models is larger than oceanographic signals of interest (Cheney and Marsh, 1981). When the altimeter-borne satellite orbits in a nonrepeating orbit, it provides a dense network of crossovers. For example, the Geodetic Satellite (Geosat) altimeter provided ~35 million crossovers over the global oceans in a period of 18 months. At each of these locations, the two crossing passes provide independent sea-level measurements at the same place but at different times. Differences of the sea surface heights at the two times, or *crossover differences*, form the basis for variability studies and can also be expressed in terms of sea-level time series.

Generation of time series from altimeter crossover differences involves a separate processing step to convert from height differences (at many combinations of times) to a time-ordered sequence of heights referred to an *arbitrary zero point* (Cheney et al., 1986). The first step is removing radial orbit error. Radial orbit error is eliminated from the reference grid by removing a quadratic trend for each pass in a simultaneous least-squares adjustment. Subsequently, crossover data are grouped into areas that are smaller relative to the horizontal scale of interest (typically a few hundred kilometers). Altimeter data obtained anywhere in the defined area can then be considered representative of the area as a whole. Each crossover difference within the area provides a measure of sea-level change between two discrete times (the times of the two intersecting passes). Once a network of crossover points is generated, the problem confines to the task of solving for a series of individual heights, which are consistent with the height differences. Least-squares techniques are usually used for computing these heights, yielding a time series of sea-level changes at any given grid point from the first altimeter pass to the last.

The task of mapping the Earth's gravity field saw great success in recent years. For example, the joint NASA-German Aerospace Centre Gravity Recovery and Climate Experiment (GRACE) mission—the latest tool for

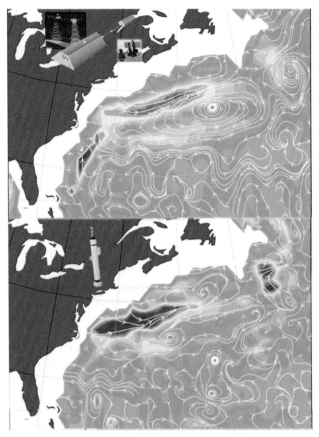

FIGURE 12.7   Ocean currents (depicted by arrows) off the east coast of the United States, 1,000 m beneath the surface. The top panel is obtained from the GRACE geoid, satellite altimetry, and ship measurements of temperature and salt. The bottom panel shows direct measurement of those currents by floats deployed from ships. Colors indicate the strength of the ocean current, with red being strongest and blue-green weakest. Areas in white have no available data. *(Sources: NASA/JPL/University of Texas Center for Space Research/GeoForschungsZentrum (GFZ) Potsdam; http://photojournal.jpl.nasa.gov/catalog/PIA04652;   http://photojournal. jpl.nasa.gov/figures/PIA04652-fig3.jpg; Ocean news, International Ocean Systems, Sept./Oct. 2003 issue, Vol. 7, Number 5, page 32.)*

scientists working to unlock the secrets of ocean circulation—released its first science product, the most accurate map yet of Earth's gravity field. GRACE senses minute variations in gravitational pull from local changes in Earth's mass by precisely measuring, to a tenth of the width of a human hair, the changes in the separation of two identical spacecrafts following the same orbit approximately 220 km apart. GRACE maps the variations from month to month, following changes imposed by the seasons, weather patterns, and short-term climate change.

The initial gravity model, created from several days of selected GRACE data, represented a major advancement in our knowledge of Earth's gravity field. Pre-GRACE models contained such large errors that many important features were obscured. GRACE brought the true state of the oceans into much sharper focus, so we can now better see ocean phenomena that have a strong impact on atmospheric weather

patterns, fisheries, and global climate change. GRACE provided a more precise definition of Earth's geoid, with cm-level precision. Scientists have studied Earth's gravity for decades, using both satellite and ground measurements that were of uneven quality. However, with just a few months' worth of the globally uniform-quality GRACE data, the accuracy of Earth's gravity model has been improved by a factor of between 10 and nearly 100, depending on the size of the gravity feature.

Figure 12.7 shows ocean currents (depicted by arrows) off the east coast of the United States, 1,000 m beneath the surface, obtained from the GRACE geoid, satellite altimetry, and ship measurements of temperature and salt, together with direct measurement of those currents by floats deployed from ships. With advances in gravity field modeling and orbit determination, satellite altimetry is expected to contribute greatly to determination of the general circulation of the oceans (the long-term mean movement of water).

12.5. EVOLUTION OF SATELLITE ALTIMETRY

Radar altimeters carried on spacecraft have provided subtle and valuable data on ocean dynamics by way of measurement of the long and short undulations of the ocean surface. The first such device was part of a payload on Skylab during a short interval in 1972 and was flown as a proof-of-concept instrument (Apel, 1982). The Skylab S-193 altimeter was the first in the series of satellite altimeters that were planned to progressively achieve the goal of detection of global circulation features through sea-level measurements. This altimeter was designed primarily for obtaining the radar measurements necessary for designing improved altimeters. This low-resolution altimeter permitted limited measurements of the sea level. Skylab operated from an altitude of 435 km and used a 100-ns uncompressed pulse of 13.9 GHz for estimation of altitude relative to the sea surface. The Skylab design was that of conventional pulse radar. The beam width of the altimeter radar was 1.5°, providing a footprint of diameter 8 km. The Skylab altimeter could achieve an accuracy of only a few meters. This accuracy was far from satisfactory for estimation of geostrophic currents and detection of mesoscale circulation phenomena such as eddies.

Subsequently, the Geodynamics Experimental Ocean Satellite-3 (GEOS-3 satellite), bearing the second of the series of satellite altimeters, was launched in 1975. This was the first globally applied altimeter system. The GEOS-3 satellite flew at an altitude of 840 km. The altimeter on board this satellite transmitted a chirp frequency of 12.5 ns derived from a 13.9 GHz pulse. The altimeter aboard the GEOS-3 satellite was the first to achieve a precision of approximately 50 cm in height measurement. This was sufficient to measure the 1- to 2-meter sea surface height difference across the Gulf Stream. In fact, the first altimetric observations of ocean dynamics phenomena took place from the GEOS-3. The GEOS-3 dataset extended over a period of three years and contained a wealth of ocean dynamics information. Using altimeter data from this satellite, Douglas and Gaborski (1979) detected a prominent cold ring, thereby establishing the effectiveness of satellite-borne altimetry in remotely detecting large-scale circulation features. Subsequently, Douglas and Cheney (1981) were able to produce a map of sea surface height variability in the Gulf Stream. Sea surface height variability of other areas such as the Gulf of Mexico and the Caribbean were also mapped in later studies. The GEOS-3 satellite demonstrated the long-term feasibility of such instruments in space for a period in excess of three years.

In an attempt to explore the dynamic ocean features more precisely, NASA launched Seasat in 1978. Seasat was, in fact, the first satellite dedicated to establishing the utility of microwave sensors for remote sensing of the oceans (Born et al., 1979). Seasat circled the earth 14 times a day at an altitude of approximately 800 km. The altimeter directly measured the radial distance (h) from the satellite to the sea surface based on the travel time ($\triangle t$) of very short pulse (3.125 ns chirp waveform) derived from a 13.5 GHz pulse. The Seasat altimeter measurements were supported by a closed-loop microprocessor range tracker and an automatic gain-control feedback loop. Because the spacecraft's longitudinal axis was oriented toward the geocenter to $\pm 0.5°$, the measured distance was very nearly in the direction defined by the geodetic vertical at the subsatellite point (Schutz et al., 1982). Microwave pulses were transmitted at a repetition rate of 1,020 Hz. The altitude information thus obtained was compressed and recorded on board the spacecraft to yield 10 data points per second. Further ground-based processing applied necessary smoothing to produce altimetric measurements at the rate of one per second. From an altitude near 800 km, it measured the satellite-to-surface separation with a precision better than 10 cm, for an overall distance accuracy of better than 12 parts per million.

The radar altimeter flown on Seasat collected approximately 1,000 orbits of continuous data around the world before it failed in orbit because of a reported short circuit in the electrical power system. However, the data collected by the Seasat altimeter represented about 90 percent of the data collected by the GEOS-3 radar altimeter during its 3.5 years of operation (Townsend, 1980). The Seasat altimeter data, together with a detailed gravimetric geoid, have been used to demonstrate the ability of satellite altimetry to detect sea surface height signatures associated with the Gulf Stream system. The presence of dynamic ocean features in the altimeter profiles was verified by comparison with standard oceanographic observations gathered from a variety of sources during the Seasat mission (Cheney, 1982).

The Seasat altimeter was a remarkable instrument that yielded information on sea surface topography and demonstrated the enormous potential of altimetry for oceanographic studies. The relatively short duration of the Seasat altimetric mission could not realize the full potential of the altimeter's "all-weather" observations of ocean fronts and eddies. A follow-on altimeter mission was expected to yield obvious benefits. Accordingly, Geosat, funded by the U.S. Navy, was launched in 1985. This was the first satellite altimeter to provide long-term global coverage. The primary purpose of this satellite was improvement of our knowledge of the marine gravity field. Geosat carried a 13.5-GHz radar altimeter that provided a continuous record of sea level along the satellite ground track. Geosat was, in fact, an improved version of the Seasat altimeter. A ground station provided information on the satellite ephermeris as well as tidal and other corrections. In April 1985, Geosat began generating a remarkable dataset that changed the way in which physical oceanographers viewed the global oceans.

During the initial 18 months of "primary mission" operation, data from Geosat was downloaded to Johns Hopkins University's Applied Physics Laboratory (APL) every 12 hours. At APL, the altimeter data were processed, enabling determination of sea-level variability and providing applications in many areas of ocean dynamics. One of the initial applications was to yield information on the ice edge on a global basis as well as oceanic fronts and eddies only from the northwest Atlantic regions encompassing the Gulf Stream. In September 1986, at the conclusion of the initial mission, Geosat was placed into a near-Seasat 17-day repeat orbit. The altimeter data collected during this Exact Repeat Mission (ERM) has been used to determine locations of ocean fronts and eddies over more of the ocean's surface than during the initial 18 months of operation. The information gathered from Geosat was routinely used in support of a variety of oceanographic parameters as well as fleet operations of naval interest (Cummings, 1988).

Although neither the Geosat nor the Seasat records were sufficient for definitive analyses of a global nature, Geosat established a new milestone in satellite oceanography by the duration and coverage of its observations. It is the first altimeter satellite to make continuous, global, multiyear measurements. During the first 18 months, Geosat accumulated 270 million sea-level observations along 200 million km of the world oceans, with a precision approaching 2 cm (Cheney et al., 1986). In a field historically limited by lack of observations, the Geosat datasets were the most extensive oceanographic datasets ever collected. These datasets offered a preview of future altimeter missions and provided a unique opportunity to gain experience in the analysis of global ocean datasets.

Altimeter technology has rapidly evolved since the first radar altimeter was flown on Skylab in 1973. Several satellite altimeter systems have been launched that have the capability to measure the range of radar altimeter signals returned from the sea surface with a variance of about ± 3 cm for sea state of about $H_s = 2$m and only slightly worse for much higher sea states. The utility of satellite altimetry for measuring ocean currents and circulation features through measurement of sea-level elevation variability has been demonstrated by a series of altimeters of increasing accuracy and precision, flown successively on several subsequent satellites.

The role of satellite altimetry in the study of global ocean circulation has been remarkable. The GEOS-3, Seasat, and Geosat experiments demonstrated that satellite altimeter technology can provide the mechanism for monitoring the oceans on a scale that appropriates the requirements of oceanographers. With the launch of the ERS-1 satellite, altimetry data began to be collected on a global scale. The first systems, carried on Seasat, Geosat, ERS-1, and ERS-2, were designed to measure week-to-week variability of currents. Topex/Poseidon, launched in 1992, was the first satellite designed to make the much more accurate measurements necessary for observing the permanent (time-averaged) surface circulation of the oceans, tides, and variability of gyre-scale currents. This was replaced by Jason in 2001. Topex/Poseidon and Jason flew over the same ground track every 9.9156 days. The great accuracy and precision of the Topex/Poseidon and Jason altimeter systems allowed them to measure the oceanic topography over ocean basins with an accuracy of ± 5 cm. This allowed them to measure:

- Changes in the mean volume of the ocean
- Seasonal heating and cooling of the ocean
- Tides
- The permanent surface geostrophic current system
- Changes in surface geostrophic currents on all scales
- Variations in the topography of equatorial current systems such as those associated with El Niño

Plotting the global distribution of time-averaged topography of the ocean surface provides a means of obtaining the global distribution of geostrophic currents at the ocean surface. The authenticity of altimetric detection of meso-scale dynamic events had been proved based on comparison with measurements obtained from satellite-tracked drifters circling the eddies during the same time when altimetric data were collected from the same location (Cheney and Marsh, 1981). Similarly, the Gulf Stream path depicted by the Seasat altimeter data closely agreed with the satellite infrared imagery during the time. These intercomparison experiments provided a high degree of confidence in the altimetric technique of remote sensing of ocean circulation features. It must, however, be accepted that although satellite altimetry has several merits, the technique does not have the capability to detect and quantify the entire range of surface circulation features. The technique is applicable to measurements of ocean circulation features of only large space scales.

Aviso has been distributing altimetric data worldwide since 1992. Consideration is now being given to altimetry missions capable of "scanning" the ocean surface to acquire data at scales of a few tens of kilometers, passing over the same spots every few days. The goal will be to monitor relatively rapid ocean variations over a period of less than 10 days at scales below 100 km. On February 28, 2013, the Indian Space Research Organization (ISRO) satellite successfully launched a CNES (France)-built Saral (Satellite with ARgos and ALtika). The altimeter (working in Ka-band, 35 GHz), named ALtika, will enable better observation of coastal zones. Land-based sea-level gauges will be used for calibration of the altimeter. In the Indian Ocean region, a network of four microwave radar gauges, which form part of a larger network, the Integrated Coastal Observation Network (ICON), for real-time monitoring of sea-level, sea-state, and surface meteorological data (Prabhudesai et al., 2010) developed by CSIR-National Institute of Oceanography, India has already been installed for this purpose.

REFERENCES

Apel, J.R., 1982. Some recent scientific results from the Seasat altimeter. Sea Technol. 23 (10), 21–27.

Barrick, D.E., Swift, C.T., 1980. The Seasat microwave instruments in historical perspective. IEEE J. Oceanic Eng. OE-5 (2), 75–79.

Biggs, A.W., Jordan, J.M., 1989. Simulation of chirp pulse spectra and effect of Doppler shifts on backscattered pulses. Quantitative Remote Sensing: An Economic Tool for the Nineties 3, 1720–1722.

Bonjean, F., Lagerloef, G.S.E., 2002. Diagnostic model and analysis of the surface currents in the tropical Pacific Ocean. J. Phys. Oceanogr. 32 (10), 2938–2954.

Born, G.H., Dunne, J.A., Lame, D.B., 1979. Seasat mission overview. Science 204, 1405–1406.

Brown, O.B., Cheney, R.E., 1983. Advances in satellite oceanography: Reviews of geophysics and space physics 21 (5), 1216–1230.

Cheney, R.E., 1982. Comparison data for Seasat altimetry in the Western North Atlantic. J. Geophys. Res. 87 (C5), 3247–3253.

Cheney, R.E., Marsh, J.G., 1981. Oceanographic evaluation of geoid surfaces in the Western North Atlantic. In: Gower, J. (Ed.), Oceanography from space. Plenum, New York, NY, USA, pp. 855–864.

Cheney, R.E., Marsh, J.G., 1981. Seasat altimeter observations of dynamic topography in the Gulf Stream region. J. Geophys. Res. 86 (C1), 473–483.

Cheney, R.E., Douglas, B.C., Sandwell, D.T., Marsh, J.G., Martin, T.V., McCarthy, J.J., 1984. Applications of satellite altimetry to oceanography and geophysics. Mar. Geophys. Res. 7, 17–32.

Cheney, R.E., Douglas, B., Agreen, R., Miller, L., Milbert, D., Porter, D., 1986. The Geosat altimeter mission: A milestone in satellite oceanography. Eos 67 (48), 1354–1355.

Cheney, R.E., Marsh, J.G., Beckley, B.D., 1983. Global mesoscale variability from repeat tracks of Seasat Altimeter data. J. Geophys. Res. 88, 4342–4354.

Cheney, R.E., Emery, W.J., Haines, B.J., Wentz, F., 1991. Recent improvements in Geosat altimeter data. Eos 72 (51), 577–580.

Clay, C.S., Medwin, H., 1977. Acoustical oceanography: Principles and applications. Wiley-Interscience, Newyork.

Cummings, T.K., 1988. Geosat: Navy applications of satellite altimetry growing. Sea Technol. 29 (11), 39–43.

Douglas, B.C., Gaborski, P.D., 1979. Observation of sea surface topography with GEOS-3 altimeter data. J. Geophys. Res. 84 (B8), 3893–3896.

Douglas, B.C., Cheney, R.E., 1981. Ocean mesoscale variability from repeat tracks of GEOS-3 altimeter data. J. Geophys. Res. 86 (C11), 10,931–10,937.

Douglas, B.C., McAdoo, D.C., Cheney, R.E., 1987. Oceanographic and geophysical applications of satellite altimetry. Rev. Geophys. 25 (5), 875–880.

Fu, L.L., Chelton, D., 2001. Large-scale ocean circulation. In: Fu, L.L., Cazenave, A. (Eds.), Satellite altimetry and earth sciences: A handbook of techniques and applications. Academic Press, pp. 133–169.

Fu, L.L., Chelton, D.B., Zlotnicki, V., 1988. Satellite altimetry: Observing ocean variability from space. Oceanography (Feature).

Hurlburt, H.E., Hogan, P.J., 2008. The Gulf Stream pathway and the impacts of the eddy-driven abyssal circulation and the Deep Western Boundary Current. Dynamics of Atmospheres and Oceans 45, 71–101.

Joseph, A., 2000. Applications of Doppler Effect in navigation and oceanography. In: Encyclopedia of Microcomputers, vol. 25. Marcel Dekker, New York, NY, USA. pp. 17–45.

Levitus, S., 1982. Climatological atlas of the world ocean. NOAA Professional Paper 13, Rockville, MD, 20852, USA.

Lorell, J., Colquitt, E., Anderle, R.J., 1982. Ionospheric correction for Seasat altimeter height measurement. J. Geophys. Res. 87 (C5), 3207–3212.

MacArthur, J.L., 1976. Design of the Seasat-A radar altimeter. Proc. Oceans '76 IEEE Conference, 10B-1–10B-8.

Marsh, J.G., Martin, T.V., 1982. The Seasat altimeter mean sea surface model. J. Geophys. Res. 87 (C5), 3269–3280.

Neumann, G., 1968. Ocean currents. Elsevier, Amsterdam- London- New York.

Powell, R.J., 1986. Relative vertical positioning using ground-level transponders with the ERS-1 altimeter. IEEE Trans. Geosci. Remote Sensing GE-24 (3), 421–425.

Powell, R.J., 1992. Measurement of mid-ocean surface levels to ±3 cm with respect to mid-continent reference points using transponders with the ERS-1 and Topex altimeters: A developing technique. Workshop Report No. 81, Intergovernmental Oceanographic Commission, 111–119.

Prabhudesai, R.G., Joseph, A., Agarwadekar, Y., Mehra, P., Vijay Kumar, K., Luis, R., 2010. Integrated Coastal Observation Network (ICON) for real-time monitoring of sea-level, sea-state, and surface meteorological data. Oceans '10: IEEE Seattle Technical Conference. September 2010.

Raney, R.K., 1998. The Delay/Doppler radar altimeter. IEEE Trans. Geosci. Remote Sensing 36 (5), 1578–1588.

Robinson, I.S., 1985. Satellite Oceanography, An introduction for oceanographers and remote sensing scientists. Ellis Horwood Ltd, Chichester.

Schutz, B.E., Tapley, B.D., Shum, C., 1982. Evaluation of the Seasat altimeter time tag bias. J. Geophys. Res. 87 (C5), 3239–3245.

Sverdrup, H.U., Johnson, M.W., Fleming, R.H., 1942. The oceans: Their physics, chemistry, and general biology. Prentice Hall, New York, NY, USA.

Tapley, B.D., Born, G.H., Parke, M.E., 1982. The Seasat altimeter data and its accuracy assessment. J. Geophys. Res. 87 (C5), 3179–3188.

Townsend, W.F., 1980. An initial assessment of the performance achieved by the Seasat-1 radar altimeter. IEEE J. Oceanic Eng. OE-5 (2), 80–92.

Ulaby, F.T., Moore, R.K., Fung, A.K., 1982. Microwave remote sensing (active and passive). In: Radar Remote Sensing and Surface Scattering and Emission Theory, vol. II. Addison-Wesley Advanced Book Program/World Science Division, Reading, Massachusetts, 1064.

Wunsch, C., 1986. Calibrating an altimeter: How many tide gauges is enough? J. Atmos. Oceanic Technol. 3, 746–754.

Wunsch, C., Gaposchkin, E.M., 1980. On using satellite altimetry to determine the general circulation of the ocean with application to geoid improvement. Rev. Geophys. 18, 725–745.

Wyrtki, K., 1977. Sea level during the 1972 El Niño. J. Phys. Oceanogr. 7 (6), 779–787.

BIBLIOGRAPHY

Bernstein, R.L., Born, G.H., Whritner, R.H., 1982. Seasat altimeter determination of ocean current variability. J. Geophys. Res. 87 (C5), 3261–3268.

Cheney, B., Miller, L., Agreen, R., Doyle, N., Lillibridge, J., 1994. Topex/Poseidon: The 2-cm solution. J. Geophys. Res. 99, 24,555–24,564.

Cheney, R.E., Marsh, J.G., 1982. Ocean current detection by satellite altimetry. Proc. Oceans '82 Conference, 409–414.

Cheney, R.E., Douglas, B.C., Miller, L., 1989. Evaluation of Geosat altimeter data with application to tropical Pacific sea-level variability. J. Geophys. Res. 94, 4737–4748.

Cheney, R.E., Marsh, J.G., Beckley, B.D., 1983. Global mesoscale variability from collinear tracks of Seasat altimeter data. J. Geophys. Res. 88 (C7), 4343–4354.

Cheney, R.E., Emery, W.J., Haines, B.J., Wentz, F., 1991. Recent improvements in Geosat altimeter data. Eos 72 (51), 577–580.

Cutting, E., Born, G.H., Frautnik, J.C., 1978. Orbit analysis for Seasat-A. J. Astronaut. Sci. XXVI, 315–342.

Delcroix, T., Boulanger, J.P., Masia, F., Menkes, C., 1994. Geosat-derived sea level and surface current anomalies in the equatorial Pacific during the 1986–1989 El Niño and La Niña. J. Geophys. Res. 99, 25,093–25,107.

Diamante, J.M., Douglas, B.C., Porter, D.L., Masterson, R.P., 1982. Tidal and geodetic observations for the Seasat altimeter calibration experiment. J. Geophys. Res. 87 (C5), 3199–3206.

Diamante, J.M., Pyle, T.E., Carter, W.E., Scherer, W.D., 1987. Global change and the measurement of absolute sea level, Progress in Oceanography. Pergamon Press.

Douglas, B.C., Cheney, R.E., 1990. Geosat: Beginning a new era in satellite oceanography. J. Geophys. Res. 95 (C3), 2833–2835.

Douglas, B.C., McAdoo, D.C., Cheney, R.E., 1987. Oceanographic and geophysical applications of satellite altimetry. Rev. Geophys. 25 (5), 875–880.

Kao, T.W., Cheney, R.E., 1982. The Gulf Stream front: A comparison between Seasat altimeter observations and theory. J. Geophys. Res. 87 (C1), 539–545.

Kolenkiewicz, R., Martin, C.F., 1982. Seasat altimeter height calibration. J. Geophys. Res. 87 (C5), 3189–3197.

Legeckis, R., 1987. Satellite observations of a western boundary current in the Bay of Bengal. J. Geophys. Res. 92 (C1 2), 12,974–12,978.

McGoogan, J.T., Miller, L.S., Brown, G.S., Hayne, G.S., 1974. The S193 radar altimeter experiment. Proc. IEEE 62 (6), 793–803.

Miller, L., Cheney, R.E., 1990. Large-scale meridional transport in the tropical Pacific Ocean during the 1986-87 El Niño. J. Geophys. Res. 95 (17), 905—920.

Miller, L., Cheney, R.E., Douglas, B.C., 1988. Geosat altimeter observations of Kelvin waves and the 1986—87 El Niño. Science 239, 52—54.

Mitchum, G.T., 1994. Comparison of Topex sea surface heights and tide gauge sea levels. J. Geophys. Res. 99 (C12), 24,541—24,553.

Parker, B.B., Cheney, R.E., Carter, W.E., 1992. NOAA global sea-level program. Sea Technol. 33 (6), 55—62.

Ponte, R.M., Lyard, F., 2002. Effects of unresolved high-frequency signals in altimeter records inferred from tide gauge data. J. Atmos. Oceanic Technol. 19 (4), 534—539.

Rossby, H.T., 1983. Eddies and the general circulation. In: Brewer, P.G. (Ed.), Oceanography; The present and future. Springer-Verlag, New York, Heidelberg, Berlin, pp. 137—161.

Schwiderski, E.W., 1991. High-precision modeling of mean sea level, ocean tides, and dynamic ocean variations with Geosat altimeter signals. In: Parker, B.B. (Ed.), Tidal Hydrodynamics. Wiley, New York, NY, USA, pp. 593—616.

Tai, C.K., White, W.B., Pazan, S.E., 1989. Geosat crossover analysis in the tropical Pacific, 2. Verification analysis of altimetric sea-level maps with expendable bathythermograph and island sea level data. J. Geophys. Res., 94,897—94,908.

Tapley, B.D., Born, G.H., Hager, H.H., Lorell, J., Parke, M.E., Diamante, J.M., Douglas, B.C., Goad, C.C., Kolenkiewicz, R., Marsh, J.G., Martin, C.F., Smith III, S.L., Townsend, W.F., Whitehead, J.A., Byrne, H.M., Fedor, L.S., Hammond, D.C., Mognard, N.M., 1979. Seasat altimeter calibration: Initial results. Science 204 (4400), 1410—1412.

Chapter 13

Conclusions

Chapter Outline

13.1. Progress in Ocean Current Measurement
Technologies 397
13.2. Moored Current Meters and Their Limitations 397
13.3. Lagrangian Measurements of Surface Currents 399
13.4. Global Observation of Sea Surface Currents
and Their Signatures through Imagery 401
13.5. Real-Time Two-Dimensional Mapping of Sea
Surface Current Vectors 402
13.6. Global Observation of Surface Geostrophic
Currents and Mesoscale Circulation Features 405

13.7. Current Profile Measurements Using Freely-Moving Sensor
Packages and ADPs 406
13.8. Evolution of Acoustic Tomography: Monitoring
Water Flow Structure from Open Ocean, Coastal
Waters, and Rivers 409
13.9. Lagrangian Measurements of Subsurface Currents 410
13.10. Comprehensive Study of Oceanic Circulation 415
References 416

Ocean-current measurements are important for a wide spectrum of applications extending from the domain of scientific research to the domain of a multitude of operational applications. Observations and experiments have been crucial to untangling the mysteries of fluid processes in the ocean. Quite often, observations guide the progress of our science. There are only very few instances in oceanography in which theory preceded observation; one such example was the prediction of southward flow under the Gulf Stream by the Stommel-Arons theory of abyssal circulation (Stommel and Arons, 1960a, b). Even in such instances, confirmation of the theory requires direct measurement.

13.1. PROGRESS IN OCEAN CURRENT MEASUREMENT TECHNOLOGIES

Over the past few decades, various methods of current measurement, through a variety of instrumentation ranging from the conceptually simple time-series position measurement of drift bottles, poles, and parachute drogues to the sophistication of current meter (CM) moorings, HF Doppler radar systems, satellite imagery, satellite altimetry, acoustic tomography, satellite-tracked surface/subsurface drifters, and remote profiling using a variety of free-falling/rising devices and acoustic Doppler techniques, have been employed. Such measurements have revealed that the ocean is populated by several small to medium-scale

circulation trajectories and patterns analogous to the weather systems in the atmosphere, and not often, some strange features and complex motions in the interior of the ocean, which are not found in the atmosphere, have been identified.

13.2. MOORED CURRENT METERS AND THEIR LIMITATIONS

Eulerian-style current measurements continue to be widely used to support offshore construction in the continental shelf and environmental projects. In such applications, regimes of interest range from the surface to several hundred meters and from concentrated, short-term measurements to sparse, long-term measurements of up to three years or more. The timeliness, reliability, and availability of the data are frequently more important to the operational user than absolute data accuracy. A detailed study in 1975 by the Current Meter Review Panel within the NOAA Office of Marine Technology (OMT) established the urgent need for measurement of currents in the mixed layers of the ocean and on the continental shelf to support marine surveying and offshore construction projects. Major obstacles to satisfying this need were the inability of available current measurement systems to operate reliably and to deliver accurate measurements under the influence of the dynamic ocean environment within these regions. The Working Conference on Current

Copyright © 2014 Elsevier Inc. All rights reserved.

Meters held at the University of Delaware in January 1978 also confirmed these observations and noted that neither the need for ocean-current data nor the number of associated problems has diminished.

General knowledge of oceanic motions increased considerably during the past few decades, primarily due to the successful use of moored current meters. The technology of mooring and maintaining chains of CMs in the deep ocean steadily improved and moorings became maintainable for over a year or more, yielding time-series measurements long enough to cover the greater part of the spectrum of oceanic motions. Fortunately, as a result of decades of design and development efforts by oceanographic technologists, laboratory-scale experimentation, flow flume experiments, and field intercomparison studies, a number of promising state-of-the-art current sensors were developed and incorporated into CMs for judicious selection by the user communities to suit their specific application needs. In some instances, modifications and adaptations of current sensors have been made to facilitate at-sea testing on subsurface arrays designed to decouple the sensors from the effect of surface waves. Some modern CMs have been designed with remote capability to allow controlling and monitoring of the status of current-measurement subsystems from the surface during short-term, at-sea testing as well as near-real-time data transmission capability. Schmitz et al. (1988) combined CMs and floats to describe mean and eddy flow in the eastern North Atlantic and found remarkable agreement of Eulerian and Lagrangian mean flows and eddy statistics.

Ocean currents, both on the surface and at mid-depth, have been measured and analyzed since the early 1900s. Several time-series records of horizontal currents and temperatures exist from fixed instruments on taut-wire moorings in the deep ocean and on continental shelf and slope locations. These measurements have shown that the energy in horizontal currents is contained mainly in low-frequency motions. Quasi-geostrophic, tidal and inertial motions account for the great bulk of the energy. Because low-frequency, energetic motions are of great interest for several studies, the temporal sampling rate is often reduced.

Unfortunately, data from fixed instruments tell a rather complicated story about vertical spatial structure. Records from adjacent instruments on an array in the vertical plane show low coherence at all but the lowest temporal frequencies (quasi-geostrophic). Webster (1968) describes the character of inertial oscillations as coherent horizontally over much greater scales (more than 3 km) than they are coherent vertically (less than 80 m). Based on examination of the extensive data bank at the Woods Hole Oceanographic Institution, it was inferred that some of the most energetic temporal frequencies are badly sampled in the vertical by typical fixed instrument spacing.

Considering further what is already known about temperature and salinity distributions, it is considered that vertical profiles of scalar or vector quantities of interest cannot be obtained with an economical number of fixed instruments. The often-observed extremely low correlation between motions at different depth levels (Webster, 1968, 1969, 1972), together with the great expense of CM chains, indicated the necessity for additional tools (say, a vertically moving sensor package providing high-resolution profiles) to effectively measure ocean currents over a variety of space and time scales. Thus, although in the past the moored CMs were useful, their limitation of poor spatial coverage is rather obvious.

Furthermore, although Eulerian-style CMs deployed from moorings have been able to measure the mean currents fairly accurately, measurements of very slow currents (such as vertical currents that are usually less than 0.001 cm/s) remain an unfulfilled dream of physical oceanographers. In the absence of suitable instrumentation for direct measurements, these small currents were computed from wind stresses or from horizontal velocity fields using the equation of continuity. Flow determination by computation only is less complete unless supported by actual measurements, at least for cross-checking. The growing emphasis is, therefore, to achieve still lower threshold, greater accuracy, and resolution so that very slow motions such as upwelling currents, which have significant biological importance, may actually be measured instead of depending entirely on mathematical calculations. Each sensor type has its own merits and limitations as regards differing range limits, thresholds, accuracy, resolution, and the like. The type of sensor to be employed in a particular situation depends on the nature of the intended study.

Although steady currents can be measured comparatively easily with sufficient reliability and accuracy, accurate measurements of mean current in the presence of complex, mixed steady, and unsteady currents containing a myriad of temporal and spatial scales pose a real challenge. The problem becomes further complicated by the lack of fixed reference points arising from the motion of the platform or the mooring that supports the CM. In estuaries and coastal environments, the current regime of which is often required to be known, the steady current will generally be disturbed by wind and wave activity. The current flow regime of the upper ocean consists of high-frequency, wave-induced spurious surface currents superimposed on long-period mean current. A surface-following float or the floating platform suspending the CM (e.g., the boat from which the CM is suspended) inevitably moves in many directions, and the CMs beneath sense such motions as well because of the physical coupling between the CM and the surface platform through the interconnecting cable. However, this limitation is absent in the case of the CMs mounted on fixed offshore platforms.

It has been reported (e.g., Gould and Sambuco, 1975) that on a mooring with a surface-following buoy, the vertical motion of the buoy in the wave field is communicated to great depths. Any CM on such a mooring is exposed to vertical motions many times stronger than the mean horizontal current. Field experiments have shown (Halpern and Pillsbury, 1976) that high-frequency surface wave motion is transmitted via mooring cable to instruments suspended at depths. It was also inferred that the CM on a mooring line would receive a good percent of the vibrations transmitted along the cable and contributes to errors in the data records. Observations have shown (Chhabra, 1977; Hogg, 1986; Hendry, 1988) that subsurface moorings could undergo considerable horizontal and hundreds of meters of vertical excursions in response to the associated horizontal drag forces, resulting in complex motions of the instrument and causing contamination of measurements. Vertical excursions associated with these horizontal movements are more problematic because horizontal currents can vary significantly with depth on the scales of the expected vertical displacements, with the result that the instrument measures a mixture of temporal and spatial variability (Hendry, 1988).

To some extent, the CMs' motion can be reduced by the use of subsurface moorings. The effect on the current measurements depends, therefore, on the kind of mooring, the kind of CM, and the environmental conditions of winds, waves, swells, and currents. Though some types of CM sensors are found to be more sensitive to current speed measurement errors due to mooring motion, contamination of current measurements by mooring-line motion is common to all types of sensors incorporated in Eulerian CMs.

Von Zweck and Saunders (1981) mathematically treated the effects of mooring motion on current spectra, independent of instruments' inherent performance characteristics, by taking into account the displacement of CMs within the oceanic current field. They found that mooring motions contaminate current measurements by redistribution of energy in frequency, wave-number-dependent attenuation and linear superposition of mooring motion. The type of contamination arising from redistribution of spectral energy to higher and lower frequencies, as well as wave-number-dependent attenuation (which is a function of the ratio of meter displacement to wavelength), are difficult to remove. This poses a serious problem in determining the true energy spectrum from the data obtained from CMs deployed on a wave-influenced mooring line.

Another source of error in measurements from moored CMs is the tilting of the CM sensor during high horizontal currents. A major difficulty involved in Eulerian measurements is that deployment and retrieval methods for such systems are both time-consuming and costly, besides the peril of losing the CM (and data) in rough seas or to shipping traffic or fishing activity in the area. Because the effects of mooring and wave motions vary with environmental conditions and are difficult to quantify, the data may lose its desired credibility for specialized applications. However, for a detailed study of a particular region on a long-term basis, ocean-current measurement in a Eulerian fashion is a necessity. Nevertheless, long-term, time-series ocean-current measurements from moorings have been essential to much of the understanding gained about ocean currents. In fact, most of the quantitative knowledge of deep currents has been gained from moored current measurements. The CM mooring with subsurface floatation has now become a standard tool for exploring the ocean and gathering statistical information on mean currents, eddy energies, and fluxes of momentum and heat. It is worth mentioning that acoustic release systems have facilitated moorings with subsurface flotation.

13.3. LAGRANGIAN MEASUREMENTS OF SURFACE CURRENTS

In oceanography, lack of systematic synoptic current observations in the Lagrangian style has been a major hindering factor in the development of synoptic current charts for the analysis and prediction of surface currents for various operational activities. Sea surface drifters of various kinds have traditionally been used to yield an estimate of sea surface currents and their trajectories. There had been several problems associated with drifter buoys (Vachon, 1977), the typical one being slippage. The effect of wind on the buoys can be reduced by keeping the amount of buoy volume above the water as small as possible, but true windage effects are not yet known. Typical results show that slippage can be as high as 1 percent of the wind speed when wave effects are included. Measurements are needed to show how well buoys with and without drogues drift relative to the water under different conditions. Another problem is the quantitative interpretation of Lagrangian measurements. In strong currents the interpretation is easier than in the mid-ocean, where mean flow may be weaker than the time-dependent flow. For statistical interpretations, large samples are required. In spite of these difficulties, the potential of the sea surface drifter techniques is large, and the entire fields of drifter technology and interpretation are very much active.

Application of Lagrangian measurements to monitor sea surface circulation greatly expanded in the 1980s and 1990s through effective utilization of the progress made in satellite technologies and global positioning systems (GPS). The global World Ocean Circulation Experiment (WOCE) took advantage of such progress and adopted the WOCE Lagrangian drifter for studies of the circulation of the surface layer of the global ocean. The WOCE-type drifters were designed for open-ocean applications. In an attempt to minimize wind-driven drag, the drogue of the drifter was sufficiently long that the current measurements represented a vertically averaged current of the 12.5- to 17.5-m water layer. Thus, the drifter was designed to record the current at ∼15 m depth and could be easily deployed because it was preprogrammed and its drogue was self-extendable. Furthermore, the WOCE drifter utilized satellite communication, which has several advantages and some disadvantages. The most obvious and highly significant advantage is global coverage. Among the disadvantages are the relatively high cost and the indirect, one-way communication with the drifter, which make the instrument expendable. In the framework of the WOCE project, over 4,000 such drifters were deployed and tracked by satellites from 1990 to 2002. Subsequently, the WOCE drifters began to be used worldwide for open-ocean Lagrangian measurement of surface circulation. Undoubtedly, these surface drifter networks have improved the description of surface circulation.

Space-based observation of offshore currents would provide added information to support coastal current observations. These measures would fulfill the currently existing need for a minimum safety level, which is being challenged through climate change. Disaster risk reduction is an essential element of climate-change adaptation.

In the recent past, the trajectory data of satellite-tracked drifters became available and made great progress with increasing long-term ocean-observing systems in response to the growing demand for climate study and global operational predictions. The World Ocean Circulation Experiment—Tropical Ocean Global Atmosphere (WOCE-TOGA) program has established the Surface Velocity Program (SVP) for monitoring ocean currents; this program arranges global deployment of Lagrangian drifters.

At the surface of the ocean, measurements are difficult for other reasons as well, primarily because sea surface lies in the wave zone. Wind waves and swell have periods of a few seconds. A 6-s-period wave will subject a surface platform to almost 1 million flexure cycles every month. It is not surprising that reliability and longevity are prime problems. Surface drifters play a continuing role in ocean research.

Note that "ARGOS" is the short form of the French satellite system known as *Advanced Research and Global Observation Satellite,* which, since 1978, offers capabilities for satellite-based position fixing of radio buoys. In the oceanographic context the platform transmit terminal (PTT) onboard a radio buoy (known as *ARGOS float*), which floats on the sea surface and freely drifts with the local sea surface current, allows periodic tracking of the radio buoy; thus allowing determination of both the trajectory of sea surface current and the current-velocity at various points along the trajectory. In contrast, "Argo"

is the short form of an international program, known as "*Array for Real-time Geostrophic Oceanography*". An *Argo float* is a robotic float (used under the *Argo* program), which autonomously ascends from a parking depth (1,000 m) to sea surface, synchronously measuring the temperature and salinity in environmental water mass, and then drifts for 10−12 h on the sea surface before it descends to its parking depth, and then freely drifts at this depth with the water motion at that depth. Subsequently it resurfaces from the parking depth to continue another cycle. This cycle repeats until the battery power depletes. Thus, whereas *Argo float* is a profiling float that is primarily used for determining subsurface current, *ARGOS float* is a freely drifting surface-float. Interestingly, the ARGOS float that is always floating on the sea surface and the Argo float that floats on the sea surface only during its 10−12 h surface-parking interval in its profiling cycle are both tracked by the ARGOS satellite system. The tracking data collected during the Argo float's 10−12 h surface-parking interval in its profiling cycle are used for determination of surface current (both trajectory and current-velocity).

Tracked by Argos satellite or relaying GPS positions by cellular phone, drifters benefit from high-tech systems already in place. Since the work of Johnson et al. (2003) and Schmidt et al. (2003), availability of low-cost handheld GPS units gave rise to the development of several surf-zone drifters at reduced expense. This enabled an increase in the number of drifters that can be deployed simultaneously, thereby further increasing the statistical confidence of surf-zone water-velocity estimates. In addition, small, inexpensive, off-the-shelf waterproof cases rated to a submersible depth of 30 m became available for most handheld GPS units, thereby reducing engineering and production time. Owing to their small size, handheld GPS units can be mounted to the exterior of a surf-zone drifter rather than in the interior, allowing variation in drifter design while reducing costs and production time. Taking into account the hazardous nature of the surf zone, inexpensive drifters relax the logistical costs and pressures of drifter recovery.

Despite the apparent difficulties of using drifters in the surf zone, their design and deployment have been surprisingly successful. Small, simple drifters have proven very robust and easy to deploy. The quality of surf-zone water-motion data (both trajectory and velocity) obtained through the GPS drifters was found to be good in terms of positioning accuracy and coverage. Both modeling and field validations have suggested that the drifter closely measures the depth-averaged and wave-averaged Eulerian current in low- to moderate-energy surf zones if drogued in an appropriate manner. Purpose-built drifters have been found to possess very significant potential as a valuable tool for surf-zone investigations and have already been used successfully by several researchers to study transient rip currents. Drifters have also been used to monitor pollution dispersion from recreational areas and study of tidal fronts. Small GPS drifters have been found to be valuable instruments in studies of circulation and dispersion in a whole range of aquatic environments in which Lagrangian measurements are scarce. These environments include the near-shore zone, small and medium-sized lakes, rivers, and estuaries. In these applications, the full capabilities of more sophisticated drifters that currently exist may not be required, and very simple, low-cost devices may be more appropriate.

13.4. GLOBAL OBSERVATION OF SEA SURFACE CURRENTS AND THEIR SIGNATURES THROUGH IMAGERY

Sea surface imagery is another useful product to examine the surface expressions of various kinds of motions in the upper ocean. For example, the ocean is rich in short- and long-lived mesoscale dynamic features. Examples include meandering currents and eddies with distinct sea surface temperature fronts, the sporadic occurrence of filaments and jets, and wind-driven coastal upwelling and downwelling. It has been found that imageries of large areas over the ocean surface can be used effectively in studies of frontal features and their impact on the mesoscale variability of large-scale ocean-current systems (e.g., the Gulf Stream and coastal boundary currents).

Upper-ocean fronts and eddies are dynamic features that importantly contribute to mesoscale variability, coupled physical-biochemical processes, and rapid changes in air-sea interaction. Such features are usually manifested by the sea surface temperature pattern and chlorophyll-*a* concentration, in thermal infrared and visible-wavelength satellite remote sensing images, respectively. In synthetic aperture radar (SAR) images they are manifested by virtue of sea surface roughness changes. Among the different components of the surface current gradient tensor, the dominant contribution is mostly related to the effects of convergence and divergence. Indeed, variations in the mean-square slope and wave breaking as well as Bragg wave modulations (via the intermediate-scale wave-breaking mechanism) are all strongly affected by such current gradients. Consequently, enhancement and suppression of surface roughness occur in the zones of convergence and divergence, respectively. Interestingly, surfactants also tend to accumulate in the zones of the current convergence to dampen short wind-driven waves. The effects of pure wave-current interaction and surfactant damping may then lead to opposite SAR signatures for current convergence areas. It is reasonable to anticipate that the resulting net effect depends on the wind conditions, the magnitude of the current convergence, and the properties of any surfactant material that is present. Satellite

observational techniques, which surpass all previous methods of synoptic surface current measurements from the open sea, promise to obtain a near-simultaneous picture of ocean surface current features.

Most of the dynamic features contribute to the complication of surface roughness modulation patterns, which are often manifested in visible-wavelength photographs and microwave images. The track of sea surface roughness contrast can be imaged in the visible-wavelength photography because the sea surface roughness variability alters the sun-glint pattern. The efficiency of visible-wavelength photography is, however, hampered during periods when clouds are present or visibility is poor. The problem is of particularly serious proportion in the cloud-belt regions within $15°$ of the equator and near the poles. Because visible-wavelength aerial photography is practicable only during daytime with a clear atmosphere, this method is generally unsuited for long-term systematic remote sensing studies of ocean surface circulation. Furthermore, the method of aerial photography is largely restricted to coastal waters. The advent of microwave imaging radar systems has removed these restrictions, thereby permitting efficient scheduling of flight operations, regardless of weather and visibility conditions. These radar systems permit obtaining round-the-clock imagery of the sea surface roughness contrasts under most weather conditions. Furthermore, radar imaging systems provide larger aerial coverage. However, as a result of relatively poor resolution arising from comparatively larger wavelength of the radar (relative to that of light used in photography), the quality of imagery obtained from radar systems does not generally match that obtained from visible-wavelength photography.

Images obtained from aircraft-borne side-looking microwave radar systems have been of great value for coastal surveillance during nights and cloudy days. However, use of aircraft for routine generation of images from large oceanic areas is impractical (costly and time-consuming). Fortunately, the fundamental equations for SAR imaging of surface current features have gradually become better known and can now be applied to simulate SAR image expressions. The spatial resolution of space-borne SARs typically ranges between a few meters and more than 100 meters, depending on the product type (that is, continuous or burst mode). Accordingly, the spatial coverage varies between approximately 100×100 km (standard image mode) to 500×500 km (wide swath or ScanSAR mode). Today three spaceborne SARs—the ERS-2, Envisat, and Radarsat-1—are available. The reasonably good resolution properties of space-borne SAR systems, as well as their independence of light and cloud conditions, together make SAR imagery a crucial source of information for a number of marine and coastal applications. SAR observations have also demonstrated their ability to routinely monitor different ocean-surface parameters, such as swell direction and amplitude (Beal et al., 1983; Hasselmann and Hasselmann, 1991). Hydrodynamic modulation of the surface roughness resulting from wave-current interaction makes possible the observation of oceanic features such as current fronts, eddies, and internal waves. Recent research activities have led to progress in more emerging SAR applications, such as Doppler-based surface current measurements and oceanic and atmospheric feature identification, which offers new insights for observing and modeling mesoscale meteo-oceanic processes.

It has been found that large-scale oceanographic features can produce corresponding signatures in SAR imagery. For example, many researchers have examined the SAR signature of the large-scale oceanic fronts such as the Gulf Stream North Wall (GSNW). The GSNW is made evident in SAR imagery by the buoyancy and air-sea momentum flux discontinuity that often accompanies it. Field observations, including hydrographic measurements, microwave imaging radar measurements, and HF radar measurements, have revealed the evolution of complicated frontal interactions between various water masses. In some instances, the water masses were found to have been separated by intersecting frontal lines configured in a manner analogous to occluded atmospheric fronts. As a result of the occlusion, the water mass with intermediate density subducts and intrudes under the most buoyant water, carrying with it strong horizontal and vertical shears and a frontal band of diverging currents, is created in the densest water mass. Model studies have suggested that in the ocean there will be an increase in hydrographic and velocity fine structure downstream of the frontal occlusion point.

SAR instruments have been used not only for imaging oceanic circulation features but also for imaging meteorological features such as cold and warm fronts of various kinds. SAR has been found to be useful in studying the evolution of convective storms from their footprints on the sea and in studying the signatures of spatially evolving atmospheric convection over the oceans. Patterns in SAR backscatter from the ocean result from corresponding modulations of the centimeter-scale, wind-induced wave state by both oceanic and atmospheric phenomena. Given the high resolution of typical SARs and their order 100–1,000 km swath widths, they are ideal devices for sensing the sea surface signatures of those phenomena over a wide range of scales. Examples of signatures of oceanic phenomena imaged by SAR include swells, internal waves, surface currents, and sea surface slicks. The signatures of atmospheric phenomena commonly imaged by SAR include convective cells, roll vortices, gravity waves, mesoscale cyclones, and synoptic-scale weather systems.

Space-borne SAR measurements have effectively been used to forecast polar mesoscale cyclones in the Bering Sea (Friedman et al., 2001). Footprints of the atmosphere can be derived from SAR images of the ocean surface.

Space-borne SAR imagery of the sea surface has effectively been used in detecting the presence and structure of the convective marine atmospheric boundary layer (Sikora et al., 1995, 1997). Analysis of several thousand RADARSAT-1 SAR images from the Gulf of Alaska and from off the east coast of North America revealed the presence of, and the mesoscale and microscale substructures associated with, synoptic-scale cold fronts, warm fronts, stationary fronts, occluded fronts, and secluded fronts. It is expected that in the near future related experimental products such as SAR-derived wind-speed datasets will have the potential to further the field of SAR meteorology.

The high-resolution properties of space-borne SAR systems, as well as their independence of light and cloud conditions, make SAR imagery a crucial source of information for a number of marine and coastal applications. Recent research activities have led to progress in more emerging SAR applications such as Doppler-based surface current measurements and oceanic and atmospheric features identification, which offers new insights for the observation and modeling of mesoscale meteo-oceanic processes. Very small gradients and low-value currents, which are inferred from sea surface heights estimated from satellite altimetry, are already routinely used in the global circulation models.

Sequences of sea surface temperatures and ocean-color data can be used either in the maximum cross-correlation method or the optical-flow method to derive absolute measurements of current vectors (Vigan et al., 2000; Yang and Parvin, 2000). However, they do not achieve the spatial resolution of SAR-based methods and they are limited by the solar-illumination and cloud-cover conditions.

13.5. REAL-TIME TWO-DIMENSIONAL MAPPING OF SEA SURFACE CURRENT VECTORS

Routine monitoring of sea surface currents and waves in the offshore and near-shore regions is of great interest to both coastal and scientific communities. Crombie's meticulously careful observation of the backscattered high-frequency (HF) radio waveband echoes from the sea (the Bragg scattering principle) led to the discovery of the unique backscattering phenomenon behind the development of radio-wave Doppler radar systems (with frequencies of 3–30 MHz and wavelengths of ~10–100 m). Bragg-resonant backscattering, which is a coherent backscattering of radio-wave energy by ocean surface waves with half the electromagnetic radar wavelength, just like X-rays are scattered in crystals, allows near-surface currents, along with wave heights and wind direction, to be measured and mapped at distances up to 200 km, depending on the transmitted frequency. Some of the transmitted HF energy is guided by the sea surface; this phenomenon allows measurements to be made beyond the normal radar horizon.

Remote sensing of near-surface currents with HF radar was demonstrated in the early 1970s by Stewart and Joy (1974). Subsequent theoretical investigations by Hasselmann (1971) and Barrick et al. (1977) strengthened Crombie's findings. Bragg waves in the HF band happen to be "short" surface gravity waves, which can be assumed to be traveling as deep-water waves, except in very shallow depths of a few meters or less. This is important because it allows information contained in the Doppler shift of Bragg peaks to be used to estimate ocean currents and other parameters. Because the ocean is generally covered by waves of many different wavelengths and directions (continuous spectrum), there are always trains of waves propagating toward and away from the radar. The current measurement by HF Doppler radars is close to a "true" surface current measurement. Because radar pulses scatter off ocean waves, the derived currents represent an integral over a depth that is proportional to the radar wavelength. Stewart and Joy (1974) show this depth (d) to be approximately $d = \lambda/(8\pi)$. Because wavelength of the resonant sea surface wave depends on the radar frequency, it is feasible to use multifrequency HF radar systems to estimate vertical shear in the top 2 m of the ocean.

At NOAA in the United States in the early 1970s, Donald Barrick, B. J. Lipa, and co-researchers played a big role in demonstrating the ability of coastal HF Doppler radar systems to map ocean surface currents. Since that time, there has been a push to develop this technology into a useful and affordable tool that would fill a big gap. Before the 1970s there was nothing else available that could map surface currents continuously over space and time. Successive advancements in technology led to the development of different kinds of devices for remote mapping sea surface current fields, sea surface waves, and winds. These include the coastal ocean dynamics applications radar (CODAR) system, SeaSonde, the ocean surface current radar (OSCR), and the Wellen radar (WERA), all of which hold a prime position in coastal-observing systems in many parts of the world. Other, more recent developments include the coastal ocean surface radar (COSRAD), Courants de Surface MEsures par Radar (COSMER), PISCES, high-frequency surface wave radar (HF-SWR), high-frequency ocean surface radar (HFOSR), multifrequency coastal radar (MCR), Ocean States Measuring and Analyzing Radar (OSMAR2000), and PortMap.

Different designs of these HF/VHF Doppler radar systems have taken full advantage of improved electronics and computer technologies and incorporated different methods to perform spatial resolution both in range and azimuth. Range resolution could be achieved by means of short pulses, frequency-modulated chirps known as frequency-modulated continuous waves (FMCW), or frequency-modulated interrupted continuous waves (FMICW). Azimuth resolution could be obtained by means of beam forming or direction finding (phase comparison).

HF/VHF Doppler radar systems provide the unique capability to continuously monitor the coastal environment far beyond the range of conventional microwave radars. The working range of Doppler radar systems depends on the attenuation of the electromagnetic wave between the transmitter and the target. Attenuation is strongly affected by the condition of the sea state (sea too rough or too calm). The attenuation also increases with decreasing seawater conductivity, increasing transmission frequency, increasing distance between the transmitter and the target, and increasing atmospheric noise and/or radio-wave interference. A compromise should be found between working range and range resolution, which decreases with decreasing frequency, because of interference from long-range radio sources (transmissions from all over the world reflected by the ionosphere).

HF Doppler radar systems based on ground-wave propagation proved highly beneficial for remotely sensing ocean surface currents and gravity waves. Coastally located HF Doppler radar networks permit synoptic time-series mapping of coastal surface current circulation over a wide area in real time, under all weather conditions, for a variety of oceanographic research purposes and operational utilities and can also provide timely and valuable information to improve safety in coastal waters. Shore-based HF and very high-frequency (VHF) radar (typically in the range 3—50 MHz) proved very useful tools for analyzing and understanding coastal sea surface water currents. Pairs of HF/VHF radar devices operate mostly from shore (except in the case of offshore platform-mounted systems) and provide a convenient way to measure surface currents over a large area.

The reason for the great importance attached to the HF/VHF Doppler radar technology is that it has the ability to map ocean surface currents in real time over large areas, revealing evolving current structures that are difficult to monitor using any other instrument. HF radar systems have now become a popular technology for studying coastal circulation processes. The high spatial and temporal resolutions make HF radar technology suitable for aiding search and rescue operations, studying pollutant and larval dispersal, and analyzing the physical forcing of coastal flows. Because HF Doppler radar systems have the advantage of being real-time and noninvasive, several countries have begun to consider shore-based systems that are capable of mapping ocean surface currents from shore to far-off distances as their national coastal ocean surface current mapping systems.

Cochin et al. (2006) demonstrated the operational feasibility of VHF Doppler radar systems in a region dominated by strong tidal currents. They also showed that extraction of tidal components can be accomplished within a large area by the same process as for *in situ* instruments. Tidal-current ellipses are particularly useful for calibrating numerical models. Doppler radar measurements are also used to increase the precision of coastal hydrodynamic models by means of data assimilation. These measurements also have an important role in terms of operational surveillance, because of their high temporal resolution (every 30 min, down to 10 min) and large-scale domain, which offer good opportunity for control and operational surveys.

The spatial coverage provided by HF/VHF Doppler radar network and its superb capability for remotely measuring ocean surface currents as far out as 200 km offshore, generating hourly maps indicating the speed and direction of the currents at a number of spatially dense locations, are ideal features for meeting the requirements of port authorities and search-and-rescue operations. Port authorities have the responsibility to mariners to provide accurate and up-to-date information to allow for the safest passage possible. Knowledge of sea surface currents and winds over a reasonably large area offshore of the ports has been recognized as an important factor in meeting this responsibility. Such data are also important for many major industries, whether it is to predict the course of an oil spill, to calculate the best path for the installation of a pipeline, or to aid in search and rescue operations. The surface current maps generated at close spatial and temporal intervals are also immensely valuable for addressing several research problems.

Today remote sensing using satellite-borne sensors as well as fixed-platform-based HF/VHF Doppler radar systems permits collection of near-real-time "snapshot pictures" of oceanic surface circulation over large areas. In terms of monitoring the ever-increasing menace of oil spills from oil tankers and oil-drilling platforms as well as conducting successful rescue operations, the capability of these devices for all-weather, near-real-time mapping of oceanic surface circulation is of immense practical utility. The land-based multiple-site HF Doppler radar system is indeed a major development that enables round-the-clock remote time-series mapping of sea surface circulation features over large areas.

While appreciating the merits of the HF/VHF Doppler radar systems, it is important to keep in mind their limitations as revealed by several intercomparison studies. Perhaps the most important among them is due to their spatial resolution, where measurements are smooth and do not provide local strong currents. Another limitation is due to the presence of spurious peaks in the Doppler spectrum. Those peaks represent energy that is stronger from the Bragg waves in the direction of the side lobes in the antenna pattern than those from the principal radar beam. This happens in some cases of wind and current conditions. Time-frequency analysis on the whole duration of the signal or unmixed processing could be applied to eliminate spurious peaks. Attenuation of the HF ground wave is strongly dependent on radio frequency and sea-water conductivity. Experimental data have confirmed the predicted decrease of propagation range with decreasing conductivity. High salinity allows optimum performance; strong attenuation prevents HF radar being used for remote sensing in freshwater lakes and over ice-covered areas. Radio interference or high sea states can limit the actual range at times. Wet and moist sandy soils enhance the ground-wave propagation, whereas dry and rocky ground reduces signal strengths.

There is a "baseline problem" that occurs where both radar sites measure the same (or nearly the same) radial component of velocity, such as along the baseline between the sites or at great distances from both sites. Generally, two radials must have an angle >30° and <150° to resolve the current vector. Typical azimuthal resolutions are ~5°. Near the coast, this gives a measurement width of ~0.5 km; the width is ~10 km at range cells 100 km offshore. Fortunately, radar backscatter use is increasing and there are several different systems in the marketplace at a range of frequencies. These have extended uses. The use of HF radar has already become commonplace; as a result, a new level of understanding of the coastal ocean is emerging.

It may be added that there is no single oceanic current measuring system "for all seasons," but HF/VHF Doppler ocean surface current radar systems are among the few that go a long way in approaching that ideal. HF/VHF Doppler radar systems offer a very interesting opportunity to retrieve the full vector of sea surface currents with a temporal sampling of a few minutes. However, the coverage area is substantially more limited than that offered by space-borne systems. Furthermore, the deployment and maintenance of such systems often induce additional costs and administrative difficulties. Nevertheless, such a system may be well suited for local measurements and validation purposes. It is quite clear that two-dimensional maps of surface currents from HF radar networks represent a useful and unique resource for the improvement of coastal ocean circulation models, particularly in the critical depth range encompassing the euphotic zone.

After a long gestation period consisting of validation studies and analyses by several groups, the operational era has arrived. Until 2010, over 13 countries had established real-time networks, defined as four or more radar systems (Barrick, 2010). In the United States, real-time data are posted to two national Websites and several individual regional ones for public use. In search-and-rescue operations, the U.S. Coast Guard incorporates these real-time maps into a short-term predictive system that narrows their search regions. NOAA and the Coast Guard use the maps for oil-spill and floating-pollutant impact studies and cleanup mitigation. Oil companies make decisions regarding safety of platform operations based on the radar outputs. Based on these data, the local coastal municipalities decide whether to close beaches due to pollution. Several groups have demonstrated how HF surface maps significantly improve numerical model forecasts, not just of surface flow but of subsurface circulation, temperature, salinity, and transport. According to Barrick (2010), one New York forecast model assimilates data in real time. Ocean research scientists now routinely accept and incorporate these data among other sensor outputs into understanding oceanic processes and their impacts on climate and biology. In recent years the evolving capability to use shore-based HF radar systems to continuously monitor vast stretches of coastal ocean surface currents has presented a new possibility for improving our understanding and monitoring capabilities in these marine environments (Paduan and Graber, 1997).

Present system and coverage capabilities of HF Doppler radar systems are quite impressive. Measurements can be made in range as short as 1 km and as long as 200 km from the shore at a resolution of $\sim 0.3-3$ km along a radial beam. Among other technologies, HF Doppler radar is now considered a new technology for tsunami early-warning systems. This is a welcome addition. In case of an approaching tsunami, a strong ocean surface current signature can be observed by the radar when the tsunami waves enter the shelf edge. HF Doppler radar measurements from offshore platforms are very useful for offshore drilling and oil prospecting applications.

But despite the availability of appropriate technology, approaches, and options for replication of good practices, there is still an unacceptable delay in implementing schemes that enable rescue of coastal population from hazards involving coastal currents. HF Doppler radar systems for routine monitoring of currents along coastal regions and around islands would also be useful for detecting the climate-induced variability in these currents. Real-time reporting of currents via the Internet would be a practical means of awareness-raising and climate-change adaptation. This approach would result in a substantial reduction of disaster losses in terms of lives and in the social, economic, and environmental assets of coastal communities.

In recent years, the utility of HF radar-derived surface velocity fields as input to data-assimilating numerical circulation models has been the focus of several studies. Pandian et al. (2010) reported an overview of recent technologies on current and wave measurement in coastal and marine applications. The potential benefits of HF radar data are large, particularly in light of the dearth of real-time observations from the marine environment. These data are also potentially important because they can cover significant portions of coastal ocean model domains. They make it possible, for the first time, to track the location and movement of mesoscale oceanic features in a fashion analogous to the superior capabilities provided by data inputs to numerical weather forecast models (Paduan and Shulman, 2004). The unique advantage of the HF radar is the ability to map the horizontal variability of currents, which is needed for several applications. Eddy dynamics, such as propagation and decay, can be studied, as can the spatial variability of tidal currents. This is a further step in research on current-wave interaction.

Finally, HF radar offers a very interesting opportunity to retrieve the full vector of sea surface currents with a temporal sampling of a few minutes. However, the coverage area is more limited than the space-borne SAR coverage. Many questions still exist, however, about the details and effectiveness of HF radar-derived surface currents as sources for data assimilation. Despite several tasks, it is clear that two-dimensional maps of surface currents from HF radar networks represent a useful and unique resource for the improvement of coastal ocean circulation models, particularly in the critical depth range encompassing the euphotic zone (Paduan and Shulman, 2004). Despite the several merits associated with HF Doppler radar systems, their application is generally restricted to coastal waters.

13.6. GLOBAL OBSERVATION OF SURFACE GEOSTROPHIC CURRENTS AND MESOSCALE CIRCULATION FEATURES

Satellite-tracked surface drifters certainly play a major role in global observation systems, providing surface measurements for calibration of satellite data, giving a direct measurement of surface currents for use with hydrographic data, and providing an interpolation between moorings. A second satellite technique is the measurement of surface topography. Because radar altimeters can measure the distance from the satellite to the sea surface with great accuracy (to better than 10 cm), one need only combine such a measurement with accurate knowledge of the geoid in order to get the surface pressure field. Accuracy in three elements—the satellite orbit, the altimeter measurement, and the geoid itself—is required. Aspects of these problems are discussed in the National Academy report *Requirements for a Dedicated Gravitational Satellite* (Committee on Geodesy, 1979). The uncertainties in the three areas are rapidly decreasing, and thus this technique may be one of the most promising for a synoptic view of surface circulation.

Measuring sea surface height anomaly provides a means for determining global surface geostrophic currents and mesoscale circulation features such as cold-core eddies, warm-core eddies, meanders, and so on. In the past, little sea-level data were available from the vast offshore regions. To circumvent this problem it was necessary to have a spatially dense set of observations made over the entire globe once every few days. Satellite altimetry provides a technique for collecting just such a dataset. A satellite altimeter is a "sea-level gauge in space" and is considered to be a key device for measuring offshore sea-level variability and deducting the just-mentioned oceanic currents and their mesoscale features through measurement of sea-level variability. To accomplish this effectively, further advances will be required in the gravity field modeling and orbit determination. Accurate computation of mesoscale variability from altimeter data, however, requires elimination of the effect of errors in the height of the satellite. Fortunately, mesoscale variability can be detected with pairs of altimeter passes that have the same track. The very long wavelength relative error between two or more repeated (collinear) passes can be effectively removed as a linear trend. The technique of collinear differences had been used effectively by several investigators to derive the global mesoscale variability from altimeter data. The dominant contributor to the random error of an altimetric satellite system used to be the long wavelength uncertainty in the orbital radius. However, calibration by a comparatively modest sea-level gauge system can drastically reduce the overall error in global estimates of large-scale oceanic variability.

Seasat was the first satellite dedicated to establishing the utility of microwave sensors for remote sensing of oceans. Satellite technology could provide the mechanism for monitoring the world ocean on a scale appropriate to oceanographers' requirement. The brief 100-day mission of Seasat in 1978 revealed for the first time how satellites carrying radar altimeters and scatterometers could provide quantitative global-scale information relevant to ocean circulation and the wind fields that force it.

Using the voluminous altimeter-derived sea-level dataset covering vast offshore regions, it has been possible to document large-scale meridional water transport in the equatorial Pacific. Similar studies of interannual variability have been conducted for the tropical Atlantic and Indian oceans as well. Subsequent to the demise of Geosat in late 1989, a few other improved versions of satellite altimeters were launched, and the large datasets derived from them have been used successfully for the study of several hitherto poorly understood oceanic phenomena, such as the warm (El Niño) and the cold (La Niña) phases of the ENSO cycle, massive quantities of surface water transport along the equator in the form of equatorial Kelvin waves, and so forth.

Very small gradients and low-value currents, which are inferred from sea surface heights estimated from satellite altimetry, are already routinely used in global circulation models. Satellite altimetry provides all-weather coverage of large-scale ocean circulation features such as gyres/eddies, western boundary currents, and the like as well as global surface geostrophic currents using surface height anomaly measurements. However, this method is unable to capture the nongeostrophic components of the surface currents, which are dominant in the tropic oceans.

13.7. CURRENT PROFILE MEASUREMENTS USING FREELY-MOVING SENSOR PACKAGES AND ADPs

The difference in horizontal velocity of neighboring depth layers in the ocean gives rise to what is known as *vertical shear of horizontal current velocity* (defined as the derivative of current with respect to depth). Although current shears are most frequently found in the upper layers of the water column, primarily because different time responses by water layers at different depths to a fluctuating wind regime induce high lateral shear, they are also found within the main thermocline and within 500 m of the seafloor.

Moored CMs can give us only a time series at a few points (depths or horizontal positions), and the moorings are not yet adequate for strong currents; features such as narrow jets may not be seen. Thus for a long time there has been interest in techniques to measure currents as a function of depth or horizontal position.

The often-observed, extremely low correlation between motions at different depth levels (Webster, 1968, 1969), together with the great expense of CM chains, indicated the necessity for additional tools (say, vertically moving sensor packages providing high-resolution profiles or appropriate remote sensing devices) to measure effectively ocean currents over a variety of space and time scales. Obviously, it would be interesting to explore how these horizontal motions on different time scales couple in the vertical. This calls for fine-resolution measurements of horizontal motions over both small and large vertical separations. Besides being suitable for directly verifying coupled thermal and circulation models in the ocean and large lakes, the dataset obtained from vertical profilers possessing fine spatial and temporal resolution should provide an insight into a number of geophysical phenomena and processes, such as the evolution of the surface mixed layer; the generation and decay of internal waves and seiches; and the turbulent transport and mixing of heat, momentum, and water masses.

Various methods have been used to determine the vertical profile of horizontal current $V(z)$. The profiler methods emphasize the vertical rather than the temporal dependence of ocean currents. In the initial years of vertical profile measurements of horizontal currents, direct-reading and -recording CMs lowered from ships had been employed. Further improvisation resulted in the use of CMs slowly lowered from a tracked buoy and freely falling CMs measuring relative velocities (e.g., CMs falling along a taut wire). Motionally induced EMF values surrounding a freely falling probe were also used to estimate velocity profiles. Subsequently, freely moving sensor packages (e.g., buoyancy-driven sensors on unattended moorings and free-falling probes horizontally advected by currents and tracked acoustically using seabed transponders) became available for high-resolution profile measurements.

The simplest profiler, in principle, is the analog of the meteorological balloon: the sinking float. The float is tracked acoustically as it sinks, and its path is differentiated to yield velocity as a function of depth. A subset of this class is the transport float, the position of which before and after a trip to the bottom is shown by dye patches at the surface. The second class is the free-fall device that has a current sensor on it, including the electromagnetic and the airfoil lift probes. We also have the class of instruments known as the cyclesonde, which consists of a CM that goes up and down a line attached to a ship, mooring, or drifting buoy. Acoustic Doppler current profilers (ADPs) outshine all the previous current profilers.

In contrast to a moored instrument, a freely sinking/rising probe has been found to possess certain advantages in that an entire profile is obtained from a single record and the measurements are not contaminated by mooring-line motions. Most profilers had ingeniously incorporated a combination of discoveries made in the fields of electronics, hydraulics, electromechanical engineering, and underwater acoustics. Ocean-current velocity profile observations have revealed considerable complexity to the vertical structure or distribution of currents in the deep sea. Until the arrival of ADPs, vertical profile measurements could not replace moored instrumentation, which was deployable for several months. The advantages of the profilers are that they can be used effectively as mobile, real-time survey tools or when detailed vertical structure measurements are required.

Examples of appropriate applications include studies of (1) the vertical or modal structure of low-frequency currents and internal waves; (2) zones of strong but spatially intermittent shear; (3) geostrophic structure and dynamic stability (e.g., Richardson numbers) based on combined density and shear measurements; (4) the velocity field in which real-time results are needed to guide further work; and (5) intense currents, such as the Gulf Stream, where moorings are difficult to deploy and maintain. No single instrument has the mobility, accuracy, and vertical resolution of all the existing profilers. There was, however, a need for a mobile, rapid, absolute velocity profiler accurate to within 1 cm/s with a vertical resolution of 10 to 50 m that yields nearly real-time results and is easy to operate at sea.

Measurement of vertical profiles of horizontal currents in the ocean was a daunting task for several years. It was only with great difficulty and good fortune that the just-mentioned profilers could be operated for more than several weeks at one site. The idea of tracking a sinking float for velocity profiles came out of World War II acoustic developments.

Successful development of ADP took several decades. In this process, a Doppler sonar system for measurement of upper ocean-current velocity was developed in the late 1970s. This sonar transmitted a narrow beam that scattered off drifting plankton and other organisms in the upper ocean. From the Doppler shift of the backscattered sound, the component of water velocity parallel to the beam could be determined to a range of 1,400 m from the transmitter with a precision of 1 cm/s. The instrument was used successfully from a floating instrument platform (FLIP). The ADP, initially used as a profiler on ships, is now used routinely in moorings. For bottom-mounted deployments, ADPs are caged to protect them from uprooting by trawling vessels.

Emergence of ADPs was a great relief to a wide spectrum of oceanographic communities who are interested in the measurement of ocean currents. The unique feature of ADPs is their capability to remotely sense the undisturbed flow velocity at multiple sections in the water column. Unlike the freely sinking current profiler probes, the ADPs can be made to "look" in any direction and their deployments may be on fixed or moving platforms or on surface, bottom, or mid-depth moorings. Furthermore, the acoustic Doppler technique permits continuous, unattended, fast profiling of currents. Remote sensing with Doppler sonar, therefore, provides an attractive alternative to all previous methods of current profile measurements.

Apart from this, in situations in which only short-term current measurements are needed, the capability of hull-mountable, downward-looking remote profilers to make measurements from moving vessels is a remarkable feature that permits large spatial coverage and virtually eliminates the tedious and time-consuming logistics in deployment and retrieval, if conventional moorings were to be used. While mounted beneath a moving vessel, the ability of shallow-water ADPs to measure the speed of the ship with respect to the bottom allows current profile measurements to be made without the motion of the ship significantly affecting the current measurements. ADPs can map the horizontal (restricted to its field of view) as well as vertical distribution of the current field and its temporal variations, yielding three-dimensional current distribution patterns. Current profiling over vertical ranges of hundreds of meters in the deep ocean was achieved by acoustic profilers mounted on top of subsurface moorings. Special precautions must be taken, however, to obtain reliable data from near the surface.

Probably ADP's most important characteristic is its ability to provide a substantial increase in the vertical density of current data obtainable from a single instrument.

Because the majority of measurements are made remotely, only a small part of the current field—immediately surrounding the instrument—is perturbed. Therefore, current measurements are made in a theoretically ideal way. An added advantage, not obtainable with older-generation current profilers, is that the acoustic system can also create profiles of acoustic backscattering strength. This information provides an additional window into the environment by which researchers can synoptically study the vertical distribution of biological scatterers such as plankton. The backscattered signal can also be useful to fisheries scientists and fishermen alike, assisting them in estimating fish population. Thus, backscattering-strength information obtained from the ADP is a valuable bonus that throws immense light on the biomass, suspended sediments, and so on. The different time responses by water layers at different depths to a fluctuating wind regime induces high lateral shear. Oceanic water-current velocity shear has great importance in turbulent mixing and dispersal effects. An interesting application of ADP in oceanography is in the study of fine structures and velocity shear in the water column.

ADP's remote sensing capability permits monitoring of current fields in logistically difficult environments such as ports and estuaries where high flow regimes, ship traffic, or fishing activity make impractical the deployment and maintenance of conventional moorings. Furthermore, bottom-mounted, moored, and ship-borne ADPs provide time-series information on the subsurface and deep-sea features of ocean currents from different horizontal layers of the ocean. In addition, use of acoustic Doppler current profilers mounted on subsurface floats can considerably reduce the number of Eulerian CMs that would have been required otherwise and provide better spatial resolution than normally achievable by conventional "single-point" Eulerian CM arrays.

Apart from the use of ADPs for studies of academic interest, offshore industry began to draw a great deal of benefit from these devices. For instance, leakage from multiphase underwater pipelines carrying oil, gas, and water in any combination produced from one or more oil wells is not uncommon in the offshore oil and gas exploration industry. The 2010 Macondo incident off the coast of Louisiana, one of the largest submarine oil spills in history, boosted the worldwide need for detection and quantification methods for leakages from offshore oil and gas installations. Large discrepancies still exist in the calculated amount of discharged oil and gas, the vertical and lateral plume dimensions and transport, and biogeochemical degradation. Whereas geochemical methods (including chemical sensors) are mainly used to identify the variable gas/fluid sources, plume distribution, and secondary degradation processes, quantification of gas fluxes and transport in the water column requires measurement using hydroacoustic devices such as ADPs. Combining measurement with modern tools such as ADPs and pipe trackers will allow for a more complete picture of the leakage.

Standard techniques for measuring sea surface gravity waves use arrays of sensors, surface-following devices, and single-point subsurface devices. These traditional methods have complementary advantages and disadvantages, and generally the sacrifice for improving data quality is more complex logistics. However, ADP changed this picture by providing the power of an array, the quality of a surface-tracking device, and the simplicity of a single-point instrument. The ADP provides the powerful method of a beam-forming array for directional wave analysis. The array can be visualized by viewing from above the ADP's multiple beams and depth cells along each beam. In the analysis, the wave orbital current fluctuations observed in successive depth cells are transformed to the surface via linear wave theory. Together with the beam data at the surface, they emulate a grid of independent sensors recording the incident wave field.

Current meter development is a vigorous field, with the number of ADPs on the market doubling every few years. Acoustic Doppler sensors have been through many evolutions. With their digital signal processors, high-performance acoustic transducers and amplifiers, and large data rates, they are truly beneficiaries of advances in high-tech fields such as digital speech processing and laptop computers, but they also have pushed certain technologies themselves. Broadband techniques have broken the sample frequency-speed, resolution-range product limit of the earlier inco-herent ADPs and permitted their application in some configurations to turbulence measurements, which require rapid sampling and high resolution of speed to resolve turbulent spectra. Broadband ADPs provide finer vertical resolution and shorter averaging times for accurate veloci-ties but require greater sophistication in their use.

ADPs are available in more frequencies, more sizes, and from more manufacturers, and they are filling specialized niches. The price is even dropping, which is the sign of a maturing technology. One of ADPs' greatest impacts has been the reduction in mooring costs for bottom-mounted instruments. Particularly in shallow water, a profile of the entire water column is possible, except for a 15-percent ambiguous region at the top where side lobe reflection from the surface interferes. But fishing activities put these bottom-mounted instruments at risk, and this danger inspired a cottage industry of trawler-proof ADP mounts. Designed to direct trawl nets over the top yet be recoverable on acoustic command, at least three designs have been developed; several have been presented at Oceans conferences (Williams, 1997).

Because the ADPs combine the required functionality to measure both waves and currents in a single compact package, there has been considerable interest in exploring their efficacy as a wave sensor. A bottom-mounted, upward-looking ADP provides a robust means of determining wave height and direction in coastal and deep waters. When equipped with a pressure sensor, the ADP yields three independent estimates of the nondirectional wave-height spectrum and hence provides an internal consistency check on the performance of the instrument. Directional spectra obtained from the ADP tend to be sharper than those from point measurements, such as pressure velocity (PUV) triplets or directional wave buoys, and because of the greater number of degrees of freedom in the measurement, the ADP can resolve complex multidirectional wave distributions.

13.8. EVOLUTION OF ACOUSTIC TOMOGRAPHY: MONITORING WATER FLOW STRUCTURE FROM OPEN OCEAN, COASTAL WATERS, AND RIVERS

Despite great success achieved in ocean-current measurement technologies, it became clear that maintaining enough moorings to monitor large-scale ocean circulation is too expensive for both equipment and logistics. One must look to other, more sophisticated methods, together with a modest number of moorings and hydrographic observations. Taking this necessity into account, Munk and Wunsch, (1979) proposed an acoustic technique of great promise. Called *ocean acoustic tomography* (OAT), after the medical procedure of producing a two-dimensional display of interior structure from exterior X-rays, the technique monitors acoustic travel time with a number of moorings. Because the number of pieces of information is the product of the number of sources, receivers, and resolvable multipath arrivals, the economics of the system is enhanced over the usual spot measurements.

The necessary precision did not appear difficult to achieve, and the main limitation at high acoustic frequencies was imposed by the effects of variable ocean fine structure (limiting horizontal scales to 1,000 km). Using geophysical inverse techniques, Munk and Wunsch showed that it should be possible to invert the system for interior changes in sound speed and, by inference, changes in geostrophic velocity associated with density variations. They concluded that such a system is achievable and that it has potential for cost-effective, large-scale monitoring of the ocean.

The acoustic tomography method, which allows remote sampling from many directions, promises a detailed study of the interior features of the large-scale oceanic circulation. OAT was found to be an important remote sensing method for making "CAT scans" of the internal temperature and current structure of large volumes of the oceans.

Testing the performance of "pure" one-way acoustic tomography was realized in 1981. The method produced useful spatially averaged profiles of sound velocity, even with the limited acoustic bandwidth then available. Although the array configuration was adequate for producing maps of mesoscale features, the poor travel-time resolution (due to the limited bandwidth) meant that large error bars prevented going much beyond pattern recognition. Initial tomographic experiments in the early 1980s demonstrated the capability of mapping the mesoscale sound velocity with an accuracy of about 1.5–2 m/s (about 0.3–0.4°C) at 700 m, which is the most energetic level. Performance worsened near the surface, and at greater depths both the error limits and the signal energies decreased. Since 1979 this technique has steadily grown from the sea-trials stage to a routine measurement tool in the oceanographer's toolbox.

Measurement of ocean currents by reciprocal sound transmission was successively carried out on scales of 10–1,000 km by Worcester (1977). Vorticity fields in the central North Pacific Ocean and in the adjacent areas of the Gulf Stream in the North Atlantic Ocean were measured with triangular and pentagonal arrays of acoustic transceivers in which reciprocal transmission was possible between a pair of stations. A reciprocal transmission system using a triangular array of the inverted echo sounder (IES) moored near the bottom was also developed for measuring depth-averaged velocities of the Kuroshio Current south of Japan. Thanks to consistent efforts expended by several oceanographers and technocrats, OAT became a well-developed technique for monitoring the mesoscale fluctuations of temperature and current fields in the open ocean.

Acoustic tomography is a subject that saw the need for intense interdisciplinary and interlaboratory collaboration. One way to speed the research efforts was the introduction of outside groups into the project. For example, in the initial phase of OAT research at Scripps and Woods Hole institutes of oceanography in the United States, the other groups helping the researchers included the Massachusetts Institute of Technology, where Wunsch and others concentrated on the data taken from the at-sea tests and performed the inversions relating to the science. Likewise, the University of Michigan provided expertise in both signal processing and underwater signal propagation; NOAA provided ship support and independent hydrographic surveys of the test areas. The private sector also showed interest, with three companies (Hydroacoustics, Inc.; Webb Research Corp.; and Gould Inc.) assisting Scripps, either as potential suppliers or through studies of prototypes. In addition, support was provided from the Underwater Sound Attachment of the Naval Research Laboratory in Orlando, Florida.

Although one could imagine that the project might become saddled with all the diverse groups, according to Wunsch one of the strengths of the project was the ability of people with diverse backgrounds to work together. The many different kinds of expertise included in the project were people who were experts in acoustic sources and receivers, getting moorings in, and knowing how to make independent quality measurements at sea.

The OAT program also was the genesis of several interesting spin-offs. For example, tomographic techniques were used to probe the subseafloor structure. They were also used to estimate sea surface roughness and internal properties of sea ice in the Arctic.

A number of studies were made to adapt deep-ocean acoustic tomography methods to coastal environments, in which shallow-water propagation and high environmental variability make the inversion significantly more difficult. Although coastal tomography remained a topic of active research, parallel developments in wireless communication technology, combined with significant advancements in computing power, opened the way to acoustically focused oceanographic sampling (AFOS). AFOS consists of a network of acoustic arrays connected to a fleet of autonomous underwater vehicles (AUV) and to a shore station using wireless local area network technology.

In contrast to the frequent use of reciprocal transmission techniques in the open ocean, no reciprocal transmission experiments were performed in the coastal ocean in OAT's initial decades. The reason was that most of the acoustic oceanographers had focused their major interest on the open ocean. Coastal acoustic tomography (CAT) was proposed as an application of OAT to the coastal sea, aiming at continuous monitoring of tidal currents in ports, bays, and semi-enclosed and inland seas without disturbing shipping traffic or fishing and marine aquaculture activities. The CAT system was developed by Hiroshima University in Japan and since 1995 has been successfully applied to measurement of current structure in coastal seas around Japan. Range-averaged water-current velocities, estimated from travel-time data obtained reciprocally, were in good agreement with the results of ADP measurement obtained along the sound transmission line. It was thus suggested that reciprocal sound transmission is applicable to velocity measurement in the inland sea having complications associated with heavy ship traffic and fishing activities.

Surprisingly, no serious CAT application experiments were carried out outside Japan. This void was filled by the CAT experiment attempted for the first time in China in 2009, with seven acoustic stations for mapping the tidal currents in the Zhitouyang Bay near the Zhoushan Island. The experiment identified the generation of a clockwise tidal vortex of diameter ~5 km in the eastern part of the bay in the transition phase from ebb to flood. It was suggested that the CAT is a powerful device for continuously mapping the horizontal tidal current structures in coastal regions in China.

Success of the CAT system led to its further enhancement and modification through technology additions to river acoustic tomography (RAT) systems with capability for river discharge measurements. The RAT technology for current measurement was successfully carried out during April through December 2009 in the upstream region of the Qiantang River in China, about 90 km from the mouth of Hangzhou Bay. Range-averaged water-current velocities, determined from the travel-time differences along the transmission line, were in good agreement with those from an ADP, producing a root-mean-square difference of 0.03 m/s. The variations of river discharge caused by tidal bores were well captured. Based on these encouraging results, it has been suggested that the RAT is a powerful instrument and that the RAT method provides a prosperous method of continuous, long-term monitoring of river discharges, even in large tidal bore-infested rivers with quite heavy shipping traffic.

Coming back to the history of open ocean-current measurements, much of classical physical oceanography was devoted to the study of the general circulation of the world oceans. Most of our early ideas about ocean circulation were based on the indirect evidence of the temperature and salinity fields and the assumption of geostrophy. The principal observational tools used in such studies were the dynamic method and water mass analysis. The latter was useful in establishing pathways, the former for determining fluxes. In deep and abyssal waters, the dynamic method, which gives only differential velocities (shear), becomes useless in the absence of reliable reference velocity information. Box models and inverse methods were developed to provide closure to the circulation in an integral sense. Clearly, it became necessary to make direct measurement of deep and abyssal flows.

13.9. LAGRANGIAN MEASUREMENTS OF SUBSURFACE CURRENTS

Considerable efforts were expended over the years in the design of subsurface drifters suitable for subsurface and abyssal current measurements. The factors that influenced the performance of a drifter as a Lagrangian current follower were wind-induced slippage and wave drag on the surface float, drag on the tether for large depths, and drag due to the finite size of the drogue.

John Swallow's (1955) pioneering development of neutrally buoyant subsurface floats provided an important new method for observing ocean subsurface current velocities, particularly low-frequency, large-scale currents in a Lagrangian perspective. With the advent of direct velocity measurements by deep floats and CMs during the 1960s and early 1970s, the necessary data for a consistent picture of ocean circulation, at least in limited areas, began to come in. Worthington's attempt to put together for the first time such a picture of circulation in the North Atlantic was based on the new direct data.

Neutrally buoyant floats were a vital tool in the exploration of the global ocean circulation and now provide a central element of the *in situ* ocean-observing system through the Argo project. The original Swallow floats were relatively thin-walled aluminum cylinders that became neutrally buoyant at a certain depth, primarily because they were less compressible than seawater. They carried acoustic sources that were tracked at relatively short range from an attending ship. The Swallow float-tracking technique proved itself to be robust and capable of application to a range of depths and geographical locations. The Western Boundary Undercurrent work really marked the transition of float use from exploration to hypothesis testing—although much more exploratory work would follow. By mid-1958, Swallow and Hamon made a further attempt in the Northeast Atlantic to use floats systematically, to extend their lives, and to compare the direct measurements with geostrophy. Rather than allowing the floats to signal continuously (their life in this mode was limited to two weeks), an internal mechanical clock programmed the transmissions for 4 h per day and thus extended the float life to 12 weeks.

The results provided several examples of closely spaced floats with very different velocities and showed no "level-of-no-motion" but rather a sheared unidirectional flow at depths between 1,500 and 4,300 m. These measurements provided a clear indication that the ocean was not behaving as theory or classical hydrography suggested it should and that the previously held notion of "level-of-no-motion" is baseless.

Apart from these deployments, floats were used by the NIO (UK) scientists in the Labrador Sea in 1962 (Swallow and Worthington, 1969), in the Norwegian Sea outflow in 1963 (Crease, 1965), and in the Somali Basin during the International Indian Ocean Expedition (Swallow and Bruce, 1966).

Until the advent of World War II, oceanography proceeded at a rather leisurely pace. The anti-submarine-warfare program during World War II forced the rapid development of underwater acoustics. Measurement of deep motions by the use of *in situ* moving floats requires some method of tracking. The great strides made in acoustics during World War II yielded the necessary technology for acoustic tracking of instruments in the ocean.

Discovery of the SOFAR channel, an acoustic wave-guide in the ocean, led to great success of tracking subsurface floats over long distances. The SOFAR channel is produced in many parts of the ocean by the combination of pressure and temperature effects on the speed of sound, which decreases with depth from the surface to about 1,000 to 1,500 m, owing to the decrease of temperature, and increases with depth below this level, owing to the increase of pressure. In the SOFAR channel a few watts of sound can be heard about 2,000 km away.

Lagrangian-style subsurface current measurement programs involving subsurface floats tracked by moored hydrophones in the SOFAR channel unfolded into two well-defined efforts; one of these was to put the sound in the water, and the other was to get the sound out. The development of the SOFAR float, based on a relatively high-power 250-Hz sound source, made possible long-range unattended tracking and operational float lifetimes of several years. The first big SOFAR floats were placed out in a triangle of listening stations at Eleuthera, Puerto Rico, and Bermuda. The range of listening was about 700 km, and the floats were successfully tracked (Rossby and Webb, 1971). This success generated enough interest that a large program could be funded as part of the MODE-1 experiment. Underlying the design of the MODE-1 experiment was the need to test the hypothesis that eddies do play an important role in the general circulation.

The weak absorption of low-frequency sound by seawater made SOFAR float tracking possible at ranges of a few thousand kilometers in those oceanic regions with an intermediate-depth sound-speed minimum (sound channel). Although developing the receiving instrumentation and the program for the data analysis were relatively straightforward, the technology involved in the development of these floats was much more demanding. Economy of power, batteries with high energy density per unit weight, and a low-frequency acoustic projector with low weight and high efficiency were all requirements where little compromise was tolerable. In particular, the projector required considerable engineering time. The requirement to efficiently project low-frequency sound made SOFAR floats large (8 m in length and 430 kg in mass) and, consequently, costly and operationally difficult. Nevertheless, Owens (1991) mapped mean flow and eddy variability over much of the western North Atlantic using a collection of SOFAR data from various regional experiments.

The isopycnal-following (i.e., following the same density surface) float was found to be an ideal tool for the investigation of a number of upper-ocean processes in frontal regions. Using clusters of these instruments along with hydrographic surveys, studies of the topographic effects (e.g., the Gulf Stream departure from the Blake Plateau south of Cape Hatteras) as well as entrainment and subduction processes in the ocean could be carried out. In addition to providing three-dimensional flow-path information, additional sensors such as oxygen, pressure, and vertical velocity could be incorporated into the isopycnal-following float. Multiple float arrays on the same density surface, together with hydrographic sampling, were expected to permit studies of relative vorticity and its variation in response to topography and curvature along the float trajectory.

Since Swallow and Worthington's (1961) measurements of absolute velocity under the Gulf Stream and Swallow's (1971) landmark discovery of what is now known as mesoscale variability, much of our understanding of ocean circulation came from floats. Early studies tended to concentrate on regions that could be relatively well sampled by tens of floats positioned by a few acoustic stations. Important contributions based on such float arrays were the pioneering study of eddy statistics by Freeland et al. (1975), the mesoscale mapping by McWilliams (1976), the discovery of small long-lived coherent eddies called *Meddies* by McDowell and Rossby (1978), and the intensive study of Gulf Stream kinematics by Bower and Rossby (1989). As more floats were deployed and more trajectories accumulated, investigators began to map large-scale, low-frequency seawater motion velocities and eddy statistics over modest areas. Rossby and Dorson (1983) estimated Eulerian mean flow in a region using floats as moving CMs (averaging together all velocity observations in specified areas) and estimated the lateral single-particle diffusivity. Combining average absolute float velocities with geostrophic shear from hydrographic sections, Richardson developed a detailed velocity section of mean flow in the Gulf Stream as it crosses 55°W, accomplishing what the pioneering study that Swallow and Worthington (1961) began. It is heartening to note that since the mid-1950s, neutrally buoyant floats have been used in various forms to explore and to discover many aspects of ocean circulation. The successful collaboration in the study of western boundary undercurrent and the exploratory work over the Iberian abyssal plain led to what is probably the best-known early use of floats: the 1960 so-called *Aries* experiment led by John Swallow. (*Aries* was a 93-foot vessel that had been donated to WHOI in 1959.) The SOFAR floats, considering their relatively short development and trial phase, were remarkably successful. The Mini-MODE system developed by Swallow and co-researchers in the early 1970s allowed up to 18 floats to be tracked simultaneously (each identified by its own frequency in the range 5.0–6.5 kHz).

The SOFAR float restriction to the western north Atlantic was removed by the ingenious development by Al Bradley and Jim Valdes of Autonomous Listening Stations (ALS). These were moored hydrophones with data loggers that recorded the signal arrival times from floats within acoustic range and were deployed on moorings with the hydrophones near the SOFAR channel axis. The use of subsurface moorings to reduce mooring cost, risk of damage, and acoustic noise meant that data were not available in real time. ALS deployments of six months to one year were typical.

Along with extending coverage in the western north Atlantic to include the Gulf Stream, the ALSs allowed SOFAR floats to be used in the eastern Atlantic. In this case, floats were deployed as a contribution to US efforts to study Meddies and to support US research in the US/Soviet Polygon Mid-Ocean Dynamics Experiment (POLYMODE; Schmitz et al., 1988). Acoustically tracked floats were ideal for this purpose, and the presence of the Mediterranean water core at around 1,000 m resulted in a double sound channel, with the deeper channel allowing tracking of floats at depths as great as 3,000 m at ranges of 1,000 km. Autonomous floats were also used to study a number of physical processes such as deep convection in the Labrador sea, diapycnal mixing in the North Atlantic Tracer Release Experiment, and subduction.

SOFAR floats were bulky, heavy, and cumbersome, and their lifetime was limited by their need to carry large battery packs while still remaining neutrally buoyant. The SOFAR floats' impediment to use in large numbers was alleviated by the development of the RAFOS acoustic tracking system in which floats listened to moored sound sources. The RAFOS float enabled greater power output and longer life. These smaller, cheaper RAFOS floats developed by Tom Rossby and his group at the University of Rhode Island recorded the signal arrivals from an array of moored sources. Acoustic receptions were processed in the RAFOS float, and at the end of a mission the float surfaced and transmitted the log of acoustic-signal arrival times through the Argos satellite system. The floats were used extensively as both traditional isobaric floats and, by the addition of a compressible element, isopycnal rather than isobaric floats. Such floats were small enough to allow successive releases from a moored near-bottom "float park" (Zenk et al., 2000).

The scientific applications of floats of various types during the 1970s, '80s, and '90s were numerous and significantly improved our understanding of the oceanic eddy fields and, to a lesser extent, the basin-scale mean circulation. The topics explored included but were not restricted to (Gould, 2005):

- The origins, dynamics, history, and distribution of discrete intense eddies
- The statistics of mesoscale eddy variability on the scale of ocean basins
- The Gulf Stream and its dynamics
- Local oceanographic phenomena, including flow interactions with topography and abyssal circulations
- Pathways of cross-equatorial flow
- Internal wave dynamics
- The processes of winter convection, subduction, and mixing

An important application of neutrally buoyant floats was to study internal waves. Several of the floats in MODE were instrumented to record pressure, temperature, and vertical water motion. As well as being applied to studies of internal waves, the technique for measuring vertical velocities was particularly applicable to the study of deep-winter convection and was used in the western Mediterranean in 1970 (Webb et al., 1970; Gascard, 1973) and subsequently in the Greenland and Labrador seas (Lherminier and Gascard, 1998).

Neutrally buoyant floats had operated only on the scale of ocean basins, and the provision of a global acoustic float-tracking network would have required far too great a level of commitment. A novel system was required to allow floats to be tracked globally. Thus, during the 1980s, Russ Davis and co-researchers jointly developed the Autonomous Lagrangian Circulation Explorer (ALACE), which was a float that would have a multiyear life and could provide useful subsurface water-motion velocity information throughout the ice-free ocean. Observing mid-depth oceanic currents is rather a challenge, especially for large-scale and near-real-time operations in the global ocean. Davis et al. (1992) discussed the mid-depth currents based on measurements by ALACE floats and successfully applied these measurements to analyze the currents in tropical and South Pacific afterward. These studies showed that the small, cheap autonomous instruments in large numbers can play a vital role in ocean observations.

If acoustic tracking was impossible globally, the only alternative was to have the floats surface periodically and be tracked by satellite. Both SOFAR and RAFOS floats had the capability of surfacing at the end of their mission by dropping a ballast weight, either on a timer or by acoustic command. What was required additionally for the ALACE was a capability to surface and then return to the parking depth repeatedly. The solution to this problem was found to be in pumping fluid from within the pressure case into an external bladder to reduce the float's density and hence drive it to the surface. Deflating the bladder would return the float to the desired depth.

In total, 1,110 ALACE-type floats were deployed in WOCE. In large measure they achieved their objectives. Although not all achieved their target five-year life, many exceeded it, and the longest-lived floats continued to operate exceeding eight years after deployment. The data were used to construct velocity fields across the entire ocean basins (e.g., Davis, 1998) and to constrain inverse calculations (e.g., Wijffels et al., 2001).

Mainly between 1991 and 1995, some 306 autonomous floats, mostly of the ALACE type, were deployed in the equatorial region and South Pacific, including the Pacific sector of the Southern Ocean. All but 23 of those floats had expired by July 2003. Between 1994 and 1996, another 228 ALACE floats were deployed in the Indian Ocean, including its Southern Ocean sector. Only 90 of those floats were still operating at the end of 2003. Based on available data the mean circulation, seasonal changes, and eddy variability in intermediate-depth flow in the Indian and South Pacific Oceans seen by these floats could be studied.

Integration of float measurements with hydrographic data yielded a three-dimensional coverage of the circulation and provided some insight into the nature of regional circulations. Because ocean *variability* is generally more energetic than *mean flow*, extracting accurate mean from observations requires substantial averaging. Thus, multiyear records are necessary to achieve useful accuracy, and for floats this requires spatial averaging or filtering. It is impossible to directly measure mean transport without observations at many depths. Combining float-measured velocity at one level with hydrography holds the promise of deducing such transport.

It may be recalled that neutrally buoyant floats had been used in various forms to explore and discover many aspects of the oceanic subsurface circulation. The demonstrated capability of P-ALACE (profiling ALACE) and similar floats for estimation of *drift current* at the parking depth and for collection of high-quality CTD data above and into the permanent thermocline, as well as the success of the numerous deployments of ALACE and P-ALACE floats in the World Ocean Circulation Experiment (WOCE), pointed the way toward their use as a tool for prolonged global-scale ocean monitoring that would complement and greatly enhance other elements such as altimetry, hydrography, XBTs, and the like. ALACE, P-ALACE, and SOLO floats are drifting instruments that measure ocean temperature and salinity. SOLO floats are very similar to P-ALACE floats but have better satellite communication and acoustic tracking capabilities. Because the floats are only about 6 feet long and 80 lbs in weight (body plus antenna), they could be deployed from any ship by one or two people without special equipment. Typically, the instruments are simply lowered over the side of the ship with a rope. In high seas, the floats can be deployed in biodegradable boxes that protect the instrumentation from rough landings. Once a float is deployed, neither ships nor people are required to obtain the measurements.

Some types of floats (e.g., SOLO) apply an offset to the pressure data on board the float and, therefore, the transmitted data need to be corrected. This has limitations when it comes to data manipulation. Therefore, as an improvement, the APEX float only transmits raw data but also notes the offset that it thinks should be given to the data. This way the user can decide how the data are manipulated. The EM-APEX float is an improvement on the APEX float that is designed to make long-duration measurements in inhospitable conditions (e.g., hurricane regions). After the floats are deployed, they move with the ambient currents and can therefore travel long distances on their own without the need of a ship or a person to handle them. Floats are programmed to come to the sea surface at regular intervals to transmit their data and geographical position to orbiting satellites. Afterward, they continue measuring ocean conditions, with at-sea missions lasting four to five years. This provides ocean-ographers with a wealth of near-real-time data, often from remote regions of the world's oceans. A deep, neutrally buoyant float that periodically pops up or reports in some acoustic mode to the surface and to a satellite would yield a kind of global coverage, as would acoustic tomography.

The international Array for Real-time Geostrophic Oceanography (Argo) program aimed at building a global array of 3,000 free-drifting profiling floats. In the upper 2,000-m ocean, an Argo float autonomously ascends from a parking depth to sea surface synchronously measuring the temperature and salinity in environmental water mass, drifts for 10–12 h on the sea surface before it descends to its parking depth, and then freely drifts at this depth. Subsequently it resurfaces from the parking depth to continue another cycle. While a float drifts on the sea surface, only its thin antenna (~2 cm) is outside of seawater and its 1-m-long cylindrical hull with ~17 cm diameter always submerges. For low wind speed, it is designed to follow water parcels well while both drifting on the surface and parking on mid-depth. With more than 5,000 floats deployed in the global oceans since the initiation of the Argo program, there has been a unique opportunity to simultaneously measure surface and mid-depth currents in the global scale in near real time.

Since 2000, the international Argo project has provided real-time monitoring of thermohaline (temperature and salinity) profiles for the global upper ocean and drift measurements at a known depth level (approximately 1,000 m) to enable estimation of thermohaline circulation at different depth levels. By late 2004, over 1,500 neutrally buoyant floats were drifting at depth throughout the global ocean. They were approximately 50 percent of the targeted final global Argo array, scheduled for completion by 2007. As of March 2007, more than 2,800 Argo-type profiling floats (Argo floats) were active in the world ocean.

Argo forms the core of the *in situ* ocean component of the Global Climate Observing System, essential for quantifying the oceans' response to climate change and to improving our understanding of, and making improved predictions about, shorter-lived climate events. The technology used for the production and operation of neutrally buoyant robotic floats such as Argo has improved greatly over recent years, thereby allowing the floats to collect more data more efficiently and reliably.

It seems that neutrally buoyant floats will remain a key element of ocean exploration and monitoring, both in the global Argo program and when used regionally to explore particular phenomena. Floats will carry a growing range of sensors. The potential for floats to act as monitors of ocean mixing through microstructure measurements has already been demonstrated. Surprisingly, floats have survived under Antarctic sea ice to download their profile data when spring arrives.

Undoubtedly, the surface current estimates from Argo trajectories provide a new means to describe the surface circulation in the global ocean in real time. Although the number of Argo float arrays is more than that of the surface drifter array (3,000 vs. 1,250) of the Surface Velocity Program (SVP), the number of surface velocity estimates from Argo float arrays is still less than that of SVP surface drifter arrays because Argo floats spend more than 90 percent of their time in the subsurface, whereas SVP surface drifters always remain at the surface. The combination of all surface trajectories can give more detailed description for global ocean circulation. Meanwhile, the preliminary results of mid-depth currents in the Pacific Ocean indicate that the trajectories of Argo floats would become one of the important sources to acquire knowledge on the specific mid-depth circulation.

Advantages of the Argo floats are ease of deployment, ability to conduct long-lived missions, and capacity to acquire and communicate data to researchers throughout the world without the direct involvement of ships or people. The Argo floats are, however, not free from limitations. For example, the datasets recorded by these floats can contain small gaps. The reasons are that ALACE floats do not acquire data when they are descending or ascending, and this scheme results in a data gap, especially when strong currents push the float a significant distance during ascent or descent. Similarly, P-ALACE and SOLO floats do not acquire data when they are descending or floating at depth, so a similar gap in data can result.

Despite these limitations, Argo marks a radical broadening of the use of floats. Seventeen countries have provided floats for the Argo array. The commitments ranged from the United States (contributing half the floats) to fewer than five floats contributed by countries such as Mauritius, Denmark, Ireland, Netherlands, New Zealand, and the Russian Federation. Many other countries assisted with float deployments and access to their Exclusive Economic Zones. Use of float data broadened, too. Operational centers are still using data from Argo in the production of ocean and climate analyses and forecasts (Gould and the Argo Science Team, 2004). So, the concept originated by John Swallow and further developed by other researchers for global application in the 1980s and 1990s grew from a rather exclusive research tool into a central element of the ocean-observing system that addresses issues of global socioeconomic significance (anthropogenic climate change, sea-level rise). According to Gould, we have arrived at our present exciting position thanks to a small number of far-sighted individuals and to a close and very productive interaction between ocean scientists and engineers. With that in mind, we can optimistically look forward to a very exciting era that was surely not envisaged by John Swallow when he scavenged the storerooms of the NIO to build the first Swallow float.

Oceanography has not yet reached meteorology's operational level because progress in numerical modeling still has to be made and because routine data assimilation is still to be developed. For assisting offshore operations, surveying companies rely on *in situ* real-time current measurements. Although providing valuable information, this approach remains costly (human and ship costs for at-sea operations, instrument deployment, etc.), time-consuming (moorings often have to be deployed for lengthy periods to gather enough data for reliable statistics to be established), and somewhat incomplete (for budget considerations, only scattered point measurements can be made). Forecasts cannot be established with *in situ* data alone.

To predict local circulation, even at relatively short time scales, a large area must be monitored synoptically with a high spatial and temporal coverage. At present, commercial companies (e.g., SAT-OCEAN) provide cost-effective solutions based on satellite imagery, allowing a reasonably accurate diagnosis and forecast of oceanic upper-layer currents over a large part of the world's oceans. This approach presents a huge potential for offshore surveying and for statistical studies of unknown oceanic provinces, as demonstrated during an assistance mission offshore South Africa (see Vigan, 2002).

Availability of a wide variety of technologies has begun to enable generation of accurate ocean environmental forecasting, which is necessary for a variety of offshore engineering operations. For example, some commercial companies (e.g., Horizon Marine Inc.) produce loop-current eddy (LCE) forecasts to prevent disruption of deep-water operations. The eddy-watch service utilizes data from oceanographic buoys, current profiles at rig locations, and all publicly available satellite observations to provide comprehensive operational monitoring and analysis of ocean currents and eddies based on numerical ocean models. It is of interest to single out the subject of satellite-tracked drifting buoys for special attention, however, because of the potential of this technique for large-scale measurements. Lagrangian floats play an important role in mixing studies as well, and several investigators are vigorously developing novel floats to study deep convection, internal wave shear, and boundary layer deepening.

13.10. COMPREHENSIVE STUDY OF OCEANIC CIRCULATION

Although it was not until much later that the successors to Seasat were launched (for example, Geosat in 1985; ERS-1 in 1991; TOPEX-Poseidon in 1992; Jason in 2001), these satellites and the development of floats, moored CMs, high-quality CTD, and tracer measurements opened up the possibility of a comprehensive study of the ocean circulation on a global scale. Several passive radiometry images

revealed visual indications of the most salient features of eddies, and the congruence between the physical and biological structure of eddies and rings was often found to be remarkable. It is conceivable that a global monitoring system for the ocean circulation could consist of a combination of several of the elements discussed: (1) satellite observations of the surface temperature and surface pressure fields; (2) direct measurements of the surface current and temperature by drifters and a modest number of moorings; and (3) deep-ocean monitoring by a combination of acoustic tomography, ship-borne hydrography, and moorings. All the elements are present now or are being tested.

Ocean-current measurement is a first-order task in ocean process research, environmental monitoring, climate studies, ship traffic control, and offshore work. Vector measurements are always harder than scalar measurements, both to make and to interpret.

Since 1965 there has been a flood of new instruments and new ideas, thanks to a few good and determined engineers and physicists. As a result, our view of the ocean has changed markedly, especially on the smaller scales, where the old instruments and techniques were essentially blind. In this endeavor, the information gained in actual field tests was crucial to the development of reliable instruments. The development of technologies in a number of areas in the 1950s and 1960s laid the foundation for rapid improvement in observing techniques, starting in the mid-1960s.

I do not want to leave the reader with the impression that the technologies for measurement of ocean currents have reached a plateau. Perhaps still better technologies will evolve in the future, with further advancements in science, technology, and sensor design. Nevertheless, the present technologies provide the opportunity to have a long-term presence in any part of the oceans, to collect continuous streams of data, and to observe processes and transmit data about them in real time. Though various elements of these real-time observations have been functional for some time, the comprehensive array and potentially integrated nature of sensor networks have only recently become possible. The datastreams from sensors within and at the junctures of atmosphere, ocean, and solid-earth realms provide the potential for the development of a quantitative understanding of ocean processes across otherwise unattainable space and time scales. This potential has fundamentally important consequences for different groups, namely, scientific researchers, commercial interests, government regulators and policy makers, teachers, the military, and the public. The solution to the problems that still remain in understanding oceanographic processes depend in great measure on the continuation of close interaction between ocean scientists and engineers.

Several techniques are developing in parallel, and each offers special advantages in certain tasks. Some technologies, specifically acoustic Doppler and radar backscatter, have expanded rapidly. We still need testing, intercomparison, and most significantly, interpreted datasets from measurements made with these techniques to gain an understanding of where they can be trusted and where not. There is no saturation of the current measurement field. Although there is maturing in specific technologies, new technologies or reinventions of older technologies spawn new cycles of development.

REFERENCES

Barrick, D.E., 2010. After 40 years, how are HF radar currents now being used? Proc. IEEE/OES/CWTM Tenth Working Conference on Current Measurement Technology, 3.

Barrick, D.E., Evans, M.W., Weber, B.L., 1977. Ocean surface currents mapped by radar. Science 198 (4313), 138−144.

Beal, R., Tilley, D., Monaldo, F., 1983. Large and small-scale spatial evolution of digitally processed ocean wave spectra from Seasat Synthetic Aperture Radar. J. Geophys. Res. 88 (C3), 1761−1778.

Bower, A.S., Rossby, T., 1989. Evidence of cross-frontal exchange processes in the Gulf Stream based on isopycnal RAFOS float data. J. Phys. Oceanogr. 19, 1177−1190.

Chhabra, N.K., 1977. Correction of vector-averaging current meter records from the MODE-1 central mooring for the effects of low-frequency mooring line motion. Deep-Sea Res. 24, 279−287.

Cochin, V., Mariette, V., Broche, P., Garello, R., 2006. Tidal current measurements using VHF radar and ADCP in the Normand Breton Gulf: Comparison of observations and numerical model. IEEE J. Ocean. Eng. 31 (4), 885−893.

Crease, J., 1965. The flow of Norwegian Sea water through the Faroe Bank Channel. Deep-Sea Res. 12, 143−150.

Davis, R.E., 1998. Preliminary results from directly measuring mid-depth circulation in the tropical and South Pacific. J. Geophys. Res. 103, 24,619−24,639.

Davis, R.E., Webb, D.C., Regier, L.A., Dufour, J., 1992. The autonomous Lagrangian circulation explorer (ALACE). J. Atmos. Oceanic Technol. 9, 264−285.

Freeland, H.J., Rhines, P.B., Rossby, T., 1975. Statistical observations of the trajectories of neutrally buoyant floats in the North Atlantic. J. Mar. Res. 33, 383−404.

Friedman, K.S., Sikora, T.D., Pichel, W.G., Clemente-Colon, P., Hufford, G., 2001. Using space-borne synthetic aperture radar to forecast polar mesoscale cyclones in the Bering Sea. Wea. Forecasting 16, 270−276.

Gascard, J.-C., 1973. Vertical motions in a region of deep water formation. Deep-Sea Res. 20, 1011−1027.

Gould, Sambuco, 1975. The effect of mooring type on measured values of ocean currents. Deep-Sea Res. 22, 55−62.

Gould, W.J., 2005. From Swallow floats to Argo: The development of neutrally buoyant floats. Deep-Sea Res. II 52, 529−543.

Halpern, D., Pillsbury, R.D., 1976. Influence of surface waves on subsurface current measurements in shallow water. Limnology Oceanogr. 21, 611−616.

Hasselmann, K., 1971. Determination of ocean wave spectra from Doppler radio return from the sea surface. Nature Physical Science 229, 16−17.

Hasselmann, K., Hasselmann, S., 1991. On the nonlinear mapping of an ocean wave spectrum into a synthetic aperture radar image spectrum and its inversion. J. Geophys. Res. 96 (C6), 10,713−10,729.

Hendry, R., 1988. A simple model of Gulf Stream thermal structure with application to the analysis of moored measurements in the presence of mooring motion. J. Atmos. Oceanic Technol. 5, 328−339.

Hogg, N.G., 1986. On the correction of temperature and velocity time series for mooring motion. J. Atmos. Oceanic Technol. 3, 204−214.

Johnson, D., Stocker, R., Head, R., Imberger, J., Pattiaratchi, C., 2003. A compact, low-cost GPS drifter for use in the oceanic nearshore zone, lakes and estuaries. J. Atmos. Oceanic Technol. 18, 1880−1884.

Lherminier, P., Gascard, J.C., 1998. Drifting isobaric float response to deep convective activity in the Greenland Sea, Academie des Sciences, Comptes Rendus. Serie 2a. Sciences de la Terre et des Planetes 326 (5), 341−346.

McDowell, S.E., Rossby, H.T., 1978. Mediterranean water: An intense mesoscale eddy off the Bahamas. Science 202, 1085−1087.

McWilliams, J.C., 1976. Maps from the Mid-Ocean Dynamics Experiment. J. Phys. Oceanogr. 6, 810−846.

Munk, J.W., Wunsch, C., 1979. Ocean acoustic tomography: a scheme for large scale monitoring. Deep-Sea Res. 26A, 123−161.

Owens, W.B., 1991. A statistical description of the mean circulation and eddy variability in the northwestern Atlantic using SOFAR floats. Prog. Oceanogr. 28, 257−303.

Paduan, J.D., Graber, H.C., 1997. Introduction to high frequency radar: Reality and myth. Oceanography 10, 36−39.

Paduan, J.D., Shulman, I., 2004. HF radar data assimilation in the Monterey Bay area. J. Geophys. Res. 109 (C07 S09), 1−17.

Rossby, H.T., Dorson, D., 1983. The Deep Drifter: A simple tool to determine average ocean currents. Deep-Sea Res. 30, 279−1288.

Rossby, T., Webb, D., 1971. The four-month drift of a Swallow float. Deep-Sea Res. 18, 1035−1039.

Schmidt, W., Woodward, B., Millikan, K., Guza, R., Raubenheimer, B., Elgar, S., 2003. A GPS-tracked surf zone drifter. J. Atmos. Oceanic Technol. 20, 1069−1075.

Schmitz, W.J., Price, J.F., Richardson, P.L., 1988. Recent moored current meter and SOFAR float observations in the eastern North Atlantic. J. Mar. Res. 46, 301−319.

Sikora, T.D., Young, G.S., Shirer, H.N., Chapman, R.D., 1997. Estimating convective atmospheric boundary layer depth from microwave radar imagery of the sea surface. J. Appl. Meteor. 36, 833−845.

Sikora, T.D., Young, G.S., Beal, R.C., Edson, J.B., 1995. Use of space-borne synthetic aperture radar imagery of the sea surface in detecting the presence and structure of the convective marine atmospheric boundary layer. Mon. Wea. Rev. 123, 3623−3632.

Stewart, R.H., Joy, J.W., 1974. HF radio measurements of surface currents. Deep Sea Res. 21, 1039−1049.

Stommel, H., Arons, A.B., 1960a. On the abyssal circulation of the world ocean—I. Stationary planetary flow patterns on a sphere. Deep-Sea Res. 6, 140−154.

Stommel, H., Arons, A.B., 1960b. On the abyssal circulation of the world ocean—II. An idealized model of the circulation pattern and amplitude in oceanic basins. Deep-Sea Res. 6, 217−233.

Swallow, J.C., 1955. A neutral-buoyancy float for measuring deep currents. Deep-Sea Res. 3, 74–81.

Swallow, J.C., Worthington, I.V., 1961. An observation of a deep countercurrent in the western North Atlantic. Deep-Sea Res. 8, 1–19.

Swallow, J.C., Bruce, J.G., 1966. Current measurements off the Somali coast during the southwest monsoon of 1964. Deep-Sea Res. 13, 861–888.

Swallow, J.C., Worthington, L.V., 1969. Deep currents in the Labrador Sea. Deep-Sea Res. 16, 77–84.

Swallow, J.C., 1971. The *Aries* current measurements in the western North Atlantic. Phil. Trans. Roy. Soc. London A 270, 451–460.

Vachon, W.A., 1977. Current measurement by Lagrangian drifting buoys: Problems and potential. Proc. Oceans '77, 46B-1–46B-7.

Vigan, X., Provost, C., Bleck, R., Courtier, P., 2000. Sea surface velocities from sea surface temperature image sequences, 1. Method and validation using primitive equation model output. J. Geophys. Res. 105 (C8), 19,499–19,514.

Von Zweck, O.H., Saunders, K.D., 1981. Contamination of current records by mooring motion. J. Geophys. Res. 86 (C3), 2071–2072.

Webb, D.C., Dorson, D.L., Voorhis, A.D., 1970. A New Instrument for the Measurement of Vertical Currents in the Ocean. Proceedings of the Institute of Electronic and Radio Engineers' Conference on "Electronic Engineering in Ocean Technology", Swansea, Sept. 1970.

WHOI Contribution No. 2518. Also printed in the Radio and Electronic Engineer 41 (9), 416–420.

Webster, F., 1968. Observations of inertial-period motions in the deep sea, Woods Hole, Massachusetts, U.S.A. Rev. Geophys. 6, 473–490.

Webster, F., 1969. On the representativeness of direct deep-sea current measurements. Prog. Oceanogr. 5, 3–15.

Webster, F., 1972. Estimates of the coherence of ocean currents over vertical distances. Deep-Sea Res. 19, 35–44.

Worcester, P.F., 1977. Reciprocal acoustic transmission in a mid-ocean environment. J. Acoustic Society of America 62, 895–905.

Wijffels, S.E., Toole, J.M., Davis, R., 2001. Revisiting the South Pacific subtropical circulation: a synthesis of world ocean circulation experiment observations along 32°S. J. Geophys. Res. 106 (C9), 19,481–19,513.

Williams 3rd, A.J., 1997. Historical overview: Current measurement technologies. Sea Technol. 38 (6), 63–69.

Yang, Q., Parvin, B., 2000. Feature-based visualization of geophysical data, Hilton Head Island, SC. Proc. Computer Vision and Pattern Recognition (CVPR). Hilton Head Island, SC, USA, 2: 276–281.

Zenk, W., Pinck, A., Becker, S., Tillier, P., 2000. The float park: a new tool for a cost-effective collection of Lagrangian time series with dual release RAFOS floats. J. Atmos. Oceanic Technol. 17, 1439–1443.

Index

Note: Page numbers with "*f*" denote figures.

A

Aanderaa current meter, 242–249
Absolute velocity profiler (AVP), 314, 315f
  free-falling, acoustically tracked, 318–320
AC, *see* Azores Current
ACC, *see* Antarctic circumpolar current
ACM, *see* Acoustic current meter
Acoustically focused oceanographic sampling
    (AFOS), 410
Acoustically tracked free-fall current velocity
    and CTD profiler (TOPS), 326–328
Acoustically tracked freely sinking pingers,
    305–308
Acoustic current meter (ACM), 282–283
Acoustic Doppler current meter, 81–82,
    291–294
Acoustic Doppler current profilers (ADCPs),
    68, 118–119, 407–409
  bottom-mounted, upward-facing, 344–353
  directional wave-measuring, 346
  horizontal, 370–371
  lowered, 367–368
  RDI WorkHorse ADCPs, 352–353
  single upward-facing, 346–350
  vertical current profiling using, 339–380
    AUV-mounted ADCPs, for current
      profiling, 368–369
    for AUV navigation, 369–371
    basic assumptions and operational issues,
      340–341
    calibration of, 371–372
    downward-facing shipboard ADCPs,
      360–362
    evaluation of, 372–374
    horizontal-facing ADCPs, 354–358
    intercomparison of, 372–374
    limitations of, 374–375
    merits of, 374–375
    operation principles, 341–344
    profiling geometries, 344–353
    subsurface moored ADCPs, 358–359
    towed ADCPs, 363–367
    trawl-resistant ADCP bottom mounts,
      353–354
Acoustic radiation, 291
Acoustic surface tracking (AST),
    347–349
Acoustic Thermometry of Ocean Climate
    (ATOC) project, 66–67

Acoustic tomography, 66–69
  coastal, 68–69, 215–226, 410
  ocean, 66–68, 201–240
  river, 69, 226–230, 409–410
Acoustic travel time (ATT) difference sensors,
    74, 281–290
Acoustic velocity meters (AVMs), 226
Acoustic wave and current (AWAC) profiler
    system, 349–350
  Platform Mount, 351–352, 351f
Active microwave radar imaging, 61–62
  of sea surface current signatures, 153–154
  technologies of, 154–166
Active radiometry, in visible-wavelength band,
    145–146
Active sensors, 139–140
ADCPs, *see* Acoustic Doppler current profilers
Advanced Research and Global Observation
    Satellite (ARGOS), 93–94, 99
Advanced very high-resolution radiometer
    (AVHRR), 60, 148–149
Aerial photography
  of surface water motion trajectories and
    patterns, 57–60
  in visible and infrared bands, 140–141
AFOS, *see* Acoustically focused
    oceanographic sampling
AGC, *see* Automatic gain control
Agulhas Current, 22–24
Agulhas retroflection, 22–24
Aircraft-borne side-looking microwave radar
    systems, 401–402
ALACE, *see* Autonomous Lagrangian
    Circulation Explorer
Almaz SAR system, 166
ALOS, 167
Ametek-Straza Doppler current profiler, 361
Amplitude reduction, 164
Antarctic circumpolar current (ACC), 1–2,
    6–7, 13–15, 14f
Antenna temperature, 152
  apparent, 150
APEX floats, 413–414
Apogean current, 36
Apparent antenna temperature, 150
Arbitrary zero point, 391
ARGOS, *see* Advanced Research and Global
    Observation Satellite
Aries, 31

Array for real-time geostrophic oceanography
    (Argo)
  floats, 6–7, 178–180, 189–198
  program, 414
AST, *see* Acoustic surface tracking
Atlantic Ocean
  Brazil Current, 17–20
  equatorial undercurrents in, 25–27
  Gulf Stream, 5, 6f, 11, 15–17, 325
  North, 17
  western boundary currents in, 15–20
ATOC, *see* Acoustic Thermometry of Ocean
    Climate project
ATT, *see* Acoustic travel time difference
    sensors
Automatic gain control (AGC), 118–119
Autonomous Lagrangian Circulation Explorer
    (ALACE), 192–195, 413–414
Autonomous underwater vehicles (AUVs),
    368, 410
  -mounted ADCPs, for current profiling,
    368–369
  navigation, acoustic Doppler current profilers
    for, 369–371
AUVs, *see* Autonomous underwater vehicles
AVHRR, *see* Advanced very high-resolution
    radiometer
AVMs, *see* Acoustic velocity meters
AVP, *see* Absolute velocity profiler
AWAC, *see* Acoustic wave and current profiler
    system
Axis-crossing counting technique, 342
Azimuth image shift, 164
Azimuth image smear, 164
Azimuth resolution enhancement of,
    160–163
Azores Current (AC), 17

B

Backscattering, 109–110, 110f
Backward-scatter laser Doppler sensor,
    78–80, 86f
BASS, *see* Benthic Acoustic Stress Sensor
BC, *see* Brazil Current
Beacon survey method, 319, 319f
Beam forming, 116–117, 117f
Benthic Acoustic Stress Sensor (BASS),
    84–85, 285–290

Biaxial dual orthogonal propeller vector-measuring current meters, 253–255
Bidirectional propeller sensor, 263
Bins, 340–342, 341f
Biplanar crossed vanes, 177–178
Black body, 146
Bottom-mounted, upward-facing ADCPs, 344–353
  current profile of, 347–350
  single upward-facing ADCP, 346–350
  surface currents–gravity wave orbital velocities combination, 345–346
  wave measurements of, 347–350
Bottom-mounted, winch-controlled vertical automatic profiling systems, 302–305
Bragg frequency, 111
Bragg resonance, 153–154
  sea surface current estimation using, 113
Brazil Current (BC), 17–20
  North, 18–20
Brightness temperature, 152
Broadband signal-processing technique, 343–344

C

Calibration
  of acoustic Doppler current profilers, 371–372
  of current meters, 255–258
Carbon fiber-reinforced plastics (CFRP), 167
Cartesian diver profiler, 315–318, 316f
CAT, see Coastal acoustic tomography
CCDs, see Charge-coupled devices
CFRP, see Carbon fiber-reinforced plastics
Challenger, 166
Charge-coupled devices (CCDs), 143
Chirp pulse, 157
Circle of inertia, 33–34
Coastal acoustic tomography (CAT), 68–69, 215–226, 410
Coastal ocean dynamics applications radar (CODAR), 62–63, 121, 403
Coastal ocean surface radar (COSRAD), 62–63, 403
Coastal Zone Color Scanner (CZCS), 142–145
CODAR, see Coastal ocean dynamics applications radar
Coherent processing, 343
Cold-core eddies, 6–7, 383–384, 383f
Collinear passes, 391
Corioli's force, 34
Coriolis parameter, 31, 385
Correlated pulse pairs, 343–344
Cosine response, 253–254
  horizontal, 253–254
  vertical, 253–254
COSMER, see Courants de Surface MEsures par Radar
COSRAD, see Coastal ocean surface radar
Courants de Surface MEsures par Radar (COSMER), 403
Cromwell Current, 27
Crossover points, 79, 391

Cryosat-2, 394
Current charts, 2, 51–52, 64
Current-driven sea surface wave transport, vector mapping based on, 62–64
Current meters, 70
  Aanderaa, 242–249
  acoustic, 282–283
  acoustic Doppler, 81–82, 291–294
  calibration of, 255–258
  electromagnetic, 269–281
  Savonius rotor, 72–74
  unidirectional impeller, 71–72
  vector-averaging, 249–250, 256–257
  vector-measuring, 254–255, 258
Current rose, 34
Cyclesonde, see Freely sinking and rising relative velocity probe
CZCS, see Coastal Zone Color Scanner

D

Data collection and location system (DCLS), 98–99
Davis Strait, 3
DCLS, see Data collection and location system
DDL, see Dispersive delay line
Dead reckoning, 53
Deep countercurrent, 16–17
Deep jets, 299–300
Delayed-mode quality-control system (DMQC), 197
Delft Object-Oriented Interferometric Software (DORIS), 166
DEM, see Digital elevation models
Differential InSAR (D-InSAR), 61
Digital elevation models (DEM), 165–166
D-InSAR, see Differential InSAR
Directional wave-measuring ADCP, 346
Direct reading current meter (DRCM), 241, 250
Dispersive delay line (DDL), 387
DMQC, see Delayed-mode quality-control system
Doppler centroid, 162–163
Doppler shift, 109–110, 110f
DORIS, see Delft Object-Oriented Interferometric Software
Dot plots, see Scatter plot
Downward-facing shipboard ADCPs, 360–362
Downward-looking ADCPs, 341
DRCM, see Direct reading current meter
Drift card, 54–55, 55f
  Duncan, 54
  Olson, 54
  trajectories of, 54, 56f
Drifter-borne Doppler transponder, 96–97
Drifter-following radar transponder, 95–96
Drifting profiling floats, 195–198
  profiling observations from polar regions, 197–198
Drifting subsurface floats motion, measurements based on, 69–70
Drift poles, 55
Dual-scatter system, 79

Duncan drift card, 54
Dwell time, 160–161

E

EAC, see East Australian Current
Early mariners' contributions to subsurface/abyssal current measurements, 64
East Auckland Current (EAUC), 20
East Australian Current (EAC), 20
East Cape Current (ECC), 20
East Madagascar Current, 22
EAUC, see East Auckland Current
Ebb current, 36
ECC, see East Cape Current
Echo intensity (EI), 340
Eckerman current meters, 71
Eddies, 5–9
  cold-core, 6–7, 383–384, 383f
  frontal, 7
  influence on fishery and weather, 11–12
  loop current, 4, 366, 415
  mesoscale, 99
  warm-core, 6–8, 383–384, 383f
EI, see Echo intensity
Ekman current profile, 32
Ekman Spiral, 31–33
Electromagnetic current meters, 269–281
Electromagnetic induction (EMI), 65, 268
Electromagnetic method, 65–66
Electromagnetic sensors, 267–281
El Niño Southern Oscillation (ESCO), 20, 28, 383
EM-APEX floats, 413–414
EMI, see Electromagnetic induction
EMVP, see Freely falling electromagnetic velocity profiler
Encoder, 251
EnKF, see Ensemble Kalman filter technique
Ensemble Kalman filter (EnKF) technique, 219, 223–224
ENVISAT, 167, 401–402
Equatorial undercurrents (EUC), 24–29, 384–385
  in Atlantic Ocean, 25–27
  in Indian Ocean, 28–29
  in Pacific Ocean, 27–28
ERAMER, 208–209
ERM, see Exact Repeat Mission
ERS-1, 61–62, 166–167, 394, 415
ERS-2, 61–62, 167, 394, 401–402
ESCO, see El Niño Southern Oscillation
Estuaries, tidal currents in, 34–36
EUC, see Equatorial undercurrents
Eulerian current meters, 68
Eulerian-style current measurement, 242, 397–418
  incorporating mechanical sensors, 241–266
    advantages of, 258–261
    biaxial dual orthogonal propeller vector-measuring current meters, 253–255
    graphical methods for displaying, 255–258
    limitations of, 258–261
    propeller rotor current meter (Plessey current meter), 250–253

Savonius rotor current meters, 242–249
vector-averaging current meters, 249–250
incorporating nonmechanical sensors, 267–296
acoustic Doppler current meter, 291–294
acoustic travel time difference sensors, 281–290
electromagnetic current meters, 269–281
electromagnetic sensors, 267–281
EVAPS, 302–305, 303f
Evolution of satellite altimetry, 392–394
Exact Repeat Mission (ERM), 393

F
Fast Fourier transform (FFT), 342
Fermat's principle, 212
FFT, *see* Fast Fourier transform
Fisheries eddies/fronts influence on, 11–12
Fixed locations at predetermined depths, measurements from, 70–82
acoustic Doppler current meter, 81–82
laser Doppler sensor, 78–81
propeller revolution registration, by mechanical counters, 71
Savonius rotor current meters, 72–74
suspended drag, 70–71
thermal sensors, for turbulent motion measurement, 77–78
ultrasonic acoustic methods, 74–77
unidirectional impeller current meters, 71–72
FLIP, *see* Floating instrument platform
Floating instrument platform (FLIP), 361–362, 407
Floats
ALACE, 192–195
APEX, 413–414
Argo, 6–7, 178–180, 189–198
drifting profiling, 195–198
EM-APEX, 413–414
MARVOR, 192
RAFOS, 188–192, 412–413
SOFAR, 183–188, 191, 204–206, 208–209
SOLO, 196, 413–414
Swallow, 69–70, 180–183, 188
Flood current, 36
Florida Straits, 201
eddy-induced sea surface temperature, 7
Flow meters, 242
FMCW, *see* Frequency-modulated continuous wave
Forward-scatter laser Doppler sensor, 78–79, 86f
Free-drop techniques, 300
Free-falling, acoustically tracked absolute velocity profiler (Pegasus), 318–320
Free-falling lift-force sensitive probes, 332–335
Free-fall probe, 330–332
Free-fall shear profiler (Yvette), 326
Freely falling, acoustically self-positioning dropsonde system (White Horse), 320–322
Freely falling electromagnetic velocity profiler (EMVP), 311–318

Freely rising acoustically tracked expendable probes (Popup), 322–325
Freely sinking and rising relative velocity probe (Cyclesonde), 308–311
operation of, 309–310
Freely sinking and wire-guided relative velocity probes, 300–302
Frequency-modulated continuous wave (FMCW), 121
Frequent altimeter calibrations, 389–390
Fringe region, 79
Frontal circulation system, 10–11
Frontal eddies, 7
Fronts
hydrographic, 10–11
influence on fishery and weather, 11–12
ocean, 10

G
GEK, *see* Geomagnetic electrokinetograph
Geodynamics Experimental Ocean Satellite-3 (GEOS-3 satellite), 392–393
Geoid errors correction, 390–391
Geomagnetic electrokinetograph (GEK), 65–66, 269
Geosat, 393
GEOS-3, 392–393
Geostrophic currents, 398, 405–407, 409, 412
Global climate change, 2–5
Global Climate Observing System, 414
Global positioning system (GPS), 208, 215–216, 400–401
-tracked drifters, 100–104
Global System for Mobile (GSM) communication, 104–105
Global Telecommunication System (GTS), 99
GOMGB, *see* Gulf of Maine and Georges Bank
GPS, *see* Global positioning system
GRACE, *see* Gravity Recovery and Climate Experiment
Gravity Recovery and Climate Experiment (GRACE), 391–392
GSM, *see* Global System for Mobile communication
GSMR, *see* Gulf Stream Meandering and Ring region
GSNW, *see* Gulf Stream North Wall
GTS, *see* Global Telecommunication System
Gulf of Maine and Georges Bank (GOMGB), 8
Gulf Stream, 15–17, 325
hydrographic fronts, 11
meandering currents in, 5, 6f
Gulf Stream Meandering and Ring (GSMR) region, 11
Gulf Stream North Wall (GSNW), 402
Gyres, 13

H
H-ADCPs, *see* Horizontal acoustic Doppler current profilers
HAMburg Shelf Ocean Model (HAMSOM), 132

HAMSOM, *see* HAMburg Shelf Ocean Model
Haro Strait, 213–214, 213f, 214f
Helmholtz coil, 269–273
Heterodyning, 75
HFOSR, *see* High-frequency ocean surface radar
HF-SWR, *see* High-frequency surface wave radar
High-frequency (HF) Doppler radar networks
remote mapping of sea surface currents using, 109–138
advantages of, 129–131
Crombie's discovery, 110–112
depth extent of, 113–114
experimental developments of, 115–118
instrumentation aspects of, 118–120
intercomparison considerations of, 126–129
operational scales development, 120–126
pulse Doppler-radar echo spectra, peculiarities of, 112–113
round-the-clock coast-observing role of, 131–132
sea surface current estimation, using Bragg resonance principle, 113
shelf-observing role of, 131–132
technological aspects of, 114–115
tsunami-induced sea surface-current jets at continental shelves, detection and monitoring of, 132–133
High-frequency (HF) Doppler radar systems, 63–64, 402–405
High-frequency ocean surface radar (HFOSR), 62–63, 403
High-frequency surface wave radar (HF-SWR), 62–63, 403
High-resolution infrared radiometer (HRIR), 60
Horizontal acoustic Doppler current profilers (H-ADCPs), 226
Horizontal current profile, 339, 362, 367, *see also* Horizontal currents, vertical profiling of
Horizontal currents, vertical profiling of, 86–88
using freely sinking/rising robes, 297–338
acoustically tracked free-fall current velocity and CTD profiler (TOPS), 326–328
acoustically tracked freely sinking pingers, 305–308
bottom-mounted, winch-controlled vertical automatic profiling systems, 302–305
free-fall probe, 330–332
free-fall shear profiler (Yvette), 326
free-falling, acoustically tracked absolute velocity profiler (Pegasus), 318–320
free-falling lift-force sensitive probes, 332–335
freely falling, acoustically self-positioning dropsonde system (White Horse), 320–322
freely falling electromagnetic velocity profiler, 311–318

Horizontal currents, vertical profiling of (*Continued*)
    freely rising acoustically tracked expendable probes (Popup), 322–325
    freely sinking and rising relative velocity probe (Cyclesonde), 308–311
    freely sinking and wire-guided relative velocity probes, 300–302
    importance of, 297–300
    limitations of, 335
    merits of, 335
    towed acoustic transducer, 329–330
Horizontal-facing ADCPs, 354–358
Horizontal-looking ADCPs, 354–357
Horizontally integrated remote measurements, using ocean acoustic tomography, 201–240
    coastal acoustic tomography, 215–226
    one-way tomography, 203–210
    river acoustic tomography, 226–230
    space-time acoustic scintillation analysis, 232–235
    from Straits, 212–215
    two-way tomography, 210–212
    vorticity, 230–232
HRIR, *see* High-resolution infrared radiometer
HY-2A, 394
Hydrographic fronts, 10–11
Hydrophones, 181
Hydrothermal vent field, 41

I

IABP, *see* International Arctic Buoy Programme
Ice floes imaging sea surface currents measurement using, 170–171
Ice thickness estimation, 350
IFOV, *see* Instantaneous field of view
Image texture, 153
Image tone, 153
Incoherent processing, 342–343
Indian Ocean
    Agulhas Current, 22–24
    equatorial undercurrents in, 28–29
    Somali Current, 24
    western boundary currents in, 22–24
Inertia Current, 33–34
Infrared bands, aerial photography in, 140–141
Infrared radiance detectors, 147–148
InSAR, *see* Interferometric synthetic aperture radar system
Instantaneous field of view (IFOV), 142–144
Integration time, 160–161
Interferometric synthetic aperture radar (InSAR) system, 61, 165–166
    differential, 61
Interferometric technique, 323
Intermediate Western Boundary Current (IWBC), 18
International Arctic Buoy Programme (IABP), 197
IWBC, *see* Intermediate Western Boundary Current

J

JAMSTEC, 197–198
Janus-Helmholtz acoustic source (JHAS), 209
Jason altimeter system, 394
    Jason-1, 394
    Jason-2, 394
JHAS, *see* Janus-Helmholtz acoustic source

K

Kalman filter method, 197
Kanmon Strait, 219–220, 222–224
Kuroshio Current, 20–22
Kuroshio Stream, 325

L

L-ADCP, *see* Lowered ADCP
Lagrangian measurements, interpretation of, 399–400
Lagrangian method, 55
Lagrangian-style subsurface current measurements, through subsurface drifters tracking, 177–200
    ALACE floats, 192–195
    drifting profiling floats, 195–198
        profiling observations from polar regions, 197–198
    moored acoustic receivers subsurface floats transmitting to, 183–188
    moored acoustic sources subsurface floats listening to, 188–192
    satellite-recovered pop-up drifters, 178–180
    surface-trackable subsurface drifters, 177–178
    swallow floats, tracked by ship-borne hydrophones, 180–183
Lagrangian-style surface current measurements, through surface drifters tracking, 93–108
    limitations of, 105
    radio buoys, 94–105
        drifter-borne Doppler transponder, 96–97
        drifter-following radar transponder, 95–96
        GPS-tracked drifters, 100–104
        telephonically tracked drifters, 104–105
        tracked by polar-orbiting satellites, 97–100
Langmuir circulation, 33, 141
Laser Doppler current meter (LDCM), 78, 80–81
Laser Doppler sensor, 78–81
Laser-induced fluorescence (LIF), 145
LCE, *see* Loop Current eddies
LDCM, *see* Laser Doppler current meter
LiDAR, *see* Light detection and ranging transmitter
LIF, *see* Laser-induced fluorescence
Light detection and ranging (LiDAR) transmitter, 145
Lomonosov Current, 25
Long antenna from small antenna synthesis of, 159–160
Longitudinal triangular ripples (LTR), 41–42
Long Range Navigation (LORAN), 177–178, 182–183, 190

Loop Current eddies (LCE), 4, 366, 415
LORAN, *see* Long Range Navigation
Lowered ADCP (L-ADCP), 367–368
LSAT-OCEAN, 415
LTR, *see* Longitudinal triangular ripples

M

MABL, *see* Marine atmospheric boundary layer
Maine Coastal Current (MCC), 8
Malvinas Current, 18
Marine atmospheric boundary layer (MABL), 140
Marine geoid, 381–382
    errors correction, 390–391
    topographic height variability independent of, null methods for obtaining, 391–392
MARVOR floats, 192
Matched filter, 387–388
Maury's Wind and Current Charts, 51, 64
MAVS, *see* Modular Acoustic Velocity Sensor
Maximum Likelihood Method (MLM), 348
MCC, *see* Maine Coastal Current
MCR, *see* Multifrequency coastal radar
Meandering currents, 5
Mechanical devices, 83
Mechanical sensors, Eulerian-style measurements incorporating, 241–266
    advantages of, 258–261
    biaxial dual orthogonal propeller vector-measuring current meters, 253–255
    graphical methods for displaying, 255–258
    limitations of, 258–261
    propeller rotor current meter (Plessey current meter), 250–253
    Savonius rotor current meters, 242–249
    vector-averaging current meters, 249–250
Meddies, 412
Medium-resolution infrared radiometer (MRIR), 60
Meridional overturning circulation (MOC), 2, 4
Mesoscale eddies, 99
METOCEAN, 197–198
Microbubbles, 340
Microwave radiometers, 149–153
    passive microwave radiometer instrumentation, 150–151
    passive microwave radiometry, 150
    scanning multichannel radiometers, 151–153
Mid-Ocean Dynamics Experiment (MODE), 186–187
MLM, *see* Maximum Likelihood Method
MOC, *see* Meridional overturning circulation
MODE, *see* Mid-Ocean Dynamics Experiment
Modular Acoustic Velocity Sensor (MAVS), 85–86, 288–291
Monostatic configuration, 341–342
Moored acoustic receivers subsurface floats transmitting to, 183–188
Moored acoustic sources subsurface floats listening to, 188–192
Moored current meters, 398–399, 406
MRIR, *see* Medium-resolution infrared radiometer

M-sequence, 216
MSS, *see* Multispectral scanner
Multifrequency coastal radar (MCR), 403
Multispectral scanner (MSS), 142−144

N

NAC, *see* North Atlantic Current
Natal Bight, 23
Natal pulses, 23
NBC, *see* North Brazil Current
Neap current, 36
Near-surface geostrophic currents, 383
NIMBUS, 97
NIP, *see* Nortek Internal Processor
Nonmechanical devices, 83−86
Nonmechanical sensors, Eulerian-style
    measurements incorporating, 267−296
    acoustic Doppler current meter, 291−294
    acoustic travel time difference sensors,
        281−290
    electromagnetic current meters, 269−281
    electromagnetic sensors, 267−281
Normalized radar cross-section, 153
Nortek Internal Processor (NIP), 352
North Atlantic Current (NAC), 17
North Brazil Current (NBC), 18−20
North Equatorial Countercurrent, 18
Null methods, for obtaining topographic
    height variability independent of
    geoid, 391−392

O

OAT, *see* Ocean acoustic tomography
Ocean acoustic tomography (OAT), 66−68,
    409−410
    horizontally integrated remote measurements
        using, 201−240
        coastal acoustic tomography, 215−226
        one-way tomography, 203−210
        river acoustic tomography, 226−230
        space-time acoustic scintillation analysis,
            232−235
        from Straits, 212−215
        two-way tomography, 210−212
        vorticity, 230−232
Ocean climate, 203−204
Ocean currents, implications of, 40−42
Ocean currents, measurement history of,
    51−92
    subsurface/abyssal current measurements,
        64−82
        early mariners' contributions, 64
        measurements from fixed locations at
            predetermined depths, 70−82
        spatially integrated measurements, 65−70
    surface current measurements, 52−64
        based on drifting surface bodies motion,
            52−57
        surface water motion trajectories and
            patterns, imaging of, 57−62
        vector mapping, based on current-driven
            sea surface wave transport, 62−64

vertical profiling, of horizontal currents,
    86−88
Ocean fronts, 10
Ocean States Measuring and Analyzing Radar
    (OSMAR2000), 403
Ocean storms, 99
Ocean surface current radar (OSCR), 62−63,
    121−123
Ocean weather, 203−204
Odden, 152−153
Olson drift card, 54
One-way tomography, 203−210
Open seas, tidal currents in, 34−36
Oscillating arm, 272
OSCR, *see* Ocean surface current radar
OSMAR2000, *see* Ocean States Measuring
    and Analyzing Radar
OTH, *see* Over-the-horizon radars
Over-speeding, 245−246
Over-the-horizon (OTH) radars,
    114−115

P

Pacific Ocean
    equatorial undercurrents in, 27−28
    western boundary currents in, 20−22
P-ALACE, 196, 413−414
Parking depth, 196
Passive radiometry
    microwave, 150
        instrumentation, 150−151
    in thermal infrared band, 146−149
    in visible-wavelength band, 142−145
Passive sensors, 139−140
PCM, *see* Profiling current meter
Pegasus, *see* Free-falling, acoustically tracked
    absolute velocity profiler
Perigean current, 36
PGS-S4 model, 382
Phytoplankton bloom, 141−142
Pings, 342
PISCES, 403
Plan position indicator (PPI), 154−155
Platform Mount AWAC profiler system,
    351−352, 351f
Platform transmit terminal (PTT), 98−99
Plessey current meter, *see* Propeller rotor
    current meter
Polarization, 114−115
Polar Ocean Profiling System (POPS),
    197−198
Polar-orbiting satellites, radio buoys tracked
    by, 97−100
Polar regions, profiling observations from,
    197−198
Polygon Mid-Ocean Dynamics Experiment
    (POLYMODE), 187−188, 412
POLYMODE, *see* Polygon Mid-Ocean
    Dynamics Experiment
POPS, *see* Polar Ocean Profiling System
Popup, *see* Freely rising acoustically tracked
    expendable probes
PortMap, 403
PPI, *see* Plan position indicator

PRF, *see* Pulse repetition frequency
Probe Recording Ocean Temperature and
    Shear (PROTAS), 330−332
Probe volume, 79
Profiling current meter (PCM), 300−302, 301f
Profiling geometries, 344−353
    bottom-mounted, upward-facing ADCPs,
        344−353
    current profile of, 350−353
    single upward-facing ADCP, 346−350
    surface currents−gravity wave orbital
        velocities combination, 345−346
    wave measurements of, 350−353
Progressive vector diagram, 258−259
Propeller revolution registration, by
    mechanical counters, 71
Propeller rotor current meter (Plessey current
    meter), 250−253
PROTAS, *see* Probe Recording Ocean
    Temperature and Shear
PTT, *see* Platform transmit terminal
Pulse Doppler-radar echo spectra, peculiarities
    of, 112−113
Pulse-limited geometry, 388
Pulse rectangle of ground-resolvable area, 156
Pulse repetition frequency (PRF), 158−159,
    161−164

R

RADARSAT-1, 167, 401−402
RADARSAT 2, 167
Radial vector currents, 120
Radial velocity, 341
    component, 120
Radiation stress, 39
Radio buoys, 94−105
    drifter-borne Doppler transponder, 96−97
    drifter-following radar transponder, 95−96
    GPS-tracked drifters, 100−104
    telephonically tracked drifters, 104−105
    tracked by polar-orbiting satellites, 97−100
Radiometers, remote detection by, 141−153
    active radiometry, in visible-wavelength band,
        145−146
    microwave radiometers, 149−153
        passive microwave radiometer
            instrumentation, 150−151
        passive microwave radiometry, 150
        scanning multichannel radiometers,
            151−153
    passive radiometry, in visible-wavelength
        band, 142−145
    passive radiometry, in thermal infrared band,
        146−149
        atmospheric effects and correction of, 148
        detectors, 147−148
        SST imaging, 148−149
Radiometry, of surface water motion
    trajectories and patterns, 60
RAFOS floats, 188−192, 412−413
Range walk, 164
RAR, *see* Real aperture radar
RAT, *see* River acoustic tomography
RDCP 600 profiler, 353

RDM, *see* Reflectivity displacement method

Real aperture radar (RAR)

active microwave imaging by, 155–158

seawater circulation features detection using, 168–170

Reciprocal tomography, *see* Two-way tomography

Reflectivity displacement method (RDM), 164

Remote mapping of sea surface currents, using HF Doppler radar networks, 109–138

advantages of, 129–131

Crombie's discovery, 110–112

depth extent of, 113–114

experimental developments of, 115–118

instrumentation aspects of, 118–120

intercomparison considerations of, 126–129

operational scales development, 120–126

coastal ocean dynamics applications radar, 121

ocean surface current radar, 121–123

SeaSonde, 123–124

special applications systems for, 126

Wellen Radar, 124–126

pulse Doppler-radar echo spectra, peculiarities of, 112–113

radial vector currents, 120

round-the-clock coast-observing role of, 131–132

sea surface current estimation, using Bragg resonance principle, 113

shelf-observing role of, 131–132

technological aspects of, 114–115

total vector currents, 120

tsunami-induced sea surface-current jets at continental shelves, detection and monitoring of, 132–133

Residual current, 261

Reynolds stress, 290

Richardson number, 302

Ridge valleys, tidal currents in, 34–36

Rings, 9–10

Rip currents, 36–40

transient, 37–38

River acoustic tomography (RAT), 69, 226–230, 410

River discharge, 226, 230

Roberts radio current meter, *see* Unidirectional impeller current meters

Rocknes, 353

Rotor pumping, 245–246

Round-the-clock coast-observing role, of HF Doppler radar networks, 131–132

S

SAR, *see* Synthetic aperture radar

Saral, 394

Satellite altimetric measurements, technological intricacies in, 386–388

Satellite altimetry, 381–396, 382f, 405–406

data error correction in, 389–392

geoid errors correction, 390–391

orbit errors correction, 389–390

topographic height variability independent of geoid, null methods for obtaining, 391–392

evolution of, 392–394

Satellite orbit errors correction, 389–390

Satellite-recovered pop-up drifters, 178–180

Satellite-tracked drifters, 400, 405

Savonius rotor current meters, 72–74, 242–249

miniature vane vector-averaging current meters, 249–250

Scanning multichannel microwave radiometers (SMMR), 151–153, 389

Scanning radiometer (SR), 60

Scattering correlation time, 340–341

Scatter plot, 258–259

Schaufelrad (paddlewheel) current meter, 71

SCM, *see* Spectrum centroid method

Seafaring Satellite (SEASAT), 61–62

Seafloor boundary layer current measurements, 82–86

Benthic Acoustic Stress Sensor, 84–85

mechanical devices, 83

Modular Acoustic Velocity Sensor, 85–86

nonmechanical devices, 83–86

SEASAT, *see* Seafaring Satellite

Seasat altimeter, 393

Seasat-A SAR system, 166

SeaSonde, 123–124, 403

Sea surface current signatures, active microwave radar imaging of, 153–154

Sea surface gravity-wave measurements, using ADCPs, 347–350

Sea surface imagery, 401

Sea surface slope, oceanic currents and features generated by, 382–385

Sea surface temperature (SST), 146, 148–149, 151

Seawater motion determination, from sea surface slope measurements, 385–386

Seawater motion signatures using remote sensors, imaging of, 139–176

aerial photography, in visible and infrared bands, 140–141

ice floes imaging sea surface currents measurement using, 170–171

radiometers, remote detection by, 141–153

active radiometry, in visible-wavelength band, 145–146

microwave radiometers, 149–153

passive radiometry, in visible-wavelength band, 142–145

passive radiometry, in thermal infrared band, 146–149

by RAR systems, 155–158

SAR systems, 158–166

azimuth resolution enhancement of, 160–163

development of, advances in, 166–168

error sources, 163–165

image data interpretation, 167–168

interferometric SAR system, 165–166

long antenna from small antenna synthesis of, 159–160

sea surface current signatures, active microwave radar imaging of, 153–154

SEC, *see* South Equatorial Current

Sensitivity time control (STC), 118–119

Shelf-observing role, of HF Doppler radar networks, 131–132

Shingles, 11

Ship-borne hydrophones swallow floats tracked by, 180–183

Shuttle Imaging Radar (SIR), 166

SIC, *see* South Indian Current

Silicon photodiodes, 143

Simrad Ultrasonic Current Meter (UCM), 282

Single upward-facing ADCP, 346–350

SIR, *see* Shuttle Imaging Radar

Skylab S-193 altimeter, 392

Slack water, 36

Sliding average method, 213

Slope current, 383

SMMR, *see* Scanning multichannel microwave radiometers

SOFAR channel, 93–94, 205, 411–413

SOFAR floats, 183–188, 191, 204–206, 208–209

SOLO floats, 196, 413–414

Somali Current, 24

SonTek/YSI-make side-looking ADCP, 354–355, 355f, 356f

South Equatorial Current (SEC), 18

South Indian Current (SIC), 7

South Pacific Current (SPC), 7

Space-time acoustic scintillation analysis, horizontally integrated current measurements using, 232–235

Spatially integrated measurements, 65–69

acoustic tomography, 66–69

electromagnetic method, 65–66

measurements based on drifting subsurface floats motion, 69–70

Spatial uncertainty, 127–128

SPC, *see* South Pacific Current

Special Sensor Microwave Imager (SSMI), 389

Spectrum centroid method (SCM), 164

Spring current, 36

SR, *see* Scanning radiometer

SSMI, *see* Special Sensor Microwave Imager

SST, *see* Sea surface temperature

STC, *see* Sensitivity time control

S-tether system, 207–208

Stick diagram, 258–259

Stochastic inverse method, 204

Straits

acoustic tomographic measurements from, 212–215

Davis Strait, 3

Haro Strait, 213–214, 213f, 214f

Florida Strait, 7, 201

Kanmon Strait, 219–220, 222–224

SUBS, *see* Subsurface, torpedo-shaped streamlined buoy

Subsurface, torpedo-shaped streamlined buoy (SUBS), 358–359

Subsurface/abyssal current measurements,
64—82
early mariners' contributions, 64
measurements from fixed locations at
predetermined depths, 70—82
acoustic Doppler current meter, 81—82
laser Doppler sensor, 78—81
propeller revolution registration by
mechanical counters, 71
Savonius rotor current meters, 72—74
suspended drag, 70—71
thermal sensors, for turbulent motion
measurement, 77—78
ultrasonic acoustic methods, 74—77
unidirectional impeller current meters,
71—72
spatially integrated measurements, 65—69
acoustic tomography, 66—69
electromagnetic method, 65—66
measurements based on drifting subsurface
floats motion, 69—70
Subsurface drifters tracking, Lagrangian-style
subsurface current measurements
through, 177—200
ALACE floats, 192—195
drifting profiling floats, 195—198
profiling observations from polar regions,
197—198
moored acoustic receivers subsurface floats
transmitting to, 183—188
moored acoustic sources subsurface floats
listening to, 188—192
satellite-recovered pop-up drifters, 178—180
surface-trackable subsurface drifters,
177—178
swallow floats, tracked by ship-borne
hydrophones, 180—183
Subsurface floats
listening to moored acoustic sources,
188—192
transmitting to moored acoustic receivers,
183—188
Subsurface moored ADCPs, 358—359
Surface current measurements, 52—64
based on drifting surface bodies motion,
52—57
surface water motion trajectories and patterns,
imaging of, 57—62
active microwave radar imaging, 61—62
aerial photography, 57—60
radiometry, 60
vector mapping, based on current-driven
surface wave transport, 62—64
Surface drifters tracking, Lagrangian-style
surface current measurements through,
93—108
Surface-trackable subsurface drifters,
177—178
Surface Velocity Program (SVP), 400, 414
Surf zone, 36—37
Suspended drag, 70—71
SVP, see Surface Velocity Program
Swallow floats, 69—70, 183, 188
tracked by ship-borne hydrophones, 180—183

Swing-around acoustic current meter, 75—76
Synthetic aperture radar (SAR), 61—62,
158—166, 401—402, 405
active microwave imaging by, 158—166
azimuth resolution enhancement of, 160—163
development of, advances in, 166—168
error sources, 163—165
image data interpretation, 167—168
interferometric, 165—166
long antenna from small antenna synthesis of,
159—160
seawater circulation features detection using,
168—170
TerraSAR, 167

T

Tasman Front, 20
Telephonically tracked drifters, 104—105
TerraSAR, 167
Texture analysis, 167
THC, see Thermohaline circulation
Thermal infrared band, passive radiometry in,
146—149
Thermal sensors, for turbulent motion
measurement, 77—78
Thermohaline circulation (THC), defined, 2
Thermohaline conveyor belt circulation, 2—5
Threshold cross-time method, 213
Tidal currents
ellipse, 34—36
in estuaries, 34—36
in open seas, 34—36
reversing, 36
in ridge valleys, 34—36
Tidal ellipse, 259—261
Topex/Poseidon, 394
TOPS, see Acoustically tracked free-fall
current velocity and CTD profiler
Total vector currents, 120
Towed acoustic transducer, 329—330
Towed ADCPs, 363—367
Trade-off triangle, 359
Transient rip currents, 37—38
Trawl-resistant ADCP bottom mounts
(TRBMs), 353—354
TRBMs, see Trawl-resistant ADCP bottom
mounts
Tsunami-induced sea surface-current jets at
continental shelves, detection and
monitoring of, 132—133
Turbulent motion measurement, using thermal
sensors, 77—78
Two-way tomography, 210—212

U

UCM, see Simrad Ultrasonic Current Meter
Ultrasonic acoustic methods, 74—77
Unidirectional impeller current meters, 71—72
Upward-looking ADCPs, 341
Upwelling, 11—12, 60

V

VACMs, see Vector-averaging current meters
Vector-averaging current meters (VACMs),
249—250, 256—257, 282—283, 322
Vector mapping, based on current-driven sea
surface wave transport, 62—64
Vector-measuring current meters (VMCMs),
254—255, 258
Vertical profile measurements, 68, see also
Vertical profiling, of horizontal
currents
Vertical profiling, using ADCPs, 339—380
bottom-mounted, upward-facing,
344—353
directional wave-measuring, 346
single upward-facing, 346—350
vertical current profiling using, 339—380
AUV-mounted ADCPs, for current
profiling, 368—369
for AUV navigation, 369—371
basic assumptions and operational issues,
340—341
calibration of, 371—372
downward-facing shipboard ADCPs,
360—362
evaluation of, 372—374
horizontal-facing ADCPs, 354—358
intercomparison of, 372—374
limitations of, 374—375
merits of, 374—375
operation principles, 341—344
profiling geometries, 344—353
subsurface moored ADCPs, 358—359
towed ADCPs, 363—367
trawl-resistant ADCP bottom mounts,
353—354
Vertical profiling, of horizontal currents,
86—88
using freely sinking/rising robes, 297—338
acoustically tracked free-fall current
velocity and CTD profiler (TOPS),
326—328
acoustically tracked freely sinking pingers,
305—308
bottom-mounted, winch-controlled vertical
automatic profiling systems, 302—305
free-falling, acoustically tracked absolute
velocity profiler (Pegasus), 318—320
free-falling lift-force sensitive probes,
332—335
free-fall probe, 330—332
free-fall shear profiler (Yvette), 326
freely falling, acoustically self-positioning
dropsonde system (White Horse),
320—322
freely falling electromagnetic velocity
profiler, 311—318
freely rising acoustically tracked
expendable probes (Popup), 322—325
freely sinking and rising relative velocity
probe (Cyclesonde), 308—311
freely sinking and wire-guided relative
velocity probes, 300—302
importance of, 297—300

Vertical profiling, of horizontal currents
    (*Continued*)
    limitations of, 335
    merits of, 335
    towed acoustic transducer, 329—330
Vertical shear of horizontal current velocity,
    297, 406
Vertical spatial structure, 398
Very high-frequency (VHF) Doppler radar
    systems, 63—64, 403—404
Very high-resolution radiometer (VHRR), 60
VHRR, *see* Very high-resolution radiometer
VIRR, *see* Visible and infrared radiometer
Viscous Lamb vortices, 370—371
Visible and infrared radiometer (VIRR), 60
Visible and infrared spin-scan radiometer
    (VISSR), 60
Visible and thermal infrared radiometer
    (VTIR), 60
Visible-wavelength aerial photography, 401
Visible wavelength bands
    active radiometry in, 145—146
    aerial photography in, 140—141
    passive radiometry in, 142—145

VISSR, *see* Visible and infrared spin-scan
    radiometer
VMCMs, *see* Vector-measuring current meters
Vorticity
    acoustic tomographic measurements of,
        230—232
    relative, 230—231
VTIR, *see* Visible and thermal infrared
    radiometer

W

Wake vortices, 11
Warm-core eddies, 6—8, 383—384, 383f
WBC, *see* Western boundary currents
Weather eddies/fronts influence on, 11—12
Wellen radar (WERA), 62—63, 124—126, 403
WERA, *see* Wellen radar
Western boundary currents (WBC), 6—7, 10,
    146, 384—385
    in Atlantic Ocean, 15—20
    in Indian Ocean, 22—24
    in Pacific Ocean, 20—22
White Horse, *see* Freely falling, acoustically
    self-positioning dropsonde system

Wind-Driven Current, 31—33
    Ekman Spiral, 31—33
    Langmuir Circulation, 33
WOCE, *see* World Ocean Circulation
    Experiment
WOCE-TOGA, *see* World Ocean Circulation
    Experiment—Tropical Ocean Global
    Atmosphere
Woods Hole, 208—209
World Ocean Circulation Experiment
    (WOCE), 195—197, 360, 413
    Lagrangian drifters, 399—400
World Ocean Circulation
    Experiment—Tropical Ocean Global
    Atmosphere (WOCE-TOGA), 400

Y

Yvette, *see* Free-fall shear profiler

Z

Zero-order kind of tomography, *see* One-way
    tomography

Printed and bound by CPI Group (UK) Ltd, Croydon, CR0 4YY

08/05/2025

01864939-0001